Supermanifolds and Superg

Mathematics and Its Applications

Volume 570

Supermanifolds and Supergroups

Basic Theory

Gijs M. Tuynman
Université de Lille I,,
Lille, France

KLUWER ACADEMIC PUBLISHERS
DORDRECHT / BOSTON / LONDON

A C.I.P. Catalogue record for this book is available from the Library of Congress.

ISBN 1-4020-2296-4 (HB)
ISBN 1-4020-2297-2 (e-book)

Published by Kluwer Academic Publishers,
P.O. Box 17, 3300 AA Dordrecht, The Netherlands.

Sold and distributed in North, Central and South America
by Kluwer Academic Publishers,
101 Philip Drive, Norwell, MA 02061, U.S.A.

In all other countries, sold and distributed
by Kluwer Academic Publishers,
P.O. Box 322, 3300 AH Dordrecht, The Netherlands.

Typeset by the author using $\mathcal{A}_{\mathcal{M}}\mathcal{S}$-TEX and the amsppt.sty macro package.
Numbering and referencing done with the acres macro package written by the author.

Gijs M. Tuynman
UMR 8524 AGAT au CNRS & UFR de Mathématiques
Université de Lille I
F.-59655 Villeneuve d'Ascq Cedex
France

e-mail : Gijs.Tuynman@univ-lille1.fr

Printed on acid-free paper

Printed in the Netherlands.

Table of Contents

Preface

This book is a self contained introduction to super differential geometry, intended for graduate students in mathematics and theoretical physics and other people who want to learn the basics about supermanifolds. It is self contained in that it only requires standard undergraduate knowledge. However, some knowledge of ordinary (non super) differential geometry will make this text much easier to read.

Various versions of super differential geometry exist, some of which are equivalent and some of which are not. The version presented here is equivalent to those that are most widely used: the H^{∞} supermanifolds of DeWitt and the sheaf theoretic approach to supermanifolds of Kostant and Leites. The approach taken here is based on an index free formalism using a graded commutative ring $\mathcal{A}$ containing the usual real numbers as well as so called anticommuting numbers. Starting with a non-standard definition of a differentiable function, valid in the real case, in the complex case and in the super case, the theory is developed as if it were ordinary differential geometry. It is shown that most constructions and theorems in ordinary differential geometry have a natural generalization to the super context. Moreover, even the proofs bear more than a superficial resemblance to their counter parts in ordinary differential geometry. The (equivalent) sheaf-theoretic approach to supermanifolds makes it manifest that the theory is "independent" of the choice of $\mathcal{A}$, but at the same time it hides the more geometric nature of the theory. The approach presented here can be seen as a theory with a parameter $\mathcal{A}$. Choosing $\mathcal{A} = \mathbf{R}$ gives ordinary differential geometry, choosing $\mathcal{A} = \bigwedge \mathbf{R}^{\infty}$ gives super differential geometry, choosing $\mathcal{A} = \mathbf{C}$ gives the theory of complex manifolds, etc. Of course, in each of these cases some small but usually superficial changes have to be made, and not all results remain true in all cases (e.g., Batchelor's theorem, which uses partitions of unity, is not valid for (super) complex manifolds). But the main body of the results is not affected by the choice of $\mathcal{A}$.

In Chapter I the general theory of graded linear algebra (graded by an arbitrary abelian group) is outlined. This plays the same role in super differential geometry as does linear algebra in ordinary differential geometry and as does commutative algebra in algebraic geometry. Since the basic ring is (in principle) not commutative, we have to make a distinction between left and right linear maps. The isomorphism between these two kinds of maps is given by the operator $\mathfrak{T}$, which will later be identified with (super) transposition of matrices.

In chapter II we specialize to $\mathbf{Z}_2$-graded linear algebra and we impose some restrictions on the ring $\mathcal{A}$. The canonical example of $\mathcal{A}$ that satisfies all conditions (the ones imposed in chapter II and also other ones imposed later on) is the exterior algebra of an infinite dimensional real vector space: $\mathcal{A} = \bigwedge \mathbf{R}^\infty$. Some of the more important points of this chapter are the following. In section 2 it is shown that any (finitely generated) free graded $\mathcal{A}$-module admits a well defined graded dimension. In section 4 the relation between matrices and linear maps is explained. The reader should really pay attention here, because there are *three* different natural ways to associate a matrix to a linear map, and these three different ways imply different ways how to multiply a matrix by an element of $\mathcal{A}$ (so as to be compatible with the multiplication of the corresponding linear map by the element of $\mathcal{A}$). It is here that we see most clearly the role of the transposition operator introduced in chapter I to relate left and right linear maps. In section 5 the graded trace is defined for any linear map (and thus for any matrix, not only the even ones), as well as its integrated version for even maps, the graded determinant or Berezinian. Finally in section 6 the body map $\mathbf{B}$ is introduced, which provides an "isomorphism" between equivalence classes of free graded $\mathcal{A}$-modules and direct sums of two real vector spaces . It is this body map which gives the link between standard linear algebra and $\mathbf{Z}_2$-graded linear algebra.

The heart of this book lies in chapter III, in which the notion of a supermanifold is developed based on a non-standard definition of differentiable functions. The key idea is expressed by the following formula, valid for functions f of class C^1 on convex domains in $\mathbf{R}^n$:

$$f(x) - f(y) = \left(\int_0^1 f'(sx + (1-s)y)\, ds \right) \cdot (x - y) .$$

If we write this as $f(x) - f(y) = g(x,y) \cdot (x - y)$, it is obvious that f is of class C^1 if and only if the function g is of class C^0. Moreover, if a g with this property exists, it is also easy to see that f', the derivative of f, is given by $f'(x) = g(x,x)$. If we now note that the formula $f(x) - f(y) = g(x,y) \cdot (x - y)$ does not involve quotients nor limits, we can apply the same definition to super functions, for which there generally do not exist quotients (because of nilpotent elements in $\mathcal{A}$), nor does the natural topology (the DeWitt topology) admit unique limits (being non Hausdorff). Based on this idea, smooth functions (C^∞) on super domains with p even coordinates and q odd coordinates are defined. It is shown, using the body map $\mathbf{B}$ defined in chapter II, that these smooth functions are in bijection with ordinary smooth real-valued functions of p real variables, multiplied by antisymmetric polynomials in q variables. This result is usually taken as the definition of smooth super functions; here it is a consequence of a more general definition, a definition which applies as well to ordinary functions as to super functions. The last two sections of chapter III are devoted to copying the standard definition of manifolds in terms of charts and transition functions to the case in which the transition functions are super smooth functions.

In chapter IV the general theory of fiber and vector bundles is developed. The first two sections deal with general fiber bundles and how to construct new ones out of given ones. The next two sections deal with vector bundles and how to generalize the construction of new $\mathcal{A}$-modules to the setting of vector bundles. In section 5 the behavior of the operation of taking sections under the various operations one can perform on $\mathcal{A}$-modules is considered. In section 6 the exterior algebra of a (dual) bundle is discussed in more detail, as well as the pull-back of sections. The main purpose of sections 5 and 6 is to provide a rigorous justification for operations everybody performs without thinking twice. In section 7 one finds a proof of part of the Serre-Swann theorem that the module of sections of a vector bundle is a finitely generated projective module over the ring of smooth functions on the base manifold. The proof of this result needs the notion of a metric on a free graded $\mathcal{A}$-module, a notion whose definition is subtly different from what one would expect. These results are not used elsewhere, but they are needed to complete the proofs of statements given in section 5. The last section in chapter IV on Batchelor's theorem merits ample attention. This theorem says that any supermanifold is "isomorphic" to an ordinary vector bundle over an ordinary manifold, or, stated differently, for any smooth supermanifold there exists an atlas in which the transition functions are of the special form: even coordinates depend on even coordinates only, and odd coordinates depend in a linear way on odd coordinates. The proof is "constructive" in that it provides an explicit algorithm to compute such an atlas given an arbitrary atlas. The quotes are needed because this algorithm requires a partition of unity on the underlying ordinary manifold.

Chapter V treats the standard machinery of differential geometry. In section 1 the tangent bundle is defined and it is shown that sections of it, called vector fields, are equivalent to derivations of the ring of smooth functions. In section 2 the tangent map is defined, which in turn gives rise to the notions of immersion and embedding. In section 3 the relationship between the tangent map and the derivative of a map are studied in more detail. It turns out that in the super case this is in general not a 1–1 correspondence. Generalizing the notion of the derivative of an $\mathcal{A}$-valued function to vector bundle valued functions, a necessary and sufficient condition is given for a vector bundle to be trivial as a vector bundle. Here one also can find an example of a vector bundle which is trivial as fiber bundle, but not as vector bundle. Sections 4 and 5 then concentrate on the notion of the flow of a vector field and the well known proposition that two vector fields commute if and only if their flows commute. For odd vector fields this amounts to saying that an odd vector field is integrable if and only if its auto commutator is zero. Section 6 treats Frobenius' theorem on integrability of subbundles of the tangent bundle, the notion of integral manifolds and the existence of leaves for a foliation. In section 7 the calculus of (exterior differential) k-forms is given, including the definition of the Lie derivative and its relation with the flow of a vector field. Finally in section 8 an elementary proof is given of the fact that the de Rham cohomology of a supermanifold is the same as that of the underlying ordinary manifold (its body).

Chapter VI treats the basic facts about super Lie groups and their associated super Lie algebras. In section 1 one finds the basic definition of a super Lie group and the construction of the associated super Lie algebra. The exponential map from the super Lie algebra to the super Lie group is defined in section 2. There one also finds the proof that it intertwines a homomorphism of super Lie groups and its induced morphism on the associated super Lie algebras. Section 3 is rather technical and computes the derivative of the exponential map. Section 4 deals with the relationship between Lie subgroups and Lie subalgebras, whereas section 5 treats homogeneous supermanifolds. Section 6 is again technical and proves that any smooth action can be transformed into a pseudo effective action. The last section gives a geometric proof that to each finite dimensional super Lie algebra corresponds a super Lie group.

Chapter VII is more advanced and discusses the general concept of a connection on a fiber bundle. Sections 1 and 8 are technical and provide the necessary theory of vector valued and vector bundle valued differential forms. In section 2 the general concept of an Ehresmann connection is introduced, as well as the more restrictive notion of FVF connection, which is an Ehresmann connection determined by the fundamental vector fields of the structure group on the typical fiber. FVF connections have nice properties: they are defined on any fiber bundle, they include the standard examples of connections such as the (principal) connection on a principal fiber bundle and linear connections on vector bundles, and they always allow parallel transport. In sections 3 and 4 the particular case of an FVF connection on a principal fiber bundle is studied, which includes the description by a connection 1-form, the exterior covariant derivative and a discussion about the curvature 2-form. In section 5 it is shown that any FVF connection can be seen as induced by an FVF connection on a principal fiber bundle. Sections 6 and 7 treat the notion of a covariant derivative on a vector bundle and prove that it is equivalent to an FVF connection. It includes the proof that the covariant derivative measures how far away a (local) section is from being horizontal. In sections 9 and 10 the covariant derivative on a vector bundle is generalized to vector bundle valued differential forms and it is shown how the exterior covariant derivative (on a principal fiber bundle), the ordinary exterior derivative of differential forms and the generalized covariant derivative (on a vector bundle) are intimately related.

This book is written in a logical order, meaning that a proof of a statement never uses future results and meaning that related subjects are put together. This is certainly not the most pedagogical way to present the subject, but it avoids the risk of circular arguments. As a consequence, the novice reader should not read this book in a linear order. For a first reading, one can easily skip sections 7 and 8 of chapter I. From chapter IV one should certainly read sections 1–3, but coming back for sections 4–6 (and then only superficially) just before starting to read section 6 of chapter V. The reader who already has a working knowledge of ordinary manifold theory need not read all sections with the same attention and at a first reading (s)he can even skip chapter IV completely.

One final word on terminology: in this introduction I have systematically used the adjective *super*. On the other hand, in the main text I never use this adjective, but rather the prefix $\mathcal{A}$. The reason to do so is that one should regard this theory not as opposed to ordinary differential geometry (super versus non-super), but more as a theory with a parameter $\mathcal{A}$, indicating over which ring it is developed.

In preparing chapters I–VI I have relied heavily on the first three chapters of F. Warner's classic "Foundations of Differentiable Manifolds and Lie Groups,", while chapter VII is based on H. Pijls' review article "The Yang-Mills equations." Other sources of inspiration have been the first volume of M. Spivak's "A Comprehensive Introduction to Differential Geometry" and "Les Tenseurs" of L. Schwartz. During the years it took me to write this book, I have benefitted from the hospitality of the following three institutions: MSRI (Berkeley, USA), CPT (Marseille, France) and LNCC (Rio de Janeiro, Brazil). Special thanks are due to P. Bongaarts for some excellent suggestions concerning chapter I and to V. Thilliez who helped me with [III.1.12]. Finally, I am convinced I got the idea for [IV.7.3] from a paper by S. Sternberg, but I can no longer find the source.

Lille, january 2004

Chapter I

$\mathfrak{A}$-graded commutative linear algebra

Linear algebra is concerned with the study of vector spaces over the real numbers (or more generally over a field) and linear maps. A standard course on linear algebra more or less starts with the introduction of the concept of a basis. Immediately afterwards one usually restricts attention to finite dimensional vector spaces. Next on the list is the concept of a subspace and with that notion one derives some elementary properties of linear maps. Then one introduces bilinear maps, with a scalar product as the most important example. This gives rise to the notions of orthogonal basis, orthogonal linear map, and orthogonal subspaces, eventually followed by a classification of quadrics. More advanced courses treat the notions of multilinear maps, tensor products, and exterior powers. Algebras, and in particular Lie algebras, are usually treated separately.

Besides analysis, these concepts in linear algebra form the basis of differential geometry. One could even say that differential geometry is the interplay between analysis and linear algebra. Algebraic geometry is closely related to differential geometry, but hardly relies on analysis; it is mainly concerned with algebraic structures. For that it needs a generalization of linear algebra in which a vector space over a field is replaced by a module over a commutative ring with unit. Commutative algebra is the theory which plays in algebraic geometry the same role as linear algebra does in differential geometry. In commutative algebra the notion of basis more or less disappears, but subspaces, tensor products, and exterior powers can still be defined.

In supergeometry one replaces the field of real numbers, not by a commutative ring, but by a graded commutative ring. Since such a ring is not commutative, commutative algebra does not apply. In this context, graded means $\mathbf{Z}_2$-graded, i.e., the ring and all modules are a direct sum of two subspaces, the even and odd parts. In this first chapter we look at an even more general situation. We denote by $\mathfrak{A}$ an arbitrary abelian group and we denote by $\mathcal{A}$ an arbitrary $\mathfrak{A}$-graded commutative ring with unit $1 \neq 0$ (i.e., a ring which splits as a direct sum of subspaces indexed by $\mathfrak{A}$ and satisfying conditions how these subspaces commute). We will show that all concepts of linear algebra that are important *for differential geometry can be generalized to $\mathfrak{A}$-graded commutative linear algebra, i.e., to the theory of $\mathfrak{A}$-graded $\mathcal{A}$-modules.*

1. $\mathfrak{A}$-GRADED COMMUTATIVE RINGS AND $\mathfrak{A}$-GRADED $\mathcal{A}$-MODULES

In this first section we give the definitions of the principal objects of this book: $\mathfrak{A}$-graded commutative algebras and $\mathfrak{A}$-graded $\mathcal{A}$-modules. $\mathfrak{A}$-graded $\mathcal{A}$-modules are a special kind of $\mathfrak{A}$-graded $\mathcal{A}$-bimodules, a fact that will greatly facilitate constructions of new $\mathfrak{A}$-graded $\mathcal{A}$-modules, one of which is discussed in this section: the $\mathfrak{A}$-graded submodule.

1.1 Definition. Given abelian groups $G_1, \ldots, G_k$ and H, a map $\phi : G_1 \times \cdots \times G_k \to H$ is called *k-additive* if for all i and for all $g_i, \widehat{g}_i \in G_i$ we have:

$$\phi(g_1, \ldots, g_{i-1}, g_i + \widehat{g}_i, g_{i+1}, \ldots, g_k) = \phi(g_1, \ldots, g_{i-1}, g_i, g_{i+1}, \ldots, g_k) + \phi(g_1, \ldots, g_{i-1}, \widehat{g}_i, g_{i+1}, \ldots, g_k) .$$

1.2 Definition. Let G be an abelian group and let G_i, $i \in I$ be a family of subgroups. One writes $G = \bigoplus_{i \in I} G_i$ if and only if for each element $g \in G$ there exist *unique* $g_i \in G_i$, only finitely many of them non-zero, such that $g = \sum_{i \in I} g_i$; it is called *the (unique) decomposition of g into G_i-components*.

1.3 Definitions. • Let $\mathcal{A}$ be a ring. A *left module over the ring $\mathcal{A}$* (or a *left $\mathcal{A}$-module*) is an abelian group E equipped with a map $m_L : \mathcal{A} \times E \to E$ that is bi-additive and satisfies

$$m_L(a, m_L(b, e)) = m_L(ab, e) .$$

This map is called *left multiplication by elements of $\mathcal{A}$*, and (as is usual) we will omit the symbol m_L if no confusion is possible and just write ae or $a \cdot e$ for $m_L(a, e)$. If $\mathcal{A}$ contains a unit $1 \neq 0$, we also require that $1 \cdot e = e$ for all $e \in E$. In a similar way, a *right $\mathcal{A}$-module* is an abelian group E equipped with a map (right multiplication) $m_R : E \times \mathcal{A} \to E$ that is bi-additive and satisfies $m_R(m_R(e, a), b) = m_R(e, ab)$. And as before, if no confusion is possible we will just write ea or $e \cdot a$ for $m_R(e, a)$. As for left $\mathcal{A}$-modules, if $\mathcal{A}$ contains a unit $1 \neq 0$, we require that $e \cdot 1 = e$ for all $e \in E$. Since $\mathcal{A}$ is in general not commutative, the notions of left and right $\mathcal{A}$-modules do not coincide.

• An *$\mathcal{A}$-bimodule* is an abelian group E which is at the same time a left and a right $\mathcal{A}$-module such that the left and right actions commute, i.e., for all $a, b \in \mathcal{A}$, $e \in E$: $m_L(a, m_R(e, b)) = m_R(m_L(a, e), b)$, which can also be written as $(ae)b = a(eb)$.

• A subset F of a left/right $\mathcal{A}$-module E is called a *submodule* if F is a subgroup with respect to the additive structure of E such that $\mathcal{A}F \subset F$. It follows that F, with the induced multiplication of $\mathcal{A}$, is itself a left/right $\mathcal{A}$-module.

1.4 Convention. Throughout this book we will denote by $\mathfrak{A}$ an arbitrary abelian group.

1.5 Definition. An $\mathfrak{A}$-*graded ring* is a ring $\mathcal{A}$ with the additional property (the $\mathfrak{A}$-grading) that there exists a family of subgroups $\mathcal{A}_\alpha, \alpha \in \mathfrak{A}$ (subgroups with respect to the additive (abelian) structure of the ring $\mathcal{A}$) such that:

(i) $\mathcal{A} = \bigoplus_{\alpha\in\mathfrak{A}} \mathcal{A}_\alpha$ and
(ii) $\forall \alpha, \beta \in \mathfrak{A} : \mathcal{A}_\alpha \cdot \mathcal{A}_\beta \subset \mathcal{A}_{\alpha+\beta}$.

The elements of $\mathcal{A}_\alpha$ are called *homogeneous elements of parity* α. For homogeneous elements the *parity map* $\varepsilon : \cup_{\alpha\in\mathfrak{A}}\mathcal{A}_\alpha \to \mathfrak{A}$ is defined by $\varepsilon(\mathcal{A}_\alpha) = \alpha$. Note that the parity map is *not* defined on the whole of $\mathcal{A}$. In analogy with the special case $\mathfrak{A} = \mathbf{Z}_2$, elements of parity 0 will always be called *even*.

1.6 Remarks. • There is an ambiguity in the definition of the parity of the zero element $0 \in \mathcal{A}$. However, we will not avoid this ambiguity since it comes in useful to say that 0 has every parity one wishes. Purists might want to exclude 0 from having a parity, but then in a lot of proofs one has to treat the zero element as a special case.

• Using the parity map, condition (ii) above can be stated as: if $a, b \in \mathcal{A}$ are homogeneous, then $\varepsilon(ab) = \varepsilon(a) + \varepsilon(b)$.

1.7 Lemma. *If an $\mathfrak{A}$-graded ring $\mathcal{A}$ has a unit $1 \neq 0$, then $\varepsilon(1) = 0$.*

Proof. Let $1 = \sum_{\alpha\in\mathfrak{A}} 1_\alpha$ be the decomposition of the unit into homogeneous components, and let $a \in \mathcal{A}$ be any homogeneous element. From the equation $a = 1 \cdot a$, the unique decomposition into homogeneous components and the fact that $\varepsilon(1_\alpha \cdot a) = \alpha + \varepsilon(a)$ we deduce that $1_0 \cdot a = a$. Since multiplication is bi-additive, it follows that $1_0 \cdot a = a$ for all $a \in \mathcal{A}$. In the same way one proves that $a \cdot 1_0 = a$ for all $a \in \mathcal{A}$. But units, if they exist, are unique and hence $1 = 1_0$. QED

1.8 Definition. An $\mathfrak{A}$-*graded commutative ring* is an $\mathfrak{A}$-graded ring $\mathcal{A}$ together with a symmetric bi-additive map $\langle\,\_\,|\,\_\,\rangle : \mathfrak{A} \times \mathfrak{A} \to \mathbf{Z}_2 \cong \{0, 1\}$ such that

$$a \in \mathcal{A}_\alpha, b \in \mathcal{A}_\beta \quad \Longrightarrow \quad a \cdot b = (-1)^{\langle\alpha|\beta\rangle} b \cdot a \ .$$

The above property is called the $\mathfrak{A}$-*graded commutativity* of $\mathcal{A}$.

1.9 Remark. A more general definition of $\mathfrak{A}$-graded commutativity is possible if $\mathcal{A}$ is an $\mathfrak{A}$-graded k-algebra [6.1] with $k = \mathbf{R}$ or $k = \mathbf{C}$. In that case one can define $\mathfrak{A}$-graded commutativity by the condition

$$a \in \mathcal{A}_\alpha, b \in \mathcal{A}_\beta \quad \Longrightarrow \quad a \cdot b = \rho(\alpha, \beta) \cdot b \cdot a \ ,$$

where $\rho : \mathfrak{A} \times \mathfrak{A} \to k^* = k \setminus \{0\}$ is a "bi-additive" map satisfying $\rho(\alpha, \beta) \cdot \rho(\beta, \alpha) = 1$ (we put bi-additive in quotes because the abelian operation in k^* is multiplication).

For instance, if $\mathfrak{A} = \mathbf{Z}_3 \times \mathbf{Z}_3 \cong \{(i,j) \mid 0 \leq i,j < 3\}$ and $k = \mathbf{C}$, we can take for ρ the map $\rho((i,j),(p,q)) = \lambda^{iq-jp}$ with λ a third root of unity. On the other hand, for $k = \mathbf{R}$ or $k = \mathbf{C}$ and $\mathfrak{A}$ cyclic, the only possibilities for ρ are of the form $\rho(\alpha, \beta) = (-1)^{\langle \alpha | \beta \rangle}$ with $\langle \alpha | \beta \rangle \in \mathbf{Z}_2$.

1.10 Convention. Throughout this text we will denote by $\mathcal{A}$ an arbitrary $\mathfrak{A}$-graded commutative ring with unit $1 \neq 0$. From time to time we will impose additional restriction on $\mathcal{A}$, but those restrictions will always be stated clearly.

1.11 Definition. An *$\mathfrak{A}$-graded left (respectively right) $\mathcal{A}$-module* is a left (respectively right) $\mathcal{A}$-module E together with a family of subgroups $E_\alpha, \alpha \in \mathfrak{A}$ (subgroups with respect to the additive (abelian) structure of E) satisfying:

(i) $E = \bigoplus_{\alpha \in \mathfrak{A}} E_\alpha$ and
(ii) $\mathcal{A}_\alpha \, E_\beta \subset E_{\alpha+\beta}$ (respectively $E_\alpha \, \mathcal{A}_\beta \subset E_{\alpha+\beta}$).

The following definitions and remarks are as for $\mathfrak{A}$-graded rings. The elements of E_α are called *homogeneous elements of parity* α. The *parity map* $\varepsilon : \cup_{\alpha \in \mathfrak{A}} V_\alpha \to \mathfrak{A}$ is defined by $\varepsilon(E_\alpha) = \alpha$. Note that the parity map is *not* defined on the whole of E and that the parity of the zero element $0 \in E$ is ambiguous. Using the parity map, condition (ii) can be stated as: for homogeneous $a \in \mathcal{A}$ and $e \in E$, $\varepsilon(ae) = \varepsilon(a) + \varepsilon(e)$. Elements of parity 0 will be called *even*.

1.12 Example. Let $\mathcal{A}$ be an $\mathfrak{A}$-graded ring and $n \in \mathbf{N}$ a natural number, then we can make $\mathcal{A}^n$ into an $\mathfrak{A}$-graded $\mathcal{A}$-bimodule. Addition and left/right multiplication by $\mathcal{A}$ is component wise, while the $\mathfrak{A}$-grading is given by $(\mathcal{A}^n)_\alpha = (\mathcal{A}_\alpha)^n$. For $n = 0$ we formally define $\mathcal{A}^0 = \{0\}$, which is trivially an $\mathfrak{A}$-graded $\mathcal{A}$-bimodule.

1.13 Definition. A subset F of an $\mathfrak{A}$-graded left/right $\mathcal{A}$-module E is called an *$\mathfrak{A}$-graded submodule* if F is a submodule of the left/right $\mathcal{A}$-module E such that F together with the subsets $F_\alpha = F \cap E_\alpha$ is itself an $\mathfrak{A}$-graded left/right $\mathcal{A}$-module.

1.14 Lemma. *Let F be a submodule of an $\mathfrak{A}$-graded left/right $\mathcal{A}$-module E. Then F is an $\mathfrak{A}$-graded submodule if and only if $\forall e \in E : e \in F \Rightarrow e_\alpha \in F$, where $e = \sum_{\alpha \in \mathfrak{A}} e_\alpha$ is the unique decomposition of e into homogeneous components in the $\mathfrak{A}$-graded left/right $\mathcal{A}$-module E.*

1.15 Lemma. *Let $\mathcal{A}$ be an $\mathfrak{A}$-graded commutative ring and let $E = \bigoplus_{\alpha\in\mathfrak{A}} E_\alpha$ be an $\mathfrak{A}$-graded left/right $\mathcal{A}$-module. Then there exists a unique right/left action of $\mathcal{A}$ on E turning E into an $\mathfrak{A}$-graded right/left $\mathcal{A}$-module with the same subsets E_α such that the left and right actions of $\mathcal{A}$ are related by the formula*

$$a \in \mathcal{A}_\alpha, e \in E_\beta \quad \Longrightarrow \quad m_R(e,a) = (-1)^{\langle\alpha|\beta\rangle} m_L(a,e) \ . \tag{1.16}$$

Proof. Let us suppose that m_L has been given, then if m_R exists, bi-additivity implies that it must be given by

$$m_R(e,a) \equiv m_R(\sum_{\beta\in\mathfrak{A}} e_\beta, \sum_{\alpha\in\mathfrak{A}} a_\alpha) = \sum_{\alpha,\beta\in\mathfrak{A}} m_R(e_\beta, a_\alpha) = \sum_{\alpha,\beta\in\mathfrak{A}} (-1)^{\langle\alpha|\beta\rangle} m_L(a_\alpha, e_\beta) \ ,$$

where $e = \sum_{\beta\in\mathfrak{A}} e_\beta$ and $a = \sum_{\alpha\in\mathfrak{A}} a_\alpha$ are the unique decompositions of these elements into homogeneous components. From this formula, the uniqueness of m_R is clear. On the other hand, bi-additivity of m_L proves that m_R defined by this formula is also bi-additive. It thus remains to show that m_R is a right action, i.e., $m_R(m_R(e,a),b) = m_R(e,ab)$. Since this relation is 3-additive in the variables $(e,a,b) \in E \times \mathcal{A} \times \mathcal{A}$, it suffices to show it for homogeneous elements:

$$\begin{aligned} m_R(m_R(e,a),b) &= (-1)^{\langle\varepsilon(e)|\varepsilon(a)\rangle} m_R(m_L(a,e),b) \\ &= (-1)^{\langle\varepsilon(e)|\varepsilon(a)\rangle} (-1)^{\langle\varepsilon(e)+\varepsilon(a)|\varepsilon(b)\rangle} m_L(b, m_L(a,e)) \\ &= (-1)^{\langle\varepsilon(e)|\varepsilon(a)+\varepsilon(b)\rangle} (-1)^{\langle\varepsilon(a)|\varepsilon(b)\rangle} m_L(ba,e) \\ &= (-1)^{\langle\varepsilon(e)|\varepsilon(a)+\varepsilon(b)\rangle} m_L(ab,e) = m_R(e,ab) \ . \end{aligned}$$

QED

1.17 Remark. The left and right actions of $\mathcal{A}$ on the $\mathfrak{A}$-graded $\mathcal{A}$-module $\mathcal{A}^n$ are related to each other by (1.16).

1.18 Lemma. *With the assumptions as in [1.15], the left and right actions of $\mathcal{A}$ on E commute, i.e., $\forall a,b \in \mathcal{A}$, $\forall e \in E : a(eb) = (ae)b$.*

Proof. The relation $a(eb) = (ae)b$ is 3-additive in (a,e,b) and it thus suffices to show it for homogeneous elements:

$$\begin{aligned} a(eb) &= a((-1)^{\langle\varepsilon(e)|\varepsilon(b)\rangle} be) = (-1)^{\langle\varepsilon(e)|\varepsilon(b)\rangle} (ab)e \\ &= (-1)^{\langle\varepsilon(e)|\varepsilon(b)\rangle} (-1)^{\langle\varepsilon(a)|\varepsilon(b)\rangle} (ba)e \\ &= (-1)^{\langle\varepsilon(e)+\varepsilon(a)|\varepsilon(b)\rangle} b(ae) = (-1)^{\langle\varepsilon(e)+\varepsilon(a)|\varepsilon(b)\rangle} (-1)^{\langle\varepsilon(ae)|\varepsilon(b)\rangle} (ae)b \\ &= (ae)b \qquad\qquad \text{because } \varepsilon(ae) = \varepsilon(a) + \varepsilon(e). \end{aligned}$$

QED

1.19 Definitions. By an $\mathfrak{A}$-*graded* $\mathcal{A}$-*module* (without the adjectives left or right) we will *always* mean an $\mathfrak{A}$-graded $\mathcal{A}$-bimodule for which the left and right actions of $\mathcal{A}$ are related by (1.16). Note that [1.15] implies that every $\mathfrak{A}$-graded left $\mathcal{A}$-module and every $\mathfrak{A}$-graded right $\mathcal{A}$-module can be turned into such an $\mathfrak{A}$-graded $\mathcal{A}$-module.

A subset F of an $\mathfrak{A}$-graded $\mathcal{A}$-module E is called an $\mathfrak{A}$-*graded submodule* of E if it is an $\mathfrak{A}$-graded submodule of the left or right $\mathcal{A}$-module E. It follows from [1.20] that this is a sensible definition and that such an F is automatically an $\mathfrak{A}$-graded $\mathcal{A}$-module itself.

1.20 Lemma. *Let E be an $\mathfrak{A}$-graded $\mathcal{A}$-module and let F be a submodule of either the $\mathfrak{A}$-graded left or the $\mathfrak{A}$-graded right $\mathcal{A}$-module E. If F is an $\mathfrak{A}$-graded submodule for this structure, it is an $\mathfrak{A}$-graded submodule for the opposite structure. It thus is itself an $\mathfrak{A}$-graded $\mathcal{A}$-module.*

Proof. Suppose F is an $\mathfrak{A}$-graded submodule for the $\mathfrak{A}$-graded left $\mathcal{A}$-module E. For $a \in \mathcal{A}$ and $f \in F$ denote by $a = \sum_{\alpha\in\mathfrak{A}} a_\alpha$ and $f = \sum_{\beta\in\mathfrak{A}} f_\beta$ their decomposition in homogeneous components. Then $fa = \sum_{\alpha,\beta\in\mathfrak{A}} f_\beta a_\alpha = \sum_{\alpha,\beta\in\mathfrak{A}} (-1)^{\langle\alpha|\beta\rangle} a_\alpha f_\beta$. This belongs to F because each f_β belongs to F [1.14]. Hence F is also a submodule of the right $\mathcal{A}$-module E. The conclusion then follows again from [1.14]. $\boxed{QED}$

1.21 Guiding principle. The guiding principle for $\mathfrak{A}$-graded linear algebra is that in any formula in which we interchange two homogeneous $\mathfrak{A}$-graded objects, a sign $(-1)^{\langle\alpha|\beta\rangle}$ appears. This (additional) sign is already visible in the definition of $\mathfrak{A}$-graded commutativity and the relation between the left and right actions of $\mathcal{A}$ on an $\mathfrak{A}$-graded $\mathcal{A}$-module. In order to adhere to this principle, we are occasionally led to change notation (e.g., [2.12]). Its advantages will be mostly notational: additional signs in equations will be "obvious" from the order in which one writes the separate terms.

1.22 Definition. Let E be a left or right $\mathcal{A}$-module, let S be any subset of E and let $\mathcal{B}$ be any subset of $\mathcal{A}$. For a left $\mathcal{A}$-module we define the subset $\mathrm{Span}_{\mathcal{B}}(S) \subset E$ by

$$\mathrm{Span}_{\mathcal{B}}(S) = \{ \sum_{i=1}^{n} a^i s_i \mid n \in \mathbf{N}, a^i \in \mathcal{B}, s_i \in S \} ;$$

for a right $\mathcal{A}$-module one just replaces $\sum_{i=1}^n a^i s_i$ in this definition by $\sum_{i=1}^n s_i a^i$. In case $\mathcal{B} = \mathcal{A}$ we will drop the subscript $\mathcal{B}$ and speak of $\mathrm{Span}(S)$. This will occur by far the most frequently, but we will occasionally need proper subsets $\mathcal{B}$ of $\mathcal{A}$. The subset $\mathrm{Span}(S)$ is obviously a submodule of E, usually called the *submodule generated by* S.

One special case should be mentioned separately. Suppose $F_i, i \in I$ is a family of submodules of E, then the submodule $\mathrm{Span}(\cup_{i\in I} F_i)$ is usually denoted as $\sum_{i\in I} F_i$ and called *the sum of the submodules* F_i. This notation is justified because obviously

$$\mathrm{Span}(\cup_{i\in I} F_i) = \{ \sum_{k=1}^{n} f_{i_k} \mid n \in \mathbf{N}, i_k \in I, f_{i_k} \in F_{i_k} \} .$$

Note that $\sum_{i\in I} F_i$ is *not* called the *direct* sum [3.1] of the submodules F_i; that is in general a completely different $\mathcal{A}$-module.

1.23 Nota Bene. If E is an $\mathfrak{A}$-graded $\mathcal{A}$-module, the notation $\mathrm{Span}_{\mathcal{B}}(S)$ is ambiguous because we have to specify whether we see E as a right or as a left $\mathcal{A}$-module. If, in either view, $\mathrm{Span}_{\mathcal{B}}(S)$ is an $\mathfrak{A}$-graded submodule, then it follows from [1.20] that it is an $\mathfrak{A}$-graded submodule of the $\mathfrak{A}$-graded $\mathcal{A}$-module E. However, if the submodule $\mathrm{Span}_{\mathcal{B}}(S)$ is not an $\mathfrak{A}$-graded submodule, then we have to specify whether we view E as a right or as a left $\mathcal{A}$-module [1.26].

1.24 Lemma. *If E is an $\mathfrak{A}$-graded $\mathcal{A}$-module and if $F_i, i \in I$ is a family of $\mathfrak{A}$-graded submodules of E, then $\sum_{i\in I} F_i$ is an $\mathfrak{A}$-graded submodule of E.*

1.25 Lemma. *If E is an $\mathfrak{A}$-graded $\mathcal{A}$-module and if $S \subset E$ consists of homogeneous elements only, then* $\mathrm{Span}(S)$ *is an $\mathfrak{A}$-graded submodule of E.*

Proof. According to [1.14] we have to prove that the homogeneous components of an element $e = \sum_{i=1}^{n} a^i s_i \in \mathrm{Span}(S)$ are itself in $\mathrm{Span}(S)$. Let $a^i = \sum_\alpha a^i_\alpha$ be the decomposition of $a^i \in \mathcal{A}$ into homogeneous components, then by assumption $a^i_\alpha s_i$ is homogeneous. Regrouping the terms in the (finite) sum $e = \sum_{i,\alpha} a^i_\alpha s_i$ according to parity immediately gives the desired result. $\boxed{QED}$

1.26 Counter example. To show that the condition of homogeneity is not superfluous, consider the $\mathfrak{A}$-graded $\mathcal{A}$-module $E = \mathcal{A}^2$ and an element $s = (\mu, \nu)$ where μ and ν are two non-zero homogeneous elements of different parity. In this case the submodule $F = \mathrm{Span}(\{s\}) = \{\,(a\mu, a\nu) \mid a \in \mathcal{A}\,\}$ of E seen as left $\mathcal{A}$-module is not an $\mathfrak{A}$-graded submodule because $F \cap \mathcal{A}^2_\alpha = \{0\}$. Note also that, had we interpreted E as a right $\mathcal{A}$-module, the submodule F would (for generic $\mathcal{A}$) have been different. This shows that for general generating subsets S we have indeed to specify whether we use the left or the right $\mathcal{A}$-module approach.

2. (Multi-) linear maps

After the introduction of $\mathfrak{A}$-graded $\mathcal{A}$-modules in §1, we introduce in this section the notion of a k-linear map, of which the more elementary notion of linear map is a special case. We show that the set of all k-linear maps $\mathrm{Map}_S(E_1, \ldots, E_k; F)$ *is not in general an $\mathfrak{A}$-graded $\mathcal{A}$-module, but that there is a natural subset* $\mathrm{Hom}_S(E_1, \ldots, E_k; F)$ *which is an $\mathfrak{A}$-graded $\mathcal{A}$-module. Two other main points of this section are that left k-linear maps* *should be seen as operating on the right rather than on the left, and that, despite their difference, there exists a natural isomorphism $\mathfrak{T}$ between left and right linear morphisms. This section ends with the definition of dual maps.*

2.1 Definition. Given $\mathfrak{A}$-graded $\mathcal{A}$-modules $E_1, \ldots, E_k$, and F, then a k-additive map $\phi : E_1 \times \cdots \times E_k \to F$ is said to be *left k-linear* if $\forall i$, $\forall e_i \in E_i$ and $\forall a \in \mathcal{A}$ we have:

$$(i) \quad \phi(e_1, \ldots, e_{i-1}, e_i a, e_{i+1}, \ldots, e_k) = \phi(e_1, \ldots, e_i, a e_{i+1}, e_{i+2}, \ldots, e_k)$$

$$(ii) \quad \phi(a e_1, e_2, \ldots, e_k) = a\phi(e_1, \ldots, e_k) \ .$$

The map ϕ is called *right k-linear* if condition (ii) is replaced by condition $(ii)'$:

$$(ii)' \quad \phi(e_1, \ldots, e_{k-1}, e_k a) = \phi(e_1, \ldots, e_{k-1}, e_k)a \ .$$

According to common usage, a 1-linear map will just be called *linear*, *bilinear* stands for 2-linear and *trilinear* for 3-linear. We denote the set of all left k-linear maps by $\mathrm{Map}_L(E_1, \ldots, E_k; F)$ and the set of all right k-linear maps by $\mathrm{Map}_R(E_1, \ldots, E_k; F)$. Since we will often discuss properties of left and right k-linear maps at the same time, we will use the notation Map_S to denote, during the whole discussion concerned, either Map_L or Map_R. This should be interpreted as that the discussion is valid for both Map_L and Map_R.

A map $\phi \in \mathrm{Map}_S(E_1, \ldots, E_k; F)$ is called *of parity* $\alpha \in \mathfrak{A}$ if

$$\forall \beta_i \in \mathfrak{A} : \phi((E_1)_{\beta_1}, \ldots, (E_k)_{\beta_k}) \subset F_{\alpha+\beta_1+\cdots+\beta_k} \ .$$

Note that a k-linear map ϕ is of parity α if and only if $\phi(e_1, \ldots, e_k)$ is homogeneous when all entries are homogeneous and such that $\varepsilon(\phi(e_1, \ldots, e_k)) = \alpha + \sum_{i=1}^k \varepsilon(e_i)$. By $\mathrm{Map}_S(E_1, \ldots, E_k; F)_\alpha \subset \mathrm{Map}_S(E_1, \ldots, E_k; F)$ we denote the subset of all maps of fixed parity α.

In the set $\mathrm{Map}_S(E_1, \ldots, E_k; F)$ we define an addition by

$$(\phi + \psi)(e_1, \ldots, e_k) = \phi(e_1, \ldots, e_k) + \psi(e_1, \ldots, e_k) \ .$$

We also define a multiplication by elements in $\mathcal{A}$; for *left* k-linear maps the *right* multiplication is defined by

$$(\phi a)(e_1, \ldots, e_k) = (\phi(e_1, \ldots, e_k))a \ ,$$

and for *right* k-linear maps the *left* multiplication is defined by

$$(a\phi)(e_1, \ldots, e_k) = a(\phi(e_1, \ldots, e_k)) \ .$$

2.2 Lemma. *With the above definitions, the set* $\mathrm{Map}_L(E_1, \ldots, E_k; F)$ *becomes a right $\mathcal{A}$-module and* $\mathrm{Map}_R(E_1, \ldots, E_k; F)$ *becomes a left $\mathcal{A}$-module.*

Proof. What we have to show is in the first place that addition and multiplication are well defined, i.e., that the result is again left/right k-linear. When that has been done, the axioms of a left/right $\mathcal{A}$-module have to be verified. All this is left to the reader. $\boxed{QED}$

2.3 Lemma. *For $\alpha, \beta \in \mathfrak{A}$ and $\phi \in \mathrm{Map}_S(E_1, \ldots, E_k; F)_\alpha$, $\psi \in \mathrm{Map}_S(F; H)_\beta$ we have $\psi \circ \phi \in \mathrm{Map}_S(E_1, \ldots, E_k; H)_{\alpha+\beta}$.*

2.4 Definition. A k-linear map $\phi \in \mathrm{Map}_S(E_1, \ldots, E_k; F)$ is called a *(homo)morphism* if it is a *finite* sum of *homogeneous* k-linear maps (see [2.6] for an example in which ϕ is an infinite sum). More precisely, we define the set $\mathrm{Hom}_S(E_1, \ldots, E_k; F)$ of all k-linear homomorphisms by:

$$\mathrm{Hom}_S(E_1, \ldots, E_k; F) = \sum_{\alpha \in \mathfrak{A}} \mathrm{Map}_S(E_1, \ldots, E_k; F)_\alpha \subset \mathrm{Map}_S(E_1, \ldots, E_k; F) .$$

In the context of morphisms, we will also use the name $\mathrm{Hom}_S(E_1, \ldots, E_k; F)_\alpha$ for the set $\mathrm{Map}_S(E_1, \ldots, E_k; F)_\alpha$ (two names for the same object!). Note that for (homo)morphisms we use the same notation as for k-linear maps: Hom_S denotes either Hom_L or Hom_R, but never both at the same time in a discussion.

In case all $\mathfrak{A}$-graded $\mathcal{A}$-modules E_i coincide, say with the $\mathfrak{A}$-graded $\mathcal{A}$-module E, we denote $\mathrm{Hom}_S(E_1, \ldots, E_k; F)$ by $\mathrm{Hom}_S(E^k; F)$. Two special cases have an alternative notation. $\mathrm{Hom}_S(E; E)$ is denoted as $\mathrm{End}_S(E)$; its elements are called *endomorphisms of E*. $\mathrm{Hom}_L(E; \mathcal{A})$ is denoted as $^*E$ and is called the *left dual of E*; $\mathrm{Hom}_R(E; \mathcal{A})$ is denoted as E^* and is called the *right dual of E*.

2.5 Lemma. *The left/right $\mathcal{A}$-module $\mathrm{Hom}_S(E_1, \ldots, E_k; F)$ together with its subsets $\mathrm{Hom}_S(E_1, \ldots, E_k; F)_\alpha$ is an $\mathfrak{A}$-graded $\mathcal{A}$-module. If the abelian group $\mathfrak{A}$ is finite, then $\mathrm{Hom}_S(E_1, \ldots, E_k; F) = \mathrm{Map}_S(E_1, \ldots, E_k; F)$.*

Proof. We give the proof for the right linear case; the left linear case is analogous. From the definition of parity it is obvious that the subsets $\mathrm{Map}_R(E_1, \ldots, E_k; F)_\alpha$ are additive subgroups that satisfy the condition

$$\mathcal{A}_\alpha \cdot \mathrm{Map}_R(E_1, \ldots, E_k; F)_\beta \subset \mathrm{Map}_R(E_1, \ldots, E_k; F)_{\alpha+\beta} .$$

It thus remains to show that each element $\phi \in \mathrm{Hom}_R(E_1, \ldots, E_k; F)$ admits a unique decomposition in homogeneous parts. That it admits a decomposition in homogeneous parts is immediate from its definition, so remains the uniqueness. Therefore suppose $\phi = \sum_{\alpha \in \mathfrak{A}} \phi_\alpha = 0$, where the maps ϕ_α are of parity α. Apply this ϕ to k homogeneous vectors $e_i \in E_i$ to obtain $\sum_{\alpha \in \mathfrak{A}} \phi_\alpha(e_1, \ldots, e_k) = 0$. But the parities $\varepsilon(\phi_\alpha(e_1, \ldots, e_k)) = \alpha + \sum_{i=1}^k \varepsilon(e_i)$ are all different. Hence for each α separately $\phi_\alpha(e_1, \ldots, e_k) = 0$ (because F is an $\mathfrak{A}$-graded $\mathcal{A}$-module). Since each ϕ_α is k-additive, $\phi_\alpha(e_1, \ldots, e_k) = 0$ for all possible choices of the e_i, not necessarily homogeneous. It follows that $\phi_\alpha = 0$, i.e., any decomposition into homogeneous components is unique.

To show the second part of the lemma, choose $\phi \in \mathrm{Map}_R(E_1, \ldots, E_k; F)$ and define the maps ϕ_α by:

$$\phi_\alpha(e_1, \ldots, e_k) = \sum_{\beta_1, \ldots, \beta_k \in \mathfrak{A}} \Big(\phi((e_1)_{\beta_1}, \ldots, (e_k)_{\beta_k}) \Big)_{\alpha+\beta_1+\cdots+\beta_k} .$$

Since ϕ is k-additive, it follows immediately that $\phi = \sum_{\alpha\in\mathfrak{A}} \phi_\alpha$, which is a finite sum by hypothesis; comparison with the condition for parity α also shows immediately that $\varepsilon(\phi_\alpha) = \alpha$. It thus remains to show that ϕ_α is k-linear. Since taking homogeneous parts is additive, the k-additivity will be obvious. To verify conditions (i) and $(ii)'$, we note that these are additive in $a \in \mathcal{A}$. We thus may assume that a is homogeneous. But for homogeneous a, these relations are obvious after a reparametrization of the appropriate dummy summation variables β_i. QED

2.6 Counter example. The finiteness condition on $\mathfrak{A}$ is not superfluous because a linear map could be an *infinite* sum of non-zero homogeneous components, in which case it is not a morphism in our sense. The following example shows that this can indeed happen. Consider the $\mathbf{Z}$-graded commutative algebra $\mathcal{A}$ defined by $\mathcal{A}_0 = \mathbf{R}$ and $\mathcal{A}_n = \{0\}$ for $n \neq 0$ with the trivial bilinear map $\langle m|n\rangle = 0$. In this case $\mathbf{Z}$-graded $\mathcal{A}$-modules are nothing more than real vector spaces. Let us consider next the full exterior algebra $E = \bigwedge X \equiv \bigoplus_{n=0}^{\infty} \bigwedge^n X$ where X is an infinite dimensional real vector space; it is a $\mathbf{Z}$-graded $\mathcal{A}$-module with grading $E_n = \{0\}$ for $n < 0$ and $E_n = \bigwedge^n X$ for $n \geq 0$. In X we choose an infinite set of independent elements $\{ x_i \in X \mid i \in \mathbf{N} \}$ and we define the linear map $\phi : E \to E$ by its restrictions to E_n: $e \in E_n = \bigwedge^n X \Rightarrow \phi(e) = e \wedge x_1 \wedge \cdots \wedge x_n$. Restricted to each E_n the map ϕ augments the parity by n, but since none of these restrictions is the zero map, there are infinitely many non-zero maps of different parities involved in the definition of ϕ. Hence ϕ is not a *finite* sum of homogeneous maps, i.e., $\phi \notin \mathrm{Hom}_S(E; E)$.

2.7 Lemma. *For $\phi \in \mathrm{Hom}_L(E_1, \ldots, E_k; F)$, $e_i \in E_i$ and $a \in \mathcal{A}$ we have the relation $(a\phi)(e_1, \ldots, e_k) = \phi(e_1, \ldots, e_{k-1}, e_k a)$. If ϕ is right k-linear instead of left k-linear, we have the relation $(\phi a)(e_1, \ldots, e_k) = \phi(ae_1, e_2, \ldots, e_k)$.*

Proof. We prove the left linear case, the right linear case being similar. The relation is additive in *all* its entries, i.e., in $e_1, \ldots, e_k$ and a but also in ϕ. We thus may assume that all entries are homogeneous. It follows that $\varepsilon(\phi(e_1, e_2, \ldots, e_k)) = \varepsilon(\phi) + \varepsilon(e_1) + \cdots + \varepsilon(e_k)$. We then compute:

$$\begin{aligned}(a\phi)(e_1, \ldots, e_k) &= (-1)^{\langle\varepsilon(a)|\varepsilon(\phi)\rangle}(\phi a)(e_1, \ldots, e_k) = (-1)^{\langle\varepsilon(a)|\varepsilon(\phi)\rangle}(\phi(e_1, \ldots, e_k))a \\ &= (-1)^{\langle\varepsilon(a)|\varepsilon(\phi)\rangle}(-1)^{\langle\varepsilon(a)|\varepsilon(\phi)+\varepsilon(e_1)+\cdots+\varepsilon(e_k)\rangle}a(\phi(e_1, \ldots, e_k)) \\ &= (-1)^{\langle\varepsilon(a)|\varepsilon(e_1)+\cdots+\varepsilon(e_k)\rangle}\phi(ae_1, e_2, \ldots, e_k) \\ &= \phi(e_1, \ldots, e_{k-1}, e_k a) \ .\end{aligned}$$

QED

2.8 Corollary. *For $\phi : E_1 \times \cdots \times E_k \to F$ we have: $\phi \in \mathrm{Hom}_L(E_1, \ldots, E_k; F)_0$ if and only if $\phi \in \mathrm{Hom}_R(E_1, \ldots, E_k; F)_0$. We are thus allowed to drop the subscript and to write $\mathrm{Hom}(E_1, \ldots, E_k; F)_0$ for both sets.*

2.9 Definitions. Given two $\mathfrak{A}$-graded $\mathcal{A}$-modules E and F, a map $\phi \in \mathrm{Hom}_S(E;F)$ is said to be *invertible* if there exists a map $\psi \in \mathrm{Hom}_S(F;E)$ such that $\phi \circ \psi = id(F)$ and $\psi \circ \phi = id(E)$. An even invertible map $\phi \in \mathrm{Hom}(E;F)_0$ is called an *isomorphism between E and F*. If there exists an isomorphism between E and F, the two $\mathfrak{A}$-graded $\mathcal{A}$-modules E and F are called *isomorphic*, denoted by $E \cong F$. An even invertible endomorphism of an $\mathfrak{A}$-graded $\mathcal{A}$-module E is called an *automorphism of* E; the set of all automorphisms of E is denoted by $\mathrm{Aut}(E)$.

We also introduce the notion of *identification* as being synonym to *even linear map.* However, use of the word identification will usually mean that the even linear map concerned will not be noted in the sequel. For instance, if $\phi : E \to F$ is an injective even linear map, we may identify E with its image and write $E \subset F$, forgetting about the actual map ϕ.

2.10 Remarks. • If a morphism $\phi \in \mathrm{Hom}_S(E;F)$ is bijective, it is elementary to show that its inverse is also a morphism, proving that invertible morphisms are the same as bijective morphisms.

• If we equip the $\mathfrak{A}$-graded $\mathcal{A}$-module $\mathrm{End}_S(E)$ with composition of maps as multiplication, it becomes an $\mathfrak{A}$-graded ring; its subset $\mathrm{Aut}(E)$ becomes a group. This follows immediately from [2.3].

2.11 Discussion. For *right* k-linear maps [2.7] leads to the following series of relations:

$$\begin{aligned}
a(\phi(e_1,\ldots,e_k)) &= (a\phi)(e_1,\ldots,e_k)\\
(\phi a)(e_1,\ldots,e_k) &= \phi(ae_1,e_2,\ldots,e_k)\\
\phi(e_1a,e_2,\ldots,e_k) &= \phi(e_1,ae_2,e_3,\ldots,e_k)\\
&\ \vdots\\
\phi(e_1,\ldots,e_{k-2},e_{k-1}a,e_k) &= \phi(e_1,\ldots,e_{k-1},ae_k)\\
\phi(e_1,\ldots,e_{k-1},e_ka) &= (\phi(e_1,\ldots,e_k))a\ .
\end{aligned}$$

These relations can be summarized by saying that for multiplication by $a \in \mathcal{A}$, the position of commas and parentheses is of no importance. The technical way to say the same is the statement that the evaluation map $\mathrm{Hom}_R(E_1,\ldots,E_k;F) \times E_1 \times \cdots \times E_k \to F$ is $(k+1)$-linear and even. Moreover, one easily verifies that the composition operator $\circ : \mathrm{Hom}_R(F;G) \times \mathrm{Hom}_R(E;F) \to \mathrm{Hom}_R(E;G)$, $(\phi,\psi) \mapsto \phi\circ\psi$ is an even bilinear map. For *left* k-linear maps the situation might seem to be not so nice. However, a small change in notation yields a similar result.

2.12 Notation. For $\phi \in \mathrm{Hom}_L(E_1,\ldots,E_k;F)$ and $e_i \in E_i$ we introduce an alternative notation for the evaluation of ϕ in the vectors e_i by

$$\iota(e_1,\ldots,e_k)\phi \equiv \phi(e_1,\ldots,e_k)\ .$$

In this notation we interpret ϕ as an operator on the right rather than on the left. It should be read as *substitution of the vectors* $e_1, \ldots, e_k$ *in the map* ϕ, or as *contraction of the vectors* $e_1, \ldots, e_k$ *with the map* ϕ, a notation well known in differential geometry. The choice of the name *contraction* is explained in [II.5.2] (which refers to [4.10]). In case $k = 1$, we will also use the notation

$$\langle\!\langle e \,\|\, \phi \rangle\!\rangle \equiv \iota(e)\phi \equiv \phi(e) \ .$$

Parallel to the contraction/evaluation operator ι we introduce an alternative notation for the composition of two left linear homomorphisms by

$$\phi \diamond \psi \equiv \psi \circ \phi \ .$$

Aficionados of categories might say that $\diamond$ is the composition operator in the opposite category; we will use it just as a different notation for the composition.

2.13 Discussion. With the alternative notation as introduced above, we obtain for *left* k-linear maps the following series of relations:

$$\begin{aligned} a(\iota(e_1, \ldots, e_k)\phi) &= \iota(ae_1, e_2, \ldots, e_k)\phi \\ \iota(e_1 a, e_2, \ldots, e_k)\phi &= \iota(e_1, ae_2, e_3, \ldots, e_k)\phi \\ &\ \ \vdots \\ \iota(e_1, \ldots, e_{k-2}, e_{k-1}a, e_k)\phi &= \iota(e_1, \ldots, e_{k-1}, ae_k)\phi \\ \iota(e_1, \ldots, e_{k-1}, e_k a)\phi &= \iota(e_1, \ldots, e_k)(a\phi) \\ \iota(e_1, \ldots, e_k)(\phi a) &= (\iota(e_1, \ldots, e_k)\phi)a \ . \end{aligned}$$

Thus, here again we see that for multiplication by $a \in \mathcal{A}$, we can ignore the position of commas and parentheses (and of course the symbol ι). As for right k-linear maps we can state this by saying that the evaluation map $\iota : E_1 \times \cdots \times E_k \times \mathrm{Hom}_L(E_1, \ldots, E_k; F)$ is an even $(k+1)$-linear map. In the same vein, the alternative composition operator $\diamond$ leads to the statement that $\diamond : \mathrm{Hom}_L(E; F) \times \mathrm{Hom}_L(F; G) \to \mathrm{Hom}_L(E; G)$ is an even bilinear map. Using the contraction operator ι and the composition operator $\diamond$ at the same time allows us to write for left linear maps:

(2.14) $$\iota(e)\phi \diamond \psi = \iota(\iota(e)\phi)\psi \qquad \text{or} \qquad \langle\!\langle e \,\|\, \phi \diamond \psi \rangle\!\rangle = \langle\!\langle \langle\!\langle e \,\|\, \phi \rangle\!\rangle \,\|\, \psi \rangle\!\rangle \ .$$

As for right linear maps, evaluation of a composite of left linear maps becomes a mere question of parentheses (and the formal evaluation operator ι, if used).

The definite advantage of our alternative notation for left linear maps is that we can adhere systematically to the guiding principle [1.21] that interchanging two objects gives an additional sign $(-1)^{\langle \alpha | \beta \rangle}$. With the alternative notation we avoid notational interchangings that do not involve additional signs (such as $(\phi a)(e) = \phi(e)a$ and $(\phi a) \circ \psi = (\phi \circ \psi)a$ for left linear maps).

2.15 Discussion/Definition. So far we have insisted on the difference between left linear and right linear morphisms, and we will continue to do so. Nevertheless there exists a natural isomorphism $\mathfrak{T} : \mathrm{Hom}_L(E; F) \to \mathrm{Hom}_R(E; F)$ between these two sets of maps. It is defined by the equation

$$(\mathfrak{T}\phi)(e) = \sum_{\alpha,\beta\in\mathfrak{A}} (-1)^{\langle\alpha|\beta\rangle} \iota(e_\alpha)\phi_\beta \equiv \sum_{\alpha,\beta\in\mathfrak{A}} (-1)^{\langle\alpha|\beta\rangle} \langle\!\langle e_\alpha \| \phi_\beta \rangle\!\rangle \,, \tag{2.16}$$

where $\phi = \sum_{\beta\in\mathfrak{A}} \phi_\beta$ and $e = \sum_{\alpha\in\mathfrak{A}} e_\alpha$ denote the decomposition of ϕ and e into their homogeneous parts. One advantage of interpreting left linear maps as acting on the right now becomes obvious: going from the left hand side to the right hand side of (2.16), we have to interchange ϕ and e, which "explains" the sign $(-1)^{\langle\alpha|\beta\rangle}$.

Several verifications have to be made, the first of which is to show that $\mathfrak{T}\phi$ is indeed right linear. Since (2.16) is obviously additive in e, it is sufficient to show the relation $(\mathfrak{T}\phi)(ea) = (\mathfrak{T}\phi)(e)a$ for homogeneous e and a. If $\varepsilon(a) = \alpha$ and $\varepsilon(e) = \gamma$, and thus $\varepsilon(ea) = \alpha + \gamma$, we compute:

$$(\mathfrak{T}\phi)(ea) = \sum_{\beta\in\mathfrak{A}} (-1)^{\langle\alpha+\gamma|\beta\rangle} \langle\!\langle ea \| \phi_\beta \rangle\!\rangle = \Big(\sum_{\beta\in\mathfrak{A}} (-1)^{\langle\gamma|\beta\rangle} \langle\!\langle e \| \phi_\beta \rangle\!\rangle\Big) a = (\mathfrak{T}\phi)(e)a \,,$$

where we have used that in the $\mathfrak{A}$-graded $\mathcal{A}$-module $\mathrm{Hom}_L(E; F)$ the right and left multiplication are related to each other by (1.16). In a similar way one proves that $\mathfrak{T}$ itself is left linear. Since an explicit expression for its inverse is given by

$$\langle\!\langle e \| \mathfrak{T}^{-1}\psi \rangle\!\rangle = \sum_{\alpha,\beta\in\mathfrak{A}} (-1)^{\langle\alpha|\beta\rangle} \psi_\beta(e_\alpha) \,,$$

and since $\mathfrak{T}$ obviously preserves parity, we conclude that it indeed is an isomorphism.

2.17 Lemma. *If ϕ and ψ are homogeneous, $\mathfrak{T}(\phi \diamond \psi) = (-1)^{\langle\varepsilon(\phi)|\varepsilon(\psi)\rangle} (\mathfrak{T}\psi) \circ (\mathfrak{T}\phi)$.*

2.18 Definition. We will call the operator $\mathfrak{T}$ *the transposition operator*, and we will call $\mathfrak{T}\phi$ *the transpose of* ϕ. These names will be justified in §II.4. However, the reader is warned that $\mathfrak{T}$ is not its own inverse: $\mathfrak{T}$ is defined on left linear maps and its inverse on right linear maps!

2.19 Discussion. If E is an $\mathfrak{A}$-graded $\mathcal{A}$-module, there exist canonical isomorphisms Φ_S : $\mathrm{Hom}_S(\mathcal{A}; E) \to E$ given by $\phi \mapsto \phi(1)$. The inverse of Φ_R is given by $\Phi_R^{-1}(e) : a \mapsto ea$. We thus find for homogeneous ϕ and a the relation

$$(\Phi_R^{-1} \circ \Phi_L)(\phi) : a \mapsto \Phi_L(\phi) \cdot a = \langle\!\langle 1 \| \phi \rangle\!\rangle \cdot a = (-1)^{\langle\phi|a\rangle} \langle\!\langle a \| \phi \rangle\!\rangle = (\mathfrak{T}\phi)(a) \,.$$

We conclude that $\Phi_R^{-1} \circ \Phi_L = \mathfrak{T} : \mathrm{Hom}_L(\mathcal{A}; E) \to \mathrm{Hom}_R(\mathcal{A}; E)$.

In the special case $E = \mathcal{A}$, the maps Φ_S provide canonical isomorphisms $\Phi_R : \mathcal{A}^* \to \mathcal{A}$ and Φ_L : ${}^*\mathcal{A} \to \mathcal{A}$. More precisely $\Phi_R^{-1}(a) : b \mapsto ab$ and $\langle\!\langle b \| \Phi_L^{-1}(a) \rangle\!\rangle = ba$, i.e., $\Phi_R^{-1}(a) = m(a, \_)$ and $\Phi_L^{-1}(a) = m(\_, a)$, where m denotes the multiplication $m : \mathcal{A} \times \mathcal{A} \to \mathcal{A}$.

2.20 Definition. For any $\phi \in \mathrm{Hom}_R(E; F)$ we define $\phi^* \in \mathrm{Hom}_L(F^*; E^*)$, called the *right dual map of* ϕ, by the formula

$$\forall \psi \in F^* : \langle\!\langle \psi \| \phi^* \rangle\!\rangle = \psi \circ \phi \,.$$

Similarly we define for any $\phi \in \mathrm{Hom}_L(E; F)$ the *left dual map* ${}^*\phi \in \mathrm{Hom}_R({}^*F; {}^*E)$ by the formula

$$\forall \psi \in {}^*F : ({}^*\phi)(\psi) = \phi \diamond \psi \,.$$

One should note that *taking the dual map* switches sides: the dual of a left linear map is right linear and vice versa.

2.21 Proposition.

(i) *If* ϕ *is surjective, its dual map is injective.*
(ii) $(\phi \circ \psi)^* = \phi^* \diamond \psi^*$ *and* ${}^*(\phi \diamond \psi) = {}^*\phi \circ {}^*\psi$.
(iii) *The map* $\mathrm{Hom}_R(E; F) \to \mathrm{Hom}_L(F^*; E^*)$, $\phi \mapsto \phi^*$ *is even and linear, as is the map* $\mathrm{Hom}_L(E; F) \to \mathrm{Hom}_R({}^*F; {}^*E)$, $\phi \mapsto {}^*\phi$.
(iv) $\forall \phi \in \mathrm{Hom}_L(E; F) : \mathfrak{T}((\mathfrak{T}\phi)^*) = \mathfrak{T} \circ {}^*\phi \circ \mathfrak{T}^{-1}$.

Proof. Properties (i), (ii), and (iii) are elementary. For (iv) one has to realize which transposition operators are involved. In $\mathfrak{T} \circ {}^*\phi \circ \mathfrak{T}^{-1}$ on the right hand side they represent the sequence $F^* \to {}^*F \to {}^*E \to E^*$. In $\mathfrak{T}((\mathfrak{T}\phi)^*)$ on the left hand side the first one represents the switch $\mathrm{Hom}_L(F^*; E^*) \to \mathrm{Hom}_R(F^*; E^*)$ and the second one the switch $\mathrm{Hom}_L(E; F) \to \mathrm{Hom}_R(E; F)$. Once one has this, the proof is elementary. QED

2.22 Remark. In [2.21-ii] we see again the advantage of the notation $\diamond$ for composition of left linear maps: we do not have to change the order of ϕ and ψ in these formulæ.

3. Direct sums, free 𝔄-graded $\mathcal{A}$-modules, and quotients

In the previous sections we have seen the construction of the 𝔄-graded $\mathcal{A}$-modules submodule and morphisms; in this section we provide three new constructions of 𝔄-graded $\mathcal{A}$-modules. In the first place the free 𝔄-graded $\mathcal{A}$-module $F(G, \varepsilon)$ on a set G of homogeneous generators whose parity is given by ε. The next construction is the direct sum of a family of 𝔄-graded $\mathcal{A}$-modules. The third construction is that of the quotient of an 𝔄-graded $\mathcal{A}$-module by an 𝔄-graded submodule.

3.1 Construction (direct sums). If $E_i, i \in I$ is a collection of 𝔄-graded $\mathcal{A}$-modules, we define their *direct sum* $E = \bigoplus_{i \in I} E_i$ as the subset of the direct product $\prod_{i \in I} E_i$

consisting of those vectors $(e_i)_{i\in I}$ with $e_i = 0$ except for finitely many indices $i \in I$ (recall that the direct sum of real vector spaces is defined exactly in this way). By defining a componentwise addition and (left) multiplication by elements of $\mathcal{A}$, E becomes a left $\mathcal{A}$-module. Finally we define the subsets E_α, $\alpha \in \mathfrak{A}$ by $E_\alpha = E \cap \prod_{i\in I}(E_i)_\alpha$. We leave it to the reader to verify that with these definitions E becomes an $\mathfrak{A}$-graded $\mathcal{A}$-module.

For each $i \in I$ we define maps $\pi_i : E \to E_i$ and $s_i : E_i \to E$ by $\pi_i((e_j)_{j\in I}) = e_i$ and $(s_i(e))_i = e$, $(s_i(e))_j = 0$ for $j \neq i$. It follows immediately that the π_i are surjective even linear maps and that the s_i are injective even linear maps, related by $\pi_i \circ s_i = id(E_i)$.

We will usually denote a general element $(e_i)_{i\in I} \in E$ by $\oplus_{i\in I} e_i$ instead of by $(e_i)_{i\in I}$, just to stress that it is not an arbitrary element of the direct product, but one with only finitely many non-zero entries. In case the index set I has a finite number n of elements, we will write $E_1 \oplus \cdots \oplus E_n$ for $E = \bigoplus_{i=1}^n E_i$, and an arbitrary element will be denoted by $e_1 \oplus \cdots \oplus e_n$. If the $\mathcal{A}$-vector spaces E_i are all equal to a given one, $\forall i : E_i = F$, the direct sum $\bigoplus_{i=1}^n E_i$ is also denoted as F^k. It is indeed the k-th power of F because for a finite index set I the direct sum equals the direct product. And if we define formally $F^0 = \{0\}$, then the equality $F^k \oplus F^\ell = F^{k+\ell}$ holds for all $k, \ell \in \mathbf{N}$.

3.2 Remark. One might ask why we do not define direct products of $\mathfrak{A}$-graded $\mathcal{A}$-modules. There are several reasons. In the first place, if both the index set I and the abelian group $\mathfrak{A}$ are infinite, one can easily find examples in which the direct product is *not* an $\mathfrak{A}$-graded $\mathcal{A}$-module, the failure being that not every element can be written as a *finite* sum of homogeneous elements. In the second place, we never need infinite direct products. And in the third place, a direct product of finitely many $\mathfrak{A}$-graded $\mathcal{A}$-modules is the same as the direct sum of these spaces.

3.3 Definition. If $F_i, i \in I$ is a family of $\mathfrak{A}$-graded submodules of a given $\mathfrak{A}$-graded $\mathcal{A}$-module E, we can consider the map $\bigoplus_i F_i \to E$ defined by $(f_i)_{i\in I} \mapsto \sum_i f_i$. Note that this map is well defined because there are only finitely many f_i's non-zero; its image is $\sum_{i\in I} F_i$. One easily verifies that this map is even and linear by definition of $\mathfrak{A}$-graded submodules. Officially $\bigoplus_i F_i$ is never a submodule of E (but $\sum_{i\in I} F_i$ is); nevertheless, we will write $E = \bigoplus_i F_i$ whenever this map is an isomorphism onto E. As for real vector spaces, this is the case if and only if every element $e \in E$ can be written in a *unique* way as $e = \sum_{i\in I} f_i$ with $f_i \in F_i$ and only finitely many of them non-zero. If I contains two elements, we will write $E = F_1 \oplus F_2$. The $\mathfrak{A}$-graded submodules F_1 and F_2 will be called *supplements* to each other.

3.4 Construction (free $\mathfrak{A}$-graded $\mathcal{A}$-modules). Let $\varepsilon : G \to \mathfrak{A}$ be a map from an abstract set G to $\mathfrak{A}$, and define $G_\alpha \subset G$ by $G_\alpha = \varepsilon^{-1}(\alpha)$. We define the space $F(G, \varepsilon)$ as the set of all maps $f : G \to \mathcal{A}$ with the property that $f(g) = 0$ for all $g \in G$ except finitely many. In $F(G, \varepsilon)$ we define an addition by $(f + f')(g) = f(g) + f'(g)$, and a (left) multiplication by elements of $\mathcal{A}$ by $(af)(g) = af(g)$. In this way $F(G, \varepsilon)$ becomes a left

$\mathcal{A}$-module. One usually identifies each element $g \in G$ with the map $\phi_g : G \to \mathcal{A}$ defined by $\phi_g(g) = 1$ and $\phi_g(h) = 0$ for $h \neq g$. It follows that each $f \in F(G, \varepsilon)$ can be written in a unique way as $f = \sum_{g \in G} f^g \phi_g \equiv \sum_{g \in G} f^g \cdot g$ where f^g is defined as $f^g = f(g)$ and where the sum is actually a finite sum by definition of $F(G, \varepsilon)$.

To make $F(G, \varepsilon)$ into an $\mathfrak{A}$-graded $\mathcal{A}$-module, we define $F(G, \varepsilon)_\alpha, \alpha \in \mathfrak{A}$ by

$$F(G, \varepsilon)_\alpha = \{ f \in F(G, \varepsilon) \mid \forall g \in G : f(g) \in \mathcal{A}_{\alpha - \varepsilon(g)} \} .$$

In words, $f = \sum_{g \in G} f^g \cdot g$ has parity α if and only if the coefficient f^g has parity $\alpha - \varepsilon(g)$. In particular the element (map) ϕ_g has parity $\varepsilon(\phi_g) = \varepsilon(g)$, justifying the use of the symbol ε for the abstract map $\varepsilon : G \to \mathfrak{A}$. Decomposing the coefficients of an arbitrary element $f \in F(G, \varepsilon)$ into homogeneous parts, it follows immediately that $F(G, \varepsilon) = \bigoplus_{\alpha \in \mathfrak{A}} F(G, \varepsilon)_\alpha$; since by construction $\mathcal{A}_\alpha F(G, \varepsilon)_\beta$ is contained in $F(G, \varepsilon)_{\alpha + \beta}$, we conclude that $F(G, \varepsilon)$ is an $\mathfrak{A}$-graded $\mathcal{A}$-module.

The $\mathfrak{A}$-graded $\mathcal{A}$-module $F(G, \varepsilon)$ is usually called *the free $\mathfrak{A}$-graded $\mathcal{A}$-module on (homogeneous) generators G with parity ε*. Using the notion of Span, we can summarize the construction of $F(G, \varepsilon)$ by saying $F(G, \varepsilon) = \mathrm{Span}(G)$. Using the notion of direct sums, we can write $F(G, \varepsilon) = \bigoplus_{g \in G} F(\{g\}, \varepsilon_g)$, where $F(\{g\}, \varepsilon_g)$ is the free $\mathfrak{A}$-graded $\mathcal{A}$-module on the single generator g of parity $\varepsilon_g = \varepsilon(g)$.

3.5 Nota Bene. We have seen that each element $f \in F(G, \varepsilon)$ admits a unique decomposition $f = \sum_{g \in G} f^g \cdot g$ with $f^g \in \mathcal{A}$. Using the induced right action of $\mathcal{A}$, it follows that there also exists a unique representation with the coefficients on the right of the ϕ_g, i.e., $f = \sum_{g \in G} g \cdot f'^g$. In general the coefficients f^g and f'^g are different; only if $\varepsilon(g) = 0$ can we be sure that $f^g = f'^g$.

For any free $\mathfrak{A}$-graded $\mathcal{A}$-module on a single homogeneous generator $F(\{g\}, \varepsilon_g)$ we can define the map $\phi : \mathcal{A} \to F(\{g\}, \varepsilon_g)$ by $a \mapsto ag$. This is a bijective linear map of parity ε_g. It is an isomorphism if and only if $\varepsilon_g = 0$. It follows that for $\varepsilon_g \neq 0$ we cannot identify (in the naive and official sense of the word) the $\mathfrak{A}$-graded $\mathcal{A}$-module $F(\{g\}, \varepsilon_g)$ with $\mathcal{A}$ because left and right multiplication in the $\mathfrak{A}$-graded $\mathcal{A}$-modules $\mathcal{A}$ and $F(\{g\}, \varepsilon_g)$ are *not* related in the same way due to the difference in parity between $1 \in \mathcal{A}$ and $g \in F(\{g\}, \varepsilon_g)$.

3.6 Corollary. *Let $F(G_i, \varepsilon_i)$, $i \in I$ be a family of free $\mathfrak{A}$-graded $\mathcal{A}$-modules on generators G_i ($i \neq j \Rightarrow G_i \cap G_j = \phi$). Then $\bigoplus_{i \in I} F(G_i, \varepsilon_i) \cong F(\cup_{i \in I} G_i, \varepsilon)$, where ε is defined as $\varepsilon|_{G_i} = \varepsilon_i$.*

3.7 Construction (quotients). Let E be an $\mathfrak{A}$-graded $\mathcal{A}$-module and let F be an $\mathfrak{A}$-graded submodule. The quotient $G = E/F$ with canonical projection $\pi : E \to G$ is defined in the sense of abelian groups, i.e., $\pi(e) = \pi(e') \Leftrightarrow e - e' \in F$. As for abelian groups, the element $\pi(e) \in G$ will also be denoted as $e \mod F$. We claim that G can be equipped with the structure of an $\mathfrak{A}$-graded $\mathcal{A}$-module. Addition and (left) multiplication by elements in $\mathcal{A}$ are defined by $\pi(e)+\pi(e') = \pi(e+e')$, $a\pi(e) = \pi(ae)$. The subgroups G_α are defined by $\pi(e) \in G_\alpha \Leftrightarrow \exists f \in F : e - f \in E_\alpha$. It follows immediately that with this grading π is an even morphism.

The only tricky point in proving that G is an $\mathfrak{A}$-graded $\mathcal{A}$-module, is in the proof that the decomposition in homogeneous components is unique. Therefore, let us suppose $\sum_{\alpha\in\mathfrak{A}} \pi(e_\alpha) = 0$ with $\pi(e_\alpha) \in G_\alpha$ (and of course only finitely many of them non-zero, which implies $\sum_{\alpha\in\mathfrak{A}} e_\alpha \in F$). By definition of G_α and the projection π we may assume that $e_\alpha \in E_\alpha$. Since these e_α are homogeneous and F is an $\mathfrak{A}$-graded submodule, we have by [1.14] that e_α belongs to F, i.e., $\pi(e_\alpha) = 0$. This proves that the decomposition into homogeneous components is unique.

3.8 Lemma. *Let E and H be $\mathfrak{A}$-graded $\mathcal{A}$-modules, F an $\mathfrak{A}$-graded submodule of E and $\phi : E \to H$ a linear map that vanishes on F, i.e., $\phi(F) = \{0\}$. Then there exists a unique induced map $\Phi : E/F \to H$ such that $\phi = \Phi \circ \pi$. If ϕ has parity α, then so has Φ; if ϕ is a homomorphism, i.e., $\phi \in \mathrm{Hom}_S(E; H)$, then so is Φ.*

3.9 Proposition. *Let E and F be $\mathfrak{A}$-graded $\mathcal{A}$-modules and let $\phi \in \mathrm{Hom}_S(E; F)$ be homogeneous. Then the following assertions hold (see also [II.3.12]).*

(i) *$\ker(\phi)$ and $\mathrm{im}(\phi)$ are $\mathfrak{A}$-graded submodules of E and F respectively.*
(ii) *If ϕ is even, there exists a canonical isomorphism $\mathrm{im}(\phi) \cong E/\ker(\phi)$.*
(iii) *If ϕ is even and if $\ker(\phi)$ admits a supplement, there exists an isomorphism $E \cong \ker(\phi) \oplus \mathrm{im}(\phi)$ which is completely determined by the choice of the supplement.*

Proof. • For any linear map it is immediate that $\ker(\phi)$ and $\mathrm{im}(\phi)$ are submodules, so we only have to check the grading. Therefore, let $f = \phi(e) \in \mathrm{im}(\phi)$, and decompose e into homogeneous components $f = \phi(\sum_\alpha e_\alpha) = \sum_\alpha \phi(e_\alpha)$. Since ϕ is homogeneous, the $\phi(e_\alpha)$ are also homogeneous. It follows that the homogeneous components of $f \in \mathrm{im}(\phi)$ are again in $\mathrm{im}(\phi)$, i.e., $\mathrm{im}(\phi)$ is a graded submodule. For $\ker(\phi)$ the reasoning is the same: if $e \in \ker(\phi)$ then $0 = \sum_\alpha \phi(e_\alpha)$. Since all $\phi(e_\alpha)$ have different parities, they must be zero, i.e., the e_α are in $\ker(\phi)$. The result then follows.

• If ϕ is even, consider the induced (even!) map $\Phi : E/\ker(\phi) \to F$. It is injective by construction and hence is an isomorphism onto $\mathrm{im}(\phi)$.

• If H is an $\mathfrak{A}$-graded submodule of E, supplement to $\ker(\phi)$, it follows that the restriction $\phi|_H : H \to F$ is injective. If ϕ is also even, it is an isomorphism $H \to \mathrm{im}(\phi)$. We thus have $E = \ker(\phi) \oplus H \cong \ker(\phi) \oplus \mathrm{im}(\phi)$. QED

4. Tensor products

In this section we introduce the construction of a tensor product of $\mathfrak{A}$-graded $\mathcal{A}$-modules. We show that forming tensor products is associative in a very nice way and we prove the *principal property of tensor products: transforming k-linear maps into linear maps; in other words, we prove that the tensor product is the solution of a universal problem. We finish with the construction of the permutation operator $\mathfrak{R}$ on multiple tensor products.*

4.1 Construction (tensor product). Let E and F be $\mathfrak{A}$-graded $\mathcal{A}$-modules and consider the set $G = (\cup_{\alpha\in\mathfrak{A}} E_\alpha \setminus \{0\}) \times (\cup_{\beta\in\mathfrak{A}} F_\beta \setminus \{0\}) \subset E \times F$, i.e., G is the product of all non-zero homogeneous elements in E and in F. On G we define a parity map $\varepsilon : G \to \mathfrak{A}$ by $\varepsilon(e, f) = \varepsilon(e) + \varepsilon(f)$, which is well defined because e and f are by assumption homogeneous. We thus can consider the free $\mathfrak{A}$-graded $\mathcal{A}$-module $F(G, \varepsilon)$ and we recall that we have identified the abstract elements $g \in G$ with the elements $\phi_g \in F(G, \varepsilon)$. With this in mind, we define the subset S of $F(G, \varepsilon)$ as the union of two subsets: $S = S_a \cup S_m$ with

$$S_a = \{\, \phi_{(e+e',f)} - \phi_{(e,f)} - \phi_{(e',f)},\quad \phi_{(e,f+f')} - \phi_{(e,f)} - \phi_{(e,f')} \mid \alpha, \beta \in \mathfrak{A}, e, e' \in E_\alpha, f, f' \in F_\beta \,\}$$

$$S_m = \{\, \phi_{(ae,f)} - a\phi_{(e,f)},\quad \phi_{(ea,f)} - \phi_{(e,af)} \mid \alpha, \beta, \gamma \in \mathfrak{A}, a \in \mathcal{A}_\alpha, e \in E_\beta, f \in F_\gamma \,\} .$$

By construction, all the elements of S are homogeneous and hence $\mathrm{Span}(S)$ is an $\mathfrak{A}$-graded submodule [1.25]. With this $\mathfrak{A}$-graded submodule we then define the $\mathfrak{A}$-graded $\mathcal{A}$-module $E \otimes F$ as the quotient $E \otimes F = F(G, \varepsilon)/\,\mathrm{Span}(S)$. This $\mathfrak{A}$-graded $\mathcal{A}$-module is called *the tensor product of E and F*.

The construction of the tensor product is not complete without the definition of the map $\chi : E \times F \to E \otimes F$:

$$\chi(e, f) = \sum_{\alpha,\beta\in\mathfrak{A}} \pi(\phi_{(e_\alpha,f_\beta)}) .$$

4.2 Lemma. *The map $\chi : E \times F \to E \otimes F$ is even and bilinear.*

Proof. If e and f are homogeneous, the sum over α and β in the definition of χ contains only one term. It follows immediately that $\varepsilon(\chi(e, f)) = \varepsilon(\pi(\phi_{(e,f)})) = \varepsilon(\phi_{(e,f)}) = \varepsilon(e) + \varepsilon(f)$, i.e., χ is even. The bi-additivity follows easily from the definition of S. Let us show for instance the additivity in e :

$$\begin{aligned}\chi(e + e', f) &= \sum_{\alpha,\beta\in\mathfrak{A}} \pi(\phi_{((e+e')_\alpha,f_\beta)}) = \sum_{\alpha,\beta\in\mathfrak{A}} \pi(\phi_{(e_\alpha+e'_\alpha,f_\beta)}) \\ &= \sum_{\alpha,\beta\in\mathfrak{A}} \pi(\phi_{(e_\alpha,f_\beta)} + \phi_{(e'_\alpha,f_\beta)}) = \sum_{\alpha,\beta\in\mathfrak{A}} \pi(\phi_{(e_\alpha,f_\beta)}) + \sum_{\alpha,\beta\in\mathfrak{A}} \pi(\phi_{(e'_\alpha,f_\beta)}) \\ &= \chi(e, f) + \chi(e', f) ,\end{aligned}$$

where we used that $\phi_{(e_\alpha+e'_\alpha,f_\beta)} - (\phi_{(e_\alpha,f_\beta)} + \phi_{(e'_\alpha,f_\beta)})$ belongs to $S_a \subset \mathrm{Span}(S) = \ker \pi$.

Once we know that χ is bi-additive, we notice that conditions (i) and (ii) of left bilinear maps are additive in all the coordinates e_i but also in a. It thus suffices to show these relations for homogeneous e_i and a. Let us show for instance condition (ii) for homogeneous a, e and f:

$$\chi(ae, f) = \pi(\phi_{(ae,f)}) = \pi(a\phi_{(e,f)}) = a\pi(\phi_{(e,f)}) = a\chi(e, f) ,$$

where we used that $\phi_{(ae,f)} - a\phi_{(e,f)}$ is in $S_m \subset \mathrm{Span}(S) = \ker \pi$. QED

4.3 Proposition. *Given any $\mathfrak{A}$-graded $\mathcal{A}$-module H and any $\psi \in \mathrm{Map}_S(E, F; H)$, there exists a unique $\Psi \in \mathrm{Map}_S(E \otimes F; H)$ such that $\psi = \Psi \circ \chi$. If ψ has parity α, then so has Ψ.*

Proof. Let us consider the case of right bilinear maps. Suppose $\psi \in \mathrm{Map}_R(E, F; H)$ has been given. Define the map $\overline{\psi} : F(G, \varepsilon) \to H$ by $\overline{\psi}(\sum_{g\in G} \phi_g c^g) = \sum_{g\in G} \psi(g)c^g$, where we have used that $G \subset E \times F$. This map is by construction right linear but, more important, it vanishes on $\mathrm{Span}(S)$ just because ψ is right bilinear. By [3.8] there exists an induced map $\Psi \in \mathrm{Map}_R(E \otimes F; H)$ such that $\overline{\psi} = \Psi \circ \pi$. If we now consider $\Psi \circ \chi \in \mathrm{Map}_R(E, F; H)$ and apply it to homogeneous e and f, we find

$$(\Psi \circ \chi)(e, f) = \Psi(\pi(\phi_{(e,f)})) = \overline{\psi}(\phi_{(e,f)}) = \psi(e, f) .$$

Since both $\Psi \circ \chi$ and ψ are bilinear, it follows that $\Psi \circ \chi = \psi$, showing existence.

To show that Ψ is unique, let Ψ_1 and Ψ_2 be two such maps. We then have two maps $\widetilde{\Psi}_i \in \mathrm{Map}_R(F(G, \varepsilon); H)$ defined by $\widetilde{\Psi}_i = \Psi_i \circ \pi$. We then compute:

$$\widetilde{\Psi}_1(\phi_g) = \Psi_1(\pi(\phi_g)) = (\Psi_1 \circ \chi)(g) = \psi(g) = \widetilde{\Psi}_2(\phi_g) ,$$

or in other words, $\widetilde{\Psi}_1 = \widetilde{\Psi}_2$. It then follows from the uniqueness in [3.8] that $\Psi_1 = \Psi_2$.

Finally, the parity claim follows from the analogous parity claim in [3.8]. For left bilinear maps, one only has to replace the definition of $\overline{\psi}$ by the formula $\langle\!\langle \sum_{g\in G} c^g \phi_g \,\|\, \overline{\psi} \rangle\!\rangle = \sum_{g\in G} c^g \iota(g)\psi$, which is left linear by construction. QED

4.4 Lemma. *Given $\mathfrak{A}$-graded $\mathcal{A}$-modules E and F, there exists up to isomorphism a unique pair $(H \equiv E \otimes F, \chi)$ with the property mentioned in [4.3]*

Proof. Suppose $(\widehat{H}, \widehat{\chi})$ is a second pair with this property. Since $\widehat{\chi}$ is bilinear and since (H, χ) has this property, there exists a $\widehat{\Phi} : H \to \widehat{H}$ with $\widehat{\chi} = \widehat{\Phi} \circ \chi$. In the same way, interchanging $(\widehat{H}, \widehat{\chi})$ and (H, χ), there exists a $\Phi : \widehat{H} \to H$ such that $\chi = \Phi \circ \widehat{\chi}$. Moreover, since χ and $\widehat{\chi}$ are even, it follows that Φ and $\widehat{\Phi}$ also are even.

Now denote $\Psi = \Phi \circ \widehat{\Phi} : H \to H$ to obtain the equality $\chi = \Psi \circ \chi$. If we view the χ on the left of this equation as arbitrary and apply the property with the pair (H, χ), we see that Ψ is the unique linear map given by [4.3]. Since the identity is also a solution, it follows that $\Psi = \Phi \circ \widehat{\Phi} = id(H)$. In exactly the same way one proves that $\widehat{\Phi} \circ \Phi = id(\widehat{H})$. We thus conclude that Φ and $\widehat{\Phi}$ are isomorphisms between H and $\widehat{H}$. This finishes the proof (modulo some small details that are left to the reader). QED

4.5 Discussion. The tensor product symbol $\otimes$ is not only used to denote the $\mathfrak{A}$-graded $\mathcal{A}$-module $E \otimes F$, it is also used to replace the map $\chi : E \times F \to E \otimes F$ in the following way:

$$e \otimes f \equiv \chi(e, f) .$$

There is no symbol, not on the left nor on the right, to tell us that it is either right or left bilinear. But this is justified because the map χ is even, and thus is left and right bilinear at the same time.

Since $F(G, \varepsilon)$ in [4.1] is generated by the elements $\phi_{(e,f)}$ with e and f homogeneous, it follows that the elements $e \otimes f$ generate $E \otimes F$ (but note that due to the quotient, a decomposition as linear combination of this kind of elements is not necessarily unique). It follows immediately that any linear map on $E \otimes F$ is completely determined by its values on elements of the form $e \otimes f$ with e and f homogeneous. Finally note that for homogeneous $e \in E$ and $f \in F$ we have $\varepsilon(e \otimes f) = \varepsilon(e) + \varepsilon(f)$.

4.6 Lemma. *Given three $\mathfrak{A}$-graded $\mathcal{A}$-modules E, F and G, there exists a canonical isomorphism (identification) between $(E \otimes F) \otimes G$ and $E \otimes (F \otimes G)$ mapping $(e \otimes f) \otimes g$ to $e \otimes (f \otimes g)$.*

Proof. Let us first distinguish the various maps χ (or $\otimes$) that intervene in the construction of these spaces: $\chi_{1,2} : E \times F \to E \otimes F$, $\chi_{12,3} : (E \otimes F) \times G \to (E \otimes F) \otimes G$, $\chi_{2,3} : F \times G \to F \otimes G$, and $\chi_{1,23} : E \times (F \otimes G) \to E \otimes (F \otimes G)$. Next consider the trilinear map $\phi \in \mathrm{Map}(E, F, G; E \otimes (F \otimes G))_0$ defined by

$$\phi(e, f, g) = \chi_{1,23}(e, \chi_{2,3}(f, g)) .$$

For a fixed element $g \in G$ we define the map $\phi_g \in \mathrm{Map}_L(E, F; E \otimes (F \otimes G))$ by $\phi_g(e, f) = \phi(e, f, g)$ (Nota Bene. In general ϕ_g is not right bilinear). Thus there exists a unique induced linear map $\overline{\phi}_g \in \mathrm{Map}_L(E \otimes F; E \otimes (F \otimes G))$ such that $\phi_g = \overline{\phi}_g \circ \chi_{1,2}$. Defining $\overline{\phi}(h, g) = \overline{\phi}_g(h)$, one can easily show that $\overline{\phi} \in \mathrm{Map}_L(E \otimes F, G; E \otimes (F \otimes G))$ is even. Thus there exists a unique induced even linear map $\Phi : (E \otimes F) \otimes G \to E \otimes (F \otimes G)$ with the property that $\overline{\phi} = \Phi \circ \chi_{12,3}$, or in other words,

$$\Phi(\chi_{12,3}(\chi_{1,2}(e, f), g)) = \chi_{1,23}(e, \chi_{2,3}(f, g)) \equiv \phi(e, f, g) .$$

In exactly the same way, starting with the trilinear map $\psi \in \mathrm{Map}(E, F, G; (E \otimes F) \otimes G)_0$ defined by $\psi(e, f, g) = \chi_{12,3}(\chi_{1,2}(e, f), g)$, one shows the existence of a unique even linear map $\Psi : E \otimes (F \otimes G) \to (E \otimes F) \otimes G$ with the property

$$\Psi(\chi_{1,23}(e, \chi_{2,3}(f, g))) = \chi_{12,3}(\chi_{1,2}(e, f), g) \equiv \psi(e, f, g) .$$

We now claim that Φ and Ψ are inverse to each other. Therefore we note that the map $\Psi \circ \Phi \in \mathrm{End}((E \otimes F) \otimes G)_0$ is a map that satisfies $(\Psi \circ \Phi)(\psi(e, f, g)) = \psi(e, f, g)$. Exactly as we showed the uniqueness of the maps Φ and Ψ, one can show the uniqueness of a map that is the identity on elements of the form $\psi(e, f, g)$. But since the identity map satisfies this property, the uniqueness proves that $\Psi \circ \Phi = id((E \otimes F) \otimes G)$. In exactly the same way one proves $\Phi \circ \Psi = id(E \otimes (F \otimes G))$. QED

4.7 Discussion. Using the identification given in [4.6] we are thus allowed to say that the operation of taking the tensor product is associative. It follows that we can write $E_1 \otimes \cdots \otimes E_k$ without using parentheses, and that we can speak of elements $e_1 \otimes \cdots \otimes e_k$ in this multiple tensor product. Note that (by an easy induction argument) the elements of the form $e_1 \otimes \cdots \otimes e_k$ with $e_i \in E_i$ homogeneous (and then $\varepsilon(e_1 \otimes \cdots \otimes e_k) = \varepsilon(e_1) + \cdots + \varepsilon(e_k)$) generate the tensor product $E_1 \otimes \cdots \otimes E_k$. It follows that a linear map defined on $E_1 \otimes \cdots \otimes E_k$ is completely known once we know its values on the elements $e_1 \otimes \cdots \otimes e_k$.

4.8 Proposition. *Given $\mathfrak{A}$-graded $\mathcal{A}$-modules $E_1, \ldots, E_k$, there exists, up to isomorphism, a unique $\mathfrak{A}$-graded $\mathcal{A}$-module $H = E_1 \otimes \cdots \otimes E_k$ and an even k-linear map $\chi \in \mathrm{Hom}(E_1, \ldots, E_k; H)_0$, $\chi(e_1, \ldots, e_k) = e_1 \otimes \cdots \otimes e_k$, with the following property. Given any $\mathfrak{A}$-graded $\mathcal{A}$-module F and any $\phi \in \mathrm{Map}_S(E_1, \ldots, E_k; F)$, there exists a unique $\Phi \in \mathrm{Map}_S(H; F)$ such that $\phi = \Phi \circ \chi$. If ϕ has parity α, then so has Φ.*

Proof. The proof of the uniqueness is a word by word copy of the same proof in the case $k = 2$, replacing bilinear by k-linear. To prove that this unique space is $E_1 \otimes \cdots \otimes E_k$, it suffices to show that it has the announced property. The easiest way to do this is by induction on k. The principle of such a proof has been used in the proof of [4.6], where essentially the case $k = 3$ has been shown. QED

4.9 Corollary. *Given $\mathfrak{A}$-graded $\mathcal{A}$-modules $E_1, \ldots, E_k$, and F, the map $\phi \mapsto \phi \circ \chi$ from $\mathrm{Map}_S(E_1 \otimes \cdots \otimes E_k; F)$ to $\mathrm{Map}_S(E_1, \ldots, E_k; F)$ is a bijection. Restricted to $\mathrm{Hom}_S(E_1 \otimes \cdots \otimes E_k; F)$ it is an isomorphism onto $\mathrm{Hom}_S(E_1, \ldots, E_k; F)$.*

4.10 Examples. • The map $\iota : E_1 \times \cdots \times E_k \times \mathrm{Hom}_L(E_1, \ldots, E_k; F) \to F$ given by $(e_1, \ldots, e_k, \phi) \mapsto \iota(e_1, \ldots, e_k)\phi$ is $(k+1)$-linear and even and thus induces an even linear map $E_1 \otimes \cdots \otimes E_k \otimes \mathrm{Hom}_L(E_1, \ldots, E_k; F) \to F$.

• The evaluation map $\mathrm{Hom}_R(E; F) \times E \to F$ given by $(\phi, e) \mapsto \phi(e)$ is even and bilinear and thus induces an even linear map $\mathrm{Hom}_R(E; F) \otimes E \to F$ which maps $\phi \otimes e$ to $\phi(e)$.

• The composition map $\diamond : \mathrm{Hom}_L(E; F) \times \mathrm{Hom}_L(F; G) \to \mathrm{Hom}_L(E; G)$ given by $(\phi, \psi) \mapsto \phi \diamond \psi \equiv \psi \circ \phi$ is even and bilinear. We thus have an induced even liner map $\mathrm{Hom}_L(E; F) \otimes \mathrm{Hom}_L(F; G) \to \mathrm{Hom}_L(E; G)$.

• Left multiplication by elements of $\mathcal{A}$ in an $\mathfrak{A}$-graded $\mathcal{A}$-module E, $m_L : \mathcal{A} \times E \to E$ is bi-additive and, by definition of multiplication, left bilinear. It thus induces a linear map $\underline{m_L} : \mathcal{A} \otimes E \to E$. Since $\underline{m_L}(1 \otimes e) = e$, the map $\underline{m_L}$ is surjective. Now any $f \in \mathcal{A} \otimes E$ can be written (in a non unique way) as $f = \sum_i \alpha_i \otimes e_i$ for homogeneous $\alpha_i \in \mathcal{A}$ and $e_i \in E$. But then we have, using the bilinearity of $\otimes$, $f = \sum_i 1 \cdot \alpha_i \otimes e_i = 1 \otimes \sum_i \alpha_i e_i$. It follows that $\underline{m_L}$ is also injective, i.e., $\underline{m_L} : \mathcal{A} \otimes E \to E$ is an isomorphism. In the same way, right multiplication $m_R : E \times \mathcal{A} \to E$ induces an isomorphism $\underline{m_R} : E \otimes \mathcal{A} \to E$.

In the sequel we will always identify $\mathcal{A} \otimes E$ and $E \otimes \mathcal{A}$ with E by the isomorphisms $\underline{m}_L$ and $\underline{m}_R$.

4.11 Notation/Definition. If $E_1, \ldots, E_k$ are $\mathfrak{A}$-graded $\mathcal{A}$-modules, we will usually denote their tensor product $E_1 \otimes \cdots \otimes E_k$ as $\bigotimes_{i=1}^k E_i$. Implicit in this notation is the order: $\bigotimes_{i=k}^1 E_i$ will denote the tensor product $E_k \otimes \cdots \otimes E_1$. If all E_i coincide with E, the k-fold tensor product is denoted as $\bigotimes^k E$, for which the above mentioned order problem does not exist. Obviously, if $k = 1$, we do not take a tensor product and $\bigotimes^1 E = E$. We also formally define $\bigotimes^0 E = \mathcal{A}$. With these definitions the equality $(\bigotimes^k E) \otimes (\bigotimes^\ell E) = \bigotimes^{k+\ell} E$ holds for all $k, \ell \in \mathbf{N}$ (for k or $\ell = 0$, use the isomorphism $\underline{m}_S$ [4.10], see also [5.8]).

4.12 Construction. For any two $\mathfrak{A}$-graded $\mathcal{A}$-modules we let $R : E \times F \to F \otimes E$ be the map defined by

$$R(e, f) = \sum_{\alpha,\beta\in\mathfrak{A}} (-1)^{\langle\alpha|\beta\rangle} f_\beta \otimes e_\alpha \ .$$

We leave it to the reader to check that R is an even bilinear map. The sign, which is in agreement with our guiding principle [1.21], will be crucial. It thus induces a linear map $\mathfrak{R} : E \otimes F \to F \otimes E$. With a slight abuse of notation, denoting the analogous map from $F \otimes E \to E \otimes F$ also by $\mathfrak{R}$, it is immediate that $\mathfrak{R} \circ \mathfrak{R}$ applied to $e \otimes f$ yields $e \otimes f$, and thus $\mathfrak{R} \circ \mathfrak{R} = id$, proving that it is an isomorphism. This canonical isomorphism $\mathfrak{R} : E \otimes F \to F \otimes E$ is called *the interchanging map of E and F*. For homogeneous elements e and f it has the property

$$\mathfrak{R}(e \otimes f) = (-1)^{\langle\varepsilon(e)|\varepsilon(f)\rangle} f \otimes e \ .$$

4.13 Discussion. We have already said that we will denote the *two* interchanging maps $E \otimes F \to F \otimes E$ and $F \otimes E \to E \otimes F$ both by $\mathfrak{R}$. However, we will employ the symbol $\mathfrak{R}$ in an even wider context. In a multiple tensor product $G \otimes E \otimes F \otimes H$ it is easy to construct an isomorphism onto $G \otimes F \otimes E \otimes H$ such that an element $g \otimes e \otimes f \otimes h$ is mapped to $g \otimes \mathfrak{R}(e \otimes f) \otimes h$. By abuse of notation we will denote this isomorphism also by $\mathfrak{R}$; we will say that it is the (interchanging) map that interchanges the neighbors E and F in such a multiple tensor product.

Let us denote by $\mathfrak{R}_{(ii+1)}$ the map that interchanges the i-th and $i+1$-st place in a multiple tensor product $E_1 \otimes \cdots \otimes E_k$. It is well known that if $\sigma \in \mathfrak{S}_k$ is a permutation of k elements, we can write it as a product of neighbor interchanges $(ii+1)$ (which permutes the elements i and $i+1$). Taking the corresponding product of isomorphisms $\mathfrak{R}_{(ii+1)}$ gives us an isomorphism $\mathfrak{R}_\sigma : E_1 \otimes \cdots \otimes E_k \to E_{\sigma^{-1}(1)} \otimes \cdots \otimes E_{\sigma^{-1}(k)}$. Obviously the target space $E_{\sigma^{-1}(1)} \otimes \cdots \otimes E_{\sigma^{-1}(k)}$ is completely determined by σ and does not depend upon the way we write σ as a product of neighbor interchanges. However, the map $\mathfrak{R}_\sigma$

itself, which is a product of maps $\mathfrak{R}_{(ii+1)}$, might quite well depend upon the way we write σ as a product of neighbor interchanges.

We are thus faced with the problem: how do we investigate whether $\mathfrak{R}_\sigma$ does or does not depend upon the way we write σ as a product of neighbor interchanges? Another way to pose this problem is to ask whether the maps $\mathfrak{R}_{(ii+1)}$ generate an action of $\mathfrak{S}_k$. Phrased this way the problem is solved in the theory of Coxeter groups [Bo, Ch IV, §1–2]. Since $\mathfrak{S}_k$ *is* a Coxeter group, the isomorphisms $\mathfrak{R}_{(ii+1)}$, $1 \leq i < k$ generate an action of $\mathfrak{S}_k$ if and only if they satisfy the relations

$$\begin{aligned} \mathfrak{R}^2_{(ii+1)} &= id & & 1 \leq i < k \, , \\ \big(\mathfrak{R}_{(ii+1)} \circ \mathfrak{R}_{(jj+1)}\big)^2 &= id & & 1 \leq i < i+1 < j < k \, , \\ \big(\mathfrak{R}_{(ii+1)} \circ \mathfrak{R}_{(i+1i+2)}\big)^3 &= id & & 1 \leq i \leq k-2 \, . \end{aligned}$$

Since the maps $\mathfrak{R}_{(ii+1)}$ are essentially all the same and of square id, the only relation that is not completely trivial in our case is the relation

$$\mathfrak{R}_{(12)} \circ \mathfrak{R}_{(23)} \circ \mathfrak{R}_{(12)} = \mathfrak{R}_{(23)} \circ \mathfrak{R}_{(12)} \circ \mathfrak{R}_{(23)} \, .$$

We leave it to the reader to verify that our interchanging maps do indeed satisfy this relation. (In mathematical physics this relation is known as the Yang-Baxter equation.)

We can summarize this discussion by saying that for each permutation $\sigma \in \mathfrak{S}_k$ we have a well defined isomorphism $\mathfrak{R}_\sigma : E_1 \otimes \cdots \otimes E_k \to E_{\sigma^{-1}(1)} \otimes \cdots \otimes E_{\sigma^{-1}(k)}$; if all e_i are homogeneous it is given by

$$\mathfrak{R}_\sigma(e_1 \otimes \cdots \otimes e_k) = \pm e_{\sigma^{-1}(1)} \otimes \cdots \otimes e_{\sigma^{-1}(k)} \, ,$$

where the sign depends on σ and the parities of the e_i. In case all the E_i are the same $\mathfrak{A}$-graded $\mathcal{A}$-module E, we obtain a representation $\mathfrak{R} : \mathfrak{S}_k \to \mathrm{Aut}(\otimes^k E)$.

4.14 Example. The fact that left and right multiplication are related to each other by (1.16) can be stated as the relations

$$\underline{m}_L = \underline{m}_R \circ \mathfrak{R} \qquad \text{or} \qquad \underline{m}_R = \underline{m}_L \circ \mathfrak{R} \, ,$$

where we interchange the factors E and $\mathcal{A}$ in the tensor product, and where $\underline{m}_S$ is as in [4.10].

5. Exterior powers

In this section the concepts of $\mathfrak{A}$-graded skew-symmetric k-linear maps and exterior products are introduced. The k-th exterior power $\bigwedge^k E$ of an $\mathfrak{A}$-graded $\mathcal{A}$-module E is defined as a quotient of the k-fold tensor product $\bigotimes^k E$; no attempt is made to identify it with an $\mathfrak{A}$-graded submodule of $\bigotimes^k E$. After we have proved the principal property of the exterior power $\bigwedge^k E$: transforming $\mathfrak{A}$-graded skew-symmetric k-linear maps into linear maps, we introduce the wedge product of two exterior powers $\bigwedge^k E$ and $\bigwedge^\ell E$, which takes values in $\bigwedge^{k+\ell} E$. This allows us to show that $\bigwedge E = \bigoplus_{k=0}^\infty \bigwedge^k E$, called the exterior algebra of E, can be given a natural structure of a $\mathbf{Z} \times \mathfrak{A}$-graded commutative ring. We thus have completed a full circle: starting with an $\mathfrak{A}$-graded commutative ring $\mathcal{A}$, we have created a $\mathbf{Z} \times \mathfrak{A}$-graded commutative ring $\bigwedge E$ via the intermediate of an $\mathfrak{A}$-graded $\mathcal{A}$-module E.

5.1 Definition. A k-linear map $\phi : E^k \to F$, where E and F are $\mathfrak{A}$-graded $\mathcal{A}$-modules, is called $\mathfrak{A}$-*graded skew-symmetric* if for all homogeneous $e_i \in E$ and all $j = 1, \ldots, k-1$ we have:

$$\begin{aligned}\phi(e_1, \ldots, e_{j-1}, e_j, e_{j+1}, e_{j+2}, \ldots, e_k)&\\ = -(-1)^{\langle \varepsilon(e_j) | \varepsilon(e_{j+1}) \rangle} \phi(e_1, \ldots, e_{j-1}, e_{j+1}, e_j, e_{j+2}, \ldots, e_k)&\ ,\end{aligned}$$

i.e., the value of ϕ changes sign if we interchange two neighboring homogeneous entries, except if both entries are odd. We will denote the set of all left/right k-linear $\mathfrak{A}$-graded skew-symmetric maps from E^k to F by $\mathrm{Map}_S^{sk}(E^k; F)$. We also define $\mathrm{Hom}_S^{sk}(E^k; F)$ as $\mathrm{Hom}_S^{sk}(E^k; F) = \mathrm{Map}_S^{sk}(E^k; F) \cap \mathrm{Hom}_S(E^k; F)$; we leave it to the reader to verify that it is an $\mathfrak{A}$-graded submodule of the $\mathfrak{A}$-graded $\mathcal{A}$-module $\mathrm{Hom}_S(E^k; F)$.

The above definition of $\mathfrak{A}$-graded skew-symmetry is the usual one, but we can give a more intrinsic definition by using the interchanging maps $\mathfrak{R}_{(jj+1)} \in \mathrm{Aut}(\bigotimes^k E)$. A k-linear map ϕ is $\mathfrak{A}$-graded skew-symmetric if and only if for all $j = 1, \ldots, k-1$ the induced linear map $\Phi : \bigotimes^k E \to F$ satisfies

$$\Phi = -\,\Phi \circ \mathfrak{R}_{(jj+1)} \qquad \text{or equivalently} \qquad \Phi \circ (id + \mathfrak{R}_{(jj+1)}) = 0\ .$$

Since the $\mathfrak{R}_{(jj+1)}$ generate the action of the permutation group $\mathfrak{S}_k$ on $\bigotimes^k E$, this condition is equivalent to the condition that for all $\sigma \in \mathfrak{S}_k$ one has $\Phi \circ \mathfrak{R}_\sigma = (-1)^\sigma \Phi$, where $(-1)^\sigma$ denotes the *signature* of the permutation σ.

If we replace this relation by $\Phi = \Phi \circ \mathfrak{R}_\sigma$, the map is called $\mathfrak{A}$-*graded symmetric*. However, the subject of $\mathfrak{A}$-graded symmetric maps will not be pursued in this book.

5.2 Construction (k-th exterior power). Let E be an $\mathfrak{A}$-graded $\mathcal{A}$-module, then we define the $\mathfrak{A}$-graded submodule N_k of $\bigotimes^k E$ as being generated by the subset T_k defined as

$$T_k = \{\, e_1 \otimes \cdots \otimes e_k + \mathfrak{R}_{(jj+1)}(e_1 \otimes \cdots \otimes e_k) \mid 1 \le j < k,\ e_i \in E \text{ homogeneous} \,\}\ .$$

Since T_k satisfies the assumptions of [1.25], $N_k = \mathrm{Span}(T_k)$ is an $\mathfrak{A}$-graded submodule of $\bigotimes^k E$. With these preparations we define the $\mathfrak{A}$-graded $\mathcal{A}$-module $\bigwedge^k E$ as the quotient

$$\textstyle\bigwedge^k E = \bigotimes^k E / N_k \ .$$

It is called *the k-th exterior power of the $\mathfrak{A}$-graded $\mathcal{A}$-module E*. Associated to this exterior power we define an even k-linear map $\omega : E^k \to \bigwedge^k E$ by $\omega = \pi \circ \chi$, i.e., as the composition of the tensor product map $\chi : E^k \to \bigotimes^k E$ with the canonical projection $\pi : \bigotimes^k E \to \bigwedge^k E$.

5.3 Nota Bene. The above definitions and constructions have no direct meaning if $k < 2$. For $k = 1$ we let the definition of $\mathfrak{A}$-graded skew-symmetry be an empty condition, i.e., we define $\mathrm{Map}_S^{sk}(E^1; F) = \mathrm{Map}_S(E; F)$. This is compatible with the definition of skew-symmetric maps as satisfying $\Phi \circ \mathfrak{R}_\sigma = (-1)^\sigma \Phi$ for $\sigma \in \mathfrak{S}_k$, because $\mathfrak{S}_1 = \{id\}$ and thus the condition of skew-symmetry reduces to the empty condition $\Phi = \Phi$. We also define $N_1 = \{0\}$ which, together with $\bigotimes^1 E = E$ [4.11], implies $\bigwedge^1 E = E$ and $\omega \equiv id(E) : E^1 \to \bigwedge^1 E$.

For $k = 0$ we define $N_0 = \{0\}$, and thus $\bigwedge^0 E = \mathcal{A}$, because $\bigotimes^0 E = \mathcal{A}$ [4.11]. We also define formally $\mathrm{Map}_S^{sk}(E^0; F) = \mathrm{Map}_S(\mathcal{A}; F)$, but no natural justification for this definition can be found and neither do we define a map $\omega : E^0 \to \bigwedge^0 E$. We thus have by definition the equalities $\mathrm{Map}_S^{sk}(E^0; F) = \mathrm{Map}_S(\bigwedge^0 E; F)$ as well as $\mathrm{Hom}_S^{sk}(E^0; F) = \mathrm{Hom}_S(\bigwedge^0 E; F)$.

5.4 Proposition. *Let E be an $\mathfrak{A}$-graded $\mathcal{A}$-module and $k \geq 1$, then $\omega : E^k \to \bigwedge^k E$ is $\mathfrak{A}$-graded skew-symmetric. Moreover, given any $\mathfrak{A}$-graded $\mathcal{A}$-module F and any map $\phi \in \mathrm{Map}_S^{sk}(E^k; F)$, there exists a unique map $\Phi \in \mathrm{Map}_S(\bigwedge^k E; F)$ such that $\phi = \Phi \circ \omega$. If ϕ has parity α, then so has Φ.*

Proof. Let $\Omega : \mathrm{Hom}(\bigotimes^k E; \bigwedge^k E)_0$ be the by ω induced linear map. In order to show the $\mathfrak{A}$-graded skew-symmetry, we have to show that $\Omega \circ (id + \mathfrak{R}_{(jj+1)}) = 0$. But Ω is the canonical projection $\pi : \bigotimes^k E \to \bigotimes^k E/N_k = \bigwedge^k E$. We thus have to show that $id + \mathfrak{R}_{(jj+1)}$ maps $\bigotimes^k E$ into N_k. But this is immediate from the definition of the generating subset T_k.

If $\phi : E^k \to F$ is an $\mathfrak{A}$-graded skew-symmetric map, then the induced linear map $\underline{\Phi} : \bigotimes^k E \to F$ satisfies the relations $\underline{\Phi} \circ (id + \mathfrak{R}_{(jj+1)}) = 0$. But this says that $\underline{\Phi}$ vanishes on T_k, hence on N_k and hence induces a unique linear map $\Phi : \bigwedge^k E \equiv \bigotimes^k E/N_k \to F$. Since both operations ϕ to $\underline{\Phi}$ to Φ preserve parity, the parity claim follows immediately.

QED

5.5 Corollary. *Given $\mathfrak{A}$-graded $\mathcal{A}$-modules E and F and $k \geq 1$, we have a bijection $\phi \mapsto \phi \circ \omega$ from $\mathrm{Map}_S(\bigwedge^k E; F)$ to $\mathrm{Map}_S^{sk}(E^k; F)$. Restricted to $\mathrm{Hom}_S(\bigwedge^k E; F)$ it is an isomorphism onto $\mathrm{Hom}_S^{sk}(E^k; F)$. For $k = 0$ we refer to [5.3].*

5.6 Lemma. *Given an $\mathfrak{A}$-graded $\mathcal{A}$-module E and $k \geq 1$, then, up to isomorphism, the $\mathfrak{A}$-graded $\mathcal{A}$-module $\bigwedge^k E$ is the unique $\mathfrak{A}$-graded $\mathcal{A}$-module enjoying the property of [5.4].*

Proof. The proof is an exact copy of the proof of [4.4]. $\boxed{QED}$

5.7 Notation. According to standard usage, one uses the wedge product symbol $\wedge$ to replace the map ω in the following way. Given k elements $e_i \in V$, one writes

$$e_1 \wedge \cdots \wedge e_k \equiv \omega(e_1, \ldots, e_k) .$$

It follows that the $\mathfrak{A}$-graded $\mathcal{A}$-module $\bigwedge^k E$ is generated by the elements $e_1 \wedge \cdots \wedge e_k = \pi(e_1 \otimes \cdots \otimes e_k)$, where the e_i run through E (one might even assume the e_i to be homogeneous [4.7]).

5.8 Construction. For $k\ell > 0$ we define the map $\Phi : \bigotimes^k E \times \bigotimes^\ell E \to \bigotimes^{k+\ell} E$ by $(K, L) \mapsto K \otimes L$. If either k or ℓ is 0, we let Φ be left/right multiplication by elements of $\mathcal{A}$ (remember that $\bigotimes^0 E = \mathcal{A}$). In this way Φ is defined for all $k, \ell \in \mathbf{N}$. Note that the induced linear map [4.3] is the canonical map which identifies $(\bigotimes^k E) \otimes (\bigotimes^\ell E)$ with $\bigotimes^{k+\ell} E$ (and thus the maps $\underline{m}_S$ are the special cases $k\ell = 0$).

We now note that both $\Phi(N_k, \bigotimes^\ell E)$ and $\Phi(\bigotimes^k E, N_\ell)$ are contained in $N_{k+\ell}$. Hence there exists a unique map $\wedge : \bigwedge^k E \times \bigwedge^\ell E \to \bigwedge^{k+\ell} E$, called the *wedge product* and also denoted by a wedge, such that the following diagram is commutative:

$$\begin{array}{ccc} \bigotimes^k E \times \bigotimes^\ell E & \xrightarrow{\ \Phi\ } & \bigotimes^{k+\ell} E \\ \pi \downarrow \quad \downarrow \pi & & \downarrow \pi \\ \bigwedge^k E \times \bigwedge^\ell E & \xrightarrow[\ \wedge\ ]{} & \bigwedge^{k+\ell} E . \end{array}$$

We leave it to the reader to verify that this map is even and bilinear (but beware: the notion of $\mathfrak{A}$-graded skew-symmetry does *not* apply). Because of the associativity of the tensor product and, more precisely, because we may use *the same* tensor product symbol throughout, we are justified in the use of the wedge product symbol throughout:

$$(e_1 \wedge \cdots \wedge e_k) \wedge (e_{k+1} \wedge \cdots \wedge e_{k+\ell}) = e_1 \wedge \cdots \wedge e_{k+\ell} ,$$

just because the wedge product is induced by the tensor product. In the particular cases k or ℓ is 0 the definition of Φ together with the fact that the projection $\pi : \bigotimes^0 E \to \bigwedge^0 E$ is the identity map $id : \mathcal{A} \to \mathcal{A}$ immediately gives, for $a \in \mathcal{A}$, the equalities

$$a \wedge e_1 \wedge \cdots \wedge e_k = ae_1 \wedge \cdots \wedge e_k \quad \text{and} \quad e_1 \wedge \cdots \wedge e_k \wedge a = e_1 \wedge \cdots \wedge e_k a .$$

5.9 Proposition. *Given an $\mathfrak{A}$-graded $\mathcal{A}$-module E, $k, \ell \in \mathbf{N}$, and $K \in \bigwedge^k E$ and $L \in \bigwedge^\ell E$ homogeneous, then*

$$K \wedge L = (-1)^{\langle \varepsilon(K)|\varepsilon(L)\rangle + k\ell} L \wedge K \in \textstyle\bigwedge^{k+\ell} E\ .$$

Proof. Since the wedge products $\bigwedge^k E \times \bigwedge^\ell E \to \bigwedge^{k+\ell} E$ and $\bigwedge^\ell E \times \bigwedge^k E \to \bigwedge^{k+\ell} E$ are in particular bi-additive, it suffices to show this equality for elements of the form $K = e_1 \wedge \cdots \wedge e_k$ and $L = f_1 \wedge \cdots \wedge f_\ell$, where the $e_i, f_j \in E$ are homogeneous. The result now follows immediately if we realize that $\varepsilon(K) = \sum_{i=1}^k \varepsilon(e_i)$, that $\varepsilon(L) = \sum_{j=1}^\ell \varepsilon(f_j)$, and that interchanging two neighboring homogeneous elements from $e_i \wedge f_j$ to $f_j \wedge e_i$ introduces the sign $(-1)^{1+\langle \varepsilon(e_i)|\varepsilon(f_j)\rangle}$. QED

5.10 Definition. The *exterior algebra* $\bigwedge E$ of an $\mathfrak{A}$-graded $\mathcal{A}$-module E is defined as the direct sum $\bigwedge E = \bigoplus_{k=0}^\infty \bigwedge^k E$. We equip $\bigwedge E$ with a $\mathbf{Z} \times \mathfrak{A}$-grading by defining for $(k, \alpha) \in \mathbf{Z} \times \mathfrak{A}$:

$$(\textstyle\bigwedge E)_{(k,\alpha)} = \begin{cases} (\bigwedge^k E)_\alpha\ , & \text{if } k \geq 0 \\ \{0\}\ , & \text{otherwise.} \end{cases}$$

We also extend the wedge product to a map $\wedge : \bigwedge E \times \bigwedge E \to \bigwedge E$, as the unique even bilinear map which, when restricted to $\bigwedge^k E \times \bigwedge^\ell E$, $k, \ell \in \mathbf{N}$, reproduces the already defined wedge product. Since taking wedge products is associative (taking tensor products is), $\bigwedge E$ equipped with the wedge product as multiplication becomes a ring. We finally define the symmetric bi-additive map $\langle\,\_\,|\,\_\,\rangle : (\mathbf{Z} \times \mathfrak{A}) \times (\mathbf{Z} \times \mathfrak{A}) \to \mathbf{Z}_2$ (abuse of notation because the same symbol is used for the map $\mathfrak{A} \times \mathfrak{A} \to \mathbf{Z}_2$) by

$$\langle (k, \alpha) | (\ell, \beta) \rangle = (k\ell \mod 2) + \langle \alpha | \beta \rangle\ .$$

It is now an immediate consequence of [5.9] that $\bigwedge E$ is a $\mathbf{Z} \times \mathfrak{A}$-graded commutative ring. Anticipating definition [6.1] of an $\mathfrak{A}$-graded $\mathcal{A}$-algebra, the fact that $\bigwedge E$ also is an $\mathfrak{A}$-graded $\mathcal{A}$-module turns it into a $\mathbf{Z} \times \mathfrak{A}$-graded commutative $\mathcal{A}$-algebra, explaining the algebra part in the name exterior algebra.

5.11 Remarks. • In terms of the more general definition [1.9] of $\mathfrak{A}$-graded commutativity, the symmetric bi-additive map on $\mathbf{Z} \times \mathfrak{A}$ becomes the product of the functions ρ for $\mathbf{Z}$, given by $\rho(k, \ell) = (-1)^{k\ell}$, and for $\mathfrak{A}$, given by $\rho(\alpha, \beta) = (-1)^{\langle \alpha|\beta\rangle}$. In other words, the function ρ for a product grading is the product of the separate functions ρ.

• In the particular case $\mathfrak{A} = \mathbf{Z}_2$ (which is the case we will use exclusively starting in chapter II), some authors define a (single) $\mathbf{Z}_2$-grading on the exterior algebra $\bigwedge E$ by

$$(\textstyle\bigwedge E)_\alpha = \displaystyle\bigoplus_{(k \mod 2) + \beta = \alpha} (\textstyle\bigwedge^k E)_\beta \quad \text{or} \quad (\textstyle\bigwedge E)_\alpha = \displaystyle\bigoplus_{(k \mod 2) + \beta + 1 = \alpha} (\textstyle\bigwedge^k E)_\beta\ .$$

However, such a $\mathbf{Z}_2$-grading is not compatible with [5.9] in the sense that there does not exist a symmetric bi-additive map on $\mathbf{Z}_2$ reproducing the sign $(-1)^{\langle \varepsilon(K)|\varepsilon(L)\rangle + k\ell}$.

6. Algebras and derivations

In the previous sections we have introduced constructions of new $\mathfrak{A}$-graded $\mathcal{A}$-modules out of given $\mathfrak{A}$-graded $\mathcal{A}$-modules. In this section we will introduce different structures on $\mathfrak{A}$-graded $\mathcal{A}$-modules. More precisely, we will introduce the notions of associative $\mathfrak{A}$-graded $\mathcal{A}$-algebra and $\mathfrak{A}$-graded $\mathcal{A}$-Lie algebra. Associated to the notion of an algebra is the notion of a derivation. It is shown that there exists a natural way to identify E^* as a collection of derivations of the exterior algebra $\bigwedge E$. This identification is the algebraic version of the contraction of a vector field with a k-form, employed systematically in differential geometry.

6.1 Definition. Let $\mathfrak{g}$ be an $\mathfrak{A}$-graded $\mathcal{A}$-module and $m : \mathfrak{g} \times \mathfrak{g} \to \mathfrak{g}$ an even bilinear map. The couple $(\mathfrak{g}, m)$ is called an *(associative) $\mathfrak{A}$-graded $\mathcal{A}$-algebra* if m seen as multiplication is associative, i.e., if $(\mathfrak{g}, +, m)$ is a ring. It is called an $\mathfrak{A}$-*graded commutative* $\mathcal{A}$-algebra if m is $\mathfrak{A}$-graded symmetric.

The couple $(\mathfrak{g}, m)$ is called an $\mathfrak{A}$-*graded $\mathcal{A}$-Lie algebra* if the map m is $\mathfrak{A}$-graded skew-symmetric and satisfies the $\mathfrak{A}$-*graded Jacobi identity*, i.e., for all homogeneous $e, f, g \in \mathfrak{g}$:

$$[e,[f,g]] = [[e,f],g] + (-1)^{\langle \varepsilon(e)|\varepsilon(f)\rangle}\, [f,[e,g]]\ ,$$

where we have written $[e,f]$ for $m(e,f)$, as we will always do for $\mathfrak{A}$-graded $\mathcal{A}$-Lie algebras. This will cause no problems concerning left/right linearity because m is even. Using the $\mathfrak{A}$-graded skew-symmetry $[e,f] = -(-1)^{\langle \varepsilon(e)|\varepsilon(f)\rangle}[f,e]$, this relation can also be written in the more symmetric form

$$(-1)^{\langle \varepsilon(g)|\varepsilon(e)\rangle}[e,[f,g]] + (-1)^{\langle \varepsilon(e)|\varepsilon(f)\rangle}[f,[g,e]] + (-1)^{\langle \varepsilon(f)|\varepsilon(g)\rangle}[g,[e,f]] = 0\ .$$

For $\mathfrak{A}$-graded $\mathcal{A}$-Lie algebras one usually calls the element $[e,f] \equiv m(e,f)$ *the bracket of e and f*.

An $\mathfrak{A}$-graded submodule F of an $\mathfrak{A}$-graded $\mathcal{A}$-Lie algebra E is called an $\mathfrak{A}$-*graded sub $\mathcal{A}$-Lie algebra of E* (or a *subalgebra* for short) if F is stable under the bracket operation $[\_,\_]$, i.e., $\forall e, f \in F : [e,f] \in F$.

6.2 Examples. • The ring $\mathcal{A}$ itself is a (rather trivial) $\mathfrak{A}$-graded commutative $\mathcal{A}$-algebra.

• If E is an $\mathfrak{A}$-graded $\mathcal{A}$-module, the $\mathfrak{A}$-graded $\mathcal{A}$-module $\mathrm{End}_S(E)$ becomes an $\mathfrak{A}$-graded $\mathcal{A}$-algebra when we take composition of maps as multiplication.

• For any $\mathfrak{A}$-graded $\mathcal{A}$-module E, the exterior algebra $\bigwedge E$ is a $\mathbf{Z} \times \mathfrak{A}$-graded commutative $\mathcal{A}$-algebra, where we give $\mathcal{A}$ a $\mathbf{Z} \times \mathfrak{A}$-grading by $\mathcal{A}_{0,\alpha} = \mathcal{A}_\alpha$ and $\mathcal{A}_{n,\alpha} = \{0\}$ whenever $n \neq 0$.

• If $(\mathfrak{g}, m)$ is an (associative) $\mathfrak{A}$-graded $\mathcal{A}$-algebra, we can introduce an even bilinear $\mathfrak{A}$-graded skew-symmetric commutator map $[\_,\_] : \mathfrak{g} \times \mathfrak{g} \to \mathfrak{g}$ by its action on homogeneous elements $e, f \in \mathfrak{g}$:

$$[e,f] = m(e,f) - (-1)^{\langle \varepsilon(e)|\varepsilon(f)\rangle} m(f,e)\ .$$

If we identify the bilinear maps m and $[\_,\_]$ with the associated linear maps $\mathfrak{g} \otimes \mathfrak{g} \to \mathfrak{g}$, the definition of the commutator $[\_,\_]$ can be written as $[\_,\_] = m - m \circ \mathfrak{R}$. We leave it to the reader to verify that $(\mathfrak{g}, [\_,\_])$ is an $\mathfrak{A}$-graded $\mathcal{A}$-Lie algebra. In particular, $(\mathfrak{g}, m)$ is $\mathfrak{A}$-graded commutative if and only if the commutator map $[\_,\_]$ is identically zero.

6.3 Definition. Applying the construction of the last example in [6.2] to the algebra $\mathfrak{g} = (\mathrm{End}_R(E), \circ)$ gives us a commutator $[\_,\_]_R$ on $\mathrm{End}_R(E)$, turning the set of right linear endomorphisms of an $\mathfrak{A}$-graded $\mathcal{A}$-module into an $\mathfrak{A}$-graded $\mathcal{A}$-Lie algebra. In the same way $\mathfrak{g} = (\mathrm{End}_L(E), \diamond)$ becomes an $\mathfrak{A}$-graded $\mathcal{A}$-Lie algebra with bracket $[\_,\_]_L$ given on homogeneous elements by

$$\begin{aligned}[\phi, \psi]_L &= \phi \diamond \psi - (-1)^{\langle \varepsilon(\phi) | \varepsilon(\psi) \rangle} \psi \diamond \phi = \psi \circ \phi - (-1)^{\langle \varepsilon(\phi) | \varepsilon(\psi) \rangle} \phi \circ \psi \\ &= -(-1)^{\langle \varepsilon(\phi) | \varepsilon(\psi) \rangle} \left(\phi \circ \psi - (-1)^{\langle \varepsilon(\phi) | \varepsilon(\psi) \rangle} \psi \circ \phi \right) .\end{aligned}$$

If no confusion is possible, we will omit the subscripts L and R in these brackets and simply write $[\_,\_]$.

6.4 Lemma. *For $\phi, \psi \in \mathrm{End}_L(E)$ we have $\mathfrak{T}([\phi, \psi]_L) = -[\mathfrak{T}\phi, \mathfrak{T}\psi]_R$.*

6.5 Remark. The above lemma is in agreement with the usual interpretation of the transpose of an endomorphism and the commutator. Moreover, written this way, no additional sign is involved according to our guiding principle [1.21], because the elements ϕ and ψ are not interchanged.

6.6 Definition. Let E be an $\mathfrak{A}$-graded $\mathcal{A}$-module and m an even bilinear map $E \times E \to E$. A *right derivation of the couple* (E, m) is a right linear endomorphism of E whose homogeneous parts ϕ satisfy for all homogeneous elements $e, f \in E$ the relations

$$\phi(m(e, f)) = m(\phi(e), f) + (-1)^{\langle \varepsilon(e) | \varepsilon(\phi) \rangle} m(e, \phi(f)) .$$

For a *left derivation*, ϕ has to be left linear and the relation has to be replaced by

$$\langle\!\langle m(e, f) \| \phi \rangle\!\rangle = m(e, \langle\!\langle f \| \phi \rangle\!\rangle) + (-1)^{\langle \varepsilon(\phi) | \varepsilon(f) \rangle} m(\langle\!\langle e \| \phi \rangle\!\rangle, f) .$$

The set of all right derivations of (E, m) is called $\mathrm{Der}_R(E, m) \subset \mathrm{End}_R(E)$, or $\mathrm{Der}_R(E)$ if the bilinear map m is understood. Similarly the left derivations are denoted by $\mathrm{Der}_L(E, m)$ or $\mathrm{Der}_L(E)$.

6.7 Lemma. *The set* $\mathrm{Der}_S(E, m)$ *is a subalgebra of* $\mathrm{End}_S(E)$ *when the latter is equipped with the commutator as bracket; in particular* $\mathrm{Der}_S(E, m)$ *is an 𝔄-graded submodule of* $\mathrm{End}_S(E)$. *Moreover,* $\mathfrak{T}\phi$ *is a right derivation if and only if* ϕ *is a left derivation.*

Proof. We leave it to the reader to verify that $\mathrm{Der}_S(E, m)$ is an 𝔄-graded submodule of $\mathrm{End}_S(E)$. Let us show that if ϕ, ψ are two homogeneous right derivations, then $[\phi, \psi]$ is a homogeneous right derivation as well. Thus let $e, f \in E$ be homogeneous and compute:

$$\begin{aligned}
[\phi,\psi]([e,f]) &= (\phi\circ\psi - (-1)^{\langle\varepsilon(\phi)|\varepsilon(\psi)\rangle}\psi\circ\phi)([e,f]) \\
&= \phi([\psi(e),f] + (-1)^{\langle\varepsilon(e)|\varepsilon(\psi)\rangle}[e,\psi(f)]) \\
&\qquad - (-1)^{\langle\varepsilon(\phi)|\varepsilon(\psi)\rangle}\,\psi([\phi(e),f] + (-1)^{\langle\varepsilon(e)|\varepsilon(\phi)\rangle}[e,\phi(f)]) \\
&= [\phi(\psi(e)),f] + (-1)^{\langle\varepsilon(\phi)|\varepsilon(\psi(e))\rangle}\,[\psi(e),\phi(f)] \\
&\qquad + (-1)^{\langle\varepsilon(e)|\varepsilon(\psi)\rangle}\,[\phi(e),\psi(f)] + (-1)^{\langle\varepsilon(e)|\varepsilon([\phi,\psi])\rangle}\,[e,\phi(\psi(f))] \\
&\qquad - (-1)^{\langle\varepsilon(\phi)|\varepsilon(\psi)\rangle}\,[\psi(\phi(e)),f] - (-1)^{\langle\varepsilon(e)|\varepsilon(\psi)\rangle}\,[\phi(e),\psi(f)] \\
&\qquad - (-1)^{\langle\varepsilon(\phi)|\varepsilon(\psi(e))\rangle}\,[\psi(e),\phi(f)] \\
&\qquad - (-1)^{\langle\varepsilon(\phi)|\varepsilon(\psi)\rangle+\langle\varepsilon(e)|\varepsilon([\phi,\psi])\rangle}\,[e,\psi(\phi(f))] \\
&= [[\phi,\psi](e),f] + (-1)^{\langle\varepsilon(e)|\varepsilon([\phi,\psi])\rangle}\,[e,[\phi,\psi](f)]\,.
\end{aligned}$$

QED

6.8 Discussion. In ungraded Lie algebras, the auto commutator $[e, e]$ of an element $e \in \mathfrak{g}$ is automatically zero, just because of the skew-symmetry. However, for 𝔄-graded $\mathcal{A}$-Lie algebras this is no longer true. Indeed, for a homogeneous element e, 𝔄-graded skew-symmetry gives us $[e, e] = -(-1)^{\langle\varepsilon(e)|\varepsilon(e)\rangle}[e, e]$, and thus $[e, e]$ necessarily equals 0 only if $\langle\varepsilon(e)|\varepsilon(e)\rangle = 0$.

This phenomenon becomes clearer in the context of derivations. If $\phi \in \mathrm{End}_S(E)$ is a derivation, its square $\phi^2 = \phi\circ\phi$ is, in general, no longer a derivation. But if $\langle\varepsilon(\phi)|\varepsilon(\phi)\rangle = 1$, it follows from the definition of the commutator that $\phi^2 = \frac{1}{2}[\phi, \phi]$ is again a derivation (see [V.1.23] for an explicit example).

6.9 Definitions. • Let $(\mathfrak{g}_1, [\,\_\,,\_\,]_1)$ and $(\mathfrak{g}_2, [\,\_\,,\_\,]_2)$ be two 𝔄-graded $\mathcal{A}$-Lie algebras and let $\phi : \mathfrak{g}_1 \to \mathfrak{g}_2$ be linear. The map ϕ is said to be a *morphism of 𝔄-graded $\mathcal{A}$-Lie algebras* if it is even and preserves brackets, i.e., $\forall e, f \in \mathfrak{g}_1 : \phi([e, f]_1) = [\phi(e), \phi(f)]_2$.

• A *left-representation* of an 𝔄-graded $\mathcal{A}$-Lie algebra $(\mathfrak{g}, [\,\_\,,\_\,])$ on an 𝔄-graded $\mathcal{A}$-module E is a morphism of 𝔄-graded $\mathcal{A}$-Lie algebras $\phi : \mathfrak{g} \to \mathrm{End}_L(E)$. *Right-representations* are defined similarly.

• For any 𝔄-graded $\mathcal{A}$-Lie algebra $(\mathfrak{g}, [\,\_\,,\_\,])$ we define maps $\mathrm{ad}_S : \mathfrak{g} \to \mathrm{End}_S(\mathfrak{g})$ by

$$\mathrm{ad}_R(e) : f \mapsto [e, f] \qquad \& \qquad \mathrm{ad}_L(e) : f \mapsto [f, e]\,.$$

The fact that the bracket $[\,\_\,,\_\,]$ is even and bilinear immediately shows that ad_S is a well defined even morphism of 𝔄-graded $\mathcal{A}$-modules. The 𝔄-graded Jacobi identity tells us that ad_R is a right-representation of $(\mathfrak{g}, [\,\_,\_\,])$ and at the same time that all $\mathrm{ad}_R(e)$ are right derivations of $(\mathfrak{g}, [\,\_,\_\,])$. The Jacobi identity plus the $\mathfrak{A}$-graded skew-symmetry tell us the same things for ad_L: it is a left-representation and $\mathrm{im}(\mathrm{ad}_L) \subset \mathrm{Der}_L(\mathfrak{g})$. These two representations are called the *(left- and right-) adjoint representations* of the $\mathfrak{A}$-graded $\mathcal{A}$-Lie algebra $(\mathfrak{g}, [\,\_,\_\,])$.

6.10 Lemma. *For any $\mathfrak{A}$-graded $\mathcal{A}$-Lie algebra $\mathfrak{g}$, the left and right adjoint representations are related by $\mathfrak{T} \circ \mathrm{ad}_L = -\,\mathrm{ad}_R$.*

Proof. This is a direct consequence of the $\mathfrak{A}$-graded skew-symmetry and [6.4]. QED

6.11 Remark. For generic $\mathfrak{A}$-graded $\mathcal{A}$-Lie algebras, the condition that a morphism of $\mathfrak{A}$-graded $\mathcal{A}$-Lie algebras preserves the brackets "implies" that it must be even, just by counting parities: $\phi([\,e,f\,]_1) = [\,\phi(e),\phi(f)\,]_2$ "implies" that we must have the equality $\varepsilon(\phi)+\varepsilon(e)+\varepsilon(f) = \varepsilon(\phi)+\varepsilon(e)+\varepsilon(\phi)+\varepsilon(f)$. However, this argument is not valid whenever the zero element is involved, e.g., if both brackets are identically zero.

6.12 Remark. We have restricted our attention to *even* bilinear maps $m : \mathfrak{g}\times\mathfrak{g} \to \mathfrak{g}$, because we will need no others. However, in the literature one also finds non-even homogeneous maps m, especially in the context of $\mathfrak{A}$-graded $\mathcal{A}$-Lie algebras. Let us give the precise definition.

An *$\mathfrak{A}$-graded $\mathcal{A}$-Lie algebra of parity* α is an $\mathfrak{A}$-graded $\mathcal{A}$-module $\mathfrak{g}$ together with a bi-additive map $[\,\_\,|m|\,\_\,] : \mathfrak{g}\times\mathfrak{g} \to \mathfrak{g}$, $(e,f) \mapsto [\,e\,|m|\,f\,]$ satisfying:

(i) for homogeneous $e,f \in \mathfrak{g}$: $\varepsilon([\,e\,|m|\,f\,]) = \varepsilon(e)+\varepsilon(f)+\alpha$ (the bracket has parity α),

(ii) for all $\lambda,\mu \in \mathcal{A}$ and $e,f \in \mathfrak{g}$: $[\,\lambda e\,|m|\,f\mu\,] = \lambda\,[\,e\,|m|\,f\,]\,\mu$ (left linearity in the first argument and right linearity in the second),

(iii) for homogeneous $e,f \in \mathfrak{g}$: $[\,e\,|m|\,f\,] = -(-1)^{\langle \varepsilon(e)+\alpha|\varepsilon(f)+\alpha\rangle}[\,f\,|m|\,e\,]$ ($\mathfrak{A}$-graded skew-symmetry of parity α),

(iv) for homogeneous $e,f,g \in \mathfrak{g}$:

$$[\,e\,|m|\,[\,f\,|m|\,g\,]\,] = [\,[\,e\,|m|\,f\,]\,|m|\,g\,] + (-1)^{\langle \varepsilon(e)+\alpha|\varepsilon(f)+\alpha\rangle}[\,f\,|m|\,[\,e\,|m|\,g\,]\,]$$

(the $\mathfrak{A}$-graded Jacobi identity of parity α).

Using (iii), (iv) can be written in the more symmetric looking form

$$\begin{aligned}
&(-1)^{\langle \varepsilon(e)+\alpha|\varepsilon(g)+\alpha\rangle}[\,[\,e\,|m|\,f\,]\,|m|\,g\,] \\
&+(-1)^{\langle \varepsilon(f)+\alpha|\varepsilon(e)+\alpha\rangle}[\,[\,f\,|m|\,g\,]\,|m|\,e\,] \\
&+(-1)^{\langle \varepsilon(g)+\alpha|\varepsilon(f)+\alpha\rangle}[\,[\,g\,|m|\,e\,]\,|m|\,f\,] = 0\,.
\end{aligned}$$

If $(\mathfrak{g}, [\,\_\,|m|\,\_\,])$ is an $\mathfrak{A}$-graded $\mathcal{A}$-Lie algebra of parity α, we can define the left and right adjoint maps $\mathrm{ad}_S : \mathfrak{g} \to \mathrm{End}_S(\mathfrak{g})$ by

$$\mathrm{ad}_R(e) : f \mapsto [\,e\,|m|\,f\,] \qquad \text{and} \qquad \mathrm{ad}_L(e) : f \mapsto [\,f\,|m|\,e\,] \,.$$

If we drop the condition that a morphism of $\mathfrak{A}$-graded $\mathcal{A}$-Lie algebras must be even (but still requiring that it be homogeneous), and if we take the parity of the bracket $[\,\_\,|m|\,\_\,]$ into account in the definition of a derivation (e.g., a right derivation should satisfy $\phi([\,e\,|m|\,f\,]) = [\,\phi(e)\,|m|\,f\,] + (-1)^{\langle \varepsilon(\phi)|\varepsilon(e)+\alpha\rangle}[\,e\,|m|\,\phi(f)\,])$, then all results given in [6.9] and [6.10] remain valid. We note that, given conditions (i) and (ii), the conditions (iii) and (iv) are the only natural ones such that [6.9] and [6.10] remain true. If one does not like conditions (i) and (ii), one can easily transform them into, for instance, a right bilinear map $\widehat{m} : \mathfrak{g} \times \mathfrak{g} \to \mathfrak{g}$ by $\widehat{m}(e, f) = (-1)^{\langle \alpha|\varepsilon(e)\rangle}[\,e\,|m|\,f\,]$. The conditions (iii) and (iv) then transform accordingly.

In a similar spirit we can define an *associative algebra of parity* α as an $\mathfrak{A}$-graded $\mathcal{A}$-module $\mathfrak{g}$ with a map $(\,\_|m|\_\,) : \mathfrak{g} \times \mathfrak{g} \to \mathfrak{g}$ satisfying (i), (ii), and

(iii)$'$ for all $e, f, g \in \mathfrak{g}$: $(\,(e|m|f)|m|g) = (e|m|(f|m|g)\,)$.

And as for the even case, one can construct an $\mathfrak{A}$-graded $\mathcal{A}$-Lie algebra of parity α out of such an associative algebra by $[\,e\,|m|\,f\,] = (e|m|f) - (-1)^{\langle \varepsilon(e)+\alpha|\varepsilon(f)+\alpha\rangle}(f|m|e)$.

If one tries to generalize even more by dropping the homogeneity condition on the bracket (or the multiplication), then the constructions become rather ugly and lose a lot of their nice properties. For instance, if one maintains that the $\mathfrak{A}$-graded Jacobi identity should tell us that $\langle\!\langle e \| \mathrm{ad}_R \rangle\!\rangle = [\,e\,|m|\,\_\,]$ is a derivation, then the symmetric formulation of the $\mathfrak{A}$-graded Jacobi identity will not hold, and $\mathrm{ad}_L(e)$ will not be a derivation.

6.13 Discussion. The conditions (i)—(iv) of an $\mathfrak{A}$-graded $\mathcal{A}$-Lie algebra of parity α become natural if we imagine that $\mathfrak{g}$ is just an $\mathfrak{A}$-graded $\mathcal{A}$-module, that we have an $\mathfrak{A}$-graded $\mathcal{A}$-Lie algebra $(\mathfrak{h}, [\,\_\,,\,\_\,])$ with even bracket (as in [6.1]), and that we have a homogeneous invertible left linear map $\phi : \mathfrak{g} \to \mathfrak{h}$ of parity α. We then can induce a bracket $[\,\_\,|m|\,\_\,]$ on $\mathfrak{g}$ by the formula

$$[\,e\,|m|\,f\,] = \phi^{-1}([\,\phi(e), \phi(f)\,]) \,.$$

(Note that we used right evaluation of ϕ instead of left evaluation as we should have done for left linear maps.) This transported bracket indeed satisfies the conditions (i)—(iv), turning $(\mathfrak{g}, [\,\_\,|m|\,\_\,])$ into an $\mathfrak{A}$-graded $\mathcal{A}$-Lie algebra of parity α and turning ϕ into a bracket preserving map.

Another way to make essentially the same observation is to define the *parity shift operation*. Given an $\mathfrak{A}$-graded $\mathcal{A}$-module E and an $\alpha \in \mathfrak{A}$, we define a new $\mathfrak{A}$-graded $\mathcal{A}$-module $E^{\flat\alpha}$ as follows: as a *left* $\mathcal{A}$-module $E^{\flat\alpha}$ is the same as E, but the $\mathfrak{A}$-grading is given by $(E^{\flat\alpha})_\beta = E_{\beta-\alpha}$, i.e., we "lower" the parity by α. With this new grading, $E^{\flat\alpha}$ will (in general) certainly *not* be equal to E as *right* $\mathcal{A}$-module. Now if $(\mathfrak{g}, [\,\_\,|m|\,\_\,])$ is an $\mathfrak{A}$-graded $\mathcal{A}$-Lie algebra of parity α, $(\mathfrak{g}^{\flat\alpha}, [\,\_\,|m|\,\_\,])$ is an $\mathfrak{A}$-graded $\mathcal{A}$-Lie algebra of parity 0, i.e., an $\mathfrak{A}$-graded $\mathcal{A}$-Lie algebra in the sense of [6.1]. The link with the previous paragraph is the observation that the identity map $\phi : \mathfrak{g} \to \mathfrak{g}^{\flat\alpha}$, $e \mapsto e$ is an invertible left linear map of parity α.

6.14 Lemma. *Let E be an $\mathfrak{A}$-graded $\mathcal{A}$-module and D a (left or right) derivation of its exterior algebra $(\bigwedge E, \wedge)$. Then $D|_{\bigwedge^0 E} = 0$ and D is uniquely determined by its restriction $D|_{\bigwedge^1 E}$.*

Proof. From the derivation property and the equation $1 \wedge 1 = 1 \cdot 1 = 1 \in \mathcal{A} \cong \bigwedge^0 E$ we deduce that $D(1) = 0$, and hence $D|_{\bigwedge^0 E} = 0$. Since an element of $\bigwedge E$ is a finite sum of terms of the form $e_1 \wedge \cdots \wedge e_k$, the second assertion follows from the derivation property of D. QED

6.15 Corollary. *If D is a derivation of $(\bigwedge E, \wedge)$ of parity (n, α) with $n < -1$, it is the zero map.*

Proof. $D(\bigwedge^1 E) \subset \bigwedge^{1+n} E$ and $\bigwedge^k E = \{0\}$ whenever $k < 0$. QED

6.16 Proposition. *There exists a unique even homomorphism of $\mathfrak{A}$-graded $\mathcal{A}$-modules $\iota : E^* \to \mathrm{Der}_R(\bigwedge E)$ such that $\iota(\phi)$ is the unique right derivation whose restriction to $\bigwedge^1 E \cong E$ is given by $e \mapsto \phi(e) \in \mathcal{A} \cong \bigwedge^0 E$. If ϕ has parity α, $\iota(\phi)$ has parity $(-1, \alpha)$ in the $\mathbf{Z} \times \mathfrak{A}$-graded $\mathcal{A}$-Lie algebra $\mathrm{Der}_R(\bigwedge E)$.*

Proof. Uniqueness is guaranteed by [6.14], so let us show existence. We construct $\iota(\phi)$ by its restrictions to $\bigwedge^k E$. To that end we define a k-linear $\mathfrak{A}$-graded skew-symmetric map $E^k \to \bigwedge^{k-1} E$ by

(6.17) $(e_1, \ldots, e_k) \mapsto$

$$\sum_{\alpha_0, \ldots, \alpha_k} \sum_{i=1}^{k} (-1)^{\langle \alpha_0 | \sum_{j=1}^{i-1} \alpha_j \rangle} e_{\alpha_1} \wedge \cdots \wedge e_{\alpha_{i-1}} \wedge \phi_{\alpha_0}(e_{\alpha_i}) e_{\alpha_{i+1}} \wedge \cdots \wedge e_{\alpha_k} ,$$

where $\phi = \sum_{\alpha_0} \phi_{\alpha_0}$ and $e_i = \sum_{\alpha_i} e_{\alpha_i}$ are the decompositions in homogeneous parts. We leave it to the reader to verify that this map is indeed k-linear and $\mathfrak{A}$-graded skew-symmetric. We thus have an induced linear map $\iota(\phi) : \bigwedge^k E \to \bigwedge^{k-1} E$. Since this map is obviously of parity α whenever ϕ is, the full map $\iota(\phi) \in \mathrm{End}_R(\bigwedge E)$ is of parity $(-1, \alpha)$ whenever ϕ is of parity α. It thus remains to show that $\iota(\phi)$ satisfies the property of a right derivation, i.e., for homogeneous ϕ and A:

$$\iota(\phi)(A \wedge B) = (\iota(\phi)A) \wedge B + (-1)^{\langle \varepsilon(\phi) | \varepsilon(A) \rangle} A \wedge \iota(\phi)B .$$

Since this relation is bilinear in A and B, it is sufficient to verify it for elements of the form $A = e_1 \wedge \cdots \wedge e_k$ and $B = e_{k+1} \wedge \cdots \wedge e_{k+\ell}$. But for these elements it is an easy consequence of the defining formula (6.17) for $\iota(\phi)$. QED

6.18 Corollary. *For any two $\phi, \psi \in E^*$ the commutator $[\,\iota(\phi), \iota(\psi)\,] \in \mathrm{Der}_R(\bigwedge E)$ is zero.*

Proof. $[\,\iota(\phi), \iota(\psi)\,]$ maps $\bigwedge^k E$ to $\bigwedge^{k-2} E$. The result now follows from [6.15]. QED

6.19 Nota Bene. If in this corollary both ϕ and ψ are homogeneous, the conclusion translates as $\iota(\phi) \circ \iota(\psi) = -(-1)^{\langle\varepsilon(\phi)|\varepsilon(\psi)\rangle}\iota(\psi) \circ \iota(\phi)$, the "extra" minus sign coming from the $\mathbf{Z}\times\mathfrak{A}$-grading: $\langle\varepsilon(\iota(\phi))|\varepsilon(\iota(\psi))\rangle = \langle(-1,\varepsilon(\phi))|(-1,\varepsilon(\psi))\rangle = 1+\langle\varepsilon(\phi)|\varepsilon(\psi)\rangle$.

6.20 Discussion. As said in [6.16], ι is an even morphism of $\mathfrak{A}$-graded $\mathcal{A}$-modules. But the target module is $\mathbf{Z} \times \mathfrak{A}$-graded, not only $\mathfrak{A}$-graded. And, just as the $\mathfrak{A}$-graded ring $\mathcal{A}$ is $\mathbf{Z} \times \mathfrak{A}$-graded (by being $\{0\}$ for $\mathbf{Z}$-parities different from 0), we can turn E^* into a $\mathbf{Z}\times\mathfrak{A}$-graded $\mathcal{A}$-module. However, here we have a choice! One obvious choice is to define $(E^*)_{n,\alpha} = \{0\}$ whenever $n \neq 0$. In that case ι becomes a morphism of $\mathbf{Z} \times \mathfrak{A}$-parity $(-1,0)$; in particular it is no longer even. Another choice is to define $(E^*)_{n,\alpha} = \{0\}$ whenever $n \neq -1$ (Nota Bene. This is a valid definition because elements of E can not be multiplied). With this definition ι is an even morphism of $\mathbf{Z} \times \mathfrak{A}$-graded $\mathcal{A}$-modules.

Both approaches have their merits: from a theoretical viewpoint the second approach is nicer because one gets even maps; moreover, the second approach concurs with the idea that elements of $E_\alpha \cong (\bigwedge^1 E)_\alpha$ have parity $(1,\alpha)$. On the other hand, the first approach is the more practical one: if ι has parity $(-1,0)$ and $\phi \in E^*$ has parity $(0,\alpha)$, then the image $\iota(\phi)$ has parity $(-1,\alpha) = (-1,0) + (0,\alpha)$.

6.21 Remark. The choice of the symbol ι for the map $E^* \to \mathrm{Der}_R(\bigwedge E)$ is motivated by the applications in differential geometry. There we mainly look at right derivations of the exterior algebra $\bigwedge {}^*E$, where E stands for the module of vector fields. In [7.26] we will show that there is a canonical identification $E \to ({}^*E)^*$, and thus the combination with [6.16] gives us a map $E \to ({}^*E)^* \to \mathrm{Der}_R(\bigwedge {}^*E)$. The image of an element $e \in E$ applied to $\phi \in \bigwedge^1 {}^*E \cong {}^*E$ evaluates exactly as $\iota(e)\phi \equiv \langle\!\langle e \| \phi \rangle\!\rangle$, justifying our choice of the symbol ι for the map $E^* \to \mathrm{Der}_R(\bigwedge E)$.

7. Identifications

In previous sections we have given constructions of $\mathfrak{A}$-graded $\mathcal{A}$-modules out of other $\mathfrak{A}$-graded $\mathcal{A}$-modules, such as homomorphisms, direct sums, tensor products, and exterior powers. Some of these constructions "commute" under certain circumstances, i.e., applying these constructions in a different order gives results that are canonically isomorphic. In this and the next section we will provide a number of such identifications, of which a summary can be found in [8.18]. In this section these identifications are introduced as maps, in the next section the technical proofs that they are isomorphisms (and the conditions needed for that) are given. All definitions of these identifications are straightforward, although we are naturally led to introduce a non-standard ordering in the definition of the tensor product of maps [7.12]. This new ordering avoids a sign in an identification *involving the exterior power [7.21], a sign that is usually present in the ungraded case [7.22].*

This section and the next one are purely technical, and their contents are not needed for the understanding of the next chapters. The reader can thus safely skip these two sections at a first reading; looking at the summary [8.18] will give him a fair idea of the contents (s)he skipped.

7.1 Discussion. In what follows we will establish identifications between various spaces. Some of these will always be isomorphisms, some will be isomorphisms only with additional conditions. In establishing an identification, we will denote it by the symbol $\mathfrak{I}$. Afterwards we will usually omit this symbol and identify the two spaces without noting the explicit identification.

7.2 Construction (sum of maps). Let F and E_i, $i \in I$ be $\mathfrak{A}$-graded $\mathcal{A}$-modules, and let $\phi_i \in \mathrm{Maps}_S(E_i; F)$, $i \in I$ be linear maps. With these we define a new linear map $\sum_{i \in I} \phi_i \in \mathrm{Maps}_S(\bigoplus_{i\in I} E_i; F)$ by the formula

$$\sum_{i\in I} \phi_i : \oplus_i e_i \mapsto \sum_{i\in I} \phi_i(e_i) \ .$$

This is well defined because only finitely many e_i are different from zero, rendering the sum a finite one. We thus obtain a map $\mathfrak{I} : \prod_{i\in I} \mathrm{Maps}_S(E_i; F) \to \mathrm{Maps}_S(\bigoplus_{i\in I} E_i; F)$, $(\phi_i)_{i\in I} \mapsto \sum_{i\in I} \phi_i$. Source and target spaces are $\mathcal{A}$-modules (left for $S = R$ and right for $S = L$) and it is elementary to show that $\mathfrak{I}$ is linear. Moreover, if all ϕ_i have parity α, so does $\sum_{i\in I} \phi_i$. On the other hand, for $\phi \in \mathrm{Maps}_S(\bigoplus_{i\in I} E_i; F)$ we can define the maps $\phi_i = \phi \circ s_i \in \mathrm{Maps}_S(E_i; F)$, where the s_i denote the canonical injections $E_i \to \bigoplus_{j\in I} E_j$ [3.1]. It is immediate that $\sum_{i\in I} \phi_i = \phi$ proving that the map $\mathfrak{I}$ is a linear bijection preserving parities.

However, we are more interested in morphisms than in just linear maps. This means that we want to look at the restriction of $\mathfrak{I}$ to the subset $\prod_{i\in I} \mathrm{Hom}_S(E_i; F)$. Moreover, since we did not define arbitrary products of $\mathfrak{A}$-graded $\mathcal{A}$-modules, we are even more interested in the restriction of $\mathfrak{I}$ to $\bigoplus_{i\in I} \mathrm{Hom}_S(E_i; F)$, which is an $\mathfrak{A}$-graded $\mathcal{A}$-module. We claim that $\mathfrak{I}$ is an even injective homomorphism from $\bigoplus_{i\in I} \mathrm{Hom}_S(E_i; F)$ into $\mathrm{Hom}_S(\bigoplus_{i\in I} E_i; F)$. Since $\mathfrak{I}$ is injective and preserves parities, it only remains to prove that its restriction takes values in $\mathrm{Hom}_S(\bigoplus_{i\in I} E_i; F)$. So let $\oplus_{i\in I}\phi_i$ be an element in $\bigoplus_{i\in I} \mathrm{Hom}_S(E_i; F)$. This means that ϕ_i is a morphism and that only finitely many of them are non-zero. Since each of the non-zero ϕ_i is a morphism, it is a finite sum of homogeneous morphisms. Since $\mathfrak{I}$ is linear and preserves parities, this means that the image is a finite sum of homogeneous morphisms, i.e., $\mathfrak{I}(\bigoplus_{i\in I} \mathrm{Hom}_S(E_i; F)) \subset \mathrm{Hom}_S(\bigoplus_{i\in I} E_i; F)$ as claimed.

For $\phi \in \mathrm{Hom}_S(\bigoplus_{i\in I} E_i; F)$ we can construct the maps $\phi_i = \phi \circ s_i$, i.e., $\mathfrak{I}^{-1}(\phi)$. Since the s_i are morphisms, $\mathfrak{I}^{-1}(\phi)$ belongs to $\prod_{i\in I} \mathrm{Hom}_S(E_i; F)$, but there is no reason to think that it should belong to $\bigoplus_{i\in I} \mathrm{Hom}_S(E_i; F)$. However, if I is finite, these sets coincide and we find that $\mathfrak{I} : \bigoplus_{i\in I} \mathrm{Hom}_S(E_i; F) \to \mathrm{Hom}_S(\bigoplus_{i\in I} E_i; F)$ is an isomorphism. If a finite set I contains n elements, we denote the map $\sum_{i=1}^{n} \phi_i$ also by $\phi_1 + \cdots + \phi_n$.

7.3 Corollary. *Let E_i, $i \in I$ be $\mathfrak{A}$-graded $\mathcal{A}$-modules. The map $(\phi_i)_{i\in I} \mapsto \sum_{i\in I} \phi_i$ provides identifications $\bigoplus_{i\in I}(E_i^*) \to (\bigoplus_{i\in I} E_i)^*$ and $\bigoplus_{i\in I}({}^*E_i) \to {}^*(\bigoplus_{i\in I} E_i)$. If I is finite, these identifications are isomorphisms.*

7.4 Nota Bene. The notation $\sum_{i=1}^{n} \phi_i \equiv \phi_1 + \cdots + \phi_n$ is ambiguous in case all E_i are the same $\mathfrak{A}$-graded $\mathcal{A}$-module E, because in that case the ϕ_i all belong to $\mathrm{Hom}(E; F)$ in which an addition is already defined. In practice this will hardly ever cause misunderstandings because it will usually be clear from the context which interpretation to choose.

7.5 Example. Let F and E_i, $i \in I$ be $\mathfrak{A}$-graded $\mathcal{A}$-modules. If $\mathfrak{A}$ is finite, the sets Map_S and Hom_S coincide [2.5]. Since for infinite I the sets $\prod_{i\in I}$ and $\bigoplus_{i\in I}$ do not coincide, it is easy to construct examples in which the map $\mathfrak{I} : \bigoplus_{i\in I} \mathrm{Hom}_S(E_i; F) \to \mathrm{Hom}_S(\bigoplus_{i\in I} E_i; F)$ is not surjective. On the other hand, it is not obvious that the image of $\prod_{i\in I} \mathrm{Hom}_S(E_i; F)$ under $\mathfrak{I}$ is not necessarily contained in $\mathrm{Hom}_S(\bigoplus_{i\in I} E_i; F)$. To show that that can happen, we recall example [2.6]. There we defined the $\mathbf{Z}$-graded commutative algebra $\mathcal{A}$ to be the trivially graded $\mathcal{A} = \mathbf{R}$ and we argued that $\mathbf{Z}$-graded $\mathcal{A}$-modules are just real vector spaces. We now choose an infinite dimensional real vector space X and we define the $\mathbf{Z}$-graded $\mathcal{A}$-module E_n by $(E_n)_0 = \bigwedge^n X$ and $(E_n)_k = \{0\}$ for $k \neq 0$. We also define the $\mathbf{Z}$-graded $\mathcal{A}$-module $F = \bigoplus_{n=0}^{\infty} \bigwedge^n X$ with grading $F_k = \{0\}$ for $k < 0$ and $F_k = \bigwedge^k X$ for $k \geq 0$. We finally define the maps $\phi_n : E_n \to F$ by $\phi_n(e) = e \in F_n$. Because of our choice of $\mathbf{Z}$-grading, the map ϕ_n has parity n. Hence the family $(\phi_n)_{n\in\mathbf{N}}$ belongs to $\prod_{n\in\mathbf{N}} \mathrm{Hom}_S(E_n; F)$. However, the map $\sum_{n\in\mathbf{N}} \phi_n$ is not a *finite* sum of homogeneous maps (because X is infinite dimensional and thus all ϕ_n are non-zero). Hence in this example the image of $\prod_{i\in I} \mathrm{Hom}_S(E_i; F)$ under $\mathfrak{I}$ is not contained in $\mathrm{Hom}_S(\bigoplus_{i\in I} E_i; F)$.

7.6 Construction (direct sum of maps). Let E and $F_i, i \in I$ be $\mathfrak{A}$-graded $\mathcal{A}$-modules. We define the identification $\mathfrak{I} : \bigoplus_{i\in I} \mathrm{Hom}_S(E; F_i) \to \mathrm{Hom}_S(E; \bigoplus_{i\in I} F_i)$ by the formula

$$\mathfrak{I}(\oplus_i \phi_i) : e \mapsto \oplus_i \phi_i(e) .$$

This is well defined because there are only finitely many maps ϕ_i that are not zero. It defines a morphism because each of the finitely many non-zero ϕ_i is itself a finite sum of homogeneous linear maps. We leave it to the reader to verify that $\mathfrak{I}$ is indeed an injective even linear map.

7.7 Construction (product of maps). Let E_i and F_i, $i \in I$ be $\mathfrak{A}$-graded $\mathcal{A}$-modules, and let $\phi_i \in \mathrm{Map}_S(E_i; F_i)$, $i \in I$ be linear maps. With these we define a new linear map $\prod_{i\in I} \phi_i \in \mathrm{Map}_S(\bigoplus_{i\in I} E_i; \bigoplus_{i\in I} F_i)$ by the formula

$$\prod_{i\in I} \phi_i : \oplus_i e_i \mapsto \oplus_i \phi_i(e_i) .$$

This is well defined because only finitely many e_i are different from zero. We thus obtain a map $\mathfrak{I} : \prod_{i\in I} \mathrm{Map}_S(E_i; F_i) \to \mathrm{Map}_S(\bigoplus_{i\in I} E_i; \bigoplus_{i\in I} F_i)$, $(\phi_i)_{i\in I} \mapsto \prod_{i\in I} \phi_i$. We leave it to the reader to show that this $\mathfrak{I}$ is injective. In case I is finite, we denote the map $\prod_{i=1}^n \phi_i$ also by $\phi_1 \times \cdots \times \phi_n$.

7.8 Lemma. *The linear map $\prod_{i\in I} \phi_i$ is injective (surjective, of parity α) if and only if all maps ϕ_i are injective (surjective, of parity α). In particular we obtain an injective map $\mathfrak{I} : \prod_{i\in I} \mathrm{Aut}(E_i) \to \mathrm{Aut}(\bigoplus_{i\in I} E_i)$, $(\phi_i)_{i\in I} \mapsto \prod_{i\in I} \phi_i$.*

7.9 Remark. If E_i and F_i, $i \in I$ are $\mathfrak{A}$-graded $\mathcal{A}$-modules, we can combine the operations of *sum of maps* and *direct sum of maps* to obtain an injective map

$$\prod_{i\in I} \bigoplus_{j\in I} \mathrm{Map}_S(E_i; F_j) \to \mathrm{Map}_S(\bigoplus_{i\in I} E_i; \bigoplus_{j\in I} F_j) .$$

Restriction to the "diagonal" $\prod_{i\in I} \mathrm{Map}_S(E_i; F_i)$ would have given us the *product of maps*. There is one subtle point in this discussion: one has to be very careful with the order in which one applies the operations *sum of maps* and *direct sum of maps,* because the operations $\bigoplus_{i\in I}$ and $\prod_{j\in I}$ do not commute in general.

7.10 Construction (homomorphisms on tensor products). Let E, F, and G be $\mathfrak{A}$-graded $\mathcal{A}$-modules. We want to show that $\mathrm{Hom}_R(E \otimes F; G)$ and $\mathrm{Hom}_R(E; \mathrm{Hom}_R(F; G))$ are isomorphic, as well as $\mathrm{Hom}_L(E \otimes F; G)$ and $\mathrm{Hom}_L(F; \mathrm{Hom}_L(E; G))$. In the right linear case we define the map $\mathfrak{I} : \mathrm{Hom}_R(E \otimes F; G) \to \mathrm{Hom}_R(E; \mathrm{Hom}_R(F; G))$ by the formula

$$\mathfrak{I}(\phi)(e) : f \mapsto \phi(e \otimes f) .$$

This is well defined because for each $e \in E$ the map $\mathfrak{I}(\phi)(e)$ is indeed right linear. Using that $e\lambda \otimes f = e \otimes \lambda f$, one can show that $\mathfrak{I}(\phi)$ is right linear. If ϕ, e, and f are homogeneous, we obviously have $\varepsilon(\mathfrak{I}(\phi)(e)(f)) = \varepsilon(\phi(e \otimes f)) = \varepsilon(\phi) + \varepsilon(e) + \varepsilon(f)$. It follows that $\mathfrak{I}$ is an even linear map. To show that it is bijective, let us construct its inverse. Given $\psi \in \mathrm{Hom}_R(E; \mathrm{Hom}_R(F; G))$, we define a right bilinear map $\widehat{\psi} : E \times F \to G$ by $\widehat{\psi}(e, f) = \psi(e)(f)$. This gives us an induced right linear map $\phi : E \otimes F \to G$. We leave it to the reader to verify that the map $\psi \mapsto \phi$ is actually the inverse to $\mathfrak{I}$. We conclude that $\mathfrak{I}$ is an isomorphism.

In the left linear case, we define $\mathfrak{I} : \mathrm{Hom}_L(E \otimes F; G) \to \mathrm{Hom}_L(F; \mathrm{Hom}_L(E; G))$ by the formula

$$\langle\!\langle f \,\|\, \mathfrak{I}(\phi) \rangle\!\rangle : e \mapsto \langle\!\langle e \otimes f \,\|\, \phi \rangle\!\rangle .$$

In the same way as for the right linear case one can show that this is an isomorphism.

7.11 Construction (homomorphisms as tensor products). Let E, F, and G be $\mathfrak{A}$-graded $\mathcal{A}$-modules. We will construct identifications $\mathfrak{J} : F \otimes \mathrm{Hom}_R(E; G) \to \mathrm{Hom}_R(E; F \otimes G)$ and $\mathfrak{J} : \mathrm{Hom}_L(E; G) \otimes F \to \mathrm{Hom}_L(E; G \otimes F)$. In the right linear case we define $\mathfrak{J}$ as follows. For $f \in F$ and $\phi \in \mathrm{Hom}_R(E; G)$ we define $\Psi(f, \phi) \in \mathrm{Hom}_R(E; F \otimes G)$ by the formula

$$\Psi(f, \phi) : e \mapsto f \otimes \phi(e) \ .$$

Obviously $\Psi(f, \phi)$ is right linear, i.e., $\Psi(f, \phi)$ is well defined. Moreover, the map $\Psi : F \times \mathrm{Hom}_R(E; G) \to \mathrm{Hom}_R(E; F \otimes G)$ is left bilinear and even. The identification $\mathfrak{J}$ is the induced even linear map $F \otimes \mathrm{Hom}_R(E; G) \to \mathrm{Hom}_R(E; F \otimes G)$. In a similar way we define the identification $\mathfrak{J} : \mathrm{Hom}_L(E; G) \otimes F \to \mathrm{Hom}_L(E; G \otimes F)$ as the even linear map induced by the bilinear map $\Psi : \mathrm{Hom}_L(E; G) \times F \to \mathrm{Hom}_L(E; G \otimes F)$ given by the formula $\Psi(\phi, f) : e \mapsto \langle\!\langle e \| \phi \rangle\!\rangle \otimes f$.

Taking $G = \mathcal{A}$ gives us an identification ${}^*E \otimes F \to \mathrm{Hom}_L(E; F)$ in which the element $\phi \otimes f$ is identified with the map

$$\langle\!\langle e \| \phi \otimes f \rangle\!\rangle = \langle\!\langle e \| \phi \rangle\!\rangle \cdot f \ .$$

In a similar way we obtain an identification $F \otimes E^* \to \mathrm{Hom}_R(E; F)$ in which $f \otimes \phi$ is identified with the map $e \mapsto f \cdot \phi(e)$.

Nota Bene: One might be tempted to think that, with these identifications, the transposition map is just the interchanging operator in disguise. But things are not so easy. Applying $\mathfrak{R}$ to $\mathrm{Hom}_L(E; F) \cong {}^*E \otimes F$ gives us the space $F \otimes {}^*E$, and we then still have to apply the transpose map $\mathfrak{T} : {}^*E \to E^*$ to get $F \otimes E^* \cong \mathrm{Hom}_R(E; F)$.

7.12 Construction (tensor product of maps). If we combine [7.10] and [7.11] we obtain in particular the following identification concerning two $\mathfrak{A}$-graded $\mathcal{A}$-modules E and F :

$$\begin{aligned} F^* \otimes E^* \cong \mathrm{Hom}_R(E; F^*) &\equiv \mathrm{Hom}_R(E; \mathrm{Hom}_R(F; \mathcal{A})) \\ &\cong \mathrm{Hom}_R(E \otimes F; \mathcal{A}) \equiv (E \otimes F)^* \ , \end{aligned}$$

which is given on elements $e \in E$, $f \in F$, $\psi \in F^*$, and $\phi \in E^*$ by

$$(\psi \otimes \phi)(e \otimes f) = \psi(\phi(e)\, f) = (-1)^{\langle \varepsilon(\psi) | \varepsilon(\phi(e)) \rangle} \phi(e)\, \psi(f) \ .$$

In order to generalize this identification to arbitrary homomorphisms, we consider the map $\Theta : \mathrm{Hom}_R(E_2; F_2) \times \mathrm{Hom}_R(E_1; F_1) \to \mathrm{Hom}_R(E_1 \otimes E_2; F_1 \otimes F_2)$ defined by

$$\Theta(\phi_2, \phi_1) : (e_1 \otimes e_2) \mapsto \sum_{\alpha, \beta, \gamma \in \mathfrak{A}} (-1)^{\langle \alpha | \beta + \gamma \rangle} \phi_{1\beta}(e_{1\gamma}) \otimes \phi_{2\alpha}(e_2) \ .$$

We leave it to the reader to verify that $\Theta(\phi_2, \phi_1)$ is indeed a well defined right linear map (the sign will be crucial), and that Θ is an even (right) bilinear map. We thus obtain an induced identification $\mathfrak{I} : \mathrm{Hom}_R(E_2; F_2) \otimes \mathrm{Hom}_R(E_1; F_1) \to \mathrm{Hom}_R(E_1 \otimes E_2; F_1 \otimes F_2)$ such that for homogeneous elements we have

$$\big(\mathfrak{I}(\phi_2 \otimes \phi_1)\big)(e_1 \otimes e_2) = (-1)^{\langle \varepsilon(\phi_2) | \varepsilon(\phi_1(e_1)) \rangle} \phi_1(e_1) \otimes \phi_2(e_2) \ .$$

We can easily generalize this by induction to two families $(E_i)_{i=1}^n$ and $(F_i)_{i=1}^n$ of $\mathfrak{A}$-graded $\mathcal{A}$-modules, to obtain an identification

$$\mathfrak{I} : \bigotimes_{i=n}^{1} \mathrm{Hom}_R(E_i; F_i) \to \mathrm{Hom}_R(\bigotimes_{i=1}^{n} E_i; \bigotimes_{i=1}^{n} F_i)$$

such that for homogeneous elements we have

$$(\mathfrak{I}(\phi_n \otimes \cdots \otimes \phi_1))(e_1 \otimes \cdots \otimes e_n) = (-1)^{\sum\limits_{i>j} \langle \varepsilon(\phi_i) | \varepsilon(\phi_j(e_j)) \rangle} \phi_1(e_1) \otimes \cdots \otimes \phi_n(e_n) \ .$$

Similarly an identification $\mathfrak{I} : \bigotimes_{i=n}^{1} \mathrm{Hom}_L(E_i; F_i) \to \mathrm{Hom}_L(\bigotimes_{i=1}^{n} E_i; \bigotimes_{i=1}^{n} F_i)$ is constructed such that for homogeneous elements we have

$$\langle\!\langle e_1 \otimes \cdots \otimes e_n \,\|\, \mathfrak{I}(\phi_n \otimes \cdots \otimes \phi_1) \rangle\!\rangle = (-1)^{\sum\limits_{i<j} \langle \varepsilon(\phi_i) | \varepsilon(\phi_j(e_j)) \rangle} \langle\!\langle e_1 \,\|\, \phi_1 \rangle\!\rangle \otimes \cdots \otimes \langle\!\langle e_n \,\|\, \phi_n \rangle\!\rangle \ .$$

Of course there is no compelling reason to use the reverse order in the tensor product $\bigotimes_{i=n}^{1} \mathrm{Hom}_L(E_i; F_i)$, except that [7.10] and [7.11] suggested it that way. There is also a kind of natural explanation: in the expression $\big(\mathfrak{I}(\phi_n \otimes \cdots \otimes \phi_1)\big)(e_1 \otimes \cdots \otimes e_n)$ the ϕ_1 is next to e_1 and is ready to be applied. Then we have to shift the result out of the way and we are ready to apply ϕ_2 to e_2, and so on. One could look at it as if two rows of dancers meet head on, forming a new row of pairs. It also is more natural in the proof of sufficient conditions for these identifications to be isomorphisms [8.12]. Moreover, it fits better when discussing duals of exterior powers [7.16] (see also [7.22]).

Taking all F_i to be $\mathcal{A}$ gives us identifications ${}^*E_n \otimes \cdots \otimes {}^*E_1 \to {}^*(E_1 \otimes \cdots \otimes E_n)$ and $E_n^* \otimes \cdots \otimes E_1^* \to (E_1 \otimes \cdots \otimes E_n)^*$.

7.13 Remark. If all $\phi_i \in {}^*E_i$ are even, and in particular also right linear, then the images $\mathfrak{I}_L(\phi_n \otimes \cdots \otimes \phi_1) \in {}^*(E_1 \otimes \cdots \otimes E_n)$ and $\mathfrak{I}_R(\phi_n \otimes \cdots \otimes \phi_1) \in (E_1 \otimes \cdots \otimes E_n)^*$ coincide (and are even), as can be verified easily.

7.14 Lemma. *If $\psi \in \mathrm{Hom}_S(E; F)$ is bijective, then $id_G \otimes \psi \in \mathrm{Hom}_S(E \otimes G; F \otimes G)$ is bijective with $id_G \otimes \psi^{-1}$ as inverse. Similarly the map $\psi \otimes id_G$ is bijective with $\psi^{-1} \otimes id_G$ as inverse.*

7.15 Construction (maps on exterior powers). Let E and F be $\mathfrak{A}$-graded $\mathcal{A}$-modules and let $\phi : E \to F$ be right/left linear. We want to investigate when this map induces a linear map $\bigwedge^k E \to \bigwedge^k F$. According to [7.12] we have a map $\phi \otimes \cdots \otimes \phi : \bigotimes^k E \to \bigotimes^k F$, which we can compose with the canonical projection onto $\bigwedge^k F$ to yield a linear map $\Phi : \bigotimes^k E \to \bigwedge^k F$. This map induces a linear map $\bigwedge^k E \to \bigwedge^k F$ if and only if the map Φ satisfies the relations $\Phi \circ \mathfrak{R}_{(ii+1)} = -\Phi$ [5.1]. It is fairly easy to see that this is the case if ϕ is even. We conclude that for an *even* linear map $\phi : E \to F$ there exists an induced even linear map $\bigwedge^k \phi : \bigwedge^k E \to \bigwedge^k F$ such that for $e_i \in E$ one has

$$\textstyle\bigwedge^k \phi : e_1 \wedge \cdots \wedge e_k \mapsto \phi(e_1) \wedge \cdots \wedge \phi(e_k) \ .$$

Since $\bigwedge^1 E = E$ [5.3], we quite naturally find that $\bigwedge^1 \phi = \phi$. For $k = 0$ we formally defined $\bigwedge^0 E = \mathcal{A}$; for maps we now formally define $\bigwedge^0 \phi = id(\mathcal{A})$.

An immediate consequence of these definitions and the definition of the exterior product [5.8] is that for $A \in \bigwedge^k E$ and $B \in \bigwedge^\ell E$ we have

$$(\textstyle\bigwedge^k \phi)(A) \wedge (\bigwedge^\ell \phi)(B) = (\bigwedge^{k+\ell} \phi)(A \wedge B) \ .$$

It should be obvious that if ϕ is bijective, i.e., an isomorphism, then $\bigwedge^k \phi$ also is bijective with inverse $\bigwedge^k (\phi^{-1})$. In categorical language one would say that $\bigwedge^k$ is a functor, but one which we only apply to *even* morphisms.

7.16 Construction (dual of exterior powers). Let E be an $\mathfrak{A}$-graded $\mathcal{A}$-module. We will construct identifications $\mathfrak{I} : \bigwedge^k (E^*) \to (\bigwedge^k E)^*$ and $\mathfrak{I} : \bigwedge^k {}^*E \to {}^*(\bigwedge^k E)$. There are two ways to construct this isomorphism, a fast way and a pedestrian way. We start with the pedestrian way for the right linear case. We denote by $\mathfrak{I}_\otimes$ the identification $\mathfrak{I}_\otimes : \bigotimes^k (E^*) \to (\bigotimes^k E)^*$ of [7.12], and we introduce the modified permutation operators $\widehat{\mathfrak{R}}_\sigma = \mathfrak{R}_{\tau\sigma\tau}$ [4.13], where $\tau \in \mathfrak{S}_k$ is the fixed permutation $\tau(i) = k+1-i$ (the one which reverses the order of k elements). For homogeneous elements e_i this implies

$$\widehat{\mathfrak{R}}_\sigma(e_k \otimes \cdots \otimes e_1) = \pm e_{\sigma^{-1}(k)} \otimes \cdots \otimes e_{\sigma^{-1}(1)} \ ,$$

where the sign is determined by the permutation σ and the parities of the e_i. We now claim that we have the following equality of maps from $\bigotimes^k E$ to $\mathcal{A}$:

$$\mathfrak{I}_\otimes(\phi_k \otimes \cdots \otimes \phi_1) \circ \mathfrak{R}_\sigma = \mathfrak{I}_\otimes(\widehat{\mathfrak{R}}_{\sigma^{-1}}(\phi_k \otimes \cdots \otimes \phi_1)) \ . \tag{7.17}$$

Since the neighbor interchanges generate $\mathfrak{S}_k$, it suffices to check this for σ of the form $\sigma = (ii+1)$. The essential computation to verify is the case $k = 2$, which is left to the reader.

We now define the skew-symmetrization operator $\mathbb{A}$ by the formula

$$\mathbb{A} = \sum_{\sigma \in \mathfrak{S}_k} (-1)^\sigma \cdot \mathfrak{R}_\sigma = \sum_{\sigma \in \mathfrak{S}_k} (-1)^\sigma \cdot \widehat{\mathfrak{R}}_{\sigma^{-1}} \ ,$$

where $(-1)^\sigma$ denotes the sign of the permutation σ. This skew-symmetrization operator has the important property that for any permutation σ we have

$$\mathbb{A} \circ \mathfrak{R}_\sigma = \mathfrak{R}_\sigma \circ \mathbb{A} = (-1)^\sigma \cdot \mathbb{A} = \widehat{\mathfrak{R}}_{\sigma^{-1}} \circ \mathbb{A} = \mathbb{A} \circ \widehat{\mathfrak{R}}_{\sigma^{-1}} \, . \tag{7.18}$$

With these preparations we can construct the identification $\mathfrak{I}$. We first define the map $\psi = \mathfrak{I}_\otimes \circ \mathbb{A} : \bigotimes^k E^* \to (\bigotimes^k E)^*$. Using (7.17) we obtain for any $\Phi \in \bigotimes^k E^*$ the equality $\psi(\Phi) = \mathfrak{I}_\otimes(\Phi) \circ \mathbb{A}$. It then follows from (7.18) that $\psi(\Phi) : \bigotimes^k E \to \mathcal{A}$ is $\mathfrak{A}$-graded skew-symmetric [5.1], and thus induces a map $\Psi(\Phi) \in (\bigwedge^k E)^*$. On the other hand, it also follows from (7.18) that $\Psi(\widehat{\mathfrak{R}}_{\sigma^{-1}}(\Phi)) = (-1)^\sigma \cdot \Psi(\Phi)$, i.e., the map $\Psi : \bigotimes^k E^* \to (\bigwedge^k E)^*$ is $\mathfrak{A}$-graded skew-symmetric. We thus obtain an induced map $\mathfrak{I} : \bigwedge^k E^* \to (\bigwedge^k E)^*$. This identification is given explicitly by the formula

$$\mathfrak{I}(\phi_k \wedge \cdots \wedge \phi_1)(e_1 \wedge \cdots \wedge e_k) = \sum_{\sigma \in \mathfrak{S}_k} (-1)^\sigma \cdot \mathfrak{I}_\otimes(\phi_k \otimes \cdots \otimes \phi_1)(\mathfrak{R}_\sigma(e_1 \otimes \cdots \otimes e_k)) \, . \tag{7.19}$$

So far the pedestrian way to define $\mathfrak{I}$. The fast way uses the concept of derivations. Recall [6.16] that for $\phi_i \in E^*$ we have defined a derivation $\iota(\phi)$ of bi-degree $(-1, \varepsilon(\phi))$ on the exterior algebra $\bigwedge E$. In particular the composite $\chi(\phi_1, \ldots, \phi_k) = \iota(\phi_1) \circ \cdots \circ \iota(\phi_k)$ restricts to a map $\bigwedge^k E \to \bigwedge^0 E = \mathcal{A}$. It follows from [6.16] and [6.18] ([6.19]) that $\chi : (E^*)^k \to (\bigwedge^k E)^*$ is k-linear and $\mathfrak{A}$-graded skew-symmetric. We thus have an induced map $\mathfrak{I} : \bigwedge^k E^* \to (\bigwedge^k E)^*$. In [7.21] we will prove that these two definitions coincide, i.e., that we are allowed to use the same symbol. That we obtain *equality* of both definitions, and not an equality up to a sign, is due to the reversed order in the tensor product in [7.12].

In the left-linear case the construction is similar. The pedestrian way uses the identification $\mathfrak{I}_\otimes : \bigotimes^k({}^*E) \to {}^*(\bigotimes^k E)$ [7.12] and the fast way uses the left-derivations equivalent of [6.16] (which gives a map ${}^*E \to \mathrm{Der}_L(\bigwedge E)$).

7.20 Example. Let $\phi_1, \phi_2 \in E^*$ and $e_1, e_2 \in E$ be homogeneous, then the identification between $\bigwedge^2 E^*$ and $(\bigwedge^2 E)^*$ gives us:

$$\begin{aligned}\mathfrak{I}(\phi_2 \wedge \phi_1)(e_1 \wedge e_2) &= \mathfrak{I}_\otimes(\phi_2 \otimes \phi_1)(e_1 \otimes e_2) - \mathfrak{I}_\otimes(\phi_2 \otimes \phi_1)\mathfrak{R}_{(12)}(e_1 \otimes e_2) \\ &= (-1)^{\langle \varepsilon(\phi_2) | \varepsilon(\phi_1(e_1)) \rangle} \phi_1(e_1)\phi_2(e_2) - \\ &\qquad (-1)^{\langle \varepsilon(\phi_1) | \varepsilon(e_1) \rangle} \phi_2(e_1)\phi_1(e_2) \, .\end{aligned}$$

Apart from the additional signs, this formula can be seen as the ordinary determinant of the matrix $\begin{pmatrix} \phi_1(e_1) & \phi_2(e_1) \\ \phi_1(e_2) & \phi_2(e_2) \end{pmatrix}$. More generally, the value of $\mathfrak{I}(\phi_1 \wedge \cdots \wedge \phi_k)(e_k \wedge \cdots \wedge e_1)$ can be seen as a generalization of $\mathrm{Det}(\phi_i(e_j))$. However, this generalization of the determinant should not be confused with the Berezinian or graded determinant to be defined in [II.5.16]. The latter is related to the group $\mathrm{Aut}(E)$, whereas the former bears no natural relation to this group.

7.21 Proposition. *Let $\phi_1, \ldots, \phi_k \in E^*$ be arbitrary, then as maps from $\bigwedge^k E$ to $\mathcal{A}$ we have the identity*

$$\mathfrak{I}(\phi_1 \wedge \cdots \wedge \phi_k) = \iota(\phi_1) \circ \cdots \circ \iota(\phi_k) ,$$

where the $\iota(\phi_i)$ are the derivations defined in [6.16].

Proof. Since both expressions are left k-linear in the ϕ_i, it suffices to check the identity for homogeneous ϕ_i. Since elements $e_1 \wedge \cdots \wedge e_k$ with $e_i \in E$ homogeneous generate $\bigwedge^k E$, it again suffices to check this identity on elements of this form. Now using the derivation property of $\iota(\phi)$, it is easy to check the following identity:

$$\iota(\phi)(e_1 \wedge \cdots \wedge e_k) = \sum_{i=1}^{k} (-1)^{(i-1)+\sum_{j<i} \langle \varepsilon(e_i) | \varepsilon(e_j) \rangle} \phi(e_i) e_1 \cdots \wedge e_{i-1} \wedge e_{i+1} \wedge \cdots \wedge e_k .$$

Using the map $id_{k-1} \otimes \phi : \bigotimes^k E \to \mathcal{A} \otimes \bigotimes^{k-1} E \cong \bigotimes^{k-1} E$ and the canonical projection $\pi : \bigotimes^k E \to \bigwedge^k E$, where we introduced the abbreviation id_k for the identity map on $\bigotimes^k E$, we can write this identity as

$$(\iota(\phi) \circ \pi)(e_1 \otimes \cdots \otimes e_k) =$$

$$(\pi \circ (id_{k-1} \otimes \phi)) \Big(\sum_{i=1}^{k} (-1)^{(i-1)+\sum_{j<i} \langle \varepsilon(e_i) | \varepsilon(e_j) \rangle} e_i \otimes e_1 \cdots \otimes e_{i-1} \otimes e_{i+1} \otimes \cdots \otimes e_k \Big) .$$

We now introduce for any $1 \le i \le j \le n$ the permutations $\sigma_i^j \in \mathfrak{S}_n$ defined as the cyclic permutation from i to j, i.e., $\sigma_i^j(p) = p$ for $p < i$ or $p > j$, $\sigma_i^j(j) = i$, and $\sigma_i^j(p) = p+1$ for $i \le p < j$. Using the definition of the operators $\mathfrak{R}_\sigma$ [4.12], [4.13] one can easily show that

$$\mathfrak{R}_{\sigma_1^i}(e_1 \otimes \cdots \otimes e_k) = \sum_{i=1}^{k} (-1)^{\sum_{j<i} \langle \varepsilon(e_i) | \varepsilon(e_j) \rangle} e_i \otimes e_1 \cdots \otimes e_{i-1} \otimes e_{i+1} \otimes \cdots \otimes e_k .$$

Using that the $e_1 \otimes \cdots \otimes e_k$ generate $\bigotimes^k E$ and that $(-1)^{\sigma_i^j} = (-1)^{j-i}$, we find

$$\iota(\phi) \circ \pi = \pi \circ (id_{k-1} \otimes \phi) \circ \Big(\sum_{i=1}^{k} (-1)^{\sigma_1^i} \mathfrak{R}_{\sigma_1^i} \Big)$$

as maps from $\bigotimes^k E$ to $\bigwedge^{k-1} E$. Composing on the left with $\iota(\psi)$ and using the same formula with k replaced by $k-1$ gives us:

$$\begin{aligned} \iota(\psi) \circ \iota(\phi) \circ \pi &= \pi \circ (id_{k-2} \otimes \psi) \circ \Big(\sum_{j=1}^{k-1} (-1)^{\sigma_1^j} \mathfrak{R}_{\sigma_1^j} \Big) \circ (id_{k-1} \otimes \phi) \circ \Big(\sum_{i=1}^{k} (-1)^{\sigma_1^i} \mathfrak{R}_{\sigma_1^i} \Big) \\ &= \pi \circ (id_{k-2} \otimes \psi \otimes \phi) \circ \Big(\sum_{j=2}^{k} \sum_{i=1}^{k} (-1)^{\sigma_2^j \cdot \sigma_1^i} \mathfrak{R}_{\sigma_2^j \cdot \sigma_1^i} \Big) . \end{aligned}$$

To obtain the second line we used that $(id_{k-2}\otimes\psi)\circ(id_{k-1}\otimes\phi) = id_{k-2}\otimes\psi\otimes\phi$ and that $\mathfrak{R}_{\sigma_1^j}\circ(id_{k-1}\otimes\phi) = (id_{k-1}\otimes\phi)\circ\mathfrak{R}_{\sigma_2^{j+1}}$ because the map $id_{k-1}\otimes\phi : \bigotimes^k E \to \bigotimes^{k-1} E$ eats up the first factor.

By induction we get the identity

$$\iota(\phi_k)\circ\cdots\circ\iota(\phi_1)\circ\pi = \pi\circ(\phi_k\otimes\cdots\otimes\phi_1)\circ\Big(\sum_{i_k=k}^{k}\cdots\sum_{i_1=1}^{k}(-1)^{\sigma_k^{i_k}\cdots\sigma_1^{i_1}}\,\mathfrak{R}_{\sigma_k^{i_k}\cdots\sigma_1^{i_1}}\Big)\ .$$

It now suffices to note that any permutation $\sigma\in\mathfrak{S}_k$ has a unique representation as $\sigma = \sigma_k^{i_k}\cdots\sigma_1^{i_1}$, where i_j is the position of $\sigma^{-1}(j)$ in the list $1,2,\ldots,k$ in which the items $\sigma^{-1}(1),\ldots,\sigma^{-1}(j-1)$ are (already) put in front. Since there are already $j-1$ elements in front which are different from $\sigma^{-1}(j)$, this position is greater or equal to j, so $\sigma_j^{i_j}$ makes sense. We conclude that

$$\iota(\phi_k)\circ\cdots\circ\iota(\phi_1)\circ\pi = (\phi_k\otimes\cdots\otimes\phi_1)\circ\mathbb{A}\ ,$$

where we have omitted the π at the right hand side because there $\pi : \bigotimes^0 E \to \bigwedge^0 E$ is the identity map $\mathcal{A}\to\mathcal{A}$. Applying this to $e_1\otimes\cdots\otimes e_k$ and comparing with (7.19) finishes the proof. QED

7.22 Remark. The fact that we have no sign at all in [7.21] is a direct consequence of the reverse order in our definition of $\mathfrak{I}_\otimes$ in [7.12]. In ordinary exterior powers (over $\mathbf{R}$), the identification is usually given as

$$\mathfrak{I}(\phi_1\wedge\cdots\wedge\phi_k)(e_1\wedge\cdots\wedge e_k) = \mathrm{Det}(\phi_i(e_j))\ ,$$

and then one gets the relation

$$\begin{aligned}(\iota(\phi_1)\circ\cdots\circ\iota(\phi_k))(e_1\wedge\cdots\wedge e_k) &= (-1)^{k(k-1)/2}\,\mathrm{Det}(\phi_i(e_j))\\ &= (-1)^{k(k-1)/2}\mathfrak{I}(\phi_1\wedge\cdots\wedge\phi_k)(e_1\wedge\cdots\wedge e_k)\ .\end{aligned}$$

7.23 Corollary. *For $\phi\in E^*$ we can define a left linear map $\mathcal{E}(\phi) : \bigwedge^k E^* \to \bigwedge^{k+1} E^*$ by $\langle\!\langle A \| \mathcal{E}(\phi)\rangle\!\rangle = A\wedge\phi$. With the right linear map $\iota(\phi) : \bigwedge^{k+1} E \to \bigwedge^k E$ defined in [6.16], we have a commutative diagram*

$$\begin{array}{ccc}\bigwedge^k E^* & \xrightarrow{\ \mathcal{E}(\phi)\ } & \bigwedge^{k+1} E^*\\ {\scriptstyle\mathfrak{I}}\big\downarrow & & \big\downarrow{\scriptstyle\mathfrak{I}}\\ (\bigwedge^k E)^* & \xrightarrow[\ \iota(\phi)^*\ ]{} & (\bigwedge^{k+1} E)^*\ ,\end{array}$$

i.e., under the identifications $\bigwedge^k E^* \cong (\bigwedge^k E)^*$ *and* $\bigwedge^{k+1} E^* \cong (\bigwedge^{k+1} E)^*$ *we may identify the right dual* $\iota(\phi)^*$ *of* $\iota(\phi)$ *with the map* $\mathcal{E}(\phi)$ *of right exterior product with* ϕ. *(A similar result holds in the left dual case.)*

Proof. For $e_1, \ldots, e_{k+1} \in E$ and $\phi_1, \ldots, \phi_k \in E^*$ we compute:

$$\begin{aligned}\Big(\iota(\mathfrak{I}(\phi_1 \wedge \cdots \wedge \phi_k))\iota(\phi)^*\Big)(e_1 \wedge \cdots \wedge e_{k+1}) &= \\ &= \mathfrak{I}(\phi_1 \wedge \cdots \wedge \phi_k)\Big(\iota(\phi)(e_1 \wedge \cdots \wedge e_{k+1})\Big) \\ &= (\iota(\phi_1) \circ \cdots \circ \iota(\phi_k))\Big(\iota(\phi)(e_1 \wedge \cdots \wedge e_{k+1})\Big) \\ &= \mathfrak{I}(\phi_1 \wedge \cdots \wedge \phi_k \wedge \phi)(e_1 \wedge \cdots \wedge e_{k+1})\ .\end{aligned}$$

Since the $e_1 \wedge \cdots \wedge e_{k+1}$ generate $\bigwedge^{k+1} E$ and the $\phi_1 \wedge \cdots \wedge \phi_k$ generate $\bigwedge^k E^*$, the result follows. QED

7.24 Discussion. So far we have introduced five constructions to create new morphisms out of a family of morphisms: $\sum_i \phi_i$, $\oplus_i \phi_i$, $\prod_i \phi_i$, $\otimes_i \phi_i$, and $\bigwedge^k \phi$. We also have constructed identifications which allow us to permute the operation *taking the dual* with the constructions *direct sum* [7.3], *tensor product* [7.12], and *exterior product* [7.16]. We now claim that these identifications also allow us to permute the operation *taking the dual map* with the five constructions of new morphisms. More precisely, one obtains identifications $(\sum_i \phi_i)^* \cong \oplus_i(\phi_i^*)$, $(\oplus_i \phi_i)^* \cong \sum_i(\phi_i^*)$, $(\prod_i \phi_i)^* \cong \prod_i(\phi_i^*)$, $(\otimes_i \phi_i)^* \cong \otimes_i(\phi_i^*)$, and $(\bigwedge^k \psi)^* \cong \bigwedge^k(\psi^*)$. We will give the exact formulation of this statement and its proof only for the case $\bigwedge^k \psi$; the other cases are similar and are left to the reader.

7.25 Proposition. *If* $\psi \in \mathrm{Hom}(E; F)_0$ *is an even morphism, we have a commutative diagram*

$$\begin{array}{ccc} \bigwedge^k {}^*F & \xrightarrow{\bigwedge^k {}^*\psi} & \bigwedge^k {}^*E \\ \mathfrak{I}\big\downarrow & & \big\downarrow\mathfrak{I} \\ {}^*(\bigwedge^k F) & \xrightarrow[{}^*(\bigwedge^k \psi)]{} & {}^*(\bigwedge^k E)\ , \end{array}$$

i.e., under the identifications $\bigwedge^k {}^*E \cong {}^*(\bigwedge^k E)$ *and* $\bigwedge^k {}^*F \cong {}^*(\bigwedge^k F)$ *we may identify* $\bigwedge^k {}^*\psi$ *with* ${}^*(\bigwedge^k \psi)$. *Similarly we may identify* $(\bigwedge^k \psi)^*$ *with* $\bigwedge^k(\psi^*)$.

Proof. For $e_1, \ldots, e_k \in E$ and $\phi_1, \ldots, \phi_k \in {}^*F$ the definitions give

$$\begin{aligned}\iota(e_1 \wedge \cdots \wedge e_k)\Big({}^*(\textstyle\bigwedge^k \psi)\mathfrak{I}(\phi_1 \wedge \cdots \phi_k)\Big) &= \iota\Big(\iota(e_1 \wedge \cdots \wedge e_k)\textstyle\bigwedge^k \psi\Big)\mathfrak{I}(\phi_1 \wedge \cdots \phi_k) \\ &= \iota\Big(\iota(e_1)\psi \wedge \cdots \wedge \iota(e_k)\psi\Big)\mathfrak{I}(\phi_1 \wedge \cdots \phi_k)\ .\end{aligned}$$

By definition of $\mathfrak{I}$, the last term is a sum (with signs) of products of terms of the form $\iota(\iota(e_i)\psi)\phi_j = \iota(e_i)({}^*\psi(\phi_j))$. Since ψ is even, the signs do not change and we find

$$\begin{aligned}\iota(e_1 \wedge \cdots \wedge e_k)\big({}^*(\textstyle\bigwedge^k \psi)\mathfrak{I}(\phi_1 \wedge \cdots \phi_k)\big) &= \iota(e_1 \wedge \cdots \wedge e_k)\mathfrak{I}\Big({}^*\psi(\phi_1) \wedge \cdots \wedge {}^*\psi(\phi_k)\Big) \\ &= \iota(e_1 \wedge \cdots \wedge e_k)\mathfrak{I}\Big(\textstyle\bigwedge^k {}^*\psi(\phi_1 \wedge \cdots \wedge \phi_k)\Big) .\end{aligned}$$

Since the $e_1 \wedge \cdots \wedge e_k$ generate $\bigwedge^k E$ and the $\phi_1 \wedge \cdots \wedge \phi_k$ generate $\bigwedge^k {}^*F$, the result follows. QED

7.26 Construction (bi-duals). For any $\mathfrak{A}$-graded $\mathcal{A}$-module E we construct an identification $\mathfrak{I}_{LR} : E \to {}^*(E^*)$ by the formula

$$\mathfrak{I}_{LR}(e) : \phi \mapsto \iota(\phi)\mathfrak{I}_{LR}(e) = \phi(e) ,$$

where $\phi : E \to \mathcal{A}$ is right linear, i.e., $\phi \in \mathrm{Hom}_R(E;\mathcal{A}) \equiv E^*$. In a similar way we define an identification $\mathfrak{I}_{RL} : E \to ({}^*E)^*$ by the formula

$$\mathfrak{I}_{RL}(e) : \phi \mapsto \mathfrak{I}_{RL}(e)(\phi) = \iota(e)\phi \equiv \langle\!\langle e \,\|\, \phi \rangle\!\rangle ,$$

where now $\phi : E \to \mathcal{A}$ is left linear, i.e., $\phi \in {}^*E$.

The advantage of identifying E with the mixed bi-duals ${}^*(E^*)$ and $({}^*E)^*$ is that we never have to change the order of terms. However, there are four "different" bi-duals. The transpositions $\mathfrak{T}_L : {}^{**}E \to ({}^*E)^*$ and $\mathfrak{T}_R : {}^*(E^*) \to E^{**}$ are isomorphisms [2.15]. Using the transposition $\mathfrak{T} : {}^*E \to E^*$ gives us an isomorphism $\mathfrak{I} : {}^*(E^*) \to ({}^*E)^*$ defined on $\psi \in {}^*(E^*)$ by

$$\mathfrak{I}(\psi) = (\mathfrak{T}_R(\psi)) \circ \mathfrak{T} = \mathfrak{T}_L(\psi \circ \mathfrak{T}) .$$

If $\psi \in {}^*(E^*)$ and $\phi \in {}^*E$ are homogeneous we have

$$\mathfrak{I}(\psi)(\phi) = (-1)^{\langle \varepsilon(\psi) | \varepsilon(\phi) \rangle} \iota(\mathfrak{T}\phi)\psi .$$

The isomorphism $\mathfrak{I}$ is compatible with the previously defined identifications $\mathfrak{I}_{LR}$ and $\mathfrak{I}_{RL}$ in the sense that $\mathfrak{I}_{RL} = \mathfrak{I} \circ \mathfrak{I}_{LR}$. The verification of this statement is left to the reader. It follows that we do not need the parentheses in the mixed bi-duals and that we may say that we have an identification $E \to {}^*E^*$.

7.27 Remark. Let $\Phi_R^E : E \to \mathrm{Hom}_R(\mathcal{A};E)$ denote the isomorphism of [2.19] characterized by $\Phi_R^E(e)(1) = e$, then we can interpret $\mathfrak{I}_{LR}(e)$ as the dual map $\Phi_R^E(e)^* : E^* \to \mathcal{A}$. Here is how it works. Let $e \in E$ and $\phi \in E^*$ be arbitrary, then we have the equality $\langle\!\langle \phi \,\|\, \Phi_R^E(e)^* \rangle\!\rangle = \phi \circ \Phi_R^E(e) \in \mathcal{A}^*$ and

$$(\langle\!\langle \phi \,\|\, \Phi_R^E(e)^* \rangle\!\rangle)(1) = \phi(e) = \langle\!\langle \phi \,\|\, \mathfrak{I}_{LR}(e) \rangle\!\rangle = \Phi_R^{\mathcal{A}}(\langle\!\langle \phi \,\|\, \mathfrak{I}_{LR}(e) \rangle\!\rangle)(1)$$

and thus $\langle\!\langle \phi \,\|\, \Phi_R^E(e)^* \rangle\!\rangle = \Phi_R^{\mathcal{A}}(\langle\!\langle \phi \,\|\, \mathfrak{I}_{LR}(e) \rangle\!\rangle)$. By abuse of notation, forgetting about the isomorphisms $\Phi_R^{\mathcal{A}}$ and Φ_R^E, we thus find $\mathfrak{I}_{LR}(e) = e^*$, where we have made the identifications $E \cong \mathrm{Hom}_R(\mathcal{A};E)$ and $\mathrm{Hom}_L(E^*;\mathcal{A}^*) \cong \mathrm{Hom}_L(E^*;\mathcal{A}) \equiv {}^*(E^*)$.

7.28 Construction (sums of tensor products). Let E be an $\mathfrak{A}$-graded $\mathcal{A}$-module and let $(F_i)_{i\in I}$ be a family of $\mathfrak{A}$-graded $\mathcal{A}$-modules. We want to show that $E \otimes (\bigoplus_{i\in I} F_i)$ and $\bigoplus_{i\in I}(E\otimes F_i)$ are isomorphic. First recall that π_i and s_i denote the canonical projections and injections between F_i and $\bigoplus_{i\in I} F_i$ [3.1]. The map $E\times(\bigoplus_{i\in I} F_i) \to \bigoplus_{i\in I}(E\otimes F_i)$ given by $(e,f)\mapsto \oplus_i(e\otimes\pi_i(f))$ is even and bilinear, and thus gives rise to an even linear map $E\otimes(\bigoplus_{i\in I} F_i)\to\bigoplus_{i\in I}(E\otimes F_i)$. On the other hand we have the even linear map $\sum_{i\in I} id\otimes s_i : \bigoplus_{i\in I}(E\otimes F_i)\to E\otimes(\bigoplus_{i\in I} F_i)$. Let us show by computation that these two maps are inverse to each other.

$$e\otimes f \mapsto \oplus_{i\in I}(e\otimes\pi_i(f)) \mapsto \sum_{i\in I} e\otimes s_i(\pi_i(f)) = e\otimes\sum_{i\in I} s_i(\pi_i(f)) = e\otimes f\ ,$$

and

$$\oplus_{i\in I}(e_i\otimes f_i)\mapsto\sum_{i\in I} e_i\otimes s_i(f_i)\mapsto\sum_{i\in I}\oplus_{j\in I}(e_i\otimes\pi_j(s_i(f_i))) = \oplus_{j\in I}(e_j\otimes f_j)\ ,$$

where we used that $\pi_j\circ s_i = 0$ whenever $i\neq j$. We conclude that we have constructed an isomorphism between $E\otimes(\bigoplus_{i\in I} F_i)$ and $\bigoplus_{i\in I}(E\otimes F_i)$. In the same way we construct an isomorphism between $(\bigoplus_{i\in I} F_i)\otimes E$ and $\bigoplus_{i\in I}(F_i\otimes E)$.

8. Isomorphisms

In §7 we have defined a number of identifications. In this section some technical proofs are given to show sufficient conditions for these identifications to be isomorphisms. It turns out that this is the case if the $\mathfrak{A}$-graded $\mathcal{A}$-modules are finitely generated and projective. The condition finitely generated and projective for ($\mathfrak{A}$-graded) modules is equivalent to the condition finite dimensional for vector spaces. At the end of this section a summary of the more interesting identifications can be found.

8.1 Definitions. • A subset G of an $\mathfrak{A}$-graded $\mathcal{A}$-module E is called a set of *generators* for E if for each $e\in E$ there exist $g_1,\dots,g_k\in G$ (a finite number!) and $a^1,\dots,a^k\in\mathcal{A}$ such that $e=\sum_i a^i g_i$. The g_i and a^i are not supposed to be unique.

• An $\mathfrak{A}$-graded $\mathcal{A}$-module E is called *finitely generated*, or *of finite type*, if there exists a finite set of generators G.

• A subset B of an $\mathfrak{A}$-graded $\mathcal{A}$-module E is called a set of *independent elements* if for all $e_1,\dots,e_k\in B$ (a finite number!) and for all $a^1,\dots,a^k\in\mathcal{A}$ one has the implication: $\sum_i a^i e_i = 0 \Rightarrow \forall i : a^i = 0$. In words: any (linear) relation between elements of B is necessarily trivial.

• A subset B of an $\mathfrak{A}$-graded $\mathcal{A}$-module E is called a *basis* for E if the elements of B are at the same time independent and generating. For vector spaces it is well known that there always exists a basis. However, for $\mathfrak{A}$-graded $\mathcal{A}$-modules the existence of a basis is no longer guaranteed. It is immediate that an $\mathfrak{A}$-graded $\mathcal{A}$-module admits a *homogeneous* basis if and only if it is (isomorphic to) a free $\mathfrak{A}$-graded $\mathcal{A}$-module $F(G,\varepsilon)$ on homogeneous generators G [3.4].

• An $\mathfrak{A}$-graded $\mathcal{A}$-module E is called *projective* if there exists an $\mathfrak{A}$-graded $\mathcal{A}$-module E' such that $E\oplus E'$ admits a *homogeneous* basis G, i.e., $E\oplus E' = F(G,\varepsilon)$. An $\mathfrak{A}$-graded $\mathcal{A}$-module that is both finitely generated and projective will be called *f.g.p*.

8.2 Remark. We have defined the notions of independence of vectors, generating sets and bases with respect to the left module structure, completely ignoring that $\mathfrak{A}$-graded $\mathcal{A}$-modules also have a (compatible) right module structure. Of course we could have developed the right module case parallel to the left module case. However, this is hardly necessary. If E is an $\mathfrak{A}$-graded $\mathcal{A}$-module, $g \in E$ homogeneous and $a \in \mathcal{A}$, we have the relation $ag = g \cdot \sum_{\alpha\in\mathfrak{A}}(-1)^{\langle\varepsilon(g)|\alpha\rangle}a_\alpha$. This shows that a (left) generating set of homogeneous elements is also generating for the right module structure. And if a set of homogeneous elements is (left) independent, it is also independent for the right module structure. Hence for homogeneous sets, there is no difference between the notions of generating and independence for the left or right module structures. Moreover, by splitting into homogeneous components, any generating set can be made homogeneous. The only possible difference between left and right module structures thus lies in non-homogeneous independent sets. Since we will not use these, we will not see any difference between our left module structure definitions and the corresponding right module equivalents.

8.3 Lemma. *An $\mathfrak{A}$-graded $\mathcal{A}$-module E is f.g.p if and only if there exists an $\mathfrak{A}$-graded $\mathcal{A}$-module E' such that $E\oplus E'$ is a free $\mathfrak{A}$-graded $\mathcal{A}$-module on a finite set of homogeneous generators.*

Proof. If $E\oplus E' = F(G,\varepsilon)$ with G a finite set, E is projective and $\pi_E(G)$ is a finite set of homogeneous generators for E, proving the if part. To prove the only if part, assume that E is projective and admits a finite number of generators. Since E is projective, there exists E' such that $E\oplus E' = F(G',\varepsilon')$ for some set of homogeneous generators G'. Our problem is that we do not know whether we can take G' to be finite. Let $G = \{g_1,\dots,g_n\}$ be a finite set of generators for E. By splitting these generators into their homogeneous parts, we may assume that all g_i are homogeneous. We thus obtain an even surjective map $\phi : F(G,\varepsilon) \to E$ defined simply by $\phi(g_i) = g_i$. Define the even map $\psi : E\oplus E' = F(G',\varepsilon') \to E$ as $\psi = id(E) + \underline{0}$ [7.2], where $\underline{0} : E' \to E$ is identically zero. Define also a map $\chi : F(G',\varepsilon') \to F(G,\varepsilon)$ as follows. For any $g' \in G'$ choose an element $f' \in F(G,\varepsilon)$ such that $\phi(f') = \psi(g')$. This is possible because ϕ is surjective. Since both ϕ and ψ are even, we may assume that f' and g' have the same parity. Now define χ by the formula

$$\chi : \sum_{g'\in G'} \lambda_{g'} g' \mapsto \sum_{g'\in G'} \lambda_{g'} f' \ .$$

This is a well defined even linear map satisfying the relation $\psi = \phi \circ \chi$. Restriction of χ to the $\mathfrak{A}$-graded submodule E gives us the relation $id(E) = \phi \circ (\chi|_E)$. By using the equality $x = \big(x - (\chi|_E)(\phi(x))\big) + (\chi|_E)(\phi(x))$ it follows easily that the $\mathfrak{A}$-graded submodules $\ker(\phi)$ and $\mathrm{im}(\chi|_E)$ are supplements. Hence by [3.9] we obtain $E \oplus \ker(\phi) = F(G, \varepsilon)$.

QED

8.4 Lemma. *If an $\mathfrak{A}$-graded $\mathcal{A}$-module E is f.g.p, then so are E^* and ${}^*E$.*

Proof. If E is a free $\mathfrak{A}$-graded $\mathcal{A}$-module on a single homogeneous generator e_o, it follows immediately that E^* is a free $\mathfrak{A}$-graded $\mathcal{A}$-module on the single homogeneous generator ϕ_o defined by $\phi_o(e_o \cdot \lambda) = \lambda$. If E is f.g.p, there exists an $\mathfrak{A}$-graded $\mathcal{A}$-module E' such that $E \oplus E' = \bigoplus_{i=1}^n E_i$, where all the E_i are free $\mathfrak{A}$-graded $\mathcal{A}$-modules on a single homogeneous generator [3.4]. Using [7.3] we thus obtain an isomorphism $E^* \oplus (E')^* \cong \bigoplus_{i=1}^n E_i^*$. Since the last one is a free $\mathfrak{A}$-graded $\mathcal{A}$-module on n generators, the result follows. The left linear case is analogous. QED

8.5 Proposition. *If E is f.g.p, the identification $\mathfrak{I} : E \to {}^*E^*$ [7.26] is an isomorphism.*

Proof. Consider the special case of ${}^*(E^*)$. If E is a free $\mathfrak{A}$-graded $\mathcal{A}$-module on a single homogeneous generator, the result follows immediately from the proof of [8.4].

If $E_1, \ldots, E_n$ are $\mathfrak{A}$-graded $\mathcal{A}$-modules, we have identifications $\mathfrak{I}_i : E_i \to {}^*(E_i^*)$ and $\mathfrak{I}_\oplus : \bigoplus_{i=1}^n E_i \to {}^*((\bigoplus_{i=1}^n E_i)^*)$. Applying the isomorphism [7.2] twice gives us the following diagram:

$$\begin{array}{ccc}
\bigoplus_{i=1}^n E_i & \xrightarrow[\prod_{i=1}^n \mathfrak{I}_i]{} & \bigoplus_{i=1}^n {}^*(E_i^*) \\
\Big\| & & \Big\downarrow \cong \\
\bigoplus_{i=1}^n E_i & \xrightarrow{\mathfrak{I}_\oplus} & {}^*((\bigoplus_{i=1}^n E_i)^*)\ .
\end{array}$$

We leave it to the reader to verify that this diagram is actually a commutative diagram. Now let E' be an $\mathfrak{A}$-graded $\mathcal{A}$-module such that $E \oplus E' = \bigoplus_{i=1}^n E_i$ where the E_i are free $\mathfrak{A}$-graded $\mathcal{A}$-modules on a single homogeneous generator. Applying the commutative diagram to the family of two: E and E', shows that $\mathfrak{I}_\oplus$ is an isomorphism if and only if both $\mathfrak{I}$ and $\mathfrak{I}'$ are isomorphisms [7.8]. Applying the commutative diagram to the finite family E_i proves that $\mathfrak{I}_\oplus$ is an isomorphism because for free $\mathfrak{A}$-graded $\mathcal{A}$-modules on a single homogeneous generator the $\mathfrak{I}_i$ are isomorphisms. It follows that $\mathfrak{I}$ is an isomorphism.

QED

8.6 Proposition. *If E is finitely generated or if the index set I is finite, then the identifications $\mathfrak{I} : \bigoplus_{i \in I} \mathrm{Hom}_S(E; F_i) \to \mathrm{Hom}_S(E; \bigoplus_{i \in I} F_i)$ [7.6] are isomorphisms.*

Proof. Since injectivity is automatic, we have to assure surjectivity. For an arbitrary ϕ in $\mathrm{Hom}_S(E; \bigoplus_{i \in I} F_i)$ we define the maps $\phi_i \in \mathrm{Hom}_S(E; F_i)$ by $\phi_i = \pi_i \circ \phi$, where

$\pi_i : \bigoplus_i F_i \to F_i$ denotes the canonical projection. It is immediate that $\mathfrak{I}(\oplus_i\phi_i) = \phi$, except for the fact that we do not know whether $(\phi_i)_{i\in I}$ lies in $\bigoplus_{i\in I} \mathrm{Hom}_S(E; F_i)$, i.e., whether only finitely many ϕ_i are non-zero. If the index set I is finite, this is obvious and we may conclude that in that case $\mathfrak{I}$ is an isomorphism. So suppose that $e_1, \ldots, e_n$ are generators for E. Since $\phi(e_j)$ lies in $\bigoplus_i F_i$, it follows that only finitely many $\phi_i(e_j) = \pi_i(\phi(e_j))$ are non-zero. Since there are only finitely many e_j, which generate E, it follows that only finitely many ϕ_i are non-zero. Hence $(\phi_i)_{i\in I} \in \bigoplus_{i\in I} \mathrm{Hom}_S(E; F_i)$, proving that $\mathfrak{I}$ is an isomorphism. QED

8.7 Proposition. *Given three $\mathfrak{A}$-graded $\mathcal{A}$-modules E, F, and G, we want to investigate the following two identifications: $\mathfrak{I} : F \otimes \mathrm{Hom}_R(E; G) \to \mathrm{Hom}_R(E; F \otimes G)$ and $\mathfrak{I} : \mathrm{Hom}_L(E; G) \otimes F \to \mathrm{Hom}_L(E; G \otimes F)$.*

(i) *If F is projective, then the $\mathfrak{I}$'s are injective.*
(ii) *If F is f.g.p, then the $\mathfrak{I}$'s are isomorphisms.*
(iii) *If E is f.g.p, then the $\mathfrak{I}$'s are isomorphisms.*

Proof. We only treat the right linear case, the left linear case being similar. Let us start with the special case in which F is a free $\mathfrak{A}$-graded $\mathcal{A}$-module on a single homogeneous generator f_o. By definition, the map $p : \mathcal{A} \to F$, $a \mapsto f_o \cdot a$ is a right linear bijection of the same parity as f_o. Using [7.14] one can show that $C : \mathrm{Hom}_R(E; F \otimes G) \to \mathrm{Hom}_R(E; \mathcal{A} \otimes G)$ defined by $C(\phi) = (p^{-1} \otimes id) \circ \phi$ is a right linear bijection. We thus obtain a map $C \circ \mathfrak{I} \circ (p \otimes id) : \mathcal{A} \otimes \mathrm{Hom}_R(E; G) \to \mathrm{Hom}_R(E; \mathcal{A} \otimes G)$. Identifying $\mathcal{A} \otimes H$ with H for any $\mathfrak{A}$-graded $\mathcal{A}$-module H, it is an elementary verification that this map is the identity on $\mathrm{Hom}_R(E; G)$. Since C and $(p \otimes id)$ are bijective, we deduce that $\mathfrak{I}$ is bijective. We conclude that $\mathfrak{I}$ is an isomorphism in case F is a free $\mathfrak{A}$-graded $\mathcal{A}$-module on a single homogeneous generator.

To prove (i) and (ii), let us fix E and G, and let us consider a family F_i, $i \in I$ of $\mathfrak{A}$-graded $\mathcal{A}$-modules. We thus have identifications $\mathfrak{I}_i : F_i \otimes \mathrm{Hom}_R(E; G) \to \mathrm{Hom}_R(E; F_i \otimes G)$ and $\mathfrak{I}_\oplus : (\bigoplus_{i\in I} F_i) \otimes \mathrm{Hom}_R(E; G) \to \mathrm{Hom}_R(E; (\bigoplus_{i\in I} F_i) \otimes G)$. According to [7.28] there exists an isomorphism $(\bigoplus_{i\in I} F_i) \otimes \mathrm{Hom}_R(E; G) \to \bigoplus_{i\in I}(F_i \otimes \mathrm{Hom}_R(E; G))$. According to [7.28] we also have an isomorphism $(\bigoplus_{i\in I} F_i) \otimes G \to \bigoplus_{i\in I}(F_i \otimes G)$. This last isomorphism combined with the identification [7.6] gives us an injective identification $\mathfrak{I}_h : \bigoplus_{i\in I} \mathrm{Hom}_R(E; F_i \otimes G) \to \mathrm{Hom}_R(E; (\bigoplus_{i\in I} F_i) \otimes G)$. We thus obtain the following diagram of maps:

$$\begin{array}{ccc}
\bigoplus_{i\in I}(F_i \otimes \mathrm{Hom}_R(E; G)) & \xrightarrow[\prod_{i\in I} \mathfrak{I}_i]{} & \bigoplus_{i\in I} \mathrm{Hom}_R(E; F_i \otimes G) \\
\cong \uparrow [7.28] & & \downarrow \mathfrak{I}_h \\
(\bigoplus_{i\in I} F_i) \otimes \mathrm{Hom}_R(E; G) & \xrightarrow[\mathfrak{I}_\oplus]{} & \mathrm{Hom}_R(E; (\bigoplus_{i\in I} F_i) \otimes G) .
\end{array}$$

We leave it to the reader to check that this diagram is commutative.

Now take a family of two: F and F'. Since the index set consists of two elements, the map $\mathfrak{I}_h$ is a bijection [7.6]. Using [7.8] we conclude that $\mathfrak{I}_\oplus$ is injective/bijective if and only if both $\mathfrak{I}$ and $\mathfrak{I}'$ are injective/bijective. If we suppose that F is projective, we can take F' such that $F \oplus F'$ is a free $\mathfrak{A}$-graded $\mathcal{A}$-module on a set of homogeneous generators, i.e., $F \oplus F' = \bigoplus_{i \in I} F_i$ where each F_i is a free $\mathfrak{A}$-graded $\mathcal{A}$-module on a single homogeneous generator. For these we know that all $\mathfrak{I}_i$ are isomorphisms. However, for this family we only know that $\mathfrak{I}_h$ is injective. We conclude that $\mathfrak{I}_\oplus$ is injective. This proves (i). If F is also finitely generated, it follows that the family F_i can be taken finite, in which case $\mathfrak{I}_h$ becomes bijective, hence $\mathfrak{I}_\oplus$ is bijective, and thus $\mathfrak{I}$ is bijective. This proves (ii).

To prove (iii) we proceed in the same way. If E is a free $\mathfrak{A}$-graded $\mathcal{A}$-module on a single homogeneous generator, one can easily establish the existence of homogeneous linear bijections between H and $\mathrm{Hom}_R(E; H)$ for any $\mathfrak{A}$-graded $\mathcal{A}$-module H. Using these bijections one then proves that $\mathfrak{I}$ is an isomorphism if E is a free $\mathfrak{A}$-graded $\mathcal{A}$-module on a single homogeneous generator. Now let us fix F and G and let us take a finite family $(E_i)_{i=1}^n$. We thus have identifications $\mathfrak{I}_i : F \otimes \mathrm{Hom}_R(E_i; G) \to \mathrm{Hom}_R(E_i; F \otimes G)$ and $\mathfrak{I}_\oplus : F \otimes \mathrm{Hom}_R(\bigoplus_{i=1}^n E_i; G) \to \mathrm{Hom}_R(\bigoplus_{i=1}^n E_i; F \otimes G)$. According to [7.2] we have an isomorphism $\bigoplus_{i=1}^n \mathrm{Hom}_R(E_i; G) \to \mathrm{Hom}_R(\bigoplus_{i=1}^n E_i; G)$. Taking the tensor product of this isomorphism with the identity on F (see [7.14]) and composing it with the isomorphism $\bigoplus_{i=1}^n (F \otimes \mathrm{Hom}_R(E_i; G)) \to F \otimes (\bigoplus_{i=1}^n \mathrm{Hom}_R(E_i; G))$ gives us an isomorphism $\mathfrak{I}_v : \bigoplus_{i=1}^n (F \otimes \mathrm{Hom}_R(E_i; G)) \to F \otimes \mathrm{Hom}_R(\bigoplus_{i=1}^n E_i; G)$. We thus obtain a diagram of maps:

$$\begin{array}{ccc}
\bigoplus_{i=1}^n (F \otimes \mathrm{Hom}_R(E_i; G)) & \xrightarrow[\prod_{i=1}^n \mathfrak{I}_i]{} & \bigoplus_{i=1}^n \mathrm{Hom}_R(E_i; F \otimes G) \\
\cong \big\downarrow \mathfrak{I}_v & & \cong \big\downarrow [7.2] \\
F \otimes \mathrm{Hom}_R(\bigoplus_{i=1}^n E_i; G) & \xrightarrow[\mathfrak{I}_\oplus]{} & \mathrm{Hom}_R(\bigoplus_{i=1}^n E_i; F \otimes G) .
\end{array}$$

As before, it is left to the reader to check that this diagram is commutative.

If we now suppose that E is f.g.p, then there exists an $\mathfrak{A}$-graded $\mathcal{A}$-module E' such that $E \oplus E' = \bigoplus_{i=1}^n E_i$, where each E_i is a free $\mathfrak{A}$-graded $\mathcal{A}$-module on a single homogeneous generator. Applying the commutative diagram to the finite family of two: E and E', shows that $\mathfrak{I}_\oplus$ is bijective if and only if both $\mathfrak{I}$ and $\mathfrak{I}'$ are bijective [7.8]. Applying it to the finite family E_i shows that $\mathfrak{I}_\oplus$ is bijective if and only if all $\mathfrak{I}_i$ are bijective. Since the E_i are free on a single homogeneous generator, the $\mathfrak{I}_i$ are isomorphisms, and we conclude that both $\mathfrak{I}$ and $\mathfrak{I}'$ are bijective. *QED*

8.8 Corollary. *Let E and F be $\mathfrak{A}$-graded $\mathcal{A}$-modules. If F is projective, the identifications ${}^*E \otimes F \to \mathrm{Hom}_L(E; F)$ and $F \otimes E^* \to \mathrm{Hom}_R(E; F)$ [7.11] are injective. If either E or F is f.g.p, these identifications are isomorphisms.*

8.9 Lemma. *If $E_i = F(G_i, \varepsilon_i)$, $i = 1, 2$ are free $\mathfrak{A}$-graded $\mathcal{A}$-modules on homogeneous generators G_i, then $E_1 \otimes E_2$ is a free $\mathfrak{A}$-graded $\mathcal{A}$-module on homogeneous generators $G_1 \times G_2$ with parity map $\varepsilon(g_1, g_2) = \varepsilon_1(g_1) + \varepsilon_2(g_2)$.*

Proof. Choosing left coordinates for E_1 and right coordinates for E_2, we define a map $\psi : E_1 \times E_2 \to F(G_1 \times G_2, \varepsilon)$ by $\psi(\sum_i \lambda_i \cdot g_{1i}, \sum_j g_{2j} \cdot \mu_j) \mapsto \sum_{i,j} \lambda_i \cdot (g_{1i}, g_{2j}) \cdot \mu_j$. We leave it to the reader to verify that ψ is even and (right) bilinear, and thus induces an identification $E_1 \otimes E_2 \to F(G_1 \times G_2, \varepsilon)$. It is easily seen that the inverse is given by the map $\sum_{i,j} \nu_{ij} \cdot (g_{1i}, g_{2j}) \mapsto \sum_{i,j} \nu_{ij} \cdot g_{1i} \otimes g_{2j}$, proving the lemma. $\boxed{QED}$

8.10 Corollary. *If E and F are projective $\mathfrak{A}$-graded $\mathcal{A}$-modules, then $E \otimes F$ is projective. If they are finitely generated, then so is $E \otimes F$.*

Proof. The elements $e \otimes f$, $e \in E$, $f \in F$ generate $E \otimes F$. If E is generated by $(e_i)_{i \in I}$ and F by $(f_j)_{j \in J}$, then $E \otimes F$ is generated by $(e_i \otimes f_j)_{i \in I, j \in J}$. We conclude that if both E and F are finitely generated, so is $E \otimes F$.

If $E \oplus E'$ and $F \oplus F'$ are free $\mathfrak{A}$-graded $\mathcal{A}$-modules on homogeneous generators, then $(E \oplus E') \otimes (F \oplus F')$ is a free $\mathfrak{A}$-graded $\mathcal{A}$-module on homogeneous generators. But by [7.28] this is isomorphic to $(E \otimes F) \oplus \big((E \otimes F') \oplus (E' \otimes F) \oplus (E' \otimes F')\big)$, showing that $E \otimes F$ is projective. $\boxed{QED}$

8.11 Discussion. If E and F are $\mathfrak{A}$-graded $\mathcal{A}$-modules, we have defined the operation of *right dual map,* which is an even linear map $^* : \mathrm{Hom}_R(E; F) \to \mathrm{Hom}_L(F^*; E^*)$. We also have identifications $F \otimes E^* \to \mathrm{Hom}_R(E; F)$, ${}^*(F^*) \otimes E^* \to \mathrm{Hom}_L(F^*; E^*)$, and $F \to {}^*(F^*)$. We leave it to the reader to verify that these identifications fit together in a commutative diagram

$$\begin{array}{ccc} F \otimes E^* & \longrightarrow & {}^*(F^*) \otimes E^* \\ \downarrow & & \downarrow \\ \mathrm{Hom}_R(E; F) & \underset{*}{\longrightarrow} & \mathrm{Hom}_L(F^*; E^*) \,. \end{array}$$

If E and F are f.g.p, we know that the three unlabeled arrows are isomorphisms, and thus that the map $^* : \mathrm{Hom}_R(E; F) \to \mathrm{Hom}_L(F^*; E^*)$ is an isomorphism as well. Obviously a similar result is true for left dual maps.

8.12 Proposition. *Let E_i and F_i, $i = 1, \ldots, n$ be two families of $\mathfrak{A}$-graded $\mathcal{A}$-modules and let $\mathfrak{J} : \bigotimes_{i=n}^{1} \mathrm{Hom}_S(E_i; F_i) \to \mathrm{Hom}_S(\bigotimes_{i=1}^{n} E_i; \bigotimes_{i=1}^{n} F_i)$ be the identification given in [7.12].*

(i) *If all E_i are f.g.p then $\mathfrak{J}$ is an isomorphism.*
(ii) *If all E_i and F_i are f.g.p, with the possible exception of a single pair (E_i, F_i), then $\mathfrak{J}$ is an isomorphism.*

Proof. We give the proof in the left linear case, the right linear case being similar. We create the following isomorphisms:

$$\begin{array}{lll} & \mathrm{Hom}_L(E_2;F_2)\otimes\mathrm{Hom}_L(E_1;F_1) & \\ \cong & \mathrm{Hom}_L(E_2;F_2\otimes\mathrm{Hom}_L(E_1;F_1)) & \text{by [8.7] with } E_2 \text{ f.g.p} \\ \cong & \mathrm{Hom}_L(E_2;\mathrm{Hom}_L(E_1;F_1)\otimes F_2) & \text{using that } \mathfrak{R} \text{ is an isomorphism} \\ \cong & \mathrm{Hom}_L(E_2;\mathrm{Hom}_L(E_1;F_1\otimes F_2)) & \text{by [8.7] with } E_1 \text{ or } F_2 \text{ f.g.p} \\ \cong & \mathrm{Hom}_L(E_1\otimes E_2;F_1\otimes F_2) & \text{by [7.10]} \end{array}$$

We leave it to the reader to trace these isomorphisms and to show that the final result is indeed the identification given in [7.12]. It follows that we have proven the proposition for $n = 2$, where we used that either E_1 and E_2 are f.g.p, or E_2 and F_2 are f.g.p. The general result follows by induction. In case (i) one uses [8.10]. In case (ii) one uses that all E_i and F_i are f.g.p for $i = 1, \ldots, n-1$. Since permuting factors in a tensor product is an isomorphism, we may indeed assume that it is the last couple (E_n, F_n) that is not f.g.p. QED

8.13 Corollary. *Let $E_1, \ldots, E_n$ be 𝔄-graded $\mathcal{A}$-modules. If all but one are f.g.p, the identification $E_n^* \otimes \cdots \otimes E_1^* \to (E_1 \otimes \cdots \otimes E_n)^*$, defined on homogeneous elements by*

$$(\phi_n \otimes \cdots \otimes \phi_1)(e_1 \otimes \cdots \otimes e_n) = (-1)^{\sum_{i>j} \langle \varepsilon(\phi_i) | \varepsilon(\phi_j(e_j)) \rangle} \cdot \phi_1(e_1) \cdots \phi_n(e_n) \ ,$$

is an isomorphism, as is the identification ${}^*E_n \otimes \cdots \otimes {}^*E_1 \to {}^*(E_1 \otimes \cdots \otimes E_n)$, defined on homogeneous elements by

$$\iota(e_1 \otimes \cdots \otimes e_n)(\phi_n \otimes \cdots \otimes \phi_1) = (-1)^{\sum_{i<j} \langle \varepsilon(\phi_i) | \varepsilon(\phi_j(e_j)) \rangle} \cdot \iota(e_1)\phi_1 \cdots \iota(e_n)\phi_n \ .$$

8.14 Proposition. *If $E = F(G, \varepsilon)$ is a free 𝔄-graded $\mathcal{A}$-module on homogeneous generators G, then $\bigwedge^k E$ is a free 𝔄-graded $\mathcal{A}$-module.*

Proof. From [8.9] we know that the set $B = \{ g_1 \otimes \cdots \otimes g_k \mid g_i \in G \}$ is a basis for $\bigotimes^k E$. We now choose a total order on G and we define the subsets $B_{g_1 \cdots g_k} \subset B$ for increasing sequences $g_1 \leq g_2 \leq \cdots \leq g_k$ by

$$B_{g_1 \cdots g_k} = \{ g_{\sigma(1)} \otimes \cdots \otimes g_{\sigma(k)} \mid \sigma \in \mathfrak{S}_k \} \ .$$

It should be obvious that the $B_{g_1 \cdots g_k}$ form a partition of B in disjunct subsets. We now introduce the modified sets $B'_{g_1 \cdots g_k}$ defined as

$$B'_{g_1 \cdots g_k} = \{ g_1 \otimes \cdots \otimes g_k - (-1)^\sigma \mathfrak{R}_\sigma(g_1 \otimes \cdots \otimes g_k) \mid \sigma \in \mathfrak{S}_k \} \ .$$

It is easy to see that the set $\{g_1 \otimes \cdots \otimes g_k\} \cup B'_{g_1 \cdots g_k}$ generates the same 𝔄-graded submodule of $\bigotimes^k E$ as $B_{g_1 \cdots g_k}$.

We now recall the construction of $\bigwedge^k E = \bigotimes^k E/N_k$ as given in [5.2]. By k-linearity, it is immediate that the $\mathfrak{A}$-graded submodule $N_k \subset \bigotimes^k E$ is also generated by the set

$$T'_k = \{\, g_1 \otimes \cdots \otimes g_k + \mathfrak{R}_{(jj+1)}(g_1 \otimes \cdots \otimes g_k) \mid 1 \leq j < k,\ g_i \in G \,\}\ .$$

An elementary but slightly tedious verification then shows that N_k is also generated by the union of all sets $B'_{g_1 \cdots g_k}$ (and recall that the index set $g_1 \ldots g_k$ is an increasing one).

Since the $B_{g_1 \ldots g_k}$ form a partition of the basis B, it follows that we have found a basis for a supplement of N_k in the form of the set of all $g_1 \otimes \cdots \otimes g_k$ subject to two conditions (i) that $g_1 \leq \cdots \leq g_k$ and (ii) that it does not belong to the $\mathfrak{A}$-graded submodule generated by $B'_{g_1 \cdots g_k}$. The projection of this basis to $\bigwedge^k E = \bigotimes^k E/N_k$ then is a homogeneous basis for $\bigwedge^k E$, proving that $\bigwedge^k E$ is a free $\mathfrak{A}$-graded $\mathcal{A}$-module. QED

8.15 Discussion. It can be shown that the condition that $g_1 \otimes \cdots \otimes g_k$ belongs to the $\mathfrak{A}$-graded submodule generated by $B'_{g_1 \cdots g_k}$ is equivalent to the existence of a permutation $\sigma \in \mathfrak{S}_k$ such that $-(-1)^\sigma \mathfrak{R}_\sigma(g_1 \otimes \cdots \otimes g_k) = g_1 \otimes \cdots \otimes g_k$. This in turn can be shown to be equivalent to the existence of an index $i < k$ such that $g_i = g_{i+1}$ and such that $(-1)^{\langle \varepsilon(g_i) | \varepsilon(g_i) \rangle} = 1$. It follows that a basis of $\bigwedge^k E$ is given by the vectors $g_1 \wedge \cdots \wedge g_k$ where the g_i form an increasing sequence such that a g_i is not repeated if $(-1)^{\langle \varepsilon(g_i) | \varepsilon(g_i) \rangle} = 1$.

8.16 Proposition. *If E is f.g.p, the identifications $\mathfrak{I} : \bigwedge^k (E^*) \to (\bigwedge^k E)^*$ and $\mathfrak{I} : \bigwedge^k ({}^*E) \to {}^*(\bigwedge^k E)$ [7.16] are isomorphisms.*

Proof. As usual we treat the right linear case, the left linear case being similar. We denote by $\pi : \bigotimes^k E \to \bigwedge^k E$ and by $\pi' : \bigotimes^k E^* \to \bigwedge^k E^*$ the canonical projections. By [2.21] the dual map $\pi^* : (\bigwedge^k E)^* \to (\bigotimes^k E)^*$ is injective. It then follows from the pedestrian way of constructing $\mathfrak{I}$ [7.16] that we have the equalities $\psi = \mathfrak{I}_\otimes \circ \mathbb{A}$, $\psi = \pi^* \circ \Psi$, and $\Psi = \mathfrak{I} \circ \pi'$. We thus obtain the equality $\pi^* \circ \mathfrak{I} \circ \pi' = \mathfrak{I}_\otimes \circ \mathbb{A}$, i.e., the commutative diagram

$$\begin{array}{ccc} \bigotimes^k E^* & \xrightarrow[\mathfrak{I}_\otimes \circ \mathbb{A}]{} & (\bigotimes^k E)^* \\ {\scriptstyle \pi'} \big\downarrow & & \big\uparrow {\scriptstyle \pi^*} \\ \bigwedge^k E^* & \xrightarrow[\mathfrak{I}]{} & (\bigwedge^k E)^*\ . \end{array}$$

The essence of the proof will be "running around this diagram." If E is f.g.p, it follows from [8.13] that $\mathfrak{I}_\otimes$ is bijective, i.e., we may write $\mathbb{A} = \mathfrak{I}_\otimes^{-1} \circ \pi^* \circ \mathfrak{I} \circ \pi'$. Since π' is $\mathfrak{A}$-graded skew-symmetric, we obviously have the equality $\pi' \circ \mathbb{A} = (k!) \cdot \pi'$. We now start running:

$$(k!) \cdot \pi' = \pi' \circ \mathbb{A} = \pi' \circ (\mathfrak{I}_\otimes^{-1} \circ \pi^* \circ \mathfrak{I} \circ \pi') = (\pi' \circ \mathfrak{I}_\otimes^{-1} \circ \pi^* \circ \mathfrak{I}) \circ \pi'\ .$$

Since π' is surjective, this implies that we have the equality $\pi' \circ \mathfrak{I}_\otimes^{-1} \circ \pi^* \circ \mathfrak{I} = (k!) \cdot id$, and thus $\mathfrak{I}$ must be injective. On the other hand, since each $\Theta \in (\bigwedge^k E)^*$ represents an

$\mathfrak{A}$-graded skew-symmetric map on E^k, it follows from (7.17) that we have the equality $\mathbb{A} \circ \mathfrak{I}_{\otimes}^{-1} \circ \pi^* = (k!) \cdot \mathfrak{I}_{\otimes}^{-1} \circ \pi^*$. We now start running again:

$$\begin{aligned}(k!) \cdot \mathfrak{I}_{\otimes}^{-1} \circ \pi^* &= \mathbb{A} \circ \mathfrak{I}_{\otimes}^{-1} \circ \pi^* = (\mathfrak{I}_{\otimes}^{-1} \circ \pi^* \circ \mathfrak{I} \circ \pi') \circ \mathfrak{I}_{\otimes}^{-1} \circ \pi^* \\ &= \mathfrak{I}_{\otimes}^{-1} \circ \pi^* \circ (\mathfrak{I} \circ \pi' \circ \mathfrak{I}_{\otimes}^{-1} \circ \pi^*) .\end{aligned}$$

Since $\mathfrak{I}_{\otimes}^{-1} \circ \pi^*$ is injective, we deduce the equality $\mathfrak{I} \circ \pi' \circ \mathfrak{I}_{\otimes}^{-1} \circ \pi^* = (k!) \cdot id$, and thus $\mathfrak{I}$ must be surjective. QED

8.17 Corollary. *If E is f.g.p, there is a natural identification* $\mathrm{Hom}_L^{sk}(E^k, \mathcal{A}) \cong \bigwedge^k {}^*E$.

8.18 Summary. The following table summarizes some of the more interesting identifications, as well as where one can find sufficient conditions for the identification to be an isomorphism (a $\cong$ indicates that it is always an isomorphism, and a $\hookrightarrow$ indicates that it is always injective). In any case, if all $\mathfrak{A}$-graded $\mathcal{A}$-modules involved are f.g.p, and if the index sets are finite, all identifications are isomorphisms.

identification			definition - isomorphism
$E \otimes (\bigoplus_{i\in I} F_i)$	$\cong$	$\bigoplus_{i\in I}(E \otimes F_i)$	[7.28] - [7.28]
$\bigoplus_{i\in I} \mathrm{Hom}_S(E_i; F)$	$\hookrightarrow$	$\mathrm{Hom}_S(\bigoplus_{i\in I} E_i; F)$	[7.2] - [7.2]
$\bigoplus_{i\in I}(E_i^*)$	$\hookrightarrow$	$(\bigoplus_{i\in I} E_i)^*$	[7.2] - [7.3]
$\bigoplus_{i\in I} \mathrm{Hom}_S(E; F_i)$	$\hookrightarrow$	$\mathrm{Hom}_S(E; \bigoplus_{i\in I} F_i)$	[7.6] - [8.6]
$\bigotimes_{i=n}^{1} \mathrm{Hom}_S(E_i; F_i)$	$\to$	$\mathrm{Hom}_S(\bigotimes_{i=1}^{n} E_i; \bigotimes_{i=1}^{n} F_i)$	[7.12] - [8.12]
$E_1^* \otimes \cdots \otimes E_n^*$	$\to$	$(E_n \otimes \cdots \otimes E_1)^*$	[7.12] - [8.13]
$\bigwedge^k (E^*)$	$\to$	$(\bigwedge^k E)^*$	[7.16] - [8.16]
E	$\to$	${}^*(E^*) \cong ({}^*E)^*$	[7.26] - [8.5]
$F \otimes E^*$	$\to$	$\mathrm{Hom}_R(E; F)$	[7.11] - [8.8]
${}^*E \otimes F$	$\to$	$\mathrm{Hom}_L(E; F)$	[7.11] - [8.8]

Chapter II

Linear algebra of free graded $\mathcal{A}$-modules

In chapter I we have studied general $\mathfrak{A}$-graded commutative algebra for an arbitrary abelian group $\mathfrak{A}$ and an arbitrary $\mathfrak{A}$-graded commutative ring $\mathcal{A}$. Starting this chapter we specialize to the case $\mathfrak{A} = \mathbf{Z}_2$ and we will abbreviate $\mathbf{Z}_2$-graded to simply graded. In order to get "close" to ordinary linear algebra, we also impose two conditions on the graded commutative ring $\mathcal{A}$: it should contain the real numbers $\mathbf{R}$ as a subring and the nilpotent elements in $\mathcal{A}$ should form a supplement to $\mathbf{R} \subset \mathcal{A}$. This allows us to define the notions of basis and graded dimension for free graded $\mathcal{A}$-modules. Except for subspaces and quotients (where one has to be a bit careful), these notions behave exactly as one would expect from ordinary linear algebra.

Special attention has to be paid to matrices associated to linear maps. Whereas in usual linear algebra there is a single natural way to associate a matrix to a linear map when a basis has been given, in graded linear algebra there are three natural ways to do so. Each of these three ways has its own advantages and disadvantages. For instance, one of them is particularly adapted to express the graded trace of a linear map. The same representation by matrices is useful to define the graded determinant, also called Berezinian, of a linear map. However, the proof that the graded determinant is a group homomorphism is a bit cumbersome.

Graded linear algebra is sometimes understood as meaning linear algebra of vector spaces (over $\mathbf{R}$) that split as a direct sum of two subspaces: the even and odd parts. In our context that would be the special case in which one takes $\mathcal{A}$ to be equal to the real numbers $\mathbf{R}$. However, it only requires a very small step to relate the general case to the special case $\mathcal{A} = \mathbf{R}$. This small step is the introduction of the notion of an equivalence class of bases. This idea allows us to reduce all discussions about free graded $\mathcal{A}$-modules to discussions about bases. And then the only difference between a generic $\mathcal{A}$ and the special case $\mathcal{A} = \mathbf{R}$ is just the choice of coordinates with respect to a basis.

1. OUR KIND OF $\mathbf{Z}_2$-GRADED ALGEBRA $\mathcal{A}$

In this section we introduce the additional conditions we will impose from now on on the ring $\mathcal{A}$: it should be a special kind of $\mathbf{Z}_2$-graded commutative $\mathbf{R}$-algebra. The basic example of such a ring is the exterior algebra of an infinite dimensional real vector space [1.2]. The most important result of this section is the lemma that says that any finite number of nilpotent elements in $\mathcal{A}$ can be "killed" multiplicatively by a non-zero nilpotent element. We also introduce matrices with entries in $\mathcal{A}$ and derive some useful properties concerning nilpotent and invertible matrices.

1.1 Definition. Starting this chapter, we will be concerned mostly with $\mathbf{Z}_2$-gradings. This being the case, we will abbreviate $\mathbf{Z}_2$-*graded* to simply *graded*. As usual elements of parity 0 will be called even, but now elements of parity 1 will be called *odd*. When speaking of graded commutativity, we will always use ordinary multiplication in $\mathbf{Z}_2$ for the bilinear symmetric map $\langle\_|\_\rangle : \mathbf{Z}_2 \times \mathbf{Z}_2 \to \mathbf{Z}_2$.

For any ($\mathbf{Z}_2$-)graded commutative ring $\mathcal{A}$ we define the set of nilpotent elements $\mathcal{N}$ by $\mathcal{N} = \{ a \in \mathcal{A} \mid \exists k \in \mathbf{N} : a^k = 0 \}$.

1.2 Examples. • The real line $\mathbf{R}$ is a graded commutative ring if we take $\mathcal{A}_0 = \mathbf{R}$ and $\mathcal{A}_1 = \{0\}$.

• The complex line $\mathbf{C}$ is a graded commutative ring if we define $\mathcal{A}_0 = \mathbf{C}$ and $\mathcal{A}_1 = \{0\}$. Had we defined $\mathcal{A}_0 = \mathbf{R}$ and $\mathcal{A}_1 = i\mathbf{R}$, the ring $\mathbf{C}$ would have been $\mathbf{Z}_2$-graded, but not graded commutative.

• Let X be a real vector space (finite or infinite dimensional), and let $\mathcal{A}$ be the exterior algebra $\mathcal{A} = \bigwedge X = \bigoplus_{k=0}^{\infty} \bigwedge^k X$. It is a graded commutative ring if we define $\mathcal{A}_0 = \bigoplus_{k=0}^{\infty} \bigwedge^{2k} X$, $\mathcal{A}_1 = \bigoplus_{k=0}^{\infty} \bigwedge^{2k+1} X$ and the wedge product as multiplication operation. We have $1 \in \bigwedge^0 X \cong \mathbf{R}$ and $\mathcal{N} = \bigoplus_{k=1}^{\infty} \bigwedge^k X$.

1.3 Lemma. $\mathcal{N} = (\mathcal{N} \cap \mathcal{A}_0) \oplus \mathcal{A}_1$ *and* $\mathcal{N}$ *is an ideal of* $\mathcal{A}$.

Proof. The proof naturally splits in four steps. (i) If $a \in \mathcal{A}_1$ then the graded commutativity implies $a \cdot a = -a \cdot a$ and hence $a^2 = 0$, i.e., $\mathcal{A}_1 \subset \mathcal{N}$. (ii) If $n \in \mathcal{N}$ then we have $0 = (n_0 + n_1)^k = n_0^k + k n_0^{k-1} n_1$ (because of the graded commutativity and (i)). Hence $n_0^{2k} = (-k n_0^{k-1} n_1)^2 = k^2 n_0^{2k-2} n_1^2 = 0$. Thus $n \in \mathcal{N} \Rightarrow n_0 \in \mathcal{N}$. This proves the first assertion.

(iii) If $n, m \in \mathcal{N}$ then $(n+m)^k = (n_0 + m_0)^k + k(n_0 + m_0)^{k-1}(n_1 + m_1)$. Since we know that n_0 and m_0 are also nilpotent, it follows from the binomial formula that $(n+m)^k = 0$ for sufficiently large k. Thus $n, m \in \mathcal{N} \Rightarrow n + m \in \mathcal{N}$. (iv) If $a \in \mathcal{A}$ and $n \in \mathcal{N}$ then $an = an_0 + an_1$. Since n_0 is even, we have $(an_0)^k = a^k n_0^k$ and thus an_0 is nilpotent. Since $(an_1)^2 = an_1 an_1 = a\,(a_0 - a_1)\, n_1^2 = 0$, an_1 is nilpotent too. Thus $a \in \mathcal{A}, n \in \mathcal{N} \Rightarrow an \in \mathcal{N}$. A similar argument applies to na. This proves the second assertion. $\boxed{QED}$

1.4 Lemma. *If $n_1, \cdots, n_N$ is a finite number of nilpotent elements in $\mathcal{A}$, then there exists a non-zero homogeneous nilpotent n such that $\forall i : nn_i = 0$.*

Proof. Without loss of generality we may assume that all n_i are non-zero. Moreover, by considering the homogeneous parts of the n_i separately, we may assume also that all n_i are homogeneous. Define $k_i \in \mathbf{N}$ such that $n_i^{k_i} \neq 0$ and $n_i^{k_i+1} = 0$. Now define $\lambda_0 = 1$ (Nota Bene. This is just formal, $\mathcal{A}$ need not have a unit) and $\lambda_i = \lambda_{i-1} n_i^{k_i}$ if $\lambda_{i-1} n_i^{k_i} \neq 0$, $\lambda_i = \lambda_{i-1}$ otherwise. Taking $n = \lambda_N$ and using the graded commutativity, we find that the product nn_i can be written as $\lambda_i n_i a$ for some $a \in \mathcal{A}$ and hence by definition of k_i we have $nn_i = 0$. QED

1.5 Definition. If $\mathcal{A}$ is a graded commutative ring, we denote by $\mathbf{B}$ the canonical projection $\mathbf{B} : \mathcal{A} \to \mathcal{A}/\mathcal{N}$ and call it the *body map*. Since $\mathcal{N}$ is an ideal in $\mathcal{A}$, $\mathcal{A}/\mathcal{N}$ is a ring and $\mathbf{B}$ a ring homomorphism. If a graded commutative ring $\mathcal{A}$ is an $\mathbf{R}$-algebra, then we have in particular that $\mathcal{A}_0, \mathcal{A}_1, \mathcal{N}$, and $\mathcal{A}/\mathcal{N}$ are vector spaces over $\mathbf{R}$ and that $\mathbf{B} : \mathcal{A} \to \mathcal{A}/\mathcal{N}$ is a linear map between vector spaces over $\mathbf{R}$.

From now on, $\mathcal{A}$ will always denote a graded commutative $\mathbf{R}$-algebra with unit such that $\mathcal{A}/\mathcal{N}$ is isomorphic (as ring) to $\mathbf{R}$. It follows that the body map $\mathbf{B} : \mathcal{A} \to \mathcal{A}/\mathcal{N} \cong \mathbf{R}$ has a canonical section $\mathbf{R} \to \mathcal{A}$ given by $r \mapsto r \cdot 1$ (which is well defined because $\mathcal{A}$ is a vector space over $\mathbf{R}$). By abuse of notation we will always identify $r \in \mathbf{R}$ with $r \cdot 1 \in \mathcal{A}$ and write $a = r + n$, when $\mathbf{B}(a) = r \in \mathbf{R}$, and $n \in \mathcal{N}$.

1.6 Examples. • The real line $\mathbf{R}$ itself verifies the conditions on $\mathcal{A}$ given in [1.5].

• The complex line $\mathbf{C}$ with $\mathbf{C}_0 = \mathbf{C}$ and $\mathbf{C}_1 = \{0\}$ is a graded commutative $\mathbf{R}$-algebra, but it does not verify the conditions of [1.5].

• If X is a vector space over $\mathbf{R}$, its exterior algebra $\bigwedge X$ is a graded commutative $\mathbf{R}$-algebra $\mathcal{A}$ verifying the condition of [1.5].

1.7 Remark. We restricted the symbol $\mathcal{A}$ to denote a graded commutative $\mathbf{R}$-algebra with unit such that $\mathcal{A}/\mathcal{N} \cong \mathbf{R}$. Although the choice of the field of real numbers is important in the next chapters, the results of this chapter remain valid if one replaces $\mathbf{R}$ by any other field of characteristic 0, e.g., $\mathbf{C}$. Moreover, most of the analytic results of the subsequent chapters also remain valid when replacing $\mathbf{R}$ by $\mathbf{C}$.

1.8 Discussion (Geometric interpretation of [1.4]). If $\mathcal{A}$ is the exterior algebra of a vector space X as in [1.2], [1.6], we can give a geometric interpretation of the element n of [1.4]. Let n_i be a non-zero nilpotent element in $\mathcal{A}$, then by definition of the exterior product as the direct sum of exterior powers of X, n_i is a finite sum of monomes, each of which is a finite wedge product of vectors in X. Since there is also a finite number of n_i's, we find a finite number of vectors in X that are involved in the definition of the n_i's.

These vectors span a finite dimensional subspace of X; the wedge product of vectors that form a basis of this finite dimensional subspace is an element $n \in \mathcal{A}$ which satisfies the requirements of the lemma.

1.9 Definitions. The set $M(m \times n, \mathcal{A})$ denotes the set of all matrices of size $m \times n$ (m being the row dimension and n the column dimension) with entries in $\mathcal{A}$. The usual matrix multiplication $M(m \times n, \mathcal{A}) \times M(n \times r, \mathcal{A}) \to M(m \times r, \mathcal{A})$ still makes sense on these sets. The body map $\mathbf{B}$ extends in a natural way to these matrices: $\mathbf{B} : M(m \times n, \mathcal{A}) \to M(m \times n, \mathbf{R})$; it is surjective and preserves matrix multiplication.

The set $M(n \times n, \mathcal{A})$ equipped with matrix multiplication is a ring with unit I_n (in terms of matrix elements $(I_n)^i_j = \delta^i_j$ with δ^i_j the Kronecker delta) for which the body map $\mathbf{B} : M(n \times n, \mathcal{A}) \to M(n \times n, \mathbf{R})$ is a surjective ring homomorphism.

For $a \in M(m \times n, \mathcal{A})$ we define the *rank of* a, denoted as $\text{rank}(a)$, as the rank of its body $\mathbf{B}a \in M(m \times n, \mathbf{R})$: $\text{rank}(a) = \text{rank}(\mathbf{B}a)$, i.e., as the number of independent rows or columns in $\mathbf{B}a$.

1.10 Lemma. *Let $a \in M(n \times n, \mathcal{A})$ such that $\mathbf{B}(a) = 0$, then a is a nilpotent matrix.*

Proof. The condition $\mathbf{B}(a) = 0$ implies that each entry of the matrix a lies in the kernel of the body map, i.e., is nilpotent. It follows that the homogeneous parts of the entries are also nilpotent. Since there are $2n^2$ homogeneous entries (i.e., a finite number), there exists a number N such that $\alpha^N = 0$ for all homogeneous entries α of a. Now the entries of the matrix a^k are sums of terms, each of which is a k-fold product of homogeneous entries of a. Hence for $k > (N-1) \cdot 2n^2$ we have $a^k = 0$. $\boxed{QED}$

1.11 Lemma. *An element $a \in M(n \times n, \mathcal{A})$ is invertible if and only if* $\text{Det}(\mathbf{B}(a)) \neq 0$.

Proof. If b is the inverse of a in $M(n \times n, \mathcal{A})$, then $\mathbf{B}(b)$ is an inverse for $\mathbf{B}(a)$ in $M(n \times n, \mathbf{R})$ and hence $\text{Det}(\mathbf{B}(a)) \neq 0$. Conversely, suppose $\text{Det}(\mathbf{B}(a)) \neq 0$ and write $a = \mathbf{B}(a) + n$ with $\mathbf{B}(n) = 0$. Writing x for the inverse of $\mathbf{B}(a)$ in $M(n \times n, \mathbf{R})$, we define $b = \sum_{k=0}^{\infty} (-xn)^k x$. This is actually a finite sum because of [1.10]; moreover, one easily verifies that b is an inverse to a in $M(n \times n, \mathcal{A})$. $\boxed{QED}$

2. Free graded $\mathcal{A}$-modules

In this section we define the graded dimension of a free graded $\mathcal{A}$-module. It consists of two integers: the number of even respectively odd vectors in a homogeneous basis. We show that this is an invariant of a (finite dimensional) free graded $\mathcal{A}$-module. More precisely, up to isomorphism there exists only one free graded $\mathcal{A}$-module of graded dimension $p|q$. These results depend crucially on the conditions we imposed on $\mathcal{A}$ in [1.5].

2.1 Definition. In this section and the subsequent ones we will be mainly interested in *free* graded $\mathcal{A}$-modules $F(G, \varepsilon)$ (on homogeneous generators G). We recall that all free graded $\mathcal{A}$-modules admit a homogeneous basis (e.g., G). We will call a free graded $\mathcal{A}$-module *finite dimensional* if it admits a finite (homogeneous) basis. A homogeneous basis $e_1, \ldots, e_n$ of a finite dimensional free graded $\mathcal{A}$-module is called *ordered* if all even vectors come first, i.e., e_i even and e_j odd implies $i < j$. A subset F of a free graded $\mathcal{A}$-module E is called a *graded subspace* if it is a graded submodule of E that in itself is (isomorphic to) a free graded $\mathcal{A}$-module.

2.2 Nota Bene. Any finite dimensional free graded $\mathcal{A}$-module is in particular f.g.p, and thus the identifications given in [I.8.18] apply.

2.3 Remark. Since $\mathcal{A}$ is in particular an $\mathbf{R}$-algebra with unit, it follows immediately that any $\mathcal{A}$-module, and thus any free graded $\mathcal{A}$-module, is also a vector space over $\mathbf{R}$. Anticipating the fact that the dimension is an invariant of a free graded $\mathcal{A}$-module, it is immediate that the dimension of a free graded $\mathcal{A}$-module E *as vector space over* $\mathbf{R}$ equals the product of the dimension of E *as free graded* $\mathcal{A}$*-module* and the dimension of $\mathcal{A}$ as vector space over $\mathbf{R}$. Since our main example for $\mathcal{A}$ is the exterior algebra of an infinite dimensional vector space, any free graded $\mathcal{A}$-module (over this $\mathcal{A}$) is infinite dimensional as vector space over $\mathbf{R}$.

2.4 Example. We know already that $\mathcal{A}$ is (trivially) a graded $\mathcal{A}$-module. The unit $1 \in \mathcal{A}$ is both generating and independent, hence $\mathcal{A}$ admits a basis. Since moreover 1 is even, $\mathcal{A}$ is a free graded $\mathcal{A}$-module. Note however that not every non-zero element is a basis for $\mathcal{A}$: if $a \neq 0$ is nilpotent, then a is not generating (1 can not be written as ba with $b \in \mathcal{A}$), nor is it independent (if $a^n \neq 0, a^{n+1} = 0$ then $a^n\, a = 0$ is a relation with a non-zero coefficient).

More generally, the graded $\mathcal{A}$-module $\mathcal{A}^n$ is a finite dimensional free graded $\mathcal{A}$-module because the vectors $(1, 0, \ldots, 0)$, $(0, 1, 0, \ldots, 0)$, $\ldots$, $(0, \ldots, 0, 1)$ are generating, independent and even.

2.5 Lemma. *Let $e_1, \ldots, e_n, f_1, \ldots, f_m$ be elements of a graded $\mathcal{A}$-module E, and suppose that there exist $a_i{}^j \in \mathcal{A}$ such that $f_i = \sum_j a_i{}^j e_j$. If the $f_1, \ldots, f_m$ are independent, then* $\operatorname{rank}(\mathbf{B}a_i{}^j)_{i=1,\ldots,m}^{j=1,\ldots,n} = m$. *In particular $m \leq n$.*

Proof. If $\operatorname{rank}(\mathbf{B}a_i{}^j) < m$, then there must exist a relation among the rows of $\mathbf{B}a_i{}^j$, i.e., there exist $r^i \in \mathbf{R}$, not all of them zero, such that for all $j : \sum_i r^i \mathbf{B}(a_i{}^j) = 0$. We then define $\mu^j = \sum_i r^i a_i{}^j \in \mathcal{A}$ and note that $\mathbf{B}(\mu^j) = 0$, i.e., that the μ^j are nilpotent. Thus, by [1.4], there exists a non-zero $n_o \in \mathcal{A}$ such that $\forall j : n_o\, \mu^j = 0$. But then $\sum_i (n_o\, r^i) f_i = \sum_{i,j} n_o(r^i a_i{}^j)\, e_j = \sum_j n_o\, \mu^j\, e_j = 0$, contradicting the independence of the f_i. We conclude that the rank of $\mathbf{B}a_i{}^j$ is m and hence $m \leq n$. QED

2.6 Proposition. *Let $e_1, \ldots, e_n$ be a basis of a graded $\mathcal{A}$-module E.*

(i) *If $f_1, \ldots, f_m$ is another basis of E, $m = n$.*
(ii) *All other bases $\{f_1, \ldots, f_n\}$ are classified by invertible matrices $a \in M(n \times n, \mathcal{A})$ with $f_i = \sum_j a_i{}^j e_j$.*
(iii) *If $f_1, \ldots, f_n$ is either generating or independent, it is a basis.*

Proof. • Since the f_i and e_j form a generating subset, there exist $a_i{}^j, b_j{}^i \in \mathcal{A}$ such that $f_i = \sum_j a_i{}^j e_j$ and $e_j = \sum_i b_j{}^i f_i$. Since the f_i and e_j are independent, it follows from [2.5] that $n \leq m$ and $m \leq n$.

• If $f_1, \ldots, f_n$ is a basis, by the preceding argument $\exists a_i{}^j \in \mathcal{A}$ such that $f_i = \sum_j a_i{}^j e_j$ and $\operatorname{rank}(\mathbf{B}a_i{}^j) = n$. Thus by [1.11] $a_i{}^j$ is invertible. On the other hand, if $f_i = \sum_j a_i{}^j e_j$ with an invertible $a_i{}^j$, then it follows easily that the f_i form a basis.

• If the f_i are independent, there exist $a_i{}^j \in \mathcal{A}$ such that $f_i = \sum_j a_i{}^j e_j$. By [2.5] we know that $\operatorname{rank}(\mathbf{B}a_i{}^j) = n$, and thus by [1.11] $a_i{}^j$ is invertible. By the preceding result the f_i form a basis. If the f_i are generating, there exist $b_j{}^i \in \mathcal{A}$ such that $e_j = \sum_i b_j{}^i f_i$. As before $b_j{}^i$ must be invertible and thus (by applying the inverse of $b_j{}^i$) the f_i form a basis. $\boxed{QED}$

2.7 Counter example. To show that not all graded $\mathcal{A}$-modules are free graded $\mathcal{A}$-modules, consider the free graded $\mathcal{A}$-module $\mathcal{A}^2$ and choose a non-zero even nilpotent element $\mu \in \mathcal{A}_0$. We then define the subset $F \subset \mathcal{A}^2$ by $F = \operatorname{Span}(\{ (1,0), (0,\mu) \})$, which is a graded submodule by [I.1.25]. We will show that F does not admit a basis, showing at the same time that not every graded $\mathcal{A}$-module is a free graded $\mathcal{A}$-module, and that not every graded submodule of a free graded $\mathcal{A}$-module is again a free graded $\mathcal{A}$-module.

Clearly $\{ (1,0), (0,\mu) \}$ is a generating set and hence a basis of F (if it exists) comprises at most 2 elements. Now let us first suppose that a single element $\{ (a, b\mu) \}$ is a basis, then there exists $\kappa \in \mathcal{A}$ with $(1,0) = \kappa(a, b\mu) = (\kappa a, \kappa b\mu)$ and hence κ must be invertible. It follows from [2.6-ii] that $(1,0)$ is a basis, which is clearly false (it is not generating). Let us suppose next that $\{ (a, b\mu), (c, d\mu) \}$ is a basis, then there exist $\kappa, \lambda \in \mathcal{A}$ with $(1,0) = \kappa(a, b\mu) + \lambda(c, d\mu) = (\kappa a + \lambda c, \kappa b\mu + \lambda d\mu)$. Since 1 is not nilpotent, it follows that not both a and c are nilpotent. Hence by applying [2.6-ii] with a suitable choice of an invertible matrix in $M(2 \times 2, \mathcal{A})$ we may assume without loss of generality that $a = 1$ and $c = 0$. But then $\mu^{k-1}(c, d\mu) = (0, d\mu^k) = (0,0)$, where k is chosen such that $\mu^{k-1} \neq 0$ and $\mu^k = 0$. Since this contradicts the independence, we conclude that F does not admit a basis in the sense of a subset that is both generating and independent.

2.8 Lemma. *Let E be a graded $\mathcal{A}$-module that admits a basis consisting of finitely many elements. Then E admits a homogeneous basis, i.e., E is a finite dimensional free graded $\mathcal{A}$-module.*

Proof. Let $\{f_1, \ldots, f_n\}$ be a basis of E and $f_i = f_{i0} + f_{i1}$ the decomposition in homogeneous parts. Since (f_i) is a basis, there exist matrices $a, b \in M(n \times n, \mathcal{A})$ such that

$f_{i0} = \sum_j a_i{}^j f_j$ and $f_{i1} = \sum_j b_i{}^j f_j$. Since the f_i are independent we have $a + b = I_n$. Decomposing $a_i{}^j = a_i{}^j{}_0 + a_i{}^j{}_1$ into homogeneous parts we find

$$f_{i0} = \sum_j (a_i{}^j{}_0 f_{j0} + a_i{}^j{}_1 f_{j1})$$

(the other two terms necessarily cancel because $f_{i0} \in E_0$). Resubstituting the expressions for f_{i0} and f_{i1} we find $\sum_k a_i{}^k f_k = \sum_{jk}(a_i{}^j{}_0 a_j{}^k + a_i{}^j{}_1 b_j{}^k) f_k$, or (since the f_i are independent) in terms of matrices:

$$a = a_0 a + a_1 b \ .$$

If we realize that $\mathbf{B}(a_i{}^j) = \mathbf{B}(a_i{}^j{}_0)$ and $\mathbf{B}(a_i{}^j{}_1) = 0$, we deduce from this equation $\mathbf{B}(a) = \mathbf{B}(a)\mathbf{B}(a)$ i.e., $\mathbf{B}(a)$ is a projection. Standard linear algebra then shows the existence of a real matrix X such that $X\mathbf{B}(a)X^{-1} = \left(\begin{smallmatrix} I_p & 0 \\ 0 & 0 \end{smallmatrix}\right)$ with $0 \le p \le n$. Since X is invertible, $g_i = \sum_j X_i{}^j f_j$ is also a basis of E. Defining the matrix $c = XaX^{-1}$, we have $g_{i0} = \sum_j c_i{}^j g_j$ and $\mathbf{B}(c) = \left(\begin{smallmatrix} I_p & 0 \\ 0 & 0 \end{smallmatrix}\right)$. It follows that $c = \left(\begin{smallmatrix} I_p + n_1 & n_2 \\ n_3 & n_4 \end{smallmatrix}\right)$, where the n_i are matrices of the appropriate size with nilpotent entries. We finally define the vectors e_i by $e_i = g_{i0}$ if $i \le p$ and $e_i = g_{i1}$ if $i > p$. It follows that $e_i = \sum_j d_i{}^j g_j$ with $d = \left(\begin{smallmatrix} I_p + n_1 & n_2 \\ -n_3 & I_{n-p} - n_4 \end{smallmatrix}\right)$. Since $\mathbf{B}(d) = I_n$, d is invertible and (e_i) is a basis, which obviously consists of homogeneous elements. QED

2.9 Proposition. *The number of even vectors in a homogeneous basis of a finite dimensional free graded $\mathcal{A}$-module E is an invariant of E.*

Proof. Let $\{ e_1, \ldots, e_n \}$ and $\{ \widehat{e}_1, \ldots, \widehat{e}_n \}$ be two ordered homogeneous bases with p and $\widehat{p}$ even elements respectively. These two are related by an invertible matrix a with

$$\widehat{e}_i = \sum_{j=1}^n a_i{}^j e_j = (\sum_{j=1}^p a_i{}^j{}_0 e_j + \sum_{j=p+1}^n a_i{}^j{}_1 e_j) + (\sum_{j=1}^p a_i{}^j{}_1 e_j + \sum_{j=p+1}^n a_i{}^j{}_0 e_j) \ .$$

Independence of the (e_i) and homogeneity of (e_i) and $(\widehat{e}_i)$ then show that $a_i{}^j$ is even if $i \le \widehat{p}$ and $j \le p$ or if $i > \widehat{p}$ and $j > p$; it is odd otherwise. It follows that $\mathbf{B}(a) = \left(\begin{smallmatrix} b & 0 \\ 0 & c \end{smallmatrix}\right)$ where b is a real matrix of size $p \times \widehat{p}$. Since such a real $n \times n$ matrix has necessarily determinant zero if $p \ne \widehat{p}$, it follows from [1.11] that $p = \widehat{p}$. QED

2.10 Definition. The *graded dimension* of a finite dimensional free graded $\mathcal{A}$-module E is a pair of integers (p, q) where p, called *the even dimension of E*, is the number of even vectors in a basis for E and q, called *the odd dimension of E*, the number of odd vectors in (the same) basis. We usually denote this as $\dim(E) = p|q$; in particular $\dim(\mathcal{A}^n) = n|0$. If the graded dimension of E is $p|q$, we define the *total dimension* of E as the sum $n = p + q$, also denoted as $\dim(E) = n$. In general, when we use the word dimension, we will mean the graded dimension, unless it is clear from the context that we mean the total version (which is notably the case when we talk about finite dimensional spaces).

2.11 Proposition. *Two finite dimensional free graded $\mathcal{A}$-modules E and F are isomorphic if and only if they have the same graded dimension.*

Proof. Let $\phi : E \to F$ be an isomorphism and B a homogeneous basis of E. Then $\phi(B)$ is a homogeneous basis for F. Since the parity is not changed by the even map ϕ, it follows that $\dim(E) = \dim(F)$. On the other hand, suppose $\dim(E) = \dim(F)$ and let (e_i) and (f_i) be *ordered* homogeneous bases of E and F respectively. Then the map ϕ defined by $\langle\!\langle \sum_i a^i e_i \,\|\, \phi \rangle\!\rangle = \sum_i a^i f_i$ is an even bijective (left) linear map, i.e., an isomorphism between E and F. QED

2.12 Remark. If one extends the definition of graded dimension to arbitrary free graded $\mathcal{A}$-modules as yielding a pair of cardinals, then [2.11] remains true; even the proof only needs cosmetic changes. Note that our definition of an isomorphism as an *even* invertible map is crucial in the invariance of the graded dimension. Had we defined an isomorphism as just any invertible map, the graded dimension would not have been an invariant.

2.13 Definition. On a graded $\mathcal{A}$-module $E = E_0 \oplus E_1$ we define an involution $\mathfrak{C} : E \to E$ by $\mathfrak{C}e \equiv \mathfrak{C}(e_0 + e_1) \equiv \mathfrak{C}(\sum_{\alpha \in \mathbf{Z}_2} e_\alpha) = \sum_{\alpha \in \mathbf{Z}_2} (-1)^{\langle \alpha | \alpha \rangle} e_\alpha \equiv e_0 - e_1$.

2.14 Lemma. *The involution $\mathfrak{C}$ is a ring homomorphism when applied to $\mathcal{A}$ itself. More generally it verifies: $\forall a, b \in \mathcal{A}\, \forall e, f \in E : \mathfrak{C}(ae + bf) = \mathfrak{C}(a)\mathfrak{C}(e) + \mathfrak{C}(b)\mathfrak{C}(f)$. If $e \in E$ is homogeneous and $a \in \mathcal{A}$, then $ae = e \cdot \mathfrak{C}^{\varepsilon(e)}(a)$, i.e., if e is even $ae = ea$ and if e is odd, $ae = e\mathfrak{C}(a)$.*

2.15 Remark. We could have defined an involution $\mathfrak{C}$ for general $\mathfrak{A}$-graded $\mathcal{A}$-modules by the formula $\mathfrak{C}(e) = \mathfrak{C}(\sum_{\alpha \in \mathfrak{A}} e_\alpha) = \sum_{\alpha \in \mathfrak{A}} (-1)^{\langle \alpha | \alpha \rangle} e_\alpha$. And then the first two parts of [2.14] remain true. However, the last part linking left and right multiplication for homogeneous elements does not hold in the general setting. And it is this part which makes this involution interesting in the $\mathbf{Z}_2$-graded case.

3. Constructions of free graded $\mathcal{A}$-modules

In this section we show that most constructions of graded $\mathcal{A}$-modules preserve the existence of a basis, i.e., if the original graded $\mathcal{A}$-modules are free, then the newly constructed graded $\mathcal{A}$-module is again free. However, this is certainly not the case for the most elementary "construction" of a graded submodule. We discuss sufficient conditions on generating subsets to generate a graded subspace, i.e., a graded submodule with a basis. For quotients, closely related to graded subspaces, the situation is slightly better: for finite *dimensional graded subspaces of a free graded $\mathcal{A}$-module, the quotient is always a free graded $\mathcal{A}$-module.*

3.1 Counter example. We have seen that the sum $\sum_{i\in I} F_i$ of graded submodules F_i of a graded $\mathcal{A}$-module E is again a graded submodule. However, if all the F_i are graded subspaces, i.e., admit a basis, there is no reason why their sum should admit a basis. An elementary counter example is given by $E = \mathcal{A}^2$ and the two graded subspaces F_1 generated by $(1, 0)$ and F_2 generated by $(1, \mu)$ where μ is an even nilpotent element. Their sum $F = F_1 + F_2$ is the graded submodule encountered in [2.7].

3.2 Proposition. *Let E be a graded $\mathcal{A}$-module and let $B \subset E$ be a set of homogeneous independent elements with $B_\alpha = B \cap E_\alpha$. Let furthermore $H_\alpha \subset \mathrm{Span}_{\mathbf{R}}(B_\alpha)$ be arbitrary. Then $F = \mathrm{Span}(H_0 \cup H_1)$ is a free graded $\mathcal{A}$-module.*

proof. By applying Zorn's Lemma, we know that there exist $G_\alpha \subset H_\alpha$ such that the elements of G_α are independent over $\mathbf{R}$ and such that every element in H_α can be written as a finite linear combination with real coefficients of elements in G_α (this can be shown along the lines of the proof that every real vector space admits a basis). Since $\mathbf{R}$ is a subring of $\mathcal{A}$, it follows that $F = \mathrm{Span}(G_0 \cup G_1)$. We now claim that the elements of $G = G_0 \cup G_1$ are independent over $\mathcal{A}$.

To prove this claim, let $\sum_{i=1}^n \mu^i g_i = 0$ be a relation between elements $g_i \in G$ with coefficients $\mu^i \in \mathcal{A}$. Since $G \subset H$, it follows that there are $e_1, \ldots, e_m \in B$ (finitely many!) and $x_i{}^j \in \mathbf{R}$, i.e., $x \in M(n \times m, \mathbf{R})$, such that $g_i = \sum_{j=1}^m x_i{}^j e_j$. Since the g_i are independent over $\mathbf{R}$, the rank of the matrix x must be n and hence $n \leq m$. Standard linear algebra then provides the existence of a matrix $y \in M(m \times n, \mathbf{R})$ with the property $\sum_{j=1}^m x_i{}^j y_j{}^k = \delta_i^k$.

With these preparations we proceed as follows: the equality $\sum_{i=1}^n \mu^i g_i = 0$ implies $\sum_{i,j} \mu^i x_i{}^j e_j = 0$ and hence by the independence of B we have $\forall j : \sum_i \mu^i x_i{}^j = 0$. We then multiply by y to obtain: $\forall k : 0 = \sum_{i,j} \mu^i x_i{}^j y_j{}^k = \sum_i \mu^i \delta_i^k = \mu^k$. Since G is an independent set of homogeneous elements and clearly generates $F = \mathrm{Span}(G)$ it only remains to show that F is a *graded* submodule. But that follows immediately from the observation $F \cap E_\alpha = \{\sum_{i=1}^n a^i g_i \mid n \in \mathbf{N}, g_i \in G, a^i \in \mathcal{A}, a^i g_i \in E_\alpha\}$. QED

3.3 Corollary. *Let E be a graded $\mathcal{A}$-module and let $B \subset E$ be a set of homogeneous independent elements. Then $F = \mathrm{Span}(B)$ is a free graded $\mathcal{A}$-module admitting B as basis.*

3.4 Definition. We will say that a graded subspace F of a free graded $\mathcal{A}$-module E admits *a supplement* if there exists another graded subspace $\widehat{F}$, called *a supplement to* F, such that $E = F \bigoplus \widehat{F}$. The last condition will be satisfied if and only if $F \cap \widehat{F} = \{0\}$ and

$F + \widehat{F} = E$. Note that the existence of a supplement to F as graded subspace is a stronger condition than the existence of a supplement to F as graded submodule [I.3.3].

3.5 Corollary. *A graded subspace F of a free graded $\mathcal{A}$-module E admits a supplement if and only if there exists a homogeneous basis B of E such that a subset of B forms a basis of F (and then the complementary subset forms a basis for a supplement).*

3.6 Proposition. *Let F be a graded subspace of a free graded $\mathcal{A}$-module E. Then F admits a supplement if and only if E/F is a free graded $\mathcal{A}$-module. If that is the case, E is isomorphic to $F \oplus (E/F)$.*

Proof. If F admits a supplement $\widehat{F}$, it follows immediately that the canonical projection restricted to $\widehat{F}$, i.e., $\pi : \widehat{F} \to E/F$ is a bijection. Whence it follows that E/F is a free graded $\mathcal{A}$-module isomorphic by π to $\widehat{F}$, and that E is isomorphic to $F \oplus (E/F)$.

On the other hand, suppose that E/F is a free graded $\mathcal{A}$-module with homogeneous basis $(x_j)_{j\in J}$. Choose homogeneous $f_j \in E$ such that $\pi(f_j) = x_j$ and let $(e_i)_{i\in I}$ be a homogeneous basis of F. We claim that $\{ e_i \mid i \in I \} \cup \{ f_j \mid j \in J \}$ is a homogeneous basis of E. First independence: let $\sum_i \lambda^i e_i + \sum_j \mu^j f_j = 0$ be a relation, then applying π we get $\sum_j \mu^j x_j = 0$ and hence $\mu^j = 0$. But then $\sum_i \lambda^i e_i = 0$ and hence $\lambda^i = 0$. Next consider $e \in E$, then $\exists \mu^j$ such that $\pi(e) = \sum_j \mu^j x_j$, but then $e - \sum_j \mu^j f_j \in F$, i.e., $\exists \lambda^i$ such that $e - \sum_j \mu^j f_j = \sum_i \lambda^i e_i$. The result now follows from [3.5]. $\boxed{QED}$

3.7 Proposition. *If F is a finite dimensional graded subspace of a free graded $\mathcal{A}$-module E, then every homogeneous basis C of F can be extended to a homogeneous basis of E.*

Proof. Let B be a homogeneous basis for E and let $C = \{f_1, \ldots, f_n\}$ be a homogeneous basis for F. Since $F \subset E$ and by definition of a basis, there exist $e_1, \ldots, e_m \in B$ (finitely many!) and $a_i{}^j \in \mathcal{A}$, i.e., $a \in M(n \times m, \mathcal{A})$, such that $f_i = \sum_j a_i{}^j e_j$. By [2.5] and the independence of the f_i we conclude that $m \geq n$.

By reshuffling the elements e_j we may assume without loss of generality that the matrix $b \in M(n \times n, \mathcal{A})$ defined by $b_i{}^j = a_i{}^j, 1 \leq i, j \leq n$ has the property that $\mathbf{B}(b)$ is non degenerate, i.e., that b is invertible [1.11]. But then the relation $f_i = \sum_j a_i{}^j e_j$ can be rewritten as $\sum_i (b^{-1})_k{}^i f_i = \sum_{i,j} (b^{-1})_k{}^i a_i{}^j f_j = e_k + \sum_{i=1}^n \sum_{j=n+1}^m (b^{-1})_k{}^i a_i{}^j f_j$. In other words, the $e_1, \ldots, e_n$ can be expressed in terms of $f_1, \ldots, f_n, e_{n+1}, \ldots, e_m$. It follows that $C \cup (B \setminus \{ e_1, \ldots, e_n \})$ is a homogeneous basis for E. $\boxed{QED}$

3.8 Corollary. *Let F be a finite dimensional graded subspace of a free graded $\mathcal{A}$-module E. Then F admits a supplement and E/F is a free graded $\mathcal{A}$-module.*

3.9 Counter example. In order to show that [3.7] cannot be extended to infinite dimensional graded subspaces without a more precise knowledge of $\mathcal{A}$, we consider two examples. First we consider $\mathcal{A} = \mathbf{R}$. Free graded $\mathcal{A}$-modules are in this case nothing more than real vector spaces split as a direct sum of two subspaces, this splitting being part of the data of the free graded $\mathcal{A}$-module (see also §6). Since the existence of bases is always guaranteed for real vector spaces, [3.7] is true for any graded subspace in the case $\mathcal{A} = \mathbf{R}$.

In the second example we use the full exterior power $\bigwedge X$ of an infinite dimensional (real) vector space X as $\mathcal{A}$. Let $\{ x_i \mid i \in \mathbf{N} \}$ be a set of independent (over $\mathbf{R}$) elements of X, and define $n_i \in \bigwedge^2 X \subset \mathcal{A}_0, i \in \mathbf{N}$ by $n_i = x_{2i} \wedge x_{2i+1}$. Obviously we have $n_i^2 = 0$, but any (finite) product of different n_i's is non-zero because of the independence of the x_i. Now let $B = \{ e_i \mid i \in \mathbf{N} \}$ be any countable abstract set and define the free graded $\mathcal{A}$-module E by $E = F(B, \varepsilon)$, where ε is constant 0 on B, i.e., E is the free graded $\mathcal{A}$-module on the even symbols in B. In E we define the even vectors $f_i, i \in \mathbf{N}$ by $f_i = e_i - n_i\, e_{i+1}$.

Our first claim is that the f_i are independent. To see this, suppose $\sum_{i=p}^q \lambda^i f_i = 0$. An elementary calculation shows

$$\sum_{i=p}^{q} \lambda^i f_i = \lambda^p e_p + \sum_{i=p+1}^{q} (\lambda^i - \lambda^{i-1} n_{i-1}) e_i - \lambda^q n_q e_{q+1} , \tag{3.10}$$

and the result follows from the fact that the e_i are independent. We conclude that the subset $F = \mathrm{Span}(\{ f_i \mid i \in \mathbf{N} \})$ is a graded subspace of E. Our second claim is that F is a proper subset of E, and more in particular, that $e_0 \notin F$. But this follows as easily from (3.10) as the independence of the f_i.

Finally suppose that there exists a basis $\widehat{B}$ for E of which a proper subset forms a basis for F. By replacing this subset by $\{ f_i \mid i \in \mathbf{N} \}$ it remains a basis for E. This is not as obvious as it might seem, but the proof (of the independence) follows along the same lines as the proof of [3.7]. Without loss of generality we thus may assume that $\{ f_i \mid i \in \mathbf{N} \} \subset \widehat{B}$. Now choose $e \in \widehat{B} \setminus \{ f_i \mid i \in \mathbf{N} \}$ arbitrarily. Since B is a basis for E, there exist $\lambda^i \in \mathcal{A}$ such that $e = \sum_{i=p}^q \lambda^i e_i$. An elementary induction argument shows the equality

$$e_i = \sum_{k=i}^{q} \Big(\prod_{\ell=i}^{k-1} n_\ell \Big) f_k + \Big(\prod_{\ell=i}^{q-1} n_\ell \Big) n_q e_{q+1} ,$$

from which we deduce the following expression for e :

$$e = \sum_{i=p}^{q} \sum_{k=i}^{q} \lambda^i \Big(\prod_{\ell=i}^{k-1} n_\ell \Big) f_k + \Big(\sum_{i=p}^{q} \lambda^i \big(\prod_{\ell=i}^{q-1} n_\ell \big) \Big) n_q e_{q+1} .$$

Multiplying this formula by n_q kills the term with e_{q+1} and we are left with a non trivial relation between e and the f_i ($n_q \neq 0$). This contradicts the independence of $\widehat{B}$. We conclude that there does not exist a basis for E of which a subset forms a basis for the graded subspace F.

3.11 Proposition. *Let E and F be free graded $\mathcal{A}$-modules and let $\phi \in \mathrm{Hom}_S(E; F)$ be homogeneous. If $\ker(\phi)$ is a graded subspace of E admitting a supplement, then $\mathrm{im}(\phi)$ is a graded subspace of F. If in addition ϕ is even, then $E \cong \ker(\phi) \oplus \mathrm{im}(\phi)$.*

proof. Let H be a supplement to $\ker(\phi)$ in E. It follows that ϕ restricted to H is injective. If C is a homogeneous basis of H, then one easily shows that $\phi(C)$ is a set of independent homogeneous elements generating $\mathrm{im}(\phi)$. In other words, $\mathrm{im}(\phi)$ is a graded subspace of F. If in addition ϕ is even, ϕ is an isomorphism between H and $\mathrm{im}(\phi)$. Since H is isomorphic to $E/\ker(\phi)$, the result follows. QED

3.12 Counter examples. The counter examples given here are intended to prove that the conditions in [I.3.9] and [3.11] cannot be weakened easily. On the other hand, they also show that these conditions, although sufficient, are not necessary.

Define $\phi : \mathcal{A}^1 \to \mathcal{A}^2$, $\psi : \mathcal{A}^2 \to \mathcal{A}^1$, and $\chi : \mathcal{A}^1 \to \mathcal{A}^1$ by $\phi(a) = (a, \mu a)$, $\psi(a, b) = b - \mu a$, and $\chi(a) = \mu a$, where $\mu \in \mathcal{A}$ is (for the moment) arbitrary. These three maps are right linear. Obviously ϕ is injective and ψ is surjective, which implies that $\ker(\phi)$ and $\mathrm{im}(\psi)$ are graded subspaces and in particular graded submodules. Moreover, $\mathrm{im}(\phi)$ and $\ker(\psi)$ are both the submodule of $\mathcal{A}^2$ generated by the vector $(1, \mu)$. However, if μ is odd, then ϕ and ψ are not homogeneous and neither $\mathrm{im}(\phi)$ nor $\ker(\psi)$ is a graded submodule.

If μ is invertible but not homogeneous, then χ is not homogeneous but $\ker(\chi)$ and $\mathrm{im}(\chi)$ are (trivially) graded subspaces. On the other hand, if μ is homogeneous, nilpotent and non-zero, then χ is homogeneous but neither $\mathrm{im}(\chi)$ nor $\ker(\chi)$ is a graded subspace (though they are graded submodules).

Let E be a (the) $\mathcal{A}$-vector space of dimension $1|1$ with homogeneous basis e, f. If we define $\phi : E \to \mathcal{A}$ by $\phi(\alpha e + \beta f) = \alpha - \beta$, then ϕ is left linear but not homogeneous. $\ker(\phi)$ is generated by $e + f$ and is a submodule but not a graded submodule: the even and odd parts of $\ker(\phi)$ do not belong to $\ker(\phi)$.

3.13 Lemma. *Suppose $E_1, \ldots, E_n$ are finite dimensional free graded $\mathcal{A}$-modules with $\dim(E_i) = p_i|q_i$, then $\bigoplus_{i=1}^n E_i$ is a finite dimensional free graded $\mathcal{A}$-module of dimension $(\sum_{i=1}^n p_i) \mid (\sum_{i=1}^n q_i)$.*

3.14 Lemma. *If E and F are finite dimensional free graded $\mathcal{A}$-modules of dimension $p|q$ and $r|s$ respectively, then $E \otimes F$ is a finite dimensional free graded $\mathcal{A}$-module of dimension $(pr + qs) \mid (ps + qr)$.*

Proof. This is an immediate consequence of [I.8.9]. QED

3.15 Lemma. *If E is a finite dimensional free graded $\mathcal{A}$-module, then the k-th exterior power $\bigwedge^k E$ is a finite dimensional free graded $\mathcal{A}$-module.*

Proof. This is an immediate consequence of [I.8.14] and [I.8.15]. QED

3.16 Nota Bene. Let E be a finite dimensional free graded $\mathcal{A}$-module of dimension $p|q$, and let $(e_i)_{i=1}^{p+q}$ be an ordered homogeneous basis of E. It follows from [I.8.15] that the set

$$B = \{\, e_{i_1} \wedge \cdots \wedge e_{i_k} \mid i_1 < \ldots i_j < i_{j+1} \leq \cdots \leq i_k \quad \& \\ \varepsilon(e_{i_1}) = \cdots = \varepsilon(e_{i_j}) = 0, \varepsilon(e_{i_{j+1}}) = \cdots = \varepsilon(e_{i_k}) = 1 \,\}$$

is a basis of $\bigwedge^k E$. An elementary computation then shows that the total dimension of $\bigwedge^k E$ is given by

$$\sum_{s=0}^{\min(k,p)} \binom{p}{s} \cdot \binom{q-1+k-s}{k-s} .$$

In this sum, the terms with even s contribute to the even dimension of $\bigwedge^k E$, terms with odd s contribute to the odd dimension.

3.17 Construction (dual basis). Let E be a finite dimensional free graded $\mathcal{A}$-module and $(e_i)_{i=1}^n$ a homogeneous basis. We define the maps $e^i \in E^*$ by $e^i(\sum_j e_j a^j) = a^i$. We leave it to the reader to verify that these $(e^i)_{i=1}^n$ are independent, generating and homogeneous of degree $\varepsilon(e^i) = \varepsilon(e_i)$. We conclude that E^* is a finite dimensional free graded $\mathcal{A}$-module of the same graded dimension as E. The basis $(e^i)_{i=1}^n$ for E^* will be called the *right dual basis to* $(e_i)_{i=1}^n$. In a similar way one shows that $^*E$ is a finite dimensional free graded $\mathcal{A}$-module of the same graded dimension as E with as basis the maps $^ie \in {}^*E$ defined by $\langle\!\langle \sum_j a^j e_j \parallel {}^ie \rangle\!\rangle = a^i$. The basis $({}^ie)_{i=1}^n$ is called the *left dual basis to* $(e_i)_{i=1}^n$.

3.18 Proposition. *If $E_1, \ldots, E_k$, and F are free graded $\mathcal{A}$-modules, of which the E_i are finite dimensional, then* $\mathrm{Hom}_S(E_1, \ldots, E_k; F)$ *is a free graded $\mathcal{A}$-module.*

Proof. According to [I.4.9] the graded $\mathcal{A}$-module $\mathrm{Hom}_S(E_1, \ldots, E_k; F)$ is isomorphic to $\mathrm{Hom}_S(E_1 \otimes \cdots \otimes E_k; F)$. According to [I.8.9] $E_1 \otimes \cdots \otimes E_k$ is a finite dimensional free graded $\mathcal{A}$-module, and thus by [I.8.8] $\mathrm{Hom}_R(E_1 \otimes \cdots \otimes E_k; F)$ is isomorphic to $F \otimes (E_1 \otimes \cdots \otimes E_k)^*$, which is a free graded $\mathcal{A}$-module by [3.17] and [I.8.9]. A similar argument holds for Hom_L. QED

3.19 Corollary. *Given free graded $\mathcal{A}$-modules E and F, E being finite dimensional, then* $\mathrm{Hom}_S^{sk}(E^k; F)$ *is a free graded $\mathcal{A}$-module.*

Proof. According to [I.5.5] there exists an isomorphism between $\mathrm{Hom}_S(\bigwedge^k E; F)$ and $\mathrm{Hom}_S^{sk}(E^k; F)$. But according to [3.18] the former is a free graded $\mathcal{A}$-module. QED

4. Linear maps and matrices

In ordinary linear algebra over a field one often identifies linear maps with matrices. Of course this requires bases, but since these are usually understood, there is no real problem. Unfortunately, for free graded $\mathcal{A}$-modules the situation is not so easy. Even after the choice of a basis, there still are three different natural ways to identify a linear map with a matrix. We discuss the merits of all three ways and we point out some of the dangers involved in this kind of identification (see also [VI.1.20]). The comparison between linear maps and matrices allows us to explain why we called the operator $\mathfrak{T}$ linking left and right linear maps the transposition operator.

4.1 Construction (linear maps and matrices). Let E and F be free graded $\mathcal{A}$-modules of dimension $n = p|q$ and $m = r|s$, and let $(e_i)_{i=1}^n$ and $(f_j)_{j=1}^m$ be homogeneous bases of E and F respectively. We denote by $({}^ie)_{i=1}^n$, $(e^i)_{i=1}^n$, $({}^jf)_{j=1}^m$, and $(f^j)_{j=1}^m$ the left and right dual bases. We know that $(f_j \otimes e^i)_{i,j=1}^{n,m}$ is a basis for $F \otimes E^*$. Since finite dimensional free graded $\mathcal{A}$-modules satisfy the conditions of [I.8.8], $F \otimes E^*$ is isomorphic to $\mathrm{Hom}_R(E;F)$. We thus are allowed to say that $(f_j \otimes e^i)_{i,j=1}^{n,m}$ is a basis for $\mathrm{Hom}_R(E;F)$. Since this basis is labeled in a natural way with two indices, coordinates of a map ϕ with respect to this basis can be seen as matrices of size $m \times n$. However, the position of the coordinates is not uniquely determined. We have three types: left coordinates ${}^\bullet M_R(\phi)^j{}_i$, right coordinates $M_R^\bullet(\phi)^j{}_i$, and middle coordinates $\dot{M}_R(\phi)^j{}_i$, determined by

$$\phi = \sum_{i,j} {}^\bullet M_R(\phi)^j{}_i \cdot f_j \otimes e^i \quad = \quad \sum_{i,j} f_j \otimes \dot{M}_R(\phi)^j{}_i \cdot e^i \quad = \quad \sum_{i,j} f_j \otimes e^i \cdot M_R^\bullet(\phi)^j{}_i \;.$$

The coefficients of these three matrices are related by ${}^\bullet M_R(\phi)^j{}_i = \mathfrak{C}^{\varepsilon(f_j)}(\dot{M}_R(\phi)^j{}_i)$ and $M_R^\bullet(\phi)^j{}_i = \mathfrak{C}^{\varepsilon(e_i)}(\dot{M}_R(\phi)^j{}_i)$.

In order to visualize these relations, we suppose that the bases of E and F are ordered, and we write any matrix $X \in M(m\times n, \mathcal{A})$ in block form $X = \begin{pmatrix} A & B \\ C & D \end{pmatrix}$ with A of size $r \times p$ and all others of the appropriate corresponding sizes. In terms of such a decomposition, the relation between the three systems of coordinates is given by

$$\dot{M}_R(\phi) = \begin{pmatrix} A & B \\ C & D \end{pmatrix} \quad \Longrightarrow \quad {}^\bullet M_R(\phi) = \begin{pmatrix} A & B \\ \mathfrak{C}(C) & \mathfrak{C}(D) \end{pmatrix} \quad , \quad M_R^\bullet(\phi) = \begin{pmatrix} A & \mathfrak{C}(B) \\ C & \mathfrak{C}(D) \end{pmatrix} \,.$$

We thus see that we have three different maps $\mathrm{Hom}_R(E;F) \to M(m \times n, \mathcal{A})$, all three "honest" coordinates, and all three obviously additive. However, their behavior under left or right multiplication by $\mathcal{A}$ is quite different: for $\phi \in \mathrm{Hom}_R(E;F)$ and $\lambda \in \mathcal{A}$ one easily verifies the following results:

$${}^\bullet M_R(\phi) = \begin{pmatrix} A & B \\ C & D \end{pmatrix} \quad \Longrightarrow \quad {}^\bullet M_R(\lambda\phi) = \begin{pmatrix} \lambda A & \lambda B \\ \lambda C & \lambda D \end{pmatrix} \qquad {}^\bullet M_R(\phi\lambda) = \begin{pmatrix} A\lambda & B\mathfrak{C}(\lambda) \\ C\mathfrak{C}(\lambda) & D\lambda \end{pmatrix}$$

$$\dot{M}_R(\phi) = \begin{pmatrix} A & B \\ C & D \end{pmatrix} \quad \Longrightarrow \quad \dot{M}_R(\lambda\phi) = \begin{pmatrix} \lambda A & \lambda B \\ \mathfrak{C}(\lambda)C & \mathfrak{C}(\lambda)D \end{pmatrix} \quad \dot{M}_R(\phi\lambda) = \begin{pmatrix} A\lambda & B\mathfrak{C}(\lambda) \\ C\lambda & D\mathfrak{C}(\lambda) \end{pmatrix}$$

$$M_R^\bullet(\phi) = \begin{pmatrix} A & B \\ C & D \end{pmatrix} \quad \Longrightarrow \quad M_R^\bullet(\lambda\phi) = \begin{pmatrix} \lambda A & \mathfrak{C}(\lambda)B \\ \mathfrak{C}(\lambda)C & \lambda D \end{pmatrix} \quad M_R^\bullet(\phi\lambda) = \begin{pmatrix} A\lambda & B\lambda \\ C\lambda & D\lambda \end{pmatrix} .$$

Since we did not (and never will!) give the set $M(m \times n, \mathcal{A})$ the structure of a graded $\mathcal{A}$-module, we cannot say that these maps are either left or right linear. Actually, saying that one of them is left (or right) linear gives $M(m \times n, \mathcal{A})$ the structure of a free graded $\mathcal{A}$-module, but each choice gives a different structure.

For left linear maps we do the same: we represent $\mathrm{Hom}_L(E;F)$ in three different ways by matrices in $M(n \times m, \mathcal{A})$. For any $\psi \in \mathrm{Hom}_L(E;F)$ we introduce left coordinates ${}^\bullet M_L(\psi)_i{}^j$, right coordinates $M_L^\bullet(\psi)_i{}^j$ and middle coordinates $\overset{\bullet}{M}_L(\psi)_i{}^j$, determined by

$$\psi = \sum_{i,j} {}^\bullet M_L(\psi)_i{}^j \cdot {}^i e \otimes f_j \quad = \quad \sum_{i,j} {}^i e \otimes \overset{\bullet}{M}_L(\psi)_i{}^j \cdot f_j \quad = \quad \sum_{i,j} {}^i e \otimes f_j \cdot M_L^\bullet(\psi)_i{}^j .$$

The visualization of the relation between these three types, as well as their behavior under multiplication by elements of $\mathcal{A}$ is *the same* as for right linear morphisms, except that in the left linear case the matrix A has size $p \times r$ (instead of $r \times p$) and the others of the appropriate corresponding sizes.

We thus obtain the maps ${}^\bullet M_S$, $\overset{\bullet}{M}_S$, and $M_S^\bullet$ which are additive bijections between $\mathrm{Hom}_S(E;F)$ and matrices. But one should never forget that these maps do depend upon the choice of bases for the free graded $\mathcal{A}$-modules E and F. Note also the position of the indices: it is always the first index that is the row index and the second one the column index. On the other hand, the height of an index depends upon the height of the corresponding basis and is more a question of custom than necessity.

4.2 Lemma. *Let E, F and H be free graded $\mathcal{A}$-modules of total dimension n, m, and ℓ respectively, each equipped with a homogeneous basis, then for $\phi_S \in \mathrm{Hom}_S(E;F)$ and $\psi_S \in \mathrm{Hom}_S(F;H)$ we have*

$$\overset{\bullet}{M}_L(\phi_L \diamond \psi_L) = \overset{\bullet}{M}_L(\phi_L) \cdot \overset{\bullet}{M}_L(\psi_L) \quad \textit{and} \quad \overset{\bullet}{M}_R(\psi_R \circ \phi_R) = \overset{\bullet}{M}_R(\psi_R) \cdot \overset{\bullet}{M}_R(\phi_R) ,$$

i.e., $\overset{\bullet}{M}_R$ and $\overset{\bullet}{M}_L$ are homomorphisms with values in $M(m \times n, \mathcal{A})$ and $M(n \times m, \mathcal{A})$ respectively.

Proof. This is an elementary verification using the definition of these matrices. The fact that $\overset{\bullet}{M}_S$ is *in between* the tensor product symbols is crucial. QED

4.3 Remark. The fact that $\overset{\bullet}{M}_L$ is a homomorphism when using the alternative composition $\diamond$ once again confirms the idea that left linear maps should be seen as operating on the right rather than on the left.

4.4 Discussion. According to [4.2], it is $\overset{\bullet}{M}_S$ that transforms the composition of homomorphisms into matrix multiplication. (We leave it to the reader to verify that the others do not.) In particular when E is the same as F, $\overset{\bullet}{M}_S$ becomes a ring isomorphism

$\overset{\bullet}{M}_S : \mathrm{End}_S(E) \to M(n \times n, \mathcal{A})$. These facts make that $\overset{\bullet}{M}_S$ is by far the most attractive matrix representation of the three types. Since the visualizations allow for an easy translation between these three types, we thus will study only the maps $\overset{\bullet}{M}_S$ in more detail.

4.5 Lemma. *Let E be a finite dimensional free graded $\mathcal{A}$-module, $(e_i)_{i=1}^n$ a homogeneous basis with the associated dual bases. Then $\mathfrak{T}({}^i e) = (-1)^{\langle \varepsilon(e_i) | \varepsilon(e_i) \rangle} \cdot e^i$.*

Proof. $\mathfrak{T}({}^i e)(e_j) = (-1)^{\langle \varepsilon(e_i) | \varepsilon(e_j) \rangle} \langle\!\langle e_j \,\|\, {}^i e \rangle\!\rangle = (-1)^{\langle \varepsilon(e_i) | \varepsilon(e_i) \rangle} \delta^i_j$. QED

4.6 Lemma. *Let E and F be free graded $\mathcal{A}$-modules, of which E is finite dimensional. Under the identification ${}^*E \otimes F \cong \mathrm{Hom}_L(E; F)$ the transposition maps $\mathfrak{T}_{\mathrm{Hom}_L(E;F)}$ and $\mathfrak{T}_{{}^*E}$ are related by $\mathfrak{T}_{\mathrm{Hom}_L(E;F)} = \mathfrak{R} \circ (\mathfrak{T}_{{}^*E} \otimes id(F)) = (id(F) \otimes \mathfrak{T}_{{}^*E}) \circ \mathfrak{R}$.*

Proof. An elementary calculation using homogeneous bases and their duals shows that all three maps send the basis elements ${}^i e \otimes f_j \in \mathrm{Hom}_L(E; F)$ to the same element $(-1)^{\langle \varepsilon(e_i) | \varepsilon(e_i) \rangle} (-1)^{\langle \varepsilon(e_i) | \varepsilon(f_j) \rangle} f_j \otimes e^i = (-1)^{\langle \varepsilon(e_i) | \varepsilon({}^i e \otimes f_j) \rangle} f_j \otimes e^i \in \mathrm{Hom}_R(E; F)$. QED

4.7 Corollary. *Let E and F be finite dimensional free graded $\mathcal{A}$-modules, both equipped with homogeneous bases, and let $\phi : E \to F$ be left linear. Then:*

$$\overset{\bullet}{M}_R(\mathfrak{T}\phi)^j{}_i = (-1)^{\langle \varepsilon(e_i) | \varepsilon(e_i) + \varepsilon(f_j) \rangle} \cdot \mathfrak{C}^{\varepsilon(e_i)+\varepsilon(f_j)}(\overset{\bullet}{M}_L(\phi)_i{}^j) .$$

Proof.

$$\begin{aligned} \mathfrak{T}(\sum_{i,j} {}^i e \otimes c_i{}^j \cdot f_j) &= \sum_{i,j} \mathfrak{C}^{\varepsilon(e_i)}(c_i{}^j) \cdot \mathfrak{T}({}^i e \otimes f_j) \\ &= \sum_{i,j} (-1)^{\langle \varepsilon(e_i) | \varepsilon(e_i) + \varepsilon(f_j) \rangle} \cdot \mathfrak{C}^{\varepsilon(e_i)}(c_i{}^j) \cdot f_j \otimes e^i \\ &= \sum_{i,j} f_j \otimes \Big((-1)^{\langle \varepsilon(e_i) | \varepsilon(e_i) + \varepsilon(f_j) \rangle} \cdot \mathfrak{C}^{\varepsilon(e_i)+\varepsilon(f_j)}(c_i{}^j) \Big) \cdot e^i . \end{aligned}$$

QED

4.8 Discussion. The result of [4.7] is quite unintelligible, except for experts in indices. So let us visualize this relation as we did for other relations in [4.1]. For a left linear map $\phi \in \mathrm{Hom}_L(E; F)$ we write:

$$\overset{\bullet}{M}_L(\phi) = \begin{pmatrix} A_L & B_L \\ C_L & D_L \end{pmatrix} \qquad \text{and} \qquad \overset{\bullet}{M}_R(\mathfrak{T}\phi) = \begin{pmatrix} A_R & B_R \\ C_R & D_R \end{pmatrix} ,$$

where A_L is of size $p \times r$, A_R of size $r \times p$, and all other submatrices of the appropriate corresponding sizes. In terms of these submatrices, [4.7] becomes

$$\begin{pmatrix} A_R & B_R \\ C_R & D_R \end{pmatrix} = \begin{pmatrix} A_L^t & -\mathfrak{C}(C_L^t) \\ \mathfrak{C}(B_L^t) & D_L^t \end{pmatrix} , \tag{4.9}$$

where the superscript ^{t} indicates the ordinary transpose of a matrix, i.e., interchanging the roles of row and column indices. If both q and s are zero, i.e., if there are no odd basis vectors, (4.9) reduces to $\overset{\bullet}{M}_R(\mathfrak{T}\phi) = \overset{\bullet}{M}_L(\phi)^t$, explaining the name *transpose* for the operator $\mathfrak{T}$. The transformation (4.9) obviously represents an internal operation on matrices given formally as $\overset{\bullet}{M}_R \circ \mathfrak{T} \circ (\overset{\bullet}{M}_L)^{-1}$. If both q and s are non-zero, this internal operation has order 4. This fact has no interpretation in terms of linear maps, essentially because applying $\mathfrak{T}$ twice does not make sense.

In case a left linear ϕ is even, it is also right linear and $\mathfrak{T}\phi = \phi$. Moreover, all entries of A_S and D_S are even and all entries of B_S and C_S are odd. We thus can apply both $\overset{\bullet}{M}_L$ and $\overset{\bullet}{M}_R$ to ϕ, and (4.9) reduces to the equality

$$\text{(4.10)} \qquad \phi = \mathfrak{T}\phi \in \mathrm{Hom}_{L/R}(E;F)_0 \quad \Longrightarrow \quad \begin{pmatrix} A_R & B_R \\ C_R & D_R \end{pmatrix} = \begin{pmatrix} A_L^t & C_L^t \\ -B_L^t & D_L^t \end{pmatrix} .$$

This operation, which for *even* ϕ relates $\overset{\bullet}{M}_L(\phi)$ to $\overset{\bullet}{M}_R(\phi)$, is sometimes called *graded transpose* or *super transpose.* However, as above, the reader is warned that this operation is not its own inverse.

4.11 Example. For a free graded $\mathcal{A}$-module E of dimension $p|q$, consider the free graded $\mathcal{A}$-module $\mathrm{Hom}_L(\mathcal{A};E)$, which we have identified with E via $\Phi_L : \phi \mapsto \langle\!\langle 1 \,\|\, \phi \rangle\!\rangle$ [I.2.19]. For $e \in E$, we will denote (in this example only) the element $\Phi_L^{-1}(e) \in \mathrm{Hom}_L(\mathcal{A};E)$ by ${}^\bullet e$; we thus have in particular $\langle\!\langle 1 \,\|\, {}^\bullet e \rangle\!\rangle = e$. After the choice of a homogeneous ordered basis $(e_i)_{i=1}^n$ of E and considering 1 to be the canonical basis of $\mathcal{A}$, we can apply the map $\overset{\bullet}{M}_L$ to obtain

$$e = \langle\!\langle 1 \,\|\, {}^\bullet e \rangle\!\rangle = \sum_i \overset{\bullet}{M}_L({}^\bullet e)^i e_i \; .$$

In other words, the map $\overset{\bullet}{M}_L$ applied to the map ${}^\bullet e$ identified with the vector e gives us the *left* coordinates of this vector with respect to the basis $(e_i)_{i=1}^n$. The case $E \cong \mathrm{Hom}_R(\mathcal{A};E)$ is similar: the map $\overset{\bullet}{M}_R$ applied to a right linear map $\Phi_R^{-1}(e) \equiv e^\bullet$ identified with the vector e gives us the *right* coordinates of this vector with respect to the basis $(e_i)_{i=1}^n$:

$$e = e^\bullet(1) = \sum_i e_i \, \overset{\bullet}{M}_R(e^\bullet)^i \; .$$

For $\psi_R \in \mathrm{Hom}_R(E;F)$ we can write $\psi_R(e) = (\psi_R \circ e^\bullet)(1)$, i.e., evaluation corresponds to composition. Similarly for $\psi_L \in \mathrm{Hom}_L(E;F)$ we have $\langle\!\langle e \,\|\, \psi_L \rangle\!\rangle = \langle\!\langle 1 \,\|\, {}^\bullet e \diamond \psi_L \rangle\!\rangle$. Since $\overset{\bullet}{M}_S$ transforms composition into matrix multiplication [4.2], we see that the sequence of numbers $\overset{\bullet}{M}_R(e^\bullet)^i$, $i = 1, \ldots, n$ should be interpreted as column vector, and that the sequence of numbers $\overset{\bullet}{M}_L({}^\bullet e)^i$, $i = 1, \ldots, n$ should be interpreted as row vector. If we now recall that $\mathfrak{T}$ relates these two identifications: $\mathfrak{T}({}^\bullet e) = e^\bullet$ [I.2.19], we see that our interpretation of $\mathfrak{T}$ as a (kind of) transpose operator is coherent with the interpretation of $\overset{\bullet}{M}_S(e)$ as row/column vectors.

But we can learn something more: in this case the visualization (4.9) reduces to the equality

$$(\overset{\bullet}{M}_R(e^{\bullet})^1, \ldots, \overset{\bullet}{M}_R(e^{\bullet})^p, \overset{\bullet}{M}_R(e^{\bullet})^{p+1}, \ldots, \overset{\bullet}{M}_R(e^{\bullet})^{p+q})$$
$$= (\overset{\bullet}{M}_L({}^{\bullet}e)^1, \ldots, \overset{\bullet}{M}_L({}^{\bullet}e)^p, \mathfrak{C}(\overset{\bullet}{M}_L({}^{\bullet}e)^{p+1}), \ldots, \mathfrak{C}(\overset{\bullet}{M}_L({}^{\bullet}e)^{p+q}))\ .$$

In this equation one can see the danger in identifying a free graded $\mathcal{A}$-module of total dimension n with $\mathcal{A}^n$ by means of the left or right coordinates: in $\mathcal{A}^n$ one does not see that left and right coordinates are quite different (see also [I.3.5]). In fact, in $\mathcal{A}^n$ the right and left coordinates are the same because its dimension is $n|0$.

4.12 Proposition. *Let E and F be finite dimensional free graded $\mathcal{A}$-modules equipped with homogeneous bases, and let ϕ_S be in $\mathrm{Hom}_S(E; F)$. Then $\overset{\bullet}{M}_R({}^*\phi_L) = \overset{\bullet}{M}_L(\phi_L)$ and $\overset{\bullet}{M}_L(\phi_R^*) = \overset{\bullet}{M}_R(\phi_R)$.*

Proof. We consider the case for ϕ_L. Let $(e_i)_{i=1}^n$ and $(f_j)_{j=1}^m$ be homogeneous bases for E and F respectively. Using the dual bases we have $\phi_L = \sum_{i,j} {}^ie \otimes \overset{\bullet}{M}_L(\phi_L)_i{}^j \cdot f_j$. Under the identification $E \cong ({}^*E)^*$ [I.7.26], [I.7.27] the vectors e_i and f_j form the right dual bases to the bases $({}^ie)_{i=1}^n$ and $({}^jf)_{j=1}^m$ of ${}^*E$ and ${}^*F$ respectively. We thus have ${}^*\phi_L = \sum_{i,j} {}^ie \otimes \overset{\bullet}{M}_R({}^*\phi_R)_i{}^j \cdot f_j \in \mathrm{Hom}_R({}^*F; {}^*E)$. The matrix elements $\overset{\bullet}{M}_R({}^*\phi_R)_i{}^j$ are given by the equation $\overset{\bullet}{M}_R({}^*\phi_R)_i{}^j = (e_i \circ {}^*\phi_L)({}^jf) = e_i(\phi_L \diamond {}^jf) = \langle\!\langle e_i \| \phi_L \diamond {}^jf \rangle\!\rangle = \overset{\bullet}{M}_L(\phi_L)_i{}^j$, where the third equality follows from the identification $E \cong ({}^*E)^*$. QED

4.13 Discussion. In standard linear algebra over a field, taking a dual map and transposing a matrix are corresponding operations. In the graded case this is no longer the case: the dual of a left linear map becomes right linear, and in general we cannot apply the same matrix representation ($\overset{\bullet}{M}_L$ or $\overset{\bullet}{M}_R$) to both. In order to be able to apply the same matrix representation, we have to apply the transposition operator $\mathfrak{T}$. However, if ϕ is even, we can apply the same representation, in which case it follows from (4.10) that the matrices of ϕ and its dual map ϕ^* are related to each other by the graded transpose.

4.14 Definitions. We denote by $M(p|q, \mathcal{A})$ the ring $M((p+q) \times (p+q), \mathcal{A})$. The reason for the notation $p|q$ rather than the obvious notation $p + q$ is because we will usually think of elements $X \in M(p|q, \mathcal{A})$ as matrices with a block structure: $X = \left(\begin{smallmatrix} A & B \\ C & D \end{smallmatrix}\right)$, where A is of size $p \times p$ and B, C, and D of the corresponding appropriate sizes. We also define $Gl(p|q, \mathcal{A})$ as the set of those invertible elements in $M(p|q, \mathcal{A})$ for which all entries of the submatrices A and D are even elements in $\mathcal{A}$ and all entries of B and C are odd. We stress that we do *not* turn $M(p|q, \mathcal{A})$ into a free graded $\mathcal{A}$-module and that these definitions are purely intended to facilitate discussions (see also [4.11]). If E is a free graded $\mathcal{A}$-module of dimension $p|q$, all three maps ${}^{\bullet}M_R$, $\overset{\bullet}{M}_R$, $M_R^{\bullet}$ are bijections from $\mathrm{End}_R(E)$ to $M(p|q, \mathcal{A})$. Declaring one of these to be an isomorphism of graded

$\mathcal{A}$-modules would equip $M(p|q, \mathcal{A})$ with the structure of a graded $\mathcal{A}$-module, but then the remaining two bijections will fail to be isomorphisms.

4.15 Notation. As we noted already, the maps $\overset{\bullet}{M}_S$ depend upon the choice of bases in the free graded $\mathcal{A}$-modules concerned. If needed we will make this dependence explicit by denoting the matrix $\overset{\bullet}{M}_S(\phi) \in M(\substack{R:m\times n\\ L:n\times m}, \mathcal{A})$ defined in [4.1] as $\overset{\bullet}{M}{}^{e,f}_S(\phi)$ to indicate that it depends upon the bases (e_i) and (f_j). In the special case when $E = F$ has dimension $p|q$ and $f_i = e_i$, we will simplify this notation to $\overset{\bullet}{M}{}^e_S : \mathrm{End}_S(E) \to M(p|q, \mathcal{A})$, i.e., we will denote $\overset{\bullet}{M}{}^{e,e}_S$ as $\overset{\bullet}{M}{}^e_S$.

4.16 Lemma. *For a given ordered homogeneous basis $(e_i)_{i=1}^{p+q}$ of a free graded $\mathcal{A}$-module E of graded dimension $p|q$, the maps $\overset{\bullet}{M}{}^e_R : (\mathrm{End}_R(E), \circ) \to M(p|q, \mathcal{A})$ and $\overset{\bullet}{M}{}^e_L : (\mathrm{End}_L(E), \diamond) \to M(p|q, \mathcal{A})$ are isomorphisms of rings. In particular, the maps $\overset{\bullet}{M}{}^e_S : \mathrm{Aut}_S(E) \to Gl(p|q, \mathcal{A})$ are isomorphisms of groups (when equipped with the corresponding composition operator).*

4.17 Lemma. *Let E and F be $\mathcal{A}$-vector spaces of dimension $p|q$ and $r|s$ respectively. Let $(e_i)_{i=1}^{p+q}$ and $(\widehat{e}_i)_{i=1}^{p+q}$ be two ordered homogeneous bases for E and let $(g_j)_{j=1}^{r+s}$ and $(\widehat{g}_j)_{j=1}^{r+s}$ be two ordered homogeneous bases for F. Let furthermore x, $\widehat{x}$, y, and $\widehat{y}$ be matrices that relate these bases by $\widehat{e}_k = \sum_i x_k{}^i\, e_i = \sum_i e_i\, \widehat{x}^i{}_k$ and $\widehat{g}_\ell = \sum_j y_\ell{}^j\, g_j = \sum_i g_j\, \widehat{y}^j{}_\ell$. Then $x, \widehat{x} \in Gl(p|q, \mathcal{A})$ and $y, \widehat{y} \in Gl(r|s, \mathcal{A})$. Moreover, for $\phi_S \in \mathrm{Hom}_S(E; F)$ we have*

$$\overset{\bullet}{M}{}^{\widehat{e},\widehat{g}}_L(\phi_L) = x \cdot \overset{\bullet}{M}{}^{e,g}_L \cdot y^{-1} \qquad \textit{and} \qquad \overset{\bullet}{M}{}^{\widehat{e},\widehat{g}}_R(\phi_R) = (\widehat{y})^{-1} \cdot \overset{\bullet}{M}{}^{e,g}_R(\phi_R) \cdot \widehat{x}\ .$$

Proof. That $x, \widehat{x} \in Gl(p|q, \mathcal{A})$ and $y, \widehat{y} \in Gl(r|s, \mathcal{A})$ is an immediate consequence of the invariance of the graded dimension and the fact that the bases are ordered. The relations between $\overset{\bullet}{M}{}^{\widehat{e},\widehat{g}}_S$ and $\overset{\bullet}{M}{}^{e,g}_S$ follow by a direct computation. QED

4.18 Definition. For future use we define the *rank of a morphism* $\phi_S \in \mathrm{Hom}_S(E; F)$, denoted as $\mathrm{rank}(\phi_S)$, by $\mathrm{rank}(\phi_S) = \mathrm{rank}(\overset{\bullet}{M}_S(\phi_S))$, where on the right hand side we use [1.9], i.e., $\mathrm{rank}(\phi_S)$ is the number of independent rows or columns in the matrix with real entries $\mathbf{B}\overset{\bullet}{M}_S(\phi_S)$. A priori this depends upon the choice of the bases, but from [4.17] we know that $\mathbf{B}\overset{\bullet}{M}{}^{\widehat{e},\widehat{g}}_L(\phi_L) = (\mathbf{B}x) \cdot (\mathbf{B}\overset{\bullet}{M}{}^{e,g}_L) \cdot (\mathbf{B}y)^{-1}$. Since $\mathbf{B}x$ and $\mathbf{B}y$ are invertible, it is a classical result of linear algebra over $\mathbf{R}$ that the rank of $\mathbf{B}\overset{\bullet}{M}{}^{\widehat{e},\widehat{g}}_L(\phi_L)$ is the same as that of $\mathbf{B}\overset{\bullet}{M}{}^{e,g}_L$, i.e., the rank of ϕ_L is a well defined number (and similarly for ϕ_R).

5. The graded trace and the graded determinant

In this section we define the notion of the graded trace on $\mathrm{End}_S(E)$ *in abstract terms before we give its expression in terms of a matrix representation [5.7]. We also prove the graded commutativity of the graded trace. We then go over to the discussion of the graded determinant for elements of* $\mathrm{Aut}(E)$. *We first define this graded determinant on matrices in* $Gl(p|q,\mathcal{A})$ *[5.16]. Some lengthy calculation are then needed to show that the graded determinant of matrices is multiplicative, which in turn is necessary to show that we can define it intrinsically on* $\mathrm{Aut}(E)$.

5.1 Construction (elementary contractions). Let E, F and G be free graded $\mathcal{A}$-modules of which E is finite dimensional. Consider the map $\gamma_E : F \times E^* \times E \times G \to F \otimes G$ given by

$$\gamma_E(f,\phi,e,g) = f \otimes (\phi(e))\, g \ .$$

It is easy to check that γ_E is even and 4-linear. We thus obtain an induced linear map $\Gamma_E : F \otimes E^* \otimes E \otimes G \to F \otimes G$. This induced map is usually called *the elementary contraction of* E^* *and* E; it satisfies the relation

$$\Gamma_E(f \otimes \phi \otimes e \otimes g) = f \otimes \phi(e)\, g \ .$$

In the same way one can define the elementary contraction of E with ${}^*E$ (note the order!) as the map ${}_E\Gamma : F \otimes E \otimes {}^*E \otimes G \to F \otimes G$ satisfying

$${}_E\Gamma(f \otimes e \otimes \phi \otimes g) = f \otimes \langle\!\langle e \| \phi \rangle\!\rangle\, g \ .$$

5.2 Examples. • For finite dimensional free graded $\mathcal{A}$-modules the examples in [I.4.10] can all be interpreted as elementary contractions. For instance, let us consider the composition $\underline{\diamond} : \mathrm{Hom}_L(E;F) \otimes \mathrm{Hom}_L(F;G) \to \mathrm{Hom}_L(E;G)$. If we make the identifications $\mathrm{Hom}_L(E;F) \cong {}^*E \otimes F$, $\mathrm{Hom}_L(F;G) \cong {}^*F \otimes G$, and $\mathrm{Hom}_L(E;G) \cong {}^*E \otimes G$, which are valid for finite dimensional free graded $\mathcal{A}$-modules, then the map $\underline{\diamond}$ becomes a map ${}^*E \otimes F \otimes {}^*F \otimes G \to {}^*E \otimes G$. In this guise, $\underline{\diamond}$ is exactly the elementary contraction ${}_F\Gamma$.

• As a slightly more elaborate version of the previous example, let us now consider the identification $\mathrm{Hom}_R(E_1;F_1) \otimes \mathrm{Hom}_R(E_2;F_2) \cong \mathrm{Hom}_R(E_2 \otimes E_1; F_2 \otimes F_1)$. Using the identification of morphisms with tensor products as before, evaluation of an element $\phi_1 \otimes \phi_2$ on an element $e_2 \otimes e_1$ can be interpreted as a map

$$F_1 \otimes E_1^* \otimes F_2 \otimes E_2^* \otimes E_2 \otimes E_1 \to F_2 \otimes F_1 \ .$$

We leave it to the reader to verify that this map decomposes as $\Gamma_{E_1} \circ \mathfrak{R}_\sigma \circ \Gamma_{E_2}$, where σ represents the permutation $(1,2,3,4) \mapsto (2,3,1,4)$.

5.3 Construction (graded trace). Let E be a finite dimensional free graded $\mathcal{A}$-module and consider the space ${}^*E \otimes E$. As we know, it is isomorphic to $\mathrm{End}_L(E)$. To this space we cannot apply directly the contraction of E with ${}^*E$, but after interchanging the factors, we can. We thus obtain an even linear map $\mathrm{gtr}_L = {}_E\Gamma \circ \mathfrak{R} : \mathrm{End}_L(E) \to \mathcal{A}$. For homogeneous elements $e \in E$ and $\phi \in {}^*E$ this map is given by

$$\mathrm{gtr}_L(\phi \otimes e) = (-1)^{\langle \varepsilon(\phi) | \varepsilon(e) \rangle} \langle\!\langle e \| \phi \rangle\!\rangle \ .$$

Another way to obtain this map is as the linear map induced by the even bilinear map ${}^*E \times E \to \mathcal{A}$ defined on homogeneous elements by $(\phi, e) \mapsto (-1)^{\langle \varepsilon(\phi) | \varepsilon(e) \rangle} \langle\!\langle e \| \phi \rangle\!\rangle$. The map gtr_L is usually called the *(left) graded trace* of a left linear endomorphism.

The *(right) graded trace* gtr_R is defined similarly: $\mathrm{gtr}_R = \Gamma_E \circ \mathfrak{R} : \mathrm{End}_R(E) \to \mathcal{A}$. For homogeneous elements e, ϕ it is given as $\mathrm{gtr}_R(e \otimes \phi) = (-1)^{\langle \varepsilon(e) | \varepsilon(\phi) \rangle} \phi(e)$.

5.4 Proposition. *For any finite dimensional free graded $\mathcal{A}$-module E we have the identities* $\mathrm{gtr}_L = \Gamma_E \circ (\mathfrak{T}_{{}^*E} \otimes id(E))$ *and* $\mathrm{gtr}_R = {}_E\Gamma \circ (id(E) \otimes \mathfrak{T}_{{}^*E}^{-1})$.

Proof. Let $(e_i)_{i=1}^n$ be a homogeneous basis of E. The first identity follows from the fact that both maps send the basis element ${}^ie \otimes e_j$ to $(-1)^{\langle \varepsilon(e_i) | \varepsilon(e_j) \rangle} \delta^i_j = (-1)^{\langle \varepsilon(e_i) | \varepsilon(e_i) \rangle} \delta^i_j$. The second identity follows from a similar argument. QED

5.5 Corollary. $\mathrm{gtr}_R \circ \mathfrak{T}_{\mathrm{End}_L(E)} = \mathrm{gtr}_L$.

5.6 Lemma. *For a finite dimensional free graded $\mathcal{A}$-module E, a homogeneous basis $(e_i)_{i=1}^n$ of E and $\phi_S \in \mathrm{End}_S(E)$ we have*

$$\mathrm{gtr}_L(\phi_L) = \sum_i (-\mathfrak{C})^{\varepsilon(e_i)} \big(\overset{\bullet}{M}{}^e_L(\phi_L)_i{}^i\big) \quad \textit{and} \quad \mathrm{gtr}_R(\phi_R) = \sum_i (-\mathfrak{C})^{\varepsilon(e_i)} \big(\overset{\bullet}{M}{}^e_R(\phi_R)^i{}_i\big) \ .$$

Proof. We prove the formula for $\mathrm{gtr}_L(\phi_L)$, the case $\mathrm{gtr}_R(\phi_R)$ being analogous. First note that $\phi_L = \sum_{i,j} {}^ie \otimes \overset{\bullet}{M}{}^e_L(\phi_L)_i{}^j \cdot e_j = \sum_{i,j} {}^ie \otimes e_j \cdot \mathfrak{C}^{\varepsilon(e_j)}(\overset{\bullet}{M}{}^e_L(\phi_L)_i{}^j)$, and thus:

$$\begin{aligned} \mathrm{gtr}_L(\phi_L) &= \sum_{i,j} \mathrm{gtr}_L({}^ie \otimes e_j) \cdot \mathfrak{C}^{\varepsilon(e_j)}(\overset{\bullet}{M}{}^e_L(\phi_L)_i{}^j) \\ &= \sum_{i,j} (-1)^{\langle \varepsilon({}^ie) | \varepsilon(e_j) \rangle} \delta^i_j \, \mathfrak{C}^{\varepsilon(e_j)}(\overset{\bullet}{M}{}^e_L(\phi_L)_i{}^j) \\ &= \sum_i (-1)^{\langle \varepsilon(e_i) | \varepsilon(e_i) \rangle} \mathfrak{C}^{\varepsilon(e_i)}(\overset{\bullet}{M}{}^e_L(\phi_L)_i{}^i) = \sum_i (-\mathfrak{C})^{\varepsilon(e_i)}(\overset{\bullet}{M}{}^e_L(\phi_L)_i{}^i) \ . \end{aligned}$$

QED

5.7 Discussion. It will be clear from the formulæ for the graded trace that in both the left and right case the operation on the corresponding matrix is the same: a sum over diagonal elements, adding a sign and applying the involution $\mathfrak{C}$ whenever e_i is odd. If $(e_i)_{i=1}^{p+q}$ is an ordered basis for E of dimension $p|q$, we write the matrices $\overset{\bullet}{M}_S(\phi_S) \in M(p|q, \mathcal{A})$ in block form as $\overset{\bullet}{M}_S(\phi_S) = \left(\begin{smallmatrix} A_S & B_S \\ C_S & D_S \end{smallmatrix}\right)$. In terms of these submatrices we find

$$\mathrm{gtr}_S(\phi_S) = \mathrm{tr}(A_S) - \mathfrak{C}(\mathrm{tr}(D_S)) ,$$

where tr denotes the ordinary trace of a matrix, i.e., the sum of the diagonal entries.

If ϕ is even, both gtr_S apply. It follows immediately from [5.5] that $\mathrm{gtr}_R(\phi) = \mathrm{gtr}_L(\phi)$. But the same result also follows from the facts that (i) for even ϕ the submatrices A and D have only even entries, (ii) the ordinary trace is invariant under ordinary transposition, and (iii) formula (4.10). For even ϕ the formula for the graded trace reduces to the formula $\mathrm{gtr}_S(\phi) = \mathrm{tr}(A) - \mathrm{tr}(D)$.

5.8 Counter example. One might be tempted to think that gtr_L and gtr_R give the same result when a map is both left and right linear. The following example shows that this need not be the case. Suppose $\mathcal{A} = \bigwedge X$, where X is a *finite* dimensional vector space over $\mathbf{R}$. Let $(x_i)_{i=1}^n$ be a basis of X and denote $0 \neq b = x_1 \wedge \cdots \wedge x_n \in \bigwedge^n X$. Consider furthermore the free graded $\mathcal{A}$-module E of dimension $0|1$ with odd basis vector e_1. Finally define the map $\phi : E \to E$ by

$$\forall a \in \mathcal{A} : \phi(e_1 a) = e_1 b a .$$

This map is obviously right linear, but is also left linear! To see this, let $\lambda \in \mathcal{A}$ be odd, then $\phi(\lambda e_1 a) = \phi(-e_1 \lambda a) = -e_1 b \lambda a = 0$, because for any odd $\lambda \in \mathcal{A}$ the product $b\lambda = 0$. Similarly $\lambda \phi(e_1 a) = 0$. And thus ϕ is also left linear. However, if n is odd, $e_1 b = -b e_1$, and thus we find $\mathrm{gtr}_R(\phi) = -\mathfrak{C}(b) = b$ and $\mathrm{gtr}_L(\phi) = -\mathfrak{C}(-b) = -b$, which is different.

5.9 Lemma. *If E and F are finite dimensional free graded $\mathcal{A}$-modules, then we have for homogeneous $\phi \in \mathrm{Hom}_S(E; F)$ and $\psi \in \mathrm{Hom}_S(F; E)$ the equality*

$$\mathrm{gtr}_S(\phi \circ \psi) = (-1)^{\langle \varepsilon(\phi) | \varepsilon(\psi) \rangle} \, \mathrm{gtr}_S(\psi \circ \phi) .$$

If $E = F$ this equality can be written as $\mathrm{gtr}_S([\phi, \psi]_S) = 0$.

Proof. In order to prove the right linear case (the left linear case being similar), we consider the free graded $\mathcal{A}$-module $\mathrm{Hom}_R(E; F) \otimes \mathrm{Hom}_R(F; E) \cong F \otimes E^* \otimes E \otimes F^*$. The equality we have to prove amounts to proving that the maps $\mathrm{gtr}_R \circ \Gamma_E$ and $\mathrm{gtr}_R \circ \Gamma_F \circ \mathfrak{R}_{(23)}$ defined on this tensor product with values in $\mathcal{A}$ are the same. But this follows from the fact that both map a basis vector $f_i \otimes e^j \otimes e_k \otimes f^\ell$ to the same value

$$(-1)^{\langle \varepsilon(f_i) | \varepsilon(f_\ell) \rangle} \, \delta_i^\ell \, \delta_j^k = (-1)^{\langle \varepsilon(f_i) + \varepsilon(e_j) | \varepsilon(e_k) + \varepsilon(f_\ell) \rangle} \, (-1)^{\langle \varepsilon(e_k) | \varepsilon(e_j) \rangle} \, \delta_i^\ell \, \delta_j^k ,$$

where the last equality follows from the Kronecker δ's. QED

5.10 Remark. Let E be a free graded $\mathcal{A}$-module of dimension $p|q$. It is an elementary exercise to prove that $\mathrm{gtr}_S : \mathrm{End}_S(E) \to \mathcal{A}$ is the unique even linear map such that (i) $\mathrm{gtr}_S(id(E)) = p - q$, and (ii) $\mathrm{gtr}_S([\phi, \psi]_S) = 0$ for all $\phi, \psi \in \mathrm{End}_S(E)$.

5.11 Definition. The ordinary determinant, defined for square matrices with real coefficients, is a polynomial function in the matrix entries. It follows that we can extend the determinant to square matrices with coefficients in any commutative ring, yielding a value in this ring. Since $\mathcal{A}_0$ is a commutative ring, we thus can extend the determinant function to $M(n \times n, \mathcal{A}_0)$. We note that the determinant so extended preserves most of its properties, in particular that it is a ring homomorphism and that it can be calculated by expansion according to a row or column.

5.12 Lemma. *Let $A \in M(n \times n, \mathcal{A}_0)$ be an invertible matrix, and let $V \in M(n \times 1, \mathcal{A}_0)$, $W \in M(1 \times n, \mathcal{A}_0)$ and $a \in \mathcal{A}_0 = M(1 \times 1, \mathcal{A}_0)$ be arbitrary. Then*

$$\mathrm{Det}\begin{pmatrix} A & V \\ W & a \end{pmatrix} = (a - WA^{-1}V) \cdot \mathrm{Det}(A) \,.$$

Proof. $\begin{pmatrix} A & V \\ W & a \end{pmatrix} = \begin{pmatrix} I_n & V \\ WA^{-1} & a \end{pmatrix}\begin{pmatrix} A & 0 \\ 0 & 1 \end{pmatrix}$ and $\mathrm{Det}\begin{pmatrix} I_n & V \\ WA^{-1} & a \end{pmatrix} = \mathrm{Det}\begin{pmatrix} I_n & V \\ 0 & a-WA^{-1}V \end{pmatrix}$. QED

5.13 Lemma. *For $V \in M(n \times 1, \mathcal{A}_1)$ and $W \in M(1 \times n, \mathcal{A}_1)$ we have*

$$\mathrm{Det}(I_n + VW) = (1 + WV)^{-1} \,.$$

Proof. First note that $VW \in M(n \times n, \mathcal{A}_0)$ and hence that the determinant function can be applied. The actual proof is by induction on n. For $n = 1$ we have $\mathrm{Det}(1 + VW) = 1 + VW = 1 - WV = (1 + WV)^{-1}$ (because $V^2 = W^2 = 0$). For $n = \widehat{n} + 1$ we define $\widehat{V}$ and $\widehat{W}$ to be the column vector consisting of the first $\widehat{n}$ entries of V respectively the row vector consisting of the first $\widehat{n}$ entries of W. With these notations we have

$$I_n + VW = \begin{pmatrix} I_{\widehat{n}} + \widehat{V}\widehat{W} & \widehat{V}W_n \\ V_n\widehat{W} & 1 + V_nW_n \end{pmatrix} .$$

We then can apply the previous lemma and the induction hypothesis; we obtain

$$\begin{aligned}
\mathrm{Det}(I_n + VW) &= (1 + V_nW_n - V_n\widehat{W}(I_{\widehat{n}} + \widehat{V}\widehat{W})^{-1}\widehat{V}W_n) \cdot \mathrm{Det}(I_{\widehat{n}} + \widehat{V}\widehat{W}) \\
&= (1 + V_nW_n - V_n\widehat{W}(\sum_{k=0}^{\infty}(-1)^k(\widehat{V}\widehat{W})^k)\widehat{V}W_n) \cdot (1 + \widehat{W}\widehat{V})^{-1} \\
&= (1 + V_n(\sum_{k=0}^{\infty}(-1)^k(\widehat{W}\widehat{V})^k)W_n) \cdot (1 + \widehat{W}\widehat{V})^{-1} \\
&= (1 - (1 + \widehat{W}\widehat{V})^{-1}W_nV_n) \cdot (1 + \widehat{W}\widehat{V})^{-1} \\
&= (1 + (1 + \widehat{W}\widehat{V})^{-1}W_nV_n)^{-1} \cdot (1 + \widehat{W}\widehat{V})^{-1} \\
&= (1 + \widehat{W}\widehat{V} + W_nV_n)^{-1} = (1 + WV)^{-1} \,.
\end{aligned}$$

QED

5.14 Corollary. *Let $A \in M(n\times n, \mathcal{A}_0)$ be an invertible matrix, and let $V \in M(n\times 1, \mathcal{A}_1)$ and $W \in M(1\times n, \mathcal{A}_1)$ be arbitrary. Then* $\mathrm{Det}(A+VW) = \mathrm{Det}(A)\cdot(1+WA^{-1}V)^{-1}$.

5.15 Lemma. *For $X \in M(p\times q, \mathcal{A}_1)$ and $Y \in M(q\times p, \mathcal{A}_1)$ we have*

$$\mathrm{Det}(I_p + XY)\cdot \mathrm{Det}(I_q + YX) = 1 .$$

Proof. The proof proceeds by induction on p; for $p=1$ it is [5.13]. If $p=\widehat{p}+1$ we define $\widehat{X} \in M(\widehat{p}\times q, \mathcal{A}_1)$, $\widehat{Y} \in M(q\times \widehat{p}, \mathcal{A}_1)$, $V \in M(q\times 1, \mathcal{A}_1)$ and $W \in M(1\times q, \mathcal{A}_1)$ such that $X = \left(\begin{smallmatrix}\widehat{X}\\ W\end{smallmatrix}\right)$ and $Y = (\widehat{Y}\ V)$. With these definitions we have

$$I_p + XY = \begin{pmatrix} I_{\widehat{p}} + \widehat{X}\widehat{Y} & \widehat{X}V \\ W\widehat{Y} & 1+WV\end{pmatrix} \qquad \text{and} \qquad I_q + YX = (I_q + \widehat{Y}\widehat{X}) + VW .$$

Using [5.12] we find

$$\begin{aligned}\mathrm{Det}(I_p + XY) &= \mathrm{Det}(I_{\widehat{p}} + \widehat{X}\widehat{Y})\cdot(1 + WV - W\widehat{Y}(I_{\widehat{p}} + \widehat{X}\widehat{Y})^{-1}\widehat{X}V) \\ &= \mathrm{Det}(I_{\widehat{p}} + \widehat{X}\widehat{Y})\cdot(1 + W(I_q + \widehat{Y}\widehat{X})^{-1}V) .\end{aligned}$$

Using [5.14] we find $\mathrm{Det}(I_q+YX) = \mathrm{Det}(I_q+\widehat{Y}\widehat{X})\cdot(1+W(I_q+\widehat{Y}\widehat{X})^{-1}V)^{-1}$. Since $\mathcal{A}_0$ is commutative the result now follows. QED

5.16 Definition. We define the *graded determinant* $\mathrm{gDet} : Gl(p|q, \mathcal{A}) \to \mathcal{A}_0$ (also called the *Berezinian*) by the following procedure. For $X = \left(\begin{smallmatrix}A & B\\ C & D\end{smallmatrix}\right) \in Gl(p|q, \mathcal{A})$ we know that the entries of A and D are even and those of B and C odd. This plus [1.11] imply that both A and D are invertible matrices. $\mathrm{gDet}(X)$ is then defined as

$$\mathrm{gDet}(X) = \mathrm{Det}(A - BD^{-1}C)\cdot(\mathrm{Det}\, D)^{-1} .$$

5.17 Proposition. *The map $\mathrm{gDet} : Gl(p|q, \mathcal{A}) \to \mathcal{A}_0$ is a homomorphism.*

Proof. Given two such matrices $X = \left(\begin{smallmatrix}A & B\\ C & D\end{smallmatrix}\right)$ and $\widehat{X} = \left(\begin{smallmatrix}\widehat{A} & \widehat{B}\\ \widehat{C} & \widehat{D}\end{smallmatrix}\right)$, we compute:

$$\begin{aligned}\mathrm{gDet}(X\widehat{X}) &= \mathrm{Det}\big(A\widehat{A} + B\widehat{C} - (A\widehat{B} + B\widehat{D})(C\widehat{B} + D\widehat{D})^{-1}(C\widehat{A} + D\widehat{C})\big) \times \\ &\qquad \mathrm{Det}(C\widehat{B} + D\widehat{D})^{-1} \\ &= \mathrm{Det}\big(A\widehat{A} + B\widehat{C} \\ &\qquad\qquad - (A\widehat{B}\widehat{D}^{-1} + B)(D^{-1}C\widehat{B}\widehat{D}^{-1} + I_q)^{-1}(D^{-1}C\widehat{A} + \widehat{C})\big) \times \\ &\qquad \mathrm{Det}(D)^{-1}\cdot \mathrm{Det}(\widehat{D})^{-1}\cdot \mathrm{Det}(D^{-1}C\widehat{B}\widehat{D}^{-1} + I_q)^{-1} .\end{aligned}$$

We introduce the matrices $Z = D^{-1}C$ and $Y = \widehat{B}\widehat{D}^{-1}$ and compute:

$$
\begin{aligned}
(A\widehat{B}\widehat{D}^{-1} &+ B)(D^{-1}C\widehat{B}\widehat{D}^{-1} + I_q)^{-1}(D^{-1}C\widehat{A} + \widehat{C}) \\
&= (AY + B)(ZY + I_q)^{-1}(Z\widehat{A} + \widehat{C}) \\
&= AY(ZY + I_q)^{-1}Z\widehat{A} + B(ZY + I_q)^{-1}Z\widehat{A} + AY(ZY + I_q)^{-1}\widehat{C} \\
&\qquad + B(ZY + I_q)^{-1}\widehat{C} \\
&= A\Big(\sum_{k=0}^{\infty}(-1)^k(YZ)^{k+1}\Big)\widehat{A} + BZ(YZ + I_p)^{-1}\widehat{A} + A(YZ + I_p)^{-1}Y\widehat{C} \\
&\qquad + B(ZY + I_q)^{-1}\widehat{C}
\end{aligned}
$$

and thus:

$$
\begin{aligned}
A\widehat{A} + B\widehat{C} &- (A\widehat{B}\widehat{D}^{-1} + B)(D^{-1}C\widehat{B}\widehat{D}^{-1} + I_q)^{-1}(D^{-1}C\widehat{A} + \widehat{C}) \\
&= A\Big(I_p - \sum_{k=0}^{\infty}(-1)^k(YZ)^{k+1}\Big)\widehat{A} - BZ(YZ + I_p)^{-1}\widehat{A} - A(YZ + I_p)^{-1}Y\widehat{C} \\
&\qquad + B\Big(I_q - \sum_{k=0}^{\infty}(-1)^k(ZY)^k\Big)\widehat{C} \\
&= A(YZ + I_p)^{-1}(\widehat{A} - Y\widehat{C}) - BZ(YZ + I_p)^{-1}\widehat{A} + BZ(YZ + I_p)^{-1}Y\widehat{C} \\
&= (A - BD^{-1}C)(YZ + I_p)^{-1}(\widehat{A} - \widehat{B}\widehat{D}^{-1}\widehat{C})
\end{aligned}
$$

We thus finally find

$$
\begin{aligned}
\mathrm{gDet}(X\widehat{X}) &= \mathrm{Det}(A - BD^{-1}C) \cdot \mathrm{Det}(YZ + I_p)^{-1} \cdot \mathrm{Det}(\widehat{A} - \widehat{B}\widehat{D}^{-1}\widehat{C}) \times \\
&\qquad \mathrm{Det}(D)^{-1} \cdot \mathrm{Det}(\widehat{D})^{-1} \cdot \mathrm{Det}(I_q + ZY)^{-1} \\
&= \mathrm{gDet}(X) \cdot \mathrm{gDet}(\widehat{X}) \cdot \big(\mathrm{Det}(YZ + I_p) \cdot \mathrm{Det}(I_q + ZY)\big)^{-1} \\
&= \mathrm{gDet}(X) \cdot \mathrm{gDet}(\widehat{X}) \qquad \text{because of [5.15]}
\end{aligned}
$$

QED

5.18 Remark. Using the decomposition $\begin{pmatrix} A & B \\ C & D \end{pmatrix} = \begin{pmatrix} A & \mathbf{0} \\ \mathbf{0} & D \end{pmatrix} \cdot \begin{pmatrix} \mathbf{1} & A^{-1}B \\ D^{-1}C & \mathbf{1} \end{pmatrix}$, [5.17], and [5.15], it is elementary to show that we have the equalities

$$
\mathrm{gDet}\begin{pmatrix} A & B \\ C & D \end{pmatrix} \equiv \frac{\mathrm{Det}(A - BD^{-1}C)}{\mathrm{Det}(D)} = \frac{\mathrm{Det}(A)}{\mathrm{Det}(D - CA^{-1}B)} .
$$

5.19 Definition. Let E be a finite dimensional free graded $\mathcal{A}$-module of dimension $p|q$, then we extend the definition of graded determinant to $\mathrm{Aut}(E)$ by the following procedure. Choose an ordered homogeneous basis (e_i) for E and define

$$
\mathrm{gDet}(\phi) = \mathrm{gDet}(\dot{M}_S^e(\phi)) .
$$

5.20 Proposition. *The value of* $\mathrm{gDet}(\phi)$ *does not depend upon the choice of* $\overset{\bullet}{M}{}^{e}_{L}$ *or* $\overset{\bullet}{M}{}^{e}_{R}$*, nor upon the choice of the basis* (e_i).

Proof. Let (e_i) be an ordered homogeneous basis of E, then, using the block matrix form $\overset{\bullet}{M}{}^{e}_{S}(\phi) = \begin{pmatrix} A_S & B_S \\ C_S & D_S \end{pmatrix}$, we have $\mathrm{gDet}(\overset{\bullet}{M}{}^{e}_{L}(\phi)) = \mathrm{Det}(A_L - B_L D_L^{-1} C_L) \cdot \mathrm{Det}(D_L)^{-1}$ and $\mathrm{gDet}(\overset{\bullet}{M}{}^{e}_{R}(\phi)) = \mathrm{Det}(A_R - B_R D_R^{-1} C_R) \cdot \mathrm{Det}(D_R)^{-1}$. But the equality of these two quantities is immediate if we use (4.9), realize that the ordinary determinant is invariant under transposition, and realize also that ${C_L}^t ({D_L}^t)^{-1} {B_L}^t = -(B_L D_L^{-1} C_L)^t$ because B and C have odd entries that produce a minus sign when interchanging. The fact that $\mathrm{gDet}(\phi)$ does not depend upon the choice of the basis is an immediate consequence of [5.17], [4.15] and [4.17]. $\boxed{QED}$

5.21 Corollary. *The map* $\mathrm{gDet} : \mathrm{Aut}(E) \to \mathcal{A}_0$ *is a well defined multiplicative homomorphism.*

5.22 Remark. The graded determinant gDet is a generalization of the ordinary determinant in the sense that it is the integrated form on the $\mathcal{A}$-Lie group $\mathrm{Aut}(E)$ of the graded trace on the $\mathcal{A}$-Lie algebra $\mathrm{End}_S(E)$ (see [III.3.14] and [VI.2.18]), just as the ordinary determinant is the integrated form of the usual trace.

6. The body of a free graded $\mathcal{A}$-module

In this section we discuss in detail the relation between a generic $\mathcal{A}$ *and the special case* $\mathcal{A} = \mathbf{R}$*. We define, in analogy with the case of the ring* $\mathcal{A}$*, the set of nilpotent vectors in a free graded* $\mathcal{A}$*-module. This allows us to extend the body map* $\mathbf{B}$ *to free graded* $\mathcal{A}$*-modules, yielding free graded* $\mathbf{R}$*-modules (i.e., a direct sum of two ordinary vector spaces over* $\mathbf{R}$*). We then introduce the notion of equivalent bases and we define an* $\mathcal{A}$*-vector space to be a free graded* $\mathcal{A}$*-module together with an equivalence class of bases. Restricting the notion of a subspace slightly, we then can prove that there is "no" difference between a generic* $\mathcal{A}$ *and the special case* $\mathcal{A} = \mathbf{R}$*. All is encoded in terms of a basis, and the only difference is what kind of coefficients one puts in front of basis vectors.*

6.1 Discussion. As has been said before, the real numbers perfectly fit the conditions imposed on $\mathcal{A}$ in [1.5], so let us start this section with a discussion what happens if we use $\mathcal{A} = \mathbf{R}$. It will be obvious that a graded $\mathbf{R}$-module is a vector space over $\mathbf{R}$ in the usual sense, but ... there is more to a graded $\mathbf{R}$-module than that. A graded $\mathbf{R}$-module splits into an even and an odd part, both of them vector spaces over $\mathbf{R}$. More precisely, if X is a graded $\mathbf{R}$-module, then it defines two subspaces X_0 and X_1 such that $X = X_0 \oplus X_1$. On the other hand, the reader can convince himself easily that, given a pair (X_0, X_1) of vector spaces over $\mathbf{R}$, the vector space $X = X_0 \oplus X_1$ is a graded $\mathbf{R}$-module when we define X_α to be the part of parity α. We conclude that graded $\mathbf{R}$-modules are nothing more nor less than vector spaces over $\mathbf{R}$ together with a splitting into a direct sum of two subspaces. In particular, a graded $\mathbf{R}$-module is always a free graded $\mathbf{R}$-module.

However, instead of taking $\mathcal{A} = \mathbf{R}$, we could have used any $\mathcal{A}$ to obtain $\mathbf{R}$, just by taking the body map $\mathbf{B} : \mathcal{A} \to \mathbf{R}$. A natural question now is, can we obtain graded $\mathbf{R}$-modules in a similar way from any graded $\mathcal{A}$-module? As we will see, the answer is positive when we restrict our attention to *free* graded $\mathcal{A}$-modules.

6.2 Definition. For a free graded $\mathcal{A}$-module E we define the set $\mathcal{N}_E$ of *nilpotent vectors* by $\mathcal{N}_E = \{\, x \in E \mid \exists a \in \mathcal{A} : a \neq 0 \;\; \& \;\; ax = 0 \,\}$.

6.3 Lemma. *$\mathcal{N}_E$ consists of those elements in E that have nilpotent coefficients with respect to any basis of E. Moreover, $\mathcal{N}_E$ is a subspace (over $\mathbf{R}$) of E, $\mathcal{A}\mathcal{N}_E \subset \mathcal{N}_E$ and $\mathcal{N}E \subset \mathcal{N}_E$.*

Proof. Let $G \subset E$ be a basis for E and let $x \in E$, then there exists $e_1, \ldots, e_n \in G$ and $a^1, \ldots, a^n \in \mathcal{A}$ such that $x = \sum_{i=1}^n a^i e_i$. If all the a^i are nilpotent, it follows from [1.4] that $x \in \mathcal{N}_E$. On the other hand suppose $x \in \mathcal{N}_E$, i.e., there exists an $a \in \mathcal{A}$, non-zero such that $ax = 0$. It follows from the properties of a basis that then $\forall i = 1, \ldots, n : aa^i = 0$. If a^i is not nilpotent, it is invertible and we deduce that $a = 0$, in contradiction with the hypothesis. This proves the first part of the lemma; the rest follows easily. $\boxed{QED}$

6.4 Counter Example. Of course the definition of nilpotent vectors can be given for any graded $\mathcal{A}$-module. And although the results of [6.3] remain valid for any graded submodule of a free graded $\mathcal{A}$-module, they are not valid in the more general setting of arbitrary graded $\mathcal{A}$-modules.

To appreciate the problem, consider the graded commutative ring $\mathcal{A}$ with $\mathcal{A}_0 = \mathbf{R}$ and $\mathcal{A}_1 = X$, where X is a vector space over $\mathbf{R}$ of dimension at least 2, and where the multiplication in $\mathcal{A}_1$ is trivial, i.e., $\mathcal{A}_1 \cdot \mathcal{A}_1 = \{0\}$. This $\mathcal{A}$ verifies the conditions given in [1.5]. In X we choose two independent vectors x and y, and we consider the submodule F of $\mathcal{A}^2$ generated by the two vectors $(x, 0)$ and $(0, y)$. Since these vectors are homogeneous (they are odd), F is a graded submodule of $\mathcal{A}^2$, and thus the quotient $E = \mathcal{A}^2/F$ is a graded $\mathcal{A}$-module. We claim that for this E the subset $\mathcal{N}_E$ is *not* a subspace (over $\mathbf{R}$) of E. If π denotes the canonical projection $\pi : \mathcal{A}^2 \to E$, the vectors $\pi(1, 0)$ and $\pi(0, 1)$ belong to $\mathcal{N}_E$ because $x \cdot \pi(1, 0) = \pi(x, 0) = 0$ and $y \cdot \pi(0, 1) = \pi(0, y) = 0$. However, we claim that $\pi(1, 1)$ does not belong to $\mathcal{N}_E$. For suppose it did, then there should exist $0 \neq a \in \mathcal{A}$ such that $a \cdot \pi(1, 1) = \pi(a, a) = 0$, i.e., $(a, a) \in F$. This implies that a must be a multiple of both x and y (remember that multiplication in $\mathcal{A}_1 = X$ is trivial), and thus must be zero. Since this contradicts the assumption $a \neq 0$, we conclude that $\pi(1, 1)$ does indeed not belong to $\mathcal{N}_E$.

6.5 Definition. We extend the notion of the body map $\mathbf{B}$ to any free graded $\mathcal{A}$-module E as being the canonical projection (in terms of real vector spaces) $\mathbf{B} : E \to E/\mathcal{N}_E$. It maps the free graded $\mathcal{A}$-module E into a vector space over $\mathbf{R}$, called *the body of* E. Note that the definition of $\mathbf{B}$ for the free graded $\mathcal{A}$-module $\mathcal{A}$ coincides with the original definition of body map just because of our requirement $\mathcal{A} = \mathbf{R} \oplus \mathcal{N}$, i.e., $\mathcal{N}_\mathcal{A} = \mathcal{N}$.

6.6 Lemma. *Let E be a free graded $\mathcal{A}$-module, then for all $a \in \mathcal{A}$ and for all $x \in E$ we have $\mathbf{B}(ax) = (\mathbf{B}a)(\mathbf{B}x)$. Moreover, if G is a set of homogeneous elements in E, then G is an independent set in E if and only if $\mathbf{B}G$ is an independent set (over $\mathbf{R}$) in $\mathbf{B}E$.*

Proof. To show the relation $\mathbf{B}(ax) = (\mathbf{B}a)(\mathbf{B}x) = \mathbf{B}((\mathbf{B}a)x)$ (the last equality because $\mathbf{B}$ is linear over $\mathbf{R}$), we have to show that $ax - (\mathbf{B}a)x = (a - \mathbf{B}a)x \in \mathcal{N}_E$. But $a - \mathbf{B}a \in \mathcal{A}$ is nilpotent, from which we deduce that indeed $ax - (\mathbf{B}a)x \in \mathcal{N}_E$.

For the second assertion, let us first assume that G is independent and let us suppose that there exist $r^1, \dots, r^n \in \mathbf{R}$ and $e_1, \dots, e_n \in G$ with $\sum_{i=1}^n r^i(\mathbf{B}e_i) = 0$, or equivalently $\sum_{i=1}^n r^i e_i \in \mathcal{N}_E$. Combining [3.3] and [3.7], we may assume that $e_1, \dots, e_n$ are elements in a homogeneous basis of E. It thus follows from [6.3] that the coefficients r^i are nilpotent. Since they are also real by hypothesis, they must be zero.

In the other direction, suppose $\mathbf{B}G$ is independent over $\mathbf{R}$ and suppose there are $a^1, \dots, a^n \in \mathcal{A}$ and $e_1, \dots, e_n \in G$ with $\sum_i a^i e_i = 0$. Applying $\mathbf{B}$ and the independence of $\mathbf{B}G$ shows $a^i \in \mathcal{N}$. Now let B be a homogeneous basis for E, then there exist $f_1, \dots, f_m \in B$ and $b_i{}^j \in \mathcal{A}$ such that $e_i = \sum_j b_i{}^j f_j$. Again applying $\mathbf{B}$ and using the independence of $\mathbf{B}G$ and $\mathbf{B}B$ (just proved!) then shows that $m \geq n$ and that the matrix $\mathbf{B}b_i{}^j$ contains an invertible $n \times n$ submatrix. Without loss of generality we may assume that it is the submatrix $(\mathbf{B}b_i{}^j)_{i,j=1}^n$. From $0 = \sum_i a^i e_i = \sum_{i,j} a^i\, b_i{}^j\, f_j$ and the independence of B we deduce that $\forall j : \sum_i a^i\, b_i{}^j = 0$. But by [1.11] the matrix $b_i{}^j, j \leq n$ is invertible, and hence $a^i = 0$, finally proving that G is an independent set. QED

6.7 Corollary. *If $E = E_0 \oplus E_1$ is a free graded $\mathcal{A}$-module, then $\mathbf{B}E = (\mathbf{B}E_0) \oplus (\mathbf{B}E_1)$ is a (free) graded $\mathbf{R}$-module of the same graded dimension; in particular $\mathbf{B}(E_\alpha) = (\mathbf{B}E)_\alpha$.*

6.8 Lemma. *Given any (left or right) linear map $\phi : E \to F$ between two free graded $\mathcal{A}$-modules, there exists a unique linear map of (free) graded $\mathbf{R}$-modules $\mathbf{B}\phi : \mathbf{B}E \to \mathbf{B}F$ making the following diagram commutative:*

$$
\begin{array}{ccc}
E & \xrightarrow{\ \phi\ } & F \\
{\scriptstyle \mathbf{B}}\big\downarrow & & \big\downarrow{\scriptstyle \mathbf{B}} \\
\mathbf{B}E & \xrightarrow[\ \mathbf{B}\phi\ ]{} & \mathbf{B}F\,.
\end{array}
$$

If $\chi : F \to G$ is a linear map, then $\mathbf{B}(\chi \circ \phi) = (\mathbf{B}\chi) \circ (\mathbf{B}\phi)$. If ϕ is homogeneous of parity α then $\mathbf{B}\phi$ is too.

Proof. If $\mathbf{B}\phi$ exists, it must be defined by $(\mathbf{B}\phi)(\mathbf{B}x) = \mathbf{B}(\phi(x))$. From this formula, linearity and uniqueness follow immediately. It thus remains to show that ϕ exists. For that it suffices to show that $\mathbf{B}(\phi(x)) = 0$ whenever $\mathbf{B}x = 0$. To see this, choose a basis for E. The vector x has nilpotent coefficients with respect to this basis, and thus by linearity of ϕ its image is a sum of terms, each a product of a vector of F and a nilpotent coefficient. From this it follows that $\phi(x)$ is a nilpotent vector in F. The equality $\mathbf{B}(\chi \circ \phi) = (\mathbf{B}\chi) \circ (\mathbf{B}\phi)$ follows from the defining equation for $\mathbf{B}$ on maps. The last statement follows immediately from the fact that $(\mathbf{B}E)_0 = \mathbf{B}(E_0)$ (and idem for F). QED

6.9 Remark. The above result can be rephrased as saying that the map $\mathbf{B}$ is a functor from the category of free graded $\mathcal{A}$-modules with its linear maps (left or right) to the category of (free) graded $\mathbf{R}$-modules with its linear maps (for which the distinction left and right does not exist).

6.10 Remark. If E is a finite dimensional free graded $\mathcal{A}$-module, $\mathrm{Hom}_S(E; F)$ is also a free graded $\mathcal{A}$-module. In that case we have two different definitions of $\mathbf{B}\phi$ for a linear map $\phi : E \to F$: the one of [6.8], but also the one as given by application of the map $\mathbf{B}$ to the free graded $\mathcal{A}$-module $\mathrm{Hom}_S(E; F)$. In the first case $\mathbf{B}\phi$ lies in $\mathrm{Hom}_S(\mathbf{B}(E); \mathbf{B}(F))$ and in the second case in $\mathbf{B}(\mathrm{Hom}_S(E; F))$. Declaring these two definitions to coincide gives us an isomorphism between these two spaces.

Actually, this is a special case of the more general fact that the body map $\mathbf{B}$ "preserves" all our constructions of free graded $\mathcal{A}$-modules, e.g., $\otimes^k(\mathbf{B}E)$ is isomorphic to $\mathbf{B}(\otimes^k E)$, where the first tensor product is that of graded $\mathbf{R}$-modules and the second that of free graded $\mathcal{A}$-modules. The proof (using homogeneous bases) of these and similar isomorphisms are left to the reader (see also [6.26] and [6.27]).

6.11 Discussion. Since any free graded $\mathcal{A}$-module E is a vector space over $\mathbf{R}$, it follows from [6.6] that $\mathbf{B} : E \to \mathbf{B}E$ is a linear map between vector spaces over $\mathbf{R}$ whose kernel is the linear subspace $\mathcal{N}_E$. However, no canonical choice for a supplement (over $\mathbf{R}$) to $\mathcal{N}_E$ presents itself. The purpose of the next definition is to create such a supplement.

6.12 Definition. Two bases $(e_i)_{i\in I}$ and $(f_j)_{j\in J}$ of a given free graded $\mathcal{A}$-module E are said to be *equivalent* if they are related to each other by real coefficients, i.e., if there exist $a_i{}^j \in \mathbf{R}$ such that $e_i = \sum_j a_i{}^j f_j$. Reflexivity and transitivity are obvious. For symmetry, suppose $f_j = \sum_i b_j{}^i e_i$ with $b_j{}^i \in \mathcal{A}$. From $f_j = \sum_{i,k} b_j{}^i a_i{}^k f_k$ it follows that $\sum_i b_j{}^i a_i{}^k = \delta_j^k$. Taking bodies, we obtain the equation $\sum_i (\mathbf{B}b_j{}^i) a_i{}^k = \delta_j^k$. Defining $\widehat{f}_j = \sum_i (\mathbf{B}b_j{}^i) e_i$, we find $\widehat{f}_j = \sum_{i,k} (\mathbf{B}b_j{}^i) a_i{}^k f_k = \sum_k \delta_j^k f_k = f_j$. From this it follows that the $b_j{}^i$ are necessarily real.

6.13 Discussion. Specifying a basis B for a free graded $\mathcal{A}$-module E automatically gives a supplement $\mathcal{R}_E$ for $\mathcal{N}_E$ (over $\mathbf{R}$) defined by $\mathcal{R}_E = \mathrm{Span}_{\mathbf{R}}(B)$. It will be obvious that an equivalent basis defines the same supplement $\mathcal{R}_E$. We thus obtain a map from the set of equivalence classes of bases to the set of supplements for $\mathcal{N}_E$. This map is injective but in general not surjective [6.15].

6.14 Lemma. *If E is a finite dimensional free graded $\mathcal{A}$-module, then any supplement $\mathcal{R}_E$ for $\mathcal{N}_E$ defines an equivalence class of bases of which $\mathcal{R}_E$ is the real span.*

Proof. By definition of a supplement, $\mathbf{B} : \mathcal{R}_E \to \mathbf{B}E$ is an isomorphism of vector spaces over $\mathbf{R}$. Let B be a homogeneous basis for the $\mathbf{R}$-vector space $\mathbf{B}E$ and denote by $\widehat{B} \subset \mathcal{R}_E$ its image under $\mathbf{B}^{-1}$. Imitating the proof of [6.6], we see that $\widehat{B}$ is independent in E (using that $\mathbf{B}\widehat{B}$ is independent but not needing that $\widehat{B}$ is homogeneous in E). Combining [6.7], [2.6-iii], and the finite dimensionality, we conclude that $\widehat{B}$ is a basis whose real span is $\mathcal{R}_E$. QED

6.15 Counter example. To show that the condition of finite dimensionality is not void, let us consider the infinite dimensional free graded $\mathcal{A}$-module E of [3.9] with its basis $\{ e_i \mid i \in \mathbf{N} \}$ and the vectors $f_i = e_i - n_i e_{i+1}$. Since the n_i are nilpotent, we have $\mathbf{B}e_i = \mathbf{B}f_i$. Since the $\mathbf{B}e_i$ form a basis of $\mathbf{B}E$, it follows that the f_i generate a supplement (over $\mathbf{R}$) for $\mathcal{N}_E$. However, the f_i do not form a basis for E as shown in [3.9].

6.16 Corollary. *For finite dimensional free graded $\mathcal{A}$-modules equivalence classes of bases can be identified with supplements $\mathcal{R}_E$ (over $\mathbf{R}$) for $\mathcal{N}_E$.*

6.17 Discussion. So far in talking about supplements for $\mathcal{N}_E$ we have interpreted E as an ordinary vector space over $\mathbf{R}$. However, we can interpret E as a graded $\mathbf{R}$-module because E_0 and E_1 are subspaces of E (over $\mathbf{R}$). Moreover, $\mathcal{N}_E$ is a graded subspace of the graded $\mathbf{R}$-module E, i.e., $\mathcal{N}_E = (\mathcal{N}_E \cap E_0) \oplus (\mathcal{N}_E \cap E_1)$. It thus is reasonable to require that a supplement $\mathcal{R}_E$ to $\mathcal{N}_E$ is a *graded* subspace. If that is the case, the map $\mathbf{B} : \mathcal{R}_E \to \mathbf{B}E$ is an isomorphism of graded $\mathbf{R}$-modules. A little reflection will show that a supplement $\mathcal{R}_E$ generated by an equivalence class of bases is a graded subspace (of E, seen as a graded $\mathbf{R}$-module) if and only if the equivalence class contains a homogeneous basis.

6.18 Counter example. To show that an equivalence class need not always contain a homogeneous basis, consider the free graded $\mathcal{A}$-module E of dimension $1|1$ with homogeneous basis $\{e, f\}$. Let $\mu \in \mathcal{N}_0$ be non-zero, then the vectors $\widehat{e} = e + \mu f$ and $\widehat{f} = f + \mu e$ form a basis for E whose equivalence class does not contain a homogeneous basis.

6.19 Definition. By an *$\mathcal{A}$-vector space E* we will mean a *finite dimensional* free graded $\mathcal{A}$-module together with an equivalence class of bases whose associated supplement $\mathcal{R}_E$ to $\mathcal{N}_E$ is a *graded* supplement (in other words, the equivalence class should contain a homogeneous basis). Moreover, when using a basis for E, it will *always* be a (homogeneous) basis within the given equivalence class.

If E is an $\mathcal{A}$-vector space, the supplement $\mathcal{R}_E$ is part of the $\mathcal{A}$-vector space structure. We thus can use the isomorphism $\mathbf{B} : \mathcal{R}_E \to \mathbf{B}E$ to identify $\mathbf{B}E$ with $\mathcal{R}_E \subset E$ and to forego the notation $\mathcal{R}_E$. This means in particular that *we see the map $\mathbf{B}$ as a projection $\mathbf{B} : E \to E$ whose kernel is $\mathcal{N}_E$ and whose image is $\mathbf{B}E \equiv \mathcal{R}_E \subset E$.* The image $\mathbf{B}E$ can be described as those points of E that have real coordinates with respect to a basis in the equivalence class.

6.20 Remark. There is no special reason to require $\mathcal{A}$-vector spaces to be finite dimensional. However, since we do not need infinite dimensional ones, our definition avoids having to say all the time that they are supposed to be finite dimensional.

6.21 Nota Bene. In case $\mathcal{A} = \mathbf{R}$ we obtain the definition of an $\mathbf{R}$-vector space. The reader should not confuse this with the notion of a vector space over $\mathbf{R}$. Since for a graded $\mathbf{R}$-module there exists only one equivalence class of bases (there are no nilpotent vectors and all linear maps are smooth), an $\mathbf{R}$-vector space is exactly the same as a graded $\mathbf{R}$-module. We conclude that an $\mathbf{R}$-vector space is the direct sum of two vector spaces over $\mathbf{R}$.

6.22 Definition. Given two $\mathcal{A}$-vector spaces E and F, a linear map $\phi : E \to F$ will be called *smooth* if and only if $\phi(\mathbf{B}E) \subset \mathbf{B}F \subset F$. Using one of the matrix representations of [4.1], ϕ is smooth if and only if all its matrix elements are real. This is most easily seen by noting that basis vectors e_i in E belong to $\mathbf{B}E$ and that the matrix elements of ϕ are (up to conjugation) the components of $\phi(e_i)$. Using [6.19] this can be stated as $\phi \in \mathbf{B}\,\mathrm{Hom}_S(E; F)$ or equivalently as $\mathbf{B}\phi = \phi$ (see also [6.10]).

6.23 Discussion/Definition. With the above definitions, we have created a new category: the category of $\mathcal{A}$-vector spaces, together with the smooth linear maps. In this category, we can still perform our constructions of new $\mathcal{A}$-vector spaces. In fact, for direct sums, tensor products, exterior powers and endomorphism spaces we have given a homogeneous basis of the new free graded $\mathcal{A}$-module in terms of homogeneous bases for the old free graded $\mathcal{A}$-modules. The equivalence class of bases for the new $\mathcal{A}$-vector space then is generated by the basis constructed from bases for the old $\mathcal{A}$-vector spaces that lie within the given equivalence classes. We leave it to the reader to check that changing the original bases within their equivalence classes does not alter the equivalence class of bases of the newly created $\mathcal{A}$-vector space. For the $\mathcal{A}$-vector space $F(G,\varepsilon)$ generated by the graded symbols G the obvious choice for a homogeneous basis is G itself.

For graded subspaces (and thus quotients) the question is more delicate since there is no obvious way to induce a basis on a graded subspace once a basis for the total $\mathcal{A}$-vector space is given, nor is there an obvious way to induce a basis on a quotient. We therefore restrict the notion of a graded subspace in our new category as being a graded subspace in the old sense with the additional restriction that *there exists a homogeneous basis for the total $\mathcal{A}$-vector space within its equivalence class such that a subset forms a basis for the graded subspace in question.* The equivalence class induced by this subset defines the equivalence class for the graded subspace (again, different bases within the equivalence class on the ambient $\mathcal{A}$-vector space induce the same equivalence class on the graded subspace).

With this definition of a graded subspace, a quotient space is well defined in our new category. If we have a homogeneous basis within the equivalence class, a subset of which defines the graded subspace, the complement induces a homogeneous basis on the quotient. And as before, changing the basis on the original space (with all the restrictions specified above) does not change the equivalence class on the quotient.

6.24 Lemma. *Let $\phi \in \mathrm{Hom}_S(E;F)$ be a smooth homogeneous linear map between $\mathcal{A}$-vector spaces. Then* $\ker(\phi)$ *and* $\mathrm{im}(\phi)$ *are graded subspaces in the sense of [6.23].*

Proof. Let $P \subset \mathbf{B}E$ and $Q \subset \mathbf{B}F$ be homogeneous bases in the corresponding equivalence classes. Left and right linear being similar, we consider the right linear case and we start with the even case. Since ϕ is smooth, the matrix representation [4.1] of ϕ (all are equal) is of the form $\phi \cong \begin{pmatrix} A & 0 \\ 0 & D \end{pmatrix}$ with A and D matrices with real entries. In particular for any $e \in P_0$, $\phi(e)$ is a linear combination with real coefficients (the matrix elements of A) of elements of Q_0. Elementary linear algebra over $\mathbf{R}$ then tells us that we can change the basis vectors in P_0 and in Q_0 by real coefficients in such a way that there exist subsets $P_0' \subset P_0$ and $Q_0' \subset Q_0$ such that P_0' is a basis for $\ker(\phi|_{\mathbf{B}E_0})$, such that Q_0' is a basis for $\mathrm{im}(\phi|_{\mathbf{B}E_0})$, and such that ϕ is a bijection between $P_0 \setminus P_0'$ and Q_0'. A similar result holds for the odd basis vectors in P_1 and Q_1. We conclude that there exists *homogeneous* bases P and Q within the corresponding equivalence classes and subsets $P' \subset P$ and $Q' \subset Q$ such that P' is a basis for $\ker(\phi|_{\mathbf{B}E})$, that Q' is a basis for $\mathrm{im}(\phi|_{\mathbf{B}E})$, and that ϕ is a bijection between $P \setminus P'$ and Q'. It follows easily that $\mathrm{Span}(P') = \ker(\phi)$ and that $\mathrm{Span}(Q') = \mathrm{im}(\phi)$. The result then follows from [3.3].

If ϕ is odd, the matrix representation [4.1] of ϕ is of the form $\phi \cong \begin{pmatrix} 0 & B \\ C & 0 \end{pmatrix}$ with B and C matrices with real entries. In particular for any $e \in P_0$, $\phi(e)$ is a linear combination with real coefficients (the matrix elements of C) of elements of Q_1. Elementary linear algebra over $\mathbf{R}$ then tells us that we can change the basis vectors in P_0 and in Q_1 by real coefficients in such a way that there exist subsets $P_0' \subset P_0$ and $Q_1' \subset Q_1$ such that P_0' is a basis for $\ker(\phi|_{\mathbf{B}E_0})$, such that Q_1' is a basis for $\mathrm{im}(\phi|_{\mathbf{B}E_0})$, and such that ϕ is a bijection between $P_0 \setminus P_0'$ and Q_1'. A similar result holds for the basis vectors in P_1 and Q_0. The rest of the proof is as for the even case. QED

6.25 Counter example. In [3.12] we have seen that the condition that ϕ be homogeneous is not superfluous. To show that smoothness is also not superfluous, we consider the map $\phi : \mathcal{A}^2 \to \mathcal{A}^2$ defined by $\phi(a,b) = (a + \mu b, \mu a + \mu^2 b)$ for some $\mu \in \mathcal{A}_0$, $\mu \notin \mathbf{R}$. This ϕ is not smooth because $\mu \notin \mathbf{R}$, $\ker phi$ is generated by $(-\mu, 1)$ and $\operatorname{im}\phi$ is generated by $(1, \mu)$. Since μ is not real, $\ker phi$ and $\operatorname{im}\phi$ are not graded subspaces in the sense of [6.23]; they are however graded subspaces in the sense of [2.1].

6.26 Remark. Let E be a finite dimensional $\mathcal{A}$-vector space with basis $(e_i)_{i=1}^n$ and F an $\mathcal{A}$-vector space with basis $(f_j)_{j\in J}$ (homogeneous, in the appropriate equivalence classes). It follows that the elements $f_j \otimes e^i$ define the equivalence class of bases for the $\mathcal{A}$-vector space $\mathrm{Hom}_R(E;F)$. A right linear map $\phi : E \to F$ is smooth if and only if its matrix elements are real numbers, i.e., its coefficients with respect to the basis $(f_j \otimes e^i)_{i=1,j\in J}^n$ are real. The analogous result holds for left linear maps. We conclude that $\phi \in \mathrm{Hom}_S(E;F)$ is smooth if and only if $\phi \in \mathbf{B}(\mathrm{Hom}_S(E;F)) \subset \mathrm{Hom}_S(E;F)$, i.e., the smooth linear maps $E \to F$ are just the maps belonging to the body of $\mathrm{Hom}_S(E;F)$.

Now recall that for a map $\phi \in \mathrm{Hom}_S(E:F)$ we have $\mathbf{B}\phi = 0$ if and only if its matrix elements are nilpotent (see the proof of [6.8]). This can be stated intrinsically either as $\phi(\mathbf{B}E) \subset \mathcal{N}_F$ or as $\phi \in \mathcal{N}_{\mathrm{Hom}_S(E;F)}$, descriptions that should be compared with the definition of smoothness: $\phi(\mathbf{B}E) \subset \mathbf{B}F$ or $\phi \in \mathbf{B}(\mathrm{Hom}_S(E;F))$.

With these preparations we can describe the isomorphism between $\mathrm{Hom}_S(\mathbf{B}E;\mathbf{B}F)$ and $\mathbf{B}(\mathrm{Hom}_S(E;F))$ (see [6.10]). The kernel of $\mathbf{B} : \mathrm{Hom}_S(E;F) \to \mathrm{Hom}_S(\mathbf{B}E;\mathbf{B}F)$ defined in [6.8] is the space $\mathcal{N}_{\mathrm{Hom}_S(E;F)}$, which is a supplement to $\mathbf{B}(\mathrm{Hom}_S(E;F))$. It follows that $\mathbf{B} : \mathbf{B}(\mathrm{Hom}_S(E;F)) \to \mathrm{Hom}_S(\mathbf{B}E;\mathbf{B}F)$ must be an isomorphism.

6.27 Discussion. On the category of $\mathcal{A}$-vector spaces and smooth maps we can still apply the body map $\mathbf{B}$ as a functor to the category of $\mathbf{R}$-vector spaces. The upshot of what we will do in the remainder of this section is that we will construct a functor $\mathbf{G}$ from the category of $\mathbf{R}$-vector spaces to our new category, which is "inverse" to $\mathbf{B}$ in the sense that it allows to show that these two categories are equivalent. However, we do not insist on this categorical language, we will merely point out some key constructions and properties that will be useful in the future. Readers versed in the language of categories will find it easy to complete the proof that these two categories are equivalent.

The fact that these two categories are equivalent immediately implies that the constructions of direct sums, free modules, and tensor products coincide, being solutions to universal problems. But we have more: our definition of graded subspace in the category of $\mathcal{A}$-vector spaces [6.23] is such that they correspond exactly with the graded subspaces in the category of $\mathbf{R}$-vector spaces.

6.28 Remark. Even with the additional data of an equivalence class of bases and requiring maps to be smooth, the only invariant of a free graded $\mathcal{A}$-module remains its graded dimension. More precisely, if E and F are two $\mathcal{A}$-vector spaces, then they have the same graded dimension if and only if there exists a *smooth* isomorphism between E and F.

6.29 Construction/Discussion. If E is an $\mathcal{A}$-vector space, one can ask whether E is completely determined by the $\mathbf{R}$-vector space $\mathbf{B}E$. We will show that the answer is positive. For an $\mathbf{R}$-vector space $X = X_0 \oplus X_1$ we define $\mathbf{G}X$ as

$$\mathbf{G}X = \mathcal{A} \otimes_{\mathbf{R}} X \ .$$

If we introduce left multiplication by $\mathcal{A}$ as $a \cdot (b \otimes x) = (ab) \otimes x$, and the grading $(\mathbf{G}X)_\alpha = (\mathcal{A}_0 \otimes_{\mathbf{R}} X_\alpha) \oplus (\mathcal{A}_1 \otimes_{\mathbf{R}} X_{1-\alpha})$, then $\mathbf{G}X$ becomes a free graded $\mathcal{A}$-module. If $(e_i)_{i\in I}$ is a basis of X, $(1 \otimes e_i)_{i\in I}$ is a basis of the free graded $\mathcal{A}$-module $\mathbf{G}X$ (use [I.7.28]). Since the equivalence class of this basis does not depend upon the choice of the original basis $(e_i)_{i\in I}$, we conclude that we can turn $\mathbf{G}X$ into an $\mathcal{A}$-vector space of the same graded dimension as the $\mathbf{R}$-vector space X.

Using [6.3] it is elementary to show that $\mathbf{B}(\mathbf{G}X) \subset \mathbf{G}X$ is just the subset of elements of the form $1 \otimes x$, i.e., we have a canonical identification between X and $\mathbf{B}(\mathbf{G}X)$. Coming back to our original question, if B is a basis for E (homogeneous, within the specified equivalence class), then $\mathbf{B}B = B$ (sic!) is a basis of $\mathbf{B}E$, and thus $1 \otimes B$ is a basis of $\mathbf{G}(\mathbf{B}E)$. We thus can identify canonically $\mathbf{G}(\mathbf{B}E)$ with E, thus answering the question in the positive.

Strictly speaking the spaces $\mathbf{B}(\mathbf{G}X)$ and X are not the same, nor are $\mathbf{G}(\mathbf{B}E)$ and E. The correct terminology is that they are canonically isomorphic. However, in the sequel we will forget such subtleties and we will write $\mathbf{B}(\mathbf{G}X) = X$ and $\mathbf{G}(\mathbf{B}E) = E$, pretending that $\mathbf{B}$ and $\mathbf{G}$ are really inverse to each other.

6.30 Lemma. *Let E and F be $\mathcal{A}$-vector spaces and let $\phi : \mathbf{B}E \to \mathbf{B}F$ be a linear map of $\mathbf{R}$-vector spaces. Then there exists a unique smooth (left or right) linear map of $\mathcal{A}$-vector spaces $\mathbf{G}\phi : E \to F$ making the following diagram commutative:*

$$\begin{array}{ccc} E & \xrightarrow{\mathbf{G}\phi} & F \\ {\scriptstyle\mathbf{B}}\downarrow & & \downarrow{\scriptstyle\mathbf{B}} \\ \mathbf{B}E & \xrightarrow[\phi]{} & \mathbf{B}F\ . \end{array}$$

Moreover, if ϕ is homogeneous of parity α, then $\mathbf{G}\phi$ is too.

Proof. $\mathbf{G}\phi$ is determined by its values on a basis. Since a basis of E is contained in $\mathbf{B}E$ (by definition of an $\mathcal{A}$-vector space), proving existence and uniqueness. Smoothness follows because this $\mathbf{G}\phi$ obviously satisfies $(\mathbf{G}\phi)(\mathbf{B}E) \subset \mathbf{B}F$. QED

6.31 Corollary. *$\mathbf{B}(\mathbf{G}\phi) = \phi$ and $\mathbf{B}$ is an isomorphism from smooth linear maps between $\mathcal{A}$-vector spaces to (smooth) linear maps between $\mathbf{R}$-vector spaces.*

Proof. The equality $\mathbf{B}(\mathbf{G}\phi) = \phi$ follows from the uniqueness in [6.8]; that $\mathbf{B}$ is an isomorphism follows from the uniqueness in [6.30]. $\boxed{QED}$

6.32 Discussion. We can summarize the above results by saying that there is no real difference between the category of $\mathcal{A}$-vector spaces (with smooth linear maps) and the category of $\mathbf{R}$-vector spaces. This fact is most easily seen when we think in terms of bases: the operations $\mathbf{B}$ and $\mathbf{G}$ do not change the basis, only the set of scalars one puts in front of a basis element. From this observation it follows immediately that all constructions that are performed using (homogeneous) bases are preserved by the functors $\mathbf{B}$ and $\mathbf{G}$. As said before, for $\mathbf{B}$ we have to be careful with graded subspaces and quotients, but for $\mathbf{G}$ there are no problems since for vector spaces over $\mathbf{R}$ one can always complete an independent set to a basis.

6.33 Nota Bene. Having claimed that there is no real difference between the categories of $\mathcal{A}$-vector spaces and $\mathbf{R}$-vector spaces (more precisely, we claim that $\mathbf{G}$ and $\mathbf{B}$ are isomorphisms of categories), we immediately have to warn the reader for a pitfall: for this statement to be true, one has to take as morphisms in the category of $\mathcal{A}$-vector spaces either left or right linear smooth maps, but not both at the same time! Whenever it is useful to consider both at the same time (and it often is), $\mathbf{B}$ and $\mathbf{G}$ are no longer isomorphisms (e.g., for every *odd* linear map between $\mathbf{R}$-vector spaces, there exist *two* smooth odd maps between $\mathcal{A}$-vector spaces: a right and a left linear one).

Chapter III

Smooth functions and $\mathcal{A}$-manifolds

For those readers who skipped or have already forgotten the first two chapters, we suggest that they imagine that the ring $\mathcal{A}$ is the exterior algebra $\mathcal{A} = \bigwedge X$ of an infinite dimensional vector space X over $\mathbf{R}$, which we usually split into an even and odd part: $\mathcal{A}_0 = \bigoplus_{k=0}^{\infty} \bigwedge^{2k} X$ and $\mathcal{A}_1 = \bigoplus_{k=0}^{\infty} \bigwedge^{2k+1} X$. We will never need these details about $\mathcal{A}$, but it fixes the ideas. Moreover, an $\mathcal{A}$-vector space E can be thought of as being a direct sum of two ordinary vector spaces over $\mathbf{R}$ (the even and odd parts) in which the coefficients with respect to a basis are replaced by elements of $\mathcal{A}$. In this interpretation the even part E_0 consists of those vectors which have even coefficients with respect to a basis of the even part and odd coefficients with respect to a basis of the odd part, while for the odd part E_1 the parities are reversed. The only thing one should remember is that for the basis vectors of the odd part it makes a difference whether one puts the coefficients on the left or on the right. Finally, we will always assume that our $\mathcal{A}$-vector spaces are finite dimensional, and that a basis is ordered in the sense that we first put the basis vectors for the even part and then those for the odd part.

The main subject of this chapter is the notion of an $\mathcal{A}$-manifold, a generalization of the notion of an ordinary manifold in which $\mathbf{R}$ is replaced by a graded commutative ring $\mathcal{A}$ and in which $\mathbf{R}^n$ is replaced by the even part E_0 of an $\mathcal{A}$-vector space E of dimension $p|q$. Since the standard approach to smooth functions cannot be copied to this more general setting, we provide an alternative definition which works in both cases. We prove that smooth functions on E_0 can be identified with ordinary smooth functions on $\mathbf{R}^p$ times skew-symmetric polynomials on $\mathbf{R}^q$, i.e., $C^\infty(E_0) \cong C^\infty(\mathbf{R}^p) \otimes \bigwedge \mathbf{R}^q$.

Once we know what smooth functions are, we define their derivatives, something which is not automatically included in our approach to smooth functions. This requires a condition on $\mathcal{A}$, but this condition is satisfied by the basic example for $\mathcal{A}$ given above. It then is elementary to generalize the implicit and inverse function theorems; even partitions of unity pose no problems. With these ingredients we then just copy the definition of ordinary manifolds in terms of charts and transition functions between charts to obtain

$\mathcal{A}$-manifolds. Product manifolds are defined easily, but the construction of a submanifold requires some care. Due to our definition of smooth functions, one cannot always construct submanifolds of lower dimension passing through an arbitrary point of the ambient manifold.

We finish this chapter by extending the body map $\mathbf{B}$ to $\mathcal{A}$-manifolds and their smooth maps. We prove that the topology of an $\mathcal{A}$-manifold M is completely determined by the topology of its body $\mathbf{B}M$ (which is a manifold in the ordinary sense of the word, or, in our terminology, an $\mathbf{R}$-manifold). This extended body map preserves all constructions, e.g., $\mathbf{B}(M \times N) = \mathbf{B}M \times \mathbf{B}N$. In terms of coordinates, the body map is just the projection onto their real part (i.e., the projection $\mathcal{A} = \bigwedge X \to \mathbf{R} = \bigwedge^0 X$ in terms of our example of $\mathcal{A}$).

1. Topology and smooth functions

In this section we provide our alternative approach to smooth functions which avoids the use of limits of difference quotients. We show that the set $C^\infty(U;\mathcal{A})$ of $\mathcal{A}$-valued smooth functions on an open set $U \subset E_0$ is a graded $\mathbf{R}$-algebra and that the set $C^\infty(U;F)$ of F-valued smooth functions (F an $\mathcal{A}$-vector space) is a free graded $C^\infty(U;\mathcal{A})$-module of the same graded dimension as F. Using the parity reversal operator, we show that the restriction to the even part of an $\mathcal{A}$-vector space is in reality not a restriction at all.

1.1 Discussion. Just as smooth functions on open sets in $\mathbf{R}^p$ are the basic ingredients for ordinary manifolds, we want smooth functions on open sets of ... to be the basic ingredients for $\mathcal{A}$-manifolds. Three questions arise: what are the open sets, what are the smooth functions and, most important for the moment, what kind of spaces to put on the dots? In trying to find reasonable answers to these questions, we will be guided by the idea that $\mathbf{R}$-manifolds, i.e., those obtained by taking $\mathcal{A} = \mathbf{R}$, should be the same as the ordinary well known manifolds.

The most naive answer to the space question would be to take any $\mathcal{A}$-vector space, but an $\mathbf{R}$-vector space X is a direct sum of two ordinary real vector spaces $X_0 \oplus X_1$ and for ordinary manifolds there is no subdivision of the (local) coordinates. A less naive answer, and the one we will adopt, is to take *the even part* E_0 of an $\mathcal{A}$-vector space E. If $\{e_1, \ldots, e_p, f_1, \ldots, f_q\}$ is a homogeneous basis of E with the e_i even and the f_i odd, then a point in E_0 has even coordinates with respect to the e_i and odd coordinates with respect to the f_i. The number of even coordinates gives us the even dimension of E and the number of odd coordinates gives us the odd dimension of E. Moreover, if we take the special case $\mathcal{A} = \mathbf{R}$, all odd coordinates are zero, and what remains is essentially $\mathbf{R}^p$, just as we wanted.

We note that we thus have a profound distinction between a generic $\mathcal{A}$ (with $\mathcal{A}_1 \neq \{0\}$) and the special case $\mathcal{A} = \mathbf{R}$. In the first case we have non trivial odd coordinates and thus retain information about q (the odd dimension), whereas in the second case we lose all information about q. Another way to see this difference is to consider the body map $\mathbf{B} : E \to \mathbf{B}E$, which provides the passage from the generic case to the special case $\mathcal{A} = \mathbf{R}$. We have seen that $\mathbf{B}E$ contains enough information to reconstruct E as $\mathbf{G}(\mathbf{B}E)$. On the other hand, $\mathbf{B}E_0$ does not contain enough information to reconstruct E_0 (and certainly not the whole of E): the odd dimension of E is missing. It is this difference that gives $\mathcal{A}$-manifolds their extra flavor.

1.2 Definition/Discussion. Let E be an $\mathcal{A}$-vector space of dimension $p|q$. We define a topology on E as the coarsest topology for which the body map $\mathbf{B} : E \to \mathbf{B}E \cong \mathbf{R}^{p+q}$ is continuous, i.e., $U \subset E$ is open if and only if $U = \mathbf{B}^{-1}(O)$ for some open subset O of $\mathbf{B}E$. Said differently, U is open if and only if $\mathbf{B}U$ is open and $U = \mathbf{B}^{-1}(\mathbf{B}U)$. Yet another way to say the same is to state that the map $\mathbf{B} : U \to \mathbf{B}U$ induces a bijection between the open sets of E and the open sets of $\mathbf{B}E \cong \mathbf{R}^{p+q}$. This topology is usually called the *DeWitt topology*. All subsets of E, and in particular E_0, will be equipped with the relative topology. Since $\mathbf{B}E = E/\mathcal{N}_E$, it follows that *open* sets $U \subset E$ are saturated with nilpotent vectors, i.e., U open implies $U + \mathcal{N}_E = U$.

For $\mathbf{R}$-vector spaces the body map $\mathbf{B}$ is the identity, so the DeWitt topology on X is the standard euclidean topology on X. In order to have a closer look at the DeWitt topology in the general case, let us choose a basis $(e_i)_{i=1}^n$ of E. With respect to this basis we consider *left* coordinates $x = \sum_i x^i\, e_i \in E$ to identify E with $\mathcal{A}^n$ as sets. This gives us an identification of $\mathbf{B}E$ with $\mathbf{R}^n$ such that the body map is given by $\mathbf{B}(x^1, \ldots, x^n) = (\mathbf{B}x^1, \ldots, \mathbf{B}x^n)$. Using this identification, a point $(x^1, \ldots, x^n)$ will be in the open set U if and only if $(\mathbf{B}x^1, \ldots, \mathbf{B}x^n)$ lies in the open set $O = \mathbf{B}U \subset \mathbf{R}^n$, i.e., the topology is completely determined by the body parts of the coordinates. If we assume furthermore that the homogeneous basis is an ordered one, a point $(x^1, \ldots, x^n)$ lies in E_0 if and only if $x^1, \ldots, x^p \in \mathcal{A}_0$ and $x^{p+1}, \ldots, x^n \in \mathcal{A}_1$. We deduce that a set $U \subset E_0$ is open if and only if there exists an open set $O \subset \mathbf{R}^p$ such that

$$(x^1, \ldots, x^n) \in U \iff (\mathbf{B}x^1, \ldots, \mathbf{B}x^p) \in O\ .$$

A particular consequence is that the induced topology on a fiber $\{(x^1, \ldots, x^p)\} \times \mathcal{A}_1^q \subset E_0$ is the indiscrete topology (either all or nothing).

Another consequence is that for the topology on E_0 we have the same characterization as for the topology on E: $U \subset E_0$ is open if and only if $\mathbf{B}U$ is open in $\mathbf{B}E_0 \cong \mathbf{R}^p$ and $U = \mathbf{B}^{-1}(\mathbf{B}U)$, where we now see $\mathbf{B}$ as the restriction $\mathbf{B} : E_0 \to \mathbf{B}E_0$. Or again, $\mathbf{B}$ induces a bijection between open sets of E_0 and open sets of $\mathbf{B}E_0 \cong \mathbf{R}^p$.

1.3 Corollary. *The DeWitt topology on an $\mathcal{A}$-vector space E and on its even part E_0 are locally connected (any neighborhood of a point contains an open connected neighborhood of that point, which is equivalent to the connected components of open sets being open* [Du, theorem V.4.2]*).*

1.4 Lemma. *Let E be an $\mathcal{A}$-vector space, then:*

(i) *the product topology of $E_0 \oplus E_1$ coincides with the topology of E; if F and $\widehat{F}$ are supplementary graded subspaces of E, then the product topology of $F \oplus \widehat{F}$ coincides with the topology of E;*

(ii) *the $\mathcal{A}$-vector space operations of addition ($E \times E \to E$) and multiplication by elements of $\mathcal{A}$ ($\mathcal{A} \times E \to E$ and $E \times \mathcal{A} \to E$) are continuous;*

(iii) *left linear maps to $\mathcal{A}$ are continuous, and in particular the coordinate projections ${}^je : x = \sum_i x^i\, e_i \mapsto x^j = \langle\!\langle x \,\|\, {}^je \rangle\!\rangle$.*

1.5 Notation. According to standard abuse of notation, we will often denote the (left) coordinate projections ${}^je : (x^1, \ldots, x^{p+q}) \mapsto x^j$ by x^j.

1.6 Discussion. So far in discussing the DeWitt topology we have treated the spaces E and $\mathbf{B}E$ as different. However, we have identified $\mathbf{B}E$ with those points in E that have real coordinates [II.6.19], i.e., we have interpreted $\mathbf{B}$ as a projection map $\mathbf{B} : E \to E$. For an arbitrary subset $U \subset E$ we thus interpret $\mathbf{B}U$ as a subset of E. Moreover, the inclusion $\mathbf{B}U \subset U$ is equivalent to the equality $\mathbf{B}U = U \cap \mathbf{B}E$, but neither condition need be true. However, if U is open in E, and thus in particular $U = \mathbf{B}^{-1}(\mathbf{B}U)$, we have $\mathbf{B}U \subset U$, and thus $\mathbf{B}U = U \cap \mathbf{B}E \subset U$. In other words, for *open* sets U the body $\mathbf{B}U$ consists of those points *in* U that have real coordinates.

1.7 Discussion. Once we have a topology, we can speak of continuous functions, and in particular we can speak of functions on open sets U of the even part of an $\mathcal{A}$-vector space E with values in an $\mathcal{A}$-vector space F. As usual we will denote these sets by $C^0(U; F)$. However, we want to speak of differentiable functions, and there we encounter a problem. The standard way to define a differentiable function is to say that the derivative should exist. Such a derivative is (usually) defined as the limit of a difference quotient. But in our context, we can not always write such a difference quotient because of the existence of nilpotent elements. Even disregarding the nilpotent elements, the DeWitt topology is *not* Hausdorff, so a limit need not be unique. In order to circumvent these problems, we will use an alternative approach to differentiable functions which does not use limits nor difference quotients. The essential idea of this approach is expressed in [1.8].

1.8 Proposition. *Let $O \subset \mathbf{R}^p$ be open, and let $f : O \to \mathbf{R}^N$ be a function. Then f is of class C^{k+1} if and only if there exists an open cover $\mathcal{U} = \{U_a \mid a \in I\}$ of O and for all $a \in I$ functions $g_{ai} : U_a^2 \to \mathbf{R}^N$, $1 \le i \le p$ of class C^k such that for all $x, y \in U_a$:*

$$f(x) - f(y) = \sum_{i=1}^{p} g_{ai}(x, y) \cdot (x^i - y^i) \,. \tag{1.9}$$

Proof. If f is of class C^{k+1}, we take any cover of O consisting of convex open sets (for instance open balls). For each convex U_a in this cover we define the functions g_{ai} by

$$(1.10) \qquad g_{ai}(x,y) = \int_0^1 \frac{\partial f}{\partial x^i}(sx + (1-s)y)\, ds \; .$$

The convexity of U_a guarantees that this integral makes sense, and since the $\partial f/\partial x^i$ are of class C^k, the g_{ai} are of class C^k as well. Computing $\sum_i g_{ai}(x,y)\cdot(x^i - y^i)$ immediately gives $f(x) - f(y)$.

If on the other hand the condition is satisfied, we compute on each open U_a the partial derivatives of f in $x \in U_a$ by

$$(1.11) \qquad \frac{\partial f}{\partial x^i}(x) = \lim_{h\to 0} \frac{f(x+he_i) - f(x)}{h} = \lim_{h\to 0} g_{ai}(x+he_i, x) = g_{ai}(x,x) \; ,$$

where $e_i = (0,\ldots,0,1,0,\ldots,0)$ denotes the i-th basis vector in $\mathbf{R}^p$. Since the g_{ai} are of class C^k, this shows that the partial derivatives of f exist everywhere and are of class C^k, i.e., f is of class C^{k+1}. QED

1.12 Remark. One can prove a stronger version of [1.8] which does not need a cover and which states that f is of class C^{k+1} if and only if there exist functions $g_i : O^2 \to \mathbf{R}^N$ of class C^k satisfying (1.9) for all $x, y \in O$. The reason we did not give this stronger result in [1.8] is that [1.8] as it is, remains true if we replace the field of real numbers $\mathbf{R}$ by the field of complex numbers $\mathbf{C}$ (and then it says that f is holomorphic if and only if the g_{ai} are continuous). On the other hand, if we replace [1.8] by the stronger result without the cover, then we no longer can go over to the complex case. The simple reason is that the stronger result needs a partition of unity argument. In fact, domains in $\mathbf{C}^p$ for which no cover is needed occur naturally in complex analysis: by a classical corollary of H. Cartan's theorem B, this is the case for any domain of holomorphy ($\equiv$ a pseudo convex domain $\equiv$ a Stein domain), see [Ra, corollary 6.26].

Let us now give the proof of the stronger result. Since the if part is proved as in [1.8], we attack the only if part. We choose a cover $\mathcal{V} = \{\, V_a \mid a \in I \,\}$ of $O \times O$ having the following property: for $V \in \mathcal{V}$ we have either (i) $\exists C$ convex open set in O such that $V = C \times C$, or (ii) $\exists j$ $(1 \le j \le n)$ such that $\forall (x,y) \in V : x^j \neq y^j$. The existence of such covers follows easily from the fact that $\mathbf{R}^p$ is Hausdorff. Given such a cover, we define on each $V_a \in \mathcal{V}$ functions $g_{ai} : V_a \to \mathbf{R}^N$ as follows. In case (i) we define g_{ai} by (1.10); in case (ii) we define $g_{aj}(x,y) = (f(x) - f(y))/(x^j - y^j)$ and for $i \neq j$ we define $g_{ai} = 0$. In both cases the functions g_{ai} are of class C^k and satisfy (1.9) for all $(x,y) \in V_a$. Now let ρ_a be a partition of unity of class C^k subordinated to the cover $\mathcal{V}$ (see [5.18] for an exact definition), and define the functions $g_i : O \times O \to \mathbf{R}^N$ by $g_i(x,y) = \sum_a \rho_a(x,y)\, g_{ai}(x,y)$. Then for any $(x,y) \in O \times O$ we have $\rho_a(x,y) \neq 0 \Rightarrow (x,y) \in V_a$, and thus $\sum_i (x^i - y^i)\, g_i(x,y) = \sum_a \rho_a(x,y) \sum_i (x^i - y^i)\, g_{ai}(x,y) = \sum_a \rho_a(x,y)\,(f(x) - f(y)) = f(x) - f(y)$. This finishes the proof.

1.13 Discussion. If $f : O \subset \mathbf{R}^p \to \mathbf{R}^N$ is smooth (of class C^∞), application of [1.8] gives us an open cover $\{U_a\}$ of O and again smooth functions $f^{(1)}_{ai} : U^2_a \to \mathbf{R}^N$ satisfying (1.9). We then can apply [1.8] again to each of the functions $f^{(1)}_{ai}$, to obtain an open cover $\{U_{ab}\}$ of U_a and smooth functions $f^{(2)}_{(ai)(bj)} : U^2_{ab} \to \mathbf{R}^N$ satisfying a condition analogous to (1.9). Continuing this procedure, we get an infinite tree of open covers $\{U_{a_1a_2\dots a_r}\}$ (i.e., $\cup_{a_r} U_{a_1a_2\dots a_r} = U_{a_1\dots a_{r-1}}$, and the index family for a_r might depend upon the sequence $a_1 \dots a_{r-1}$) and an infinite tree of smooth functions $f^{(r)}_{(a_1i_1)\dots(a_ri_r)} : U^2_{a_1\dots a_r} \to \mathbf{R}^N$, $1 \le i_j \le p\,2^{j-1}$ satisfying for all $x, y \in U_{a_1\dots a_rb}$:

$$(1.14)\quad f^{(r)}_{(a_1i_1)\dots(a_ri_r)}(x) - f^{(r)}_{(a_1i_1)\dots(a_ri_r)}(y) = \sum_{j=1}^{p\,2^r}(x^j - y^j)\cdot f^{(r+1)}_{(a_1i_1)\dots(a_ri_r)(bj)}(x,y)\ .$$

If on the other hand we have such a tree of covers and functions (without any differentiability assumption) satisfying (1.14), we can argue as follows. If for a fixed value of r all functions $f^{(r)}_{(a_1i_1)\dots(a_ri_r)}$ are continuous, then by [1.8] all functions $f^{(r-1)}_{(a_1i_1)\dots(a_{r-1}i_{r-1})}$ are of class C^1. Continuing this argument we conclude that the initial function $f \equiv f^{(0)}$ is of class C^r. Varying the value of r proves [1.15].

1.15 Proposition. *Let $f : O \subset \mathbf{R}^p \to \mathbf{R}^N$ be a function. Then f is smooth if and only if there exists an infinite tree of open covers $\{U_{a_1\dots a_r}\}$ and continuous functions $f^{(r)}_{(a_1i_1)\dots(a_ri_r)}$ as described in [1.13] satisfying (1.14).*

1.16 Definitions. With the above preparations concerning smooth functions on $\mathbf{R}^p$, we now return to the question how to define smooth functions on open subsets $U \subset E_0$ of the even part of some $\mathcal{A}$-vector space E. We start with some preparatory definitions.

• Given a continuous function $f : U \to F$ with $U \subset E_0$ open and E and F $\mathcal{A}$-vector spaces, a *smooth tree associated to* f will be an infinite tree of open covers $\{U_{a_1a_2\dots a_r}\}$ and continuous functions $f^{(r)}_{(a_1i_1)\dots(a_ri_r)} : U^2_{a_1\dots a_r} \to F$ satisfying for all $x, y \in U_{a_1\dots a_rb}$:

$$f^{(r)}_{(a_1i_1)\dots(a_ri_r)}(x) - f^{(r)}_{(a_1i_1)\dots(a_ri_r)}(y) = \sum_{j=1}^{2^r\,\dim(E)} (x^j - y^j)\cdot f^{(r+1)}_{(a_1i_1)\dots(a_ri_r)(bj)}(x,y)\ .$$

• A symbol $\mathcal{F}$ is called a *smooth system of F-valued functions* if for all E and for all $U \subset E_0$ open, we have a set $\mathcal{F}(U;F) \subset C^0(U;F)$ of continuous functions from U to F such that for any $f \in \mathcal{F}(U;F)$ the following two conditions are satisfied:

(A1) there exists a family $U_a \subset U$ of open subsets covering U and $n = \dim E$ functions $\phi_{a,i} \in \mathcal{F}(U^2_a;F)$ such that

$$\forall x,y \in U_a \quad : \quad f(x) - f(y) = \sum_{i=1}^{n}(x-y)^i\cdot\phi_{a,i}(x,y)\ ,$$

where the x^i are the coordinates of x with respect to some basis of E (homogeneous, in the specified equivalence class);

(A2) $f(\mathbf{B}U) \subset \mathbf{B}F$.

1.17 Remark. In case $\mathcal{A} = \mathbf{R}$, condition (A2) is completely superfluous. We added it for exactly the same reason we added a similar condition for *smooth* linear maps in [II.6.22]: to be able to prove that the body map $\mathbf{B}$ is a bijection on certain sets of functions (see [2.16]).

1.18 Remark. We have restricted our attention to functions defined on open sets of the *even* part of an $\mathcal{A}$-vector space. However, one can define as easily differentiable functions on open sets of the entire $\mathcal{A}$-vector space. We also have restricted our attention to infinitely often differentiable functions, but functions of class C^k can be defined in a similar way. For more details the reader is referred to [Tu2].

1.19 Example. Let $f : U \subset E_0 \to F$ be a continuous function, and suppose there exists a smooth tree of functions associated to f. We then can construct a smooth system $\mathcal{F}_f$ (of F-valued functions) as follows.

$$\mathcal{F}_f(U^2_{a_1a_2\dots a_r}; F) = \{\, f^{(r)}_{(a_1i_1)\dots(a_ri_r)} \mid 1 \le i_j \le 2^{j-1}\dim(E)\,\}\ ,$$

and if O is not of the form $U^2_{a_1a_2\dots a_r}$, then $\mathcal{F}_f(O; F) = \phi$, with the exception of the first term of the tree $\mathcal{F}_f(U; F) = \{f\}$. The definition of a smooth tree immediately proves that $\mathcal{F}_f$ is indeed a smooth system.

1.20 Definition. We can order smooth systems by inclusion, i.e., $\mathcal{F}_1 \le \mathcal{F}_2$ if and only if for all U : $\mathcal{F}_1(U; F) \subset \mathcal{F}_2(U; F)$. It is also easy to prove that the union $\mathcal{F}_\cup$ of two smooth systems $\mathcal{F}_1$ and $\mathcal{F}_2$, defined by $\mathcal{F}_\cup(U; F) = \mathcal{F}_1(U; F) \cup \mathcal{F}_2(U; F)$, is again a smooth system. It thus makes sense to speak about the *maximal* smooth system. Which brings us to our final definition: C^∞ *is the maximal smooth system of* F*-valued functions.* Elements $f \in C^\infty(U; F)$ will be called *smooth functions* (on U, with values in F).

More generally, if D is an arbitrary subset of F, we define the sets $C^\infty(U; D)$ by

$$C^\infty(U; D) = \{\, f \in C^\infty(U; F) \mid f(U) \subset D\,\}\ .$$

The special cases $D = F_\alpha$, $\alpha \in \mathbf{Z}_2$ are also denoted as $C^\infty(U; F_\alpha) = C^\infty(U; F)_\alpha$. Note that this definition is quite different from the definition of $\mathrm{Map}_S(E; F)_\alpha$, although both concern sets of maps.

1.21 Proposition. *In the case $\mathcal{A} = \mathbf{R}$ the above definition of $C^\infty(U; F)$ is equivalent to the usual definition of infinitely often differentiable functions.*

Proof. If $f : U \to F$ is infinitely often differentiable, we can construct a smooth tree associated to f by [1.15]. As in [1.19] we then can construct a smooth system $\mathcal{F}_f$. By maximality of C^∞, $\mathcal{F}_f$ is included in C^∞, and thus $f \in C^\infty(U; F)$. On the other hand, if $f \in C^\infty(U; F)$, then by repeatedly applying property (A1) we can construct a smooth tree associated to f. And then by [1.15] the function f is infinitely often differentiable. QED

1.22 Discussion. We have, intentionally, swept one slightly disturbing aspect of the proof of [1.21] under the rug. It concerns the number of functions to be used in (A1). In case $\mathcal{A} = \mathbf{R}$, an $\mathbf{R}$-vector space X of dimension $p|q$ is a direct sum of two ordinary vector spaces $X = X_0 \oplus X_1$ of dimensions p and q respectively. Functions of the even part $X_0 = X_0 \oplus \{0\}$ are obviously functions of p real coordinates, not of $n = p+q$ coordinates. Now in [1.15] the smooth tree is constructed with p functions at each stage, whereas (A1) requires $n = p + q$ functions. This problem is easily solved by taking the zero functions for the missing ones. Since the additional q coordinates are identically zero, these terms do not contribute to the summation in (A1). On the other hand, when constructing the smooth tree from (A1), we get at each stage $n = p + q$ functions, where we need only p. This is even simpler to solve: we just neglect the functions we do not need before we apply [1.15].

In view of formula (1.11), we will want to define partial derivatives $\partial_i f(x)$ as $\phi_{a,i}(x, x)$, i.e., as the diagonal of $\phi_{a,i}i$. Although the functions $\phi_{a,i}$ are not unique, we learn from (1.11) that the diagonal of $\phi_{a,i}$ is independent of the choice for $\phi_{a,i}$. On the other hand, the above considerations tell us that for $\mathcal{A} = \mathbf{R}$, we can choose the functions $\phi_{a,i}$ with $i > p$ completely arbitrary, and thus the partial derivatives $\partial_i f$ are also arbitrary for $i > p$. There is no contradiction, because (1.11) was given only for $i \leq p$; and no one would try to define partial derivatives $\partial_i f$ for $i > p$ if the function f depends only upon p coordinates. But it shows that we have to be careful when we want to define partial derivatives for a more general $\mathcal{A}$. And indeed, we will see that the structure of $\mathcal{A}$ is crucial in the definition of partial derivatives by means of (1.11) (see [3.1] and [V.1.8]). But before we can address the question of partial derivatives, we need to analyze the consequences of our definition of smooth functions in more detail.

1.23 Lemma. *The sets $C^\infty(U; F)$ enjoy the following properties:*

(a) *All constant maps with real image belong to $C^\infty(U; \mathcal{A})$, as do the coordinate projections ${}^je$.*
(b) *For $f \in C^\infty(U; \mathcal{A})$ and $\alpha \in \mathbf{Z}_2$ define $g : U \to \mathcal{A}_\alpha \subset \mathcal{A}$ by $g(x) = (f(x))_\alpha$. Then $g \in C^\infty(U; \mathcal{A})_\alpha \subset C^\infty(U; \mathcal{A})$.*
(c) *If $\lambda \in \mathbf{R}$ and $f \in C^\infty(U; \mathcal{A})$ then $\lambda \cdot f \in C^\infty(U; \mathcal{A})$.*
(d) *If $f, g \in C^\infty(U; \mathcal{A})$, then $f + g \in C^\infty(U; \mathcal{A})$.*
(e) *If property (A1) holds for some homogeneous basis, it holds for all homogeneous bases.*

(f) *For* $f \in C^\infty(U;\mathcal{A})$ *and* E' *an* $\mathcal{A}$*-vector space we define* $g : U \times E'_0 \to \mathcal{A}$ *by* $g(x,x') = f(x)$. *Then* $g \in C^\infty(U \times E'_0;\mathcal{A})$.

(g) *Suppose* $E = E' \oplus E''$ *for two* $\mathcal{A}$*-vector spaces* E' *and* E'' *and suppose that* $U' \subset U \cap E'_0$ *is open in* E'_0. *Let* $x'' \in \mathbf{B}E'' \cap U$ *and* $f \in C^\infty(U;F)$ *be arbitrary. If* $g : U' \to F$ *is defined by* $g(x') = f(x',x'')$, *then* $g \in C^\infty(U';F)$.

(h) *If* $f,g \in C^\infty(U;\mathcal{A})$ *then* $f \cdot g \in C^\infty(U;\mathcal{A})$.

(i) $f \in C^\infty(U;F)$ *if and only if for all* j : ${}^j\epsilon \circ f \in C^\infty(U;\mathcal{A})$, *where* ${}^j\epsilon$ *are the (left-linear) coordinate functions on* F.

(j) *Let* $\widehat{U}$ *be an open set of* F_0, $f \in C^\infty(U;F)$ *and* $g \in C^\infty(\widehat{U};G)$. *If* $\mathrm{im}(f) \subset \widehat{U}$, *then* $g \circ f \in C^\infty(U;G)$.

(k) *Let* $U_i \subset U$ *be a family of open sets covering* U. *If* $f : U \to F$ *is such that* $\forall i : f|_{U_i} \in C^\infty(U_i;F)$, *then* $f \in C^\infty(U;F)$.

Proof. In this proof we leave it to the reader to verify, whenever necessary, that condition (A2) is satisfied and that the constructed functions are continuous; the conditions in the lemma assure that there will be no problem. The basic line of proof will be the construction of a smooth tree of (continuous) functions satisfying (A1) by means of induction to the level. As in [1.19] we then can form a smooth system, from which we can conclude that it must be included in C^∞ because the latter is maximal. We will abbreviate this procedure by saying that we use the recursion argument.

• To prove (a), first note that constant functions and the ${}^je$ are continuous [1.4]. For any constant function f we can choose the functions $\phi_i \equiv 0$. Hence by maximality all constant functions belong to $C^\infty(U;F)$. For the ${}^je$ one can choose $\phi_j \equiv 1$, all others constant 0 (with respect to the same basis as used for the coordinate projections). Note that in both cases the cover $\{U_a\}$ consists of the single element U itself.

• To prove (b), note that the i-th coordinate of $x \in E_0$ has parity $\varepsilon(e_i)$. Using the same cover $\{U_a\}$ as for f, we define the functions $\chi_{a,i} = (\phi_{a,i})_{\alpha+\varepsilon(e_i)}$. It follows that $g(y) = g(x) + \sum_i (y-x)^i \cdot \chi_{a,i}(y,x)$. By [1.4] the $\chi_{a,i}$ are continuous; the rest follows by the recursion argument.

• For (c), multiply the $\phi_{a,i}$'s by λ and use the recursion argument.

• For (d) and (e), let f and g be as in (d). First note that, by taking pairwise intersections of elements of the original covers for f and g, we may assume that the same cover $\{U_a\}$ serves for both f and g. Denote by (e_i) and ϕ_i the basis and functions in (A1) for f and denote by (ϵ_i) and χ_i the same ingredients for g. Finally denote by z^i the (left!) coordinates of the vector $x - y$ with respect to the basis (e_i) and by ζ^i its coordinates with respect to the basis (ϵ_i). It follows that there exists a matrix $\lambda_i{}^j \in \mathbf{R}$ such that $\zeta^j = \sum_i z^i \cdot \lambda_i{}^j$ and hence we have

$$\begin{aligned} f(x) + g(x) &= f(y) + \sum_i z^i \cdot \phi_i(x,y) + g(y) + \sum_j \zeta^j \cdot \chi_j(x,y) \\ &= f(y) + g(y) + \sum_i z^i \cdot \Big(\phi_i(x,y) + \sum_j \lambda_i{}^j \cdot \chi_j(x,y)\Big) \,. \end{aligned}$$

We conclude the proof of (d) by noting that $\lambda_i{}^j \cdot \chi_j(x,y)$ is smooth by (c) and by applying the recursion argument. The proof of (e) follows by taking $f \equiv 0$.

• To prove (f), note that

$$g(x,x') - g(y,y') = f(x) - f(y) = \sum_i (x-y)^i \cdot \phi_{a,i}(x,y)$$
$$= \sum_i (x-y)^i \cdot \chi_{a,i}(x,x',y,y') ,$$

where the functions $\chi_{a,i} : (U_a \times E'_0)^2 \to \mathcal{A}$ are defined by $\chi_{a,i}(x,x',y,y') = \phi_{a,i}(x,y)$. We then apply the recursion argument after having added zeros for the functions $\chi_{a,i}$ for coordinates in E'_0.

• For (g), let $(e_i)_{i=1}^m$ be a basis for E' and let $(e_i)_{i=m+1}^n$ be a basis for E'', making $(e_i)_{i=1}^n$ a basis for E. If we define $\chi_i(x',y') = \phi_i((x',x''),(y',x''))$, $i \leq m$ on the subset $U'_a = U' \cap U_a$, then obviously $g(x') = g(y') + \sum_{i=1}^m (x'-y')^i \cdot \chi_i(x',y')$. The result then follows by the recursion argument.

• For (h), let $\phi_{a,i}$ and $\chi_{a,i}$ be the functions of property (A1) for the functions f and g respectively. As before, we may assume that these functions share the same cover $\{U_a\}$. By [1.4] the function $f \cdot g$ is continuous. By (b) and (d) we may assume that the images of f and g are homogeneous. Using the functions $\phi_{a,i}$ and $\chi_{a,i}$ we obtain

$$(f \cdot g)(x) - (f \cdot g)(y) = \sum_{i=1}^n (x-y)^i \cdot \Big(\phi_{a,i}(x,y) \cdot g(y)$$
$$+ (-1)^{\langle \varepsilon(e_i) | \varepsilon(f(x)) \rangle} \cdot f(x) \cdot \chi_{a,i}(x,y)\Big) .$$

The result now follows if we apply (c), (d), (f), and the recursion argument.

• To prove (i), suppose first that f is smooth. Using the left-linearity of the ${}^j\epsilon$ we obtain $({}^j\epsilon \circ f)(x) = ({}^j\epsilon \circ f)(y) + \sum_i (x-y)^i \cdot ({}^j\epsilon \circ \phi_{a,i})(x,y)$. The recursion argument then shows that the ${}^j\epsilon \circ f$ are smooth.

For the other implication, suppose that the ${}^j\epsilon \circ f$ are smooth. We then have to prove that the function $f = \sum_j ({}^j\epsilon \circ f) \cdot \epsilon_j$ is smooth, where (ϵ_j) denotes the basis of F dual to the coordinate functions ${}^j\epsilon$. By hypothesis each of the functions ${}^j\epsilon \circ f$ has locally defined functions $\phi^j_{a,i}$ as in (A1). As before, by taking multiple intersections when necessary, we may assume that the domains of definition of these functions coincide. We then define the functions $\phi_{a,i} : U_a^2 \to F$ by $\phi_{a,i}(x,y) = \sum_j \phi^j_{a,i}(x,y) \cdot \epsilon_j$. By definition of the topology, these functions are continuous. Moreover, they satisfy $\phi^j_{a,i} = {}^j\epsilon \circ \phi_{a,i}$. We then conclude by the recursion argument.

• To prove (j), let $\phi_{a,i}$ and $\chi_{b,j}$ be the functions of property (A1) for the functions f and g respectively. By taking intersections with the sets $g^{-1}(\widehat{U}_b)$ if necessary, we may assume that each $f(U_a)$ is contained in some $\widehat{U}_b$. We then compute:

$$(g \circ f)(x) = (g \circ f)(y) + \sum_j (f(x) - f(y))^j \cdot \chi_{b,j}(f(x), f(y))$$
$$= (g \circ f)(y) + \sum_i (x-y)^i \cdot \Big(\sum_j ({}^j\epsilon \circ \phi_{a,i})(x,y) \cdot \chi_{b,j}(f(x), f(y))\Big) .$$

The result then follows by applying (d), (h), (i), and the recursion argument.

• Finally, for (k) let $U_{i,a}$ be the families of open sets as defined by property (A1) for each U_i. Since the U_i cover U, it follows that $\cup_{i,a} U_{i,a} = U$, proving that f is smooth. QED

1.24 Summary. *We can reformulate most of [1.23] in the following more manageable form.*

(i) *Being a smooth function is a local property, stable under composition and fixing of variables (to real values).*
(ii) *The set* $C^\infty(U;\mathcal{A}) = C^\infty(U;\mathcal{A})_0 \oplus C^\infty(U;\mathcal{A})_1$ *is a graded commutative* **R**-*algebra with unit under pointwise addition and multiplication of functions.*
(iii) *For any* $\mathcal{A}$-*vector space* F, $C^\infty(U;F) = C^\infty(U;F)_0 \oplus C^\infty(U;F)_1$ *is a free graded* $C^\infty(U;\mathcal{A})$-*module of the same graded dimension as* F.

Proof. The graded commutativity of $C^\infty(U;\mathcal{A})$ follows from the graded commutativity of $\mathcal{A}$ and the fact that multiplication in $C^\infty(U;\mathcal{A})$ is pointwise. That $C^\infty(U;F)$ is a free module of the same dimension as F is an immediate consequence of the fact that we can write $f = \sum_j ({}^j\epsilon \circ f)\cdot \epsilon_j$ for an element $f \in C^\infty(U;F)$ and that we can interpret ϵ_j as a constant function with real image on U. QED

1.25 Discussion. If $f : U \to F$ is a smooth function, we get from property (A1) the functions $\phi_{a,i} : U_a^2 \to F$. We can combine these functions into a single function $\phi_a : U_a^2 \to \mathrm{Hom}_L(E;F)$ defined by $\phi_a(x,y) = \sum_i {}^i e \otimes \phi_{a,i}(x,y)$. It is an immediate consequence of [1.23] that ϕ is smooth. It follows that for each function $f \in C^\infty(U;F)$ there exist functions $\phi_a \in C^\infty(U_a^2;\mathrm{Hom}_L(E;F))$ such that

$$\forall x,y \in U_a : f(x) - f(y) = \langle\!\langle\, x - y \,\|\, \phi_a(x,y) \,\rangle\!\rangle \ .$$

In this way we obtain a definition of smooth functions which is obviously independent of a basis. It should be noted that we have made a choice here to take *left linear* endomorphisms, a choice already present in (A1) by using left coordinate functions and writing the coordinates $(y-x)^i$ to the left of the $\phi_{a,i}$.

In analogy with ordinary real valued C^∞ functions (see (1.11)) we will (want to) define partial derivatives in our graded setting by the functions $\phi_{a,i}(x,x)$. It follows that we need that these functions are uniquely determined by the function f. We will see in [3.1] that this is not automatically true, but that it requires a condition on $\mathcal{A}$.

1.26 Construction. The reader might have got the impression that only the even part of an $\mathcal{A}$-vector space can be the domain of smooth functions (in the sense of [1.20]). However, any $\mathcal{A}$-vector space E can be interpreted as the even part of an $\mathcal{A}$-vector space $E^\sharp$. Before we give the definition of $E^\sharp$, let us have a quick look at what we need. A vector $x \in E$ is even only if it has homogeneous coordinates x^i of the correct parity with respect to a homogeneous basis $(e_i)_{i=1}^n$. But a generic vector does not have homogeneous coordinates. On the other hand, if we split the coordinates x^i into their homogeneous components $x^i = (x^i)_0 + (x^i)_1$, we get homogeneous coordinates, but twice as many and as many even as odd ones. The $\mathcal{A}$-vector space $E^\sharp$ thus should have twice the dimension of E and with the same number of even and odd basis vectors.

To prepare the definition of $E^\sharp$, we recall the parity shift operation introduced in [I.6.13]. For any $\alpha \in \mathbf{Z}_2$ we define the $\mathcal{A}$-vector space $E^{\flat\alpha}$ in the following way: as a *left* $\mathcal{A}$-module it is the same as E, but the $\mathbf{Z}_2$-grading is given by $(E^{\flat\alpha})_\beta = E_{\beta-\alpha}$. Since $\mathbf{Z}_2$ has only two elements, we have $E^{\flat 0}$, which is obviously the same as E, and $E^{\flat 1}$, which has its parities reversed with respect to E : $(E^{\flat 1})_0 = E_1$ and $(E^{\flat 1})_1 = E_0$. In the literature the operation $^{\flat 1}$ is also called the parity reversal operation. As a *right* $\mathcal{A}$-module $E^{\flat 1}$ is certainly not the same as E, but the (set theoretical) identity map $\widehat{id} : E^{\flat 1} \to E$ is an odd left linear bijection. Using this bijection, we can transport the structure of an $\mathcal{A}$-vector space from E to $E^{\flat 1}$: if $(e_i)_{i=1}^n$ is a basis in the prescribed equivalence class for E, the vectors $\overline{e}_i = \widehat{id}^{-1}(e_i)$ define an equivalence class of bases on $E^{\flat 1}$, independent of the choice of the basis (e_i) in its equivalence class. With these preparations, we define the $\mathcal{A}$-vector space $E^\sharp$ as $E^\sharp = E \oplus E^{\flat 1}$. And then indeed $E_0^\sharp = E_0 \oplus E_0^{\flat 1} = E_0 \oplus E_1 \cong E$, but the last identification is not an identification of $\mathcal{A}$-vector spaces.

In order to get a better idea of what this space $E^\sharp$ is, we define the (left linear) projection $\pi : E^\sharp \to E$ as $\pi = id + \widehat{id}$. We also define the map $\varphi : E \to E^\sharp$ as $\varphi : x \mapsto id(x_0) \oplus \widehat{id}^{-1}(x_1)$, where $x = x_0 + x_1$ is the decomposition into homogeneous parts of $x \in E$. If $(e_i)_{i=1}^n$ is a homogeneous basis of E, then the vectors $\overline{e}_i = \widehat{id}^{-1}(e_i)$ form a basis for $E^{\flat 1}$ (see above) and thus $(e_i, \overline{e}_i)_{i=1}^n$ is a homogeneous basis of $E^\sharp$. Since $\varepsilon(\overline{e}_i) = 1 - \varepsilon(e_i)$, the dimension of $E^\sharp$ is $n|n$. In terms of these bases, the maps π and φ take the form

$$\pi : y = \sum_i (y^i e_i + \overline{y}^i \,\overline{e}_i) \mapsto \sum_i (y^i + \overline{y}^i) e_i$$

$$\varphi : x = \sum_i x^i \cdot e_i \mapsto \sum_i \big((x^i)_{\varepsilon(e_i)} \cdot e_i + (x^i)_{1-\varepsilon(e_i)} \cdot \overline{e}_i \big) .$$

Since $y \in E^\sharp$ is even if and only if we have $\varepsilon(y^i) = \varepsilon(e_i)$ and $\varepsilon(\overline{y}^i) = \varepsilon(\overline{e}_i) = 1 - \varepsilon(e_i)$, it follows easily that φ is injective onto $E_0^\sharp$ and that its inverse is given by the restriction of π to $E_0^\sharp$. We leave it to the reader to verify that $\varphi : E \to E_0^\sharp$ is a homeomorphism, linear over $\mathcal{A}_0$, but certainly *not* linear over $\mathcal{A}_1$.

It is with this φ that we identify E as the even part of the $\mathcal{A}$-vector space $E^\sharp$. In terms of coordinates, each point $x \in E$ has n "coordinates" $x^i \in \mathcal{A}$, i.e., n even and n odd "coordinates" $(x^i)_0$ and $(x^i)_1$, in perfect accordance with the dimension of $E^\sharp$.

1.27 Discussion. Let $\psi : E \to F$ be a (left or right) linear map between $\mathcal{A}$-vector spaces. Identifying E with $E_0^\sharp$ we can interpret this map as a map $\psi : E_0^\sharp \to F$. We thus have two notions of smoothness: smooth as a linear map $E \to F$ in the sense of [II.6.22] or smooth as a map $E_0^\sharp \to F$ in the sense of [1.20]. We leave it to the reader to verify that these two notions coincide.

Instead of looking at $E^\sharp$, we also could have restricted ψ to E_0. It is a direct corollary of [1.23-a] that if $\psi \in \mathrm{Hom}_S(E; F)$ is a smooth linear map in the sense of [II.6.22], then its restriction to E_0 is smooth in the sense of [1.20], i.e., $\psi|_{E_0} \in C^\infty(E_0; F)$. Moreover, if $\psi \in \mathrm{Hom}_S(E;F)_\alpha$ is smooth *and* homogeneous of parity α, then $\psi|_{E_0} \in C^\infty(E_0;F)_\alpha$ is also homogeneous of the same parity.

On the other hand, for a linear map $\psi \in \mathrm{Hom}_L(E;F)$, knowledge of the restriction $\psi|_{E_0}$ does not necessarily imply knowledge of ψ itself. For instance, the linear map ϕ defined in [II.5.8] satisfies the condition $\phi|_{E_0} = 0$, which is smooth. However, ϕ itself is neither smooth nor the zero map. However, as we will show in [2.31], if ψ is a member of a *smooth* family of linear maps and if $\mathcal{A}$ satisfies a condition we will need anyway, then ψ is completely determined by its restriction to E_0.

2. The structure of smooth functions

In this section we analyze in more detail the structure of smooth functions. More precisely, if U is an open set in E_0 with E an $\mathcal{A}$-vector space of dimension $p|q$, we show that $C^\infty(U;F)$ is isomorphic as graded commutative $\mathbf{R}$-algebra with $C^\infty(O;\mathbf{B}F) \otimes \bigwedge \mathbf{R}^q$, where $O = \mathbf{B}U$ is open in $\mathbf{R}^p$. For this result to be true, the ring $\mathcal{A}$ has to be big enough in the sense that there should exist q odd elements whose product is non-zero.

2.1 Lemma. *Let $f \in C^\infty(U;F)$, $x_o \in U$, and $1 \le j \le n$ be arbitrary. Then there exist a decreasing sequence of open sets $x_o \in U_k \subset U$ and functions $f^{(k)} \in C^\infty(U_k;F)$ with the following property. If $N \in \mathbf{N}$ and $a \in \mathcal{A}_{\varepsilon(e_j)}$ are such that $a^N = 0$, then for all $x \in U_N$:*

$$f(x^1,\ldots,x^{j-1},x^j + a, x^{j+1},\ldots,x^n) = \sum_{k=0}^{\infty} a^k \cdot f^{(k)}(x) \ .$$

Proof. We introduce the $\mathcal{A}$-vector space $\widehat{E} = E \oplus \mathrm{Span}(e_j)$, i.e., adding one more coordinate to E of the same parity as e_j. By $\pi : \widehat{E} \to E$ we denote the canonical projection, by $d : E \to \widehat{E}$ the diagonal injection $d(x) = (x, x^j)$, and by $D : \widehat{E} \to \widehat{E}$ the map $D(y) = (y^1,\ldots,y^n,y^j)$. We will say that a subset $\widehat{U} \subset \widehat{E}$ is diagonal if $y \in \widehat{U}$ implies $D(y) \in \widehat{U}$.

Now suppose that $g^{(k)}$ is a smooth function on $\widehat{U}_k$, where $\widehat{U}_k$ is a diagonal open subset of $\widehat{E}$ containing $d(x_o)$. Then there exist an open subset $d(x_o) \in \widehat{U}_{k+1} \subset \widehat{U}_k$ and smooth functions χ_i^{k+1} on $\widehat{U}_{k+1}^2$ such that for $y, z \in \widehat{U}_{k+1}$:

$$g^{(k)}(y) = g^{(k)}(z) + \sum_{i=1}^{n+1} (y-z)^i \cdot \chi_i^{k+1}(y,z) \ .$$

Using [1.4-i] and shrinking $\widehat{U}_{k+1}$ if necessary, we may assume that $\widehat{U}_{k+1}$ is diagonal. We thus can write for all $y \in \widehat{U}_{k+1}$:

$$g^{(k)}(y) = g^{(k)}(D(y)) + (y^{n+1} - y^j) \cdot \chi_{n+1}^{k+1}(y, D(y)) \ .$$

Since the set of $2n+2$ coordinates of $(y, D(y))$ only contains $n+1$ different entries, namely those of y, we can apply [1.23-g,i] to deduce that the function $g^{(k+1)}(y) = \chi_{n+1}^{k+1}(y, D(y))$ is smooth on $\widehat{U}_{k+1}$.

Starting from a $g^{(0)}$ and an open diagonal set $\widehat{U}_0$, both to be defined later, we thus create by induction a decreasing sequence of open diagonal sets $\widehat{U}_k$ and smooth functions $g^{(k)}$ on them such that for all $y \in \widehat{U}_N$:

$$g^{(0)}(y) = \sum_{k=0}^{N-1} (y^{n+1} - y^j)^k \cdot g^{(k)}(D(y)) + (y^{n+1} - y^j)^N \cdot g^N(y) . \tag{2.2}$$

To get us started, note that, again by [1.4-i], there exists an open diagonal set $\widehat{U}_0 \subset \widehat{E}$ such that $g^{(0)}(y) = f(y^1, \ldots, y^{j-1}, y^{n+1}, y^{j+1}, \ldots y^n)$ is a well defined smooth function on it. We finally define $U_k = \pi(\widehat{U}_k)$ and $f^{(k)}(x) = g^{(k)}(d(x))$, which is a well defined smooth function because $\widehat{U}_k$ is diagonal. Since open sets are saturated with nilpotent vectors, if $x \in U_N$, then not only $d(x) \in \widehat{U}_N$, but also $y = (x, x^j + a) \in \widehat{U}_N$. The result then follows immediately from (2.2) and the fact that $f^{(k)}(x) = g^{(k)}(D(y))$. $\boxed{QED}$

2.3 Remark. The above result is a kind of Taylor expansion in a single variable but with a nilpotent increment. As such, the functions $f^{(k)}$ look, up to a factor $k!$, like higher order (partial) derivatives. In [2.27] we will improve this result by showing that there exist functions $f^{(k)}$ on the whole of U, independent of the point x_o. In [3.7] we will show that, under suitable conditions on $\mathcal{A}$, the $(k!) \cdot f^{(k)}$ are indeed the higher order (partial) derivatives of f.

2.4 Notation. The result of [2.1] is particularly useful when e_j is odd, i.e., when the coordinate x^j is itself nilpotent (of order 2). It is often useful to make the distinction between coordinates with respect to even and odd basis vectors visually more explicit. This will be done by introducing the following notational conventions. In the first place, general elements of an $\mathcal{A}$-vector space E will be denoted by lower case roman letters (as we did always). Coordinates with respect to even basis vectors will be denoted by the same lower case roman letter with a superscript, and the coordinates with respect to odd basis vectors will be denoted by the corresponding greek letter with a superscript. More precisely, we will usually assume that our homogeneous basis of E is an ordered one, and then we write

$$x = \sum_{i=1}^{p} x^i \cdot e_i + \sum_{i=1}^{q} \xi^i \cdot e_{p+i} \cong (x^1, \ldots, x^p, \xi^1, \ldots, \xi^q) .$$

In particular, $x \in E_0$ if and only if the coordinates $x^i \in \mathcal{A}_0$ and $\xi^i \in \mathcal{A}_1$. We will call the x^i the *even coordinates* and the ξ^j the *odd coordinates* of the point $x \in E_0$ (for which $(\xi^i)^2 = 0$).

2.5 Proposition. *For any* $f \in C^\infty(U;F)$ *there exist open subsets* $U_a \subset U$ *with* $\cup_a U_a = U$ *and functions* $g_{a,i_1\dots i_k} \in C^\infty(U_a;F)$, $0 \le k \le q$ *and* $1 \le i_1 < i_2 < \cdots < i_k \le q$ *such that for* $x \in U_a$ *:*

$$f(x^1,\dots,x^p,\xi^1,\dots,\xi^q) = \sum_{k=0}^{q} \sum_{1\le i_1<\dots<i_k\le q} \xi^{i_1}\cdots\xi^{i_k}\cdot g_{a,i_1\dots i_k}(x^1,\dots,x^p,0,\dots,0)\ .$$

Proof. A direct consequence of [2.1] using induction on the number of odd coordinates.

QED

2.6 Discussion. Up till now, in our analysis of smooth functions, we only used property (A1), never property (A2). But [2.5] is as far as one can reasonably go using only (A1). Using [1.23-f,g] we can ignore the zero's in the functions $g_{a,i_1\dots i_k}$ of [2.5], and thus we are led to the study of smooth functions that depend only on even variables. Using (A2) we will show that these are in one to one correspondence with ordinary smooth functions of real variables.

2.7 Definition. If $f \in C^\infty(U;F)$ is a smooth function, then by (A2) it follows that $f(\mathbf{B}U) \subset \mathbf{B}F$. We thus can define a function $\mathbf{B}f : \mathbf{B}U \to \mathbf{B}F$ by $\mathbf{B}f = f|_{\mathbf{B}U}$. But then the fact that f is smooth and [1.21] show that $\mathbf{B}f$ is an ordinary smooth function on $\mathbf{B}U \subset \mathbf{B}E_0 \cong \mathbf{R}^p$. Using [2.1] it follows that $f(x) - f(\mathbf{B}x)$ is a nilpotent vector, and thus that $\mathbf{B}(f(x)) = f(\mathbf{B}x)$. We thus can characterize the function $\mathbf{B}f \in C^\infty(\mathbf{B}U;\mathbf{B}F)$ as the unique function which makes the diagram

$$\begin{array}{ccc} U & \xrightarrow{\ f\ } & F \\ {\scriptstyle\mathbf{B}}\big\downarrow & & \big\downarrow{\scriptstyle\mathbf{B}} \\ \mathbf{B}U & \xrightarrow[\mathbf{B}f=f|_{\mathbf{B}U}]{} & \mathbf{B}F \end{array}$$

commutative (uniqueness follows from the surjectivity of $\mathbf{B} : U \to \mathbf{B}U$ because U is open).

2.8 Lemma. *The map* $\mathbf{B} : C^\infty(U;\mathcal{A}) \to C^\infty(\mathbf{B}U;\mathbf{R})$ *is a morphism of graded commutative* $\mathbf{R}$*-algebras with unit. For any* $\mathcal{A}$*-vector space* F, $\mathbf{B} : C^\infty(U;F) \to C^\infty(\mathbf{B}U;\mathbf{B}F)$ *is a morphism from graded modules over* $C^\infty(U;\mathcal{A})$ *to graded modules over* $C^\infty(\mathbf{B}U;\mathbf{R})$.

2.9 Definition. Let E be an $\mathcal{A}$-vector space and $\{e_1,\dots,e_p,f_1,\dots,f_q\}$ an ordered homogeneous basis such that the e_i are even and the f_i odd. Any other *ordered homogeneous* basis (in the equivalence class) is related to this basis by a matrix with real entries of the form $\left(\begin{smallmatrix} a & 0 \\ 0 & b \end{smallmatrix}\right)$. It follows that the graded subspace $E^{w.o.d} = \text{Span}(\{e_1, \ldots, e_p\})$ is independent of the choice of a homogeneous basis (in its equivalence class!). We thus can associate to an $\mathcal{A}$-vector space E two graded "subspaces" : its body $\mathbf{B}E$, and $E^{w.o.d}$, called *the part of E without odd dimensions*. (Nota Bene. A point in $E^{w.o.d}$ can have odd coordinates.) The superscripts $^{w.o.d}$ (*without odd dimension*) will be applied more generally to any subset A of an $\mathcal{A}$-vector space E, and will be defined as

$$A^{w.o.d} = A \cap E^{w.o.d} \ .$$

Note that the definition of $E^{w.o.d}$ implies that $\mathbf{B}(E^{w.o.d}) = \mathbf{B}E_0 = (\mathbf{B}E)^{w.o.d}$.

An $\mathcal{A}$-vector space E satisfies $E = E^{w.o.d}$ if and only if it admits a homogeneous basis without odd elements; such an $\mathcal{A}$-vector space will naturally be called *an $\mathcal{A}$-vector space without odd dimensions*.

2.10 Discussion/Definition. In §II.6 we have seen that the $\mathbf{R}$-vector space $\mathbf{B}E$ completely determines the $\mathcal{A}$-vector space E. In [1.1] we have argued that the same is not true for the even part: $\mathbf{B}E_0$ does not determine E_0. If we now insist that we want to reconstruct an open subset $U \subset E_0$ given the open subset $\mathbf{B}U \subset \mathbf{B}E_0$ in the even part of the $\mathbf{R}$-vector space $X = \mathbf{B}E$, the best we can do is to suppose that $X_1 = \{0\}$, since we have no data concerning X_1.

Let us formalize this idea by means of a map $\mathbf{G}$ that applies to open subsets O of the even part of an $\mathbf{R}$-vector space X. This $\mathbf{G}$ can be seen as an extension of the map $\mathbf{G}$ defined on $\mathcal{A}$-vector spaces in [II.6.29], although it is not a very straightforward extension. We define $\mathbf{G}O$ as the unique open subset of the $\mathcal{A}$-vector space *without odd dimensions* $E = \mathbf{G}(X_0 \oplus \{0\}_1)$ such that $\mathbf{B}(\mathbf{G}O) = O$. In other words, $\mathbf{G}O = \mathbf{B}^{-1}(O)$, with $\mathbf{B} : E \to \mathbf{B}(\mathbf{G}(X_0 \oplus \{0\}_1)) = X_0$.

Now suppose that U is an open subset in the even part of an arbitrary $\mathcal{A}$-vector space E, what can we say about $\mathbf{G}(\mathbf{B}U)$? The set $\mathbf{B}U$ is open in $\mathbf{B}E_0$ and hence $\mathbf{G}(\mathbf{B}U)$ is open in $(\mathbf{G}(\mathbf{B}E_0 \oplus \{0\}_1))_0$. But this is just $E^{w.o.d}$. We conclude that $\mathbf{G}(\mathbf{B}U) = U \cap E^{w.o.d} \equiv U^{w.o.d}$.

2.11 Construction. Let E be an $\mathcal{A}$-vector space without odd dimensions, U an open subset of E_0 and $f \in C^\infty(\mathbf{B}U; \mathbf{B}F)$. By definition of open sets, $\mathbf{B}U = U \cap \mathbf{B}E$, and thus every point $x \in U$ decomposes as $x = \mathbf{B}x + n$ with $\mathbf{B}x \in \mathbf{B}U$ and $n \in \mathcal{N}_E$. We now define the function $\mathbf{G}f : U \to F$ by the formula

$$(\mathbf{G}f)(x) = \sum_{k=0}^{\infty} \frac{1}{k!} \big((D^k f)(\mathbf{B}x)\big)(n, \ldots, n) \ . \tag{2.12}$$

In this formula $D^k f$ is the k-th order derivative of f (i.e., the "matrix" of all k-th order partial derivatives) seen as a map from $\mathbf{B}U$ to the space of k-linear symmetric maps from $\mathbf{B}E$ to $\mathbf{B}F \subset F$. Applying such a map to k vectors $n \in E_0$ makes sense because all the coordinates of n are even ($E = E^{w.o.d}$ and $U \subset E_0$) and thus commute. Finally the infinite sum is actually a finite sum because the coordinates of n are nilpotent.

2.13 Remark. Let E be an $\mathcal{A}$-vector space without odd dimensions, then $\mathbf{B}E_0 = \mathbf{B}E$. Hence, if $f \in C^\infty(\mathbf{B}E_0, \mathbf{B}F)$ is linear, then the map $\mathbf{G}f : E_0 \to F$ constructed in [2.11] is the same as the restriction to E_0 of the map $\mathbf{G}f : E \to F$ constructed in [II.6.30].

2.14 Lemma. *For $f \in C^\infty(\mathbf{B}U; \mathbf{R})$ and $g, h \in C^\infty(\mathbf{B}U; \mathbf{B}F)$ we have the equalities $\mathbf{G}(g + h) = (\mathbf{G}g) + (\mathbf{G}h)$ and $\mathbf{G}(f \cdot g) = (\mathbf{G}f) \cdot (\mathbf{G}g)$.*

Proof. The first equality is obvious, the second follows from the formula

$$(D^k(f \cdot g)(x))(v, \ldots, v) = \sum_{\ell=0}^{k} \binom{k}{\ell} (D^\ell f(x))(v, \ldots, v) \cdot (D^{k-\ell} g(x))(v, \ldots, v) ,$$

valid for ordinary C^∞-functions. QED

2.15 Lemma. *If E is an $\mathcal{A}$-vector space without odd dimensions and $f \in C^\infty(\mathbf{B}U; \mathbf{B}F)$, then $\mathbf{G}f \in C^\infty(U; F)$ and $\mathbf{B}(\mathbf{G}f) = f$. Moreover, if f is homogeneous of degree α, then so is $\mathbf{G}f$.*

Proof. With the usual abuse of notation (writing the same symbol for the coordinate and the coordinate function), we denote by x^i the i-th coordinate *function* on E and by $(\mathbf{B}x)^i$ the i-th coordinate *function* on $\mathbf{B}E$. It is straightforward to show that $\mathbf{B}(x^i) = (\mathbf{B}x)^i$ *as functions* and that $\mathbf{G}(\mathbf{B}x^i) = x^i$.

Since f is smooth, there exist local functions $\phi_{a,i} \in C^\infty((\mathbf{B}U_a)^2; \mathbf{B}F)$ such that

$$f(\mathbf{B}y) - f(\mathbf{B}x) = \sum_i (\mathbf{B}y^i - \mathbf{B}x^i) \cdot \phi_{a,i}(\mathbf{B}y, \mathbf{B}x) .$$

Note that by definition of the DeWitt topology we have $U_a = \mathbf{B}^{-1}(\mathbf{B}U_a)$, justifying the definition of the domain of definition of $\phi_{a,i}$ in terms of open subsets on E. Applying $\mathbf{G}$ to both sides of the defining equality for $\phi_{a,i}$ (seeing them as functions on $(\mathbf{B}U_a)^2$) and using [2.14] gives us

$$(\mathbf{G}f)(y) - (\mathbf{G}f)(x) = \sum_i (y^i - x^i) \cdot (\mathbf{G}\phi_{a,i})(y, x) .$$

The recursion argument (see the proof of [1.23]) then proves that $\mathbf{G}f \in C^\infty(U; F)$. The second assertion is immediate because $\big((D^k f)(\mathbf{B}x)\big)(n, \ldots, n)$ is nilpotent for $k > 0$.

For the homogeneity statement, note that f is homogeneous of degree α if and only if f takes values in $\mathbf{B}F_\alpha$. Since the n in [2.11] are even, the result $((D^k f)(\mathbf{B}x))(n, \ldots, n)$ remains in F_α. QED

2.16 Proposition. *Let E be an $\mathcal{A}$-vector space without odd dimensions and U an open subset of E_0. Then $\mathbf{B} : C^\infty(U; F) \to C^\infty(\mathbf{B}U, \mathbf{B}F)$ is an isomorphism.*

Proof. Surjectivity is assured by [2.15]. To prove injectivity, let $f \in C^\infty(U; F)$ be such that $\mathbf{B}f \equiv 0$, i.e., $f|_{\mathbf{B}U}$ is identically zero, and let $x_o \in \mathbf{B}U \subset U$. Now look at the proof of [2.1] and suppose that $g^{(k)}$ is zero on $\mathbf{B}\widehat{U}_k$. It immediately follows that $g^{(k+1)}$ is zero on $\mathbf{B}\widehat{U}_{k+1}$, provided $\mathbf{B}y^{n+1} \neq \mathbf{B}y^j$. By continuity of $g^{(k+1)}$ and the fact that it takes *real* values on $\mathbf{B}\widehat{U}_{k+1}$, it follows that $g^{(k+1)}$ is zero on the whole of $\mathbf{B}\widehat{U}_{k+1}$. This last conclusion would not have been justified, had we not known that $g^{(k+1)}$ takes real values on $\mathbf{B}\widehat{U}_{k+1}$: the non-Hausdorff character allows for non-unique limits! We conclude that $f^{(k)}|_{\mathbf{B}U_k} \equiv 0$ for all k. If there were only one even coordinate, the expansion of [2.1] would prove that $f \equiv 0$. For more than one even coordinate we apply induction on their number n. We fix $x^n = r^n + a^n$ and apply the expansion of [2.1] with $j = n$. By the previous argument we have $\mathbf{B}f^{(k)} \equiv 0$. Since in the expansion of [2.1] the functions $f^{(k)}$ only appear with the real valued n-th coordinate r^n, we may by [1.23-g] interpret the $f^{(k)}$ as smooth functions of $n-1$ coordinates. By the (unstated) induction hypothesis we conclude that $f^{(k)} \equiv 0$, and thus $f \equiv 0$. $\boxed{QED}$

2.17 Remark. In §II.6 we showed that $\mathbf{B}$ is an isomorphism when applied to smooth *linear* maps. Here we show that $\mathbf{B}$ is an isomorphism when applied to arbitrary smooth functions on an open subset of the even part of an $\mathcal{A}$-vector space without odd dimensions. Said differently, as long as the domain of definition only contains *even* coordinates, there is no difference between the general case and the special case $\mathcal{A} = \mathbf{R}$. The restriction to $\mathcal{A}$-vector spaces *without odd dimensions* is necessary because we take as domain of definition *not* (an open subset of) an $\mathcal{A}$-vector space, but only (an open subset of) its even part. The same phenomenon is also reflected in the fact that in reconstructing U from $\mathbf{B}U$ by means of $\mathbf{G}$ we only recover the part without odd dimensions, i.e., $\mathbf{G}(\mathbf{B}U) = U$ only if E has no odd dimensions.

2.18 Notation. Up until now we have carefully distinguished an ordinary real valued smooth function of p real variables and its corresponding smooth function $\mathbf{G}f$ of p even variables. However, [2.16] has shown that there is no real distinction between these two objects. Therefore we will in the future use the same symbol for both objects. We thus might start with an ordinary real valued smooth function f and use even coordinates as arguments, meaning that we take $\mathbf{G}f$. Or we might start with a smooth function of even variables, plug in real coordinates (a point in $\mathbf{B}E$) and claim that we have an ordinary real valued function, this being justified by either taking $\mathbf{B}f$ or using (A2).

2.19 Discussion. Let E be an $\mathcal{A}$-vector space of dimension $p|q$ and let U be an open subset of E_0. Combining [2.5] and [2.16], we see that we have a nearly complete control over the elements of $C^\infty(U; F)$. The functions $g_{a,i_1\dots i_k} \in C^\infty(U_a; F)$ are only used with all odd coordinates being zero. This means that we only need their restrictions to $U_a^{w.o.d}$, but (by [2.16]) these are uniquely determined by ordinary real valued smooth functions $f_{a,i_1\dots i_k} \in C^\infty(\mathbf{B}U_a; \mathbf{B}F)$ as $g_{a,i_1\dots i_k}(x^1, \dots, x^p, 0, \dots, 0) = f_{a,i_1\dots i_k}(x^1, \dots, x^p)$ (remember, we do not use the symbol $\mathbf{G}$ any longer). However, our control is not complete because (i) this description of f is only local on U_a, not on U, and (ii) it is not clear whether the function f determines the local functions $f_{a,i_1\dots i_k}$ uniquely. In order to make our control complete, we need to formulate conditions on $\mathcal{A}$, conditions that will also be useful in defining the partial derivatives using the diagonal of the functions $\phi_{a,i}$ as in (1.11).

2.20 Definition. Let $\mathcal{A}$ be a graded commutative $\mathbf{R}$-algebra and let $n \in \mathbf{N}^*$ be a non-zero natural number. We say that $\mathcal{A}$ satisfies (C [n]) if

$$\text{(C [}n\text{])} \qquad \exists a_1, \dots, a_n \in \mathcal{A}_1 \quad : \quad a_1 \cdots a_n \neq 0 \,.$$

We will say that $\mathcal{A}$ satisfies (C [∞]) if it satisfies (C [n]) for all $n \in \mathbf{N}^*$. Obviously (C [∞]) and (C [$n{+}1$]) imply (C [n]). In the standard example $\mathcal{A} = \bigwedge X$ with X a vector space over $\mathbf{R}$, (C [n]) is satisfied if and only if $\dim X \geq n$, a condition valid also for $n = \infty$.

2.21 Proposition. *Let E be an $\mathcal{A}$-vector space of dimension $p|q$, let f be in $C^\infty(U; F)$, and let $f_{i_1\dots i_k} \in C^\infty(\mathbf{B}U, \mathbf{B}F)$ be such that $f(x) = \sum_k \sum_{i_j} \xi^{i_1} \cdots \xi^{i_k} \cdot f_{i_1\dots i_k}(x^1, \dots, x^p)$. Then the functions $f_{i_1\dots i_k}$ are unique if and only if $\mathcal{A}$ satisfies* (C [k]).

Proof. We start by observing that the separate additive terms $\xi^{i_1} \cdots \xi^{i_k} \cdot f_{i_1\dots i_k}(x^1, \dots, x^p)$ are always uniquely determined by f : a simple induction argument with respect to k and taking all coordinates $\xi^j = 0$ whenever $j \notin \{i_1, \dots, i_k\}$ suffices. Now suppose $\xi^{i_1} \cdots \xi^{i_k} \cdot f_{i_1\dots i_k}(x^1, \dots, x^p)$ to be identically zero and take the coordinates $x^1, \dots, x^p$ to be real. It follows that $\xi^{i_1} \cdots \xi^{i_k} \cdot f_{i_1\dots i_k}(x^1, \dots, x^p) = 0$, which implies that either $\xi^{i_1} \cdots \xi^{i_k}$ or $f_{i_1\dots i_k}(x^1, \dots, x^p)$ is zero (because the latter is real, and thus can be inverted if non-zero). If (C [k]) is satisfied, we can choose $\xi^{i_1} \cdots \xi^{i_k}$ to be non-zero, implying that $f_{i_1\dots i_k}$ must be zero. On the other hand, if (C [k]) is not satisfied, the term $\xi^{i_1} \cdots \xi^{i_k}$ is always zero, leaving the function $f_{i_1\dots i_k}$ completely undetermined. QED

2.22 Corollary. *Let E be an $\mathcal{A}$-vector space of dimension $p|q$ and let $U \subset E_0$ be open. Then $C^\infty(U; F)$ is in bijection with $C^\infty(\mathbf{B}U; \mathbf{B}F)^d$, i.e., with d copies of $C^\infty(\mathbf{B}U; \mathbf{B}F)$, where d is determined as follows.*

(i) *If $\mathcal{A}_1 = \{0\}$, then $d = 1$.*
(ii) *If $1 \leq \ell < q$ is such that $\mathcal{A}$ satisfies* (C [ℓ]) *but not* (C [$\ell{+}1$]), *then $d = \sum_{j=0}^{\ell} \binom{q}{j}$.*
(iii) *If $\mathcal{A}$ satisfies* (C [q]), *then $d = 2^q$.*

Proof. • If $\mathcal{A}_1 = \{0\}$, it follows from [2.5] that $f(x,\xi) = g_{a,\emptyset}(x, 0)$, which proves that $f(x,\xi) = f(x, 0)$, i.e., f is a function of the even coordinates only. And then the conclusion follows from [2.16].

• Let us now suppose that $\mathcal{A}$ satisfies (C [k]). From [2.5] we obtain $\binom{q}{k}$ local smooth functions $g_{a,i_1\dots i_k}$ ($\binom{q}{k}$ = the number of increasing sequences $1 \leq i_1 < \cdots < i_k \leq q$). Since $f_{a,i_1\dots i_k}(x^1, \dots, x^p) = g_{a,i_1\dots i_k}(x^1, \dots, x^p, 0, \dots, 0)$ depends only upon even coordinates, it follows that $f_{a,i_1\dots i_k} \in C^\infty(\mathbf{B}U; \mathbf{B}F)$. But then we can combine [2.5] and

[2.21] to conclude that the functions $f_{a,i_1\dots i_k} \in C^\infty(\mathbf{B}U;\mathbf{B}F)$ are uniquely determined by f. Hence they must coincide on overlaps $U_a \cap U_b$. It thus follows from [1.23-k] that we have global functions $f_{i_1\dots i_k} \in C^\infty(\mathbf{B}U;\mathbf{B}F)$ such that $f_{a,i_1\dots i_k} = f_{i_1\dots i_k}|_{U_a}$. Since (C [$k$]) implies (C [$k-1$]), the same conclusion holds for all lower order functions. On the other hand, if $\mathcal{A}$ does not satisfy (C [k]), the terms $\xi^{i_1}\cdots\xi^{i_k}\cdot g_{a,i_1\dots i_k}(x,0)$ do not contribute to f. The conclusions of (ii) and (iii) follow. $\boxed{QED}$

2.23 Discussion. With [2.22] we have complete control over smooth functions. If the $\mathcal{A}$-vector space E has dimension $p|q$ and if $U \subset E_0$ is open, then there is a one to one correspondence between smooth functions $f \in C^\infty(U,F)$ and collections of functions $f_{i_1\dots i_k} \in C^\infty(\mathbf{B}U,\mathbf{B}F)$ given by

(2.24)
$$f(x^1,\dots,x^p,\xi^1,\dots,\xi^q) = \sum_{k=0}^{\min(q,\ell)} \sum_{1\le i_1<\cdots<i_k\le q} \xi^{i_1}\cdots\xi^{i_k}\cdot f_{i_1\dots i_k}(x^1,\dots,x^p)\ ,$$

where ℓ is the maximal value such that $\mathcal{A}$ satisfies (C [ℓ]) (if $\mathcal{A}_1 = \{0\}$ we take $\ell = 0$).

2.25 Remark. If one takes into account how the functions $f_{i_1\dots i_k}$ in (2.24) behave under a change of the odd basis vectors, then it is not hard to show that $C^\infty(U;F)$ is bijective with

$$C^\infty(U;F) \cong C^\infty(\mathbf{B}U;\mathbf{B}F) \otimes \Big(\bigoplus_{k=0}^{\min(q,\ell)} \textstyle\bigwedge^k(\mathbf{R}^q)\Big)\ .$$

It is not surprising that the exterior power $\bigwedge^k(\mathbf{R}^q)$ appears because a k-fold product of the q odd variables $\xi^1,\dots,\xi^q$ is k-linear and skew-symmetric in these variables. If $\ell \ge q$ we have the identification $C^\infty(U;F) \cong C^\infty(\mathbf{B}U;\mathbf{B}F)\otimes\bigwedge\mathbf{R}^q$, which is sometimes taken as definition of smooth functions of p even variables and q odd variables. Since starting in [3.3] we will assume that $\mathcal{A}$ satisfies (C [∞]) (and thus $\infty = \ell \ge q$), we obtain this identification as a consequence of a more basic definition of smooth functions.

2.26 Examples. • According to [II.1.11] and [I.2.9], the group $\mathrm{Aut}(E)$ is an open subset of $\mathrm{End}_S(E)_0$. We claim that taking inverses $Inv : \mathrm{Aut}(E) \to \mathrm{End}_S(E)_0$ is smooth (see also [VI.1.6]). To prove this, it suffices to show that Inv is of the form (2.24) when using left-coordinates ${}^\bullet M_S$ on $\mathrm{Aut}(E)$. On the other hand, calculations are much simpler when using the coordinates $\overset{\bullet}{M}_S$ because these preserve composition (see [II.4.2]), and in particular $\overset{\bullet}{M}_S(\phi^{-1}) = (\overset{\bullet}{M}_S(\phi))^{-1}$. Writing $\overset{\bullet}{M}_S(\phi) = \left(\begin{smallmatrix} A & B\\ C & D\end{smallmatrix}\right)$, A and D are invertible matrices with even entries, and B and C have odd entries. We thus can write :

$$\begin{aligned}\begin{pmatrix} A & B\\ C & D\end{pmatrix}^{-1} &= \begin{pmatrix} A^{-1} & \mathbf{0}\\ \mathbf{0} & D^{-1}\end{pmatrix}\cdot\left[I_{p+q} - \begin{pmatrix} \mathbf{0} & -BD^{-1}\\ -CA^{-1} & \mathbf{0}\end{pmatrix}\right]^{-1}\\ &= \begin{pmatrix} A^{-1} & \mathbf{0}\\ \mathbf{0} & D^{-1}\end{pmatrix}\cdot\sum_{k=0}^{\infty}\begin{pmatrix} \mathbf{0} & -BD^{-1}\\ -CA^{-1} & \mathbf{0}\end{pmatrix}^k\ ,\end{aligned}$$

where the infinite sum is actually finite because B and C are nilpotent (it breaks off after the $2pq$th term). We now note that (fortunately) the left coordinates ${}^{\bullet}M_S(\phi)$ only differ by a sign from the coordinates $\overset{\bullet}{M}_S(\phi)$ (see [II.4.1]). We thus are allowed to use the entries of A, B, C, and D as coordinates on $\mathrm{Aut}(E)$.

Since taking the inverse of a matrix with real entries is smooth, the maps $A \mapsto A^{-1}$ and $D \mapsto D^{-1}$ are smooth in the even coordinates of A and D. It thus follows that the matrix entries of $\overset{\bullet}{M}_S(\phi)^{-1}$ are smooth functions of the even coordinates and polynomials in the odd coordinates. In other words, $Inv : \mathrm{Aut}(E) \to \mathrm{Aut}(E) \subset \mathrm{End}_S(E)_0$ is smooth.

• The ordinary determinant function is a polynomial in its entries. It thus follows immediately from the definition of the graded determinant and the arguments given in the previous example that $\mathrm{gDet} : \mathrm{Aut}(E) \to \mathcal{A}_0$ is smooth.

2.27 Proposition. *Let $f \in C^\infty(U; F)$ and $1 \leq j \leq n$ be arbitrary. Then there exist functions $f^{(k)} \in C^\infty(U; F)$, $f^{(0)} = f$ such that $\forall x \in U$, $\forall a \in \mathcal{N} \cap \mathcal{A}_{\varepsilon(e_j)}$:*

$$f(x^1, \ldots, x^{j-1}, x^j + a, x^{j+1}, \ldots, x^n) = \sum_{k=0}^{\infty} a^k \cdot f^{(k)}(x) \,. \tag{2.28}$$

Proof. We will define the $f^{(k)}$ by means of (2.24) and [2.16]. If $\varepsilon(e_j) = 1$, i.e., if the j-th coordinate is odd, we define $f^{(1)}_{i_1 \ldots i_k} = 0$ whenever $j \in \{i_1, \ldots, i_k\}$. If $j \notin \{i_1, \ldots, i_k\}$, we define $f^{(1)}_{i_1 \ldots i_k} = \pm f_{j_0 \ldots j_k}$, where $\{j_0, \ldots, j_k\} = \{j, i_1, \ldots, i_k\}$ and where the sign is determined by the position of j within the (increasing) sequence $i_1 \ldots i_k$: if j is in m-th position, the sign is $(-1)^{m-1}$. (In particular $f^{(1)}_\phi = f_j$.) For $k > 1$ we define $f^{(k)} \equiv 0$. The result then follows from (2.24).

If $\varepsilon(e_j) = 0$, we first note that by using [2.21] and (2.24) we can reduce the problem to functions of even coordinates only. For a fixed sequence $i_1, \ldots, i_\ell$ we then define $f^{(k)}_{i_1 \ldots i_\ell}$ by $f^{(k)}_{i_1 \ldots i_\ell} = k! \cdot (\partial_{x^j})^k f_{i_1 \ldots i_\ell}$. This makes sense because functions of even coordinates only can be interpreted as ordinary infinitely often differentiable functions. And then a simple but tedious computation using definition (2.12) completes the proof. *QED*

2.29 Remark. [2.27] is similar to [2.1], but stronger in that the $f^{(k)}$ are defined on the whole of U (and do not depend on the point x_o). We will see in [3.7] that if $\mathcal{A}$ satisfies (C [$q+1$]), then we can interpret the functions $k! \cdot f^{(k)}$ as the k-th order partial derivatives of f with respect to the j-th coordinate. In this way we have a convergent Taylor series expansion for nilpotent increments.

The obvious question which comes to mind when seeing this result is: couldn't we prove [2.27] directly, without the intermediate result [2.1]? The answer is no, and the reason is the presence of a cover in property (A1) of smooth functions. The argument given in [1.12] can be used to show that in the real case this cover is not needed, in which case one could prove [2.27] directly. We did not follow this approach because it would not be valid in a complexified setting. In the approach we used here, we are sure that [2.27] is valid in a complexified setting as well (i.e., a setting in which we consider $\mathcal{A}$ as a graded commutative **C**-algebra such that $\mathcal{A}/\mathcal{N} \cong \mathbf{C}$, see §II.1).

2.30 Discussion. In [1.27] we raised the question whether a linear map is determined by its restriction to the even part of an $\mathcal{A}$-vector space. We now are able to prove that under suitable circumstances the restriction of a smooth family of linear maps to the even part of an $\mathcal{A}$-vector space does indeed determine this family completely.

2.31 Proposition. *Let E, F, and G be $\mathcal{A}$-vector spaces, let $U \subset G_0$ be open, and let $\psi : U \to \mathrm{Hom}_S(E;F)$ be a smooth map. Suppose furthermore that $\mathcal{A}$ satisfies* (C [q+1]) *with q the odd dimension of G. If $\forall g \in U : \psi(g)|_{E_0} = 0$, then $\forall g \in U : \psi(g) = 0$. In other words, $\psi(g)|_{E_0}$ uniquely determines $\psi(g)$.*

Proof. We treat the right linear case, the left linear case being similar. Let (e_i) and (f_j) be bases of E and F respectively (in their equivalence classes!). Since the map $\mathrm{Hom}_R(E;F) \times E \to F$, $(\psi, e) \mapsto \psi(e)$ is smooth, as are the maps f^j, we conclude that the maps $\psi^j : U \times E \to \mathcal{A}$, $(g,e) \mapsto f^j(\psi(g)(e))$ are smooth. Evaluating ψ^j on a basis vector e_i gives us the middle matrix element $\psi^j(g)(e_i) = \overset{\bullet}{M}_R(\psi(g))^j{}_i$. Knowledge of these matrix elements completely determines ψ. If e_i is an even basis vector, our assumption implies that $\psi^j(g)(e_i) = \overset{\bullet}{M}_R(\psi(g))^j{}_i = 0$, i.e., all the corresponding matrix elements are zero. Now suppose that e_i is an odd basis vector. By [1.23-g] the map $\phi : U \times \mathcal{A}_1 \to \mathcal{A}$, $(g, \eta) \mapsto \psi^j(g)(e_i\eta)$ is smooth. By assumption it is identically zero and by linearity it is given by $\phi(g,\eta) = \psi^j(g)(e_i) \cdot \eta$. Since $U \times \mathcal{A}_1$ contains $q+1$ odd coordinates and since $\mathcal{A}$ satisfies (C [q+1]), we deduce from [2.21] that $\psi^j(g)(e_i)$ is zero. We conclude that all matrix elements of $\psi(g)$ are zero, i.e., $\psi(g)$ is zero. QED

3. Derivatives and the inverse function theorem

The first part of this section is devoted to the definition of the (partial) derivative(s) of a smooth function. This requires a condition on $\mathcal{A}$ slightly stronger than the one mentioned in §2. The second part is devoted to the classical theorems of differential calculus: the inverse function theorem, the implicit function theorem, invariance of graded dimension, and the canonical form of a smooth function of maximal rank.

3.1 Proposition. *Let E be an $\mathcal{A}$-vector space of dimension $p|q$, let f belong to $C^\infty(U;F)$, let $\phi_{a,i}$ be as in property* (A1) *and let $g_i \in C^\infty(U;F)$ be defined by $g_i(x) = \phi_{a,i}(x,x)$. For $i \le p$, g_i is uniquely determined by f, independent of the index a; for $i > p$ it is uniquely determined if and only if $\mathcal{A}$ satisfies* (C [q+1]).

Proof. Fix i and let $\phi_{a,i}$ and $\phi_{b,i}$ be two possibilities. Let $x, y \in U_a \cap U_b$ be such that $x^j \neq y^j \Rightarrow j = i$, i.e., x and y differ only in the i-th coordinate. It follows that $(y^i - x^i) \cdot (\phi_{a,i}(y,x) - \phi_{b,i}(y,x)) = 0$. Denoting by h the difference $h = y^i - x^i$, it follows that the function $\phi_{a,i}(y,x) - \phi_{b,i}(y,x)$ is a function $\Delta(x,h)$ of the $p+q+1$ variables (x,h). We thus are given the equality $h \cdot \Delta(x,h) = 0$, while we want to prove that $\Delta(x,0) = 0$. Once we have that, smoothness of g_i on U follows from the fact that it is smooth on all U_a and [1.23-k]

Consider first the case $i \leq p$, i.e., h is an even coordinate. According to expansion (2.24) there exist smooth functions (of even variables only) $\chi_{i_1,\ldots,i_k}$ defined in a neighborhood of $(x,0)$ such that

$$\Delta(x,h) = \sum_k \sum_{i_j} \xi^{i_1} \cdots \xi^{i_k} \cdot (\chi_{i_1 \ldots i_k})(x^1, \ldots, x^p, h) .$$

Taking real values for $x^1, \ldots, x^p$ and h, we deduce (as in [2.21]) from $h \cdot \Delta(x,h) = 0$ that $\xi^{i_1} \cdots \xi^{i_k} \cdot h \cdot \chi_{i_1 \ldots i_k}(x^1, \ldots, x^p, h)$ must be zero. Since the coordinates $(\xi^{i_1}, \ldots, \xi^{i_k})$ and $(x^1, \ldots, x^p, h)$ are independent, it follows that either $\xi^{i_1} \cdots \xi^{i_k}$ or $h \cdot \chi_{i_1 \ldots i_k}(x^1, \ldots, x^p, h)$ must be identically zero. If the former, the term $\xi^{i_1} \cdots \xi^{i_k} \cdot \chi_{i_1 \ldots i_k}(x^1, \ldots, x^p, h)$ does not contribute to $\Delta(x,h)$. If the latter, continuity of $\chi_{i_1 \ldots i_k}$ (and the fact that it is real valued) implies that $h \cdot \chi_{i_1 \ldots i_k}(x^1, \ldots, x^p, h)$ is identically zero if and only if $\chi_{i_1 \ldots i_k}(x^1, \ldots, x^p, h)$ is identically zero. We conclude that $\Delta(x,h) = 0$ and in particular $\Delta(x,0) = 0$. This shows that g_i is uniquely defined by f.

Next let $i > p$, i.e., h is an odd coordinate. Again using expansion (2.24), there exist smooth functions $\chi_{i_1,\ldots,i_k}$ and $\psi_{i_1,\ldots,i_k}$ defined in a neighborhood of $(x,0)$ such that

$$\Delta(x,h) = \sum_k \sum_{i_j} \xi^{i_1} \cdots \xi^{i_k} \cdot \Big((\chi_{i_1 \ldots i_k})(x^1, \ldots, x^p) + h \cdot (\psi_{i_1 \ldots i_k})(x^1, \ldots, x^p) \Big) .$$

Since $h^2 = 0$, $h \cdot \Delta(x,h) = 0$ implies that either $h \cdot \xi^{i_1} \cdots \xi^{i_k}$ or $\chi_{i_1 \ldots i_k}(x^1, \ldots, x^p)$ must be identically zero. If (C [$q+1$]) holds, the first is impossible, so $\chi_{i_1 \ldots i_k}$ is identically zero, proving that $\Delta(x,0) = 0$, i.e., that g_i is uniquely determined by f. On the other hand, if (C [$q+1$]) does not hold, we can change $\phi_{a,i}$ by adding $\xi^1 \cdots \xi^k$, where k is chosen such that this product is not identically zero, but any $k+1$-fold product of odd elements is zero (in particular if $k = 0$, i.e., $\mathcal{A}_1 = \{0\}$, we add the constant function 1). By hypothesis such a $k \leq q$ exists. This does not invalidate (A1) but changes g_i also by this non-zero amount, showing that g_i is not uniquely determined by f. QED

3.2 Remark. As in the proof of [2.16], the fact that we can reduce functions of even variables to real valued functions of real variables is essential. Otherwise, the non-Hausdorff character of the DeWitt topology would invalidate the continuity argument.

3.3 Definition/Convention. As suggested already several times, we want to define the partial derivatives of a smooth function f by the diagonal of the functions $\phi_{a,i}$ from

(A1), i.e., by the functions g_i from [3.1]. In analogy with ordinary smooth functions and knowing that g_i depends only upon f, we will denote these partial derivatives by $g_i \equiv \partial_i f \equiv \partial f/\partial x^i$, $1 \le i \le n = p + q$. If we want to make an explicit distinction between even and odd coordinates we will use the notation $\partial_{x^i} f$ or $\partial f/\partial x^i$ for $i \le p$ and $\partial_{\xi^{i-p}} f$ or $\partial f/\partial \xi^{i-p}$ for $i > p$.

According to [3.1] the partial derivatives $\partial_i f$ are always defined for $i \le p$, i.e., for the even directions. On the other hand, the partial derivatives in the odd directions exist only if $\mathcal{A}$ satisfies (C[$q+1$]). This immediately raises the question what happens for $\mathcal{A} = \mathbf{R}$ (or more generally when $\mathcal{A}_1 = \{0\}$), because then (C[$q+1$]) is certainly not satisfied. Of course this is no problem if there are no odd coordinates present, i.e., when E is without odd dimensions. But in all other cases it poses a problem. Thinking of the case $\mathcal{A} = \mathbf{R}$, in which we know that there ought to be only p partial derivatives, not $p + q$, we can "solve" this dilemma by just ignoring the undefined partial derivatives in the odd directions. That this is reasonable to do is confirmed in [V.1.8], where we show that if $\mathcal{A}_1 = \{0\}$, then we do not lose anything by ignoring the $\partial_i f$ for $i > p$. For all other cases, i.e., when $q > 0$, $\mathcal{A}_1 \ne \{0\}$, and (C[$q+1$]) not satisfied, we do have a problem. *For that reason we will assume throughout the rest of this book that $\mathcal{A}$ satisfies* (C[∞]). Most of the time this is stronger than strictly needed, but it avoids changing (the condition on) $\mathcal{A}$ every time we change the odd dimension. It also guarantees that a smooth function f is always represented in (2.24) by the maximal number (only depending upon the odd dimension) of ordinary smooth functions $f_{i_1\dots i_k}$.

Of course, with this convention we do not cover the case $\mathcal{A} = \mathbf{R}$. However, most of the statements, including their proofs, remain valid for $\mathcal{A} = \mathbf{R}$. It is only occasionally that there will be a difference in treatment, but these we will point out in separate remarks.

3.4 Discussion. Let us now for the last time distinguish between an ordinary smooth (vector valued) function f of p real coordinates, and the (smooth) function $\mathbf{G}f$ of p even coordinates. It follows immediately from the proof of [1.21] that $\partial_i \mathbf{G}f = \mathbf{G}(\partial_i f)$, which justifies at the same time our use of the symbol ∂_i for this operation, as well as our identification of f with $\mathbf{G}f$. With this knowledge, the reader should be able to convince himself of the correctness of the following result: if a function $f \in C^\infty(U;F)$ is explicitly given by 2^q functions $f_{i_1\dots i_k} \in C^\infty(\mathbf{B}U, \mathbf{B}F)$ (expansion (2.24)), then the partial derivatives are given by

$$\frac{\partial f}{\partial x^\ell}(x) = \sum_{k=0}^{q} \sum_{1\le i_1<\dots<i_k\le q} \xi^{i_1}\cdots\xi^{i_k}\cdot\Big(\frac{\partial f_{i_1\dots i_k}}{\partial x^\ell}\Big)(x^1,\dots,x^p)\,,$$

(3.5)

$$\frac{\partial f}{\partial \xi^\ell}(x) = \sum_{k=0}^{q} \sum_{1\le i_1<\dots<i_k\le q} \sum_{j=1}^{k} (-1)^{j-1}\cdot\xi^{i_1}\cdots\xi^{i_{j-1}}\cdot\delta_\ell^{i_j}\cdot\xi^{i_{j+1}}\cdots\xi^{i_k}\cdot f_{i_1\dots i_k}(x^1,\dots,x^p)\,.$$

In the last formula, the $\delta_\ell^{i_j}$ is the Kronecker delta, always zero except if $i_j = \ell$ when it is 1. The sign $(-1)^{j-1}$ in the second formula appears when we put the coordinate ξ^{i_j} in front of the rest.

3.6 Proposition. *The maps* ∂_i *belong to* $\mathrm{Der}_R(C^\infty(U;\mathcal{A}))$*, they have parity* $\varepsilon(e_i)$*, and they commute:* $[\partial_i, \partial_j] = 0$.

Proof. We will use the proof of [1.23] to prove the first two statements. From part (b) it follows that the parity of ∂_i is the same as that of e_i. From parts (c,d) it follows that it is linear over $\mathbf{R}$. (Nota Bene. $C^\infty(U;\mathcal{A})$ is an $\mathbf{R}$-algebra and as such right and left linear are the same, the difference becomes apparent in the derivation property.) For homogeneous f, part (h) shows that $\partial_i(f\cdot g) = (\partial_i f)\cdot g + (-1)^{\langle\varepsilon(e_i)|\varepsilon(f)\rangle}\cdot f\cdot(\partial_i g)$, proving that ∂_i is a right-linear derivation of parity $\varepsilon(e_i)$.

For the last statement we use (3.5). If x^i and x^j are both even coordinates, the equality $\partial_i\circ\partial_j = \partial_j\circ\partial_i$ follows from the same equality for ordinary real valued smooth functions and (3.5). If x^i and x^j have different parities, the equality $\partial_i\circ\partial_j = \partial_j\circ\partial_i$ follows immediately from (3.5). Finally, if x^i and x^j are both odd coordinates, the equality $[\partial_i,\partial_j]\equiv\partial_i\circ\partial_j + \partial_j\circ\partial_i = 0$ follows from (3.5) by an elementary calculation. QED

3.7 Discussion. In §2 we have given two variants of an expansion that looked like a convergent Taylor series with a nilpotent increment. Let us show that it indeed is a Taylor series (up to factors $k!$). And remember, our convention that $\mathcal{A}$ satisfies (C[∞]) guarantees that the partial derivatives exist for any odd dimension. Looking at the proof of [2.27] and using that for an even coordinate x^j we have $\partial_j\mathbf{G}f = \mathbf{G}(\partial_j f)$, we see immediately that the functions $k!\cdot f^{(k)}$ in [2.27] are exactly the k-th order partial derivatives with respect to an even x^j. The same result follows from the proof of [2.27] for odd coordinates, once one realizes that for odd coordinates $(\partial_j)^2 = 0$ [3.6]. We thus are allowed to write (finally) for homogeneous nilpotent a of the correct parity:

$$f(x^1,\ldots,x^{j-1},x^j+a,x^{j+1},\ldots,x^n) = \sum_{k=0}^{\infty}\frac{a^k}{k!}\cdot((\partial_j)^k f)(x) = \sum_{k=0}^{\infty}\frac{a^k}{k!}\cdot\frac{\partial^k f}{(\partial x^j)^k}(x)\ .$$

In case there are no odd coordinates and if we apply this expansion repeatedly to all (even) coordinates, we obviously recover definition (2.12) of $\mathbf{G}f$. But also in the case of odd coordinates have we already obtained such an expansion! Using (3.5) we have an alternative way to describe the functions $f_{i_1\ldots i_k}$ appearing in the expansion (2.24): they are given as partial derivatives of f with respect to odd variables:

$$f_{i_1\ldots i_k}(x) = \left(\partial_{\xi^{i_k}}\cdots\partial_{\xi^{i_1}} f\right)(x,\xi=0)\ .$$

It is obviously a circular argument to think that we could *define* the functions $f_{i_1\ldots i_k}$ in terms of derivatives. We needed them in order to be able to show that (partial) derivatives exist. On the other hand, the expansion (2.24) now can be written in the form of a Taylor expansion:

$$\begin{aligned} f(x,\xi) &= \sum_{k=0}^{q} \sum_{1\leq i_1<\cdots<i_k\leq q} \xi^{i_1}\cdots\xi^{i_k}\cdot\left(\partial_{\xi^{i_k}}\cdots\partial_{\xi^{i_1}} f\right)(x,\xi=0) \\ &= \sum_{i_1,\ldots,i_q=0}^{1} (\xi^1)^{i_1}\cdot(\xi^2)^{i_2}\cdots(\xi^q)^{i_q}\cdot\left((\partial_{\xi^q})^{i_q}\cdots(\partial_{\xi^1})^{i_1} f\right)(x,\xi=0)\ . \end{aligned}$$

3.8 Discussion/Definition. Let E and F be $\mathcal{A}$-vector spaces with bases (e_i) and (ϵ_j) respectively, and let $f : U \subset E_0 \to F$ be smooth. Using the left coordinate functions ${}^j\epsilon$ on F, we now introduce the functions $f^j : U \to \mathcal{A}$ by $f^j = {}^j\epsilon \circ f \equiv f \diamond {}^j\epsilon$. Obviously we have for $x \in U$ the equalities $\langle\!\langle f(x) \,\|\, {}^j\epsilon \rangle\!\rangle = f^j(x)$ and $f(x) = \sum_j f^j(x)\epsilon_j$. According to [1.23-i] f is smooth if and only if all f^j are. Moreover, from the proof of [1.23-i] we deduce that $(\partial_i({}^j\epsilon \circ f))(x) = \langle\!\langle (\partial_i f)(x) \,\|\, {}^j\epsilon \rangle\!\rangle$. This property can also be stated in the following equivalent ways:

$$(3.9)\quad (\partial_i f)^j = \partial_i(f^j) \qquad \text{or} \qquad \partial_i f = \textstyle\sum_j (\partial_i f^j)\epsilon_j \qquad \text{or} \qquad \left(\frac{\partial f}{\partial x^i}\right)^j = \frac{\partial f^j}{\partial x^i}\ ,$$

where x^i now denotes any of the $p+q$ even or odd coordinates on E_0. If we denote by y^j the (even and odd) coordinate functions $y^j \equiv {}^j\epsilon$ on F_0, then the proof of [1.23-j] shows that we have the chain rule for partial derivatives in the following form (and remember, for this to make sense we have to assume that $f(U) \subset F_0$):

$$\partial_i(g\circ f)(x) = \sum\nolimits_j\ (\partial_i f^j)(x)\cdot(\partial_j g)(f(x))\ ,$$

or equivalently:

$$(3.10)\qquad \frac{\partial(f\diamond g)}{\partial x^i}(x) = \sum\nolimits_j \frac{\partial f^j}{\partial x^i}(x)\cdot\frac{\partial g}{\partial y^j}(f(x))\ .$$

The order in these formulæ is important and reflects our choice to write the coordinates on the left in (A1).

For $f \in C^\infty(U;F)$ we define *the Jacobian* $Jac(f) \in C^\infty(U;\mathrm{Hom}_L(E;F))$ by

$$Jac(f)(x) = \textstyle\sum_i {}^i e \otimes \partial_i f(x) = \sum_{i,j} {}^i e \otimes \partial_i f^j(x)\cdot\epsilon_j\ .$$

It follows from the proof of [1.23-e] that this expression is independent of the choice of a basis for E. With respect to the bases (e_i) for E and (ϵ_j) for F, the matrix $\dot{M}_L(Jac(f)(x))_i{}^j$ [II.4.1] of this left-linear morphism is $(\partial_i f^j)(x)$. Since ∂_i has the same parity as ${}^i e$, it follows immediately that for homogeneous $f \in C^\infty(U;F)_\alpha$, the Jacobian has the same parity as f, i.e., $Jac(f) \in C^\infty(U; \mathrm{Hom}_L(E; F))_\alpha$. More explicitly, for a vector $v = \sum_i v^i\, e_i$ we have

$$(3.11) \qquad \langle\!\langle v \,\|\, Jac(f)(x) \rangle\!\rangle = \sum\nolimits_i \, v^i \frac{\partial f}{\partial x^i}(x) = \sum\nolimits_{i,j} \, v^i \frac{\partial f^j}{\partial x^i}(x)\, \epsilon_j \ .$$

In terms of the Jacobian, the chain rule can be stated as

$$Jac(g \circ f)(x) = Jac(g)(f(x)) \circ Jac(f)(x) \ ,$$

or equivalently, and preferably because $Jac(f)$ is *left* linear, as

$$Jac(f \diamond g)(x) = Jac(f)(x) \diamond Jac(g)(f(x)) \ .$$

3.12 Remark. In the special case $\mathcal{A} = \mathbf{R}$, we decided to ignore the (non-existing) partial derivatives in the odd directions. We thus have only the p partial derivatives in the even directions. Hence the Jacobian is an element of $\mathrm{Hom}_L(E_0; F)$ instead of $\mathrm{Hom}_L(E; F)$. Moreover, if f is an even function, i.e., taking values in F_0, the Jacobian should be seen as taking values in $\mathrm{Hom}_L(E_0; F_0)$. We are forced to this decision if we want (3.10) to make sense, because the ∂_{y^j} only exist for the even directions. In a certain sense this is the natural thing to do, because for even f all components f^j (and thus the $\partial_i f^j$) are zero for $j > p$.

3.13 Discussion. As an application of the chain rule, we consider four $\mathcal{A}$-vector spaces E, F, G, and H, open subsets $U \subset E_0$, $V \subset F_0$, and $W \subset (E \times G)_0 \equiv E_0 \times G_0$, and smooth maps $g : V \to G_0$, and $f : W \to H$. With these we define the map $h : E_0 \times V \to E_0 \times G_0$ by $h(x, y) = (x, g(y))$. If $\forall x \in U \ \forall y \in V : (x, g(y)) \in W$, we can consider the composite function $k : U \times V \to G$ defined by $k = f \circ h$, i.e., $k(x, y) = f(x, g(y))$. If we denote by (x^i) the (even and odd) coordinates on E_0, they can be seen either as part of a system of coordinates on $E_0 \times F_0$ or as part of a system of coordinates on $E_0 \times G_0$. If we see them as coordinates on $E_0 \times F_0$ and if we denote by h^j one of the coordinates of h in $E_0 \times G_0$, then by definition of h the partial derivative $\partial_{x^i} h^j$ is zero unless h^j is the coordinate corresponding to x^i, in which case it is 1. The chain rule then tells us that

$$k(x, y) = f(x, g(y)) \qquad \Longrightarrow \qquad \frac{\partial k}{\partial x^i}(x, y) = \frac{\partial f}{\partial x^i}(x, g(y)) \ .$$

Stated in words, this equation tells us that *taking partial derivatives with respect to some coordinates* and *substituting functions for other coordinates* are commuting operations. This applies in particular to the case in which g is a constant functions (thus taking values in $\mathbf{B}G_0$): *taking partial derivatives with respect to some coordinates* and *fixing some coordinates to real values* are commuting operations. This particular case can also be deduced directly from the proof of [1.23-g].

3.14 Examples. • Associated to a smooth $\phi \in \mathrm{Hom}_L(E;F)$ we define $f : E_0 \to F$ by $f \equiv \phi|_{E_0} \in C^\infty(E_0;F)$. For $x, y \in E_0$ we compute: $f(y) - f(x) = \langle\!\langle y - x \| \phi \rangle\!\rangle$. It follows that $Jac(f) \in C^\infty(E_0; \mathrm{Hom}_L(E;F))$ is the constant map ϕ, i.e., $\forall x \in E_0$: $Jac(f)(x) = \phi$.

• If $\phi \in \mathrm{Hom}_R(E;F)$ is smooth and $f = \phi|_{E_0} \in C^\infty(E_0;F)$, a similar computation shows that $Jac(f)$ is the constant map $\mathfrak{T}^{-1}\phi \in \mathrm{Hom}_L(E;F)$.

• Consider the graded determinant gDet $\in C^\infty(\mathrm{Aut}(E))$, where we view $\mathrm{Aut}(E)$ as an open set in $\mathrm{End}_S(E)_0$. In order to compute $Jac(\mathrm{gDet})(id(E))$ we apply [3.13]. A simple calculation using the matrix expression for gDet shows that the partial derivatives with respect to non-diagonal entries are zero (at $id(E)$!). The remaining partial derivatives are ± 1 in such a way that we have $Jac(\mathrm{gDet})(id(E)) = \mathrm{gtr}_S \in \mathrm{Hom}_L(\mathrm{End}_S(E);\mathcal{A})$.

3.15 Lemma. *Let E and F be $\mathcal{A}$-vector spaces, $U \subset E_0$ open, $V \subset F_0$ open and connected, and $f : U \times V \to \mathcal{A}$ a smooth function. We denote by (y^i) the even and odd coordinates on V. There exists a (unique, smooth) function $g : U \to \mathcal{A}$ such that $f = g \circ \pi_U$ if and only if for all i $\partial_{y^i} f = 0$.*

Proof. Let us denote by x even and by ξ the odd coordinates on U, and let us suppose that the dimension of E is $p|q$ and that of F $r|s$. If y^i is an odd coordinate, it follows from (2.24) that f is of the form $f = f_0 y^i f_1$ with f_0 and f_1 function on $U \times V$ independent of the coordinate y^i. Since $\partial_{y^i} f = f_1$, it follows from $\partial_{y^i} f = 0$ that $f_1 = 0$, i.e., $f = f_0$ is a function independent of the coordinate y^i. Repeating the argument for all odd coordinates y^i proves that f does not depend upon the odd coordinates, i.e., f is of the form

$$f(x, \xi, y^1, \dots, y^{r+s}) = \sum_{k=0}^{q} \sum_{1 \le i_1 < \cdots < i_k \le q} \xi^{i_1} \cdots \xi^{i_k} \cdot f_{i_1 \dots i_k}(x, y^1, \dots, y^r) \, .$$

Now if y^i is an even coordinate ($i \le r$), the condition $\partial_{y^i} f = 0$ implies, using (3.5) and [2.21], that $\partial_{y^i} f_{i_1 \dots i_k} = 0$ for all $i_1, \dots, i_k$. These equations we can see as equations of ordinary smooth $\mathbf{R}$-values functions. But then the connectedness of $\mathbf{B}V$ implies that the functions $f_{i_1 \dots i_k}$ are independent of the coordinates y^i. *QED*

3.16 Corollary. *Let $U \subset E_0$ be an open connected set and $f : U \to \mathcal{A}$ a smooth function. Then f is constant if and only if $\partial_i f = 0$ for all $1 \le i \le \dim E$.*

3.17 Discussion. To prepare the inverse function theorem, let E and F be $\mathcal{A}$-vector spaces of dimension $p|q$ and $r|s$ respectively, let $U \subset E_0$ be open and let $f : U \to F_0$ be a smooth function. We denote by (x, ξ) the even and odd coordinates of a point in U with respect to an ordered basis and by (y, η) the coordinates of a point in F_0. It follows from [2.23] that there exist real valued functions $g^i_{i_1\ldots i_k}$ and $h^i_{i_1\ldots i_k}$ such that

$$y^i = \sum_{k=0}^{q} \sum_{1\leq i_1<\cdots<i_k\leq q} g^i_{i_1\ldots i_k}(x)\cdot\xi^{i_1}\cdots\xi^{i_k}$$

$$\eta^i = \sum_{k=0}^{q} \sum_{1\leq i_1<\cdots<i_k\leq q} h^i_{i_1\ldots i_k}(x)\cdot\xi^{i_1}\cdots\xi^{i_k}\ .$$

However, y^i is an even coordinate and thus the expansion cannot contain terms with an odd number of ξ. Likewise, the expansion of η^i cannot contain terms with an even number of ξ. It follows that we can write

$$y^i = \sum_{k=0}^{[q/2]} \sum_{1\leq i_1<\cdots<i_{2k}\leq q} g^i_{i_1\ldots i_{2k}}(x)\cdot\xi^{i_1}\cdots\xi^{i_{2k}}$$

$$\eta^i = \sum_{k=0}^{[q/2]} \sum_{1\leq i_1<\cdots<i_{2k+1}\leq q} h^i_{i_1\ldots i_{2k+1}}(x)\cdot\xi^{i_1}\cdots\xi^{i_{2k+1}}\ .$$

These expansions are rather awkward, especially when one needs to write them often. We therefore introduce the following notational shorthand. In the first place, we will not write the superscript i (in x^i, y^i, ξ^i, or η^i) to denote which coordinate is concerned. In the second place we will abbreviate any product of k factors ξ as ξ^k (and like wise for the η). At the same time we will omit the summation sign over indices $i_1,\ldots,i_k$ as well as these indices in the coefficient functions g and h. We thus will write

$$y = \sum_{i\geq 0} f_{2i}(x)\cdot\xi^{2i} \quad , \quad \eta = \sum_{j\geq 0} f_{2j+1}(x)\cdot\xi^{2j+1}\ .$$

Here we have used the symbol f again for the coefficient functions $g^i_{i_1,\ldots,i_k}$ and $h^i_{i_1,\ldots,i_k}$. We also have added a subscript to indicate with which power of ξ they are associated. No confusion between y and η is possible, since y only contains even powers of ξ and η only odd powers. On the other hand, these simplifications are easily misunderstood, so we will use them only when strictly needed.

Taking all ξ to be zero, we obtain that $y = f(x,0) = f_0(x)$ and $\eta = 0$. In other words, f_0 is the real valued function $f_0 = \mathbf{B}f$. On the other hand, $f_1(x)$ represents the matrix valued real function $x \mapsto h^i_{i_1}(x)$ of size $s \times q$.

3.18 Lemma. *Let $f : U \subset E_0 \to F_0$ be a smooth function as in the above discussion, and let V be an open set in F_0. Then the following two properties are equivalent.*

(i) *$f(U) = V$ and there exists a function $g \in C^\infty(V;E)$ such that $g\circ f = id(U)$ and $f\circ g = id(V)$.*

(ii) *$f_0(\mathbf{B}U) = \mathbf{B}V$ and there exist smooth functions g_0 and g_1 defined on $\mathbf{B}V$ such that $g_0\circ f_0 = id(\mathbf{B}U)$, $f_0\circ g_0 = id(\mathbf{B}V)$ and such that for all $x\in\mathbf{B}U$ and all $y\in\mathbf{B}V$: $g_1(f_0(x))\cdot f_1(x) = I_q$ and $f_1(g_0(y))\cdot g_1(y) = I_s$.*

Here I_n denotes the $n\times n$ identity matrix. In other words, the function f admits a smooth inverse if and only if $f_0 = \mathbf{B}f$ admits a smooth inverse and f_1 is everywhere invertible.

Proof. If we assume (i), we have $y = \sum_{i\geq 0} f_{2i}(x)\cdot\xi^{2i}$ and $\eta = \sum_{j\geq 0} f_{2j+1}(x)\cdot\xi^{2j+1}$, but also $x = \sum_{k\geq 0} g_{2k}(y)\cdot\eta^{2k}$ and $\xi = \sum_{\ell\geq 0} g_{2\ell+1}(y)\cdot\eta^{2\ell+1}$. Substituting the first relations in the second ones we obtain

$$\begin{aligned} x &= \sum_{k\geq 0} g_{2k}\Big(\sum_{i\geq 0} f_{2i}(x)\cdot\xi^{2i}\Big)\cdot\Big(\sum_{j\geq 0} f_{2j+1}(x)\cdot\xi^{2j+1}\Big)^{2k} \\ &= \sum_{\ell\geq 0} g_0^{(\ell)}(f_0(x))\cdot\Big(\sum_{i>0} f_{2i}(x)\cdot\xi^{2i}\Big)^{\ell} + \\ &\qquad \sum_{k>0} g_{2k}\Big(\sum_{i\geq 0} f_{2i}(x)\cdot\xi^{2i}\Big)\cdot\Big(\sum_{j\geq 0} f_{2j+1}(x)\cdot\xi^{2j+1}\Big)^{2k}. \end{aligned} \tag{3.19}$$

To obtain the second line, we have applied [2.27] on g_0 and the fact that $\sum_{i>0} f_{2i}(x)\cdot\xi^{2i}$ is nilpotent. In a similar way we obtain for ξ the equation

$$\begin{aligned} \xi &= \sum_{\ell\geq 0} g_{2\ell+1}\Big(\sum_{i\geq 0} f_{2i}(x)\cdot\xi^{2i}\Big)\cdot\Big(\sum_{j\geq 0} f_{2j+1}(x)\cdot\xi^{2j+1}\Big)^{2\ell+1} \\ &= \sum_{k\geq 0} g_1^{(k)}(f_0(x))\cdot\Big(\sum_{i>0} f_{2i}(x)\cdot\xi^{2i}\Big)^{k}\cdot\Big(\sum_{j\geq 0} f_{2j+1}(x)\cdot\xi^{2j+1}\Big) + \\ &\qquad \sum_{\ell>0} g_{2\ell+1}\Big(\sum_{i\geq 0} f_{2i}(x)\cdot\xi^{2i}\Big)\cdot\Big(\sum_{j\geq 0} f_{2j+1}(x)\cdot\xi^{2j+1}\Big)^{2\ell+1}. \end{aligned} \tag{3.20}$$

Since the coefficients of the separate powers of ξ in an expansion are uniquely determined [2.21], one deduces from (3.19) in order 0 that $g_0\circ f_0 = id(\mathbf{B}U)$. The order 1 term in (3.20) is given by $\xi = g_1(f_0(x))\cdot f_1(x)\cdot\xi$, and thus by the same argument one obtains $g_1(f_0(x))\cdot f_1(x) = I_q$. The rest of property (ii) follows by interchanging the roles of f and g.

To prove the implication (ii) $\Rightarrow$ (i), we define the function g by defining the smooth functions g_i by induction on i using (3.19) and (3.20). By assumption the functions g_0 and g_1 exist. Now consider in (3.19) the terms with ξ^{2n}. This gives us the equation

$$0 = g_{2n}(f_0(x))\cdot(f_1(x)\cdot\xi)^{2n} + \text{other terms}\ , \tag{3.21}$$

where the other terms only contain functions g_{2k} with $k < n$. To define g_{2n} we now proceed as follows. In (3.21) we replace each occurrence of x by $g_0(y)$ and each occurrence of ξ by $g_1(y)\cdot\eta$. Finally we leave out all the η's. Since $f_0(g_0(y)) = y$, we obtain a definition of $g_{2n}(y)$ in terms of the functions g_1 and $g_{2k}\cdot\eta^{2k}$ with $k < n$, which are defined by the induction assumption.

With a similar argument using (3.20) we obtain the functions $g_{2\ell+1}$. We thus obtain a function g that satisfies *by construction* (and property (ii)) the condition $g\circ f = id(U)$. Interchanging the roles of f and g gives the existence of a smooth function h with $h_0 = f_0$ and $h_1 = f_1$ such that $h\circ g = id(V)$. We deduce that g is both injective and surjective and that hence $h = f$, proving property (i). QED

3.22 Remark. In the special case $\mathcal{A} = \mathbf{R}$, there are no (non-zero) odd coordinates. Hence an even function f is uniquely represented by the function f_0 in the shorthand of [3.17]. Part (ii) of [3.18] should thus be replaced by omitting all references to f_1 and g_1.

3.23 Theorem (inverse function theorem). *Let $O \subset E_0$ be open, $f \in C^\infty(O; F)_0$, and $m \in O$ arbitrary. Then the Jacobian $Jac(f)(m)$ is invertible if and only if there exists an open neighborhood U of m such that (i) $V = f(U)$ is open in F_0 and (ii) $f : U \to V$ admits a smooth inverse.*

Proof. Let us start analyzing the invertibility of the Jacobian: $Jac(f)(m)$ is invertible if and only if its matrix $\partial_i f^j(m)$ is invertible, which is invertible if and only if its body $\mathbf{B}(\partial_i f^j(m))$ is invertible [II.1.11]. Using an ordered basis (as we usually do when splitting coordinates in even and odd ones) and applying our knowledge of how partial derivatives work on expanded functions, we obtain $m = (x, \xi)$ and $\mathbf{B}(\partial_i f^j(m))$ in block form as

$$\mathbf{B}(\partial_i f^j(x,\xi)) = \begin{pmatrix} (\partial f_0^k / \partial x^\ell)(x) & \mathbf{0} \\ \mathbf{0} & f_1(x) \end{pmatrix} .$$

If this matrix is invertible, it follows that both $(\partial f_0^k / \partial x^\ell)(x)$ and $f_1(x)$ are invertible matrices. Application of the standard inverse function theorem to the function f_0 shows that there exists an open neighborhood $O_0 \subset \mathbf{B}E_0$ of x and an open subset $P_0 \subset \mathbf{B}F_0$ such that $f_0 : O_0 \to P_0$ admits a smooth inverse. Smoothness of f_1 implies that there exists an open neighborhood $O_1 \subset \mathbf{B}E_0$ of x on which the matrix f_1 is everywhere invertible. We thus conclude that condition (ii) of [3.18] is fulfilled on the open neighborhood $U = \mathbf{B}^{-1}(O_0 \cap O_1)$ and the open set $V = \mathbf{B}^{-1}(f_0(O_0 \cap O_1)) \subset F_0$. It follows that f admits a smooth inverse on U.

On the other hand, if f admits a smooth inverse on U, we infer from condition (ii) of [3.18] and the standard inverse function theorem that $(\partial f_0^k / \partial x^\ell)(x)$ and $f_1(x)$ are invertible on $\mathbf{B}U$, i.e., that the Jacobian matrix is invertible. *QED*

3.24 Remark. In the special case $\mathcal{A} = \mathbf{R}$, the statement of the inverse function theorem is still valid. However, the Jacobian should be restricted to the even-even part as argued in [3.12]. This is also "obvious" from the proof because f_1 does not exist (is identically zero).

3.25 Corollary (invariance of dimension). *If $f \in C^\infty(U; F)_0$ admits a smooth inverse, the graded dimensions of $E \supset U$ and F are the same.*

3.26 Remark. In the special case $\mathcal{A} = \mathbf{R}$, the invariance of dimension only concerns the even dimension. This follows from the fact that in that case the Jacobian takes its values in $\mathrm{Hom}_L(E_0; F_0)$ [3.12], in which the odd dimensions have disappeared.

3.27 Theorem (implicit function theorem). *Let E and F be $\mathcal{A}$-vector spaces, U open in $(E \oplus F)_0$ and $f \in C^\infty(U; F)_0$. Denoting by x the coordinates (even and odd) in E and by y the coordinates in F, we suppose that $(x_o, y_o) \in \mathbf{B}U$. If the square matrix $(\partial f^i/\partial y^j)(x_o, y_o)$ of partial derivatives with respect to the coordinates in F_0 is invertible, then there exist open sets $x_o \in V \subset E_0$ and $y_o \in W \subset F_0$ with $V \oplus W \subset U$, and a smooth function $g : V \to W$ with $g(x_o) = y_o$ such that*

$$\forall (x, y) \in V \oplus W : f(x, y) = f(x_o, y_o) \iff y = g(x) \ .$$

Proof. Define the map $\Phi \in C^\infty(U; E \oplus F)_0$ by $\Phi(x, y) = (x, f(x, y))$. By hypothesis the Jacobian of this map in the point (x_o, y_o) is invertible. Hence there exists an open set of the form $V \oplus W$ on which Φ admits a smooth inverse Ψ. This inverse must have the form $\Psi(x, y) = (x, \psi(x, y))$ with $f(x, \psi(x, y)) = y$ and $\psi(x, f(x, y)) = y$. We then define the function $g : V \to W$ by

$$g(x) = \psi(x, f(x_o, y_o)) \ .$$

Since $f(x_o, y_o)$ has real coordinates, g is smooth by [1.23-g]. It obviously satisfies $g(x_o) = \psi(x_o, f(x_o, y_o)) = y_o$ and $f(x, g(x)) = f(x, \psi(x, f(x_o, y_o))) = f(x_o, y_o)$. Finally, if $f(x, y) = f(x_o, y_o)$, then $y = \psi(x, f(x, y)) = \psi(x, f(x_o, y_o)) = g(x)$. *QED*

3.28 Definition. For $f \in C^\infty(U; F_0)$ we define *the rank of f at $m \in U$*, denoted by $\mathrm{rank}(f)(m)$, as the rank [II.4.18] of its Jacobian $\mathit{Jac}(f)(m)$, i.e., as the rank of the morphism of $\mathbf{R}$-vector spaces $\mathbf{B}(\mathit{Jac}(f)(m))$.

3.29 Lemma. *Let E and F be finite dimensional $\mathcal{A}$-vector spaces and $\phi \in \mathrm{Hom}_S(E; F)$. Then ϕ is injective if and only if $\mathrm{rank}\,\phi = \dim E$, which is the case if and only if $\mathbf{B}\phi$ is injective.*

Proof. Right and left linear being similar, we will treat the right linear case. It is well known from linear algebra over $\mathbf{R}$ that a linear map is injective if and only if its rank is equal to the dimension of the source space. This proves the equivalence between $\mathrm{rank}\,\phi = \dim E$ and $\mathbf{B}\phi$ injective because $\mathrm{rank}\,\phi$ is just defined as the (ordinary) rank of $\mathbf{B}\phi$ [II.1.9], [II.4.18].

Now assume that ϕ is injective and that $v \in \mathbf{B}E$ is such that $(\mathbf{B}\phi)(v) = 0$. Since $v = \mathbf{B}v$, it follows that $\mathbf{B}(\phi(v)) = 0$, i.e., $\phi(v)$ is a nilpotent vector. Hence there exists $\lambda \in \mathcal{A}$, $\lambda \neq 0$ such that $(\phi(v))\lambda = 0$. By right linearity and injectivity of ϕ we conclude that $v\lambda = 0$. But v has real coefficients with respect to a basis and $\lambda \neq 0$, and thus all these real coefficients must be zero, i.e., $v = 0$. This proves that $\mathbf{B}\phi$ is injective.

Finally assume that $\mathbf{B}\phi$ is injective. Then $n = \dim E \leq \dim F = m$, and thus, with respect to bases of E and F, the matrix $\dot{M}_R(\phi)$ has the form $\binom{X}{Y}$ with X a $n \times n$ matrix and Y a $(m-n) \times n$ matrix. Up to a permutation of the basis vectors for F we may assume that $\mathbf{B}X$ is invertible because $\dot{M}_R(\mathbf{B}\phi) = \binom{\mathbf{B}X}{\mathbf{B}Y}$ is injective. But then X is invertible by [II.1.11]. We thus can define the right linear map $\psi : F \to E$ by $\dot{M}_R(\psi) = (X^{-1}\ \mathbf{0})$. It follows that $\psi \circ \phi = id_E$ and thus that ϕ is injective. *QED*

3.30 Proposition. *Let E and F be $\mathcal{A}$-vector spaces of total dimension n and p respectively and suppose that $f \in C^\infty(U; F_0)$ has rank k at $m \in U$. Then there exist open sets $V \subset U$, $m \in V$ and $V' \subset E_0$ and a smooth map $g : V \to V'$ with smooth inverse such that*

$$f \circ g^{-1} : (x^1, \ldots, x^n) \mapsto (y^1, \ldots, y^p) = (x^1, \ldots, x^k, f^{k+1}(g^{-1}(x)), \ldots f^p(g^{-1}(x))) ,$$

with respect to given bases of E and F (up to a permutation of the basis vectors).

In case $k = n$, which is equivalent to Jac$(f)(m)$ being injective [3.29], there exists an open set $W \subset F_0$ and a smooth map $h : W \to W$ with smooth inverse such that

$$h \circ f \circ g^{-1} : (x^1, \ldots, x^n) \mapsto (y^1, \ldots, y^p) = (x^1, \ldots, x^n, 0, \ldots, 0) .$$

Proof. The condition $\mathrm{rank}(f)(m) = k$ implies that there exists an invertible square submatrix of size $k \times k$ of the Jacobian $Jac(f)(m)$. By permuting the coordinates we may assume that $(\partial f^i/\partial x^j)(m), 1 \le i, j \le k$ is invertible. Now define $g : U \to E_0$ by

$$g(x^1, \ldots, x^n) = (f^1(x), \ldots, f^k(x), x^{k+1}, \ldots, x^n) .$$

Our assumption implies that the Jacobian of g at m is invertible, and thus, by the inverse function theorem, there exist open set $V \subset U$ and $V' \subset E_0$ such that $g|_V : V \to V'$ admits a smooth inverse. This $g|_V$ satisfies the requirements.

If $k = n$, and thus $p \ge n$, we define $W = \{ (y^1, \ldots, y^p) \in F_0 \mid (y^1, \ldots, y^n) \in V' \}$ and the smooth map $h : W \to W$ by

$$h^i(y^1, \ldots, y^p) = \begin{cases} y^i & 1 \le i \le n \\ y^i - f^i(g^{-1}(y^1, \ldots, y^n)) & n < i \le p \end{cases}$$ QED

3.31 Counter examples. For ordinary real-valued smooth functions the condition $k = n$ can easily be weakened to the condition that the rank of f equals k on the whole of V without affecting the result [Spi, Ch2 Th9]. In the graded case the situation is far more delicate as can be seen in the following counter examples.

- Let E be an $\mathcal{A}$-vector space of dimension $0|2$ with coordinates (ξ, η). The map $f \in C^\infty(E_0; \mathcal{A}_0)$, defined by

$$f(\xi, \eta) = \xi\eta ,$$

has rank 0 on E_0. However, it is not a constant map.

- Let F be an $\mathcal{A}$-vector space of dimension $1|2$ with coordinates (x, ξ, η). The map $f \in C^\infty(F_0; F_0)$, defined by

$$f(x, \xi, \eta) = (\xi\eta, \xi, 0) ,$$

has rank 1 on F_0. However, an elementary calculation shows that it can not be transformed (not even locally) into the form $(z^1, z^2, z^3) \mapsto (z^1, 0, 0)$.

4. $\mathcal{A}$-MANIFOLDS

In this section we copy the definition of manifolds in terms of charts and (smooth) transition functions to obtain the definition of $\mathcal{A}$-manifolds. The definition of a smooth function between $\mathcal{A}$-manifolds is standard and the extension of the body map to $\mathcal{A}$-manifolds and their smooth functions is straightforward. There is only one subtlety involved: since the DeWitt topology is not Hausdorff, we cannot require directly that an $\mathcal{A}$-manifold should be in particular a second countable Hausdorff topological space. We thus are "obliged" to make a slight detour via proto $\mathcal{A}$-manifolds where we forego these topological conditions.

4.1 Definitions. Let M be a topological space; a *chart* for M is a triplet $\varphi : U \to O$, where φ is a homeomorphism from an open subset U of M onto an open subset O of the *even* part of some $\mathcal{A}$-vector space E. When no confusion is possible, we will often denote a chart by the open set $U \subset M$ or by the pair (U, φ) instead of by the full triplet.

Two charts $\varphi_a : U_a \to O_a \subset (E_a)_0$ and $\varphi_b : U_b \to O_b \subset (E_b)_0$ for M are said to be *compatible* if the two maps $\varphi_a \circ \varphi_b^{-1}$ and $\varphi_b \circ \varphi_a^{-1}$ are smooth. More precisely:

$$\varphi_a \circ \varphi_b^{-1} \in C^\infty(\varphi_b(U_a \cap U_b); E_a)_0$$
$$\varphi_b \circ \varphi_a^{-1} \in C^\infty(\varphi_a(U_a \cap U_b); E_b)_0 \ .$$

A *smooth $\mathcal{A}$-structure* for a topological space M is a collection of charts $\mathcal{S} = \{ \varphi_a : U_a \to O_a \mid a \in I \}$ satisfying three conditions:

(SS1) $\cup_{a \in I} U_a = M$,

(SS2) any two charts in $\mathcal{S}$ are compatible and,

(SS3) if $\varphi : U \to O$ is a chart for M that is compatible with all elements of $\mathcal{S}$, then this chart belongs to $\mathcal{S}$.

A *proto $\mathcal{A}$-manifold* is a topological space M together with the prescription of a smooth $\mathcal{A}$-structure $\mathcal{S}$; an *atlas* for a proto $\mathcal{A}$-manifold is a subset $\mathcal{U} \subset \mathcal{S}$ satisfying properties (SS1) and (SS2), but not necessarily (SS3).

4.2 Discussion/Definition. The attentive reader might have noticed that the above definitions do not really specify the $\mathcal{A}$-vector space in which the open sets O_a lie. This means in particular that the different O_a might lie in $\mathcal{A}$-vector space of different graded dimensions. However, if $m \in M$ lies in two different charts U_a and U_b, the even smooth maps $\varphi_a \circ \varphi_b^{-1}$ and $\varphi_b \circ \varphi_a^{-1}$ are inverse to each other. It follows from the invariance of the graded dimension [3.25] that O_a and O_b lie in $\mathcal{A}$-vector spaces of the same graded dimension. We thus may conclude that the graded dimension is a locally constant function on M, i.e., the connected components of M have a well defined graded dimension. We will usually assume that all connected components have the same graded dimension. Since the graded dimension is the only invariant of an $\mathcal{A}$-vector space E, we will say that a proto $\mathcal{A}$-manifold M is *modeled* on the $\mathcal{A}$-vector space E if its graded dimension is the same as that of E, i.e., if all the charts can be thought of as lying in E_0. If M is modeled on an $\mathcal{A}$-vector space of dimension $p|q$, we will say that the *(graded) dimension of M* is $p|q$.

4.3 Remark. Of course, for proto $\mathbf{R}$-manifolds the invariance of graded dimension [3.25] should be replaced by the invariance of the even dimension [3.26]. Thus for proto $\mathbf{R}$-manifolds it is the even dimension of the $\mathbf{R}$-vector space on which a chart is modeled which is the invariant, the odd dimension does not matter (see also [2.10]).

4.4 Lemma. *Given a topological space M, a collection of charts $\mathcal{U} = \{ U_a \mid a \in I \}$ satisfying* (SS1) *and* (SS2) *defines a unique smooth $\mathcal{A}$-structure for M for which $\mathcal{U}$ is an atlas.*

Proof. To prove the existence, define $\mathcal{S}$ to be the collection of all charts for M that are compatible with all charts in $\mathcal{U}$. Since (SS2) is valid for $\mathcal{U}$, it is included in $\mathcal{S}$, and hence (SS1) is satisfied. To prove (SS2), let $\varphi : U \to O$ and $\psi : V \to Q$ be two elements of $\mathcal{S}$. Notice that $\{ \psi(U \cap V \cap U_a) \mid a \in I \}$ is an open cover of $\psi(U \cap V)$ because $\mathcal{U}$ satisfies (SS1). Since on $\psi(U \cap V \cap U_a)$ we have $\varphi \circ \psi^{-1} = \varphi \circ \varphi_a^{-1} \circ \varphi_a \circ \psi^{-1}$, we deduce that $\varphi \circ \psi^{-1}$ restricted to $\psi(U \cap V \cap U_a)$ is smooth as composition of smooth maps. Hence by [1.23-k], $\varphi \circ \psi^{-1}$ is smooth. Similarly, $\psi \circ \varphi^{-1}$ is smooth and hence these two charts are compatible. Finally, to prove (SS3), let U be a chart compatible with all elements of $\mathcal{S}$. Since $\mathcal{U} \subset \mathcal{S}$, U is compatible with all elements of $\mathcal{U}$ and hence belongs to $\mathcal{S}$.

To prove uniqueness, let $\mathcal{S}'$ be another smooth $\mathcal{A}$-structure for M containing $\mathcal{U}$. Any chart in $\mathcal{S}'$ is obviously compatible with all elements in $\mathcal{U}$ and hence belongs to $\mathcal{S}$, proving $\mathcal{S}' \subset \mathcal{S}$. On the other hand, if U is a chart in $\mathcal{S} \supset \mathcal{S}'$, then it is compatible with all elements in $\mathcal{S}'$ and hence belongs to $\mathcal{S}'$ because $\mathcal{S}'$ satisfies (SS3). *QED*

4.5 Conventions. From now on we will adopt the following conventions. In the first place, when talking about proto $\mathcal{A}$-manifolds, we will never specify explicitly a smooth $\mathcal{A}$-structure, at most we will specify an atlas for it (which defines the smooth $\mathcal{A}$-structure completely according to [4.4]). In the second place, when we talk about charts, this will always mean a chart *in the smooth $\mathcal{A}$-structure*, i.e., compatible with all charts in an atlas.

4.6 Corollary. *Every set of charts covering a proto $\mathcal{A}$-manifold is an atlas.*

4.7 Lemma. *Let $\varphi : U \to O$ be a chart of a proto $\mathcal{A}$-manifold M and let V be an open subset of U. Then $\varphi|_V : V \to \varphi(V)$ is a chart for M.*

4.8 Examples. • For any $\mathcal{A}$-vector space E, its even part E_0 is a proto $\mathcal{A}$-manifold when equipped with the trivial atlas $id : E_0 \to E_0$. In particular $\mathcal{A}_0$ is a proto $\mathcal{A}$-manifold. But $\mathcal{A}_1$ also is a proto $\mathcal{A}$-manifold. Consider the $\mathcal{A}$-vector space E of dimension $0|1$, i.e., generated by a single odd basis vector e. The standard map $\mathcal{A}_1 \to E, a \to a \cdot e$ defines a homeomorphism between $\mathcal{A}_1$ and E_0, i.e., an atlas for $\mathcal{A}_1$.

• More generally, every $\mathcal{A}$-vector space E can be given the structure of a proto $\mathcal{A}$-manifold by identifying it with with the even part of the $\mathcal{A}$-vector space $E^\sharp$.

• If M is a proto $\mathcal{A}$-manifold, any open subset of M is again a proto $\mathcal{A}$-manifold in the obvious way: by restricting an atlas to the open subset in question.

4.9 Construction. Let M be a proto $\mathcal{A}$-manifold and $\mathcal{U} = \{ \varphi_a : U_a \to O_a \mid a \in I \}$ an atlas. We then define open subsets $O_{ab} \subset O_a$ and smooth maps $\varphi_{ba} : O_{ab} \to O_{ba}$ by $O_{ab} = \varphi_a(U_a \cap U_b)$ and $\varphi_{ba} = \varphi_b \circ \varphi_a^{-1}$. It is easy to verify that the open subsets and maps $O_{ab} \subset O_a, \varphi_{ab}$ have the properties

$$(4.10) \qquad \begin{aligned} \forall a \in I \quad &: \quad \varphi_{aa} = id(O_a) \ , \\ \forall a, b, c \in I \quad &: \quad \text{on } O_{ab} \cap O_{ac} \ : \ \varphi_{ca} = \varphi_{cb} \circ \varphi_{ba} \ . \end{aligned}$$

What we intend to do is to show that these ingredients completely determine the proto $\mathcal{A}$-manifold M. Let us be more precise.

Suppose we are given the following data: (i) a collection $\{ O_a \mid a \in I \}$ where the O_a are open in the even part of some $\mathcal{A}$-vector space, (ii) for all $a, b \in I$ an open subset $O_{ab} \subset O_a$ and (iii) for all $a, b \in I$ a smooth map $\varphi_{ba} : O_{ab} \to O_{ba}$. Suppose furthermore that these data satisfy (4.10). We then can (and will) construct a proto $\mathcal{A}$-manifold M and an atlas $\mathcal{U}$ such that these data appear as in the previous paragraph.

In order to carry out this construction, first note that implicit in (4.10) are the conditions that (i) $O_{aa} = O_a$ and (ii) $\forall a, b, c \in I : \varphi_{ba}(O_{ab} \cap O_{ac}) \subset O_{bc}$, because else the formulæ would not make sense. Next we consider the disjoint union $\amalg_{a \in I} O_a$ of all the O_a, on which we define an equivalence relation by

$$x \sim y \quad \Longleftrightarrow \quad \exists a, b \in I : x \in O_{ab}, y \in O_{ba}, y = \varphi_{ba}(x) \ .$$

This is an equivalence relation because of (4.10). We finally define the topological space M as the quotient $M = \coprod_{a \in I} O_a / \sim$. Denoting by π the canonical projection $\pi : \coprod_{a \in I} O_a \to M$, we define the open subsets U_a of M by $U_a = \pi(O_a)$. Since no two different points in a single O_a are equivalent, $\pi : O_a \to U_a$ is a homeomorphism. Hence we obtain charts $\varphi_a : U_a \to O_a$ by taking $\varphi_a = \pi^{-1}$. It follows easily that $\varphi_a(U_a \cap U_b) = O_{ab}$ and $\varphi_a \circ \varphi_b^{-1} = \varphi_{ab}$, from which one can deduce that the collection of these charts for M satisfies (SS1) and (SS2), and hence defines a smooth $\mathcal{A}$-structure on M.

4.11 Construction. If E is an $\mathcal{A}$-vector space, we have defined a distinguished subset $\mathbf{B}E$, consisting of those points in E that have real coordinates (with respect to a basis in the equivalence class), and a canonical projection $\mathbf{B} : E \to \mathbf{B}E$. Using these ingredients, we will define similar things for proto $\mathcal{A}$-manifolds. In the first place we define the body $\mathbf{B}M$ of a proto $\mathcal{A}$-manifold M by

$$\mathbf{B}M = \{ m \in M \mid \exists \text{ chart } \varphi : U \to O : m \in U \text{ and } \varphi(m) \in \mathbf{B}O \} \ ,$$

i.e., $m \in \mathbf{B}M$ if and only if it has real coordinates in a chart. This is well defined because if m has real coordinates in a chart $\varphi_a : U_a \to O_a$, then its coordinates in a chart $\varphi_b : U_b \to O_b$ are given by $(\varphi_b \circ \varphi_a^{-1})(\varphi_a(m))$. Since $\varphi_b \circ \varphi_a^{-1}$ is smooth, the image of a point with real coordinates has again real coordinates. We conclude that if a point m has real coordinates in some chart, it has real coordinates in all charts.

In a similar spirit we define the canonical projection $\mathbf{B} : M \to \mathbf{B}M$ by

$$\varphi : U \to O \text{ a chart} \quad \Longrightarrow \quad \mathbf{B}|_U = \varphi^{-1} \circ \mathbf{B} \circ \varphi \,,$$

where $\mathbf{B}$ is the previously defined body projection $\mathbf{B} : O \subset E \to \mathbf{B}O \subset \mathbf{B}E$. Again this definition is independent of the choice of the chart $\varphi : U \to O$ because smooth maps preserve points with real coordinates. A quick glance at the examples [4.8] shows that this new definition of $\mathbf{B}$ coincides with the previously defined one, i.e., in case the proto $\mathcal{A}$-manifold is an open subset of the even part of an $\mathcal{A}$-vector space.

4.12 Lemma. *Let M be a proto $\mathcal{A}$-manifold and $U \subset M$ an arbitrary subset. Then U is open if and only if $\mathbf{B}U$ is open in $\mathbf{B}M$ equipped with the induced topology and $\mathbf{B}^{-1}(\mathbf{B}U) = U$, in which case we have $\mathbf{B}U = U \cap \mathbf{B}M \subset U$.*

Proof. For a chart $\varphi_a : U_a \to O_a$ we have $\mathbf{B}|_{U_a} = \varphi_a^{-1} \circ \mathbf{B} \circ \varphi_a$ and hence $\mathbf{B}U_a \subset U_a$. Now let us suppose that for arbitrary $m \in M$ we have $\mathbf{B}m \in U_a$. If m lies in the chart $\varphi_b : U_b \to O_b$, it follows that $\mathbf{B}m \in U_a \cap U_b$, hence $\mathbf{B}\varphi_b(m) = \varphi_b(\mathbf{B}m) \in O_{ba}$ [4.9]. By the DeWitt topology we then also have $\varphi_b(m) \in O_{ba}$ and thus $m \in U_a \cap U_b$. Together with $\mathbf{B}U_a \subset U_a$ we thus have $\mathbf{B}U_a = U_a \cap \mathbf{B}M$ and $\mathbf{B}^{-1}(\mathbf{B}U_a) = U_a$. Since any open U can be covered by charts (use [4.7] to restrict charts to U), the same holds for U and in particular $\mathbf{B}U = U \cap \mathbf{B}M$ is open in $\mathbf{B}M$ equipped with the induced topology.

On the other hand, assume that $\mathbf{B}U$ is open in $\mathbf{B}M$ and that $\mathbf{B}^{-1}(\mathbf{B}U) = U$. Then in particular there exists an open $V \subset M$ such that $\mathbf{B}U = \mathbf{B}M \cap V = \mathbf{B}V$. But then $U = \mathbf{B}^{-1}(\mathbf{B}U) = \mathbf{B}^{-1}(\mathbf{B}V) = V$ and thus U is open. $\boxed{QED}$

4.13 Corollary. *The subset $\mathbf{B}M$ of a proto $\mathcal{A}$-manifold with its induced topology inherits a natural structure of a proto $\mathbf{R}$-manifold. In particular, if M has dimension $p|q$, then $\mathbf{B}M$ has dimension p.*

Proof. If $\varphi_a : U_a \to O_a$ and $\varphi_b : U_b \to O_b$ are two (compatible) charts for M, then in the first place $\varphi_a : \mathbf{B}U_a \to \mathbf{B}O_a$ and $\varphi_b : \mathbf{B}U_b \to \mathbf{B}O_b$ are two charts for $\mathbf{B}M$. Since the restriction of a smooth map to real coordinates is a smooth map in the ordinary sense [2.7], it follows that these two charts are compatible. Hence the smooth $\mathcal{A}$-structure on M induces a smooth $\mathbf{R}$-structure on $\mathbf{B}M$.

If M is modeled on an $\mathcal{A}$-vector space E, $\mathbf{B}M$ is modeled on $\mathbf{B}E_0$. In particular, if M has dimension $p|q$, then $\mathbf{B}M$ has dimension p. The fact that the odd dimension disappears is because the charts of M lie in the *even* part of E, and thus the odd coordinates become zero when taking the body map. $\boxed{QED}$

4.14 Definition. In the literature, a manifold is usually defined as a second countable (i.e., a countable basis for the topology) Hausdorff topological space equipped with a smooth $\mathbf{R}$-structure. Sometimes the condition *second countable* is replaced by *paracompact*. These conditions are added besides the smooth structure condition in order to get rid of some pathological examples and in order to have partitions of unity. Most of the general theory can be developed without these conditions, but not much is gained by doing so.

We thus define an *$\mathcal{A}$-manifold* as a proto $\mathcal{A}$-manifold M for which its body $\mathbf{B}M$ is a second countable Hausdorff topological space. It follows immediately that $\mathbf{B}M$ is an $\mathbf{R}$-manifold in the usual sense. Although the condition of second countability could have been imposed directly on M, the Hausdorff condition can not because the DeWitt topology is not Hausdorff.

4.15 Definition. A map $f : M \to N$ between two $\mathcal{A}$-manifolds is *smooth* if for all charts (U, φ) of M and all charts (V, ψ) of N, one has

$$\psi \circ f \circ \varphi^{-1} \text{ is smooth on } \varphi(U \cap f^{-1}(V)) .$$

Note that this condition is empty whenever $U \cap f^{-1}(V)$ is the empty set. Since smooth maps are in particular continuous, it follows that smooth maps between $\mathcal{A}$-manifolds are continuous. The set of all smooth maps from M to N is denoted by $C^\infty(M; N)$.

A map $f : M \to N$ is a *diffeomorphism* if it is a smooth bijection whose inverse is also smooth. If that is the case, the $\mathcal{A}$-manifolds M and N are said to be *diffeomorphic*.

4.16 Nota Bene. If the map $f : M \to N$ is smooth and if $\phi : U \to O \subset E_0$ and $\psi : V \to P \subset F_0$ are charts, then the map $\psi \circ f \circ \phi^{-1} : \phi(U \cap f^{-1}(V)) \subset E_0 \to F$ is an *even* smooth map. In other words, smooth maps between $\mathcal{A}$-manifolds are always even in the sense of [1.20].

4.17 Lemma. *Let $f : M \to N$ be a map, $\mathcal{U}$ an atlas for M, and $\mathcal{V}$ a set of charts on N covering $f(M)$. If f satisfies the condition given in [4.15] for all charts $U \in \mathcal{U}$ and $V \in \mathcal{V}$, then f is smooth.*

Proof. Denote by (U, φ) and (V, ψ) arbitrary charts and by (U_a, φ_a) and (V_b, ψ_b) charts in $\mathcal{U}$ and $\mathcal{V}$ respectively. On the set $\varphi(U_a \cap U \cap f^{-1}(V_b \cap V))$ we have the equality $\psi \circ f \circ \varphi^{-1} = (\psi \circ \psi_b^{-1}) \circ (\psi_b \circ f \circ \varphi_a^{-1}) \circ (\varphi_a \circ \varphi^{-1})$, which is smooth by definition of charts and the hypothesis of the lemma. Since the sets $U_a \cap U \cap f^{-1}(V_b \cap V)$ cover the set $U \cap f^{-1}(V)$ (the U_a cover M and the V_b cover $f(M)$), we conclude by [1.23-k]. QED

4.18 Corollary. *Let $f : M \to N$ be a map between $\mathcal{A}$-manifolds, let $\mathcal{U}$ be an atlas for M. If we define the maps f_a by $f_a = f|_{U_a}$, where $U_a \in \mathcal{U}$ is seen as an $\mathcal{A}$-manifold [4.8], then the map f is smooth if and only if all the maps f_a are smooth.*

4.19 Lemma. *Let L, M, and N be $\mathcal{A}$-manifolds, and let $f : L \to M$ and $g : M \to N$ be smooth maps. Then $g \circ f : L \to N$ is smooth.*

Proof. Let (U, φ), (V, ψ), and (W, χ) be charts for L, M, and N respectively. What we have to show is that the map $\chi \circ (g \circ f) \circ \varphi^{-1}$ is smooth on $\varphi(U \cap (g \circ f)^{-1}(W))$. On $\varphi(U \cap (g \circ f)^{-1}(W) \cap f^{-1}(V))$ we have $\chi \circ (g \circ f) \circ \varphi^{-1} = (\chi \circ g \circ \chi^{-1}) \circ (\chi \circ f \circ \varphi^{-1})$, which is smooth since f and g are smooth. By varying V, we cover $\varphi(U \cap (g \circ f)^{-1}(W))$ by sets of the form $\varphi(U \cap (g \circ f)^{-1}(W) \cap f^{-1}(V))$. Using either [4.18] or [1.23-k] we conclude that $\chi \circ (g \circ f) \circ \varphi^{-1}$ is smooth. QED

4.20 Discussion. If M is an $\mathcal{A}$-manifold and F an $\mathcal{A}$-vector space, we could have defined smooth maps $f : M \to F$ as those maps such that for all charts (U, φ) of M the map $f \circ \varphi^{-1}$ is smooth. This makes sense because it is a map from an open subset of the even part of some $\mathcal{A}$-vector space to the $\mathcal{A}$-vector space F. On the other hand, we can interpret F as an $\mathcal{A}$-manifold as in [4.8] and apply [4.15]. Since F has a single chart $F \to F_0^\sharp$ it follows from [1.23-b,i] that these two definitions coincide. Now let $O \subset E_0$ and $U \subset M$ be open. The above discussion gives us the following characterization: a map $\phi : U \to O$ is a chart [4.1] if and only if ϕ is a diffeomorphism [4.15].

4.21 Lemma. *Let M be an $\mathcal{A}$-manifold and F an $\mathcal{A}$-vector space. The set of all $\mathcal{A}$-valued smooth functions $C^\infty(M) \equiv C^\infty(M; \mathcal{A})$ is a graded $\mathbf{R}$-algebra and $C^\infty(M; F)$ is a free graded $C^\infty(M)$-module of the same graded dimension as F.*

4.22 Proposition. *Given a smooth map $f : M \to N$ between two $\mathcal{A}$-manifolds, the restriction of f to $\mathbf{B}M \subset M$ is the unique smooth map $\mathbf{B}f : \mathbf{B}M \to \mathbf{B}N$ between $\mathbf{R}$-manifolds such that the following diagram is commutative:*

$$\begin{array}{ccc} M & \xrightarrow{\ f\ } & N \\ {\scriptstyle \mathbf{B}}\downarrow & & \downarrow{\scriptstyle \mathbf{B}} \\ \mathbf{B}M & \xrightarrow[\mathbf{B}f = f|_{\mathbf{B}M}]{} & \mathbf{B}N\,. \end{array}$$

Proof. Uniqueness of $\mathbf{B}f$ is immediate from the surjectivity of $\mathbf{B}$. Since f is smooth, the image of a point with real coordinates must have real coordinates, hence $f(\mathbf{B}M) \subset \mathbf{B}N$. Smoothness as a map between $\mathbf{R}$-manifolds follows from [2.7]. QED

4.23 Lemma. *Let M be an $\mathcal{A}$-manifold and let F be an $\mathcal{A}$-vector space. Then the map $\mathbf{B} : C^\infty(M; \mathcal{A}) \to C^\infty(\mathbf{B}M; \mathbf{R})$ is a morphism of graded $\mathbf{R}$-algebras with unit and*

the map $\mathbf{B} : C^\infty(M;F) \to C^\infty(\mathbf{B}M;\mathbf{B}F)$ *is a morphism from graded modules over* $C^\infty(M;\mathcal{A})$ *to graded modules over* $C^\infty(\mathbf{B}M;\mathbf{R})$.

5. Constructions of $\mathcal{A}$-manifolds

This section treats subjects directly related to the creation of $\mathcal{A}$*-manifolds. The definitions of a direct product and of a submanifold are straightforward generalizations of the usual definitions. The implicit function theorem allows us in the standard way to define submanifolds of* E_0 *by equations (a more general version can be found in [V.2.10]). The only point in which submanifolds of* $\mathcal{A}$*-manifolds differ from their ordinary counterparts is that it is not possible to find submanifolds of lower dimension through an arbitrary point of the ambient* $\mathcal{A}$*-manifold. We define and prove the existence of partitions of unity, and we finish this section by extending the body map* $\mathbf{B}$ *and the "inverse" map* $\mathbf{G}$ *to* $\mathcal{A}$*-manifolds. As for the even part* E_0 *of an* $\mathcal{A}$*-vector space* E*,* $\mathbf{G}$ *can not reconstruct the odd dimension of an* $\mathcal{A}$*-manifold. More precisely:* $\mathbf{G}(\mathbf{B}M)$ *reconstructs the even part of* M*, omitting the odd coordinates completely.*

5.1 Definition. A subset $N \subset M$ of an $\mathcal{A}$-manifold M modeled on an $\mathcal{A}$-vector space E is a *submanifold* if there exists for each point $m_o \in N$ a chart $\varphi : U \to O \subset E_0$ containing m_o and a graded subspace F of E such that

$$\varphi(U \cap N) = O \cap F_0 \equiv O \cap F .$$

5.2 Discussion. If F is a graded subspace of the $\mathcal{A}$-vector space E, there exists a homogeneous basis $(e_i)_{i=1}^n$ such that a subset is a basis for F. Without loss of generality we may assume that F is generated by $(e_i)_{i=1}^k$. It follows that the submanifold N is locally described by the $n-k$ equations

$$U \cap N = \{\, m \in U \mid \forall k < i \leq n : {}^i\!e(\varphi(m)) = 0 \,\} .$$

5.3 Nota Bene. The definition of a submanifold looks quite innocuous and completely similar to the ordinary definition for $\mathbf{R}$-manifolds. However, one should realize that, contrary to the ordinary case, one can not guarantee that through every point of an $\mathcal{A}$-manifold passes a submanifold of lower dimension. This is due to the fact that we require that F is a graded subspace of E, and not just any graded submodule.

To show the problem, let E be the $\mathcal{A}$-vector space of dimension $0|2$ with basis e_1, e_2. We take $M = E_0 \cong \mathcal{A}_1^2$ the $\mathcal{A}$-manifold of dimension $0|2$. Any graded subspace of dimension $0|1$ of E must be generated by a vector of the form $f = ae_1 + be_2$ with $a, b \in \mathbf{R}$. Now let (ξ_1, ξ_2) be a point of M and suppose that it belongs to a submanifold N of M of dimension $0|1$. Since there is just one single chart possible, this means that $(\xi_1, \xi_2) = \xi_1 e_1 + \xi_2 e_2$ must belong to the subspace generated by $ae_1 + be_2$, i.e., there exists $\lambda \in \mathcal{A}_1$ such that $\xi_1 = \lambda a$ and $\xi_2 = \lambda b$. If neither ξ_1 nor ξ_2 is zero, this implies that neither a nor b can be zero. Since they are real, we deduce that $\xi_2 = \xi_1 b/a$. And if $\mathcal{A}_1$ is not too small, it is easy to find ξ_1 and ξ_2 that are not a real multiple of each other. Hence there does not exists a submanifold of dimension $0|1$ through such a point.

5.4 Proposition. *Let N be a submanifold of an $\mathcal{A}$-manifold M equipped with the induced topology. Then there exists a (canonically defined) smooth $\mathcal{A}$-structure on N making it an $\mathcal{A}$-manifold for which the canonical injection $i : N \to M$ is smooth. Moreover, $\mathbf{B}N$ is a submanifold of $\mathbf{B}M$ in the sense of $\mathbf{R}$-manifolds.*

Proof. Let $\varphi : U \to O \subset E_0$ be a chart as specified in the definition of submanifold and let F be the graded subspace of E such that $\varphi(U \cap N) = O \cap F_0$. By definition of induced topology, the set $V = U \cap N$ is open in N, $P = O \cap F_0$ is open in F_0 and the map $\psi = \varphi|_V : V \to P$ is a homeomorphism. In other words, $\psi : V \to P$ is a chart for N. If we denote by $\mathcal{V}$ the collection of all such charts for N, it automatically satisfies condition (SS1) of a smooth $\mathcal{A}$-structure.

To show that this $\mathcal{V}$ also satisfies condition (SS2), let $\psi_a : V_a \to P_a \subset (F_a)_0$ and $\psi_b : V_b \to P_b \subset (F_b)_0$ be two charts of N associated to charts $\varphi_a : U_a \to O_a \subset (E_a)_0$ and $\varphi_b : U_b \to O_b \subset (E_b)_0$ of M. The map $\varphi_b \circ \varphi_a^{-1} : \varphi_a(U_a \cap U_b) \to (E_b)_0$ is smooth by definition of charts, so by [1.23-g] its restriction to $\varphi_a(U_a \cap U_b) \cap (F_a)_0$ is smooth. But $\varphi_a(U_a \cap U_b) \cap (F_a)_0 = \varphi_a(U_a \cap U_b \cap N) = \psi_a(V_a \cap V_b)$, hence $\varphi_a^{-1}(\psi_a(V_a \cap V_b)) \subset N$ and thus $\varphi_b \circ \varphi_a^{-1}(\psi_a(V_a \cap V_b)) \subset (F_b)_0$. It follows from [1.23-i] that $\varphi_b \circ \varphi_a^{-1} : \psi_a(V_a \cap V_b) \to (F_b)_0$ is smooth. But with these restrictions, the map $\varphi_b \circ \varphi_a^{-1}$ is exactly the map $\psi_b \circ \psi_a^{-1}$, showing that $\mathcal{V}$ satisfies also condition (SS2), i.e., $\mathcal{V}$ is an atlas for a smooth $\mathcal{A}$-structure on N by [4.4].

Finally $\mathbf{B}N$ is a second countable Hausdorff space because it is a subset of $\mathbf{B}M$ which has these properties. In terms of the given atlas for N it is immediate from [4.17] that the canonical injection is smooth.

To prove the statement about $\mathbf{B}N$, we apply the body map to the definition of a submanifold. Since for open sets we have $\mathbf{B}^{-1}(\mathbf{B}U) = U$ and $\mathbf{B}^{-1}(\mathbf{B}O) = O$ we have $\mathbf{B}(U \cap N) = \mathbf{B}U \cap \mathbf{B}N$ and $\mathbf{B}(O \cap F_0) = \mathbf{B}O \cap \mathbf{B}F_0$. Since the $\varphi : \mathbf{B}U \to \mathbf{B}O$ are the charts for $\mathbf{B}M$, it follows immediately that $\mathbf{B}N$ is a submanifold of $\mathbf{B}M$ in the sense of $\mathbf{R}$-manifolds. *QED*

5.5 Remark. We will show in [V.2.17] that the smooth $\mathcal{A}$-structure on a submanifold $N \subset M$ given in [5.4] is the *unique* smooth $\mathcal{A}$-structure such that the canonical injection $i : N \to M$ is an embedding.

5.6 Lemma. *Let M be an $\mathcal{A}$-manifold, $N \subset M$ a submanifold, and $L \subset N$ a submanifold of the $\mathcal{A}$-manifold N. Then $L \subset M$ is a submanifold of M.*

Proof. Let M be modeled on E, $F \subset E$ a graded subspace on which N is modeled, and $H \subset F$ a graded subspace on which L is modeled. Let $\psi : V \to P$ be a chart for N such that $\psi(V \cap L) = P \cap H_0$. By taking smaller charts we may assume that ψ is a chart induced from a chart $\phi : U \to O$ for M, i.e., $V = U \cap N$, $P = O \cap F_0$, and $\psi = \phi|_V$. It follows that $\phi(U \cap L) = O \cap H_0$. $\boxed{QED}$

5.7 Lemma. *Let N be a submanifold of M and $f : M \to L$ a smooth map. Then $f|_N : N \to L$ is smooth.*

Proof. Let $\phi : U \to O$ be a chart for M such that $U \cap N = V$, $P = O \cap F_0$ and $\psi = \phi|_V : V \to P$ is a chart for N. Let furthermore $\chi : W \to Q$ be a chart for L. According to [4.17] it suffices to prove that $\chi \circ f|_N \circ \psi^{-1} : \psi(V \cap (f|_N)^{-1}(W)) \to Q$ is smooth. But this is the map $\chi \circ f \circ \phi^{-1}$ restricted to $\phi(U \cap f^{-1}(W)) \cap F_0$, which is smooth by [1.23-g]. $\boxed{QED}$

5.8 Lemma. *Let M and L be $\mathcal{A}$-manifolds, let $N \subset M$ be a submanifold, and let $f : L \to M$ be a smooth map. If $f(L) \subset N$, then the map $f : L \to N$ is smooth when N is equipped with the smooth $\mathcal{A}$-structure of [5.4]*

Proof. Let us denote the map from L to the $\mathcal{A}$-manifold N by f_o in order to distinguish it from the initial map to M. (Nota Bene: If f_o is smooth, automatically $f = i \circ f_o$ is smooth.) Let $\phi : U \to O$ be a chart for M such that $U \cap N = V$, $P = O \cap F_0$ and $\psi = \phi|_V : V \to P$ is a chart for N. Since F is a graded subspace of the finite dimensional $\mathcal{A}$-vector space E, there exists a supplement [II.3.7], [II.6.23], and thus a smooth projection $\pi : E \to F$. Since $f(L) \subset N$, obviously the restriction of f_o to $f^{-1}(U)$ is given as $f_o = \pi \circ f$. Since the $f^{-1}(U)$ cover L, f_o is smooth. $\boxed{QED}$

5.9 Example. For any $\mathcal{A}$-manifold M modeled on an $\mathcal{A}$-vector space E there exists a (canonically defined) submanifold $M^{w.o.d}$ modeled on $E^{w.o.d}$. Let $\{\, U_a \mid a \in I \,\}$ be an atlas for M and define the set $M^{w.o.d}$ by $M^{w.o.d} = \cup_{a \in I} \varphi_a^{-1}(O_a \cap E^{w.o.d})$. To prove that it is a submanifold, we claim that the charts U_a satisfy the conditions of [5.1]. We first compute $\varphi_b(\varphi_a^{-1}(O_a \cap E^{w.o.d}) \cap U_b) = (\varphi_b \circ \varphi_a^{-1})(\varphi_a(U_b \cap U_a) \cap E^{w.o.d})$. Since $\varphi_b \circ \varphi_a^{-1}$ is a smooth map, it follows from [3.17] that $(\varphi_b \circ \varphi_a^{-1})(\varphi_a(U_b \cap U_a) \cap E^{w.o.d})$ is contained in $\varphi_b(U_b \cap U_a) \cap E^{w.o.d}$. Hence $\varphi_b(M^{w.o.d} \cap U_b) = O_b \cap E^{w.o.d}$.

The submanifold $M^{w.o.d}$ can also be described as consisting of those points in M for which the odd coordinates in some chart (and hence in all charts) are zero. It follows that we have the inclusions $\mathbf{B}M \subset M^{w.o.d} \subset M$. Moreover, $\mathbf{B}(M^{w.o.d}) = \mathbf{B}M \subset M$. Given that we have a (canonical) projection $\mathbf{B} : M \to \mathbf{B}M$, one might be tempted to think that there also exists a natural projection $M \to M^{w.o.d}$. That this is a false belief (at least when the odd dimension of M is greater than 1) is shown by the next example.

Consider the $\mathcal{A}$-manifold $M = E_0$ with coordinates (x, ξ, η), where E is an $\mathcal{A}$-vector space of dimension $1|2$. The map $f : (x, \xi, \eta) \mapsto (x + \xi\eta, \xi, \eta)$ is a diffeomorphism. However, projection onto the even coordinate is not invariant under this diffeomorphism (but projection onto the body is).

Let $f : M \to N$ be a smooth map between $\mathcal{A}$-manifolds. If the odd coordinates of $m \in M$ are zero, i.e., $m \in M^{w.o.d}$, the odd coordinates of $f(m)$ are also zero (if all odd coordinates are zero, the parity of the image of a smooth function is necessarily even). We thus have an induced map $f|_{M^{w.o.d}} : M^{w.o.d} \to N^{w.o.d}$. Since $\mathbf{B}M \subset M^{w.o.d}$ and $\mathbf{B}N \subset N^{w.o.d}$, we can combine this into a commutative diagram

$$\begin{array}{ccccc} \mathbf{B}M & \subset & M^{w.o.d} & \subset & M \\ {\scriptstyle f|_{\mathbf{B}M} \equiv \mathbf{B}f}\downarrow & & \downarrow{\scriptstyle f|_{M^{w.o.d}}} & & \downarrow{\scriptstyle f} \\ \mathbf{B}N & \subset & N^{w.o.d} & \subset & N\,. \end{array}$$

5.10 Example (submanifolds defined by equations). Let $f : E_0 \to F$ be a smooth map and $M = f^{-1}(0) \subset E_0$. Suppose that for all $m \in M$: $\operatorname{rank}(f)(m) = \dim F = k$ (implying that $k = \dim F \leq \dim E = n$). Let $g : V \to V'$ be as in [3.30], i.e., such that $f \circ g^{-1}(x^1, \ldots, x^n) = (x^1, \ldots, x^k)$. Since g and g^{-1} are smooth, $g : V \to V'$ is a chart for E_0. Moreover, $g(V \cap M) = V' \cap \operatorname{Span}(e_{k+1}, \ldots, e_n)$. We conclude that M is a submanifold of E_0, and thus in particular that M is an $\mathcal{A}$-manifold.

5.11 Construction. Given two $\mathcal{A}$-manifolds M and N, we equip the cartesian product $M \times N$ with smooth $\mathcal{A}$-structure by the following prescription. Let $\mathcal{U}$ be an atlas for M and $\mathcal{V}$ one for N. Now define an atlas $\mathcal{W}$ for $M \times N$ by

$$\mathcal{W} = \{\, \varphi_a \times \psi_b : U_a \times V_b \to O_a \times P_b \mid a \in I, b \in J \,\}\,.$$

In this construction, the map $\varphi_a \times \psi_b$ is defined by $(\varphi_a \times \psi_b)(x, y) = (\varphi_a(x), \psi_b(y))$; that it is smooth is a direct consequence of [1.23-f,i]. We leave it to the reader to verify that $\mathcal{W}$ indeed satisfies (SS1) and (SS2), and thus defines a smooth $\mathcal{A}$-structure for $M \times N$. Since the product of two second countable Hausdorff spaces is again second countable and Hausdorff, it follows that $M \times N$ is an $\mathcal{A}$-manifold, called the *direct product* of the $\mathcal{A}$-manifolds M and N. If M is modeled on E and N on F, then $M \times N$ is modeled on $E \oplus F$. Associated with this direct product, we have two canonical projections $\pi_M : M \times N \to M$ and $\pi_N : M \times N \to N$; it is easy to show that these two maps are smooth maps between $\mathcal{A}$-manifolds.

Applying the body map $\mathbf{B}$ to the direct product $M \times N$, it is easy to see that we have $\mathbf{B}(M \times N) = (\mathbf{B}M) \times (\mathbf{B}N)$. Strictly speaking this is not an equality but a canonical diffeomorphism. However, as usual we will forego such subtleties and pretend that it actually *is* an equality.

5.12 Lemma. *Let L, M, and N be $\mathcal{A}$-manifolds, and let $f : L \to M$ and $g : L \to N$ be maps. Then the map $(f, g) : L \to M \times N$ defined by $(f, g)(x) = (f(x), g(x))$ is smooth if and only if $f = \pi_M \circ (f, g)$ and $g = \pi_N \circ (f, g)$ are smooth.*

Proof. If (f, g) is smooth, $f = \pi_M \circ (f, g)$ and $g = \pi_N \circ (f, g)$ are smooth by [4.19]. So suppose f and g are smooth. Denote by $\chi : W \to Q$ a chart of L and use the notation of [5.11]. We have to show that $(\varphi_a \times \psi_b) \circ (f, g) \circ \chi^{-1} : Q \to O_a \times P_b$ is smooth. But $((\varphi_a \times \psi_b) \circ (f, g) \circ \chi^{-1})(x) = ((\varphi_a \circ f \circ \chi^{-1})(x), (\psi_a \circ g \circ \chi^{-1})(x))$, which is smooth by the hypothesis and [1.23-i]. QED

5.13 Lemma. *If $f : M \to N$ is a smooth map between $\mathcal{A}$-manifolds, then its graph $Gr(f) = \{ (m, n) \in M \times N \mid n = f(m) \}$ is a submanifold of $M \times N$. Moreover, the projection $\pi_M : Gr(f) \to M$ is a diffeomorphism.*

Proof. Let $(m, f(m)) \in Gr(f)$ be arbitrary, let $\psi : V \to P \subset F$ be a chart of N containing $f(m)$ and let $\phi : U \to O \subset E$ be a chart of M containing m such that $f(U) \subset V$. Such charts exist by continuity of f. We thus have obtained a chart $\phi \times \psi : U \times V \to O \times P$ of $M \times N$. Since f is smooth, the map $O \times F \to O \times F$, $(x, y) \mapsto (x, y - \psi(f(\phi^{-1}(x))))$ is a diffeomorphism. Denote by Q the image of $O \times P$ under this map. It follows that $\Phi : U \times V \to Q$, $(m, n) \mapsto (\phi(m), \psi(n) - \psi(f(m)))$ is a chart for $M \times N$. Obviously $\Phi(M \times N \cap Gr(f) = Q \cap (E \oplus \{0\})$, proving that $Gr(f)$ is a submanifold of $M \times N$. For the second part, note that $\widehat{f} : M \to M \times N$ defined by $\widehat{f}(m) = (m, f(m))$ is smooth by [5.12]; by [5.8] we can see $\widehat{f}$ as a smooth map $M \to Gr(f)$. Obviously $\pi_M \circ \widehat{f} = id(M)$ and $\widehat{f} \circ (\pi_M|_{Gr(f)}) = id(Gr(f))$. QED

5.14 Lemma. *If $f : M \to N$ and $g : K \to L$ are smooth maps, then so is the map $f \times g : M \times K \to N \times L$ defined by $(f \times g)(m, k) = (f(m), g(k))$.*

Proof. By [5.12] it suffices to show that $\pi_N \circ (f \times g)$ and $\pi_L \circ (f \times g)$ are smooth. But $(\pi_N \circ (f \times g))(m, k) = f(m)$, which is smooth essentially by [1.23-f]. A similar argument for the other component finishes the proof. QED

5.15 Definitions. If $f : M \to \mathcal{A}$ is a continuous map, the set $\text{supp}(f) \subset M$, called *the support of f*, is defined as the closure of the set of those points where $f(m)$ is non-zero. In other words, $m \notin \text{supp}(f)$ if and only if there exists an open set $U \ni m$ such that $f|_U$ is identically zero.

A collection of subsets $U_\alpha \subset M$, $\alpha \in J$ is called *locally finite* if for each point $m \in M$ there exists a neighborhood of m that meets only a finite number of U_α's; in formula:

$$\forall m \in M \, \exists U \ni m \text{ open } \exists \alpha_1, \ldots, \alpha_n \in J : \alpha \notin \{\alpha_1, \ldots, \alpha_n\} \Rightarrow U \cap U_\alpha = \phi \, .$$

5.16 Lemma. *Let $U_o \subset M$ be open and $f_o : U_o \to \mathcal{A}$ smooth. If $\operatorname{supp}(f_o) \subset U_o$ is closed in M, then $f : M \to \mathcal{A}$ defined by $f|_{U_o} = f_o$, $f \equiv 0$ outside U_o, is a smooth function on M.*

Proof. Let $\mathcal{U} = \{ U_a \mid a \in I \}$ be an atlas of M containing U_o. Define a new atlas $\widehat{\mathcal{U}} = \{ \widehat{U}_a \mid a \in I \}$ by $\widehat{U}_o = U_o$ and $\widehat{U}_a = U_a \cap (M \setminus \operatorname{supp}(f_o))$ if $a \neq o$. This is indeed an atlas because a subset of a chart is a chart; they cover M because $\operatorname{supp}(f_o) \subset U_o$. It follows that for $a \neq o$ we have $\widehat{U}_a \cap \operatorname{supp}(f_o) = \phi$ and hence $f|_{\widehat{U}_a} \equiv 0$. Smoothness of f now follows from [4.18]. QED

5.17 Lemma. *Let $f_\alpha : M \to \mathcal{A}$, $\alpha \in J$ be a collection of smooth maps such that the collection $\operatorname{supp}(f_\alpha)$, $\alpha \in J$ is locally finite. Then $f = \sum_{\alpha \in J} f_\alpha$ is a well defined smooth function on M.*

Proof. Let $\mathcal{U}$ be an atlas of M such that each element $U \in \mathcal{U}$ meets only finitely many $\operatorname{supp}(f_\alpha)$'s. It follows that $f|_U = \sum_{i=1}^n f_{\alpha_i}|_U$, where the α_i are the only indices with $U \cap \operatorname{supp}(f_{\alpha_i}) \neq \phi$. As a finite sum of smooth functions, $f|_U$ is smooth. The result now follows from [4.18]. QED

5.18 Definition. Let $\mathcal{V} = \{ V_\alpha \mid \alpha \in J \}$ be a collection of open sets on M that cover M. *A partition of unity subordinated to the open cover* $\mathcal{V}$ is a collection of smooth functions $\rho_\alpha : M \to \mathcal{A}_0$, $\alpha \in J$ satisfying the following conditions.

(i) $\operatorname{supp}(\rho_\alpha) \subset V_\alpha$,
(ii) the collection $\{\operatorname{supp}(\rho_\alpha)\}$ is locally finite,
(iii) $\forall m \in M : \mathbf{B}(\rho_\alpha(m)) \in [0, 1]$, and
(iv) $\sum_{\alpha \in J} \rho_\alpha = 1$.

5.19 Corollary. *Let $\mathcal{V} = \{ V_\alpha \mid \alpha \in J \}$ be an open cover of M, $f_\alpha \in C^\infty(V_\alpha)$, and $\{ \rho_\alpha \mid \alpha \in J \}$ a partition of unity subordinated to $\mathcal{V}$. Then $f = \sum_{\alpha \in J} \rho_\alpha f_\alpha$ is a well defined smooth function on M.*

Proof. An elementary combination of [5.16] and [5.17]. QED

5.20 Proposition. *Let $\mathcal{V}$ be an open cover of an $\mathcal{A}$-manifold M. Then there exists a partition of unity subordinated to $\mathcal{V}$.*

Proof. We proceed in three steps. The first step is to *admit* the existence of partitions of unity for ordinary manifolds. The reader is referred to [Spi, Ch 2] for proofs. The second step is to prove the existence in case $\mathcal{V}$ is an atlas, and the third step is to prove the general case.

We thus start with the second step and assume that $\mathcal{U} = \{ \varphi_a : U_a \to O_a \mid a \in I \}$ is an atlas for M. It follows that $\{\mathbf{B}U_a\}$ is an open cover of $\mathbf{B}M$. By the first step there exists a partition of unity $\sigma_a : \mathbf{B}M \to [0,1]$ subordinated to this cover. Let U_o be a fixed (but otherwise arbitrary) chart in the atlas. It follows that the map $\sigma_o \circ (\mathbf{B}\varphi_o)^{-1} : \mathbf{B}O_o \to [0,1]$ is a smooth map. Applying $\mathbf{G}$ we obtain a smooth map $\mathbf{G}(\sigma_o \circ (\mathbf{B}\varphi_o)^{-1}) : O_o^{w.o.d} \to \mathcal{A}_0$. We can interpret this function as the first term in the expansion (2.24) of a smooth function on O_a, the other terms being identically zero. In other words, we obtain a smooth function $\hat{\rho}_o = \mathbf{G}(\sigma_o \circ (\mathbf{B}\varphi_o)^{-1}) \circ \varphi_o : U_o \to \mathcal{A}_0$, a function independent of the odd coordinates in O_o. It follows directly from the construction that we have $\mathbf{B} \circ \hat{\rho}_o = \sigma_o \circ \mathbf{B}$. Moreover, the definition of $\mathbf{G}$ by means of derivatives shows that $\mathbf{B}m \notin \mathrm{supp}(\sigma_o)$ implies $m \notin \mathrm{supp}(\hat{\rho}_o)$, i.e., $\mathrm{supp}(\hat{\rho}_o) \subset \mathbf{B}^{-1}(\mathrm{supp}(\sigma_o))$. A priori the set $\mathrm{supp}(\hat{\rho}_o)$ is closed in U_o but not necessarily in M. However, since $\mathrm{supp}(\sigma_o)$ is closed in $\mathbf{B}M$, the above inclusion shows that $\mathrm{supp}(\hat{\rho}_o) \subset U_o$ is indeed closed in M. By [5.16] we may extend $\hat{\rho}_o$ to the whole of M by zero outside of U_o without losing smoothness.

We thus have created smooth functions $\hat{\rho}_a : M \to \mathcal{A}_0$ associated to the cover (atlas) $\mathcal{U}$ which manifestly satisfy conditions (i) and (iii) of a partition of unity. It also satisfies condition (ii) because $\mathrm{supp}(\hat{\rho}_a) \subset \mathbf{B}^{-1}(\mathrm{supp}(\sigma_a))$ and because the σ_a are a partition of unity. We deduce by [5.17] that $\hat{\rho} = \sum_{a\in I} \hat{\rho}_a : M \to \mathcal{A}_0$ is a smooth function, which obviously satisfies $\mathbf{B} \circ \hat{\rho} \equiv 1$. (Nota Bene. $\hat{\rho}$ itself need not be identically 1.) Since inversion is smooth, the functions $\rho_a = \hat{\rho}_a / \hat{\rho}$ are smooth. The collection of these maps is a partition of unity subordinated to the cover (atlas) $\mathcal{U}$. Note that, just as the $\hat{\rho}_a$, the ρ_a satisfy the relation $\mathbf{B} \circ \rho_a = \sigma_a \circ \mathbf{B}$, i.e., $\mathbf{B}\rho_a = \sigma_a$.

To prove the third step, let $\mathcal{V} = \{ V_\alpha \mid \alpha \in J \}$ be an arbitrary open cover of M and let $\mathcal{U} = \{ U_a \mid a \in I \}$ be an atlas. It follows that $\{ U_{a\alpha} = U_a \cap V_\alpha \mid a \in I, \alpha \in J \}$ is an atlas. Hence by the previous construction there exists a partition of unity $\rho_{a\alpha}$ subordinated to this atlas. We thus can define the smooth functions $\rho_\alpha = \sum_{a\in I} \rho_{a\alpha}$. Since the $\mathrm{supp}(\rho_{a\alpha})$ are locally finite, the inclusion $\mathrm{supp}(\rho_\alpha) \subset \cup_{a\in I} \mathrm{supp}(\rho_{a\alpha})$ follows. From this we deduce that the $\mathrm{supp}(\rho_\alpha)$ are locally finite, thus proving property (ii) of a partition of unity. The other three properties being obvious, we have shown the existence of a partition of unity subordinated to the (arbitrary) open cover $\mathcal{V}$. QED

5.21 Lemma. *Let m be a point in an open set U of an $\mathcal{A}$-manifold M. Then there exists a $\rho \in C^\infty(M)_0$ and an open neighborhood $V \subset U$ of m such that* $\mathrm{supp}(\rho) \subset U$ *and such that $\rho|_V \equiv 1$. We will call such a function a* plateau function around m in U.

Proof. Let $\varphi_o : U_o \to O_o$ be a chart of M such that $m \in U_o \subset U$. Since the topology on M is induced by the topology of $\mathbf{B}M$ [4.12], and since any $\mathbf{R}^p$ is normal, there exists an open subset V of M such that $m \in V \subset \overline{V} \subset U_o \subset U$, where overlining here means the closure. Finally let $\{\rho, \tau\}$ be a partition of unity subordinated to the open cover $\{U, M \setminus \overline{V}\}$. QED

5.22 Definition. In §2 we have seen that, under suitable circumstances, the body map $\mathbf{B}$ admits an "inverse" $\mathbf{G}$. We will now show that under similar conditions the same is true for $\mathcal{A}$-manifolds. To that end we need the notion of an *$\mathcal{A}$-manifold without odd dimensions*, which is an $\mathcal{A}$-manifold modeled on an $\mathcal{A}$-vector space without odd dimensions. It should be obvious that M is an $\mathcal{A}$-manifold without odd dimensions if and only if $M = M^{w.o.d}$.

5.23 Construction. Let M be an $\mathbf{R}$-manifold modeled on an $\mathbf{R}$-vector space X. Since the odd part X_1 does not play any role in the construction of M, we may assume that X is an $\mathbf{R}$-vector space without odd dimensions, i.e., $X_1 = \{0\}$. Denoting by E the $\mathcal{A}$-vector space without odd dimensions $E = \mathbf{G}X = \mathbf{G}(X_0 \oplus \{0\}_1)$, we will construct an $\mathcal{A}$-manifold without odd dimensions $\mathbf{G}M$ modeled on E such that $\mathbf{B}(\mathbf{G}M) = M$; moreover, for any $\mathcal{A}$-manifold without odd dimensions N we will have $\mathbf{G}(\mathbf{B}N) = N$.

Let $\mathcal{U} = \{\varphi_a : U_a \to O_a \mid a \in I\}$ an atlas for M. We then consider the subsets $O_{ab} = \varphi_a(U_a \cap U_b) \subset O_a$ and the maps $\varphi_{ba} = \varphi_b \circ \varphi_a^{-1}$, to which we apply the map $\mathbf{G}$ (see §2) to obtain open sets $\mathbf{G}O_a$ in E_0, open subsets $\mathbf{G}O_{ab} \subset \mathbf{G}O_a$, and smooth maps $\mathbf{G}\varphi_{ba}$. These new data satisfy exactly the conditions of [4.9] and thus define a proto $\mathcal{A}$-manifold $\mathbf{G}M$.

If we apply the body map $\mathbf{B}$ to the data $\mathbf{G}O_{ab} \subset \mathbf{G}O_a$ and $\mathbf{G}\varphi_{ba}$, we recover exactly the data needed to reconstruct M, showing that $\mathbf{B}(\mathbf{G}M) = M$. Since M is supposed to be second countable and Hausdorff, $\mathbf{G}M$ is an $\mathcal{A}$-manifold. By the same token we see that for an $\mathcal{A}$-manifold without odd dimensions N we have $\mathbf{G}(\mathbf{B}N) = N$. However, if N is *not* "without odd dimensions," we still can construct $\mathbf{G}(\mathbf{B}N)$, which *is* an $\mathcal{A}$-manifold without odd dimensions. In the $\mathcal{A}$-manifold $\mathbf{G}(\mathbf{B}N)$ we recover the even coordinates of N, but not the odd coordinates. We leave it to the reader to convince himself that $\mathbf{G}(\mathbf{B}N)$ "is" the submanifold $N^{w.o.d}$ defined in [5.9].

5.24 Lemma. *Let M and N be $\mathcal{A}$-manifolds without odd dimensions and $f : \mathbf{B}M \to \mathbf{B}N$ a smooth map. Then there exists a smooth map $\mathbf{G}f : M \to N$ such that the following diagram is commutative:*

$$\begin{array}{ccc} M & \xrightarrow{\mathbf{G}f} & N \\ {\scriptstyle\mathbf{B}}\downarrow & & \downarrow{\scriptstyle\mathbf{B}} \\ \mathbf{B}M & \xrightarrow[f]{} & \mathbf{B}N\,. \end{array}$$

Proof. Let $\mathcal{U} = \{\varphi_a : U_a \to O_a \mid a \in I\}$ and $\mathcal{V} = \{\psi_b : V_b \to P_b \mid b \in J\}$ be atlases for M and N. By taking intersections with $f^{-1}(\mathbf{B}V_b)$ if necessary, we may assume that for each $U_a \in \mathcal{U}$ there exists a $V_b \in \mathcal{V}$ such that $f(\mathbf{B}U_a) \subset \mathbf{B}V_b$. We then define maps $\mathbf{G}f_a : U_a \to N$ by

$$\mathbf{G}f_a = \psi_b^{-1} \circ \mathbf{G}\big((\mathbf{B}\psi_b) \circ f \circ (\mathbf{B}\varphi_a)^{-1}\big) \circ \varphi_a\ ,$$

where the $\mathbf{G}$ on the right hand side is defined in [2.11]. In order to apply $\mathbf{G}$, it is essential that M is without odd dimensions. By definition of smooth maps, the $\mathbf{G}f_a$ are smooth maps *on* U_a.

Let us show that the $\mathbf{G}f_a$ coincide on intersections. For $m \in U_a \cap U_{a'}$ we have by definition of $\mathbf{B}$ on $\mathcal{A}$-manifolds [4.11] that $\mathbf{B}(\mathbf{G}f_a(m)) = f(\mathbf{B}m) = \mathbf{B}(\mathbf{G}f_{a'}(m))$. Hence, if $f_a(\mathbf{B}U_a) \subset \mathbf{B}V_b$ it follows that $\mathbf{G}f_a(m) \in V_b$, but also $\mathbf{G}f_{a'}(m) \in \mathbf{B}^{-1}(\mathbf{B}V_b) = V_b$. We thus calculate on $U_a \cap U_{a'}$:

$$\begin{aligned}
\mathbf{G}f_{a'} &= \psi_b^{-1} \circ \psi_b \circ \psi_{b'}^{-1} \circ \mathbf{G}((\mathbf{B}\psi_{b'}) \circ f \circ (\mathbf{B}\varphi_{a'})^{-1}) \circ \varphi_{a'} \circ \varphi_a^{-1} \circ \varphi_a \\
&= \psi_b^{-1} \circ \mathbf{G}(\mathbf{B}\psi_b \circ \mathbf{B}\psi_{b'}^{-1}) \circ \mathbf{G}(\mathbf{B}\psi_{b'} \circ f \circ \mathbf{B}\varphi_{a'}^{-1}) \circ \mathbf{G}(\mathbf{B}\varphi_{a'} \circ \mathbf{B}\varphi_a^{-1}) \circ \varphi_a \\
&= \psi_b^{-1} \circ \mathbf{G}(\mathbf{B}\psi_b \circ f \circ \mathbf{B}\varphi_a^{-1}) \circ \varphi_a \\
&= \mathbf{G}f_a \ .
\end{aligned}$$

In going from the first line to the second we have used the definition of the maps $\mathbf{B}\varphi$ and $\mathbf{B}\psi$ and that N and M are without odd dimensions.

We conclude that the $\mathbf{G}f_a$ define a global function $\mathbf{G}f : M \to N$. [4.18] then tells us that $\mathbf{G}f$ is a smooth map. It follows immediately from the equality $\mathbf{B}(\mathbf{G}f_a(m)) = f(\mathbf{B}m)$ that $\mathbf{B} \circ \mathbf{G}f_a = f|_{\mathbf{B}U_a} \circ \mathbf{B}$, proving commutativity of our diagram. QED

5.25 Proposition. *Let M be an $\mathcal{A}$-manifold without odd dimensions and $f : \mathbf{B}M \to \mathbf{B}N$ a smooth map. Then there exists a unique smooth map $\mathbf{G}f : M \to N$ such that the following diagram is commutative:*

$$\begin{array}{ccc}
M & \xrightarrow{\mathbf{G}f} & N \\
\mathbf{B}\big\downarrow & & \big\downarrow\mathbf{B} \\
\mathbf{B}M & \xrightarrow[f]{} & \mathbf{B}N \, .
\end{array}$$

Proof. Consider the canonical injection $i : \mathbf{GB}N \cong N^{w.o.d} \to N$. Applying the body map gives us the identity on $\mathbf{B}N$, i.e., $\mathbf{B}i = id(\mathbf{B}N)$. According to [5.24] there exists a smooth map $\mathbf{G}_1 f : M \to \mathbf{GB}N$ such that $\mathbf{B} \circ \mathbf{G}_1 f = f \circ \mathbf{B}$. It follows immediately that $\mathbf{G}f = i \circ \mathbf{G}_1 f : M \to N$ satisfies our requirements.

To prove uniqueness, note first that uniqueness in [4.22] forces $\mathbf{B}(\mathbf{G}f) = f$. So suppose $g, h : M \to N$ are two solutions, i.e., $\mathbf{B}g = \mathbf{B}h = f$, and let $m \in M$ be arbitrary. Let $\varphi : U \to O$ be a chart for M containing m, and let $\psi : V \to P$ be a chart for N containing $g(m)$. Since $\mathbf{B}(g(m)) = (\mathbf{B}g)(\mathbf{B}m) = f(\mathbf{B}m) = (\mathbf{B}h)(\mathbf{B}m) = \mathbf{B}(h(m)) \in \mathbf{B}V$, it follows that $h(m) \in V = \mathbf{B}^{-1}(\mathbf{B}V)$. Using the definition of $\mathbf{B}$ on $\mathcal{A}$-manifolds and smooth maps, we then compute $\mathbf{B}B(\psi \circ g \circ \varphi^{-1}) = \mathbf{B}\psi \circ f \circ \mathbf{B}\varphi^{-1} = \mathbf{B}(\psi \circ h \circ \varphi^{-1})$ as maps from $\mathbf{B}O$ to $\mathbf{B}P$. Since O is an open subset in the even part of an $\mathcal{A}$-vector space *without odd dimensions*, it follows from [2.16] that $\psi \circ g \circ \varphi^{-1} = \psi \circ h \circ \varphi^{-1}$, and in particular that $g(m) = h(m)$. Since m was arbitrary, we conclude $g = h$. QED

5.26 Corollary. *If M is an $\mathcal{A}$-manifold without odd dimensions, then the body map $\mathbf{B} : C^\infty(M; N) \to C^\infty(\mathbf{B}M; \mathbf{B}N)$ is bijective.*

5.27 Remark. Looking at the commutative diagram at the end of [5.9], we now can identify $f|_{M^{w.o.d}}$ as the map $\mathbf{GB}f = \mathbf{G}(f|_{\mathbf{B}M})$.

5.28 Discussion. In previous sections we have defined the body map $\mathbf{B}$ on subsets of $\mathcal{A}$-vector spaces and on smooth maps. The image was the corresponding $\mathbf{R}$-object. In this section we have extended the notion of the body map $\mathbf{B}$ to $\mathcal{A}$-manifolds and smooth maps between them. Obviously the extended definition of $\mathbf{B}$ coincides with the old one in case the old and new apply, i.e., when an $\mathcal{A}$-manifold is just an open subset of the even part of an $\mathcal{A}$-vector space. Moreover, the body map on $\mathcal{A}$-manifolds preserves the notions of product and submanifold as defined in §4. More precisely, if N is a submanifold of M, then $\mathbf{B}N$ is a submanifold of the $\mathbf{R}$-manifold $\mathbf{B}M$; if M and N are $\mathcal{A}$-manifolds, then $\mathbf{B}(M \times N) = (\mathbf{B}M) \times (\mathbf{B}N)$ as $\mathbf{R}$-manifold.

Chapter IV

Bundles

In the next chapter we will, among others, introduce the tangent bundle of an $\mathcal{A}$-manifold M as well as differential forms and their exterior derivatives. These definitions require some knowledge about vector bundles and about the relation between operations on vector bundles and the corresponding operations on the modules of their sections. The purpose of this chapter is to provide these notions and to put them in a wider context.

Let us be more precise (with some liberties). A vector bundle is a natural generalization of a vector space in the sense that it is literally a bunch (bundle) of vector spaces. The set which indexes the various vector spaces (the fibers) is called the base manifold. The various operations one can perform to create new vector spaces out of old ones, such as the direct sum, tensor product, etcetera, can also be performed on vector bundles by doing it in each fiber separately. On the other hand, one can form the set of sections of a vector bundle, meaning the set of functions which assign a point in each of the vector spaces of the bundle. This set is in a natural way a module over the set of functions on the base manifold. It follows that we can perform the operations of direct sum, tensor product, etcetera, also on these modules. In this context there is one important question for differential geometry: how to interpret a k-form. On the one hand a k-form is defined as an element of

$$\textit{sections of } \mathrm{Hom}_L^{sk}(\,(\textit{tangent bundle})^k, \mathcal{A})\ ,$$

while on the other hand, especially for the definition of the exterior derivative, it is seen as an element of

$$\mathrm{Hom}_L^{sk}(\,(\textit{sections of the tangent bundle})^k, (\textit{functions on the base manifold}\,)\,)\ .$$

In the first point of view we first apply the construction $E \to {}^*(\bigwedge^k E)$ to the tangent bundle and then we take sections, whereas in the second point of view we first take sections and then apply this construction to the resulting module. The obvious question is whether *these two points of view coincide; the answer is positive. Once this question is stated this way, it is natural to ask whether a similar result holds for the other constructions.*

It turns out that for direct sums and quotients the results are almost trivially true. For spaces of homomorphisms it requires some work to prove that a homomorphism of sections depends only pointwise on the values of the initial sections. For tensor products and exterior powers though, the proof is much harder. For tensor products and exterior powers we need to know a part of the Serre-Swann theorem which states that the space of sections of a vector bundle is a finitely generated and projective module over the ring of functions on the base manifold. In order to prove this, we need the notion of a metric on an $\mathcal{A}$-manifold in order to define an orthogonal supplement to a subbundle. For $\mathcal{A}$-vector spaces the notion of a metric is subtly different from what one would expect, and the proof of the existence of an orthogonal complement uses the specific properties of graded smooth functions.

The last section of this chapter treats Batchelor's theorem. This theorem could have been stated and proved already at the end of chapter III. The reason it is included in this chapter is that the interpretation of this theorem requires the notion of a vector bundle, albeit in the classical sense of $\mathbf{R}$*-manifolds.*

1. Fiber bundles

In this section we introduce locally trivial fiber bundles with structure group G and typical fiber F over an $\mathcal{A}$-manifold M as well as their morphisms and sections. We show that such a bundle is characterized by its transition functions with respect to a trivializing atlas.

1.1 Definitions. • An *$\mathcal{A}$-Lie group* is an $\mathcal{A}$-manifold G that admits at the same time the structure of a group for which the multiplication is smooth. More precisely, denoting by $m : G \times G \to G$ the multiplication in G, m should be a smooth map from the $\mathcal{A}$-manifold $G \times G$ to G. As is customary, we will usually denote the multiplication by a dot or just nothing, i.e., $m(g, h) = g \cdot h = gh$. It will be shown in [VI.1.6] that in an $\mathcal{A}$-Lie group the map $Inv : G \to G, g \mapsto g^{-1}$ is automatically smooth.

• Let G and H be $\mathcal{A}$-Lie groups. A map $\rho : G \to H$ is called a *morphism of $\mathcal{A}$-Lie groups* if it is smooth and a homomorphism of (abstract) groups.

• Let M be an $\mathcal{A}$-manifold and G an $\mathcal{A}$-Lie group. A *smooth (left) action of G on M* is a smooth map $\Phi : G \times M \to M$ satisfying

(i) $\forall m \in M : \Phi(id_G, m) = m$ (where id_G denotes the identity element in G),
(ii) $\forall g, h \in G\, \forall m \in M : \Phi(g \cdot h, m) = \Phi(g, \Phi(h, m))$.

For a fixed $g \in G$ we will denote the map $m \mapsto \Phi(g, m)$ by Φ_g. If no confusion is possible, we will denote the image $\Phi(g, m)$ in the following shortened ways: as $\Phi_g(m)$, as $g(m)$, or just as gm.

• In a similar way we define a *smooth right action* by switching the arguments of Φ. More precisely, it is a smooth map $\Phi : M \times G \to M$ such that $\Phi(m, id_G) = m$ and $\Phi(m, g \cdot h) = \Phi(\Phi(m, g), h)$. As for left actions, we denote the map $m \mapsto \Phi(m, g)$ by Φ_g, but evaluation will be denoted on the left, i.e., $\Phi(m, g) = \langle\!\langle m \| \Phi_g \rangle\!\rangle$. Again as for left actions, if no confusion is possible, we will use the shortened form $\Phi(m, g) = mg$.

• An action of G on M is said to be *effective* if $\Phi_g = id(M)$ implies that $g = id_G \in G$.

• An action of G on M is *pseudo effective* if for all $\mathcal{A}$-manifolds N and all *smooth* maps $\psi : N \to G$ we have: $\big[\forall n \in N : \Phi_{\psi(n)} = id(M)\big] \implies \big[\forall n \in N : \psi(n) = id(G)\big]$.

1.2 Nota Bene. If Φ is a smooth (left or right) action of an $\mathcal{A}$-Lie group G on an $\mathcal{A}$-manifold M, it is in particular continuous. It follows that all maps $\Phi_g : M \to M$ are homeomorphisms. If $g \in G$ has real coordinates, Φ_g is a diffeomorphism, but if g does not have real coordinates, there is no reason to think that Φ_g is smooth. See [VI.1.3] for more details.

1.3 Discussion/Example. Let $G = \mathrm{Aut}(E)$ be the group of automorphisms of a finite dimensional $\mathcal{A}$-vector space E. On this group we have defined *two* different group structures: the usual composition $\circ$ and the opposite composition $\diamond$. The usual composition $\circ$ corresponds to the point of view that elements of $\mathrm{Aut}(E)$ are right linear, i.e., $\mathrm{Aut}(E) \subset \mathrm{End}_R(E)$. Moreover, with the composition $\circ$, the action of $\mathrm{Aut}(E)$ on E is a *left* action. On the other hand, the opposite composition $\diamond$ corresponds to the point of view that elements of $\mathrm{Aut}(E)$ are (also) left linear, i.e., $\mathrm{Aut}(E) \subset \mathrm{End}_L(E)$ (for left linear morphisms we have argued repeatedly that its natural composition law is the opposite $\diamond$ defined by $\phi \diamond \psi = \psi \circ \phi$). And when equipped with $\diamond$, the action of $\mathrm{Aut}(E)$ on E becomes a *right* action. Indeed, seen as *left* linear maps we have $\langle\!\langle v \| \phi \diamond \psi \rangle\!\rangle = \langle\!\langle \langle\!\langle v \| \phi \rangle\!\rangle \| \psi \rangle\!\rangle$. It follows that the choice for one of these group structures on $\mathrm{Aut}(E)$ depends upon whether we want to see it as having a right or left action on E. Since traditionally one prefers left actions over right actions, we will adhere to this choice. This means that we will use the standard composition law $\circ$ and that we will view $\mathrm{Aut}(E)$ as being a subset of the right linear endomorphisms $\mathrm{End}_R(E)$.

But we can say more about $\mathrm{Aut}(E)$. It has a canonical structure of an $\mathcal{A}$-manifold, being an open subset of the even part of the $\mathcal{A}$-vector space $\mathrm{End}_R(E)$ [III.2.26]. Taking a basis for E it is immediate from the matrix representation of $\mathrm{Aut}(E)$ that the multiplication in $\mathrm{Aut}(E)$ is smooth, proving that it is an $\mathcal{A}$-Lie group. With a similar argument one can show that the evaluation $\Phi : \mathrm{Aut}(E) \times E \to E$, $(\phi, e) \mapsto \phi(e)$ is an effective smooth (left) action of $\mathrm{Aut}(E)$ on $E = E_0^\sharp$.

1.4 Remark. It is immediate that an effective action is pseudo effective. Moreover, in the context of $\mathbf{R}$-manifolds the converse is also true: if $\Phi_g = id(M)$, we can take the constant map $\psi(n) = g$ to obtain that $g = id(G)$. However, in the context of $\mathcal{A}$-manifolds this argument no longer holds: not any constant map is smooth!

1.5 Example. Let E be a finite dimensional $\mathcal{A}$-vector space and E_α one of its homogeneous parts. Since $\mathrm{Aut}(E)$ consists of *even* morphisms, its action on E preserves E_α. We claim that the left action of $\mathrm{Aut}(E)$ on E_α is pseudo effective. The case of E_0 is a direct consequence of [III.2.31] and the fact that we required $\mathcal{A}$ to satisfy (C[∞]) (in order to be assured that smooth functions have partial derivatives). If $\psi : N \to \mathrm{Aut}(E)$ is a smooth map such that all $\psi(n)$ act as the identity on E_0, all $\psi(n) - id(E)$ act as the zero map on E_0. Hence by [III.2.31] they are the zero maps, i.e., all $\psi(n) = id(E)$. For E_1, it suffices to transform this into the previous case by noting that $E_1 = E_0^{\flat 1}$ (see [III.1.26]) and that the map $id : \mathrm{Aut}(E) \to \mathrm{Aut}(E^{\flat 1})$ which sends a map to the same map (E and $E^{\flat 1}$ are the same as sets) is smooth. Thus if $\psi : N \to \mathrm{Aut}(E)$ is a smooth map such that all $\psi(n)$ act as the identity on E_1, the smooth map $\overline{\psi} = id \circ \psi : N \to \mathrm{Aut}(E^{\flat 1})$ is such that all $\overline{\psi}(n)$ act as the identity on $E_0^{\flat 1}$. Hence they are all the identity.

1.6 Discussion. The reason we introduce the notion of pseudo effective actions is that it is difficult to prove (if possible at all) that a given action is effective. The problem is that in order to prove effectiveness, one has to disentangle the coordinates of the $\mathcal{A}$-Lie group from those of the $\mathcal{A}$-manifold it is acting on. Since $\mathcal{A}$ usually contains zero divisors, separating these coordinates is hard. On the other hand, for pseudo effective actions we do not have separate elements that act as the identity, but smooth families of elements acting as the identity. Separating the coordinates of the $\mathcal{A}$-Lie group from those of the $\mathcal{A}$-manifold then is provided by [III.2.21].†

1.7 Definition. Let B, M, and F be $\mathcal{A}$-manifolds, let G be an $\mathcal{A}$-Lie group with a smooth and pseudo effective left action on F, and let $p : B \to M$ be a smooth surjective map. A *local trivializing F-chart* (often shortened to a *trivializing chart*) for p is a chart U for the $\mathcal{A}$-manifold M together with a diffeomorphism $\psi : p^{-1}(U) \to U \times F$ such that $\pi_U \circ \psi = p$, where π_U is the canonical projection on U.

To any pair of local trivializing F-charts (U_a, ψ_a) and (U_b, ψ_b) is associated a map $\widetilde{\psi}_{ba} : U_a \cap U_b \times F \to F$, defined by the equation $(\psi_b \circ \psi_a^{-1})(m, f) = (m, \widetilde{\psi}_{ba}(m, f))$. For a fixed $m \in U_a \cap U_b$, the map $\widetilde{\psi}_{ba}(m, \_)$ is a map from F to F admitting an inverse $\widetilde{\psi}_{ab}(m, \_)$, i.e., the $\widetilde{\psi}_{ba}(m, \_)$ are bijections. Two local trivializing F-charts (U_a, ψ_a) and (U_b, ψ_b) are said to be *G-compatible* (or compatible if no confusion is possible) if there exist *smooth* maps $\psi_{ba}, \psi_{ab} : U_a \cap U_b \to G$ such that

$$m \in U_a \cap U_b \quad \Rightarrow \quad \widetilde{\psi}_{ba}(m, f) = \psi_{ba}(m)(f) \quad \& \quad \widetilde{\psi}_{ab}(m, f) = \psi_{ab}(m)(f) .$$

Pseudo effectiveness of the G-action on F implies that the maps ψ_{ba} and ψ_{ab}, if they exist, are unique. These maps are called *the transition functions between the trivializing*

†If $\mathcal{A}$ is the exterior algebra of a vector space X over $\mathbf{R}$, the fact that we assume that $\mathcal{A}$ satisfies (C[∞]) implies that X has to be infinite dimensional. In that case it is not hard to show that the action of $\mathrm{Aut}(E)$ on E_α [1.5] is actually effective (which it is *not* if X is *finite* dimensional; see [II.5.8] for the essential idea for a counter example). I conjecture that this is general: if $\mathcal{A}$ is the exterior algebra of an infinite dimensional vector space X over $\mathbf{R}$ then any pseudo effective action is actually effective.

charts U_a and U_b of the bundle B, or simply *transition functions* if the rest is clear from the context.

A *fiber bundle structure on B with typical fiber F and structure group G* is a collection of local trivializing F-charts $\mathcal{S} = \{ (U_a, \psi_a) \mid a \in I \}$ satisfying three conditions:

(FB1) $\cup_{a \in I} U_a = M$,
(FB2) any two charts in $\mathcal{S}$ are G-compatible, and
(FB3) if (U, ψ) is a local trivializing F-chart that is G-compatible with all elements of $\mathcal{S}$, then $(U, \psi) \in \mathcal{S}$.

A *locally trivial fiber bundle over M with typical fiber F and structure group G* (or in short, a fiber bundle over M) is a smooth surjective map $p : B \to M$ together with a fiber bundle structure on B with typical fiber F and structure group G. A *trivializing atlas* for a fiber bundle over M is a subset $\mathcal{U} \subset \mathcal{S}$ that satisfies properties (FB1) and (FB2), but not necessarily (FB3). For any point $m \in M$ the set $p^{-1}(m) \subset B$ is called *the fiber over m*; it is usually denoted by $p^{-1}(m) \equiv B_m \equiv B|_m$. Since every point m lies in a local trivializing F-chart (U, ψ), we have a homeomorphism $\psi : B_m \to F$, explaining the name *typical fiber*. If m has real coordinates, this homeomorphism is a diffeomorphism.

1.8 Remarks. • For two local trivializing F-charts to be compatible, we required the existence of smooth transition functions ψ_{ba} and ψ_{ab}. However, existence of one of them implies the existence of the other. Note first that for a fixed $m \in U_a \cap U_b$ the map $\widetilde{\psi}_{ba}(m, \_)$ is a bijection on F with inverse $\widetilde{\psi}_{ab}(m, \_)$. Existence of ψ_{ba} implies that $\widetilde{\psi}_{ba}(m, \_) = \Phi(\psi_{ba}(m), \_)$. Since $\Phi(\psi_{ba}(m)^{-1}, \_)$ is its inverse, it follows that the smooth map $\psi_{ab} : U_a \cap U_b \to G$ defined by $\psi_{ab}(m) = \psi_{ba}(m)^{-1}$ satisfies the requirement (remember that inversion in an $\mathcal{A}$-Lie group is smooth).

• In the definition of a trivializing chart we required $U \subset M$ to be a *chart* for M, i.e., a coordinate chart for the smooth structure on M. One usually only requires U to be open. The restriction to charts is not really a restriction and avoids awkward phrasing later on.

• We have defined fiber bundles using structure groups. In this we have followed [Hi] and [Ste]. Other authors, such as [GHV] and [Hu], do not use structure groups.

1.9 Nota Bene. Note that, due to (FB1), an atlas $\mathcal{U} = \{ (U_a, \psi_a) \mid a \in I \}$ of trivializing charts for a bundle $p : B \to M$ is at the same time an atlas for the smooth structure for M. This justifies to a certain extent the abuse of notation for the word atlas.

1.10 Lemma. *Given a smooth surjective map $p : B \to M$, any collection of trivializing F-charts $\mathcal{U} = \{ (U_a, \psi_a) \mid a \in I \}$ satisfying* (FB1) *and* (FB2) *defines a unique fiber bundle structure on B with typical fiber F and structure group G such that $\mathcal{U}$ is a trivializing atlas.*

Proof. Define $\mathcal{S}$ as the collection of all local trivializing F-charts that are G-compatible with all elements in $\mathcal{U}$. The only not completely trivial part of the proof is to show property

(FB2). So let (U_i, ψ_i), $i = 1, 2$ be two charts in $\mathcal{S}$, for which we compute

$$m \in U_1 \cap U_2 \cap U_a \quad \Longrightarrow \quad \begin{cases} (m, \widetilde{\psi}_{21}(m, f)) = (\psi_2 \circ \psi_1^{-1})(m, f) \\ \qquad = (\psi_2 \circ \psi_a^{-1} \circ \psi_a \circ \psi_1^{-1})(m, f) \\ \qquad = (m, \psi_{2a}(m)(\psi_{a1}(m)(f))) \,. \end{cases}$$

From this we deduce that $\psi_{21}(m) = \psi_{2a}(m) \cdot \psi_{a1}(m)$ exists and is smooth on $U_1 \cap U_2 \cap U_a$. Pseudo effectiveness of the action shows that it is the unique solution on $U_1 \cap U_2 \cap U_a$. Since the U_a cover (by (FB1)) the manifold M, we can glue them together to a solution on $U_1 \cap U_2$. It then follows from [III.4.18] that (U_1, ψ_1) and (U_2, ψ_2) are G-compatible. QED

1.11 Remark. The fact that the structure group G acts pseudo effectively on the typical fiber F is used in the proof of [1.10]. In §VI.6 we will show that if we have a smooth left action of an $\mathcal{A}$-Lie group G on an $\mathcal{A}$-manifold F, then there exists a (proper) normal $\mathcal{A}$-Lie subgroup $G^{\psi o} \subset G$ acting as the identity on F such that $G/G^{\psi o}$ is an $\mathcal{A}$-Lie group and such that the induced action on F is pseudo effective. This means that the condition that G acts pseudo effectively on F is not restrictive at all: any action can be transformed in a pseudo effective action.

1.12 Terminology. Just as for $\mathcal{A}$-manifolds, we will never mention a fiber bundle structure explicitly, at the most we will specify a trivializing atlas for it. And just as for manifolds, a trivializing chart will always mean an element in the fiber bundle structure.

1.13 Discussion/Definition. Let $p : B \to M$ be a fiber bundle with typical fiber F and let N be a submanifold of M. We then can consider the subset $B|_N = p^{-1}(N) \subset B$ and the restriction $p_N : B|_N \to N$ of the map p to this subset. We claim that $B|_N$ is a submanifold of B and that $p_N : B|_N \to N$ is a fiber bundle with typical fiber F (and the same structure group as B). This bundle is called *the restriction of B to the submanifold N*. An important class of examples of this kind are submanifolds N that are open subsets in M.

To prove our claim, let $\phi : U \to O$ be a chart as in [III.5.1] with $\phi(U \cap N) = O \cap \widehat{E}$, where $\widehat{E}$ is a graded subspace of E. By taking intersections with trivializing charts for B, we may assume that there exists a diffeomorphism $\psi : p^{-1}(U) \to U \times F$ in the fiber bundle structure. Now let $\chi : V \to P$ be any chart for F, then $(\phi \times \chi) \circ \psi : \psi^{-1}(U \times V) \to O \times P$ is a chart for B with $((\phi \times \chi) \circ \psi)(\psi^{-1}(U \times V) \cap B|_N) = (O \cap \widehat{E}) \times P$. This proves that $B|_N$ is a submanifold of B. Finally, the charts $U \cap N$ for N together with the maps $\psi|_{p^{-1}(U \cap N)}$ form a trivializing atlas for the fiber bundle $p_N : B|_N \to N$; the details are left to the reader.

1.14 Definition. Let $p : B \to M$ and $q : C \to N$ be two fiber bundles. A smooth map $h : B \to C$ is a *fiber bundle map* if there exists a smooth map $g : M \to N$ such that the following diagram is commutative:

$$\begin{array}{ccc} B & \xrightarrow{\ h\ } & C \\ {\scriptstyle p}\big\downarrow & & \big\downarrow{\scriptstyle q} \\ M & \xrightarrow[\ g\ ]{} & N\,. \end{array}$$

Since p is surjective, such a map g, if it exists, is unique. If the base spaces are the same, we will as a rule require that g is the identity on M.

1.15 Remarks. • If a smooth map $h : B \to C$ induces a set theoretical map $g : M \to N$ making the diagram commutative, then g must be smooth. This can be seen by the following argument. Let (U, ψ) be a trivializing chart for $p : B \to M$ and choose a point $f \in \mathbf{B}F \subset F$. Commutativity of the diagram implies that $g|_U$ is given by $m \mapsto q(h(\psi^{-1}(m, f)))$. Since f has real coordinates, this is smooth.

• Some authors use the name *strong bundle map* for those fiber bundle maps between fiber bundles with the same base space that induce the identity on the base space. Since we will hardly ever use fiber bundle maps over the same base space that do *not* induce the identity on the base space, we do not introduce this distinction, but will specify it explicitly when needed.

1.16 Definitions. • An *isomorphism of fiber bundles* is diffeomorphism $h : B \to C$ of $\mathcal{A}$-manifolds between two fiber bundles $p : B \to M$ and $q : C \to M$ over the same base space that is at the same time a fiber bundle map (which by our convention induces the identity on M). Two fiber bundles B and C over the same base space are said to be *isomorphic* if there exists a fiber bundle isomorphism $h : B \to C$.

• Let M and F be $\mathcal{A}$-manifolds, let $B = M \times F$ be the direct product and consider the projection $p = \pi_M : B = M \times F \to M$. For any chart U of M we define the local trivializing F-chart (U, ψ) for p by $\psi = id : p^{-1}(U) \to U \times F$. Taking $G = \{id_F\}$, the set of all such local trivializing F-charts (U, ψ) are G-compatible and thus form a trivializing atlas for B. This turns $p : B \to M$ into a fiber bundle over M with typical fiber F and structure group $G = \{id_F\}$, called *the trivial bundle over M with fiber F*.

• A fiber bundle $p : B \to M$ with typical fiber F is said to be *trivial* if B is isomorphic to the trivial bundle $M \times F$.

1.17 Nota Bene. Let $p : B \to M$ be a fiber bundle with typical fiber F and structure group G. Suppose furthermore that $h : B \to M \times F$ is an isomorphism of fiber bundles showing that B is trivial. If $U \subset M$ is any chart of M, then $h|_{p^{-1}(U)} : p^{-1}(U) \to U \times F$ is a trivializing F-chart for B. However, it is by no means guaranteed that this trivializing chart is G-compatible with the trivializing charts of B that define the bundle structure of B. A rather trivial (counter) example is the following. Let $B = M \times F$ be the trivial bundle and let H, $H \neq \{id\}$, be an $\mathcal{A}$-Lie group acting pseudo effectively on F. Let furthermore $\varphi : M \to H$ be a smooth map such that $\forall m \in M : \varphi(m) \neq id$. Then the map $h : B \to M \times F$, $h(m, f) = (m, \varphi(m) \cdot f)$ is a fiber bundle isomorphism. But no restriction $h|_{p^{-1}(U)} : p^{-1}(U) \to U \times F$ is G-compatible with the fiber bundle structure on $B = M \times F$, simply because $G = \{id\}$ is too small. In [V.3.17] a less trivial counter example to this phenomenon can be found. There $G \cong \mathbf{Z}_2$ but it is still too small for the actual global trivialization to be G-compatible.

1.18 Discussion. We now can easily explain the name *locally trivial fiber bundle*. A local trivializing chart (U, ψ) for the fiber bundle B is a chart U of M such that the restriction $B|_U$ of B to U is trivial, the fiber bundle isomorphism being given by ψ. It follows that a locally trivial fiber bundle is just that: every point admits a neighborhood (chart) U such that the restriction of the bundle to that chart is trivial.

1.19 Definition. Let $p : B \to M$ be a fiber bundle and N a submanifold of M. *A local (smooth) section of B over N* is a (smooth) map $s : N \to B$ such that $p \circ s = id(N)$. We will denote the set of all smooth sections of B over N by $\Gamma_N(B)$; in case $N = M$ we omit the subscript and write $\Gamma(B) \equiv \Gamma_M(B)$. From the definition of the restriction of B to N it follows immediately that we can write $\Gamma_N(B) = \Gamma(B|_N)$. The reader should be aware that there is no guarantee that the set $\Gamma_N(B)$ is not empty.

1.20 Discussion. Let $p : B \to M$ be a fiber bundle with typical fiber F and suppose that $U \subset M$ is an open set such that $B|_U$ is trivial. This implies the existence of a diffeomorphism $\psi : B|_U = p^{-1}(U) \to U \times F$ such that $p_U = \pi_U \circ \psi$. Associated to any local section $s : U \to B$ we then have a map $s_\psi : U \to F$ defined by $s_\psi = \pi_F \circ \psi \circ s$, i.e.,

$$(\psi \circ s)(m) = (m, s_\psi(m)) .$$

Since ψ is a diffeomorphism, the map $s \mapsto s_\psi$ is a bijection between local sections $s : U \to B$ and functions $s_\psi : U \to F$. Moreover, s is smooth if and only if s_ψ is smooth, i.e., if B is trivial over U, then $\Gamma_U(B) \cong C^\infty(U; F)$.

Changing the context slightly, let $s : M \to B$ be a section and $\mathcal{U} = \{(U_a, \psi_a)\}$ a trivializing atlas for B. We thus obtain local functions $s_a \equiv s_{\psi_a} : U_a \to F$ as above. It is an immediate consequence of [III.4.18] that s is smooth if and only if all functions s_a are smooth. Moreover, the local functions s_a satisfy the equations

$$m \in U_a \cap U_b \quad \Rightarrow \quad s_b(m) = \psi_{ba}(m)(s_a(m)) . \tag{1.21}$$

On the other hand, if we have a set of local functions $s_a : U_a \to F$ satisfying these equations, they define a global section $s : M \to B$ by the equation $s(m) = \psi_a^{-1}(m, s_a(m))$;

(1.21) guarantees that s is well defined on intersections $U_a \cap U_b$. We conclude that sections $s : M \to B$ are in 1-1 correspondence with sets of local functions $s_a : U_a \to F$ associated to the trivializing atlas $\mathcal{U}$ and satisfying (1.21). Moreover, s is smooth (i.e., $s \in \Gamma(B)$) if and only if all s_a are smooth.

1.22 Corollary. *Let $p : B \to M$ be a fiber bundle with typical fiber F. If B is trivial, then any fiber bundle isomorphism $h : B \to M \times F$ induces a bijection $\Gamma(B) \to C^\infty(M; F)$ given by $s \mapsto \pi_F \circ h \circ s$.*

1.23 Lemma. *Let $p : B \to M$ be a fiber bundle, U an open set in M and $s \in \Gamma_U(B)$ a local smooth section. Then the image $s(U)$ is a submanifold of B and $p : s(U) \to U$ is a diffeomorphism.*

Proof. Since being a submanifold is a local property, we may assume without loss of generality that U is a trivializing chart. If $\psi : \pi^{-1}(U) \to U \times F$ is a trivializing diffeomorphism, it transforms the image $s(U)$ into the graph $Gr(s_\psi)$. The result then follows from [III.5.13]. $\boxed{QED}$

1.24 Construction. Let $p : B \to M$ be a bundle with typical fiber F and structure group G, and let $\mathcal{U}$ be a trivializing atlas. It is easy to show, using the pseudo effectiveness of the action, that the collection of maps ψ_{ba} has the following properties:

$$(1.25)\qquad \begin{aligned} &\forall m \in U_a \qquad \psi_{aa}(m) = id_G \text{ the identity element in } G\ , \\ &m \in U_a \cap U_b \cap U_c \quad \Rightarrow \quad \psi_{cb}(m) \cdot \psi_{ba}(m) = \psi_{ca}(m)\ . \end{aligned}$$

What we will do is show that the collection $\{U_a, \psi_{ba}\}$ completely determines the fiber bundle B.

To make this more precise, let M and F be $\mathcal{A}$-manifolds and let G be an $\mathcal{A}$-Lie group with a smooth action on F. Suppose also that we have an atlas $\mathcal{U} = \{\, \varphi_a : U_a \to O_a \,\}$ of M and smooth maps $\psi_{ba} : U_a \cap U_b \to G$ satisfying (1.25). With these data, we claim, we can (re)construct a bundle $p : B \to M$ with typical fiber F and structure group G such that the maps ψ_{ba} appear as described above.

To define the $\mathcal{A}$-manifold B and the surjection $p : B \to M$ we will use [III.4.9]. The ingredients for this construction are the open sets $\widehat{O}_a = O_a \times F$, the subsets $\widehat{O}_{ab} = O_{ab} \times F$ and the transition functions $\widehat{\varphi}_{ba}$ defined by $\widehat{\varphi}_{ba}(x, f) = (\varphi_{ba}(x), \psi_{ba}(\varphi_a^{-1}(x))(f))$. To realize that these ingredients satisfy (III.4.10), remember first that $O_{ab} = \varphi_a(U_a \cap U_b)$ and that $\varphi_{ba} = \varphi_b \circ \varphi_a^{-1}$. It follows immediately that the maps $\widehat{\varphi}_{ba}$ are smooth; that they also satisfy (III.4.10) follows from (1.25). According to [III.4.9] we thus may conclude that B is a well defined $\mathcal{A}$-manifold. The projection $p : B \to M$ we define in local charts by $p : (x, f) \in \widehat{O}_a = O_a \times F \mapsto x \in O_a$. It should be obvious from the definition of the maps $\widehat{\varphi}_{ba}$ that this is a well defined, smooth surjective map as required.

It thus remains to exhibit the existence of an atlas of trivializing charts satisfying (FB1) and (FB2). We claim that the atlas $\{ (U_a, \psi_a) \mid a \in I \}$ with $\psi_a = (\varphi_a^{-1} \times id(F)) \circ \widehat{\varphi}_a$ will do. But this should be obvious from the commutative diagram

$$\begin{array}{ccccc} p^{-1}(U_a) = \widehat{U}_a & \xrightarrow{\widehat{\varphi}_a} & \widehat{O}_a = O_a \times F & \xrightarrow{\varphi_a^{-1} \times id(F)} & U_a \times F \\ {\scriptstyle p}\downarrow & & \downarrow{\scriptstyle \pi_{O_a}} & & \downarrow{\scriptstyle \pi_{U_a}} \\ U_a & \xrightarrow{\varphi_a} & O_a & \xrightarrow{\varphi_a^{-1}} & U_a \end{array}$$

and the definition of the transition functions $\widehat{\varphi}_{ba}$ for the $\mathcal{A}$-manifold B (and recall that the charts φ_a and $\widehat{\varphi}_a$ are recovered/defined as in [III.4.9]).

We thus have shown that the data $\{U_a, \psi_{ba}\}$ determine a bundle $p : B \to M$ with typical fiber F and structure group G. To show that this bundle is "unique," we proceed as follows. Suppose $p : B \to M$ and $\widehat{p} : \widehat{B} \to M$ are two bundles with typical fiber F and structure group G. Suppose furthermore that $\{(U_a, \psi_a)\}$ is a trivializing atlas for B and that $\{(U_a, \widehat{\psi}_a)\}$ is one for $\widehat{B}$, i.e., the same charts on M but different trivializing maps (by taking intersections we can always accomplish this). If we suppose that the transition functions ψ_{ba} and $\widehat{\psi}_{ba}$ are the same (i.e., that $\psi_b \circ \psi_a^{-1} = \widehat{\psi}_b \circ \widehat{\psi}_a^{-1}$ as diffeomorphisms of $(U_a \cap U_b) \times F$), then we can define an isomorphism of fiber bundles $h : B \to \widehat{B}$ by $h|_{p^{-1}(U_a)} = \widehat{\psi}_a^{-1} \circ \psi_a$. The fact that the transition functions are the same guarantees that this is well defined. We conclude that B and $\widehat{B}$ are isomorphic fiber bundles. And thus we have proven that the data $\{U_a, \psi_{ba}\}$ determine the bundle $p : B \to M$ uniquely up to isomorphisms.

1.26 Remark. Attentive readers will object that the above construction is faulty because we cannot apply [III.4.9]. They are right because F is not an open set in the even part of some $\mathcal{A}$-vector space. However, taking an atlas for the $\mathcal{A}$-manifold F and replacing the open sets $\widehat{O}_a$ by direct products of a chart O_a and a chart for F will make the construction a valid one. The details are left to the reader.

1.27 Discussion/Definition. What we have shown in the above construction is that giving a bundle B over an $\mathcal{A}$-manifold with typical fiber F and structure group G is completely equivalent to giving an atlas $\mathcal{U}$ and transition functions $\psi_{ba} : U_a \cap U_b \to G$ satisfying (1.25). It thus follows that such a set of transition functions defines the bundle $p : B \to M$. It is this way that we will define and/or construct most of our bundles.

2. Constructions of fiber bundles

In this section we discuss various constructions of new fiber bundles out of old ones: associated bundles, pull-back bundles, and product bundles. We also define the notion of a principal fiber bundle and we show that all fiber bundles can be seen as associated to a principal fiber bundle.

2.1 Construction (associated bundles). Let $p : B \to M$ be a bundle with typical fiber F and structure group G. Let furthermore H be another $\mathcal{A}$-Lie group with a pseudo effective smooth action on an $\mathcal{A}$-manifold E. Now suppose we have an $\mathcal{A}$-Lie group morphism $\rho : G \to H$. Associated to ρ and B we will construct a new bundle over M with typical fiber E and structure group H, which we will denote as $p^\rho : B^{\rho,E} \to M$, and which is called an *associated bundle*, associated to B by the "representation" ρ. To that end, let $\mathcal{U} = \{(U_a, \psi_a) \mid a \in I\}$ be a trivializing atlas for B with the associated set of transition functions ψ_{ba}. It is easy to verify that the functions $\rho \circ \psi_{ba}$ satisfy the requirements of [1.24]. We thus obtain a bundle $p^\rho : B^{\rho,E} \to M$ with typical fiber E and structure group H defined by these $\rho \circ \psi_{ba}$. Moreover, the construction is such that we automatically have a trivializing atlas $\mathcal{U}' = \{(U_a, \psi'_a) \mid a \in I\}$ for the new bundle $B^{\rho,E}$ with the same trivializing charts U_a as $\mathcal{U}$ and whose associated transition functions are the $\rho \circ \psi_{ba}$. (Note that this implies that for each trivializing chart (U, ψ) for B we have a corresponding trivializing chart (U, ψ') for $B^{\rho,E}$, just by adding (U, ψ) to $\mathcal{U}$.)

There are two ways to assure that the result does not depend upon the choice of a trivializing atlas. The first is to take the whole fiber bundle structure as trivializing atlas (the biggest possible). The second is to show that a different choice of trivializing atlas leads to an isomorphic fiber bundle. The details of this are left to the reader.

2.2 Construction (pull-back bundle). Let $q : C \to N$ be a fiber bundle with typical fiber F and structure group G and let $g : M \to N$ be a smooth map. We will construct a fiber bundle $p \equiv g^*q : B \equiv g^*C \to M$ with typical fiber F and structure group G, as well as a fiber bundle map $\breve{g} : g^*C \to C$ inducing g. We thus will have a commutative diagram

$$\begin{array}{ccc} B \equiv g^*C & \xrightarrow{\ \breve{g}\ } & C \\ {\scriptstyle p\equiv g^*q}\big\downarrow & & \big\downarrow{\scriptstyle q} \\ M & \xrightarrow[\ g\ ]{} & N\,. \end{array}$$

Moreover, the construction will be such that the restriction $\breve{g} : p^{-1}(m) \to q^{-1}(g(m))$ is a diffeomorphism. This fiber bundle will be called *the pull-back of C over g*.

Let us start with the abstract set-theoretic definition. Consider the direct product $M \times C$ and the subset $g^*C = \{\,(m, c) \in M \times C \mid g(m) = q(c)\,\}$. We claim that the surjection $p \equiv g^*q = \pi_M : g^*C \subset M \times C \to M$ is the fiber bundle we are looking for and that the map $\breve{g} = \pi_C : g^*C \subset M \times C \to C$ is the bundle map. What remains to be shown is that

g^*C is a fiber bundle over M in the sense of $\mathcal{A}$-manifolds and that all maps involved are smooth.

Let $\mathcal{V} = \{(V_\alpha, \chi_\alpha) \mid \alpha \in I\}$ be a trivializing atlas for C and let $\mathcal{U} = \{U_a \mid a \in J\}$ be an atlas for M such that for each $U_a \in \mathcal{U}$ there exists $V_{g(a)} \in \mathcal{V}$ such that $g(U_a) \subset V_{g(a)}$ (we use the same symbol g to denote the map on indices). We now define the map $\psi_a : p^{-1}(U_a) \subset M \times C \to U_a \times F$ by

$$\psi_a : (m, c) \mapsto (m, \pi_F(\chi_{g(a)}(c))) \;,$$

where $\chi_{g(a)} : q^{-1}(V_{g(a)}) \to V_{g(a)} \times F$ is the local trivialization. It is elementary to check that ψ_a is a bijection. Moreover, one also can check that we have

$$\psi_b \circ \psi_a^{-1} : (m, f) \mapsto \big(m, \chi_{g(a)g(b)}(g(m))(f)\big) \;,$$

where $\chi_{\alpha\beta}$ denotes the transition function with respect to the atlas $\mathcal{V}$. In other words, $B \equiv g^*C$ is a fiber bundle with transition functions $\psi_{ba} = \chi_{g(b)g(a)} \circ g$ (use [1.24]). Finally, there are two ways to show that $\breve{g}$ is smooth. The first is to note that g^*C is a submanifold of $M \times C$ and then applying [III.5.7]. The second is to note that on the open set $p^{-1}(U_a)$ the map $\breve{g}$ is defined by

$$\breve{g} = \chi_{g(a)}^{-1} \circ (g \times id(F)) \circ \psi_a \;.$$

In other words, in the trivializations (u_a, ψ_a) and $(V_{g(a)}, \chi_{g(a)})$ the map $\breve{g}$ is given as $(m, f) \mapsto (g(m), f)$. The result then follows from [III.4.18]. This also shows that $\breve{g}$ is a diffeomorphism when restricted to a fiber: in the given local trivializations it is the identity on F.

2.3 Example. Let $p : B \to M$ be a fiber bundle and N a submanifold of M with canonical injection $i : N \to M$. It is immediate from the definition of a pull-back bundle that i^*B is isomorphic to the restricted bundle $B|_N = p^{-1}(N)$, and that $\breve{i} : i^*B \to B$ corresponds to the canonical injection of $B|_N$ in B. In other words, restriction of a bundle to a submanifold is a particular case of a pull-back bundle.

2.4 Proposition. *Let $h : B \to C$ be a fiber bundle map between the fiber bundles $p : B \to M$ and $q : C \to N$, inducing the map $g : M \to N$. Then there exists a unique fiber bundle map $h_{ind} : B \to g^*C$ such that $h = h_{ind} \circ \breve{g}$, i.e., we have a commutative diagram*

$$\begin{array}{ccccc}
 & & \xrightarrow{\quad h \quad} & & \\
B & \xrightarrow[h_{ind}]{} & g^*C & \xrightarrow[\breve{g}]{} & C \\
{\scriptstyle p}\big\downarrow & & \big\downarrow{\scriptstyle g^*q} & & \big\downarrow{\scriptstyle q} \\
M & \xrightarrow[id(M)]{} & M & \xrightarrow[g]{} & N \;.
\end{array}$$

Proof. Let us start with uniqueness. If $b \in B$ then, by our definition of fiber bundle map, $h_{ind}(b)$ must lie in $(g^*q)^{-1}(p(b))$. Since $\breve{g}$ is a bijection from $(g^*q)^{-1}(p(b))$ to $q^{-1}(g(p(b)))$, the condition $h = h_{ind} \circ \breve{g}$ implies the sought for uniqueness.

To show the existence of h_{ind}, we define the map $h_{ind} : B \to g^*C$ by the formula $h_{ind}(b) = (p(b), h(b)) \in g^*C \subset M \times C$. The image lies within g^*C because h is a fiber bundle map. To verify that h_{ind} is smooth one uses the local charts defined in the construction of the smooth structure of g^*C; this is left to the reader. $\boxed{QED}$

2.5 Corollary. *The map $h \mapsto h_{ind}$ is a 1-1 correspondence between fiber bundle maps $h : B \to C$ that induce a given map $g : M \to N$ and fiber bundle maps $h_{ind} : B \to g^*C$ (which induce by definition the identity on M).*

2.6 Discussion. The above corollary shows that we can always transform a question concerning a fiber bundle map between two fiber bundles over different spaces into a question concerning fiber bundle maps over the same base space. This is another justification not to consider very extensively fiber bundle maps between fiber bundles with the same base space that do not induce the identity.

2.7 Construction (product bundles). Let $p : B \to M$ and $q : C \to M$ be two bundles over M with typical fiber F and structure group G respectively E and H. Let $\{ (U_a, \psi_{ba}) \}$ respectively $\{ (U_a, \chi_{ba}) \}$ be the atlas with the transition functions defining the bundle B respectively C. Indeed, by taking, if necessary, pairwise intersections, we may assume that both atlases have the same set of charts U_a. We now define the smooth maps $\vartheta_{ba} : U_a \cap U_b \to G \times H$ by $\vartheta_{ba}(m) = (\psi_{ba}(m), \chi_{ba}(m))$; they obviously satisfy (1.25). Moreover, there exists a natural smooth action Ψ of $G \times H$ on $F \times E$ defined by $\Psi(g,h)(f,e) = (\Phi_G(g)(f), \Phi_H(h)(e))$. We thus obtain a new bundle over M with typical fiber $F \times E$ and structure group $G \times H$. This bundle is denoted by $B \times_M C$ and called *the product bundle over M*.

2.8 Discussion. In the construction of product bundles we did not prove that the result does not depend upon the choice of the trivializing atlases. One way to prove the independence is to show that a different choice leads to an isomorphic bundle. But there is a nicer way, using pull-back bundles.

Consider $q : C \to M$ as a fiber bundle with typical fiber E and consider $p : B \to M$ as a smooth map. We thus can form the pull-back bundle $p^*q : p^*C \to B$ (also with typical fiber E) and the smooth map $\breve{p} : p^*C \to C$. A glance at the definition of p^*C and the maps p^*q and $\breve{p}$ shows that it is rather symmetric in p and q; in particular $\breve{p} = q^*p$ and

$$p^*C = q^*B = \{ (b,c) \in B \times C \mid p(b) = q(c) \} \subset B \times C \ .$$

We leave it to the reader to convince himself that the product bundle $B \times_M C$ is (isomorphic to) the pull-back bundle $p^*C = q^*B$ with projection $p \circ p^*q = q \circ q^*p$, i.e., we have a commutative diagram

$$\begin{array}{ccc} B \times_M C & \xrightarrow{q^*p} & C \\ {\scriptstyle p^*q}\downarrow & \searrow & \downarrow{\scriptstyle q} \\ B & \xrightarrow[p]{} & M\,. \end{array}$$

We conclude that not only is $B \times_M C$ a fiber bundle over M (with typical fiber $F \times E$), but also a fiber bundle over B with typical fiber E and a fiber bundle over C with typical fiber F.

2.9 Discussion. So far we did not pay much attention to the structure group G of a fiber bundle $p : B \to M$. In fact, the only role it plays is in restricting the possible compatible local trivializing charts. Two reactions are possible: take the smallest one so as to restrict as far as possible the number of compatible charts. And at the other extreme: take the biggest one in order to have the maximum number of compatible charts. To describe this idea in more detail, we suppose that $i : G \to H$ is an injective $\mathcal{A}$-Lie group morphism and that Φ_H is a pseudo effective smooth action of H on F. We then immediately have an induced pseudo effective smooth action Φ_G of G on $F : \Phi_G(g, f) = \Phi_H(i(g), f)$.

We now first consider the case in which G (and not H) is the structure group of a fiber bundle $p : B \to M$ over M with typical fiber F, i.e., the transition functions ψ_{ba} (relative to a trivializing atlas $\mathcal{U}$) take their values in G. By combining with i, we get smooth maps $\psi'_{ba} = i \circ \psi_{ba}$ with values in H. It follows that we can say that the trivializing charts in $\mathcal{U}$ are H-compatible. By [1.10] it follows that we can say that $p : B \to M$ admits the structure of a fiber bundle with typical fiber F and structure group H. We say that we have *enlarged the structure group to* H. This change of viewpoint from structure group G to structure group H does not affect the manifold structure of either B or M, nor does it affect the smooth map p. The only thing which changes is that the number of charts in the fiber bundle structure increases. We conclude that the only function of the structure group is, as suggested before, to reduce the number of charts in a fiber bundle structure.

In order to explore this idea further, we change our viewpoint and we assume that it is H, not G, which is the structure group of the fiber bundle. If one can find a trivializing atlas such that the ψ_{ba} take values in $i(G) \subset H$, it follows from [VI.4.11] that the maps $\widehat{\psi}_{ba} = i^{-1} \circ \psi_{ba} : U_a \cap U_b \to G$ are smooth. (Nota Bene. This is a non trivial result, valid for injective $\mathcal{A}$-Lie group morphisms, but not for general injective maps [V.2.19].) We deduce that the charts in this special atlas are G-compatible, and thus that B admits the structure of a fiber bundle with typical fiber F and structure group G. We say that we can *reduce the structure group from* H *to* G. The reader should note that reduction of the structure group is the inverse procedure of enlarging the structure group.

An example of reduction of the structure group and its importance is the tangent bundle of an $\mathbf{R}$-manifold: its natural structure group is $Gl(n)$, but it can always be reduced to the orthogonal group $O(n)$; it can be reduced to $SO(n)$ if and only if M is orientable.

2.10 Remarks. • The process of enlarging the structure group can be interpreted as a special case of an associated bundle. The map $i : G \to H$ is a smooth group homomorphism, and G and H have pseudo effective smooth actions on F. We thus can apply the construction of associated bundles. Of course, in this case the result is the same bundle, seen as an $\mathcal{A}$-manifold.

• As an application of reduction of the structure group, one could say that a fiber bundle is trivial if and only if its structure group can be reduced to the identity.

2.11 Definition. A *principal fiber bundle over M with structure group G* is a fiber bundle $p : P \to M$ with typical fiber G and structure group G whose pseudo effective smooth left action on the typical fiber is the multiplication $m : G \times G \to G$ (which is effective).

2.12 Lemma. *Let $p : P \to M$ be a principal fiber bundle with structure group G. There exists a canonically defined effective smooth right action Φ of G on P with the property that each local trivialization (U, ψ) is equivariant for the natural right action of G on $U \times G$, i.e., for $v \in p^{-1}(U)$, $m \in U$, and $g, h \in G$:*

$$\psi(v) = (m, g) \qquad \Longrightarrow \qquad \psi(\Phi(v, h)) = (m, gh) \ .$$

As a consequence, the orbits of the G-action are exactly the inverse images $p^{-1}(m)$, $m \in M$, i.e., M can be identified with the orbit space P/G.

Proof. We use the equivariance condition to *define* the action Φ on any local trivialization (U, ψ) by

$$\Phi(v, g) = \psi^{-1}(\psi(v) \cdot g) \ ,$$

where on $U \times G$ we use the canonical right action of G given as $(m, h) \cdot g = (m, hg)$. In order to show that this definition is independent of the chosen trivialization, we choose two local trivializations (U_a, ψ_a) and (U_b, ψ_b). According to the definition of a general fiber bundle we have a transition function $\psi_{ba} : U_a \cap U_b \to G$. According to the definition of a principal fiber bundle this means that we have $(\psi_b \circ \psi_a^{-1})(m, h) = (m, \psi_{ba}(m) \cdot h)$. Using the canonical left action of G on any product $U \times G$ defined as $g \cdot (m, h) = (m, gh)$, we can write this as $(\psi_b \circ \psi_a^{-1})(m, h) = \psi_{ba}(m) \cdot (m, h)$. We then compute:

$$\begin{aligned}
\psi_b^{-1}(\psi_b(v) \cdot g) &= \psi_b^{-1}([((\psi_b \circ \psi_a^{-1}) \circ \psi_a)(v)] \cdot g) = \psi_b^{-1}([\psi_{ba}(m) \cdot \psi_a(v)] \cdot g) \\
&= \psi_b^{-1}(\psi_{ba}(m) \cdot [\psi_a(v) \cdot g]) = (\psi_b^{-1} \circ (\psi_b \circ \psi_a^{-1}))(\psi_a(v) \cdot g) \\
&= \psi_a^{-1}(\psi_a(v) \cdot g) \ ,
\end{aligned}$$

where we introduced $m = p(v) = \pi_1(\psi_a(v)) = \pi_1(\psi_b(v))$ with π_1 the projection on the first factor and where we used that multiplication in G is associative. QED

2.13 Discussion/Notation. Looking at the proof of [2.12], one sees that we have transported the natural right action of G on the (local) product $U \times G$ to $p^{-1}(U)$ and that the form of the transition functions of a principal fiber bundle guarantee that this local definition is compatible on overlaps and thus extends globally. We will encounter a similar phenomenon in [3.2] for vector bundles. Once we have defined the right action of G on P, we can say that each local trivialization intertwines the (global) right action of G on P with the (local) right action of G on $U \times G$. As is usual, we will denote $\Phi(p, g)$, the right action of $g \in G$ on $b \in P$, also as $\Phi_g(p)$ or as pg or $p \cdot g$ if no confusion is possible.

2.14 Lemma. *Let $q : B \to M$ be a fiber bundle with typical fiber F and structure group G, whose action on F is denoted by Φ. Then there exists a canonically defined principal fiber bundle $p : P \to M$ with structure group G, namely $P = B^{id,G}$, such that $q : B \to M$ is the associated bundle to P by the action Φ of G on F, i.e., $B = P^{id,G}$.*

Proof. This is actually playing around with associated bundles. The $\mathcal{A}$-Lie group G has a pseudo effective smooth action on both F and G. We thus can apply the construction of an associated bundle to $q : B \to M$ and the identity $\rho = id : G \to G$. Obviously the result is a principal fiber bundle $q^{id} : B^{id,G} \to M$ with structure group G, which we now denote as $p : P \to M$. Changing the roles of the typical fibers F and G, we can apply the construction of an associated bundle to $p : P \to M$ and the identity $\rho = id : G \to G$. We then recover the original fiber bundle $q : B \to M$. $\boxed{QED}$

2.15 Remark. One could say that [2.14] tells us that principal fiber bundles are a kind of universal fiber bundle in the sense that all other fiber bundles can be obtained from them by the procedure of associated bundles. More properties of principal fiber bundles can be found in chapter VII, especially in [VII.3.1]–[VII.3.6] and [VII.5.2]–[VII.5.4].

3. Vector bundles and sections

In this section we specialize to vector bundles and their morphisms. We discuss the notion of a vector subbundle in detail and show that under suitable circumstances the kernel of a vector bundle morphism is a vector subbundle. In between we show that the set of all smooth sections of a vector bundle is a graded $C^\infty(M)$-module and that local sections can be glued together to a global section using a partition of unity.

3.1 Definition. A *vector bundle over* M is a fiber bundle $p : B \to M$ whose typical fiber $E \cong E_0^\sharp$ is an $\mathcal{A}$-vector space and whose structure group is (can be enlarged to) $\mathrm{Aut}(E)$. The *rank of (the vector bundle)* B is the graded dimension of the typical fiber E.

3.2 Discussion. Let $p : B \to M$ be a vector bundle with typical fiber E and let (U_a, ψ_a) and (U_b, ψ_b) be two trivializing charts for B. For $m \in U_a \cap U_b$ we thus have two homeomorphisms $\psi_a : B_m \to \{m\} \times E$ and $\psi_b : B_m \to \{m\} \times E$. Since E is a (finite dimensional) free graded $\mathcal{A}$-module, we can transport this structure to B_m. Since $\psi_b \circ \psi_a$ induces an automorphism on E by definition of the structure group of a vector bundle, we can use either ψ_a or ψ_b without running into contradictions. We conclude that the fibers of a vector bundle are in a canonical way (finite dimensional) free graded $\mathcal{A}$-modules.

We have been very careful in saying that the fibers are free graded $\mathcal{A}$-modules, and not $\mathcal{A}$-vector spaces. In fact, the transition functions $\psi_{ba}(m)$ belong to $\mathrm{Aut}(E)$, but are not necessarily smooth in the sense of [II.6.22], [III.1.27]. This means that an equivalence class of bases is not preserved by the transition functions. Which in turn means that we can not transport the structure of an $\mathcal{A}$-vector space to B_m without running into contradictions. Of course $\psi_{ba}(m)$ is smooth when m has real coordinates (i.e., $m \in \mathbf{B}M$), hence for those points we can endow B_m with the structure of an $\mathcal{A}$-vector space. But not all points in M have real coordinates.

3.3 Nota Bene. If B is a vector bundle, it is in particular an $\mathcal{A}$-manifold modeled on the even part of an $\mathcal{A}$-vector space and as such all points are even. On the other hand, each point lies in a certain fiber, which is a free graded $\mathcal{A}$-module. Seen this way, a point can be even or odd (or neither if it is a mixture). This change of viewpoint corresponds of course to the fact that the grading of E gets lost in the identification with $E_0^\sharp$. The fact that we define a vector bundle as a fiber bundle with typical fiber an $\mathcal{A}$-vector space instead of the even part of an $\mathcal{A}$-vector space will turn out to be of importance for some of the constructions of vector bundles (see [4.12]).

3.4 Remark. Let $p : B \to M$ be a vector bundle with typical fiber E and let $\mathcal{U}$ be a trivializing atlas for B. Contrary to the discussion in [1.24], where we were rather imprecise, in the case of the *vector* bundle B the sets $U_a \times E$ *do* form an atlas for B, just because the canonical identification $E \to E_0^\sharp$ is an atlas for E.

3.5 Definitions. Let $p : B \to M$ and $q : C \to N$ be vector bundles with typical fibers E and F respectively. A fiber bundle map $h : B \to C$ inducing a map $g : M \to N$ [1.14] is called a *(left linear) vector bundle morphism* if the restriction to any fiber is left-linear, i.e.,

$$h|_{B_m} \in \mathrm{Hom}_L(B_m; C_{g(m)}) \ .$$

The map h is said to be *of parity* α if all linear maps $h|_{B_m}$ are of parity α. Similar definitions hold for right linear vector bundle morphisms.

A *(vector bundle) isomorphism* is an *even* vector bundle morphism $h : B \to C$ which is at the same time an isomorphism of fiber bundles. In view of our definition of an isomorphism of fiber bundles [1.16], this requires in particular that B and C are vector bundles over the same base space M, and that g is the identity on M.

The vector bundle B is said to be *trivial* if it is isomorphic as vector bundle to the trivial vector bundle $M \times E$. One should be aware that this is stronger than being trivial as a fiber bundle. The difference is that for a vector bundle not only fibers should be mapped diffeomorphically to fibers, but also the $\mathcal{A}$-vector space structure should be preserved. In [V.3.14] and subsequent items we will discuss this difference in more detail.

3.6 Discussion. If $h : B \to C$ is a vector bundle isomorphism, it follows that its restriction to a fiber B_m is an isomorphism of free graded $\mathcal{A}$-modules. On the other hand, if $h : B \to C$ is an even vector bundle morphism (inducing the identity on the base space) such that the restriction to each fiber B_m is an isomorphism of free graded $\mathcal{A}$-modules, then it is not hard to show that h is a vector bundle isomorphism (using that taking the inverse in $\mathrm{Aut}(E)$ is smooth).

If $h : B \to M \times E$ is a vector bundle isomorphism, then we know that B is trivial (as a vector bundle). If $U \subset M$ is a chart for M, then $(U, h|_{p^{-1}(U)})$ is a local trivializing E-chart for B. Contrary to the general case as described in [1.17], this trivializing chart *is* $\mathrm{Aut}(E)$-compatible with the trivializing charts in the vector bundle structure of B. The reason is that the restriction of h to each fiber is, by definition, linear, even and invertible, and the structure group $\mathrm{Aut}(E)$ of B contains all even invertible linear maps.

3.7 Discussion. The fact that fibers of a vector bundle $p : B \to M$ are free graded $\mathcal{A}$-modules has some important consequences. A first one is that the set of (smooth) sections $\Gamma(B)$ is never empty; it always contains the element $\underline{0}$ called *the zero section* defined by $\underline{0}(m) = 0 \in B_m$. Since in a trivializing chart (U, ψ) we have $(\psi \circ \underline{0})(m) = (m, 0)$, it follows that $\underline{0}$ is a smooth map, i.e., indeed $\underline{0} \in \Gamma(B)$. We obtain in particular a canonical submanifold $\underline{0}(M) \subset B$ which is diffeomorphic to M. One often identifies M with this submanifold of B.

Another consequence is that the set $\Gamma(B)$ can be given the structure of a graded $C^\infty(M)$-module in the following way. Let $s, t \in \Gamma(B)$ be two smooth sections and $f \in C^\infty(M)$ a smooth map. We define the section $s + f \cdot t$ by

$$(s + f \cdot t)(m) = s(m) + f(m) \cdot t(m) \ .$$

If (U, ψ) is a trivializing chart, it follows that $(s + f \cdot t)_\psi$ is given by $s_\psi + f|_U \cdot t_\psi$ [1.20]. It thus follows from [III.1.24] and [1.20] that $s + f \cdot t$ is smooth, i.e., $(s + f \cdot t) \in \Gamma(B)$. We will say that a section $s \in \Gamma(B)$ *has parity* α if $s(m) \in p^{-1}(m)$ has parity α for all $m \in M$. (Nota Bene. We use the parity of E, not that of $E_0^\sharp$!) It follows from [III.1.23-b] that $\Gamma(B)$ is a graded $C^\infty(M)$-module.

If $p : B \to M$ is a trivial vector bundle over M with typical fiber E, we know that $\Gamma(B)$ is in 1-1 correspondence with $C^\infty(M; E)$ [1.20], [1.22]. Moreover, we now know that *both* sets are graded $C^\infty(M)$-modules. It is easy to check that the correspondence preserves this structure (see also [3.6]), i.e., $\Gamma(B)$ and $C^\infty(M; E)$ are isomorphic as graded $C^\infty(M)$-modules. But, as in [1.22], this isomorphism depends upon the trivialization.

However, in most cases there will be a natural choice for a trivialization, which thus gives a natural identification between sections and functions. A particular example is the trivial vector bundle $B = M \times \mathcal{A}$ with typical fiber $\mathcal{A}$, in which case $\Gamma(B) \cong C^\infty(M)$.

3.8 Definition. Let $p : B \to M$ be a vector bundle and $s \in \Gamma(B)$. The set $\operatorname{supp}(s) \subset M$, called *the support of* s, is defined as the closure of the set of those points where $s(m)$ is non-zero (cf. [III.5.15]).

3.9 Lemma. *Let $p : B \to M$ be a vector bundle.*

(i) *Let $U_o \subset M$ be open and $s_o \in \Gamma_{U_o}(B)$ a local section. If $\operatorname{supp}(s_o) \subset U_o$ is closed in M, then $s : M \to B$ defined by $s|_{U_o} = s_o$, $s \equiv 0$ outside U_o, is a smooth global section of B.*

(ii) *Let $s_\alpha \in \Gamma(B)$, $\alpha \in J$ be a collection of global sections such that the collection $\operatorname{supp}(s_\alpha)$, $\alpha \in J$ is locally finite. Then $s = \sum_{\alpha\in J} s_\alpha$ is a well defined smooth global section of B.*

(iii) *Let $\mathcal{V} = \{ V_\alpha \mid \alpha \in J \}$ be an open cover of M, $s_\alpha \in \Gamma_{V_\alpha}(B)$, and $\{ \rho_\alpha \mid \alpha \in J \}$ a partition of unity subordinated to $\mathcal{V}$. Then $s = \sum_{\alpha\in J} \rho_\alpha s_\alpha$ is a well defined smooth global section of B.*

Proof. Left to the reader as an elementary combination of the arguments in [1.20], [3.7], [III.5.16], [III.5.17], and [III.5.19]. QED

3.10 Proposition. *Let $p : B \to M$ be a vector bundle of rank n and typical fiber E, and let $U \subset M$ be open. Then the following three properties are equivalent.*

(i) *B is trivial over U, i.e., $p : p^{-1}(U) \equiv B_U \to U$ is the trivial bundle.*

(ii) *There exist n homogeneous smooth sections $s_i \in \Gamma_U(B)$ such that $(s_i(m))_{i=1}^n$ is a basis for B_m for all $m \in U$.*

(iii) *There exist n homogeneous smooth sections $s_i \in \Gamma_U(B)$ such that $(s_i)_{i=1}^n$ is a basis for the graded $C^\infty(U)$-module $\Gamma_U(B)$.*

If $\psi : p^{-1}(U) \to U \times E$ is a (potential) trivialization and if $(e_i)_{i=1}^n$ is a homogeneous basis of E, then the link between (i), (ii), and (iii) is given by $\psi^{-1}(m, \sum_i x^i\, e_i) = \sum_i x^i s_i(m)$.

Proof. • Let $(e_i)_{i=1}^n$ be a homogeneous basis of E, fixed during this proof. Now assume (i) and let $\psi : p^{-1}(U) \to U \times E$ be a trivialization. We define the homogeneous sections $s_i \in \Gamma_U(B)$ by $s_i(m) = \psi^{-1}(m, e_i)$. These sections obviously satisfy (ii) and (iii), proving the implications (i) ⇒ (ii) and (i) ⇒ (iii).

• To prove (ii) ⇒ (i), we first note that the graded dimension of a free graded $\mathcal{A}$-module is an invariant and thus that we may assume that the parity of s_i is the same as that of e_i. From this it follows that the map $\chi : U \times E \to p^{-1}(U)$ defined by

$$\chi(m, \textstyle\sum_i x^i e_i) = \sum_i x^i s_i(m)$$

is a smooth bijection whose restriction to each fiber is even. We intend to prove that its inverse $\chi^{-1} : p^{-1}(U) \to U \times E$ is also smooth and a (local) vector bundle morphism. To that end, let $\psi : p^{-1}(V) \to V \times E$ be a trivializing chart such that $V \subset U$. Since the s_i are smooth, there exist smooth functions $\sigma_i{}^j \in C^\infty(V)$ such that for all $m \in V$:

$$\psi(s_i(m)) = (m, \textstyle\sum_j \sigma_i{}^j(m)e_j) \ .$$

It follows that $\psi \circ \chi$ restricted to $V \times E$ is given by $(m, \sum_i x^i e_i) \mapsto (m, \sum_{i,j} x^i \sigma_i{}^j(m)e_j)$. Since the $s_i(m)$ form a basis for B_m, the matrix $\sigma_i{}^j(m)$ must be invertible. Since taking the inverse of a matrix is smooth [III.2.26], it follows that the inverse map $\chi^{-1} \circ \psi^{-1}$: $(m, \sum_i x^i e_i) \mapsto (m, \sum_{i,j} x^i (\sigma^{-1})_i{}^j(m)e_j)$ is also smooth. Since we can cover U by such charts V, this proves that χ^{-1} is smooth and at the same time that it is a (local) vector bundle morphism (because it relates to ψ by an automorphism of E). This proves that χ^{-1} is a vector bundle trivialization of $p^{-1}(U)$.

• Finally to prove (iii) $\Rightarrow$ (ii), let (V, ψ) be a trivializing chart with $V \subset U$ and let ρ be a plateau function around $m_o \in V$ in V [III.5.21]. With this we define the smooth sections $t_i \in \Gamma_V(B)$ by $t_i(m) = \rho(m)\psi^{-1}(m, e_i)$. Since $\mathrm{supp}(\rho) \subset V \subset U$, we may assume that the t_i belong to $\Gamma_U(B)$. By hypothesis there exist functions $\sigma_i{}^j \in C^\infty(U)$ such that $t_i = \sum_i \sigma_i{}^j s_j$. Writing $\psi(s_j(m_o)) = (m_o, \sum_k s_j{}^k e_k)$ and using the fact that $\psi(t_i(m_o)) = (m_o, e_i)$, it follows that $e_i = \sum_{j,k} \sigma_i{}^j(m_o)\, s_j{}^k e_k$. Hence $\sigma_i{}^j(m_o)$ is invertible, and thus the $s_i(m_o)$ form a basis of B_{m_o}. In particular there are as many even elements among the s_i as among the e_i (invariance of the graded dimension). QED

3.11 Definition. Let $p : B \to M$ be a vector bundle with typical fiber E and $U \subset M$ open. A set of homogeneous local sections $s_i \in \Gamma_U(B)$, $i = 1, \ldots, \dim(E)$, is called a *set of local trivializing sections of* B if it satisfies either [3.10-ii] or [3.10-iii].

3.12 Definition. Let $p : B \to M$ be a vector bundle with typical fiber E and let F be a graded subspace of E in the sense of $\mathcal{A}$-vector spaces. A *subbundle with typical fiber F of the vector bundle B* is a subset $C \subset B$ for which there exists a trivializing atlas $\{ U_a, \psi_a \}$ of B such that $\psi_a(p^{-1}(U_a) \cap C) = U_a \times F$. Such an atlas is called *an atlas adapted to the subbundle.* If C is a subbundle of B, it follows in particular that for all $m \in M$ the set $C_m = C \cap B_m$ is a graded subspace of the free graded $\mathcal{A}$-module B_m isomorphic to F [II.2.1].

3.13 Discussion. Let C be a subbundle of a vector bundle B with typical fiber E. Let $F \subset E$ and let $\mathcal{U} = \{ U_a, \psi_a \}$ be as specified by the definition of subbundle. Since $U_a \times F = (U_a \times E) \cap (U_a \times F)$ it should be obvious that C is also a submanifold of B. If we denote by $q : C \to M$ the restriction $q = p|_C$, it follows that $q^{-1}(U_a) = \psi_a^{-1}(U_a \times F)$, i.e., that $(U_a, \psi_a|_{q^{-1}(U_a)})$ is a local trivializing F-chart for the projection $q : C \to M$. Since the transition functions ψ_{ba} are even linear maps, the same holds true for the transition functions associated to the trivializing F-charts for $q : C \to M$. We conclude that $q : C \to M$ is a vector bundle with typical fiber F, justifying its name.

The existence of a subbundle tells us also something about the vector bundle $p : B \to M$ itself. Denote by $G \subset \mathrm{Aut}(E)$ the subgroup of automorphisms of E that preserve the graded subspace F. If $(e_i)_{i=1}^n$ is a basis of E such that $(e_i)_{i=1}^k$ is a basis for F, the matrix of an automorphism g of E has a block form $g = \left(\begin{smallmatrix} a & b \\ c & d \end{smallmatrix}\right)$ such that g belongs to G if and only if $c = 0$. Hence $G \subset \mathrm{Aut}(E)$ is an $\mathcal{A}$-Lie group on its own and the canonical injection $G \to \mathrm{Aut}(E)$ is an $\mathcal{A}$-Lie group morphism. Since $\mathcal{U}$ is adapted to the subbundle, the transition functions ψ_{ba} associated to this atlas take their values in G. In other words, the structure group of B can be reduced from $\mathrm{Aut}(E)$ to G.

3.14 Proposition. *Let $p : B \to M$ be a vector bundle with typical fiber E, let $C \subset B$ be a subset, let F be a graded subspace of E, and let $(e_i)_{i=1}^n$ be a basis of E such that $(e_i)_{i=1}^k$ is a basis of F.*

If for all $m_o \in M$ there exists a chart $U \ni m_o$ and k homogeneous sections $s_i \in \Gamma_U(B)$ of the same parity as e_i such that

$$(3.15) \qquad \forall m \in U : (s_i(m))_{i=1}^k \text{ is a basis of } C_m = C \cap B_m \ ,$$

then C is a subbundle of B with typical fiber F. Nota Bene: (3.15) requires/implies that all $C_m = C \cap B_m$ be graded subspaces [II.2.1] of B_m (in the sense of free graded $\mathcal{A}$-modules).

On the other hand, if C is a subbundle of B with typical fiber F, then for all $m_o \in M$ there exists a chart $U \ni m_o$ and a set of local trivializing sections (n in total!) $s_i \in \Gamma_U(B)$ of the same parity as e_i satisfying (3.15).

Proof. If (U, ψ) is a trivializing chart adapted to the subbundle, it follows that the sections $s_i(m) = \psi^{-1}(m, e_i)$ satisfy the condition, proving the second part. For the first part, let U and $(s_i)_{i=1}^k$ be as stated. Let (V, ψ) be a trivializing chart $m_o \in V \subset U$, and let the sections $\underline{e}_i \in \Gamma_V(B)$ be defined as $\underline{e}_i(m) = \psi^{-1}(m, e_i)$. We thus can write $s_i(m) = \sum_{j=1}^n \sigma_i{}^j(m)\, \underline{e}_j$ for some smooth functions $\sigma_i{}^j \in C^\infty(V)$. Since ϕ is injective, the rank of the $n \times k$ matrix $\sigma_i{}^j$ must be k on V. Reordering the e_i if necessary, we may assume that $\left((s_i(m_o))_{i=1}^k, (\underline{e}_i(m_o))_{i=k+1}^n\right)$ constitutes a basis of B_{m_o}. By continuity (shrinking V if necessary) we may assume that $\left((s_i(m))_{i=1}^k, (\underline{e}_i(m))_{i=k+1}^n\right)$ constitutes a basis of B_m for all $m \in V$. It then follows from [3.10] that the sections $\left((s_i)_{i=1}^k, (\underline{e}_i)_{i=k+1}^n\right)$ on the chart V define a local trivialization in a neighborhood of m_o adapted to the subbundle. QED

3.16 Discussion. The interest of [3.14] is that it allows us to define subbundles of a vector bundle $p : B \to M$ in terms of local sections. According to [3.14] it suffices to specify an atlas $\mathcal{U}$ and for each $U \in \mathcal{U}$ local homogeneous sections $s_1, \ldots, s_k \in \Gamma_U(B)$ such that (i) the number of even sections is the same for every U, and (ii) the graded subspace of B_m generated by $(s_i(m))_{i=1}^k$ has total dimension k and is independent of $U \ni m$. If M is connected, even the requirement that the number of even sections is the same for every U can be omitted because the graded dimension is an invariant of a free graded $\mathcal{A}$-module.

Another consequence of [3.14] is that it says more or less that, if the local sections $s_1, \ldots, s_k \in \Gamma_U(B)$ generate C, then we can find sections $s_{k+1}, \ldots, s_n \in \Gamma_U(B)$ such that $(s_i)_{i=1}^n$ is a set of local trivializing sections of B. This follows by the second part of [3.14]. The only catch is that one might have to take a smaller U to find the $s_{k+1}, \ldots, s_n$.

3.17 Lemma. *Let E and F be $\mathcal{A}$-vector spaces, Q an $\mathcal{A}$-manifold, and let $\psi : Q \times E \to F$ be a smooth map such that for each $q \in Q$ the map $v \mapsto \psi(q, v)$ is right linear. Then there exists a unique smooth map $\psi : Q \to \mathrm{End}_R(E; F)$ (abuse of notation justified by the result) such that $\psi(q, e) = \psi(q)(e)$.*

Proof. Let $(e_i)_{i=1}^n$, $(f_j)_{j=1}^m$ be bases of E and F, and let $(e^i)_{i=1}^n$, $(f^j)_{j=1}^m$ be the associated right dual bases. We then define functions $\psi^j{}_i : Q \to \mathcal{A}$ by $\psi^j{}_i(q) = f^j(\psi(q, e_i))$. These functions are smooth because the maps $f^j : F \to \mathcal{A}$ are smooth and because the e_i have real coordinates. But then we can form the smooth function $\psi : Q \to \mathrm{End}_R(E; F)$ by $\psi(q) = \sum_{i,j} f_j \otimes \psi^j{}_i(g) \cdot e^i$. Since each $\psi(q, \_)$ is right linear, it is immediate that $\psi(q, e) = \psi(q)(e)$ for all $e \in E$. QED

3.18 Proposition. *Let $p : B \to M$ and $q : C \to N$ be vector bundles with typical fibers E and F respectively, and let $h : B \to C$ be a homogeneous left or right linear vector bundle morphism inducing a map $g : M \to N$ such that $h|_{B_m} : B_m \to C_{g(m)}$ is surjective for all $m \in M$. Then*

$$\ker(h) = \cup_{m \in M} \ker(h|_{B_m} : B_m \to C_{g(m)})$$

is a subbundle of B with $\mathrm{rank}(\ker(h)) = \mathrm{rank}(B) - \mathrm{rank}(C)$.

Proof. We will prove the right linear case, the left linear case being similar. Our strategy will be to apply [3.14] to obtain the desired result. For $m_o \in M$, let $W \ni g(m_o)$ be a trivializing chart for the bundle C with trivialization $\chi : q^{-1}(W) \to W \times F$ and let $V \ni m_o$ be a trivializing chart for B with trivialization $\psi : p^{-1}(V) \to V \times E$ such that $g(V) \subset W$ (such charts exist). The map $\chi \circ h \circ \psi^{-1} : V \times E \to W \times F$ is of the form $(m, e) \mapsto (g(m), \widehat{h}(m, e))$. The fact that h is a right linear vector bundle morphism translates as the fact that each map $\widehat{h}(m, \_) : E \to F$ is right linear. According to [3.17] this implies that the matrix elements of these right linear maps are smooth functions. More precisely, if $(e_i)_{i=1}^n$ is a basis of E and $(f_j)_{j=1}^k$ a basis of F, then there exist smooth functions $h^j{}_i \in C^\infty(V)$ such that

$$(\chi \circ h \circ \psi^{-1})(m, \sum_{i=1}^n e_i\, x^i) = (g(m), \sum_{i=1}^n \sum_{j=1}^k f_j\, h^j{}_i(m)\, x^i) \ .$$

Since $h|_{B_m}$ is surjective, the rank [II.1.9] of the $n \times k$ matrix $h^j{}_i(m)$ must be k. By reordering the basis (e_i), we may assume that the $k \times k$ matrix $(h^j{}_i(m_o))_{i,j=1}^k$ is invertible.

Thus there exists a chart $m_o \in U \subset V$ such that $(h^j{}_i(m))_{i,j=1}^k$ is invertible for all $m \in U$. Since inversion of matrices is smooth, there exist smooth functions $g^p{}_j \in C^\infty(U)$ such that the matrix $(g^p{}_j(m))_{p,j=1}^k$ is the inverse of the matrix $(h^j{}_i(m))_{i,j=1}^k$.

With these ingredients we define the sections $s_i \in \Gamma_U(B)$, $k < i \leq n$ by

$$s_i(m) = \psi^{-1}(m, e_i - \sum_{p,j=1}^{k} e_p\, g^p{}_j(m)\, h^j{}_i(m)) .$$

By elementary linear algebra the $(s_i(m))_{i=k+1}^n$ form a basis for $\ker(h|_{B_m}) = \ker(h) \cap B_m$. They are homogeneous because h is homogeneous. We conclude by [3.14] that $\ker(h)$ is a subbundle of B. Since $n = \operatorname{rank} B$ and $k = \operatorname{rank} C$, we automatically have $\operatorname{rank}(\ker(h)) = n - k = \operatorname{rank} B - \operatorname{rank} C$. QED

3.19 Counter example. As in [III.3.30], the condition that each $h|_{B_m}$ is surjective can not be weakened easily, as can be seen in the following example. Let F be an $\mathcal{A}$-vector space of dimension $0|2$, let $M = N = F_0$ with coordinates (ξ, η), and let $B = C = M \times \mathcal{A}$ be the trivial bundle. Then the map $h : B \to C$, $(\xi, \eta, a) \mapsto (\xi, \eta, \xi\eta a)$ is a right linear vector bundle morphism. But $\ker(h)$ is not a subbundle. Even more, except for $\xi\eta = 0$, $\ker(h|_{B_{(\xi,\eta)}})$ is not a graded subspace of $B_{(\xi,\eta)}$ (though it is a graded submodule).

4. Constructions of vector bundles

We show that the various operations one can perform on $\mathcal{A}$-vector spaces (quotients, direct sums, tensor products, exterior powers, homomorphisms) have their analogon for vector bundles. Even the constructions of new linear maps out of old ones have their counter part in vector bundle morphisms. As a byproduct of these constructions, we explain why we took the full $\mathcal{A}$-vector space as typical fiber for vector bundles instead of only its even part.

4.1 Discussion. Let $p : B \to M$ and $q : C \to M$ be vector bundles over the same base space with typical fibers E and F. For $m \in M$ this implies that B_m is isomorphic to E and C_m to F. Since these spaces are free graded $\mathcal{A}$-modules, we can perform various operations on these fibers, e.g., taking the direct sum $B_m \oplus C_m$, or the tensor product $B_m \otimes C_m$. Whence the obvious question: can we construct in a natural way out of B and C a vector bundle $r : D \to M$ whose typical fiber is, say, $E \oplus F$ such that we have a natural identification $D_m \cong B_m \oplus C_m$.

The obvious way to do this is to define $D = \cup_{m\in M}(B_m \oplus C_m)$. The only draw-back is that we have to indicate the structure of an $\mathcal{A}$-manifold on this D. In the next constructions we will do exactly that, although in a roundabout way. We will define vector bundles with the appropriate typical fiber, and we will show once (in [4.5] for the direct sum case) how this constructed vector bundle relates to D.

4.2 Construction (quotient bundle). Let $p : B \to M$ be a vector bundle with typical fiber E, C a subbundle with typical fiber F, and $\mathcal{U}$ a trivializing atlas adapted to the subbundle C. Let G be the subgroup of $\mathrm{Aut}(E)$ that leaves the graded subspace $F \subset E$ invariant. Since G leaves F invariant, it induces an action on the quotient E/F, i.e., there exists a group morphism $\rho : G \to \mathrm{Aut}(E/F)$. Using a basis as in [3.13] shows that ρ is smooth (it is given by $\begin{pmatrix} a & b \\ 0 & d \end{pmatrix} \mapsto d$). We thus can construct the associated vector bundle with typical fiber E/F. It is called the *quotient bundle of B by C* and denoted by B/C.

4.3 Construction (direct sums). Let $p_i : B_i \to M$, $i = 1, \ldots, n$ be a finite family of vector bundles over M with typical fibers E_i. Let $\{ U_a, \psi_{i,ba} \}$ be an atlas and transition functions defining the bundle B_i. Indeed, by taking n-fold intersections we may assume that the charts U_a are the same for all bundles. Since the maps $\psi_{i,ba} : U_a \cap U_b \to \mathrm{Aut}(E_i)$ are smooth, it follows that the map $\prod_i \psi_{i,ba} : U_a \cap U_b \to \prod_i \mathrm{Aut}(E_i) \subset \mathrm{Aut}(\oplus_i E_i)$ defined as $(\prod_i \psi_{i,ba})(m) = \prod_i \psi_{i,ba}(m)$ [I.7.7] is smooth. From the definition of $\prod_i \psi_{i,ba}(m)$ we deduce the equality

$$(\prod_i \psi_{i,cb}(m)) \circ (\prod_i \psi_{i,ba}(m)) = \prod_i (\psi_{i,cb}(m) \circ \psi_{i,ba}(m)) = \prod_i \psi_{i,ca}(m)$$

on the triple intersection $U_a \cap U_b \cap U_c$. Hence $\{ U_a, \prod_i \psi_{i,ba} \}$ defines a vector bundle over M with typical fiber $\oplus_i E_i$ whose structure group can be reduced to $\prod_i \mathrm{Aut}(E_i)$. This bundle, denoted by $\oplus_i B_i$, is called the *direct sum bundle of the vector bundles B_i*; it is also called the *Whitney sum of the B_i*. In case there are only two vector bundles involved, one writes $B_1 \oplus B_2$ for this direct sum bundle.

4.4 Remarks. • The definition of a direct sum of two vector bundles is essentially the same as the definition of a direct product bundle. Strictly speaking we should have indexed the $\oplus$ in $B_1 \oplus B_2$ with an M to distinguish this sum from the direct sum of $\mathcal{A}$-vector spaces. However, a vector bundle will hardly ever be at the same time an $\mathcal{A}$-vector space, so no confusion is possible, since a direct sum of two $\mathcal{A}$-manifolds is not defined.

• Since we can reduce the structure group of $\oplus_i B_i$ to $\prod_i \mathrm{Aut}(E_i)$, it should be fairly obvious that each separate bundle B_i can be seen as a subbundle of $\oplus_i B_i$.

4.5 Discussion. Let $p : B \to M$ and $q : C \to M$ be vector bundles over the same base space with typical fibers E and F. It follows that $B \oplus C$ is a vector bundle with typical fiber $E \oplus F$. In [4.1] we suggested the space $D = \cup_{m \in M} (B_m \oplus C_m)$, with the obvious projection $r : D \to M$, as a candidate for this vector bundle. Let us show that we can identify these two spaces, i.e., that we can equip D with the structure of a vector bundle over M with the same transition functions as the bundle $B \oplus C$.

Let $\mathcal{U}$ be a trivializing atlas for both bundles at the same time with diffeomorphisms $\psi_a : p^{-1}(U_a) \to U_a \times E$ and $\chi_a : q^{-1}(U_a) \to U_a \times F$ that induce the transition functions ψ_{ba} and χ_{ba}. By definition of the structure of a free graded $\mathcal{A}$-module on each fiber, the maps $\psi_a(m) : p^{-1}(m) \to E$ and $\chi_a(m) : q^{-1}(m) \to F$ are isomorphisms. We thus obtain an isomorphism

$$\psi_a(m) \times \chi_a(m) : B_m \oplus C_m \to E \oplus F \ .$$

Varying m over U_a gives us a bijection between $r^{-1}(U_a) = \cup_{m \in U_a}(B_m \oplus C_m) \subset D$ and $U_a \times (E \oplus F)$. Declaring this to be a diffeomorphism equips $r^{-1}(U_a)$ with the structure of an $\mathcal{A}$-manifold, compatible with the projection r. It remains to show that these structures coincide on overlaps $U_a \cap U_b$. But on overlaps we have (by definition of the transition functions) the equality $(\psi_b(m) \times \chi_b(m)) \circ (\psi_a(m) \times \chi_a(m))^{-1} = \psi_{ba} \times \chi_{ba}$, which is a smooth function. It thus follows at the same time that the smooth structures coincide on overlaps *and* that the transition functions of the vector bundle D (yes, we just proved it is one!) are exactly the same as those of $B \oplus C$.

In exactly the same spirit one should interpret the fibers of B/C (assuming of course that C is a subbundle of B) as being the quotient of the corresponding fibers in B and C :

$$(B/C)_m \cong B_m/C_m \ .$$

Similar remarks apply to the forthcoming constructions of tensor product bundles, exterior powers and homomorphism bundles.

4.6 Definition. Let $p : D \to M$ be a vector bundle with typical fiber E and let B and C be subbundles with typical fibers F_1 and F_2 respectively. We will say that B and C are *supplements* if $\forall m \in M : B_m \oplus C_m = D_m$, i.e., if $\forall m \in M :$ B_m and C_m are supplements to each other in the sense of free graded $\mathcal{A}$-modules [II.2.1], [II.3.4].

4.7 Proposition. *Let $p : D \to M$, B, and C be as in [4.6]. If B and C are supplements, then*

(i) *if $(s_i)_{i=1}^k \in \Gamma_U(D)$ is a set of local trivializing sections for B, if $(t_j)_{j=1}^\ell \in \Gamma_U(D)$ is a set of local trivializing sections for C, then (s_i, t_j) is a set of local trivializing sections for D, and*

(ii) *$B \oplus C$ is isomorphic as a vector bundle to D.*

Proof. For (i) it suffices to note that by [3.14] $\forall m \in U$ $(s_i(m))$ is a basis for B_m, and that $(t_j(m))$ is a basis of C_m, and thus $(s_i(m), t_j(m))$ is a basis of D_m because $B_m \oplus C_m = D_m$ (which means that the map $(b, c) \mapsto b + c$, $B_m \oplus C_m \to D_m$ is an isomorphism). The announced result then follows again by [3.14].

For (ii) we use the interpretation [4.5] of $B \oplus C = \cup_{m \in M} B_m \oplus C_m$ and define (set theoretically) the isomorphism $h : B \oplus C \to D$ by $h|_{B_m \oplus C_m} : B_m \oplus C_m \to D_m$,

$(b, c) \mapsto b + c$. Since B and C are supplements, h is an even fiber bundle map, an isomorphism when restricted to each fiber, and inducing the identity on M. According to [3.6] it thus suffices to show that h is smooth in order to prove the result. To prove that h is smooth, we choose local trivializing sections $(s_i) \in \Gamma_U(B) \subset \Gamma_U(D)$ and $(t_j) \in \Gamma_U(C) \subset \Gamma_U(D)$ around a point $m_o \in M$ (by shrinking we can always assume that we have the same U). We then denote by ψ the local trivialization of B associated to the (s_i) [3.10], by χ the local trivialization of C associated to the (t_j), and by ϕ the local trivialization of D associated to (s_i, t_j). Using the notation $\cup_{m \in M} \psi(m) \times \chi(m)$ as in [4.5], we have a commutative diagram

$$\begin{array}{ccc}
r^{-1}(U) = \bigcup_{m \in M} B_m \oplus C_m & \xrightarrow{\ h\ } & \bigcup_{m \in M} D_m = p^{-1}(U) \\
{\scriptstyle \cup_{m\in M} \psi(m)\times\chi(m)} \Big\downarrow & & \Big\downarrow {\scriptstyle \cup_{m\in M} \phi(m) \equiv \phi} \\
U \times (F_1 \oplus F_2) & \longrightarrow & U \times E \\
(m, (f_1, f_2)) & \longrightarrow & (m, f_1 + f_2) \ ,
\end{array}$$

just because of our choice of ϕ. Since we know that all maps except h are diffeomorphisms, we conclude that h is smooth on $r^{-1}(U)$. Since the U cover M (by varying m_o), we conclude. QED

4.8 Remark. If B and C are subbundles of a vector bundle $p : D \to M$, one might think that, in analogy with sums of graded subspaces, we could define a sum bundle as $\cup_{m \in M}(B_m + C_m)$. However, this is a very unstable definition because there is no easy way to guarantee that $B_m + C_m \subset D_m$ has a graded dimension independent of $m \in M$. And if this graded dimension is not constant, we do not get a vector bundle in our sense.

4.9 Construction (tensor products). Let $p_i : B_i \to M, i = 1, \ldots, n$ be a finite family of vector bundles over M with typical fibers E_i. Let $\{ U_a, \psi_{i,ba} \}$ be an atlas and transition functions defining the bundle B_i. Since the maps $\psi_{i,ba} : U_a \cap U_b \to \mathrm{Aut}(E_i)$ are smooth, it follows that the map $\otimes_{i=n}^{1} \psi_{i,ba} : U_a \cap U_b \to \otimes_{i=n}^{1} \mathrm{Aut}(E_i) \subset \mathrm{Aut}(\otimes_{i=1}^{n} E_i)$ defined as $(\otimes_{i=n}^{1} \psi_{i,ba})(m) = \otimes_{i=n}^{1} \psi_{i,ba}(m)$ [I.7.12] is smooth. (Nota Bene. We only have an inclusion $\otimes_{i=n}^{1} \mathrm{Aut}(E_i) \subset \mathrm{Aut}(\otimes_{i=1}^{n} E_i)$ because the even part of a tensor product is not the tensor product of the even parts.) From the definition of $\otimes_{i=n}^{1} \psi_{i,ba}(m)$ and the fact that the $\psi_{i,ba}(m)$ are even, we deduce the equality

$$(\otimes_{i=n}^{1} \psi_{i,cb}(m)) \circ (\otimes_{i=n}^{1} \psi_{i,ba}(m)) = \otimes_{i=n}^{1} (\psi_{i,cb}(m) \circ \psi_{i,ba}(m)) = \otimes_{i=n}^{1} \psi_{i,ca}(m)$$

on the triple intersection $U_a \cap U_b \cap U_c$. Hence $\{ U_a, \otimes_{i=n}^{1} \psi_{i,ba} \}$ defines a vector bundle over M with typical fiber $\otimes_{i=1}^{n} E_i$. This bundle, denoted by $\otimes_{i=1}^{n} B_i$ (or by $B_1 \otimes \cdots \otimes B_n$ to make the order even more explicit), is called the *tensor product bundle of the vector bundles* B_i.

4.10 Construction (exterior powers). Let $p : B \to M$ be a vector bundle with typical fiber E, let $\{ U_a, \psi_{ba} \}$ be an atlas and transition functions defining the bundle B and let k be a positive integer. Since the map $\psi_{ba} : U_a \cap U_b \to \mathrm{Aut}(E)$ is smooth, it follows that the map $\bigwedge^k \psi_{ba} : U_a \cap U_b \to \mathrm{Aut}(\bigwedge^k E)$ defined as $(\bigwedge^k \psi_{ba})(m) = \bigwedge^k \psi_{ba}(m)$ [I.7.15] is smooth. From the definition of $\bigwedge^k \psi_{ba}(m)$ we deduce the equality

$$(\textstyle\bigwedge^k \psi_{cb}(m)) \circ (\bigwedge^k \psi_{ba}(m)) = \bigwedge^k (\psi_{cb}(m) \circ \psi_{ba}(m)) = \bigwedge^k \psi_{ca}(m)$$

on the triple intersection $U_a \cap U_b \cap U_c$. It thus follows that $\{ U_a, \bigwedge^k \psi_{ba} \}$ defines a vector bundle over M with typical fiber $\bigwedge^k E$. This bundle, denoted by $\bigwedge^k B$, is called the *k-th exterior power of the vector bundle B*. According to [I.7.15] we have $\bigwedge^1 \psi_{ba}(m) = \psi_{ba}(m)$ and $\bigwedge^0 \psi_{ba}(m) = id(\mathcal{A})$. Together with $\bigwedge^1 E = E$ and $\bigwedge^0 E = \mathcal{A}$ this gives $\bigwedge^1 B = B$ and $\bigwedge^0 B = M \times \mathcal{A}$.

4.11 Definition/Discussion. Let $p : B \to M$ be a vector bundle with typical fiber E. Since elements of $\mathrm{Aut}(E)$ are even, they preserve E_α ($\alpha \in \mathbf{Z}_2$); we thus obtain a smooth action of $\mathrm{Aut}(E)$ on the $\mathcal{A}$-manifold E_α. Since the action of $\mathrm{Aut}(E)$ on E_α is pseudo effective [1.5], we can apply the construction of an associated bundle with the identity map $\rho = id(\mathrm{Aut}(E))$, but with an action on different $\mathcal{A}$-manifolds. The resulting *fiber* bundle will be denoted by $p : B_\alpha \to M$, called *the degree-α part of B*. Another way to define B_α is as a submanifold of B by $B_\alpha = \{ b \in B \mid b \in p^{-1}(p(b))_\alpha \}$. That it is a submanifold of B follows immediately from the fact that E_α is a submanifold of $E \cong E_0^\sharp$. That it is indeed the same as the associated bundle defined previously is left to the reader.

In order to study more carefully the above situation, let us denote by ψ_{ba} the transition functions of the bundle B, and by $\psi_{\alpha,ba}$ those of the bundle B_α. These maps are all essentially the same, they act on different spaces. Since $E = E_0 \times E_1$ in terms of $\mathcal{A}$-manifolds, it is natural to look at the transition functions $\psi_{0,ba} \times \psi_{1,ba}$ who act on $E_0 \times E_1 \cong E$, as do the ψ_{ba}. An elementary calculation shows that these two sets of transition functions are the same. We conclude that the product bundle $B_0 \times_M B_1$ is isomorphic to the original bundle B.

Although the fiber bundle B_α is not a vector bundle, it resembles it very closely. The abelian group E_α is stable under multiplication by elements of $\mathcal{A}_0$, i.e., E_α is an $\mathcal{A}_0$-module. The action of $\mathrm{Aut}(E)$ on E_α preserves this structure and thus the fibers of B_α have the structure of an $\mathcal{A}_0$-module. Copying the discussion for vector bundles shows that $\Gamma(B_\alpha)$ is a $C^\infty(M)_0$-module. More precisely, one can show that we have an identification $\Gamma(B_\alpha) = \Gamma(B)_\alpha$, a result in complete agreement with the corresponding fact for smooth functions $C^\infty(M;F)_\alpha = C^\infty(M;F_\alpha)$, and a result which is rather obvious once we see B_α as a submanifold of B. It then follows that we can apply (adapt) the results of [3.9] to sections of B_α, in particular that local sections of B_α can be glued together to a global section by means of a partition of unity.

4.12 Discussion. Let us now explain why we defined vector bundles as fiber bundles with as typical fiber a full $\mathcal{A}$-vector space E (seen as the even part of $E^\sharp$), instead of the even part of an $\mathcal{A}$-vector space. The reason is that we could not have performed all the constructions of vector bundles as we did, in particular not the tensor product. Let us show the problem.

As usual we take $\mathcal{A} = \bigwedge X$ where X is an infinite dimensional vector space over $\mathbf{R}$ with basis $(e_i)_{i \in \mathbf{N}}$. We then consider the $\mathcal{A}$-vector space E of dimension $0|1$, for which we have a natural identification $E \otimes E \cong \mathcal{A}$ (if $b \in E$ is a basis, we can identify $b \otimes b$ with $1 \in \mathcal{A}$). If we consider only E_0, it will not be an $\mathcal{A}$-module, but only an $\mathcal{A}_0$-module. Hence we cannot take the tensor product over $\mathcal{A}$, but only over $\mathcal{A}_0$. The interesting question is whether this tensor product $E_0 \otimes_{\mathcal{A}_0} E_0$ is in a natural way the even part of an $\mathcal{A}$-vector space, and more in particular whether it is the even part of $E \otimes_{\mathcal{A}} E$.

As a set E_0 is isomorphic to $\mathcal{A}_1$, and by definition of $\mathcal{A}$, the $e_i \in \bigwedge^1 X \subset \mathcal{A}_1$ generate $\mathcal{A}_1$ as an $\mathcal{A}_0$-module. It follows that the elements $e_i \otimes e_j$ generate the tensor product $E_0 \otimes_{\mathcal{A}_0} E_0$ (over $\mathcal{A}_0$). As said before, $E \otimes_{\mathcal{A}} E \cong \mathcal{A}$, and in particular $(E \otimes_{\mathcal{A}} E)_0 \cong \mathcal{A}_0$ is a free $\mathcal{A}_0$-module on one (even) generator. Now suppose that $E_0 \otimes_{\mathcal{A}_0} E_0$ is a free $\mathcal{A}_0$-module on one generator $g \in E_0 \otimes_{\mathcal{A}_0} E_0$. Since the $e_i \otimes e_j$ generate $E_0 \otimes_{\mathcal{A}_0} E_0$, there exist $\lambda_{ij} \in \mathcal{A}_0$ (finitely many!) such that $g = \sum_{ij} \lambda_{ij} e_i \otimes e_j$. If we now take $n \in \mathcal{A}_0$ to be the (exterior) product of all different e_i's appearing in this sum (adding one not in that sum if there is an odd number of different e_i's), then n is not zero (X is infinite dimensional), but ng *is* zero.

The conclusion of these computations is that in general the $\mathcal{A}_0$-modules $(E \otimes_{\mathcal{A}} F)_0$ and $E_0 \otimes_{\mathcal{A}_0} F_0$ are not the same. On the other hand, if B and C are vector bundles with typical fibers E_0 and F_0 respectively, it is quite natural to require from a *tensor product of bundles* that the typical fiber of $B \otimes C$ is $(E \otimes F)_0$. Since this cannot be done by means of the tensor product of $\mathcal{A}_0$-modules, we use the full $\mathcal{A}$-vector space E as fiber, take the tensor product in the category of $\mathcal{A}$-vector spaces, and restrict later to the even (or odd) part.

4.13 Discussion. Of course all our definitions and constructions of fiber bundles and vector bundles are also valid in the context of $\mathbf{R}$-manifolds, i.e., in the special case $\mathcal{A} = \mathbf{R}$. Moreover, the body map $\mathbf{B}$ maps fiber/vector bundles in the category of $\mathcal{A}$-manifolds to the corresponding objects in the category of $\mathbf{R}$-manifolds. But in the category of $\mathbf{R}$-manifolds something special happens with vector bundles. If $X = X_0 \oplus X_1$ is an $\mathbf{R}$-vector space of dimension $p|q$, its separate parts X_α can also be seen as $\mathbf{R}$-vector spaces: X_0 can be seen as the $\mathbf{R}$-vector space $X_0 \oplus \{0\}$ of dimension $p|0$, and X_1 can be seen as the $\mathbf{R}$-vector space $\{0\} \oplus X_1$ of dimension $0|q$. Since automorphisms of X are even, they respect this decomposition, and it is not hard to see that we have an identification $\mathrm{Aut}(X) \cong \mathrm{Aut}(X_0) \times \mathrm{Aut}(X_1)$ (regarding X_α as an $\mathbf{R}$-vector space). From these observations it follows that if $p : B \to M$ is a vector bundle in the category of $\mathbf{R}$-manifolds, then the B_α are also vector bundles in this category. We are thus able to write $B \cong B_0 \times_M B_1 \cong B_0 \oplus B_1$, i.e., in the category of $\mathbf{R}$-manifolds the homogeneous parts of a vector bundle are vector bundles in their own right and a vector bundle *is* the direct sum of its even and odd part. We said that $\mathbf{R}$-manifolds are nothing more (nor less) than manifolds in the classical sense. We now see that the same is not true for vector bundles: vector bundles in the category of **R**-manifolds are not vector bundles in the classical sense, they are a direct sum of two vector bundles in the classical sense.

4.14 Discussion. Let $p : B \to M$ and $q : C \to N$ be vector bundles with typical fibers E and F respectively, and let $h : B \to C$ be a right linear vector bundle morphism, inducing the map $g : M \to N$ on the base spaces. Let furthermore $\mathcal{V}$ be a trivializing atlas for C with transition functions χ_{dc} and $\mathcal{U}$ a trivializing atlas for B with transition functions ψ_{ba} such that for all $U_a \in \mathcal{U}$ there exists a $V_{g(a)} \in \mathcal{V}$ such that $g(U_a) \subset V_{g(a)}$. (As in [2.2] we use the same name for the function on the base space as on the indices.) By definition of a fiber bundle morphism there exist smooth functions $h_a : U_a \times E \to F$ determined by $(\chi_{g(a)} \circ h \circ \psi_a^{-1})(m, e) = (g(m), h_a(m, e))$. Since h is a vector bundle morphism, the maps $h_a(m, e)$ are right linear in e for fixed m. According to [3.17] we can interpret them as smooth functions $h_a : U_a \to \mathrm{Hom}_R(E; F)$. Since h is globally defined, the h_a are related on overlaps $U_a \cap U_b$ by

$$h_b(m) = \chi_{g(b)g(a)}(g(m)) \circ h_a(m) \circ \psi_{ba}(m)^{-1} \ . \tag{4.15}$$

On the other hand, given a set of smooth functions $h_a : U_a \to \mathrm{Hom}_R(E; F)$ verifying (4.15), we can construct a right linear vector bundle morphism $h : B \to C$ inducing the map $g : M \to N$ as follows. The restriction of f to $p^{-1}(U_a)$ is defined as the smooth map

$$f|_{p^{-1}(U_a)} = \chi_{g(a)}^{-1} \circ (g \times h_a) \circ \psi_a \ ,$$

where $g \times h_a$ denotes the map $U_a \times E \to V_{g(a)} \times F$, $(m, e) \mapsto (g(m), h_a(m)(e))$. Using the condition (4.15) it is elementary to show that these local expressions of f coincide on overlaps. In this way we obtain a bijection between right linear vector bundle morphisms and sets of smooth local functions $h_a : U_a \to \mathrm{Hom}_R(E; F)$ verifying (4.15). And obviously a similar result is valid in the left linear context.

4.16 Construction (Hom-**bundles).** Let $p : B \to M$ and $q : C \to M$ be two vector bundles with typical fibers E and F respectively. Let $\mathcal{U}$ be an atlas of trivializing charts for both bundles and denote by $\psi_{ba} : U_a \cap U_b \to \mathrm{Aut}(E)$ and $\chi_{ba} : U_a \cap U_b \to \mathrm{Aut}(F)$ their transition functions. For a fixed $m \in U_a \cap U_b$ we obtain an automorphism $H_{ba}(m)$ of $\mathrm{Hom}_L(E; F)$ defined by

$$H_{ba}(m) : A \mapsto \chi_{ba}(m) \circ A \circ \psi_{ba}(m)^{-1} \equiv \psi_{ba}(m)^{-1} \diamond A \diamond \chi_{ba}(m) \ . \tag{4.17}$$

Using [I.7.12], the identification $\mathrm{Hom}_L(E; F) \cong {}^*E \otimes F$, and that $\psi_{ba}(m)$ and $\chi_{ba}(m)$ are even (and thus left *and* right linear), we can identify $H_{ba}(m)$ with $\chi_{ba}(m) \otimes {}^*\psi_{ba}(m)^{-1}$. We thus obtain smooth maps $H_{ba}(m) : U_a \cap U_b \to \mathrm{Aut}(\mathrm{Hom}_L(E; F)) \cong \mathrm{Aut}({}^*E \otimes F)$. It is by now elementary to show that $H_{cb}(m) \circ H_{ba}(m) = H_{ca}(m)$ on the triple intersection $U_a \cap U_b \cap U_c$. It thus follows that $\{ U_a, H_{ba} \}$ defines a vector bundle over M with typical fiber $\mathrm{Hom}_L(E; F)$. This bundle is denoted by $\mathrm{Hom}_L(B; C)$ and called the *bundle of (left)* *homomorphisms from* B *to* C. In the same way we obtain a vector bundle $\mathrm{Hom}_R(B;C)$ with typical fiber $\mathrm{Hom}_R(E;F)$: the transition functions H_{ba} are given by the same expression (4.17), but in the identification $\mathrm{Hom}_R(E;F) \cong F \otimes E^*$ they are given as $H_{ba}(m) \cong (\psi_{ba}(m)^{-1})^* \otimes \chi_{ba}(m)$.

If C is the trivial bundle $C = M \times \mathcal{A}$, the typical fiber of the bundle $\mathrm{Hom}_L(B;C)$ is $\mathrm{Hom}_L(E;\mathcal{A}) \cong {}^*E$. One usually denotes this bundle $\mathrm{Hom}_L(B; M \times \mathcal{A})$ by ${}^*B$ and calls it the *(left) dual bundle of* B. Since for the trivial bundle $M \times \mathcal{A}$ all transition functions χ_{ba} are the identity on $\mathcal{A}$, it follows that the transition functions of the dual bundle are given by ${}^*\psi_{ba}^{-1} \otimes id(\mathcal{A}) \cong {}^*\psi_{ba}^{-1}$. In a similar way one defines the right dual bundle B^*.

4.18 Discussion. Let $p : B \to M$ and $q : C \to M$ be two vector bundles with typical fibers E and F respectively. The name *bundle of homomorphisms* for the vector bundle $\mathrm{Hom}_L(B;C)$ suggests that there might be a relation with vector bundle homomorphisms $h : B \to C$ (which by convention induce the identity on M). We will show that there is indeed such a relation: global smooth sections s of $\mathrm{Hom}_L(B;C)$ are in bijection with left linear vector bundle morphisms $h : B \to C$.

Given a trivializing atlas, we have seen in [1.20] that smooth sections s of $\mathrm{Hom}_L(B;C)$ are in 1-1 correspondence with families of smooth maps $s_a : U_a \to \mathrm{Hom}_L(E;F)$ satisfying the relations (1.21) with the transition functions H_{ba}. But these conditions are exactly the conditions (4.15) (in the left linear version), once one realizes that g is the identity and that we use the same trivializing atlas for both bundles. We thus obtain a left linear vector bundle morphism $h : B \to C$. We leave it to the reader to prove that the correspondence $s \mapsto h$ is a bijection. At the intrinsic level this correspondence is very easy to understand: a section $s \in \Gamma(\mathrm{Hom}_L(B;C))$ determines, for every $m \in M$, an element $s(m) \in \mathrm{Hom}_L(B;C)_m \cong \mathrm{Hom}_L(B_m;C_m)$, and a left linear vector bundle morphism determines a left linear map $h|_{B_m} : B_m \to C_m$. The given identification $s \mapsto h$ is nothing more than saying that these two are the same: $s(m) = h|_{B_m}$.

4.19 Proposition. *Let* $p : B \to M$ *and* $q : C \to N$ *be vector bundles and let* $h : B \to C$ *be a (left/right linear) vector bundle morphism inducing the map* $g : M \to N$. *Then* $g^*q : g^*C \to M$ *is a vector bundle with the same typical fiber as* C *[2.2],* $\breve{g} : g^*C \to C$ *is an even vector bundle morphism, and* $h_{ind} : B \to g^*C$ *is a (left/right linear) vector bundle morphism [2.4].*

4.20 Proposition. *Let* $g : M \to N$ *be a smooth map between* $\mathcal{A}$*-manifolds. The operation of pull-back bundle, which transforms a fiber bundle over* N *into a fiber bundle over* M *with the same typical fiber, commutes with the constructions of vector bundles given above. E.g., if* $p : B \to N$ *and* $q : C \to N$ *are vector bundles over* N, *then the vector bundles* $g^*(B \oplus C)$ *and* $(g^*B) \oplus (g^*C)$ *are the same (isomorphic), as are* $g^*(B \otimes C)$ *and* $(g^*B) \otimes (g^*C)$.

Proof. Let $\mathcal{V}$ be a trivializing atlas on N for the bundles over N that are involved, and let $\mathcal{U}$ be an atlas of M such that for each $U_a \in \mathcal{U}$ there exists a $V_{g(a)} \in \mathcal{V}$ such that $g(U_a) \subset V_{g(a)}$. In [2.2] it was shown that the $\psi_{ba} = \chi_{g(b)g(a)} \circ g$ form the transition functions of the pull-back bundle over M when the χ_{dc} are the transition functions of the bundle over N. Let us take the tensor product of two bundles as typical example, with transition functions $\chi_{1,dc}$ and $\chi_{2,dc}$. The transition functions of $g^*(B \otimes C)$ are given by $(\chi_{1,g(b)g(a)} \otimes \chi_{2,g(b)g(a)}) \circ g$, which is the same as $(\chi_{1,g(b)g(a)} \circ g) \otimes (\chi_{2,g(b)g(a)} \circ g)$, which are the transition functions of the bundle $(g^*B) \otimes (g^*C)$. QED

4.21 Discussion. This result is not really surprising when we recall the general idea behind the vector bundle constructions. Let us again take the tensor product as typical example. The tensor product bundle $B \otimes C$ is the bundle whose fibers are just the tensor products of the original fibers: $(B \otimes C)_m \cong B_m \otimes C_m$. On the other hand, the pull-back bundle g^*B is a bundle whose fiber $(g^*B)_m$ is exactly the fiber $B_{g(m)}$. We thus see that all these constructions are done fiberwise, making it easy to understand the commutativity stated in [4.20].

4.22 Nota Bene. Restricting a bundle $p : B \to M$ to a submanifold $N \subset M$ is a particular instance of a pull-back. It follows that restriction commutes with the vector bundle constructions. This fact will be frequently used in chapters V and VI without further mentioning.

4.23 Proposition. *Let $p : B \to M$ and $q : C \to M$ be vector bundles over the same base space with typical fibers E and F respectively. Let ψ_{ba} and χ_{ba} be the transition functions with respect to a joint trivializing atlas $\mathcal{U}$. Finally, let $\mathfrak{I} \in \mathrm{BHom}(E; F)$ be a smooth morphism of $\mathcal{A}$-vector spaces. If $\mathfrak{I}$ intertwines ψ_{ba} and χ_{ba}, i.e., if for each $m \in M$ we have a commutative diagram*

$$\begin{array}{ccc} E & \xrightarrow{\ \mathfrak{I}\ } & F \\ {\scriptstyle \psi_{ba}(m)}\downarrow & & \downarrow{\scriptstyle \chi_{ba}(m)} \\ E & \xrightarrow{\ \mathfrak{I}\ } & F\,, \end{array}$$

then it induces a morphism of vector bundles $\widetilde{\mathfrak{I}} : B \to C$. If $\mathfrak{I}$ is of parity α, $\widetilde{\mathfrak{I}}$ is too; if $\mathfrak{I}$ is an isomorphism, $\widetilde{\mathfrak{I}}$ is an isomorphism of vector bundles.

Proof. For each $U_a \in \mathcal{U}$ we define the map $h_a : U_a \to \mathrm{Hom}_S(E; F)$ as being the constant $\mathfrak{I}$. By commutativity of the given diagram, these functions h_a verify (4.15) with $g = id(M)$. Hence they define a vector bundle morphism $\widetilde{\mathfrak{I}}$. The remaining assertions are left to the reader. QED

4.24 Nota Bene. A trivializing atlas is never unique, and neither is a joint trivializing atlas. It follows that the transition functions ψ_{ba} and χ_{ba} in [4.23] are not unique. The condition in [4.23] thus can be read as a search for a joint trivializing atlas such that $\mathfrak{I}$ intertwines the transition functions. However, in most applications a natural candidate for a joint trivializing atlas is available which satisfies the conditions. But the reader should be aware that the induced identification $\widetilde{\mathfrak{I}}$ depends upon the chosen joint trivializing atlas.

4.25 Discussion. Using [4.23] we can construct many (iso)morphisms of vector bundles. Since all morphisms between various constructions of $\mathcal{A}$-vector spaces as given in §I.7 (see [I.8.18]) intertwine the relevant transition functions, we obtain, among others, the following isomorphisms (finite dimensional $\mathcal{A}$-vector spaces are f.g.p!) of vector bundles: $\mathrm{Hom}_R(B;C) \cong C \otimes B^*$, $^*(B \oplus C) \cong {}^*B \oplus {}^*C$, etcetera. We also obtain an even morphism $\mathrm{Hom}_R(B;C) \otimes B \to C$, which can be interpreted as fiberwise evaluation. Even the interchanging map $\mathfrak{R}$ satisfies the requirements, so we obtain a vector bundle interchanging map $\mathfrak{R} : B \otimes C \to C \otimes B$ (abuse of notation).

4.26 Construction. In §I.7 we have given a number of constructions of new linear maps out of a family of linear maps: the sum [I.7.2], the direct sum [I.7.6], the product [I.7.7], the tensor product [I.7.12], and the exterior power [I.7.15]. We will show that we can obtain, under suitable hypotheses, similar results for vector bundle morphisms. The basic idea is easily explained. Let the B_i be vector bundles over M, let the C_i be vector bundles over N, and let h_i be a vector bundle morphism between the bundles B_i and C_i. If all vector bundle morphisms involved induce the same map g between the base spaces, they all map fibers above a point $m \in M$ to fibers above $g(m) \in N$. Since the restrictions are linear, we can perform our constructions of new linear maps on these restrictions. This will certainly define an abstract map. Using [4.14] one can show that the new maps are vector bundle morphisms inducing g. We will state the precise result only for exterior powers; the other cases are similar and are left to the reader (cf. [I.7.24]).

4.27 Proposition. *Let $p : B \to M$ be a vector bundle over M, let $q : C \to N$ be a vector bundle over N, and let $h : B \to C$ be an even (left) linear vector bundle morphism inducing $g : M \to N$. Then there exists a canonically defined even linear vector bundle morphism $\bigwedge^k h : \bigwedge^k B \to \bigwedge^k C$ inducing g.*

Proof. Let $\mathcal{V}$ be a trivializing atlas for C with transition functions χ_{dc}, and let $\mathcal{U}$ be a trivializing atlas for B with transition functions ψ_{ba} such that the conditions of [4.14] are verified, i.e., we have local smooth maps $h_a : U_a \to \mathrm{Hom}_L(E;F)_0$ satisfying (4.15), where E is the typical fiber of B and F the typical fiber of C. Now recall that the transition functions of $\bigwedge^k B$ are given by $\bigwedge^k \psi_{ba}$, and those of $\bigwedge^k C$ by $\bigwedge^k \chi_{dc}$. One then can prove easily the equality $\bigwedge^k h_b(m) = (\bigwedge^k \chi_{g(b)g(a)})(g(m)) \circ (\bigwedge^k h_a(m)) \circ (\bigwedge^k \psi_{ba})(m)^{-1}$. As in [4.10] the maps $m \mapsto \bigwedge^k h_a(m)$ are smooth. This proves that the local maps $\bigwedge^k h_a : U_a \to \mathrm{Hom}_L(\bigwedge^k E; \bigwedge^k F)_0$ define a left linear even vector bundle morphism, which we denote by $\bigwedge^k h$. We leave it to the reader to verify that it coincides with the map described intuitively in [4.26]. $\boxed{QED}$

4.28 Remark. For $k = 0$ the map $\bigwedge^0 h : \bigwedge^0 B \to \bigwedge^0 C$ reduces to the "trivial" map $M \times \mathcal{A} \to N \times \mathcal{A}$, $(m, a) \mapsto (g(m), a)$. See also [I.7.15].

5. Operations on sections and on vector bundles

In §4 we discussed various constructions of vector bundles. But in §3 we showed that the set of all sections of a vector bundle is a graded $C^\infty(M)$-module, and thus we can perform analogous constructions on the modules of sections. In this section we show that there is a correspondence between these two sets of constructions: they commute with the operation of taking sections. For our purpose the most important result of this section is [5.14], which says that sections of the k-th exterior power of the left dual bundle of a vector bundle $B \to M$ can be interpreted as graded skew-symmetric k-linear (over $C^\infty(M)$) maps from sections of B to $C^\infty(M)$.

5.1 Discussion. If $p_i : B_i \to M$ are vector bundles over M with typical fibers E_i, we know that the sets $\Gamma(B_i)$ are graded $C^\infty(M)$-modules. Since $C^\infty(M)$ is a graded $\mathbf{R}$-algebra, we can perform our constructions of direct sums, tensor products, exterior powers, etcetera on these modules. It is thus natural to ask whether these operations "commute" with the corresponding operations on the vector bundles, e.g., whether $\Gamma(B_1 \oplus B_2)$ is isomorphic to $\Gamma(B_1) \oplus \Gamma(B_2)$, where the second direct sum is in the category of graded $C^\infty(M)$-modules. The answer to this general question is positive, although the proofs have an increasing complexity when we go from direct sums via morphisms to tensor products. To be more precise, we will construct new sections out of old ones, and these constructions will all be pointwise, i.e., the new section at $m \in M$ will depend only on the values of the old sections at m. The first problem we then have to solve is whether the new section if smooth when all the old ones are, and whether the correspondence so obtained is an even morphism of graded $C^\infty(M)$-modules. But the "real" problems come when we want to prove that these identifications are isomorphisms.

5.2 Construction. Let $p_i : B_i \to M$, $i = 1, \ldots, n$ be a finite number of vector bundles with typical fibers E_i. If s_i is a section of B_i, we define a section $\mathfrak{I}(\oplus_i s_i)$ of $\bigoplus_i B_i$ by

$$\mathfrak{I}(\oplus_i s_i)(m) = \oplus_i s_i(m) \in \textstyle\bigoplus_i (B_i)_m \cong (\bigoplus_i B_i)_m \ ,$$

where we have used that the fibers of $\oplus_i B_i$ are canonically isomorphic to the direct sum of the corresponding fibers in B_i. To see that this is indeed a smooth section, it suffices to verify it on trivializing charts. If $\mathcal{U}$ is a trivializing atlas, the sections s_i are represented by functions $s_{i,a} : U_a \to E_i$, and the section $\mathfrak{I}(\oplus_i s_i)$ by the function $U_a \to \bigoplus_i B_i$, $m \mapsto \oplus_i s_i(m)$. One could take this as a definition of $\mathfrak{I}(\oplus_i s_i)$, of course after having verified that these local functions glue together to form a global section [1.20] (which they do by definition of the transition functions of $\bigoplus_i B_i$). Since the $s_{i,a}$ are smooth, their direct sum is smooth as well, proving that $\mathfrak{I}(\oplus_i s_i)$ is a smooth section of $\bigoplus_i B_i$. Moreover, it is an elementary exercise (left to the reader) to show that $\mathfrak{I}$ thus defined is an even morphism of graded $C^\infty(M)$-modules.

5.3 Proposition. *$\mathfrak{I} : \bigoplus_i \Gamma(B_i) \to \Gamma(\bigoplus_i B_i)$ is an isomorphism of graded $C^\infty(M)$-modules.*

Proof. If $\mathfrak{I}(\oplus_i s_i) = 0$, then in all charts U_a we have $\oplus_i s_{i,a}(m) = 0$, implying that the $s_{i,a}$ are identically zero on U_a, i.e., the s_i are zero. We conclude that $\mathfrak{I}$ is injective.

If a section u of $\bigoplus_i B_i$ is represented by local functions $u_a : U_a \to \bigoplus_i E_i = \prod_i E_i$, we can project on the separate factors to obtain smooth functions $s_{i,a} : U_a \to E_i$ such that $u_a(m) = (s_{i,a}(m))_{i=1}^n = \oplus_i s_{i,a}(m)$. By definition of the transition functions of $\bigoplus_i B_i$, these functions glue together to form global sections $s_i \in \Gamma(B_i)$. Obviously $\mathfrak{I}(\oplus_i s_i) = u$, proving surjectivity. QED

5.4 Construction. Let $p : B \to M$ be a vector bundle with typical fiber E and let C be a subbundle with typical fiber F. If s is a section of B, we define a section $\mathfrak{I}(s)$ of B/C by

$$\mathfrak{I}(s)(m) = \pi(s(m)) \in B_m/C_m \cong (B/C)_m ,$$

where π denotes the canonical projection of B_m on B_m/C_m, and where we have used that the fibers of B/C are canonically isomorphic to the quotient of the corresponding fibers in B and C. To see that this is indeed a smooth section, it suffices (as for direct sums) to verify it on trivializing charts. If $\mathcal{U}$ is a trivializing atlas for B adapted to the subbundle C, the section s is represented by the function $s_a : U_a \to E$ and the section $\mathfrak{I}(s)$ by the function $\mathfrak{I}(s)_a : U_a \to E/F$, $m \mapsto \pi(s_a(m))$ (abuse of notation: this π denotes the canonical projection $E \to E/F$). (As for direct sums, we could take this as a definition of $\mathfrak{I}(s)$.) In a suitable basis for E [3.13] the last map is given as ignoring the last coordinates of $s_a(m)$. Since s_a is smooth, $\mathfrak{I}(s)_a$ is smooth as well, proving that $\mathfrak{I}(s)$ is a smooth section of B/C. Again it is straightforward to prove that $\mathfrak{I}$ is an even morphism of graded $C^\infty(M)$-modules.

5.5 Proposition. *The map $\mathfrak{I} : \Gamma(B) \to \Gamma(B/C)$ induces an isomorphism of graded $C^\infty(M)$-modules between $\Gamma(B)/\Gamma(C)$ and $\Gamma(B/C)$.*

Proof. First note that, since C is a subbundle of B, each section of C automatically is a section of B, i.e., that $\Gamma(C)$ is a graded submodule of $\Gamma(B)$. To prove injectivity, it thus suffices to show that $\ker(\mathfrak{I}) = \Gamma(C)$. But $\mathfrak{I}(s) = 0$ if and only if for all $m \in U_a$ we have $0 = \mathfrak{I}(s)_a(m) = \pi(s_a(m))$. This is equivalent to saying that s_a takes its values in F, i.e., that s is a section of the subbundle C.

To prove surjectivity, let $u : M \to B/C$ be a global smooth section, represented by local functions $u_a : U_a \to E/F$. Since F is a graded subspace of E, it follows that there exist smooth functions $\tau_a : U_a \to E$ such that $u_a = \pi \circ \tau_a$. These local functions τ_a represent local sections $t_a \in \Gamma_{U_a}(B)$, but there is no reason to assume that they satisfy condition (1.21). We thus choose a partition of unity ρ_a subordinated to the open cover $\mathcal{U}$ and define the global section $s = \sum_a \rho_a t_a$ [3.9]. We claim that $\mathfrak{I}(s) = u$, finishing the proof. For $m \in M$ we compute: $\mathfrak{I}(s)(m) = \pi(s(m)) = \sum_a \rho_a(m)\pi(t_a(m)) = \sum_a \rho_a(m)u(m) = u(m)$. The last equality follows because $\sum_a \rho_a(m) = 1$ and the next to last equality follows because by construction all local sections t_a (whenever defined) induce u when taken modulo C_m. QED

5.6 Discussion/Notation. As said in [5.1], the constructed identifications are all pointwise, and they work also if the initial sections are not smooth. To be more precise, let us denote by $\Gamma^{-1}(B)$ the set of all sections $s : M \to B$ of a vector bundle $p : B \to M$, and let us denote by $C^{-1}(M) = \Gamma^{-1}(m \times \mathcal{A})$ the set of all $\mathcal{A}$-valued functions on M. It follows immediately that $\Gamma^{-1}(B)$ is a graded $C^{-1}(M)$-module. If we now look at [5.2], it is obvious that the identification $\mathfrak{I} : \bigoplus_i \Gamma^{-1}(B_i) \to \Gamma^{-1}(\bigoplus_i B_i)$ is an even morphism of graded $C^{-1}(M)$-modules. It is also not hard to show that it is an isomorphism (just copy the proof of [5.3]). Since $C^\infty(M)$ is a subset of $C^{-1}(M)$, we can see this identification as a morphism of graded $C^\infty(M)$-modules. [5.3] then tells us that when we restrict to *smooth* sections $\Gamma(B_i) \subset \Gamma^{-1}(B)$, we get an isomorphism onto the smooth sections $\Gamma(\bigoplus_i B_i) \subset \Gamma^{-1}(\bigoplus_i B_i)$. Obviously, a similar remark holds for quotients.

5.7 Construction. Morphisms can be left or right k-linear and may eventually be graded skew-symmetric. We will take the right k-linear graded skew-symmetric maps as a typical example, leaving the other cases (such as the isomorphism $\Gamma(\mathrm{Hom}_S(B_1, \dots, B_k; C)) \cong \mathrm{Hom}_S(\Gamma(B_1), \dots, \Gamma(B_k); \Gamma(C))$) to the reader. We thus suppose that $p : B \to M$ and $q : C \to M$ are vector bundles with typical fibers E and F respectively, and we want to construct an isomorphism between $\Gamma(\mathrm{Hom}_R^{sk}(B^k; C))$ and $\mathrm{Hom}_R^{sk}(\Gamma(B)^k; \Gamma(C))$. For a section ϕ of $\mathrm{Hom}_R^{sk}(B^k; C)$ we define $\mathfrak{I}(\phi) \in \mathrm{Hom}_R^{sk}(\Gamma(B)^k; \Gamma(C))$ as follows. For $s_i \in \Gamma(B)$ the section $\mathfrak{I}(\phi)(s_1, \dots, s_k)$ of C is defined by

$$\mathfrak{I}(\phi)(s_1, \dots, s_k)(m) = \phi(m)(s_1(m), \dots, s_k(m)) \ .$$

Since ϕ and the s_i are smooth sections, $\mathfrak{I}(\phi)(s_1, \dots, s_k)$ is also smooth. It is straightforward to show that $\mathfrak{I}(\phi)$ is k-linear over $C^\infty(M)$ and graded skew-symmetric. At a higher level, $\mathfrak{I}$ is left-linear over $C^\infty(M)$ in ϕ and even. Before we can prove that $\mathfrak{I}$ is an isomorphism, we need some preparations. As in [5.6] we denote by $\Gamma^{-1}(B)$ the set of all sections of B and by $C^{-1}(M)$ the set of all $\mathcal{A}$-valued functions on M.

5.8 Lemma. *Let $p : B \to M$ be a vector bundle with typical fiber E, and let $q : C \to M$ be a vector bundle.*

(i) *If $\Phi : \Gamma(B)^k \to \Gamma^{-1}(C)$ is a right k-linear graded skew-symmetric morphism of graded $C^\infty(M)$-modules, then for sections $s_i \in \Gamma(B)$ and $m \in M$ arbitrary, the value $\Phi(s_1, \ldots, s_k)(m) \in C_m$ depends only upon the values $s_i(m)$.*

(ii) *If in (i) we replace $\Gamma(B)$ by $\Gamma^{-1}(B)$ and $C^\infty(M)$ by $C^{-1}(M)$, then the conclusion still holds.*

(iii) *If $\Phi : \Gamma(B)^k \to \Gamma^{-1}(C)$ is a right k-linear graded skew-symmetric morphism of graded $C^\infty(M)$-modules, then there exists a unique right k-linear morphism of graded $C^{-1}(M)$-modules $\Psi : \Gamma^{-1}(B)^k \to \Gamma^{-1}(C)$ such that $\Psi|_{\Gamma(B)^k} = \Phi$.*

Proof. • The proof of (i) breaks into two steps. Let s_i and s_i' be two sets of sections that coincide on an open set $U \ni m$. If ρ is a plateau function around m in U [III.5.21], we have the global equalities $\rho \cdot s_i = \rho \cdot s_i'$ (because $\mathrm{supp}(\rho) \subset U$). Since ρ is even, $\rho(m) = 1$, and Φ is k-linear over $C^\infty(M)$, we compute:

$$\begin{aligned}\Phi(s_1, \ldots, s_k)(m) &= \rho(m)^k \cdot \Phi(s_1, \ldots, s_k)(m) = \Phi(\rho s_1, \ldots, \rho s_k)(m) \\ &= \Phi(\rho s_1', \ldots, \rho s_k')(m) = \Phi(s_1', \ldots, s_k')(m) \ .\end{aligned}$$

This proves that the value of $\Phi(s_1, \ldots, s_k)(m)$ does not depend upon the particular s_i, provided they coincide on an open neighborhood of m.

To prove that it only depends upon the values $s_i(m)$ we assume that s_i and s_i' are two sets of sections such that $s_i(m) = s_i'(m)$. Let (U, ψ) be a trivializing chart for B, let ρ be a plateau function around m in U, and let $V \subset U$ be an open set such that $\rho|_V \equiv 1$. If $(e_j)_{j=1}^n$ is a basis of E, we define local sections $\underline{e}_j \in \Gamma_U(B)$ by $\underline{e}_j(m') = \psi^{-1}(m', e_j)$. Hence there exist smooth functions $s_i^j \in C^\infty(U)$ such that $(\rho s_i)|_U = \sum_j \underline{e}_j \cdot s_i^j$. Since the support of ρ is contained in U, the support of the s_i^j (which is by definition contained in U) is contained in the support of ρ, and thus $\mathrm{supp}(s_i^j)$ is closed in M, hence the s_i^j can be regarded as smooth functions on M. By the same argument $\rho \underline{e}_j$ can be seen as a global smooth section of B [3.9]. We thus obtain the equality of *global* sections

$$\rho^2 s_i = \textstyle\sum_j \rho\, \underline{e}_j \cdot s_i^j \in \Gamma(B) \ .$$

Using the k-linearity of Φ we compute

$$\begin{aligned}\Phi(s_1, \ldots, s_k)(m) &= \Phi(\rho^2 s_1, \ldots, \rho^2 s_k)(m) \qquad \text{by the first step} \\ &= \sum \Phi(\rho\, \underline{e}_{i_1}, \ldots, \rho\, \underline{e}_{i_k})(m) \cdot \mathcal{E}_{i_1 \ldots i_k}(s_1^{i_1}, \ldots, s_k^{i_k})(m) \ ,\end{aligned}$$

where $\mathcal{E}_{i_1 \ldots i_k}(x_1, \ldots, x_k)$ is a function on $\mathcal{A}^k$ obtained by shifting the coefficients $s_j^{i_j}$ (from $\rho s_j = \sum_{i_j} \underline{e}_{i_j} s_j^{i_j}$) out of Φ by k-linearity. These functions can be inductively defined by $\mathcal{E}_i(x) = x$ and $\mathcal{E}_{i_1, \ldots, i_{k+1}}(x_1, \ldots, x_{k+1}) = \mathfrak{C}^\alpha(x_1) \cdot \mathcal{E}_{i_2, \ldots, i_{k+1}}(x_2, \ldots, x_{k+1})$ with $\alpha = \varepsilon(e_{i_2}) + \cdots + \varepsilon(e_{i_{k+1}})$. Obviously the value of $\mathcal{E}_{i_1 \ldots i_k}(s_1^{i_1}, \ldots, s_k^{i_k})$ at m depends only upon the values of the $s_j^{i_j}$ at m. Since the same is true for the sections s_i', it follows from the equality $s_i(m) = s_i'(m)$ that $\Phi(s_1, \ldots, s_k)(m) = \Phi(s_1', \ldots, s_k')(m)$ as claimed.

• To prove (ii), we can copy the proof of (i), except for the fact that the functions s_i^j need not be smooth. However, there is a much shorter proof. Suppose that s_i and s_i' are sections that coincide at m. Then we can define the (non-continuous) function ρ by $\rho(m) = 1$ and $\rho(m') = 0$ for all $m' \neq m$. Then $\rho s_i = \rho s_i'$ and hence linearity over $C^{-1}(M)$ proves (ii).

• To prove (iii) we start with uniqueness. So we suppose that Ψ exists, and that s_i are arbitrary sections. If (U, ψ), V, ρ, and $s^j{}_i$ are as above (and note, the $s^j{}_i$ need not be smooth), we know that

$$\begin{aligned}\Psi(s_1, \ldots, s_k)(m) &= \sum \Psi(\rho\, \underline{e}_{i_1}, \ldots, \rho\, \underline{e}_{i_k})(m) \cdot \mathcal{E}_{i_1 \ldots i_k}(s_1^{i_1}, \ldots, s_k^{i_k}) \\ &= \sum \Phi(\rho\, \underline{e}_{i_1}, \ldots, \rho\, \underline{e}_{i_k})(m) \cdot \mathcal{E}_{i_1 \ldots i_k}(s_1^{i_1}, \ldots, s_k^{i_k}) ,\end{aligned}$$

where the last equality is by the assumption $\Psi|_{\Gamma(B)^k} = \Phi$. This proves that the element $\Psi(s_1, \ldots, s_k)(m)$ is completely determined by Φ, hence it must be unique.

On the other hand, we can use this formula to define Ψ. More precisely, for sections $s_i : M \to B$ and a point $m \in M$ we define $\Psi(s_1, \ldots, s_k)(m)$ by

$$\Psi(s_1, \ldots, s_k)(m) = \sum \Phi(\rho\, \underline{e}_{i_1}, \ldots, \rho\, \underline{e}_{i_k})(m) \cdot \mathcal{E}_{i_1 \ldots i_k}(s_1^{i_1}, \ldots, s_k^{i_k}) .$$

It remains to show that this is well defined, right k-linear over $C^{-1}(M)$, and graded skew-symmetric. To that end we note that by (i) the right hand side does not depend upon the specific choice for ρ, and hence neither does it depend upon U or V. An elementary but tedious computation using the explicit expression for $\mathcal{E}_{i_1 \ldots i_k}$ shows that it is also independent of the choice of the trivializing sections $\underline{e}_i$ (two different choices are related by a matrix of *smooth* functions, which can be brought inside Φ). Another tedious verification shows that this formula is right k-linear over $C^{-1}(M)$ and graded skew-symmetric (because Ψ is). $\boxed{QED}$

5.9 Proposition. *$\mathfrak{I} : \Gamma(\mathrm{Hom}_R^{sk}(B^k; C)) \to \mathrm{Hom}_R^{sk}(\Gamma(B)^k; \Gamma(C))$ is an isomorphism of graded $C^\infty(M)$-modules.*

Proof. Since we know that it is an even morphism of graded $C^\infty(M)$-modules, it suffices to show that it is bijective. We will do this by constructing its inverse. This means that for $\Phi \in \mathrm{Hom}_R^{sk}(\Gamma(B)^k; \Gamma(C))$ we have to define a ϕ such that $\mathfrak{I}(\phi) = \Phi$. For $m \in M$ the value of $\phi(m)$ should be a right k-linear graded skew-symmetric map $B_m^k \to C_m$. We can define this map by providing the images $\phi(m)(b_1, \ldots, b_k)$ with $b_i \in B_m$. We define these by

$$\phi(m)(b_1, \ldots, b_k) = \Phi(s_1, \ldots, s_k)(m) , \tag{5.10}$$

where the $s_i \in \Gamma(B)$ are such that $s_i(m) = b_i$. It follows from [5.8-i] that this is well defined. The only problem is that there need not exist *smooth* sections satisfying the condition $s_i(m) = b_i$ (if m has real coordinates, $s_i(m)$ has real coordinates, but b_i need not have real coordinates). However, in order to define a k-linear graded skew-symmetric map, we need not define it on the whole of B_m^k, but only on a generating subset, and even a basis would suffice. Since the values $s_i(m)$ certainly generate B_m when s_i varies over the smooth sections of B, (5.10) indeed defines a unique k-linear graded skew-symmetric map on B_m. We do not obtain a contradiction (the values $s_i(m)$ are certainly over-complete) because Φ itself is k-linear and graded skew-symmetric.

To prove that this $\phi : M \to \mathrm{Hom}_R^{sk}(B^k; C)$, $m \mapsto \phi(m) \in \mathrm{Hom}_R^{sk}(B_m^k; C_m)$ is smooth, it suffices to show that in a trivializing chart (U, ψ) the associated function $\phi_\psi : U \to \mathrm{Hom}_R^{sk}(E^k; F))$ is smooth. Now this function is smooth if and only if its coefficients with respect to a basis are smooth. If we use (U, ψ), V, and ρ as in he proof of [5.8], these coefficients are given by $\phi_\psi(m)(e_{i_1}, \ldots, e_{i_k})$ (or more precisely, by the coefficients of this vector with respect to a basis of F), and these are given on the neighborhood V (see above) by the formula

$$\phi_\psi(m)(e_{i_1}, \ldots, e_{i_k}) = (\pi_F \circ \psi)(\Phi(\rho\, \underline{e}_{i_1}, \ldots, \rho\, \underline{e}_{i_k})(m)) \;,$$

which is manifestly smooth. QED

5.11 Discussion. Let us introduce some names for spaces of right k-linear graded skew-symmetric morphisms: $\mathrm{Hom}_R^{sk}(\Gamma^{-1}(B)^k; \Gamma^{-1}(C))$ will contain those that are linear over $C^{-1}(M)$, $\mathrm{Hom}_R^{sk}(\Gamma(B)^k; \Gamma^{-1}(C))$ and $\mathrm{Hom}_R^{sk}(\Gamma(B)^k; \Gamma(C))$ those that are linear over $C^\infty(M)$. For any $\Psi \in \mathrm{Hom}_R^{sk}(\Gamma^{-1}(B)^k; \Gamma^{-1}(C))$ we can take its restriction to $\Gamma(B)^k$: $\Phi = \Psi|_{\Gamma(B)^k} \in \mathrm{Hom}_R^{sk}(\Gamma(B)^k; \Gamma^{-1}(C))$. [5.8-iii] shows that the map $\Psi \mapsto \Phi$ is a bijection. On the other hand, $\mathrm{Hom}_R^{sk}(\Gamma(B)^k; \Gamma(C))$ is obviously a graded submodule of $\mathrm{Hom}_R^{sk}(\Gamma(B)^k; \Gamma^{-1}(C))$. We conclude that we can interpret $\mathrm{Hom}_R^{sk}(\Gamma(B)^k; \Gamma(C))$ as a graded submodule of $\mathrm{Hom}_R^{sk}(\Gamma^{-1}(B)^k; \Gamma^{-1}(C))$. But the reader should be aware that the first space concerns morphisms linear over $C^\infty(M)$ and that the second one concerns morphisms linear over $C^{-1}(M)$.

Looking at the definition of $\mathfrak{I}$ in [5.7], it is obvious that it defines an identification $\Gamma^{-1}(\mathrm{Hom}_R^{sk}(B^k; C)) \to \mathrm{Hom}_R^{sk}(\Gamma^{-1}(B)^k; \Gamma^{-1}(C))$. Looking at the proof of [5.9], it is not hard to show that this is an isomorphism of graded $C^{-1}(M)$-modules. The result of [5.9] can now be interpreted as saying that if we restrict attention to smooth sections in the source space, then we get an isomorphism onto the graded submodule $\mathrm{Hom}_R^{sk}(\Gamma(B)^k; \Gamma(C)) \subset \mathrm{Hom}_R^{sk}(\Gamma^{-1}(B)^k; \Gamma^{-1}(C))$. An interesting consequence is the following. Suppose we have an arbitrary section ϕ of $\mathrm{Hom}_R^{sk}(B^k; C)$ and suppose that under the identification $\mathfrak{I}$ it maps $\Gamma(B)^k$ to $\Gamma(C)$, i.e., it maps smooth sections of B to smooth sections of C. Then ϕ must be smooth.

5.12 Definition. If $h : B \to C$ is a (left or right linear) vector bundle morphism of bundles over the same base space, we define an induced map $h_* : \Gamma(B) \to \Gamma(C)$ by $h_*(s) = h \circ s$. For a section $s \in \Gamma(B)$, the section $h_*(s) \in \Gamma(C)$ is usually called the *push forward of*

s by h. If h is left/right linear of parity α, the same is true for h_* (which "obliges" us to write $\langle\!\langle\, s \,\|\, h_* \rangle\!\rangle = s \diamond h$ in the left linear case). Also, if h is an isomorphism, so is h_*. This applies especially to the isomorphisms given in [4.25], which gives, for example, an isomorphism $\Gamma(\mathrm{Hom}_L(B;C)) \cong \Gamma({}^*B \otimes C)$.

5.13 Discussion. Let $p : B \to M$ and $q : C \to M$ be vector bundles over M. In [4.18] we argued that sections of $\mathrm{Hom}_L(B;C)$ are in bijection with left linear vector bundle morphisms $B \to C$. With the identification $\Gamma(\mathrm{Hom}_L(B;C)) \cong \mathrm{Hom}_L(\Gamma(B);\Gamma(C))$ and the push forward we can describe this bijection in a roundabout way.

For a left linear vector bundle morphism $h : B \to C$ we obtain a left linear morphism of graded $C^\infty(M)$-modules $h_* \in \mathrm{Hom}_L(\Gamma(B);\Gamma(C))$. Under the identification of this space with $\Gamma(\mathrm{Hom}_L(B;C))$, we obtain an element $\mathfrak{I}^{-1}(h_*) \in \Gamma(\mathrm{Hom}_L(B;C))$. We leave it to the reader to verify that this is exactly the section of $\mathrm{Hom}_L(B;C)$ we have defined in [4.18]. An interesting consequence of all these isomorphisms is that any morphism $\Gamma(B) \to \Gamma(C)$ of graded $C^\infty(M)$-modules is necessarily of the form h_* for some vector bundle morphism $h : B \to C$.

5.14 Discussion. Using the bundle isomorphisms $\mathrm{Hom}_L^{sk}(B^k;C) \cong \mathrm{Hom}_L(\bigwedge^k B;C) \cong {}^*(\bigwedge^k B) \otimes C \cong \bigwedge^k({}^*B) \otimes C$, the induced isomorphisms (by push forwards) on the modules of sections, and [5.9], we obtain an isomorphism between $\Gamma(\bigwedge^k({}^*B) \otimes C)$ and $\mathrm{Hom}_L^{sk}(\Gamma(B)^k;\Gamma(C))$. If we take for C the trivial bundle $C = M \times \mathcal{A}$, we thus obtain the isomorphism

$$\Gamma(\textstyle\bigwedge^k {}^*B) \cong \mathrm{Hom}_L^{sk}(\Gamma(B)^k; C^\infty(M)) \ , \tag{5.15}$$

and in particular for $k = 1$:

$$\Gamma({}^*B) \cong {}^*\Gamma(B) \qquad \text{and similarly} \qquad \Gamma(B^*) \cong \Gamma(B)^* \ .$$

One of the consequences of [5.9] for these particular examples is the following (see also [5.11]). Let $\omega : M \to \bigwedge^k {}^*B$ be any section and let $s_i : M \to B$ be smooth sections. If the function

$$m \mapsto \iota(s_1(m), \dots, s_k(m))\omega(m)$$

is smooth for all possible choices of the s_i, then ω must be smooth.

5.16 Construction. Let $P : B \to M$ and $q : C \to M$ be vector bundles with typical fibers E and F respectively. If s is a smooth section of B and t a smooth section of C, we define a section $\mathfrak{I}(s \otimes t)$ of $B \otimes C$ by

$$\mathfrak{I}(s \otimes t)(m) = s(m) \otimes t(m) \in B_m \otimes C_m \cong (B \otimes C)_m \ ,$$

where we have used that the fibers of $B \otimes C$ are canonically isomorphic to the tensor product of the corresponding fibers in B and C. Using arguments similar to those used for direct sums and a quotient, one can easily show that this defines indeed a smooth section and that $\mathfrak{I}$ thus defined is an even morphism of graded $C^\infty(M)$-modules.

5.17 Proposition. *$\mathfrak{I} : \Gamma(B) \otimes \Gamma(C) \to \Gamma(B \otimes C)$ is an isomorphism of graded $C^\infty(M)$-modules.*

Proof. An elementary but tedious verification shows that we have the following commutative diagram:

$$\begin{array}{ccccc}
\Gamma(B)\otimes\Gamma(C) & \longrightarrow & & & \Gamma(B\otimes C) \\
{\scriptstyle [I.8.5]}\big\downarrow{\scriptstyle \cong} & & & & {\scriptstyle \cong}\big\uparrow{\scriptstyle [I.8.5]} \\
\Gamma({}^*(B^*))\otimes\Gamma(C) & & & & \Gamma({}^*(B^*)\otimes C) \\
{\scriptstyle [5.14]}\big\downarrow{\scriptstyle \cong} & & & & {\scriptstyle \cong}\big\uparrow{\scriptstyle [I.8.8]} \\
{}^*\Gamma(B^*)\otimes\Gamma(C) & \underset{[I.7.11]}{\longrightarrow} & \mathrm{Hom}_L(\Gamma(B^*);\Gamma(C)) & \underset{[5.9]}{\overset{\cong}{\longrightarrow}} & \Gamma(\mathrm{Hom}_L(B^*;C)) ,
\end{array}$$

in which two arrows are not yet known to be isomorphisms. In [7.25] we will prove that for any vector bundle $p : B \to M$ the graded $C^\infty(M)$-module $\Gamma(B)$ is f.g.p, and thus by [I.8.8] the identification ${}^*\Gamma(B^*) \otimes \Gamma(C) \to \mathrm{Hom}_L(\Gamma(B^*); \Gamma(C))$ is an isomorphism. Hence the last arrow must also be an isomorphism. QED

5.18 Discussion. If we have a finite family $p_i : B_i \to M$ of vector bundles with typical fibers E_i, it is by now obvious how to construct an identification $\mathfrak{I} : \otimes_i \Gamma(B_i) \to \Gamma(\otimes_i B_i)$: if the s_i are sections of B_i, $\mathfrak{I}(s_1 \otimes \cdots \otimes s_k)$ is defined by

$$\mathfrak{I}(s_1 \otimes \cdots \otimes s_k)(m) = s_1(m) \otimes \cdots \otimes s_k(m) \in \otimes_i (B_i)_m \cong (\otimes_i B_i)_m .$$

Since we know that this identification is an isomorphism of graded $C^\infty(M)$-modules when $k = 2$, we can use induction and associativity of the tensor product to prove that it is an isomorphism for all $k \in \mathbf{N}$.

5.19 Construction. Let $p : B \to M$ be a vector bundle with typical fiber E. If $s_1, \ldots, s_k$ are smooth sections of B, we construct a section $\mathfrak{I}(s_1 \wedge \cdots \wedge s_k)$ of $\bigwedge^k B$ by

$$\mathfrak{I}(s_1 \wedge \cdots \wedge s_k)(m) = s_1(m) \wedge \cdots \wedge s_k(m) \in \textstyle\bigwedge^k B_m \cong (\bigwedge^k B)_m .$$

By now standard arguments show that this defines a smooth section and that the map $\mathfrak{I} : \bigwedge^k \Gamma(B) \to \Gamma(\bigwedge^k B)$ is an even morphism of graded $C^\infty(M)$-modules.

5.20 Proposition. *$\mathfrak{I} : \bigwedge^k \Gamma(B) \to \Gamma(\bigwedge^k B)$ is an isomorphism of graded $C^\infty(M)$-modules.*

Proof. A rather tedious verification shows that the map $\mathfrak{I}$ can also be written as the following series of isomorphisms/identifications:

$$\begin{aligned}\textstyle\bigwedge^k \Gamma(B) &\overset{[I.8.5]}{\cong} \textstyle\bigwedge^k \Gamma({}^*(B^*)) \overset{[5.14]}{\cong} \bigwedge^k {}^*\Gamma(B^*) \overset{[I.7.16]}{\longrightarrow} {}^*(\bigwedge^k \Gamma(B^*)) \\ &\overset{[I.5.5]}{\cong} \mathrm{Hom}_L^{sk}(\Gamma(B^*)^k; \Gamma(M \times \mathcal{A})) \overset{[5.9]}{\cong} \Gamma(\mathrm{Hom}_L^{sk}((B^*)^k; M \times \mathcal{A})) \\ &\overset{[I.5.5]}{\cong} \Gamma({}^*(\textstyle\bigwedge^k B^*)) \overset{[I.8.16]}{\cong} \Gamma(\bigwedge^k {}^*(B^*)) \overset{[I.8.5]}{\cong} \Gamma(\bigwedge^k B)\,.\end{aligned}$$

In [7.25] we will show that $\Gamma(B)$ is f.g.p [7.25], and thus by [I.8.16] the identification $\bigwedge^k {}^*\Gamma(B^*) \to {}^*(\bigwedge^k \Gamma(B^*))$ is an isomorphism. QED

6. The pull-back of a section

In this section we introduce operations which are crucial for differential geometry: the exterior or wedge product of sections and the contraction of a section of a bundle with a section of the k-th exterior power of its dual bundle. We also introduce the notion of the pull-back of a section and we show how this relates to the various operations discussed before.

6.1 Discussion. Let $p : B \to M$ be a vector bundle with typical fiber E. In [I.5.8] we defined the wedge product as a bilinear map $\bigwedge^k E \times \bigwedge^\ell E \to \bigwedge^{k+\ell} E$, factoring in a linear map $\bigwedge^k E \otimes \bigwedge^\ell E \to \bigwedge^{k+\ell} E$. It is straightforward to show that this map intertwines the corresponding transition functions [4.23], giving us an even bundle morphism $\bigwedge^k B \otimes \bigwedge^\ell B \to \bigwedge^{k+\ell} B$. Now taking the push forward of this morphism acting on sections, we can form the even bilinear morphism of graded $C^\infty(M)$-modules

$$\mathfrak{I}_\wedge : \Gamma(\textstyle\bigwedge^k B) \times \Gamma(\bigwedge^\ell B) \to \Gamma(\bigwedge^k B) \otimes \Gamma(\bigwedge^\ell B) \overset{[5.16]}{\longrightarrow} \Gamma(\bigwedge^k B \otimes \bigwedge^\ell B) \to \Gamma(\bigwedge^{k+\ell} B)\,.$$

Tracing the various morphisms immediately shows that this map is given by the pointwise formula $\mathfrak{I}_\wedge(\alpha, \beta)(m) = (\alpha(m)) \wedge (\beta(m))$. In view of this pointwise identification we will denote this bilinear map $\mathfrak{I}_\wedge : \Gamma(\bigwedge^k B) \times \Gamma(\bigwedge^\ell B) \to \Gamma(\bigwedge^{k+\ell} B)$ also by just a wedge product: $\mathfrak{I}_\wedge(\alpha, \beta) \equiv \alpha \wedge \beta$. Since the original map for $\mathcal{A}$-vector spaces (or better, family of maps, since they depend upon the powers k and ℓ) satisfies the relation $A \wedge B = (-1)^{\langle \varepsilon(A), \varepsilon(B)\rangle + k\ell} B \wedge A$, it follows immediately from the fact that this identification is pointwise that the same is true for the induced map on sections as described above. It follows that the graded $C^\infty(M)$-module $\bigoplus_{k \in \mathbf{Z}} \Gamma(\bigwedge^k B)$ can be given the structure of a ($\mathbf{Z} \times \mathbf{Z}_2$-graded) $C^\infty(M)$-algebra. Combined with the identifications $\bigwedge^k \Gamma(B) \to \Gamma(\bigwedge^k B)$, which is also pointwise, we thus obtain a morphism of graded commutative $C^\infty(M)$-algebras

$$\mathfrak{I} : \bigwedge\Gamma(B) \equiv \bigoplus_{k=0}^{\infty} \bigwedge^k \Gamma(B) \to \bigoplus_{k=0}^{\infty} \Gamma(\bigwedge^k B) ,$$

where [5.20] shows that it is in fact an isomorphism. Note that we carefully did *not* write $\Gamma(\bigoplus_k \bigwedge^k B)$, as $\bigoplus_k \bigwedge^k B$ is not (because of the infinite sum) a vector bundle in our sense.

For $\mathfrak{A}$-graded $\mathcal{A}$-modules E we have defined in [I.6.16] (see also [I.6.21]) a contraction operator $\iota : E \otimes \bigwedge^k {}^*E \to \bigwedge^{k-1} {}^*E$ such that, for a fixed $e \in E$, the operator $\iota(e)$ is a left derivation of $\bigwedge {}^*E$. It is elementary to verify that this map ι satisfies the conditions of [4.23] and thus induces a morphism of the corresponding bundles. Taking sections, we can construct the morphism

$$\iota : \Gamma(B) \times \Gamma(\bigwedge^k {}^*B) \to \Gamma(B) \otimes \Gamma(\bigwedge^k {}^*B) \to \Gamma(B \otimes \bigwedge^k {}^*B) \to \Gamma(\bigwedge^{k-1} {}^*B) .$$

There are two justifications for using the same symbol ι for this morphism on sections. In the first place, as the reader can check, this operation is given for $s \in \Gamma(B)$ and $\Phi \in \Gamma(\bigwedge^k {}^*B)$ by the pointwise formula

$$\iota(s)\Phi \equiv \iota(s \otimes \Phi) : m \mapsto \iota(s(m))\Phi(m) \in \bigwedge^{k-1} {}^*B_m .$$

It follows easily that we obtain an even morphism of graded $C^\infty(M)$-modules

$$\iota : \Gamma(B) \to \mathrm{Der}_L(\bigoplus_{k=0}^{\infty} \Gamma(\bigwedge^k {}^*B)) .$$

On the other hand, we could have applied [I.6.16] directly to the graded $C^\infty(M)$-module ${}^*\Gamma(B)$ itself and have obtained a contraction operator $\iota : \Gamma(B) \to \mathrm{Der}_L(\bigwedge^*\Gamma(B))$. We leave it (again) to the reader to show that these two contraction operators coincide under the identification $\bigwedge {}^*\Gamma(B) \cong \bigwedge \Gamma({}^*B) \to \bigoplus_{k=0}^{\infty} \Gamma(\bigwedge^k {}^*B)$, i.e., that we have a commutative diagram

$$\begin{array}{ccc} \bigwedge^*\Gamma(B) & \xrightarrow{\iota(s)} & \bigwedge^*\Gamma(B) \\ \downarrow & & \downarrow \\ \bigoplus_{k=0}^{\infty} \Gamma(\bigwedge^k {}^*B) & \xrightarrow{\iota(s)} & \bigoplus_{k=0}^{\infty} \Gamma(\bigwedge^k {}^*B) . \end{array}$$

Now let $s_1, \ldots, s_k \in \Gamma(B)$ be arbitrary, then we obtain, by composition, a map

$$\iota(s_k) \circ \cdots \circ \iota(s_1) \quad : \quad \Gamma(\bigwedge^k {}^*B) \longrightarrow C^\infty(M) .$$

On the other hand we have the evaluation map

$$\iota(s_k, \ldots, s_1) \ : \ \mathrm{Hom}_L^{sk}(\Gamma(B); C^\infty(M)) \quad \longrightarrow \quad C^\infty(M) .$$

We leave it to the reader to check (as always using that all identifications are pointwise) that these two maps coincide under the identification $\Gamma(\bigwedge^k {}^*B) \cong \mathrm{Hom}_L^{sk}(\Gamma(B); C^\infty(M))$ (5.15) (see also [I.7.21]).

6.2 Discussion. Since all our identifications are pointwise, they commute with restrictions to submanifolds (and thus in particular with restrictions to open subsets). Let us elucidate this by a few examples, in which $N \subset M$ is a submanifold. Let B and C be vector bundles over M and let $s \in \Gamma(B)$ and $t \in \Gamma(C)$ be smooth sections. Then the identification in [5.2] satisfies

$$\mathfrak{I}(s|_N \oplus t|_N) = (\mathfrak{I}(s \oplus t))|_N ,$$

obviously because for $m \in N$ we have $s|_N(m) = s(m)$ and $t|_N(m) = t(m)$. In a similar way, if $\phi \in \Gamma(\mathrm{Hom}_R^{sk}(B^k; C)) \cong \Gamma(\bigwedge^k B^* \otimes C)$ and $s_i \in \Gamma(B)$ are smooth sections, then, using the identification $\mathfrak{I}$ from [5.7], we have

$$\mathfrak{I}(\phi|_N)(s_1|_N, \ldots, s_k|_N) = (\mathfrak{I}(\phi)(s_1, \ldots, s_k))|_N .$$

As a final example, consider $\phi \in \Gamma(\bigwedge^k B^*)$ and $s \in \Gamma(B)$ smooth, then

$$(\iota(s)\mathfrak{I}(\phi))|_N = \iota(s|_N)\mathfrak{I}(\phi|_N) .$$

6.3 Definition. So far we have always considered bundles over the same base space, but now we will look at bundles over different base spaces. Let $p : B \to M$ and $q : C \to N$ be vector bundles and let $h : B \to C$ be a right linear vector bundle morphism inducing the map $g : M \to N$ on the base spaces. For each $m \in M$ we have a right linear map $h_m = h|_{B_m} : B_m \to C_{g(m)}$, and thus we can form its right dual $h_m^* : C_{g(m)}^* \to B_m^*$. However, the collection of all h_m^* does *not* define a vector bundle map $C^* \to B^*$ unless g is bijective: it is only defined on fibers of the form $C_{g(m)}$, missing fibers of C^* if g is not surjective, it could be multiply defined if g is not injective, and it does not induce a well defined map $N \to M$ (unless g^{-1} exists and is smooth).

On the other hand, if we take a section $\tau : N \to C^*$ of the right dual bundle of C, we can form a section $\langle\!\langle \tau \| h^* \rangle\!\rangle : M \to B^*$ of the right dual of B by the formula

$$\langle\!\langle \tau \| h^* \rangle\!\rangle(m) = \langle\!\langle \tau(g(m)) \| h_m^* \rangle\!\rangle = \tau(g(m)) \circ h_m : B_m \to C_{g(m)} \to \mathcal{A} , \tag{6.4}$$

or equivalently by the commutative diagram

$$\begin{array}{ccc} B_m^* \ni \langle\!\langle \tau(g(m)) \| h_m^* \rangle\!\rangle & \xleftarrow{\;h_m^*\;} & \tau(g(m)) \in C_{g(m)}^* \\ {\scriptstyle \langle\!\langle \tau \| h^* \rangle\!\rangle}\big\uparrow & & \big\uparrow{\scriptstyle \tau} \\ m & \xrightarrow[g]{} & g(m) . \end{array}$$

The section $\langle\!\langle \tau \| h^* \rangle\!\rangle$ is called *the pull-back of* τ *by* h. By varying τ we obtain a map $h^* : \Gamma^{-1}(C^*) \to \Gamma^{-1}(B^*)$ which is called *the pull-back map*. It follows easily from the definition that h^* is additive and verifies for any $\phi : N \to \mathcal{A}$:

$$\langle\!\langle \phi \cdot \tau \| h^* \rangle\!\rangle = (\phi \circ g) \cdot \langle\!\langle \tau \| h^* \rangle\!\rangle .$$

This looks like a left linear map (the reason why we wrote it as $\langle\!\langle \tau \| h^* \rangle\!\rangle$), but not quite: $\Gamma^{-1}(C^*)$ is a graded module over $C^{-1}(N)$, while $\Gamma^{-1}(B^*)$ is a graded module over $C^{-1}(M)$. This is also apparent from the fact that on the right we have to multiply by $\phi \circ g$ instead of just by g.

We now show that if the section τ is smooth, i.e., $\tau \in \Gamma(C^*)$, then the pull-back $\langle\!\langle \tau \| h^* \rangle\!\rangle$ is also smooth. For that we evaluate on a smooth section $s \in \Gamma(B)$ according to [5.14], which gives

$$\langle\!\langle \tau \| h^* \rangle\!\rangle(s) : m \mapsto (\langle\!\langle \tau \| h^* \rangle\!\rangle(m))(s(m)) = \tau(g(m))(h(s(m))) .$$

Since this is smooth when s and τ are smooth (g and h are smooth from the beginning), we conclude by [5.14] that $\langle\!\langle \tau \| h^* \rangle\!\rangle$ is smooth. We thus have shown that h^* maps $\Gamma(C^*)$ to $\Gamma(B^*)$.

If h is left linear instead of right linear, we can construct in a similar way a pull-back map ${}^*h : \Gamma^{-1}({}^*C) \to \Gamma^{-1}({}^*B)$ by the formula

$${}^*h(\tau)(m) = {}^*(h_m)(\tau(g(m))) = h_m \diamond \tau(g(m)) = \tau(g(m)) \circ h_m : B_m \to \mathcal{A} ,$$

or equivalently by its action on a section $s \in \Gamma^{-1}(B)$:

$$\langle\!\langle s \| {}^*h(\tau) \rangle\!\rangle : m \mapsto \langle\!\langle \langle\!\langle s(m) \| h \rangle\!\rangle \| \tau(g(m)) \rangle\!\rangle .$$

It can be shown easily that ${}^*h$ is additive and verifies for any $\phi : N \to \mathcal{A}$:

$${}^*h(\sigma \cdot \phi) = ({}^*h(\sigma)) \cdot (\phi \circ g) .$$

And just as for the right linear case, ${}^*h$ maps smooth sections to smooth sections, i.e., ${}^*h : \Gamma({}^*C) \to \Gamma({}^*B)$.

It should be noted that, although h^* and ${}^*h$ have many properties that resemble the properties of right and left dual maps, some properties are *not* shared. In particular a bi-dual for vector bundle morphisms does not exists, simply because h^* and ${}^*h$ are not morphisms of modules over the *same* ring. The fact that we used a notation as if they were dual maps will never leads to confusion because a vector bundle morphism simply is not a morphism of modules.

6.5 Remark. Using the construction of pull-back bundles, the pull-back map can be decomposed into a standard pull-back map to the pull-back bundle, and a part involving bundles over the same base space. In the case of pull-back bundles, one can "pull back" any section, not only those of the dual bundle (restriction is the typical example). And then, when one is in bundles over the same base space, one can apply the dual map. Although such a picture is aesthetically very pleasing, we will never need it, so we leave it to the interested reader to pursue this line of thought on his own.

6.6 Discussion. In [6.2] we argued by examples that all identifications commuted with restriction. For the pull-back map, although also pointwise in a certain sense, this is not precisely the case because the bundles are not over the same base space. Nevertheless, some kind of commutativity still holds for pull-back maps. To be more precise, let $p : B \to M$ and $q : C \to N$ be vector bundles and let $h : B \to C$ be a right linear vector bundle morphism inducing $g : M \to N$. Now suppose that $\widehat{M} \subset M$ is a submanifold. If we denote (slight abuse of notation) by $h|_{\widehat{M}}$ the restriction of h to $B|_{\widehat{M}}$ (the restriction of the bundle B to $\widehat{M}$), then it follows immediately from the definition (6.4) that for any section $\tau \in \Gamma(C^*)$ we have

$$(h|_{\widehat{M}})^* \tau = (h^* \tau)|_{\widehat{M}} \ .$$

On the other hand, if $\widehat{N} \subset N$ is a submanifold such that $g(M) \subset \widehat{N}$, then again a direct verification shows that

$$h^*(\tau|_{\widehat{N}}) = h^* \tau \ ,$$

where on the left we interpret h as a right linear vector bundle morphism from B to $C|_{\widehat{N}}$ (which is allowed because $g(M) \subset \widehat{N}$; if one really wants to be precise, one should use [V.2.17] and [V.2.18] to show that this map is smooth). This result is due, of course, to the fact that the pull-back map is essentially the collection of dual maps h_m^* defined on fibers of the form $C_{g(m)}$. It is in these two senses that the pull-back map commutes with restrictions.

6.7 Discussion. There are two interesting special cases to consider for the pull-back map. The first concerns the case when the vector bundles B and C are over the same base space M. So let $h : B \to C$ be a right linear vector bundle morphism (which induces the identity on M!), i.e., for each $m \in M$ we have a right linear map $B_m \to C_m$. We thus can form the right dual map $(h_m)^* : (C_m)^* \to (B_m)^*$. The collection of all these right dual maps forms a left linear vector bundle morphism $f : C^* \to B^*$. A careful look at the definition of the pull-back shows that $\langle\!\langle \sigma \,\|\, h^* \rangle\!\rangle$ is the composition of $\sigma \in \Gamma(C^*)$ with this f. Comparing this with the definition of the push forward map [5.12], we see that in case the bundles are over the same base space, the pull-back map is exactly the same as the push forward of the dual map.

Another way to see this phenomenon is to play around with the various identifications. For $\mathcal{A}$-vector spaces E and F we can identify $\mathrm{Hom}_R(E; F)$ with $\mathrm{Hom}_L(F^*; E^*)$ via $\mathrm{Hom}_R(E; F) \cong F \otimes E^* \cong {}^*(F^*) \otimes E^* \cong \mathrm{Hom}_L(F^*; E^*)$, an identification which is explicitly given by taking the right dual. In [4.18] we proved that right linear vector bundle morphisms are in bijection with sections of $\mathrm{Hom}_R(B; C)$. Since our identifications remain valid at the vector bundle level [4.25], we thus have an identification of a right linear vector bundle morphism $B \to C$ with a left linear vector bundle morphism $C^* \to B^*$. And again the isomorphism is given by taking the right dual, in agreement with the previous discussion.

The second interesting case is when both bundles are the trivial bundle with fiber $\mathcal{A}$. A right linear vector bundle morphism $h : M \times \mathcal{A} \to N \times \mathcal{A}$ which induces the map

$g : M \to N$ must necessarily be of the form $h(m, a) = (g(m), f(m)a)$ for some smooth function $f : M \to \mathcal{A}$. On the other hand, sections of $M \times \mathcal{A}$ and its dual are just functions on M, the same being true for N. Moreover, evaluation of a dual "function" on a normal "function" is just pointwise multiplication of those functions. An elementary computation then shows that

$$\langle\!\langle \sigma \,\|\, h^* \rangle\!\rangle = (\sigma \circ g) \cdot f \ ,$$

where we identified sections with functions. In the applications of this result, the function f, representing the right linear vector bundle morphism h, usually is identically 1, i.e., $h(m, a) = (g(m), a)$, in which case the formula simplifies to $\langle\!\langle \sigma \,\|\, h^* \rangle\!\rangle = \sigma \circ g$.

With a similar argument a left linear vector bundle morphism is given as $h(m, a) = (g(m), af(m))$, and the pull-back map is given by ${}^*h(\sigma) = f \cdot (\sigma \circ g) \equiv f \cdot (g \diamond \sigma)$.

6.8 Discussion. Once we know of the existence of the pull-back map, we might recall that we also have constructions of new morphisms out of families of old ones, both for morphisms of graded $C^\infty(M)$-modules (in §I.7) and for morphisms of vector bundles (see [4.26] and [4.27]). The natural question is: how do these constructions behave under taking the pull-back map? The answer is that they behave as one would expect (see also [I.7.24] and [I.7.25]), except that this expected behavior requires some of the identifications we met before. [6.9] gives the precise result for exterior powers; similar results for the other constructions are left to the reader.

6.9 Proposition. *Let $p : B \to M$ be a vector bundle over M, let $q : C \to N$ be a vector bundle over N, and let $h : B \to C$ be an even (left) linear vector bundle morphism inducing $g : M \to N$. Then the map ${}^*(\bigwedge^k h)$ corresponds to the map $\bigwedge^k ({}^*h)$ under the identifications $\mathfrak{I}_B : \bigwedge^k \Gamma({}^*B) \to \Gamma({}^*(\bigwedge^k B))$ and $\mathfrak{I}_C : \bigwedge^k \Gamma({}^*C) \to \Gamma({}^*(\bigwedge^k C))$, i.e., we have a commutative diagram*

$$\begin{array}{ccc} \bigwedge^k \Gamma({}^*C) & \xrightarrow{\bigwedge^k ({}^*h)} & \bigwedge^k \Gamma({}^*B) \\ {\scriptstyle \mathfrak{I}_C}\big\downarrow & & \big\downarrow{\scriptstyle \mathfrak{I}_B} \\ \Gamma({}^*(\bigwedge^k C)) & \xrightarrow[{}^*(\bigwedge^k h)]{} & \Gamma({}^*(\bigwedge^k B)) \ . \end{array}$$

Proof. Let us first start to analyze in more detail the identification $\mathfrak{I}_B$, which decomposes into two parts. The first part is the map $\mathfrak{I} : \bigwedge^k \Gamma({}^*B) \to \Gamma(\bigwedge^k {}^*B)$, which sends $\sigma_k \wedge \cdots \wedge \sigma_1 \in \bigwedge^k \Gamma({}^*B)$ to the section $\mathfrak{I}(\sigma_k \wedge \cdots \wedge \sigma_1)$ given in any point $m \in M$ by $\mathfrak{I}(\sigma_k \wedge \cdots \wedge \sigma_1)(m) = \sigma_k(m) \wedge \cdots \wedge \sigma_1(m)$. We then use (I.7.19) to identify this section with a section of ${}^*(\bigwedge^k B)$. More precisely, we have

$$\mathfrak{I}_B(\sigma_k \wedge \cdots \wedge \sigma_1)(m) = \mathfrak{I}_\wedge(\sigma_k(m) \wedge \cdots \wedge \sigma_1(m)) \in {}^*(\textstyle\bigwedge^k B_m) \ ,$$

where $\mathfrak{I}_\wedge$ is the map $\mathfrak{I}$ defined in [I.7.16]. We added the subscript $\wedge$ to distinguish this map from the other identifications.

We are now ready to prove the commutativity of the diagram. For $\tau_i \in \Gamma({}^*C)$ we have

$$\begin{aligned}(\mathfrak{I}_B \circ \textstyle\bigwedge^k ({}^*h))(\tau_k \wedge \cdots \wedge \tau_1)(m) &= \mathfrak{I}_\wedge({}^*h(\tau_k)(m) \wedge \cdots \wedge {}^*h(\tau_1)(m)) \\ &= \mathfrak{I}_\wedge(\tau_k(g(m)) \circ h_m \wedge \cdots \wedge \tau_1(g(m)) \circ h_m) \\ &= \mathfrak{I}_\wedge(\tau_k(g(m)) \wedge \cdots \wedge \tau_1(g(m))) \circ \textstyle\bigwedge^k h_m \\ &= \mathfrak{I}_C(\tau_k \wedge \cdots \wedge \tau_1)(g(m)) \circ (\textstyle\bigwedge^k h)_m \\ &= ({}^*(\textstyle\bigwedge^k h) \circ \mathfrak{I}_C)(\tau_k \wedge \cdots \wedge \tau_1) . \qquad \boxed{QED}\end{aligned}$$

6.10 Discussion. Let $p : B \to M$ be a vector bundle over M, let $q : C \to N$ be a vector bundle over N, and let $h : B \to C$ be an even (left) linear vector bundle morphism inducing $g : M \to N$. Since the bundles ${}^*\bigwedge^k B$ and $\bigwedge^k {}^*B$ are isomorphic (see [4.25]), we can interpret the pull-back map ${}^*(\bigwedge^k h)$ as a map

$${}^*(\textstyle\bigwedge^k h) : \Gamma(\textstyle\bigwedge^k {}^*C) \to \Gamma(\textstyle\bigwedge^k {}^*B) .$$

Now let $\alpha \in \Gamma(\bigwedge^k {}^*C)$ and $\beta \in \Gamma(\bigwedge^\ell {}^*C)$ be arbitrary, then we can form the wedge product $\alpha \wedge \beta \in \Gamma(\bigwedge^{k+\ell} {}^*C)$ according to [6.1]. Using that the identifications $\mathfrak{I}_B$ and $\mathfrak{I}_C$ of [6.9] are isomorphisms, it follows immediately from [I.7.15] and [6.9] that we have the equality

$${}^*(\textstyle\bigwedge^{k+\ell} h)(\alpha \wedge \beta) = {}^*(\textstyle\bigwedge^k h)(\alpha) \wedge {}^*(\textstyle\bigwedge^\ell h)(\beta) .$$

An other way to prove this equality without using that $\mathfrak{I}_B$ and $\mathfrak{I}_C$ are isomorphisms is the following pointwise argument. In [6.1] we have seen that $(\alpha \wedge \beta)_{g(m)} = \alpha_{g(m)} \wedge \beta_{g(m)}$, and in [6.3] that $({}^*(\bigwedge^k h)\alpha)(m) = ({}^*(\bigwedge^k h)_m)(\alpha_{g(m)}) = ({}^*(\bigwedge^k h_m))(\alpha_{g(m)})$. And then we obtain immediately

$$\begin{aligned}({}^*(\textstyle\bigwedge^{k+\ell} h)(\alpha \wedge \beta))(m) &= {}^*(\textstyle\bigwedge^{k+\ell} h_m)(\alpha \wedge \beta)_{g(m)} \\ &\overset{[I.7.25]}{=} (\textstyle\bigwedge^{k+\ell} {}^*h_m)(\alpha_{g(m)} \wedge \beta_{g(m)}) \\ &\overset{[I.7.15]}{=} (\textstyle\bigwedge^k {}^*h_m)\alpha_{g(m)} \wedge (\textstyle\bigwedge^\ell {}^*h_m)\beta_{g(m)} \\ &= ({}^*(\textstyle\bigwedge^k h)(\alpha) \wedge {}^*(\textstyle\bigwedge^\ell h)(\beta))(m) ,\end{aligned}$$

where in the first line we have used the identification of $\bigwedge^{k+\ell} {}^*C_{g(m)}$ with ${}^*\bigwedge^{k+\ell} C_{g(m)}$.

7. Metrics on vector bundles

In this section we introduce the notion of a (pseudo) metric on an $\mathcal{A}$-vector space and we show its canonical form. We then use a metric to define the notion of an orthogonal complement to a subspace of an $\mathcal{A}$-vector space. These results can be implemented on vector bundles to show that a metric on a vector bundle can be used to define a supplement to a given vector subbundle. This in turn we use to show that the module of sections of a vector bundle is a finitely generated projective module over the ring of smooth functions on the base space.

7.1 Definition. We define the complexified ring $\mathcal{A}^{\mathbf{C}}$ as the $\mathcal{A}$-vector space $\mathcal{A}^2$ with multiplication $(a,b) \cdot (a',b') = (aa' - bb', ab' + ba')$. Introducing as usual the notation $1 \equiv (1,0), i \equiv (0,1) \in \mathcal{A}^{\mathbf{C}}$, an element $(a,b) \in \mathcal{A}^{\mathbf{C}}$ can be written as $(a,b) = a + ib$. On $\mathcal{A}^{\mathbf{C}}$ we define an involution (complex conjugation) by

$$\overline{a + ib} = a - ib \ .$$

A direct verification shows that this involution is a ring morphism. Moreover, the body map $\mathbf{B} : \mathcal{A}^{\mathbf{C}} \equiv \mathcal{A}^2 \to \mathbf{R}^2$ can be interpreted as a surjective ring morphism $\mathbf{B} : \mathcal{A}^{\mathbf{C}} \to \mathbf{C}$, which respects complex conjugation.

7.2 Remark. In the (mainly physics) literature one can find a different definition of complex conjugation, which involves an internal operation on $\mathcal{A}$ (changing the order in the exterior powers of $\mathcal{A} \cong \bigwedge X$). In our context this internal operation has a disadvantage: either it is not a smooth map on $\mathcal{A}$ in the sense of [III.1.20], or, redefining $\mathcal{A}$ as the subset of $\mathcal{A}^{\mathbf{C}}$ on which complex conjugation *is* the identity, $\mathcal{A}$ is not a ring.

7.3 Definition. Let E be a free graded $\mathcal{A}$-module. An even bilinear map $g : E \times E \to \mathcal{A}^{\mathbf{C}}$ is called a *pseudo metric*† *on* E if it satisfies:

(i) for all homogeneous $e, f \in E : \overline{g(e,f)} = (-1)^{\langle \varepsilon(e) | \varepsilon(f) \rangle} g(f,e)$;
(ii) $\mathbf{B}g : \mathbf{B}E \times \mathbf{B}E \to \mathbf{C}$ is non-degenerate;
(iii) for all homogeneous $e, f \in E : \mathbf{B}(\overline{g(e,f)}) = (-1)^{\langle \varepsilon(e) | \varepsilon(f) \rangle} \mathbf{B}(g(e,f))$.

7.4 Discussion. Let us investigate the conditions in more detail. The first one is the natural definition of symmetry (of a hermitian metric), the sign appears because we changed the order of the elements e and f. The second condition is the natural definition of non-degeneracy. The third one is a strange one: the sign is not justified by a change of order of the elements e and f; *apart from this sign,* it says that $\mathbf{B}g$ should take its values in $\mathbf{R}$. Using condition (i), condition (iii) can be rephrased as $\mathbf{B}g(e,f) = \mathbf{B}g(f,e)$, which says that $\mathbf{B}g$ is real symmetric, whereas condition (i) says that g itself should be hermitian symmetric. And of course the strangeness did not disappear: e and f are interchanged *without* an additional sign.

7.5 Lemma. *Let E be a finite dimensional free graded $\mathcal{A}$-module of dimension $n = p|q$, and let $(e_i)_{i=1}^n$ be an ordered homogeneous basis. If $g : E \times E \to \mathcal{A}^{\mathbf{C}}$ is a pseudo metric*

---

†Our free graded $\mathcal{A}$-modules are essentially "real" in the sense that they are in particular vector spaces over $\mathbf{R}$, but not (necessarily) over $\mathbf{C}$. It thus will be strange to see a definition of a (pseudo) metric in the sense of vector spaces over $\mathbf{C}$. However, I challenge the reader to find a "natural" definition of a "real" metric such that a linear combination of metrics with positive coefficients is again a metric.

on E, it defines a matrix g_{ij} in $M(p|q, \mathcal{A}^{\mathbf{C}})$ by

$$g_{ij} = g(e_i, e_j) \cong \begin{pmatrix} P & Q \\ R & T \end{pmatrix} ,$$

with P of size $p \times p$ and T of size $q \times q$. The matrices P, Q, R, and T verify $P_{ij} = \overline{P_{ji}}$, $T_{ij} = -\overline{T_{ji}}$, $R_{ij} = \overline{Q_{ji}}$. Moreover, $\mathbf{B}P$ and $-i\mathbf{B}T$ are non-degenerate symmetric $\mathbf{R}$-valued matrices.

Conversely, given matrices P, Q, R, and T verifying these properties, then the bilinear map $g : E \times E \to \mathcal{A}^{\mathbf{C}}$ defined by

$$g(\sum_i x^i \cdot e_i, \sum_j e_j \cdot y^j) = \sum_{ij} x^i \cdot g_{ij} \cdot y^j$$

(note the order in the coefficients!) is a pseudo metric on E.

Proof. The properties of P, Q, R, and T follow immediately from the first condition on a pseudo metric. The properties of $\mathbf{B}P$ and $-i\mathbf{B}T$ follow from this and the last two conditions. The converse is a direct (tedious) verification, using that the coefficients x^i and y^j are $\mathcal{A}$-valued and not $\mathcal{A}^{\mathbf{C}}$-valued. And note that among the properties of the matrices P, Q, R, and T is the property that the entries of P and T are even and that the entries of Q and R are odd (g represents an even bilinear map). QED

7.6 Discussion. By definition, a pseudo metric on an $\mathcal{A}$-vector space E can be seen as an element of $\mathrm{Hom}(E \otimes E; \mathcal{A}^{\mathbf{C}})_0$. Using our identifications [I.8.8], this space is isomorphic either to $\left({}^*(E \otimes E) \otimes \mathcal{A}^{\mathbf{C}}\right)_0$ (when we see it as left linear), or to $\left(\mathcal{A}^{\mathbf{C}} \otimes (E \otimes E)^*\right)_0$ (when we see it as right linear). Using the identifications ${}^*(E \otimes E) \cong {}^*E \otimes {}^*E$ and $(E \otimes E)^* \cong E^* \otimes E^*$ and using the dual bases $({}^i e)$ of ${}^*E$ and (e^i) of E^*, we find that a pseudo metric is represented by

$$g = \sum_{ij} g_{ij} \otimes e^j \otimes e^i = \sum_{ij} {}^j e \otimes {}^i e \otimes g_{ij} ,$$

where we interpret the g_{ij} as elements of the free graded $\mathcal{A}$-module $\mathcal{A}^{\mathbf{C}}$. The order of the indices of g_{ij} in these formulæ is a consequence of our definition of the identification given in [I.7.12]. That we obtain the same coefficients in both expressions is a direct consequence of the fact that g is even; it can also be understood in terms of the identity $\mathfrak{T}({}^i e) = (-1)^{\langle \varepsilon(e_i)|\varepsilon(e_i)\rangle} e^i$ [II.4.5].

Let us now investigate in more detail how the matrix elements of g depend upon the basis (e_i) of E. So let (e_i) and (f_j) be two (ordered homogeneous) bases of E related to each other by a matrix $\lambda^i{}_j \cong \left(\begin{smallmatrix} A & B \\ C & D \end{smallmatrix}\right) \in M(p|q, \mathcal{A})$ as $f_j = \sum_i e_i \, \lambda^i{}_j$ (note the choice of order). Invariance of graded dimension tells us that the entries of A and D are in $\mathcal{A}_0$ and those of B and C in $\mathcal{A}_1$. If we denote by $\mathrm{mat}^e(g)$ and $\mathrm{mat}^f(g)$ the matrices of g with respect to the bases e and f respectively, then an elementary calculation shows

$$\mathrm{mat}^f(g) = \begin{pmatrix} A^t & -C^t \\ B^t & D^t \end{pmatrix} \cdot \mathrm{mat}^e(g) \cdot \begin{pmatrix} A & B \\ C & D \end{pmatrix} , \tag{7.7}$$

where (as usual) the superscript t denotes the ordinary transpose of a matrix, i.e., interchanging the roles of row and column index.

7.8 Proposition. *Let g be a pseudo metric on a finite dimensional free graded $\mathcal{A}$-module E, then there exists a basis in which the matrix representation of g is given as*

$$\mathrm{mat}(g) = \begin{pmatrix} S_0 + iX & \mathbf{0} \\ \mathbf{0} & iS_1 + Y \end{pmatrix} ,$$

where S_0 and S_1 are diagonal matrices with only ± 1 on the diagonal, and where X and Y are skew-symmetric matrices (i.e., whose transpose is their own negative) with entries in $\mathcal{N}_0 \subset \mathcal{A}_0$.

Proof. If M is a symmetric non-degenerate matrix with entries in $\mathbf{R}$, a standard result of linear algebra tells us that there exists an invertible matrix T with real entries such that $T^t \cdot M \cdot T$ is a diagonal matrix with only ± 1 on the diagonal. Applying this to a pseudo metric g, it follows from [7.5] that there exists a basis f, linked to the basis e by a real valued λ (i.e., A and D real valued, B and C zero) such that

$$\mathrm{mat}^f(g) = \begin{pmatrix} S_0 + X & Z \\ \overline{Z}^t & iS_1 + Y \end{pmatrix} , \tag{7.9}$$

where S_0 and S_1 are diagonal matrices with only ± 1 on the diagonal (and thus $S_i^{-1} = S_i$), and where X, Y and Z are matrices with nilpotent entries (in $\mathcal{A}^{\mathbf{C}}$). The change of basis from e to f can be obtained by the Gram-Schmidt orthogonalization process on $\mathbf{B}g$. Our goal is to prove that we can find a basis in which Z is zero and in which X and Y have a special form. The idea is to apply a Gram-Schmidt type of procedure. The problem is that a change of basis has to be done with $\mathcal{A}$-valued matrices, and that $\mathrm{mat}(g)$ is $\mathcal{A}^{\mathbf{C}}$-valued. And indeed, if we were allowed to use $\mathcal{A}^{\mathbf{C}}$-valued matrices λ to change a basis, we could easily eliminate X, Y, and Z completely.

In order to obtain our goal, we define for any square matrix N with even nilpotent entries a new matrix $(\mathbf{1} + N)^{-1/2}$ as a power series in N with coefficients defined by the power series expansion of the real valued function $x \mapsto (1+x)^{-1/2}$. This is well defined because a matrix with nilpotent entries is nilpotent and thus any power series is finite. We also introduce three special types of coordinate changes:

$$\text{type I} : \quad \begin{pmatrix} A & B \\ C & D \end{pmatrix} = \begin{pmatrix} A & \mathbf{0} \\ \mathbf{0} & D \end{pmatrix} , \qquad \mathbf{B}A = \mathbf{1} , \quad \mathbf{B}D = \mathbf{1} ,$$

$$\text{type II} : \quad \begin{pmatrix} A & B \\ C & D \end{pmatrix} = \begin{pmatrix} id & B \\ \mathbf{0} & id \end{pmatrix} ,$$

$$\text{type III} : \quad \begin{pmatrix} A & B \\ C & D \end{pmatrix} = \begin{pmatrix} id & \mathbf{0} \\ C & id \end{pmatrix} .$$

The last ingredient we will need are the subsets $N^{(k)} \subset \mathcal{A}^{\mathbf{C}}$ which consist of all finite sums/differences of products of at least k entries of the matrices X, Y, and Z in (7.9). Since the entries of these matrices are nilpotent (and because there are only finitely many of them), it follows that there is a $k \in \mathbf{N}$ such that $N^{(k)} = \{0\}$. Although we will frequently change the basis while keeping the general form (7.9) of $\mathrm{mat}(g)$, the subsets $N^{(k)}$ are defined once and for all in terms of the initial (sub)matrices X, Y, and Z.

We will denote by Re and Im the operations that give the real and imaginary part of an element in $\mathcal{A}^{\mathbf{C}}$; we will say that X (respectively Y or Z) is homogeneous if either $\mathrm{Re}(X)$ or $\mathrm{Im}(X)$ is zero (and remember that the entries of these matrices are already homogeneous with respect to the dichotomy even-odd). It should be clear that a priori these matrices are not homogeneous. We now start our analysis by noting that initially the matrix Z has entries in $\mathcal{N}^{(1)}$. Our strategy will be to change bases such that the matrix Z will belong to $\mathcal{N}^{(k)}$ with steadily increasing k. To simplify notations, we will say that Z has order k if its entries belong to $\mathcal{N}^{(k)}$.

Our first claim is that we may assume that $\mathrm{Re}(X)$ and $\mathrm{Im}(Y)$ are zero (and thus X and Y are homogeneous). This result can be accomplished by a type I change, using $A = S_0 \cdot (\mathbf{1} + \mathrm{Re}(X) \cdot S_0)^{-1/2}$ and $D = S_1 \cdot (\mathbf{1} + \mathrm{Im}(Y) \cdot S_1)^{-1/2}$. In order to prove the result, one should use that products of power series in $\mathrm{Re}(X) \cdot S_0$ commute, that $S_0 \cdot (\mathrm{Re}(X) \cdot S_0)^\ell = (S_0 \cdot \mathrm{Re}(X))^\ell \cdot S_0$, and that S_0 and $\mathrm{Re}(X)$ are symmetric (i.e., the same as their transpose). Apart from the result on X and Y, it should be noted that Z transforms to $A^t Z D$, and thus that the order of Z and the eventual homogeneity of Z do not change under this change of basis.

The next claim is that, if $\mathrm{Re}(X)$ is zero, then we may assume that $\mathrm{Re}(Z)$ is zero. To obtain this result, we perform a type II change with $B = -S_0 \cdot \mathrm{Re}(Z)$, which transforms Z to $Z - \mathrm{Re}(Z) - X S_0 \,\mathrm{Re}(Z)$. This change of basis will destroy the homogeneity of X and Y. Moreover, if $\mathrm{Im}(Z)$ were zero before, it will be non-zero after, but of at least one order higher than before.

The third claim is that, if $\mathrm{Im}(Y)$ is zero, then we may assume that $\mathrm{Im}(Z)$ is zero. This can be obtained by a type III change with $C = S_1 \cdot (\mathrm{Im}(Z))^t$, which transforms Z to $Z - i\,\mathrm{Im}(Z) - \mathrm{Im}(Z) S_1 Y$. As before, this destroys the homogeneity of X and Y. And as before, if $\mathrm{Re}(Z)$ were zero before, it will be non-zero after, but of at least one order higher than before.

We now are in business for our induction. Applying the first and second argument, we may assume that $\mathrm{Re}(Z)$ is zero. If we then apply, in this order, the first, the third, the first, and finally the second argument, we again have a Z with $\mathrm{Re}(Z)$ zero, but with its order increased by at least 2. Since $\mathcal{N}^{(k)} = \{0\}$ if k is sufficiently large, it follows that after a finite number of basis changes, we obtain a matrix representation of the pseudo metric g of the form (7.9) with Z being zero. Applying once more the first argument, we also may assume that $\mathrm{Re}(X)$ and $\mathrm{Im}(Y)$ are zero. QED

7.10 Discussion. Once we have the notion of a pseudo metric g on a finite dimensional free graded $\mathcal{A}$-module E, we would like to define the notion of the orthogonal $F^\perp$ of a graded subspace $F \subset E$. A naive definition would be $F^\perp = \{\, e \in E \mid \forall f \in F : g(e,f) = 0 \,\}$. It is easy to show that this $F^\perp$ is a graded submodule of E, but unfortunately, even in simple cases, it need not be a graded subspace. For instance, consider $E = \mathcal{A}^2$ of dimension $2|0$, let $F \subset E$ be the graded subspace generated by $(1,0)$, and let the pseudo metric g be given in the basis $(1,0)$, $(0,1)$ by the matrix $\begin{pmatrix} 1 & in \\ -in & 1 \end{pmatrix}$ for some even non-zero nilpotent $n \in \mathcal{N}$. An immediate computation learns that $F^\perp = \{\, (0,b) \mid b \in \mathcal{A}, bn = 0 \,\}$. This is indeed a graded submodule, but it is not itself a free graded $\mathcal{A}$-module, i.e., it is not a graded subspace.

Since we still want to create new graded subspaces out of old ones by means of the notion of orthogonal, we need another definition of orthogonal. Before we can state, in [7.15], a definition that works, we need some preparations.

7.11 Definition. Let g be a pseudo metric on E. A basis $(e_i)_{i=1}^n$ is called *an orthonormal basis with respect to* g if the matrix of g with respect to the basis e has the form given in [7.8].

7.12 Lemma. *Let g be a pseudo metric on a finite dimensional free graded $\mathcal{A}$-module E, and let $(e_i)_{i=1}^n$ be an orthonormal basis with respect to g. If $\left(\begin{smallmatrix} A & B \\ C & D \end{smallmatrix}\right)$ describes a change of basis such that the new basis is again orthonormal, then necessarily B and C are zero.*

Proof. If the old and the new basis are orthonormal, then especially the Z part of $\mathrm{mat}(g)$ in the new basis must be zero. According to (7.7), this is given by the equation

$$A^t \cdot (S_0 + iX) \cdot B - C^t \cdot (iS_1 + Y) \cdot D = \mathbf{0} .$$

Splitting this equation into its real and imaginary parts, we obtain the two equations $A^t \cdot S_0 \cdot B = C^t \cdot Y \cdot D$ and $A^t \cdot X \cdot B = C^t \cdot S_1 \cdot D$. Since $\left(\begin{smallmatrix} A & B \\ C & D \end{smallmatrix}\right)$ describes a change of basis, it is invertible. Since it is also even, this implies that A and D must be invertible. We then obtain the equation $C^t = A^t \cdot X \cdot B \cdot D^{-1} \cdot S_1$ and thus

$$B = (S_0 \cdot X) \cdot B \cdot (D^{-1} \cdot S_1 \cdot Y \cdot D) .$$

Iterating this equation gives us the equality

$$B = (S_0 \cdot X)^k \cdot B \cdot (D^{-1} \cdot S_1 \cdot Y \cdot D)^k . \tag{7.13}$$

Since X and thus $S_0 \cdot X$ is nilpotent, there is a $k \in \mathbf{N}$ such that $(S_0 \cdot X)^k = \mathbf{0}$, i.e., B is zero. The equation for C^t then tells us that C also is zero. QED

7.14 Definition. A pseudo metric g on a finite dimensional free graded $\mathcal{A}$-module E is called a *metric on* E if the restriction of $\mathbf{B}g$ to $\mathbf{B}E_0$ and the restriction of $-i\mathbf{B}g$ to $\mathbf{B}E_1$ are positive definite. This definition makes sense because in terms of a basis, these restrictions are represented by the matrices $\mathbf{B}P$ and $-i\mathbf{B}T$ [7.5]. Since these are symmetric and $\mathbf{R}$-valued, we can require that they are positive definite. Moreover, under a change of basis these matrices change to $\mathbf{B}A^t \cdot \mathbf{B}P \cdot \mathbf{B}A$ and $-i\mathbf{B}D^t \cdot \mathbf{B}T \cdot \mathbf{B}D$ (7.7), and thus the notion of being positive definite is independent of the choice of a basis. Note that a pseudo metric g is a metric if and only if for an orthonormal basis the matrices S_0 and S_1 of [7.8] are the identity.

7.15 Definition. Let g be a metric on a finite dimensional free graded $\mathcal{A}$-module E and let F be a graded subspace of E. It is immediate that the restriction of g to F defines a metric on F (the same is in general not true for a pseudo metric for which the restriction to a graded subspace might be degenerate). Let $(e_i)_{i=1}^k$ be an orthonormal basis of F. We define the *orthogonal complement* $F^\perp \subset E$ by

$$F^\perp = \left\{ e \in E \;\middle|\; \begin{array}{l} \varepsilon(e_i) = 0 \Rightarrow \mathrm{Re}(g(e, e_i)) = 0 \\ \varepsilon(e_i) = 1 \Rightarrow \mathrm{Im}(g(e, e_i)) = 0 \end{array} \right\} . \tag{7.16}$$

7.17 Lemma. *The subset $F^\perp \subset E$ is a graded subspace of E, independent of the orthonormal basis $(e_i)_{i=1}^k$ of F. Moreover, it is a supplement to F.*

Proof. It is immediate from the definition of $F^\perp$ that it is a graded submodule of E. According to [7.12] two orthonormal bases are related in such a way that even (respectively odd) vectors in the new basis are a linear combination of the even (respectively odd) vectors in the old basis. This implies that $F^\perp$ is independent of the specific orthonormal basis for F. To finish the proof, it thus suffices to prove that we can extend the basis $(e_i)_{i=1}^k$ for F to a basis $(e_i)_{i=1}^n$ for E such that $(e_i)_{i=k+1}^n$ is a basis for $F^\perp$.

Since F is a graded subspace of E, we can extend the orthonormal basis for F to a basis for E [II.3.7]. Let this basis of E be ordered in such a way that within the subsets $(e_i)_{i=1}^k$ and $(e_i)_{i=k+1}^n$ the even ones come before the odd ones. In this basis, the matrix of g looks like

$$\mathrm{mat}(g) = \begin{pmatrix} \mathbf{1} + i\,* & \mathbf{0} & * + i\,* & * + i\,* \\ \mathbf{0} & i\mathbf{1} + * & * + i\,* & * + i\,* \\ * + i\,* & * + i\,* & ** + i\,** & ** + i\,** \\ * + i\,* & * + i\,* & ** + i\,** & ** + i\,** \end{pmatrix} ,$$

where $**$ denotes an arbitrary matrix with entries in $\mathcal{A}$ and where $*$ denotes an arbitrary matrix with entries in $\mathcal{N} \subset \mathcal{A}$. We now modify the vectors $e_{i>k}$ by the formula

$$\forall i > k : \widehat{e}_i = e_i - \sum_{j=1}^k \mathrm{Re}\left(\frac{g(e_i, e_j)}{g(e_j, e_j)}\right) e_j \, . \tag{7.18}$$

It is obvious that $e_1, \ldots, e_k, \widehat{e}_{k+1}, \ldots, \widehat{e}_n$ still form a basis for E and then (using that $g(e_j, e_j) = 1$ for $j \leq k$ and e_j even and $g(e_j, e_j) = i$ for $j \leq k$ and e_j odd) a direct computation shows that in this basis g looks like

$$\mathrm{mat}(g) = \begin{pmatrix} \mathbf{1} + i\,* & \mathbf{0} & i\,* & i\,* \\ \mathbf{0} & i\mathbf{1} + * & * & * \\ i\,* & * & ** + i\,** & ** + i\,** \\ i\,* & * & ** + i\,** & ** + i\,** \end{pmatrix} . \tag{7.19}$$

From this matrix form of g it is immediate that the $(\widehat{e}_i)_{i=k+1}^n$ form a basis for $F^\perp$. QED

7.20 Definition. Let $p : B \to M$ be a vector bundle over an $\mathcal{A}$-manifold M. A *(pseudo) metric on* B is a smooth section g of the bundle $\underline{\mathcal{A}}^{\mathbf{C}} \otimes (B \otimes B)^*$ (where $\underline{\mathcal{A}}^{\mathbf{C}}$ denotes the trivial bundle $M \times \mathcal{A}^{\mathbf{C}} \to M$), such that for each $m \in M$ the value $g(m)$ is a (pseudo) metric on the finite dimensional free graded $\mathcal{A}$-module $B_m = p^{-1}(m)$.

7.21 Proposition. *Let $p : B \to M$ be a vector bundle over M with typical fiber E, then there exists a metric on B.*

Proof. Let E be the typical fiber of B, let $\mathcal{U} = \{ U_\alpha \mid \alpha \in J \}$ be a trivializing atlas for the bundle $\underline{\mathcal{A}}^{\mathbf{C}} \otimes (B \otimes B)^*$, and let $\{\rho_\alpha\}$ be a partition of unity subordinated to $\mathcal{U}$. According to [1.20] we have an identification $\Gamma_{U_\alpha}(\underline{\mathcal{A}}^{\mathbf{C}} \otimes (B \otimes B)^*) \cong C^\infty(U_\alpha; \mathcal{A}^{\mathbf{C}} \otimes (E \otimes E)^*)$. Moreover, on the finite dimensional $\mathcal{A}$-vector space E we have a more or less obvious choice g_o for a metric: the one represented by the matrix $\begin{pmatrix} \mathbf{1} & \mathbf{0} \\ \mathbf{0} & i\mathbf{1} \end{pmatrix}$. Since the metric g_o has real coordinates, the constant local section $g_\alpha : U_\alpha \to \mathcal{A}^{\mathbf{C}} \otimes (E \otimes E)^*$ given by $g_\alpha(m) = g_o$ is smooth.

We thus can form the global section $g = \sum_\alpha \rho_\alpha \cdot g_\alpha$ [3.9-iii]. Since the ρ_α take their values in $\mathcal{A}_0$, not $\mathcal{A}_0^{\mathbf{C}}$, conditions (i) and (iii) of a pseudo metric are automatically satisfied on each B_m. Remains to prove that the restrictions $\mathbf{B}g$ to $\mathbf{B}(B_m)_0$ and $-i\mathbf{B}g$ to $\mathbf{B}(B_m)_1$ are positive definite. But for each $m \in M$, $\mathbf{B}g$ is a finite sum of terms $(\mathbf{B}\rho_\alpha) \cdot (\mathbf{B}g_\alpha)$ (the ρ_α are a partition of unity, thus only finitely many of them are non-zero at a time). Since each of the g_α is a metric, a finite sum with positive coefficients of positive definite forms is again positive definite. And thus $g(m)$ is a metric. QED

7.22 Proposition. *Let $p : B \to M$ be a vector bundle, let g be a metric on B, and let $C \subset B$ be a subbundle [3.12]. Then there exists a well defined subbundle $C^\perp \subset B$ which is a supplement to C [4.6].*

Proof. For each $m \in M$, $g(m)$ is a metric on the finite dimensional free graded $\mathcal{A}$-module B_m and C_m is a graded subspace of B_m. We thus can define $C^\perp$ *as a set* by $(C^\perp)_m = (C_m)^\perp$. We "only" have to show that this subset $C^\perp$ is a subbundle and a supplement to C. To prove that, we will reconsider the proofs of [7.8] and [7.17] in order to check that all basis changes depend in a smooth way on the point m. In doing so, we will use a different kind of degree than that used in [7.8]. If $f : U \to \mathcal{A}$ is a smooth function, we know that f decomposes as $f = \sum_i f_i \, \xi^i$ in the notational shorthand of [III.3.17]. We will say that f has degree d if $f = \sum_{i \geq d} f_i \, \xi^i$. From this definition it is obvious that if f has degree d_1 and g has degree d_2, then their product $f \cdot g$ has degree $d_1 + d_2$. And then the details.

Since C is a subbundle of B, there exists a trivializing atlas $\mathcal{U}$ and, for each $U \in \mathcal{U}$, there exist local trivializing sections $s_1, \ldots, s_n \in \Gamma_U(B)$ [3.14] such that $s_1, \ldots, s_k$ are local trivializing sections for C. Following [7.8] we want to modify the local sections $s_1, \ldots, s_k$ so that the $\hat{s}_1(m), \ldots, \hat{s}_k(m)$ form an orthonormal basis of C_m for all $m \in U$. The first step to take is to get them into the form (7.9). This is done by a standard Gram-Schmidt process applied to $\mathbf{B}g(m)$, which is described by an upper triangular matrix whose entries are rational functions in the matrix entries of $\mathbf{B}g(m)$. It follows that we can perform this operation on our smooth sections $s_1, \ldots, s_k$ and obtain new smooth trivializing sections for C. Since in our notational shorthand $\mathbf{B}g = g_0$, it follows that g_0 is a constant diagonal matrix with 1 or i on the diagonal. It follows that the entries of the matrices X and Y are smooth functions on U of order at least 2, and the entries of Z have at least order 1. Hence the power series definition of a type I coordinate change necessarily breaks of after the $q/2$-th term (q being the odd dimension of M). This implies that all three types of basis changes preserve smoothness of the local sections. Moreover, a type II or III change will increase the degree of (the entries of) Z by at least 1, and thus after q steps we have found local sections $(\hat{s}_i)_{i=1}^k$ such that they form an orthonormal basis at each C_m.

Up till now we have not touched the local sections $(s_i)_{i=k+1}^n$ of B. We now modify these sections according to (7.18), i.e., we replace the vectors e_i in (7.18) by the elements $s_i(m)$. Since the functions $m \mapsto \mathrm{Re}\big(g(m)(s_i(m), \hat{s}_j(m))/g(m)(\hat{s}_j(m), \hat{s}_j(m))\big)$ are smooth, the new sections $(\hat{s}_i)_{i=k+1}^n$ are again smooth local sections of B. According to [7.17], they form a set of local trivializing sections for $C^\perp$. We conclude that $C^\perp$ is indeed a subbundle and a supplement to C. QED

7.23 Theorem. [GHV, Ch I.1, Thm 1] *Let M be an $\mathbf{R}$-manifold and $\mathcal{U}$ an open cover of M. Then there exists a $d \in \mathbf{N}$ and an open cover $\mathcal{V} = \{\, V_{ij} \mid i, j \in \mathbf{N}, j \leq d \,\}$ such that*

(i) *for all $V \in \mathcal{V}$ there is an $U \in \mathcal{U}$ such that $V \subset U$,*
(ii) $i \neq i' \Longrightarrow V_{ij} \cap V_{i'j} = \phi$.

7.24 Corollary. *Let $p : B \to M$ a vector bundle over an $\mathcal{A}$-manifold M. Then there exists a finite number of global smooth homogeneous sections $s_1, \ldots, s_r \in \Gamma(B)$ such that for all $m \in M$ the $s_1(m), \ldots, s_r(m)$ generate the free graded $\mathcal{A}$-module B_m.*

Proof. Let $\mathcal{U} = \{\, U_\alpha \mid \alpha \in J \,\}$ be a trivializing atlas for B. It follows that $\mathbf{B}\mathcal{U}$ is an open cover of the $\mathbf{R}$-manifold $\mathbf{B}M$. Since $\mathbf{B}$ is a bijection between open sets in M and open sets in $\mathbf{B}M$, there exists an open cover $\mathcal{V} = \{\, V_{ij} \mid i, j \in \mathbf{N}, j \leq d \,\}$ of M such that $\mathbf{B}\mathcal{V}$ has the properties mentioned in [7.23] with respect to $\mathbf{B}\mathcal{U}$. But then $\mathcal{V}$ has these properties with respect to $\mathcal{U}$. Now define the open sets $W_1, \ldots, W_d$ as $W_j = \cup_{i \in \mathbf{N}} V_{ij}$. Since $\mathcal{U}$ is a trivializing atlas, there exist local trivializing sections $s_1^\alpha, \ldots, s_n^\alpha \in \Gamma_{U_\alpha}(B)$. By [7.23-i] there exists for each (ij) at least one $\alpha(ij)$ such that $V_{ij} \subset U_{\alpha(ij)}$. We thus can define the local trivializing sections $s_1^j, \ldots, s_n^j \in \Gamma_{W_j}(B)$ by

$$s_k^j \text{ restricted to } V_{ij} = s_k^{\alpha(ij)} \text{ restricted to } V_{ij} \,.$$

This is a correct definition because the V_{ij} are disjoint for fixed j by [7.23-ii]. Finally, let ρ_j be a partition of unity subordinated to the (finite) open cover $W_1, \ldots, W_d$. The global sections $\rho_j \cdot s_k^j$ are homogeneous, finite in number (there are $r = nd$ of them) and at each point they generate B_m because there is at least one $\mathbf{B}\rho_j(m)$ different from zero, and then the $\rho_j \cdot s_k^j$ with this j form a basis of B_m. QED

7.25 Theorem. *Let $p : B \to M$ be a vector bundle over an $\mathcal{A}$-manifold M. Then $\Gamma(B)$ is a f.g.p graded $C^\infty(M)$-module.*

Proof. Let $s_1, \ldots, s_r \in \Gamma(B)$ be homogeneous global sections as in [7.24] and let F be the $\mathcal{A}$-vector space generated by the abstract elements s_i. To distinguish the abstract basis of F from the actual sections, we will denote the former as s_i^F. We then define a vector bundle morphism $\phi : M \times F \to B$ from the trivial bundle $M \times F$ to B by

$$\phi(m, \textstyle\sum_i x^i\, s_i^F) = \textstyle\sum_i x^i\, s_i(m) \; .$$

Since the $s_i(m)$ generate the fiber B_m, it follows easily that ϕ is a smooth surjective even bundle morphism. By [3.18] we deduce that $C = \ker(\phi)$ is a subbundle of $M \times F$. By [7.22] we can find a subbundle $C^\perp$ of B such that $C \oplus C^\perp = M \times F$. By definition of C, the restriction of ϕ to $C^\perp$ is a vector bundle isomorphism between $C^\perp$ and B. We thus obtain the sequence of isomorphisms

$$\Gamma(C) \oplus \Gamma(B) \cong \Gamma(C) \oplus \Gamma(C^\perp) \overset{[5.3]}{\cong} \Gamma(C \oplus C^\perp) \cong \Gamma(M \times F) \cong C^\infty(M; F) \; ,$$

from which we deduce that $\Gamma(B)$ is f.g.p (because $C^\infty(M; F)$ is a finitely generated free graded $C^\infty(M)$-module). QED

8. Batchelor's theorem

In this section we prove Batchelor's theorem which says that for any $\mathcal{A}$-manifold one can find an atlas such that the transition functions have a special form: even coordinates in one chart depend only upon the even coordinates in another chart and odd coordinates in one chart depend linearly on the odd coordinates in another chart (and on the even coordinates in the other chart). This can be rephrased as saying that an $\mathcal{A}$-manifold is a vector bundle over an $\mathbf{R}$-manifold in the ordinary sense and that smooth functions correspond to sections of the exterior algebra of this ordinary vector bundle.

8.1 Remark. Theorem [8.2] could have been stated in §III.4 and proved in §III.5. However, the interpretation [8.8] requires the notion of a vector bundle, so we postponed it till now.

8.2 Theorem (Batchelor [Ba]**).** *If M is an $\mathcal{A}$-manifold, then there exists an atlas $\mathcal{U} = \{ \varphi_a : U_a \to O_a \mid a \in I \}$ such that the transition functions $(x_b, \xi_b) = (\varphi_b \circ \varphi_a^{-1})(x_a, \xi_a)$ are of the form*

$$x_b^i = f^i(x_a)$$
$$\xi_b^j = \sum_k g_k^j(x_a) \cdot \xi_a^k \; .$$

Proof. In this proof we will use extensively [III.2.23], [III.2.27] and the notational shorthand introduced in [III.3.17]. The general idea of the proof is that given any atlas $\mathcal{U} = \{\varphi_a : U_a \to O_a \mid a \in I\}$, we will find a new atlas $\widehat{\mathcal{U}}$ with the same sets U_a, O_a and O_{ab}, but with different maps $\widehat{\varphi}_a$ such that the transition functions are given as claimed in the theorem. We divide the proof in two steps. In the first step we eliminate by induction the ξ-dependence in the transition functions of the even coordinates. In the second step we eliminate the higher powers of ξ from the transition functions of the odd coordinates.

To start, we thus assume that all transition functions of the even coordinates are given by

$$x_b = \varphi_{ba,0}(x_a) + \sum_{i\geq k} \varphi_{ba,2i}(x_a) \cdot (\xi_a)^{2i} . \tag{8.3}$$

The induction starts trivially at $k = 1$; we will show that we can increase k by 1, thus eliminating the ξ dependence. The idea is to change the coordinates x such that they absorb the term with $i = k$. Since $\varphi_{ba}^{-1} = \varphi_{ab}$, it follows from [III.3.18] that $\varphi_{ba,0} : O_{ab}^{w.o.d} \to O_{ba}^{w.o.d}$ has $\varphi_{ab,0}$ as smooth inverse. We apply $\varphi_{ab,0}$ to (8.3) and we invoke [III.2.27], using that $\sum_{i\geq k} \varphi_{ba,2i}(x_a)\cdot(\xi_a)^{2i}$ is nilpotent. We deduce the existence of smooth functions $\chi_{ba,2i}$, $i \geq k$ defined on $O_{ab}^{w.o.d}$ such that

$$\varphi_{ab,0}(x_b) = x_a + \sum_{i\geq k} \chi_{ba,2i}(x_a) \cdot (\xi_a)^{2i} ,$$

The term of order $2k$ interests us most and can be computed as

$$\chi_{ba,2k}(x_a) \cdot (\xi_a)^{2k} = \varphi_{ab,0}^{(1)}(\varphi_{ba,0}(x_a)) \cdot \varphi_{ba,2k}(x_a) \cdot (\xi_a)^{2k} ,$$

where $\varphi_{ab,0}^{(1)}$ is (the equivalent of) the function $f^{(1)}$ in (III.2.28). We next compute a relation between various χs, needed to prove that we indeed will increment k by 1. For a point $(x_a, \xi_a) \in O_{ab} \cap O_{ac}$ one has

$$x_c = \varphi_{ca,0}\big(x_a + \sum_{i\geq k} \chi_{ca,2i}(x_a) \cdot (\xi_a)^{2i}\big) = \varphi_{cb,0}\Big(\varphi_{ba,0}\big(x_a + \sum_{i\geq k} \chi_{ca,2i}(x_a) \cdot (\xi_a)^{2i}\big)\Big)$$

$$x_c = \varphi_{cb,0}\big(x_b(x_a, \xi_a) + \sum_{i\geq k} \chi_{cb,2i}(x_b(x_a, \xi_a)) \cdot (\xi_b(x_a, \xi_a))^{2i}\big) .$$

Applying $\varphi_{bc,0}$, substituting (8.3) for $x_b(x_a, \xi_a)$, expanding into powers of ξ_a and retaining only $(\xi_a)^{2k}$, we obtain the equality

$$\begin{aligned} \chi_{cb,2k}(\varphi_{ba,0}(x_a)) \cdot (\varphi_{ba,1}(x_a) \cdot \xi_a)^{2k} + \varphi_{ba,2k}(x_a) \cdot (\xi_a)^{2k} \\ = \varphi_{ba,0}^{(1)}(x_a) \cdot \chi_{ca,2k}(x_a) \cdot (\xi_a)^{2k} . \end{aligned} \tag{8.4}$$

With these preparations, we can introduce a coordinate change that eliminates the $(\xi)^{2k}$ in all transition functions at once. To do so, let $\hat{\rho}_a$ be a partition of unity subordinated to the cover $\mathcal{U}$, and define the functions $\rho_a : O_a \to \mathcal{A}_0$ as $\rho_a = \hat{\rho}_a \circ \varphi_a^{-1}$. We then note that if σ is a smooth function defined on O_{ab}, then $\rho_b \cdot \sigma$ is a smooth function defined on O_a ($\mathrm{supp}(\rho_b|_{O_{ab}}) \subset O_{ab}$ is closed in O_a). We thus can define the diffeomorphisms $\psi_a : O_a \to O_a$, $(x_a, \xi_a) \mapsto (y_a, \xi_a)$ by

$$y_a = x_a + \sum_c \rho_c(\varphi_{ca,0}(x_a)) \cdot \chi_{ca,2k}(x_a) \cdot (\xi_a)^{2k} \,. \tag{8.5}$$

That ψ_a is indeed a diffeomorphism follows from [III.3.18]. With these diffeomorphisms we create new charts $\widehat{\varphi}_a : U_a \to O_a$ for M by $\widehat{\varphi}_a = \psi_a \circ \varphi_a$; they also form an atlas for M. To see how the transition functions $\widehat{\varphi}_{ba} = \widehat{\varphi}_b \circ \widehat{\varphi}_a^{-1}$ look like, we have to calculate ψ_a^{-1}. Expanding x_a in terms of functions of y_a and powers of ξ_a and substituting this in (8.5) shows that this map is given as

$$x_a = y_a - \sum_c \rho_c(\varphi_{ca,0}(y_a)) \cdot \chi_{ca,2k}(y_a) \cdot (\xi_a)^{2k} + h.o.t. \,,$$

where $h.o.t.$ stands for *higher order terms in* ξ_a. We now compute the transition function $\widehat{\varphi}_{ba}$:

$$\begin{aligned}
y_b &= x_b + \sum_c \rho_c(\varphi_{cb,0}(x_b)) \cdot \chi_{cb,2k}(x_b) \cdot (\xi_b)^{2k} \\
&= \varphi_{ba,0}(x_a) + \varphi_{ba,2k}(x_a) \cdot (\xi_a)^{2k} \\
&\quad + \sum_c \rho_c(\varphi_{cb,0}(\varphi_{ba,0}(x_a))) \cdot \chi_{cb,2k}(\varphi_{ba,0}(x_a)) \cdot (\varphi_{ba,1}(x_a) \cdot \xi_a)^{2k} + h.o.t. \\
&= \varphi_{ba,0}(y_a) - \varphi_{ba,0}^{(1)}(y_a) \sum_c \rho_c(\varphi_{ca,0}(y_a)) \cdot \chi_{ca,2k}(y_a) \cdot (\xi_a)^{2k} + \varphi_{ba,2k}(y_a) \cdot (\xi_a)^{2k} \\
&\quad + \sum_c \rho_c(\varphi_{ca,0}(y_a)) \cdot \chi_{cb,2k}(\varphi_{ba,0}(y_a)) \cdot (\varphi_{ba,1}(y_a) \cdot \xi_a)^{2k} + h.o.t. \\
&= \varphi_{ba,0}(y_a) + \sum_c \rho_c(\varphi_{ca,0}(y_a)) \cdot \Big(\chi_{cb,2k}(\varphi_{ba,0}(y_a)) \cdot (\varphi_{ba,1}(y_a) \cdot \xi_a)^{2k} \\
&\quad + \varphi_{ba,2k}(y_a) \cdot (\xi_a)^{2k} - \varphi_{ba,0}^{(1)}(y_a) \cdot \chi_{ca,2k}(y_a) \cdot (\xi_a)^{2k} \Big) + h.o.t. \,.
\end{aligned}$$

In view of (8.4), we may conclude that we have indeed incremented k by 1. By induction we thus may conclude that there exists an atlas in which we have eliminated the ξ-dependence of the x transition functions.

To eliminate the higher powers of ξ from the transition functions of the odd coordinates, we proceed again by induction, assuming now that the transition functions are given by

$$\begin{aligned}
x_b &= \varphi_{ba,0}(x_a) \,, \\
\xi_b &= \varphi_{ba,1}(x_a) \cdot \xi_a + \sum_{i \geq k} \varphi_{ba,2i+1}(x_a) \cdot (\xi_a)^{2i+1} \,.
\end{aligned} \tag{8.6}$$

Since we know that the matrix $\varphi_{ba,1}$ is invertible [III.3.18], we can find smooth functions $\chi_{ba,2i+1}$, $i \geq k$ defined on $O_{ab}^{w.o.d}$ such that

$$\xi_b = \varphi_{ba,1}(x_a) \cdot \Big(\xi_a + \sum_{i\geq k} \chi_{ba,2i+1}(x_a) \cdot (\xi_a)^{2i+1}\Big) .$$

The leading term is easily computed as

$$\chi_{ba,2k+1}(x_a) \cdot (\xi_a)^{2k+1} = \varphi_{ab,1}(\varphi_{ba,0}(x_a)) \cdot \varphi_{ba,2k+1}(x_a) \cdot (\xi_a)^{2k+1} .$$

We then compute a compatibility relation between various χ on $O_{ab} \cap O_{ac}$:

$$\xi_c = \varphi_{ca,1}(x_a) \cdot (\xi_a + \sum_{i\geq k} \chi_{ca,2i+1}(x_a) \cdot (\xi_a)^{2i+1})$$

$$\xi_c = \varphi_{cb,1}(\varphi_{ba,0}(x_a)) \cdot (\xi_b(x_a, \xi_a) + \sum_{i\geq k} \chi_{cb,2i+1}(\varphi_{ba,0}(x_a)) \cdot (\xi_b(x_a, \xi_a))^{2i+1}) .$$

Substituting (8.6) for $\xi_b(x_a, \xi_a)$, multiplying by $\varphi_{bc,1}(\varphi_{cb,0}(\varphi_{ba,0}(x_a)))$, expanding into powers of ξ_a and retaining $(\xi_a)^{2k+1}$, one obtains the compatibility relation

$$\begin{aligned} &\chi_{cb,2k+1}(\varphi_{ba,0}(x_a)) \cdot (\varphi_{ba,1}(x_a) \cdot \eta_a)^{2k+1} + \varphi_{ba,2k+1}(x_a) \cdot (\eta_a)^{2k+1} \\ &\qquad = \varphi_{ba,1}(x_a) \chi_{ca,2k+1}(x_a) \cdot (\eta_a)^{2k+1} . \end{aligned} \tag{8.7}$$

As for the x coordinates, we now use the partition of unity $\hat{\rho}$ and define the diffeomorphism $\psi_a : O_a \to O_a$, $(x_a, \xi_a) \mapsto (x_a, \eta_a)$ by

$$\eta_a = \xi_a + \sum_c \rho_c(\varphi_{ca,0}(x_a)) \cdot \chi_{ca,2k+1}(x_a) \cdot (\xi_a)^{2k+1} .$$

As for the x coordinates, it is easy to check that the inverse relation is given by

$$\xi_a = \eta_a - \sum_c \rho_c(\varphi_{ca,0}(x_a)) \cdot \chi_{ca,2k+1}(x_a) \cdot (\eta_a)^{2k+1} + h.o.t. .$$

Still following the line of argument used for the x coordinates, we define new maps $\widehat{\varphi}_a : U_a \to O_a$ by $\widehat{\varphi}_a = \psi_a \circ \varphi_a$. These maps define a new atlas for which we calculate the transition function $(x_a, \eta_a) \mapsto (x_b, \eta_b) = (\varphi_{ba,0}(x_a), \eta_b(x_a, \eta_a)$:

$$\begin{aligned} \eta_b &= \xi_b + \sum_c \rho_c(\varphi_{cb,0}(x_b)) \cdot \chi_{cb,2k+1}(x_b) \cdot (\xi_b)^{2k+1} \\ &= \varphi_{ba,1}(x_a) \cdot \xi_a + \varphi_{ba,2k+1}(x_a) \cdot (\xi_a)^{2k+1} \\ &\qquad + \sum_c \rho_c(\varphi_{cb,0}(\varphi_{ba,0}(x_a)) \cdot \chi_{cb,2k+1}(\varphi_{ba,0}(x_a)) \cdot (\varphi_{ba,1}(x_a) \cdot \xi_a)^{2k+1} + h.o.t. \\ &= \varphi_{ba,1}(x_a) \cdot \eta_a - \varphi_{ba,1}(x_a) \cdot \sum_c \rho_c(\varphi_{ca,0}(x_a)) \cdot \chi_{ca,2k+1}(x_a) \cdot (\eta_a)^{2k+1} \\ &\qquad + \sum_c \rho_c(\varphi_{ca,0}(x_a) \cdot \chi_{cb,2k+1}(\varphi_{ba,0}(x_a)) \cdot (\varphi_{ba,1}(x_a) \cdot \eta_a)^{2k+1} \\ &\qquad + \varphi_{ba,2k+1}(x_a) \cdot (\eta_a)^{2k+1} + h.o.t. \\ &= \varphi_{ba,1}(x_a) \cdot \eta_a + \sum_c \rho_c(\varphi_{ca,0}(x_a)) \cdot \Big(\chi_{cb,2k+1}(\varphi_{ba,0}(x_a)) \cdot (\varphi_{ba,1}(x_a) \cdot \eta_a)^{2k+1} \\ &\qquad + \varphi_{ba,2k+1}(x_a) \cdot (\eta_a)^{2k+1} - \varphi_{ba,1}(x_a) \chi_{ca,2k+1}(x_a) \cdot (\eta_a)^{2k+1}\Big) + h.o.t. . \end{aligned}$$

In view of (8.7), this means that we have incremented k by 1, and thus by induction we can assume that we have eliminated all higher powers of the odd coordinates in the transition functions of the odd coordinates. $\boxed{QED}$

8.8 Discussion. Batchelor's theorem has a ramification which transcends the practical applications of it. Recall that the functions f^i and g^j_k can be interpreted as real valued smooth functions on $\mathbf{B}O_{ab}$. More in particular, the f^i are the components of the transition function $\mathbf{B}\varphi_{ba} : \mathbf{B}O_{ab} \to \mathbf{B}O_{ba}$. The functions g^j_k we can interpret as components of a matrix valued function ψ_{ba} from $\mathbf{B}O_{ab}$ to real valued matrices of size $q \times q$, where q is the number of odd coordinates of the $\mathcal{A}$-manifold M. Changing slightly our viewpoint, the functions $f^i \cong \mathbf{B}\varphi_{ba}$ are the coordinate changes in the $\mathbf{R}$-manifold $\mathbf{B}M$ and the functions $g^j_k \cong \psi_{ba}$ become matrix valued functions on $\mathbf{B}U_a \cap \mathbf{B}U_b \subset \mathbf{B}M$. But we have more: since the ψ_{ba} are derived from the original φ_a, they satisfy the condition

$$m \in \mathbf{B}U_a \cap \mathbf{B}U_b \cap \mathbf{B}U_c \quad \Longrightarrow \quad \psi_{cb}(m) \cdot \psi_{ba}(m) = \psi_{ca}(m) .$$

In view of [1.24] this means that we have constructed a vector bundle $\pi : \mathbf{V}M \to \mathbf{B}M$ with typical fiber $\mathbf{R}^q \cong \mathbf{R}^q \oplus \{0\}$, all within the category of $\mathbf{R}$-manifolds.

On the other hand, let $\pi : \mathbf{V}M \to \mathbf{B}M$ be a vector bundle over $\mathbf{B}M$ (of dimension p) with typical fiber $\mathbf{R}^q \cong \mathbf{R}^q \oplus \{0\}$. Such a vector bundle can be described by coordinate changes $\mathbf{B}\varphi_{ba}$ for $\mathbf{B}M$ and transition functions ψ_{ba}. If E is an $\mathcal{A}$-vector space of dimension $p|q$, we can define $U_a = \mathbf{B}^{-1}(\mathbf{B}U_a) \subset E_0$, and then the pairs $(\mathbf{B}\varphi_{ba}, \psi_{ba})$ define coordinate changes for an $\mathcal{A}$-manifold M modeled on E, coordinate changes that already have the property given in Batchelor's theorem. Obviously this construction is the "inverse" of the construction that associates to an $\mathcal{A}$-manifold M a vector bundle $\pi : \mathbf{V}M \to \mathbf{B}M$ in the category of $\mathbf{R}$-manifolds.

But there is more to this correspondence. Associated to M we have the $\mathbf{R}$-algebra $C^\infty(M)$ of smooth $\mathcal{A}$-valued functions on M. Given the vector bundle $\pi : \mathbf{V}M \to \mathbf{B}M$, we can construct the exterior algebra bundle $\bigwedge \mathbf{V}M \to \mathbf{B}M$, the sections of which also form an $\mathbf{R}$-algebra $\Gamma(\bigwedge \mathbf{V}M)$. We claim that these two $\mathbf{R}$-algebras are isomorphic. If f is a smooth function on M, then for any chart $\varphi_a : U_a \to O_a$ we obtain a smooth function $f \circ \varphi_a^{-1} : O_a \to \mathcal{A}$ which can be written as

$$(f \circ \varphi_a^{-1})(x, \xi) = \sum_{k=0}^{q} \sum_{1 \le i_1 < \cdots < i_k \le q} \xi^{i_1} \cdots \xi^{i_k} \cdot f_{i_1 \ldots i_k}(x) .$$

On the other hand, any section $s : \mathbf{B}M \to \bigwedge \mathbf{V}M$ is represented locally by a function $s_a : \mathbf{B}O_a \to \bigwedge \mathbf{R}^q$. If $(v_i)_{i=1}^q$ is a basis of $\mathbf{R}^q$, such a function can be written as

$$s_a(x) = \sum_{k=0}^{q} \sum_{1 \le i_1 < \cdots < i_k \le q} v_{i_1} \wedge \cdots \wedge v_{i_k} \cdot s_{i_1 \ldots i_k}(x) .$$

We thus can identify the local function $f \circ \varphi_a$ with the local section s_a. Using an atlas *as in Batchelor's theorem [8.2]*, it is elementary to show that these local identifications glue together to a global identification.

We can summarize this discussion by saying that any $\mathcal{A}$-manifold can be described as a vector bundle over an ordinary $\mathbf{R}$-manifold, and that smooth functions on the $\mathcal{A}$-manifold are just the sections of the exterior power of the ordinary vector bundle. In other words, $\mathcal{A}$-manifolds can be studied entirely in the category of $\mathbf{R}$-manifolds.

8.9 Discussion. In [8.8] we have been a little vague about the kind of correspondence we have constructed between $\mathcal{A}$-manifolds and (ordinary) vector bundles. In the first place, the construction of the vector bundle $\mathbf{V}M \to \mathbf{B}M$ depends upon the choice of the atlas having the properties of Batchelor's theorem [8.2]. It is elementary to show that another atlas will give us another vector bundle, but one that is isomorphic to the first. An intrinsic map thus should associate to an $\mathcal{A}$-manifold an isomorphism class of vector bundles over $\mathbf{B}M$. Going the other way, the $\mathcal{A}$-manifold associated to a vector bundle depends upon the actual vector bundle, but its diffeomorphism class is independent. We thus can conclude that we have a bijection between diffeomorphism classes of vector bundles over $\mathbf{R}$-manifolds and diffeomorphism classes of $\mathcal{A}$-manifolds.

The natural reflex is to say that this bijection is functorial. However, there is a "small" problem: a functor not only maps objects to objects, but also morphisms to morphisms. And as soon as the odd dimension is greater than 1, there seems to be no natural way to associate a morphism of vector bundles to a morphism of $\mathcal{A}$-manifolds.

Chapter V

The tangent space

In this chapter we introduce the standard machinery of differential geometry of $\mathcal{A}$-manifolds. Differential geometry can be seen as the interplay between analysis and geometry, or as the passage from local to global. The first step in this interplay is the definition of the tangent bundle of an $\mathcal{A}$-manifold and the interpretation that the module of sections of the tangent bundle is isomorphic to the module of left derivations of the ring of smooth functions on the $\mathcal{A}$-manifold. The next step is the generalization of the derivative of a function, which leads to the definition of the tangent map, defined on the tangent bundle. Yet another passage from local to global is the definition of the flow of a vector field, which generalizes the (local) existence of solutions of differential equations. Another globalization is Frobenius' theorem which discusses the integrability of a distribution (roughly speaking a system of vector fields) and the existence of leaves (maximal integral manifolds). Dual to these results concerning the tangent bundle is the theory of differential forms, i.e., sections of (the k-th exterior power of the dual of) the tangent bundle. Here one finds the exterior derivative, the Lie derivative, and the de Rham cohomology. However, we make no attempt to integrate differential forms. Using the flow of a vector field, we also have a kind of reverse passage from global to local. If we let the flow of a vector field X act on a vector field Y, we can derive with respect to the flow parameter and we obtain the commutator $[X, Y]$. If we let the flow act on a differential form ω, then the derivative with respect to the flow parameter yields the Lie derivative $\mathcal{L}(X)\omega$. With a few exceptions, all results as well as their proofs are standard. The first exception concerns the integration of vector fields. Only homogeneous vector fields are considered for integration, and we show that only those having auto-commutator zero admit a flow. The second exception concerns Frobenius' theorem. We show (as expected) that a subbundle of the tangent bundle is involutive (a foliation) if and only if it is locally generated by some of the ∂_i. Due to our approach to smooth functions (constant functions must take real values), this does not imply that through every point passes an integral manifold (leaf of the foliation). The existence of leaves is only guaranteed through points that have real coordinates.

1. Derivations and the tangent bundle

In this section we define the tangent bundle TM of an $\mathcal{A}$-manifold M directly in terms of transition functions with respect to an atlas. We then show that the smooth sections of TM (vector fields) correspond to derivations of the ring of smooth functions $C^\infty(M)$. This correspondence is used to define the commutator of two vector fields.

1.1 Definition. Let M be an $\mathcal{A}$-manifold modeled on an $\mathcal{A}$-vector space E and let $\mathcal{U} = \{\varphi_a : U_a \to O_a \mid a \in I\}$ be the atlas of *all* charts for M (i.e., the differentiable structure). Using notation as in [III.4.9] we have the *even* functions $\varphi_{ba} : O_{ab} \to O_{ba}$, and associated to them their Jacobians $Jac(\varphi_{ba}) : O_{ab} \to \mathrm{Aut}(E)$. And recall that in terms of a basis (e_i) of E the Jacobian is given as $Jac(\varphi_{ba})(x) = \sum_{i,j} {}^i e \otimes \left(\partial_i \varphi_{ba}^j(x)\right) e_j$.

We thus can define the functions $\psi_{ba} : U_a \cap U_b \to \mathrm{Aut}(E)$ by $\psi_{ba} = Jac(\varphi_{ba}) \circ \varphi_a$. It follows from the chain rule that these functions satisfy the conditions of [IV.1.24]. The vector bundle so obtained is called *the tangent bundle* $\pi : TM \to M$; its typical fiber is E.

For a point $m \in M$ the fiber of TM above m is usually denoted by $T_m M$, instead of by $(TM)_m$. Moreover, if $X \in \Gamma_U(TM)$ is a local section of TM, the value $X(m)$ is also denoted by X_m or by $X|_m$.

1.2 Discussion. It is obvious that we do not really need the whole differentiable structure of M to define the tangent bundle TM, an atlas is sufficient. The reason to use the whole differentiable structure is to have a definition which is manifestly independent of the choice of an atlas.

Since an atlas is sufficient, it follows immediately that if $O \subset E_0$ is open, its tangent bundle is just the trivial bundle $TO = O \times E$ with projection $p = \pi_O : O \times E \to O$. More generally, if $U \subset M$ is an open subset of an $\mathcal{A}$-manifold, the induced smooth structure on U consists of all intersections $U_a \cap U$ for charts U_a of M. It follows that the tangent bundle TU of U is (isomorphic to) the restriction of the vector bundle TM to the open set $U \subset M$, and thus $\Gamma_U(TM) = \Gamma(TU)$ (see also [IV.1.13] and [IV.1.19]).

1.3 Discussion. Let M be an $\mathcal{A}$-manifold and consider the graded commutative $\mathbf{R}$-algebra $C^\infty(M)$ and the $\mathbf{R}$-vector space of its right linear derivations $\mathrm{Der}_R(C^\infty(M))$. We can turn $\mathrm{Der}_R(C^\infty(M))$ into a graded $C^\infty(M)$-module by the following procedure. For $f \in C^\infty(M)$ and $D \in \mathrm{Der}_R(C^\infty(M))$ we define $f \cdot D$ by

$$\forall g \in C^\infty(M) : (f \cdot D)(g) = f \cdot (D(g)) \ .$$

We leave it to the reader to verify that this indeed turns $\mathrm{Der}_R(C^\infty(M))$ into a graded $C^\infty(M)$-module. What we intend to do is to show that there is a natural identification between the $C^\infty(M)$-modules $\mathrm{Der}_R(C^\infty(M))$ and $\Gamma(TM)$.

1.4 Proposition. *Let O be open in M, $f, g \in C^\infty(O)$, and $D \in \mathrm{Der}_R(O)$. If U is open in O and $f|_U = g|_U$, then $(Df)|_U = (Dg)|_U$.*

Proof. Without loss of generality we may assume that D is homogeneous and that f and g are homogeneous of the same parity. For any $m \in U$, let ρ be a plateau function around m in U [III.5.21]. It follows that $\rho \cdot (f - g) = 0$. Since D is a derivation, we obtain

$$\begin{aligned} 0 &= D(\rho \cdot (f - g))(m) \\ &= (D\rho)(m) \cdot (f(m) - g(m)) + (-1)^{\langle \varepsilon(D)|\varepsilon(\rho)\rangle} \rho(m) \cdot ((Df)(m) - (Dg)(m)) \ . \end{aligned}$$

Since $f(m) = g(m)$ and $\rho(m) = 1$ we deduce that $(Df)(m) = (Dg)(m)$. QED

1.5 Proposition. *Let $U \subset O \subset M$ be open and $D \in \mathrm{Der}_R(C^\infty(O))$. Then there exists a unique $D_U \in \mathrm{Der}_R(C^\infty(U))$ such that $\forall f \in C^\infty(O) : (Df)|_U = D_U(f|_U)$.*

Proof. The main problem of this proposition is that not every smooth function on U need be the restriction of a smooth function on O. So let $g \in C^\infty(U)$ and $m \in U$ be arbitrary and let ρ be a plateau function around m in U [III.5.21]. It follows that ρg is a well defined smooth function on O. We now define $D_U g$ by

$$(D_U g)(m) = D(\rho g)(m) \ . \tag{1.6}$$

To see that this is well defined, independent of the choice for ρ, suppose ρ' has the same properties as ρ. It follows that ρg and $\rho' g$ coincide in a neighborhood of m, and thus by [1.4] $D(\rho g)(m) = D(\rho' g)(m)$. By the same argument we have that $(Df)|_U = D_U(f|_U)$ for any $f \in C^\infty(O)$.

Uniqueness of D_U follows in the same way. Suppose D_U exists. The functions g and ρg coincide in a neighborhood of m, and thus by [1.4] $D_U(\rho g)(m) = D_U(g)(m)$. But then by $D_U(f|_U) = (Df)|_U$ we have $D_U(\rho g)(m) = D(\rho g)(m)$, and thus D_U must be defined by (1.6).

Finally to prove that D_U is a derivation, suppose that D and $f, g \in C^\infty(U)$ are homogeneous. The functions $fg\rho$ and $fg\rho^2$ coincide on a neighborhood of m and thus

$$\begin{aligned} D_U(fg)(m) &= D(fg\rho)(m) = D((f\rho)(g\rho))(m) \\ &= D(f\rho)(m) \cdot (g\rho)(m) + (-1)^{\langle \varepsilon(D)|\varepsilon(f)\rangle} (f\rho)(m) \cdot D(g\rho)(m) \\ &= (D_U f)(m) \cdot g(m) + (-1)^{\langle \varepsilon(D)|\varepsilon(f)\rangle} f(m) \cdot (D_U g)(m) \ . \end{aligned}$$

QED

1.7 Corollary. *Let $V \subset U \subset O \subset M$ be open and $D \in \mathrm{Der}_R(C^\infty(O))$. Then $D_V = (D_U)_V$.*

1.8 Proposition. *Let E be an $\mathcal{A}$-vector space of dimension $p|q$ and let $O \subset E_0$ be open. If $\mathcal{A}$ satisfies* (C [q+1]) *(or if $q = 0$), then $\mathrm{Der}_R(C^\infty(O))$ is a free graded $C^\infty(O)$-module generated by the ∂_i, $1 \le i \le p+q$. If $\mathcal{A}_1 = \{0\}$, then $\mathrm{Der}_R(C^\infty(O))$ is a free graded $C^\infty(O)$-module generated by the ∂_i, $1 \le i \le p$.*

Proof. Let us first assume that $\mathcal{A}$ satisfies (C [q+1]), then we know from [III.3.3] and [III.3.6] that all ∂_i, $1 \le i \le p+q$ exist and are derivations.

For the independence of the (∂_i), suppose that $D = \sum_i D^i \cdot \partial_i = 0 \in \mathrm{Der}_R(C^\infty(O))$ with $D^i \in C^\infty(O)$. Denoting by (x^i) the even and odd coordinate functions associated to a basis (e_i) of E, we obtain $0 = Dx^i = D^i$, i.e., the (∂_i) are indeed independent.

To prove that the (∂_i) generate $\mathrm{Der}_R(C^\infty(O))$, let $D \in \mathrm{Der}_R(C^\infty(O))$ be arbitrary, for which we define functions $D^i \in C^\infty(O)$ by $D^i = Dx^i$. We claim that $D = \sum_{i=1}^{p+q} D^i \cdot \partial_i$. To prove the equality we change notation and denote by x the even coordinates of a point and by ξ the odd ones. Let $f = \sum \xi^{i_1} \cdots \xi^{i_k} \cdot f_{i_1 \ldots i_k}(x)$ be an arbitrary smooth function on O. Since D and $\sum_{i=1}^{p+q} D^i \cdot \partial_i$ are both derivations, it is immediate that $D(\xi^{i_1} \cdots \xi^{i_k}) = (\sum_{i=1}^{p+q} D^i \cdot \partial_i)(\xi^{i_1} \cdots \xi^{i_k})$. To show the equality, it thus suffices to show that both derivations act the same on functions of even coordinates only.

To do so, note first that the identity $1 \cdot 1 = 1$ and the fact that a derivation is $\mathbf{R}$-linear imply that $Df = 0$ whenever f is a constant function (with a real value!). Now let f depend upon even coordinates only and let $x_0 \in \mathbf{B}O$ be arbitrary, then (by condition (A1) in [III.1.20]) there exist an open $U \subset O$ and local functions ϕ_i depending upon even coordinates only such that $\forall x \in U : f(x) = f(x_o) + \sum_{i=1}^{p} (x^i - x_o^i) \cdot \phi_i(x, x_o)$. Since x_o has real coordinates, $f(x_o)$ is a real constant and $\phi_i(x, x_o)$ is a smooth function of x on U. Thus, using [1.5], we find

$$
\begin{aligned}
(Df)(x_o, \xi) &= (D_U f)(x_o, \xi) \\
&= \Big(D_U \big(\sum_{i=1}^{p} (x^i - x_o^i) \cdot \phi_i(x, x_o) \big) \Big)(x_o, \xi) \\
&= \sum_{i=1}^{p} (D_U x^i)(x_o, \xi) \cdot \phi_i(x_o, x_o) + \sum_{i=1}^{p} (x_o^i - x_o^i) \cdot (D_U \phi_i)(x_o, \xi) \\
&= \sum_{i=1}^{p} D^i(x_o, \xi) \cdot (\partial_i f)(x_o, \xi) = \Big(\sum_{i=1}^{p+q} D^i \cdot \partial_i f \Big)(x_o, \xi) ,
\end{aligned}
$$

where the last equality follows from the fact that f does not depend upon the odd coordinates, i.e., $\partial_i f = 0$ whenever $i > p$. Since a smooth function is completely determined by its values in points with real even coordinates and arbitrary odd coordinates, we have shown that $Df = \sum_{i=1}^{p+q} D^i \cdot \partial_i f$ also holds for functions of even coordinates only.

If we now assume that $\mathcal{A}_1 = \{0\}$, then again according to [III.3.3] and [III.3.6] the ∂_i, $1 \le i \le p$ exist and are derivations. Their independence is proven as for the previous case. That they are generating is also proved in the previous case, once one realizes that, when $\mathcal{A}_1 = \{0\}$, any smooth function on O automatically is a function of the even coordinates only. QED

1.9 Remark. The fact that we obtain $D = \sum_i D^i \cdot \partial_i$ in that order is a consequence of the fact that we write $x - x_o$ to the *left* of $\phi(x_o, x)$ in the definition of smooth functions and the fact that we use *right* derivations.

1.10 Discussion. With [1.8] we have proved the difference between the generic case in which $\mathcal{A}$ satisfies (C [q+1]) (with q the odd dimension) and the special case $\mathcal{A} = \mathbf{R}$ (or more generally $\mathcal{A}_1 = \{0\}$). In the first case $\mathrm{Der}_R(C^\infty(O))$ has dimension $p|q$, whereas in the second case it has dimension $p \cong p|0$. This confirms the idea that in the real case $\mathcal{A} = \mathbf{R}$ we do not have partial derivatives in the odd directions. Of course there is a kind of no man's land in between where $\mathcal{A}_1 \neq \{0\}$ but where $\mathcal{A}$ does not satisfy (C [q+1]). In order to avoid that we enter this no man's land (by using too large odd dimensions), we required $\mathcal{A}$ to satisfy (C [∞]).

1.11 Remark. An alternative approach to the tangent bundle could be to use [1.8] as a kind of definition in the following sense. For $O \subset E_0$ an open subset, its tangent space TO is defined as $TO = O \times F$ with F an $\mathcal{A}$-vector space of the same graded dimension as $\mathrm{Der}_R(C^\infty(O))$. This requires that $\mathrm{Der}_R(C^\infty(O))$ is a free graded $C^\infty(O)$-module, which in turn puts a condition on $\mathcal{A}$. If $\mathcal{A} = \mathbf{R}$ or if $\mathcal{A}$ satisfies (C [∞]) it follows from [1.8] that this is the case for all possible dimensions of E. We could go one step further and define the *differential dimension* of O as the (graded) dimension of $\mathrm{Der}_R(C^\infty(O))$, provided of course that $\mathrm{Der}_R(C^\infty(O))$ is a free graded $C^\infty(O)$-module. In this way the even part E_0 of an $\mathbf{R}$-vector space E of dimension $p|q$ would have differential dimension $p|0$, in accordance with the fact that E_0 is completely determined by p even (real) coordinates.

1.12 Discussion. Our next purpose is to construct an isomorphism between $\Gamma(TM)$ and $\mathrm{Der}_R(C^\infty(M))$. Let $\{\varphi_a : U_a \to O_a \mid a \in I\}$ be an atlas and $f : U_a \to \mathcal{A}$ arbitrary. Since by definition f is smooth if and only if $f \circ \varphi_a$ is smooth on O_a, we deduce from [1.8] that $\mathrm{Der}_R(C^\infty(U_a))$ is a free graded $C^\infty(U_a)$-module generated by the derivations ∂_{ai} defined by

$$\partial_{ai} f = (\partial_i (f \circ \varphi_a^{-1})) \circ \varphi_a \ . \tag{1.13}$$

Defining the (local) coordinate functions x_a^i on U_a by $x_a^i = x^i \circ \varphi_a \equiv {}^i e \circ \varphi_a$, it follows immediately that $\partial_{ai}(x_a^j) = \delta_i^j$. This shows that the derivations ∂_{ai} can be interpreted as the partial derivatives $\partial_{x_a^i}$ with respect to the coordinates x_a^i on U_a.

To construct the isomorphism, let $D \in \mathrm{Der}_R(C^\infty(M))$ be arbitrary. On each chart we have $D_a \equiv D_{U_a} = \sum_i D_a^i \cdot \partial_{ai}$ with $D_a^i \in C^\infty(U_a)$ defined by $D_a^i = D_a(x_a^i)$. We thus can construct a family of functions $s_a : U_a \to E$ by

$$s_a(m) = \sum_i D_a^i(m) \cdot e_i \ .$$

We now want to apply the result of [IV.1.20] to deduce that this family defines a global section $s_D \in \Gamma(TM)$. Therefore we have to prove that $\langle\!\langle s_a(m) \| \psi_{ba}(m) \rangle\!\rangle = s_b(m)$ for $m \in U_a \cap U_b$ Although $\psi_{ba}(m)$ is even, we use the left linear character because the Jacobian is left linear in general. Substituting the explicit expressions, this amounts to proving that $\sum_i D_a^i(m) \cdot (\partial_i \varphi_{ba}^j)(\varphi_a(m)) = D_b^j(m)$ (use that the Jacobian is given by $Jac(\varphi_{ba})(x) = \sum_{i,j} {}^i e \otimes \big(\partial_i \varphi_{ba}^j(x)\big)\, e_j$ [III.3.8]). We thus compute on $U_a \cap U_b$:

$$\begin{aligned} D_b^j(m) &= D_b(x^j \circ \varphi_b)(m) = D_a(x^j \circ \varphi_b)(m) \\ &= \sum\nolimits_i (D_a x_a^i)(m) \cdot \partial_{ai}(x^j \circ \varphi_b)(m) \\ &= \sum\nolimits_i D_a^i(m) \cdot \partial_i(x^j \circ \varphi_b \circ \varphi_a^{-1})(\varphi_a(m)) \\ &= \sum\nolimits_i D_a^i(m) \cdot (\partial_i \varphi_{ba}^j)(\varphi_a(m)) \; . \end{aligned}$$

In the first line we have used that on $U_a \cap U_b$ the derivations D_a and D_b both coincide with $D_{U_a \cap U_b}$; to obtain the second line we applied $D_a = \sum_i D_a^i \cdot \partial_{ai}$. Since this is exactly what we needed to prove, we conclude that we have defined a map $D \mapsto s_D$, $\mathrm{Der}_R(C^\infty(M)) \to \Gamma(TM)$. Since the parity of the ∂_{ai} is the same as of the e_i, it should be rather obvious from the construction that this map is an even morphism of $C^\infty(M)$-modules. It thus remains to show that it admits an inverse.

If $s \in \Gamma(TM)$ is a global section, it defines, according to [IV.1.20], a family of maps $s_a : U_a \to E$, $s_a(m) = \sum_i s_a^i(m) \cdot e_i$ satisfying $\langle\!\langle s_a(m) \| \psi_{ba}(m) \rangle\!\rangle = s_b(m)$, i.e., $\sum_i s_a^i(m) \cdot (\partial_i \varphi_{ba}^j)(\varphi_a(m)) = s_b^j(m)$, on intersections $U_a \cap U_b$. With these, we define a derivation $D_s \in \mathrm{Der}_R(C^\infty(M))$ by the formula

$$(1.14) \qquad (D_s f)|_{U_a} = \sum\nolimits_i s_a^i \cdot \partial_{ai}(f|_{U_a}) \; .$$

Since the ∂_{ai} belong to $\mathrm{Der}_R(C^\infty(U_a))$, the derivation property of D_s follows immediately. It only remains to be shown that $D_s f$ is a globally well-defined smooth function, i.e., we have to show that these local expressions coincide on $U_a \cap U_b$. We therefore compute

$$\begin{aligned} (\sum\nolimits_i s_a^i \cdot \partial_{ai} f)(m) &= \sum\nolimits_i s_a^i(m) \cdot \partial_i(f \circ \varphi_a^{-1})(\varphi_a(m)) \\ &= \sum\nolimits_i s_a^i(m) \cdot \partial_i(f \circ \varphi_b^{-1} \circ \varphi_{ba})(\varphi_a(m)) \\ &= \sum\nolimits_{i,j} s_a^i(m) \cdot (\partial_i \varphi_{ba}^j)(\varphi_a(m)) \cdot \partial_j(f \circ \varphi_b^{-1})(\varphi_{ba}(\varphi_a(m))) \\ &= \sum\nolimits_j s_b^j(m) \cdot \partial_j(f \circ \varphi_b^{-1})(\varphi_b(m)) = (\sum\nolimits_j s_b^j \cdot \partial_{bj} f)(m) \; . \end{aligned}$$

To obtain the third line we have applied the chain rule (beware the order!). This calculation shows that D_s is a well defined derivation. Since it is rather obvious that $s \mapsto D_s$ is inverse to $D \mapsto s_D$, we have proved [1.15].

1.15 Proposition. *The map $D \mapsto s_D$ from $\mathrm{Der}_R(C^\infty(M))$ to $\Gamma(TM)$ and the map $s \mapsto D_s$ in the other direction are isomorphisms of graded $C^\infty(M)$-modules, inverse to each other.*

1.16 Discussion/Definition. Sections of TM are usually called *vector fields on* M; we thus have proved that there is an isomorphism between *smooth* vector fields on M (elements of $\Gamma(TM)$) and derivations of $C^\infty(M)$. Given that we can interpret smooth vector fields as derivations, it is customary to introduce some notation to reflect that fact. Let $\psi_a : \pi^{-1}(U_a) \to U_a \times E$ denote the formal local trivialization of $\pi^{-1}(U_a) \subset TM$. Then we define the elements $\partial_{ai}|_m \equiv \partial_{x_a^i}|_m \in T_mM$ by

$$\partial_{ai}|_m = \frac{\partial}{\partial x_a^i}|_m = \psi_a^{-1}(m, e_i) \ .$$

Since the free graded $\mathcal{A}$-module structure on T_mM is defined by means of ψ_a, it follows that the vectors $\partial_{ai}|_m$ form a basis of T_mM. It also follows that the coefficients of a vector $X \in T_mM$ with respect to the basis $\partial_{ai}|_m$ are exactly the same as the coefficients of the vector $\psi_a(X)$ with respect to the basis e_i of E. We thus will often confuse the vectors $\partial_{ai}|_m$ in T_mM with the vectors e_i in the corresponding trivialization. As a particular application, it follows that for any local section $X : U_a \to TM|_{U_a}$, not necessarily a smooth one, there exist functions $X_a^i : U_a \to \mathcal{A}$ such that $X(m) = \sum_i X_a^i(m) \cdot \partial_{ai}|_m$.

For the moment the $\frac{\partial}{\partial x_a^i}|_m$ (or $\partial_{ai}|_m$ or $\partial_{x_a^i}|_m$) are just formal symbols, but when a vector field is seen as a derivation, they are exactly the partial derivations ∂_{ai} with respect to the (local) coordinates $x_a^i \equiv {}^ie \circ \varphi_a$ defined in (1.13), as follows immediately from (1.14).

Now let (U_a, ψ_a) and (U_b, ψ_b) be two trivializing charts, then there exist transition functions $\psi_b \circ \psi_a^{-1} : U_a \times E \to U_b \times E$, $(m, v) \mapsto (m, \langle\!\langle v \| \mathit{Jac}(\varphi_{ba})(\varphi_a(m)) \rangle\!\rangle)$. In particular

$$\begin{aligned} \langle\!\langle e_i \| \mathit{Jac}(\varphi_{ba})(\varphi_a(m)) \rangle\!\rangle &= \sum\nolimits_j (\partial_i \varphi_{ba}^j)(\varphi_a(m))\, e_j \qquad\qquad \text{(see [III.3.11])} \\ &= \sum\nolimits_j \partial_i({}^je \circ \varphi_b \circ \varphi_a^{-1})(\varphi_a(m))\, e_j = \sum\nolimits_j (\partial_{ai} x_b^j)(m) \cdot e_j \ . \end{aligned}$$

It follows that for the abstract vectors $\partial_{ai}|_m, \partial_{bj}|_m \in T_mM$ we find the relation

$$\frac{\partial}{\partial x_a^i}|_m = \sum_j \frac{\partial x_b^j}{\partial x_a^i}(m) \cdot \frac{\partial}{\partial x_b^j}|_m \ .$$

1.17 Remark. In the literature one can also find another approach to define the tangent space which encodes in a certain way the derivation property directly. The idea is to consider, for a fixed $m \in M$, maps $\delta_m : C^\infty(M) \to \mathcal{A}$ satisfying the derivation property $\delta_m(f \cdot g) = (\delta_m(f)) \cdot g(m) + (-1)^{\langle \varepsilon(\delta_m) | \varepsilon(f) \rangle} f(m) \cdot (\delta_m(g))$. As in [1.4] one can show that if f and g coincide in a neighborhood U of m, then $\delta_m f = \delta_m g$. For a fixed $m \in M$ the set of all such δ_m is a graded $\mathcal{A}$-module which is taken as the definition of the tangent space at $m \in M$. One then has to show that it indeed defines a vector bundle.

In our approach, each $X_m \in T_mM$ defines a map δ_m on smooth functions defined in a neighborhood U of m by $f \mapsto \sum_i X^i\, (\partial_{x_a^i} f)(m)$, where we used a local coordinate system (x_a^i) around m in which the tangent vector takes the form $X_m = \sum_i X^i \partial_{ai}|_m$ (1.14). This satisfies the condition for a map δ_m because the partial derivative operators ∂_{ai} are derivations [III.3.6].

1.18 Discussion. In [III.1.24] and [III.4.21] we have seen that $C^\infty(M)$ is a graded $\mathbf{R}$-algebra. From §I.6 we then learn that $\mathrm{Der}_R(C^\infty(M))$ is a graded $\mathbf{R}$-Lie algebra with even bracket $[X,Y] = X \circ Y - (-1)^{\langle \varepsilon(X) | \varepsilon(Y) \rangle} \cdot Y \circ X$. By the isomorphism between $\mathrm{Der}_R(C^\infty(M))$ and $\Gamma(TM)$ we can transport this bracket to vector fields. Although $\mathrm{Der}_R(C^\infty(M))$ is also a graded $C^\infty(M)$-module [1.3], it is not a graded $C^\infty(M)$-Lie algebra. The relation between the module structure and the Lie algebra structure is expressed in [1.19]

1.19 Lemma. *For all homogeneous $f \in C^\infty(M)$, $X, Y \in \mathrm{Der}_R(C^\infty(M))$ we have*

$$[X, f \cdot Y] = (Xf) \cdot Y + (-1)^{\langle \varepsilon(X) | \varepsilon(f) \rangle} f \cdot [X,Y] .$$

Proof. A derivation is determined by its action on functions g, so we compute

$$\begin{aligned} [X, f \cdot Y](g) &= X((fY)(g)) - (-1)^{\langle \varepsilon(X) | \varepsilon(fY) \rangle} (fY) \cdot Xg) \\ &= X(f \cdot Yg) - (-1)^{\langle \varepsilon(X) | \varepsilon(f) \rangle + \langle \varepsilon(X) | \varepsilon(Y) \rangle} f \cdot Y(Xg) \\ &= (Xf) \cdot (Yg) + (-1)^{\langle \varepsilon(X) | \varepsilon(f) \rangle} f \cdot X(Yg) \\ &\qquad - (-1)^{\langle \varepsilon(X) | \varepsilon(f) \rangle + \langle \varepsilon(X) | \varepsilon(Y) \rangle} f \cdot Y(Xg) \\ &= ((Xf) \cdot Y + (-1)^{\langle \varepsilon(X) | \varepsilon(f) \rangle} [X,Y])(g) . \end{aligned}$$

QED

1.20 Lemma. *Let X and Y be vector fields on M, and let $U \subset M$ be open. Then $X|_U \in \mathrm{Der}_R(C^\infty(U))$ and $[X,Y]|_U = [X|_U, Y|_U]$.*

Proof. Since TU is the restriction of TM to U, we have $X|_U \in \Gamma_U(TM) \cong \Gamma(TU) \cong \mathrm{Der}_R(C^\infty(U))$. Uniqueness in [1.5] proves the second part. QED

1.21 Lemma. *If $X|_{U_a} = \sum_i X_a^i \cdot \partial_{ai}$ and $Y|_{U_a} = \sum_i Y_a^i \cdot \partial_{ai}$ are the local expressions of homogeneous vector fields X and Y, the local expression of $[X,Y]$ is given by*

$$[X,Y]|_{U_a} = \sum_{i,j} \Big(X_a^i \cdot (\partial_{ai} Y_a^j) - (-1)^{\langle \varepsilon(X) | \varepsilon(Y) \rangle} \cdot Y_a^i \cdot (\partial_{ai} X_a^j) \Big) \cdot \partial_{aj} . \tag{1.22}$$

Proof. Since on U_a we have $[X,Y]|_{U_a} = \sum_j ([X,Y]|_{U_a} x_a^j) \cdot \partial_{aj}$, we compute

$$\begin{aligned} [X|_{U_a}, Y|_{U_a}] x_a^j &= X|_{U_a}(Y|_{U_a} x_a^j) - (-1)^{\langle \varepsilon(X) | \varepsilon(Y) \rangle} \cdot Y|_{U_a}(X|_{U_a} x_a^j) \\ &= X|_{U_a}(Y_a^j) - (-1)^{\langle \varepsilon(X) | \varepsilon(Y) \rangle} \cdot Y|_{U_a}(X_a^j) \\ &= \sum_i \Big(X_a^i \cdot (\partial_{ai} Y_a^j) - (-1)^{\langle \varepsilon(X) | \varepsilon(Y) \rangle} \cdot Y_a^i \cdot (\partial_{ai} X_a^j) \Big) . \end{aligned}$$

QED

1.23 Nota Bene. If X is an even vector field, obviously $[X, X] = 0$, but the operation $f \mapsto X(X(f))$ is not a derivation of $C^\infty(M)$ (unless $X = 0$). On the other hand, if X is an odd vector field, we have (globally) $[X, X]f = 2X(Xf)$, and the local expression for $[X, X]$ gives us:

$$[X, X]|_{U_a} = 2\sum_{i,j}\Big(X_a^i \cdot (\partial_{ai} X_a^j)\Big) \cdot \partial_{aj} \ .$$

This is not necessarily zero, but it shows that the operation $f \mapsto X(X(f))$ *is* a derivation of $C^\infty(M)$ (see also [I.6.8]).

As an example in which $[X, X]$ is not zero, consider an $\mathcal{A}$-vector space E of dimension $0|2$, and use (odd) coordinates (ξ^1, ξ^2) on the $\mathcal{A}$-manifold $M = E_0$. The odd vector field $X(\xi^1, \xi^2) = \partial_{\xi^1} + \xi^1\xi^2 \cdot \partial_{\xi^2}$ on M has auto-commutator $[X, X] = 2\xi^2 \cdot \partial_{\xi^2}$, which is manifestly not the zero vector field.

1.24 Definition. Let M be an $\mathcal{A}$-manifold and $X \in \Gamma(TM) \cong \mathrm{Der}_R(C^\infty(M))$ a smooth vector field on M, i.e., X acts on smooth functions $f : M \to \mathcal{A}$. If F is an $\mathcal{A}$-vector space, we can extend this action to functions $f : M \to F$ in the following way. Let $(\epsilon_j)_{j=1}^n$ be a homogeneous basis for F, then associated to a function $f : M \to F$ we have its component functions $f^j = f \diamond {}^j\epsilon : M \to \mathcal{A}$, i.e., $f(m) = \sum_j f^j(m)\, \epsilon_j$. Moreover, f is smooth if and only if all f^j are smooth. We now define the function $Xf : M \to F$ as $(Xf)(m) = \sum_j (Xf^j)(m)\, \epsilon_j$. This is a straightforward generalization of (III.3.9). Moreover, it is independent of the choice of the basis (in its equivalence class) because the vector field X acts as the zero operator on (real) constants (the matrix elements of a basis change).

2. The tangent map and some standard applications

In this section we define the tangent map $Tf : TM \to TN$ associated to a smooth map $f : M \to N$. Once we have the tangent map, we can define immersions and embeddings and their relation to submanifolds, as well as some of their elementary properties. We also show the behavior of the tangent map with respect to the product of $\mathcal{A}$-manifolds and smooth maps between them, which provides some useful identifications. At the end of this section we define the notion of vector fields related by a smooth map and we show that this relation preserves brackets.

2.1 Construction. Let M be an $\mathcal{A}$-manifold modeled on E, N an $\mathcal{A}$-manifold modeled on F and let $f : M \to N$ be a smooth map. As before (especially in [III.3.8]) we denote by (e_i) a basis of E and by (ϵ_j) a basis of F. Our purpose is to construct a smooth even (left linear) vector bundle morphism $Tf : TM \to TN$ which induces f. Let $\mathcal{V} = \{\,\chi_\alpha : V_\alpha \to P_\alpha\,\}$ be an atlas for N and $\mathcal{U} = \{\,\varphi_a : U_a \to O_a\,\}$ an atlas for M such that for each $U_a \in \mathcal{U}$ there exists a $V_\alpha \in \mathcal{V}$ such that $f(U_a) \subset V_\alpha$. Using [IV.4.14] we define the map Tf by the set of maps $h_a : U_a \to \mathrm{Hom}_L(E; F)$,

$$h_a = Jac(\chi_\alpha \circ f \circ \varphi_a^{-1}) \circ \varphi_a \ .$$

That this set satisfies the conditions (IV.4.15) is a direct consequence of the chain rule. And thus Tf is a well defined left linear (even) vector bundle map, called the *tangent map*. In terms of charts $O_a \times E$ and $P_\alpha \times F$ for M and N respectively, the tangent map takes the form

$$\begin{aligned} Tf : O_a \times E &\to P_{f(a)} \times F \\ (x, v) &\mapsto \big((\chi_\alpha \circ f \circ \varphi_a^{-1})(x)\ ,\ \langle\!\langle v \,\|\, Jac(\chi_\alpha \circ f \circ \varphi_a^{-1})(x)\rangle\!\rangle\big) \ . \end{aligned}$$

Now that we know how Tf is defined, let us determine its expression in terms of the basis $\partial_{ai}|_m$ for $T_m M$ introduced in [1.16]. Let U_a and V_α be charts around m and $f(m)$ respectively as above, then we know that a vector $X_m = \sum_i X_a^i\, \partial_{ai}|_m \in T_m M$ corresponds to the vector $v = \sum_i X_a^i\, e_i$ in the local trivialization $U_a \times E$. The image $\langle\!\langle X_m \,\|\, Tf \rangle\!\rangle$ then is represented in the trivialization $V_\alpha \times F$ by

$$\langle\!\langle v \,\|\, Jac(\chi_\alpha \circ f \circ \varphi_a^{-1})(\varphi_a(m))\rangle\!\rangle = \sum_{i,j} X_a^i \cdot \frac{\partial(\varphi_a^{-1} \diamond f \diamond \chi_\alpha \diamond {}^j\epsilon)}{\partial x^i}(\varphi_a(m)) \cdot \epsilon_j \in F \ ,$$

where we used the $\overset{\bullet}{M}_L$-matrix representation of the Jacobian as determined in [III.3.8].

If we denote by $y_\alpha^j = {}^j\epsilon \circ \chi_\alpha \equiv \chi_\alpha \diamond {}^j\epsilon$ the (even and odd) coordinates on V_α, it follows that $f_\alpha^j = {}^j\epsilon \circ \chi_\alpha \circ f \equiv f \diamond \chi_\alpha \diamond {}^j\epsilon$ are the local coordinates of f in the chart V_α, and that $\epsilon_j \in F$ corresponds to $\partial_{\alpha j}|_{f(m)} \in T_{f(m)}N$. Using (1.13), we thus find

$$\langle\!\langle \sum_i X_a^i \frac{\partial}{\partial x_a^i}|_m \,\|\, Tf \rangle\!\rangle = \sum_{i,j} X_a^i \cdot \frac{\partial f_\alpha^j}{\partial x_a^i}(m) \cdot \frac{\partial}{\partial y_\alpha^j}|_{f(m)} \ .$$

Now this is the expression in the special preselected charts U_a and V_α of the chosen atlases, so how does it look in arbitrary charts U_b around m and V_β around $f(m)$? Using the results of [1.16], we find

$$\langle\!\langle \sum_i X_b^i \frac{\partial}{\partial x_b^i}|_m \,\|\, Tf \rangle\!\rangle = \sum_{i,j,k,\ell} X_b^i \cdot \frac{\partial x_a^j}{\partial x_b^i}(m) \cdot \frac{\partial f_\alpha^k}{\partial x_a^j}(m) \cdot \frac{\partial y_\beta^\ell}{\partial y_\alpha^k}(f(m)) \cdot \frac{\partial}{\partial y_\beta^\ell} \ .$$

It looks like we can apply the chain rule to combine these three matrices into the single one $(\partial_{x_b^i} f_\beta^\ell)(m)$, except for the fact that the arguments do not really conform to the conditions of the chain rule (twice the same argument m is suspicious). So let us have a more detailed look, using the formal definitions for the derivations ∂_{ai} etcetera:

$$\begin{aligned} &\sum_{j,k} \frac{\partial x_a^j}{\partial x_b^i}(m) \cdot \frac{\partial f_\alpha^k}{\partial x_a^j}(m) \cdot \frac{\partial y_\beta^\ell}{\partial y_\alpha^k}(f(m)) = \\ &\qquad \sum_{j,k} \frac{\partial(\varphi_b^{-1} \diamond \varphi_a \diamond {}^j e)}{\partial x^i}(\varphi_b(m)) \cdot \frac{\partial(\varphi_a^{-1} \diamond f \diamond \chi_\alpha \diamond {}^k\epsilon)}{\partial x^j}(\varphi_a(m)) \\ &\qquad\qquad \cdot \frac{\partial(\chi_\alpha^{-1} \diamond \chi_\beta \diamond {}^\ell\epsilon)}{\partial y^k}(\chi_\alpha(f(m))) \ . \end{aligned}$$

Written this way, we have the official partial derivatives ∂_{x^i} and ∂_{y^j} of §III.3 and here the chain rule does make perfect sense and gives us indeed

$$\frac{\partial(\varphi_b^{-1} \diamond f \diamond \chi_\beta \diamond {}^\ell\epsilon)}{\partial x^i}(\varphi_b(m)) = \frac{\partial f_\beta^\ell}{\partial x_b^i}(m) \ .$$

We thus obtain at the same time that our intuitive feeling about the chain rule was correct, and that we have for the tangent map the general expression

$$\langle\!\langle \sum_i X_b^i \frac{\partial}{\partial x_b^i}|_m \,\|\, Tf \rangle\!\rangle = \sum_{i,j} X_b^i \cdot \frac{\partial f_\beta^j}{\partial x_b^i}(m) \cdot \frac{\partial}{\partial y_\beta^j}|_{f(m)} \ , \tag{2.2}$$

valid in any chart U_b around m and V_β around $f(m)$. Since the formal definition of the tangent map is rather awkward to use, it is (2.2) that is used in practice as if it were the definition. It also justifies to a certain extent the fact that we defined the tangent map as being left linear: the form of (2.2) is $\langle\!\langle X\,\partial \| Tf \rangle\!\rangle = X\,(\partial f)\,\partial$, i.e., we do not have to change any order.

2.3 Remark. Comparing the definition of the tangent map with the definition of the transition functions for the tangent bundle shows that the latter are exactly the tangent maps of the transition functions between the charts.

2.4 Notation. If $f : M \to N$ is a smooth map between $\mathcal{A}$-manifolds, and $m \in M$ a point, we have a left linear map $Tf|_{T_mM} : T_mM \to T_{f(m)}N$. It is customary to denote this map by T_mf, i.e., $T_mf \equiv Tf|_{T_mM}$. According to the definition of Tf, the left linear map T_mf is essentially the Jacobian $Jacf(m)$.

If X is a vector field on M, X_m belongs to T_mM. With the notations we have introduced, we thus can write $\langle\!\langle X_m \| T_mf \rangle\!\rangle = \langle\!\langle X \| T_mf \rangle\!\rangle = \langle\!\langle X_m \| Tf \rangle\!\rangle \in T_{f(m)}N$. However, the notation $\langle\!\langle X \| Tf \rangle\!\rangle$ does not indicate a point in TN, it denotes the map $M \to TN$, $m \mapsto \langle\!\langle X \| Tf \rangle\!\rangle(m) = \langle\!\langle X_m \| T_mf \rangle\!\rangle$, a map which we can also denote as $X \diamond Tf$ or as $Tf \circ X$.

2.5 Nota Bene. If $f : M \to N$ is a smooth map between $\mathcal{A}$-manifolds, then for $m \in M$ we have defined the left linear map $T_mf : T_mM \to T_{f(m)}N$. But this map is also *even* and thus *right linear as well.* Hence for any $X \in T_mM$ we may write $T_mf(X)$ for $\langle\!\langle X \| T_mf \rangle\!\rangle$. The left linear notation has the advantage that we do not have to interchange the symbols X and f in computing the result (see (2.2)), whereas the right linear notation has the advantage of tradition which says that the function symbol should be written before its argument upon which it acts. As it turns out, the left linear notation is used quite systematically in this chapter, whereas in the next two chapters the right linear notation is slightly favored.

2.6 Proposition. *Let $f : M \to N$ and $g : L \to M$ be smooth maps between $\mathcal{A}$-manifolds, then $T(f \circ g) = (Tf) \circ (Tg)$.*

Proof. According to the construction of the tangent map, it suffices to verify this formula on a local chart. But on a local chart it is just the chain rule. QED

2.7 Remark. With the notation $T_m f$ for the restriction of the tangent map to the fiber $T_m M$, the above chain rule for tangent maps takes the form $T_\ell(f \circ g) = T_{g(\ell)} f \circ T_\ell g$. If we want to stress the left linear character of the tangent maps, we could write this as $T_\ell(g \diamond f) = T_\ell g \diamond T_{g(\ell)} f$ (see also [III.3.8]).

2.8 Corollary. *If $f : M \to N$ is a diffeomorphism, $Tf : TM \to TN$ is a diffeomorphism.*

2.9 Definition. Let $f : M \to L$ be a smooth map between $\mathcal{A}$-manifolds, let $\ell \in \mathbf{B}L$ be arbitrary, and define $N = f^{-1}(\ell) \subset M$. The point $\ell \in \mathbf{B}L$ is called *a regular value of f* if for all $m \in N$ the (left) linear map $T_m f : T_m M \to T_\ell L$ is surjective.

2.10 Lemma. *If $\ell \in \mathbf{B}L$ is a regular value for $f : M \to L$, then $N = f^{-1}(\ell)$ is a submanifold of M (see also [III.5.10]).*

Proof. Suppose M is modeled on the $\mathcal{A}$-vector space E of total dimension n and L on the $\mathcal{A}$-vector space F of total dimension $k \leq n$. Let $\psi : V \to P \subset F_0$ be a coordinate chart around $\ell \in V$ and let $\phi : U \to O \subset E_0$ be a chart around some point $m \in N$ such that $f(U) \subset V$ (such charts exists by continuity of f). By assumption the map $\psi \circ f \circ \phi^{-1} : O \to P$ has rank $\dim L$ at $\phi(m)$. According to [III.3.30] there exists an open set $O' \subset O$ containing m and a diffeomorphism $g : O' \to O'' \subset E_0$ such that $(\psi \circ f \circ \phi^{-1} \circ g^{-1})(x^1, \ldots, x^n) = (x^1, \ldots, x^k)$. Since g is a diffeomorphism, $g \circ \phi : U' = \phi^{-1}(O') \to O''$ is a coordinate chart around m. Since ℓ has real coordinates, translation over $\psi(\ell)$ in F_0 is smooth. We thus may assume without loss of generality that $\psi(\ell) = (0, \ldots, 0)$. Hence $N \cap U'$ is described by the equations $x^1 = \cdots = x^k = 0$ with $(x^1, \ldots, x^n)$ the coordinates on U'. Since $m \in N$ was arbitrary, we conclude that N is a submanifold of M. QED

2.11 Definition. A smooth map $f : N \to M$ is called an *immersion* if for all $n \in N$ the map $T_n f : T_n N \to T_{f(n)} M$ is injective. It follows that f is an injective immersion if and only if $Tf : TN \to TM$ is injective. An injective immersion is called an *embedding* if it is a homeomorphism onto its image.

2.12 Lemma. *If $f : N \to M$ is an injective immersion, it is an embedding if and only if $\mathbf{B}f : \mathbf{B}N \to \mathbf{B}M$ is an embedding.*

Proof. If f is an injective immersion, then obviously $\mathbf{B}f = f|_{\mathbf{B}N} : \mathbf{B}N \to \mathbf{B}M$ is an injective immersion of $\mathbf{R}$-manifolds (because $\mathbf{B}Tf = Tf|_{\mathbf{B}TM}$). We thus have to verify that the induced topology on $f(N)$ coincides with that of N if and only if the induced topology of $f(\mathbf{B}N)$ coincides with that of $\mathbf{B}N$. But that is a direct consequence of the fact that $\mathbf{B}$ induces a bijection between open sets of M (respectively N) and those of $\mathbf{B}M$ (respectively $\mathbf{B}N$) [III.4.12]. $\boxed{QED}$

2.13 Remark. At first sight [2.12] looks natural. On second thoughts however, it has a strange aspect: the map $f : \mathbf{B}N \to \mathbf{B}M$ ignores everything about the odd coordinates and nevertheless it determines a result on $f : N \to M$. The secret is that the odd coordinates constitute a vector bundle over the body [IV.8.8]. The condition that f is an injective immersion then says more or less that the odd coordinates of N form a kind of subbundle of the odd coordinates of M. The problems of an embedding are thus concentrated in the body parts.

2.14 Lemma. *Let M and N be $\mathcal{A}$-manifolds and let $i : N \to M$ be a smooth map. If for some $n \in N$ the map $T_n i$ is injective, then there exists a neighborhood U of n such that $f|_U : U \to M$ is an embedding. If $T_n i$ is bijective (which implies* $\dim M = \dim N$*), then there exists a neighborhood U of n and a neighborhood V of $i(n)$ such that $i : U \to V$ is a diffeomorphism. In particular, if i is an immersion between $\mathcal{A}$-manifolds of the same graded dimension, it is a local diffeomorphism and an open map. If i is also injective, it is a diffeomorphism onto its (open) image $i(N) \subset M$.*

Proof. • If $T_n i$ is injective, it is of maximal rank, and then the first part follows from [III.3.30-ii].

• If $T_n i$ is bijective, we deduce from the inverse function theorem [III.3.23] the existence of U and V with the required properties.

• If i is an immersion, $T_n i$ is injective for all $n \in N$. But since N and M are supposed to have the same graded dimension, it must be bijective [II.2.6-iii]. Hence we can apply the previous result to show that i is an open map and a local diffeomorphism. If i is also injective, the second result follows immediately. $\boxed{QED}$

2.15 Nota Bene. The first part of [2.14] is a purely *local* result. It is by no means true that an immersion (not even an injective immersion) is automatically a global embedding, although it is so locally. An explicit counter example is given in [2.19].

2.16 Proposition. *Let N be a connected submanifold of M and $i : N \to M$ the canonical injection. Then i is an embedding and $Ti(TN)$ is a subbundle of $TM|_N$.*

Conversely, if $i : N \to M$ is an embedding, then $i(N)$ is a submanifold of M and the canonical injection $i : N \to i(N)$ is a diffeomorphism when $i(N)$ is given the $\mathcal{A}$-manifold structure of [III.5.4].

Proof. Let us use the atlas for the $\mathcal{A}$-manifold N introduced in [III.5.4] with charts $\psi_a : V_a \to P_a \subset F_0$ induced from charts on M $\varphi_a : U_a \to O_a \subset E_0$ by $\psi_a = \varphi_a|_{V_a}$, $V_a = U_a \cap N$, and $P_a = O_a \cap F$. Since N is connected, we may assume that the graded subspace $F \subset E$ of definition [III.5.1] is the same for all these charts. We already know that $i : N \to M$ is injective and a smooth homeomorphism onto its image. In terms of the given charts Ti is given as $Ti : P_a \times F \to O_a \times E$, $(x, v) \mapsto (x, v)$, and thus i is also an immersion, proving that i is an embedding.

Denoting by $\pi_N : TN \to N$ and $\pi_M : TM \to M$ the canonical projections, it follows from the commutative diagram $i \circ \pi_N = \pi_M \circ Ti$ that $Ti(TN) \subset TM|_N$. In terms of the above charts, $TM|_N$ has charts $P_a \times E$ and $Ti(TN)$ is the subset described as $P_a \times F$. It follows that $Ti(TN)$ is a subbundle of $TM|_N$ and that the given atlas is adapted to the subbundle [IV.3.12].

Conversely, let $i : N \to M$ be an embedding and let $n_o \in N$ be arbitrary. If U_o is a chart of M around $m_o = i(n_o)$, $i^{-1}(U_o)$ is open in N and thus there exists a chart $\psi : V \to P$ of N around n_o with $V \subset i^{-1}(U_o)$. Since i is an embedding, there exists a chart $\varphi : U \to O$, $U \subset U_o$ around m_o such that $i(V) = i(N) \cap U$. Again because i is an embedding, the map $f = \varphi \circ i \circ \psi^{-1} : P \to O$ has rank $n = \dim F$ at $\psi(n_o) \in P$. Applying the second part of [III.3.30] we conclude that there exists a chart in which $i(N)$ is described by equations of the type ${}^{i}e(\varphi(m)) = 0$ [III.5.2], and thus $i(N)$ is a submanifold. Using the same charts, it follows immediately that $i : N \to i(N)$ is a bijective immersion when $i(N)$ is equipped with the $\mathcal{A}$-manifold structure given in [III.5.4]. The result now follows from [2.14]. $\boxed{QED}$

2.17 Corollary. *Let $N \subset M$ be a submanifold. Then the $\mathcal{A}$-manifold structure on N given in [III.5.4] is the unique structure such that the canonical injection $i : N \to M$ is an embedding.*

Proof. Let us denote by N' the set N endowed with another $\mathcal{A}$-manifold structure. If the canonical injection $i' : N' \to M$ is an embedding, it follows from the second part of [2.16] that the identity map $j : N' \to N$ is a diffeomorphism, i.e., that $N' = N$ as an $\mathcal{A}$-manifold. $\boxed{QED}$

2.18 Corollary. *Let M, N, and L be $\mathcal{A}$-manifolds, let $i : N \to M$ be an embedding, and let $f : L \to M$ be a smooth map such that $f(L) \subset i(N)$. Then there exists a unique smooth map $f_o : L \to N$ such that $i \circ f_o = f$.*

Proof. A direct combination of [III.5.8] and the second part of [2.16]. $\boxed{QED}$

2.19 Counter example. The condition that $i : N \to M$ is an embedding in [2.18] can not be weakened easily to injective immersion, as can be seen from the following example. Let $N = L = \mathcal{A}_0$, let $M = \mathcal{A}_0^2$, and define the maps $i : N \to M$ and $f : L \to M$ by $i(t) = (t(1+t^4)^{-1}, t^3(1+t^4)^{-1})$ and $f(x) = (x^3(1+x^4)^{-1}, x(1+x^4)^{-1})$. Both have as their image the figure eight, and their parameters are related by $t = x^{-1}$, which shows that f can not be factorized through N. On the other hand, we will see in [6.8] a situation in which the embedding condition is not needed.

2.20 Discussion. The direct product $M \times N$ of two $\mathcal{A}$-manifolds M and N is defined as the set of ordered pairs. When doing computations with tangent vectors it is often very useful to be able to see tangent vectors to $M \times N$ also as ordered pairs: of tangent vectors to M and N respectively. This boils down to a canonical identification $T(M \times N) \cong TM \times TN$ given in [2.21]. Once we know that tangent vectors to $M \times N$ can be seen as pairs of tangent vectors, it becomes interesting to see how a (special kind of) tangent map behaves on these pairs. This is done in [2.22] and [2.23], yielding identifications $T(f, g) \cong (Tf, Tg)$ and $T(f \times g) \cong Tf \times Tg$. All these identifications will be used in the sequel without further mentioning.

2.21 Proposition. *For $\mathcal{A}$-manifolds M and N, the map $(T\pi_M, T\pi_N)$ from $T(M \times N)$ to $(TM) \times (TN)$ given by $X \mapsto (\langle\!\langle X \| T\pi_M \rangle\!\rangle, \langle\!\langle X \| T\pi_N \rangle\!\rangle)$ is an isomorphism of vector bundles.*

Proof. Using the notation of [2.1], $\{U_a \times E\}$ is an atlas for TM, $\{V_\alpha \times F\}$ one for TN, and $\{U_a \times V_\alpha\}$ one for $M \times N$. It follows that $\{(U_a \times V_\alpha) \times (E \times F)\}$ is an atlas for $T(M \times N)$ and that $\{(U_a \times E) \times (V_\alpha \times F)\}$ is an atlas for $(TM) \times (TN)$. With respect to these two atlases, the map $(T\pi_M, T\pi_N)$ is given by $((m, n), (v, w)) \mapsto ((m, v), (n, w))$. From this observation the result follows. Those who want to go into the details should use the fact that for the transition functions $\varphi_{ba} \times \chi_{\beta\alpha}$ of $M \times N$ [III.5.11] the Jacobian is given by $Jac(\varphi_{ba} \times \phi_{\beta\alpha}) = Jac(\varphi_{ba}) \times Jac(\phi_{\beta\alpha}) \in \mathrm{End}_L(E \times F) \equiv \mathrm{End}_L(E \oplus F)$. QED

2.22 Corollary. *Let $f : L \to M$ and $g : L \to N$ be smooth maps between $\mathcal{A}$-manifolds L, M, and N. In the identification $T(M \times N) \cong TM \times TN$ of [2.21] the map $T(f, g)$ (see [III.5.12]) is given as (Tf, Tg).*

Proof. To get the expression of $T(f, g)$ in terms of the given identification, we just have to compose with the identification map $(T\pi_M, T\pi_N)$. But $(T\pi_M, T\pi_N) \circ T(f, g) = (T\pi_M \circ T(f, g), T\pi_N \circ T(f, g)) = (T(\pi_M \circ (f, g)), T(\pi_N \circ (f, g)) = (Tf, Tg)$. QED

2.23 Corollary. *Let $f : M \to N$ and $g : K \to L$ be smooth maps between $\mathcal{A}$-manifolds. In the identifications $T(M \times K) \cong TM \times TK$ and $T(N \times L) \cong TN \times TL$ of [2.21], the tangent map of $f \times g : M \times K \to N \times L$ (see [III.5.14]) is given as $T(f \times g) \cong Tf \times Tg$, i.e., we have a commutative diagram*

$$\begin{array}{ccc} T(M \times K) & \xrightarrow{T(f \times g)} & T(N \times L) \\ {\scriptstyle (T\pi_M, T\pi_K)}\big\downarrow & & \big\downarrow{\scriptstyle (T\pi_N, T\pi_L)} \\ TM \times TK & \xrightarrow[Tf \times Tg]{} & TN \times TL\,. \end{array}$$

Proof. Since the identification maps are bijections, it suffices to verify that the maps $(Tf \times Tg) \circ (T\pi_M, T\pi_K)$ and $(T\pi_N, T\pi_L) \circ T(f \times g)$ are the same. But that is immediate from the equalities $\pi_N \circ f \times g = f \circ \pi_M$ and $\pi_L \circ f \times g = g \circ \pi_K$. $\boxed{QED}$

2.24 Construction. If $X : M \to TM$ and $Y : N \to TN$ are vector fields, we have the product map $X \times Y : M \times N \to TM \times TN$ defined by $(X \times Y)(m,n) = (X(m), Y(n))$. Using the identification $TM \times TN \cong T(M \times N)$ of [2.21], the map $X \times Y$ becomes a vector field on $M \times N$. If Y is the zero vector field $Y = \underline{0}$, the vector field $X \times \underline{0}$ can be interpreted as the vector field X extended to the product $M \times N$ (by being independent of N). Using that we can add vector fields, the general case can be written as $X \times Y = X \times \underline{0} + \underline{0} \times Y$. If it is clear from the context that the vector field X on M has to be seen as being extended to $M \times N$ as $X \times \underline{0}$, one abbreviates $X \times \underline{0}$ as just X. With this shorthand, one thus can write $X \times Y = X + Y$.

2.25 Definition. If $f : M \to N$ is a smooth map and X and Y vector fields such that $Tf \circ X = Y \circ f$, i.e., such that we have the commutative diagram

$$\begin{array}{ccc} TM & \xrightarrow{\;Tf\;} & TN \\ {\scriptstyle X}\uparrow & & \uparrow{\scriptstyle Y} \\ M & \xrightarrow[\;f\;]{} & N\,, \end{array} \tag{2.26}$$

then we will say that *X and Y are related by f*. Sometimes the vector field Y is called the push forward of X and denoted as $Y = f_* X$. We will *not* do so because it suggests that Y is determined by X, which is not the case. Even the existence of a vector field Y that is related to X by f is not guaranteed.

2.27 Lemma. *Let $f : M \to N$ be smooth map, X a vector field on M and Y a vector field on N. The vector fields X and Y are related by f if and only if for all smooth functions $g \in C^\infty(N)$ we have the equality $(Yg) \circ f = X(g \circ f)$.*

Proof. Let V_α be a chart for N and U_a a chart for M such that $f(U_a) \subset V_\alpha$. Using the local expressions for the vector fields $X|_{U_a} = \sum_i X_a^i \cdot \partial_{ai}$ and $Y|_{V_\alpha} = \sum_j Y_\alpha^j \cdot \partial_{\alpha j}$, commutativity of the diagram (2.26) translates as

$$Y_\alpha^j(f(m)) = \sum_i X_a^i(m) \cdot \frac{\partial(y_\alpha^j \circ f)}{\partial x_a^i}(m)\,, \tag{2.28}$$

where the x_a^i and y_α^j denote the local coordinates on U_a and V_α. Commutativity of the diagram and the chain rule then imply for $g \in C^\infty(N)$ that $(Yg)(f(m)) = \sum_j Y_\alpha^j(f(m)) \cdot (\partial_{\alpha j} g)(f(m)) = \sum_i X_a^i(m) \cdot (\partial_{ai}(g \circ f))(m) = (X(g \circ f))(m)$.

To prove the equivalence in the other direction, we restrict the domain of definition to U_a and substitute for g one of the local coordinates y_α^j to obtain immediately (2.28). $\boxed{QED}$

2.29 Proposition. *Let $f : M \to N$ be smooth map, X_i vector fields on M and Y_i vector fields on N $(i = 1, 2)$. If X_i and Y_i are related by f, then $[X_1, X_2]$ and $[Y_1, Y_2]$ are related by f.*

Proof. Since the relation is additive in both X_i and Y_i, we may suppose that these vector fields are homogeneous. Since Tf is even, it follows that X_i is of the same parity as Y_i. We now apply [2.27] and compute

$$\begin{aligned}[X_1, X_2](g \circ f) &= X_1(X_2(g \circ f)) - (-1)^{\langle \varepsilon(X_1) | \varepsilon(X_2) \rangle} \cdot X_2(X_1(g \circ f)) \\ &= X_1((Y_2 g) \circ f) - (-1)^{\langle \varepsilon(X_1) | \varepsilon(X_2) \rangle} \cdot X_2((Y_1 g) \circ f) \\ &= (Y_1(Y_2 g) - (-1)^{\langle \varepsilon(Y_1) | \varepsilon(Y_2) \rangle} \cdot Y_2(Y_1 g)) \circ f = ([Y_1, Y_2] g) \circ f \,. \end{aligned}$$ QED

3. Advanced properties of the tangent map

In this section we discuss in some detail the relation between the tangent map Tf and the derivative of f for maps taking values in an $\mathcal{A}$-vector space. Using the tangent map, we generalize the notion of the (directional) derivative of a map with values in an $\mathcal{A}$-vector space to maps with values in a vector bundle. This allows us to show that a vector bundle is trivial as a vector bundle if and only if either its even or its odd part is trivial as a fiber bundle. We then define the tangent map of a map which is itself not necessarily smooth, but which can be seen as being part of a smooth family of maps. As a first application of this generalized notion of tangent map we prove a generalization of the fact that a function is constant if and only if its derivative is zero.

3.1 Discussion. Let M be an $\mathcal{A}$-manifold, $f : M \to \mathcal{A}$ a smooth map and X a smooth vector field on M. With these ingredients we can define a new smooth map $Xf : M \to \mathcal{A}$ as well as a smooth map $X \diamond Tf : M \to T\mathcal{A}$. A superficial glance at the definitions of these maps shows that they look very much alike. On the other hand, the notion $X \diamond Tf$ also makes sense for a smooth map $f : M \to N$ to an arbitrary $\mathcal{A}$-manifold N, whereas Xf is no longer defined in such a general setting. However, in particular cases we will need, and we will still can indeed, make sense out of Xf. In order to prepare this generalization of the action of a vector field on a function and in order to study in more detail the link between Xf and $X \diamond Tf$, we start with studying (2.2) in more detail. So let E and F be $\mathcal{A}$-vector spaces, let $O \subset E_0$ be open, and let $f : O \to F_0$ be a smooth map. The tangent spaces of O and F_0 are the trivial bundles $TO = O \times E$ and $TF_0 = F_0 \times F$, and thus the tangent map is a map $Tf : O \times E \to F_0 \times F$. Let (x^i) be the coordinates of a point $x \in O$ with respect to a basis (e_i) of E, and let v^i be the coordinates of a (tangent) vector $v \in E$ with respect to the associated basis (∂_{x^i}). Similarly we denote by y^j the coordinates of a point $y \in F_0$ with respect to a basis (ϵ_j) and by w^j the coordinates of a (tangent) vector $w \in F$ with respect to the associated basis (∂_{y^j}). Now let X be a vector field on O, which we can identify with a map $X : O \to E$, i.e., $X(x) = \sum_i X^i(x)\partial_{x^i}$ with (smooth) functions $X^i : O \to \mathcal{A}$. According to (2.2), the map $X \diamond Tf$ is given by

$$(X \diamond Tf)(x) = \langle\!\langle X(x) \| Tf \rangle\!\rangle = \sum_j \Big(\sum_i X^i(x) \frac{\partial f^j}{\partial x^i}(x) \Big) \frac{\partial}{\partial y^j}|_{f(x)} \ .$$

In this expression, $f^j = {}^j\epsilon \circ f$ is a homogeneous $\mathcal{A}$-valued function of parity $\varepsilon(y^j)$, and $\sum_i X^i(x)\partial_{x^i} f^j(x)$ is the function $(Xf^j)(x)$, i.e., the action of the derivation X on the function f^j. If we now remember that the basis (∂_{y^j}) is essentially the same as the basis (ϵ_j) [1.16], it becomes very tempting to write the right hand side as $(Xf)(x)$, in which the action of X on the F_0-valued function f should be interpreted component wise [1.24]. If we are careful, this can indeed be done.

3.2 Lemma. *Let M be an $\mathcal{A}$-manifold, F an $\mathcal{A}$-vector space, $f : M \to F_0$ a smooth map, and X a smooth vector field on M. In the trivialization $TF_0 \cong F_0 \times F$, the map $X \diamond Tf : M \to TF_0$ is given as $(X \diamond Tf)(m) = (f(m), (Xf)(m)) \in F_0 \times F \cong TF_0$, i.e., $X \diamond Tf \cong (f, Xf)$.*

Proof. In [3.1] this was proved in case M is an open subset of E_0. The generalization to arbitrary $\mathcal{A}$-manifolds follows immediately from (1.14). $\boxed{QED}$

3.3 Example. Let $\phi \in \mathbf{B}\,\mathrm{Hom}_L(E; F)$ be a smooth linear map. We define the function $f \in C^\infty(E_0; F_0)$ as $f = \phi|_{E_0}$. According to [III.3.14] the Jacobian of f is ϕ. In the trivializations $TE_0 = E_0 \times E$ and $TF_0 = F_0 \times F$, the tangent map Tf is given by

$$Tf : (x, v) \in E_0 \times E \mapsto (f(x), \phi(v)) \in F_0 \times F \ ,$$

i.e., $T(\phi|_{E_0}) = (\phi \times \phi)|_{E_0 \times E}$ [III.5.14].

3.4 Discussion. In [3.2] we saw that if we know the map $f : M \to F_0$, then the map $X \diamond Tf$ is completely determined in a very natural way by the map $Xf : M \to F$. But Xf makes sense also when f takes values in the full $\mathcal{A}$-vector space F, not only in its even part. One thus might think that again Xf determines the map $X \diamond Tf$. To investigate this idea, we suppose that $f : M \to F$ is a smooth map. As an $\mathcal{A}$-manifold, F is modeled on the $\mathcal{A}$-vector space $F^\sharp$ with basis $(\epsilon_j, \overline{\epsilon}_j)$ [III.1.26]. We thus can write

$$f(m) = \sum_j f^j(m)\epsilon_j = \sum_j \Big(f^j_{\varepsilon(\epsilon_j)}(m)\epsilon_j + f^j_{1-\varepsilon(\epsilon_j)}\overline{\epsilon}_j \Big) \ ,$$

where f^j_α denotes the homogeneous part of parity α of the $\mathcal{A}$-valued function f^j. In terms of the homogeneous functions $f_\alpha : M \to F_\alpha$ these functions are given by the formula

$$f^j_{\varepsilon(\epsilon_j)} = \langle\!\langle f_0 \| {}^j\epsilon \rangle\!\rangle \qquad \text{and} \qquad f^j_{1-\varepsilon(\epsilon_j)} = \langle\!\langle f_1 \| {}^j\epsilon \rangle\!\rangle \ .$$

Denoting by $(y^j, \overline{y}^j)$ the coordinates with respect to the basis $(\epsilon_j, \overline{\epsilon}_j)$, we find

$$\langle\!\langle X(m) \| Tf \rangle\!\rangle = \sum_j \Big(\langle\!\langle (Xf_0)(m) \| {}^j\epsilon \rangle\!\rangle \frac{\partial}{\partial y^j}|_{f(m)} + \langle\!\langle (Xf_1)(m) \| {}^j\epsilon \rangle\!\rangle \frac{\partial}{\partial \overline{y}^j}|_{f(m)} \Big) ,$$

whereas $(Xf)(m)$ is given as $\sum_j (Xf^j)(m)\epsilon_j$. Obviously these two objects are not the same. Still, there is a link.

In order to investigate this link in detail, we introduce, in analogy with the projection $\pi : F^\sharp \to F$ defined in [III.1.26], the left linear map $\pi_\partial : TF = F \times F^\sharp \to F$ defined as

$$\pi_\partial(v, \sum_j (w^j \frac{\partial}{\partial y^j} + \overline{w}^j \frac{\partial}{\partial \overline{y}^j})) = \sum_j (w^j + \overline{w}^j)\epsilon_j .$$

It then follows immediately that $\pi_\partial \langle\!\langle X(m) \| Tf \rangle\!\rangle = (Xf)(m)$. Although π_∂ is not injective, there are several situations in which knowledge of Xf is sufficient to reconstruct $\langle\!\langle X(m) \| Tf \rangle\!\rangle$. In the first place if f is an even function, in which case all $\langle\!\langle f_1 \| {}^j\epsilon \rangle\!\rangle$ are zero, and thus $\langle\!\langle X(m) \| Tf \rangle\!\rangle = \sum_j \langle\!\langle (Xf)(m) \| {}^j\epsilon \rangle\!\rangle \, \partial_{y^j}$. It also happens if f is an odd function (with all $\langle\!\langle f_0 \| {}^j\epsilon \rangle\!\rangle$ zero), in which case we have $\langle\!\langle X(m) \| Tf \rangle\!\rangle = \sum_j \langle\!\langle (Xf)(m) \| {}^j\epsilon \rangle\!\rangle \, \partial_{\overline{y}^j}$. A third case is when X is a homogeneous vector field of parity $\varepsilon(X)$. In that case we have

$$\langle\!\langle X(m) \| Tf \rangle\!\rangle = \sum_j (\langle\!\langle (Xf)_{\varepsilon(X)}(m) \| {}^j\epsilon \rangle\!\rangle \, \partial_{y^j} + \langle\!\langle (Xf)_{1-\varepsilon(X)}(m) \| {}^j\epsilon \rangle\!\rangle \, \partial_{\overline{y}^j}) .$$

The conclusion is that if either X or f is homogeneous of known parity, then Xf determines the vector $\langle\!\langle X(m) \| Tf \rangle\!\rangle$ completely, i.e., we have proven [3.5]. That these conditions cannot easily be weakened can be seen in the following example. Let $f : \mathcal{A} \to \mathcal{A}$ be the map $f(x, \xi) = x + \xi$, i.e., the identity, and let X be the vector field $X = \partial_x - \partial_\xi$. Then obviously $Xf = 0$, but $\langle\!\langle X \| Tf \rangle\!\rangle = X \neq 0$.

3.5 Lemma. *Let M be an $\mathcal{A}$-manifold, F an $\mathcal{A}$-vector space, $f : M \to F$ a smooth map, and X a smooth vector field on M. Then the map $Xf : M \to F$ is the same as the map $X \diamond Tf \diamond \pi_\partial : M \to TM \to TF \to F$. Moreover, if either X or f is homogeneous, then $X \diamond Tf$ is completely determined by Xf.*

3.6 Discussion. A little thought will reveal that there is no hope to define something like Xf for a map $f : M \to B$ with values in an arbitrary $\mathcal{A}$-manifold B. However, if B is a vector bundle $p : B \to N$ over some $\mathcal{A}$-manifold N and if the vector field X behaves properly with respect to the bundle structure, then indeed can one define a new function $Xf : M \to B$. The proper behavior of X is that it does not see the base manifold N, but only the fibers of B, which are $\mathcal{A}$-vector spaces.

The idea is as follows. Since any vector bundle is locally trivial, we assume for simplicity that $B = N \times F$ is the trivial bundle with fiber the $\mathcal{A}$-vector space F. Any map $f : M \to B$ decomposes as a couple $f = (g, h)$ with $g = \pi_N \circ f : M \to N$ and $h = \pi_F \circ f : M \to F$ (see [III.5.12]). If X is a vector field on M, we still do not know what Xg should be, but Xh is defined. So the idea is that if X is such that $X \diamond Tg \equiv X \diamond T(f \diamond \pi_N)$ is the zero map, then we can define Xf as being the couple (g, Xh). If N consists of a single point, the first entry in the product $N \times F$ is always the same, so we can ignore it. If we do so, f becomes the map h and Xf becomes the map Xh, i.e., we recover the old definition of the action of the vector field X on the F-valued function h. As an example of the setting in which we will need this generalization, we mention the case $M = \mathcal{A}_0 \times N$, $B = TN$ the tangent bundle, $f : M \to B$ of the form $f(t, n) \in T_nN$, and $X = \partial_t$. Since $f(t, n) \in T_nN$ for all $t \in \mathcal{A}_0$, $X \diamond T(f \diamond \pi) = 0$ (with $\pi : B = TN \to N$ the bundle projection), and thus we will be able to define $\partial_t f : \mathcal{A}_0 \times N \to TN$. In order to make all this more precise, we need to generalize the map $\pi_\partial : TF \to F$ for an $\mathcal{A}$-vector space F to the vector bundle setting.

3.7 Construction. Let $p : B \to M$ be a vector bundle with typical fiber F over an $\mathcal{A}$-manifold N modeled on the $\mathcal{A}$-vector space E. We thus have the smooth vector bundle map $Tp : TB \to TN$. Since T_bp is surjective for all $b \in B$, we can apply [IV.3.18] to conclude that $\ker(Tp)$ is a subbundle of TB over B. We intend to construct a surjective vector bundle map $\pi_\partial : \ker(Tp) \to B$ such that the special case with N a single point reduces to the previously defined π_∂.

The tangent space of a local trivialization $U \times F$ of the vector bundle B is given by $T(U \times F) = TU \times TF = U \times E \times F \times F^\sharp$. Since p is described as $p(n, v) = n$, the map Tp is described as $Tp(n, e, v, f) = (n, e)$. The subbundle $\ker(Tp)$ is thus described as the points $(n, 0, v, f)$. Two (local) trivializations $U_a \times F$ and $U_b \times F$ are related (on $U_a \cap U_b$) by $(n, v_a) \mapsto (n, \psi_{ba}(n)(v_a))$, with $\psi_{ba}(n) \in \mathrm{Aut}(F)$. This implies that the associated trivializations of TB are related by

$$(n, e, v, f) \mapsto (n, \widehat{e}(n, v, e, f), \psi_{ba}(n)(v), \widehat{\psi_{ba}(n)}(f)) \ , \tag{3.8}$$

where $\widehat{\psi_{ba}(n)} \in \mathrm{Aut}(F^\sharp)$ is the natural extension of ψ to $F^\sharp$ defined as follows. Let (f_i) be a basis for F, and let $(f_i, \overline{f}_i)$ be the associated basis of $F^\sharp$. Then for $\psi \in \mathrm{Aut}(F)$ we define $\widehat{\psi} \in \mathrm{Aut}(F^\sharp)$ by

$$\psi(f_i) = \sum\nolimits_j \psi_i{}^j\, f_j \quad \Longrightarrow \quad \widehat{\psi}(f_i) = \sum\nolimits_j \psi_i{}^j\, f_j \quad \text{and} \quad \widehat{\psi}(\overline{f}_i) = \sum\nolimits_j \psi_i{}^j\, \overline{f}_j \ .$$

This is indeed even, and it preserves the graded subspaces F and $F^{\flat 1}$ of $F^\sharp = F \oplus F^{\flat 1}$ [III.1.26]. If, as before, π denotes the (left linear) projection $F^\sharp \to F$ of [III.1.26], it is obvious from the definition of $\widehat{\psi}$ that $\pi \circ \widehat{\psi} = \psi \circ \pi$. The tangent vector $\widehat{e}(n, e, v, f)$ in (3.8) depends linearly on e and f; it picks up tangent vectors in the direction of e because $\psi_{ba}(n)v$ depends upon n.

We now are ready to define the left linear map $\pi_\partial : \ker(Tp) \subset TB \to B$. On the local trivialization $U \times \{0\} \times F \times F^\sharp$ we define it by

$$\pi_\partial : (n, 0, v, f) \mapsto (n, \pi_\partial(v, f)) = (n, \pi(f)) \ . \tag{3.9}$$

Our discussion on how two different trivializations are related proves that this is a globally well defined smooth map. And obviously, if M consists of a single point we recover the initial definition of π_∂.

3.10 Definition. Let $p : B \to N$ be a vector bundle with typical fiber F, let $f : M \to B$ be a smooth map, and let X be a smooth vector field on M such that $X \diamond Tf$ belongs to $\ker(Tp)$, i.e., for all $m \in M$: $\langle\!\langle X(m) \| Tf \diamond Tp \rangle\!\rangle = 0$. Then we define the map $Xf : M \to B$ by

$$Xf = X \diamond Tf \diamond \pi_\partial \ .$$

This makes sense because $X \diamond Tf$ takes values in $\ker(Tp)$ which is the domain of definition of π_∂. According to [3.5] this definition coincides with the old one in case N reduces to a single point, i.e., $B = F$ an $\mathcal{A}$-vector space.

3.11 Discussion. In order to see that [3.10] fulfills our idea of [3.6], we consider a local trivialization $U \times F$ of the bundle $p : B \to N$. In this trivialization the map f restricted to a suitably chosen $V \subset M$ is given as $f(m) = (p(f(m)), h(m))$ for some smooth map $h : V \to F$. In other words, in the trivialization $U \times F$ the map f is given as the couple $(p \circ f, h)$. According to [2.22] the tangent map Tf is given as $(T(f \diamond p), Th)$. Since $X \diamond Tf$ takes values in $\ker Tp$, we have $X \diamond Tf = (\underline{0}, X \diamond Th)$. And then (3.9) and [3.5] tell us that on the trivialization $U \times F$ we have

$$Xf \equiv X \diamond Tf \diamond \pi_\partial = (f \diamond p, X \diamond Th \diamond \pi_\partial) \equiv (f \diamond p, Xh) \ . \tag{3.12}$$

We conclude that the formal definition of Xf in the bundle setting indeed gives us what we expected in [3.6].

3.13 Lemma. *Let $f : M \to B$ be a smooth map, $p : B \to N$ a vector bundle, and X a vector field on M such that $X \diamond Tf$ takes values in* $\ker Tp$. *If either X or f is homogeneous (f taking values in B_α), then the map $Xf = X \diamond Tf \diamond \pi_\partial$ completely determines the map $X \diamond Tf$. This means in particular that for all $m \in M$: $(Xf)(m) = 0$ if and only if $(X \diamond Tf)(m) \equiv \langle\!\langle X_m \| Tf \rangle\!\rangle = 0$.*

Proof. This is an immediate consequence of [3.5] and the local expression (3.12) for Xf. QED

3.14 Discussion. We now have sufficient material to attack the question of triviality of vector bundles. Let $p : B \to M$ be a vector bundle with typical fiber the $\mathcal{A}$-vector space F. If we suppose that B is trivial as a vector bundle, we have a vector bundle isomorphism $h : B \to M \times F$. Since the restriction of h to a fiber B_m is an even linear map, it preserves the parity, and thus the restriction of h to the subbundle B_α [IV.4.11] gives us a diffeomorphism $h : B_\alpha \to M \times F_\alpha$. We conclude that the subbundles B_α are trivial as fiber bundles.

3.15 Proposition. *Let $p : B \to M$ be a vector bundle with typical fiber F. Then B is trivial as a vector bundle if and only if either B_0 or B_1 is trivial as a fiber bundle.*

Proof. In view of [3.14] we only have to prove that if B_α is trivial, then B is trivial as a vector bundle. For simplicity we only treat the case $\alpha = 0$, the case $\alpha = 1$ being similar. So let us suppose that $g : M \times F_0 \to B_0$ is a fiber bundle isomorphism. This implies that the restriction of g to a fiber $\{m\} \times F_0$ is a homeomorphism onto the fiber $(B_m)_0$, but there is nothing that says that this should be linear (in whatever sense). Let (x^i) be coordinates on F_0 with respect to a basis (f_i) for F. Then we define vector fields V_i on $M \times F_0$ by $V_i(m, x) = \partial_{x^i}$. In terms of a local trivialization $U \times F$ of B, the map g is described by $g(m, x) = (m, \widehat{g}(m, x))$ with $\widehat{g} : U \times F_0 \to F_0$ a smooth map. It follows that $V_i \diamond Tg$ is given in this trivialization as

$$\langle\!\langle V_i|_{(m,x)} \,\|\, Tg \rangle\!\rangle = \sum_j \frac{\partial \widehat{g}^j}{\partial x^i}(m, x) \frac{\partial}{\partial x^j} . \tag{3.16}$$

Since B_0 is a submanifold of B, its tangent bundle TB_0 is a subbundle of $TB|_{B_0}$ [2.16]. From (3.16) it is obvious that $\langle\!\langle V_i|_{(m,x)} \,\|\, Tg \rangle\!\rangle$ lies in $\ker Tp$ [3.7]. We thus can use the map π_∂ to define the smooth sections $s_i : M \to B$ by

$$s_i(m) = (\pi_\partial \circ Tg \circ V_i)(m, 0) \in B_m ,$$

i.e., $s_i(m) = (V_i g)(m, 0)$ [3.10]. We claim that these sections satisfy the condition of [IV.3.10-ii], and thus that B is trivial as a vector bundle by [IV.3.10-i]. First we note that the vector fields V_i are homogeneous and everywhere independent, and thus, because Tg is a diffeomorphism, the tangent vectors $(Tg \circ V_i)(m, 0)$ are homogeneous (of the same parity as f_i) and independent vectors in $T_{g(m,0)}B_0 \subset T_{g(m,0)}B$. Since g takes values in B_0, the vectors $(Tg \circ V_i)(m, 0)$ do not have components in the directions $\partial_{\overline{x}^i}$ associated to the overlined vectors in $F^\sharp$. But the projection $\pi : F^\sharp \to F$ is an even bijection when restricted to the graded subspace generated by the non-overlined basis vectors (see [III.1.26]). It follows that the vectors $(\pi_\partial \circ Tg \circ V_i)(m, 0)$ are independent and homogeneous of parity $\varepsilon(f_i)$. Since B_m is isomorphic to the $\mathcal{A}$-vector space F, an independent set of the correct number of vectors is also generating, i.e., a basis. We conclude that the sections s_i satisfy [IV.3.10-ii], and thus that B is trivial as a vector bundle by [IV.3.10-i].

In the odd case ($\alpha = 1$), the only difference worth mentioning is that there the vectors $(Tg \circ V_i)(m, 0)$ lie completely in the graded subspace generated by the overlined vectors. And then the restriction of $\pi : F^\sharp \to F$ to that graded subspace reverses the parity of these vectors. QED

3.17 Counter example. Let $p : B \to M$ be a vector bundle with typical fiber F. If B is trivial as a vector bundle, we have seen that then B_α is trivial as a fiber bundle. Applying the body map, we deduce that $\mathbf{B}B \to \mathbf{B}M$ is trivial as an $\mathbf{R}$-vector bundle and that $\mathbf{B}B_\alpha \to \mathbf{B}M$ are also trivial as a fiber bundle.

Now let F be of dimension $1|1$ with basis f_0, f_1 (f_α of parity α), and let $M = \mathbf{G}S^1$ be the $\mathcal{A}$-manifold without odd dimensions whose underlying $\mathbf{R}$-manifold is the circle.

We cover M by two charts $U_\pm$ given by $U_+ \cong \{\theta \in \mathcal{A}_0 \mid 0 < \mathbf{B}\theta < 2\pi\}$ and $U_- \cong \{\theta \in \mathcal{A}_0 \mid -\pi < \mathbf{B}\theta < \pi\}$. The change of charts $\varphi_{-+} : U_- \to U_+$ is given by $\theta \mapsto \theta$ if $\mathbf{B}\theta > 0$ and $\theta \mapsto \theta + 2\pi$ is $\mathbf{B}\theta < 0$. Over M we define the vector bundle $p : B \to M$ with typical fiber F by the transition functions ψ_{-+} given by $\theta \mapsto \left(\begin{smallmatrix} 1 & 0 \\ 0 & 1 \end{smallmatrix}\right)$ if $\mathbf{B}\theta > 0$ and $\theta \mapsto \left(\begin{smallmatrix} -1 & 0 \\ 0 & -1 \end{smallmatrix}\right)$ if $\mathbf{B}\theta < 0$.

This vector bundle is essentially twice (in the f_0 direction and in the f_1 direction) the Möbius bundle over the circle. We claim that this bundle is trivial as a fiber bundle, but not trivial as a vector bundle. To prove that it is not trivial as a vector bundle, we consider the fiber bundle $\mathbf{B}B_0 \to \mathbf{B}M$. This is a fiber bundle over the circle S^1 with typical fiber $\mathbf{R} \cong \mathbf{B}F_0$ given by the transition functions $\widehat{\psi}_{-+}(\theta) = \operatorname{sign}\theta \in \operatorname{Aut}(\mathbf{R})$, i.e., the Möbius bundle, which is not trivial. Hence B can not be trivial as a vector bundle. On the other hand, let us denote by (x, ξ, y, η) the even and odd coordinates in $F \cong F_0^\sharp$ according to $(x + \xi)f_0 + (y + \eta)f_1 = xf_0 + \eta f_1 + \xi\overline{f}_0 + y\overline{f}_1$. Then we can define a fiber bundle isomorphism $h : M \times F \to B$ in the trivialization U_+ by

$$\begin{aligned} h(\theta, x, \xi, y, \eta) = (\theta, &\cos(\theta/2)x + \sin(\theta/2)y, \cos(\theta/2)\xi + \sin(\theta/2)\eta, \\ &- \sin(\theta/2)x + \cos(\theta/2)y, - \sin(\theta/2)\xi + \cos(\theta/2)\eta) \ . \end{aligned}$$

We leave it to the reader to show that this can indeed be extended to a global smooth diffeomorphism. Since it obviously respects the fibers, it is a trivialization of B as a fiber bundle.

This example shows that a vector bundle which is trivial as fiber bundle need not be trivial as a vector bundle. The obstruction lies in the fact that the group $\operatorname{Aut}(F)$ is more or less incapable of mixing even and odd basis vectors. This is seen most clearly when we look at a matrix representation of $\mathbf{B}\operatorname{Aut}(F)$, which decomposes as $Gl(p, \mathbf{R}) \times Gl(q, \mathbf{R})$ if the dimension of F is $p|q$.

3.18 Remark. In the special case $\mathcal{A} = \mathbf{R}$, a vector bundle $p : B \to M$ is trivial as a vector bundle if and only if *both* B_0 and B_1 are trivial as fiber bundles. The reason is that for generic $\mathcal{A}$ (satisfying (C[∞])) the subbundle B_0 still contains information about B_1 in the form of the odd coordinates with respect to the odd directions, while for $\mathcal{A} = \mathbf{R}$ one loses this information. This is confirmed by the fact that for $\mathcal{A} = \mathbf{R}$ the group $Gl(p|q, \mathbf{R})$ decomposes as the direct product $Gl(p, \mathbf{R}) \times Gl(q, \mathbf{R})$, whereas for generic $\mathcal{A}$ there is a cross-over between the even and odd directions in the form of odd matrix elements.

3.19 Discussion. Let M, N, and Q be three $\mathcal{A}$-manifolds and $f : M \times N \to Q$ a smooth map. For a fixed $m \in M$ we define the map $f_m : N \to Q$ by $f_m(n) = f(m, n)$. According to [III.1.23-g], if m has real coordinates, then f_m is smooth. On the other hand, if m does not have real coordinates, there is no reason to assume that f_m will be smooth (in general it will not be). Thus we can not speak about the tangent map Tf_m for an arbitrary fixed $m \in M$. Since it is often desirable to do so anyway, we will circumvent this problem as follows. We define the *generalized tangent map* $Tf_m : TN \to TQ$ by

$$Tf_m : Y_n \in T_nN \mapsto \langle\!\langle (\underline{0}_m, Y_n) \| Tf \rangle\!\rangle \ .$$

Since Tf is even and left linear on any fiber $T_{(m,n)}M \times N$, Tf_m is even and left linear on any fiber T_nN. But for general m it will not be a smooth map. However, if we do not look at a fixed value of m, but see it as a variable, we get a map $M \times TN \to TQ$. Using the identification $T(M \times N) \cong TM \times TN$ [2.21], we can see this map as the composition of the smooth maps $\underline{0} \times id(TN) : M \times TN \to TM \times TN$ and Tf. It follows that the *family* of maps Tf_m with varying $m \in M$ is smooth.

In a similar way we can fix a point $n \in N$ and define $\widehat{f}_n : M \to Q$ by $\widehat{f}_n(m) = f(m,n)$. And then we define the generalized tangent tangent map $T\widehat{f}_n$ by

$$T\widehat{f}_n : X_m \mapsto \langle\!\langle (X_m, \underline{0}_n) \| Tf \rangle\!\rangle \ .$$

As for the maps f_m, this is an even (left) linear map when restricted to a fiber T_mM, but $T\widehat{f}_n$ need not be smooth. However, when we see n as a variable, we get a smooth map $TM \times N \to TQ$ (the restriction of Tf to TM times the zero section), which *is* smooth.

More generally, let $O \subset M \times N$ be an open set and $f : O \to Q$ a smooth map. For a fixed $m \in M$ we define the set $O_m \subset N$ by $O \cap \{m\} \times N = \{m\} \times O_m$. It is an open subset of N on which we have a map $f_m : O_m \to Q$ defined by $f_m(n) = f(m,n)$. As before, if m does not have real coordinates there is no reason to assume that f_m is smooth, but we can define a generalized tangent map $Tf_m : TO_m \to TQ$ by the formula $\langle\!\langle Y_n \| Tf_m \rangle\!\rangle = \langle\!\langle (\underline{0}_m, Y_n) \| Tf \rangle\!\rangle$. To show that the family of tangent maps Tf_m is smooth, we need a more careful discussion about the domain of definition of this family. Since we will not need it, we will not go into these details.

3.20 Lemma. *Let M, N, and Q be $\mathcal{A}$-manifolds of which N is connected, and let $f : M \times N \to Q$ be smooth. Then the following three assertions are equivalent.*

(i) *There exists a smooth map $g : M \to Q$ such that $f = g \circ \pi_M$.*
(ii) *There exists a collection of vector fields X on N such that at each $n \in N$ the X_n generate T_nN and such that $(\underline{0} \times X) \diamond Tf = 0$.*
(iii) $\forall (m,n) \in M \times N : T_nf_m = 0$.

In the case Q is an $\mathcal{A}$-vector space, the condition $(\underline{0} \times X) \diamond Tf = 0$ in (ii) can be replaced by $(\underline{0} \times X)f = 0$.

Proof. • (ii) $\Leftrightarrow$ (iii) By definition we have $\langle\!\langle (\underline{0} \times X)_{(m,n)} \| Tf \rangle\!\rangle = \langle\!\langle X_n \| T_nf_m \rangle\!\rangle$, hence if $T_nf_m = 0$ then $(\underline{0} \times X) \diamond Tf = 0$. On the other hand, if the X_n generate T_nN, it follows that (ii) implies that T_nf_m is zero on a generating set, i.e., $T_nf_m = 0$.

• (i) $\Leftrightarrow$ (ii) Since $Tf = Tg \circ T\pi_M$ and $(\underline{0} \times X) \diamond T\pi_M = 0$, we have immediately the implication (i) $\Rightarrow$ (ii). To prove the reverse implication we proceed as follows. For each $(m,n) \in M \times N$ we choose a chart $W_{m,n}$ of Q around $f(m,n)$ and then, by continuity of f, charts $U_{m,n}$ of M around m and $V_{m,n}$ of N around n, $V_{m,n}$ connected, such that $f(U_{m,n} \times V_{m,n}) \subset W_{m,n}$. It follows that $\mathcal{U} = \{(U_{m,n} \times V_{m,n} \mid (m,n) \in M \times N\}$ is an open cover of $M \times N$ with the property that for each $U \times V \in \mathcal{U}$ there exists a chart W such that $f(U \times V) \subset W$.

We now first prove a local result. For $U \times V \in \mathcal{U}$ and $W \subset Q$ such that $f(U \times V) \subset W$, we choose coordinates x^i on U, y^j on V, and z^k on W. For any vector field X on N its local expression is given as $X|_V = \sum_j X^j(y)\,\partial_{y^j}$. And then the condition $(\underline{0} \times X) \diamond Tf = 0$ becomes

$$\sum_j X^j(y) \cdot \frac{\partial f^k}{\partial y^j}(x,y) \cdot \frac{\partial}{\partial z^k} = 0 .$$

Since we have this equation for vector fields X that generate at each point the tangent space, we conclude that we have $(\partial_{y^j} f^k)(x,y) = 0$. By [III.3.15] we conclude that there exist functions $g^k_{U\times V} : U \to \mathcal{A}$ such that $f^k|_{U\times V} = g^k_{U\times V} \circ \pi_U$. Combining the functions $g^k_{U\times V}$ into a single function $g_{U\times V} : U \to W$, we conclude that for each $U \times V \in \mathcal{U}$ there exists a function $g_{U\times V} : U \to Q$ such that $f|_{U\times V} = g_{U\times V} \circ \pi_M|_{U\times V}$.

Once we have the local result, we have to glue it together to get the global result. For each $m \in M$ we define $\mathcal{U}_m = \{U \times V \in |m \in U\}$. Obviously $\mathcal{U}_m$ covers $\{m\} \times N$ which is connected. For $U_1 \times V_1, U_2 \times V_2 \in \mathcal{U}_m$ such that $V_1 \cap V_2 \neq \phi$ we must have $g_{U_1\times V_1}(m) = g_{U_2\times V_2}(m)$, and thus by connectedness of N, there exists $g(m) \in Q$ such that $\forall U \times V \in \mathcal{U}_m : g_{U\times V}(m) = g(m)$.† In this way we obtain a function $g : M \to Q$. But for any $U \times V \in \mathcal{U}$ and any $m \in U$ we have $g_{U\times V}(m) = g(m)$ and thus g is smooth. Moreover, $g|_U \circ \pi_M|_{U\times V} = g_{U\times V} \circ \pi_M|_{U\times V} = f|_{U\times V}$ and thus $g \circ \pi_M = f$.

• If Q is an $\mathcal{A}$-vector space and if (ii) holds, then by [3.5] obviously $(\underline{0} \times X)f = 0$. On the other hand, if $(\underline{0} \times X)f = 0$ holds for vector fields that span the tangent space of N at each point, we can assume that it holds for *all* vector fields X on N. It thus holds in particular for all homogeneous vector fields, which span TN. For these we can apply [3.5] in the other direction to conclude that $(\underline{0} \times Y) \diamond Tf = 0$ for a set of vector fields on N which generates TN (i.e., all homogeneous vector fields). But then it also holds for the original set of vector fields X. QED

3.21 Corollary. *Let N and Q be $\mathcal{A}$-manifolds, N connected, and let $f : N \to Q$ be smooth. then the following three assertions are equivalent.*

(i) *f is constant.*
(ii) *$Tf = 0$.*
(iii) *There exists a collection of vector fields X on N such that at each $n \in N$ the X_n generate T_nN and such that $X \diamond Tf = 0$.*

In the case Q is an $\mathcal{A}$-vector space, the condition $X \diamond Tf = 0$ in (ii) can be replaced by $Xf = 0$.

---

†This is an often neglected exercise in point set topology. Let X be a connected topological space, $\mathcal{U}$ an open cover of X, and $x_o \in X$ an arbitrary point. If $C(x_o)$ denotes the set of $x \in X$ for which there exists a finite sequence $U_0, \ldots, U_n \in \mathcal{U}$ such that $x_o \in U_0$, $x \in U_n$, and $U_i \cap U_{i+1} \neq \phi$, then $C(x_o) = X$.

4. Integration of vector fields

In this section we investigate how to integrate homogeneous vector fields: even vector fields should have an even time parameter, odd vector fields an odd one. It turns out that for even vector fields the results are the same as in the ordinary case: there exists a flow on an open connected subset of time times the manifold, and this flow has the usual properties of a 1-parameter group of local diffeomorphisms. For odd vector fields a flow exists if and only if its auto commutator is zero. If that is the case the flow is defined on the whole of the (odd) time times the manifold.

4.1 Definitions. An *interval* is a connected subset I of $\mathcal{A}_\alpha$; it is called *even* if $\alpha = 0$, *odd* if $\alpha = 1$.

Let $I \subset \mathcal{A}_\alpha$ be an open interval, $t_o \in \mathbf{B}I$ a real number, and U an open set in M. An open subset W of $I \times U \subset \mathcal{A}_\alpha \times M$ is called *a wave in $I \times U$ around t_o* if $\{t_o\} \times U \subset W$ and if $W \cap \mathcal{A}_\alpha \times \{m\}$ is an open interval for all $m \in U$. In particular, a wave is connected if U is. A *border point for W* is a point $(t_1, m_1) \in (\mathbf{B}I) \times U \setminus W$ such that $J \times \{m_1\} \cap W \neq \phi$ for any open interval $t_1 \in J \subset \mathcal{A}_\alpha$. Existence of a border point requires that W is not the whole of $I \times U$.

4.2 Lemma. *There exists only one odd open interval: $\mathcal{A}_1$ itself. An even open interval I is of the form $I = \{\, a \in \mathcal{A}_0 \mid r_\ell < \mathbf{B}a < r_r \,\}$ for some $r_\ell, r_r \in \mathbf{R} \cup \{-\infty, \infty\}$.*

4.3 Definitions. Let X be a homogeneous vector field of parity α on M. A *local flow for X* is a smooth map $\phi : W \to M$ defined on a wave W in $I \times U$ ($U \subset M$, I an interval) around $t_o \in I$ satisfying the condition

$$T\phi \circ \partial_t = X \circ \phi \qquad \text{or equivalently} \qquad \partial_t \diamond T\phi = \phi \diamond X \;. \tag{4.4}$$

In this condition t denotes the (canonical) coordinate on $\mathcal{A}_\alpha$ and ∂_t is the shorthand for the vector field on $\mathcal{A}_\alpha \times M$ officially denoted by $\partial_t \times \underline{0}$ [2.24]. The coordinate t is usually called the time coordinate. Condition (4.4) can be stated in words as saying that ∂_t and X are related by ϕ; it makes sense because both X and ∂_t are of parity α.

An *initial condition at time t_o* for a local flow $\phi : W \to M$ for X is a smooth function $f : U \to M$ such that $\phi(t_o, m) = f(m)$ for all $m \in U$. Note that we need that t_o is real to be sure that $\phi(t_o, \_)$ is a smooth function on $\{t_o\} \times U$.

The flow $\phi_X : W_X \to M$ (without the adjective local) for the homogeneous vector field X is a local flow defined on a maximal wave W_X in $\mathcal{A}_\alpha \times M$ around 0 and satisfying the initial condition $id(M)$ at time $t = 0$.

A homogeneous vector field X is said to be *integrable* if it admits a flow. We will show that this is the case if and only if $[X, X] = 0$, a condition automatically satisfied for even vector fields.

4.5 Discussion. We start our investigations for *even* vector fields X only and thus assume that we have an *even* time parameter $t \in \mathcal{A}_0$. To see more clearly what condition (4.4) amounts to, we assume that we have a local flow $\phi : W \to M$ defined on a wave in $I \times U$ around t_o. We assume furthermore that U is a chart and that there exists a chart V such that $\phi(W) \subset V$. Let (x, ξ) be local (even and odd) coordinates on U and let (y, η) be local coordinates on V. Using the notational shorthand introduced in [III.3.17], we write $(y, \eta) = \phi(t, x, \xi)$ with

$$y = \sum_i \phi_{2i}(t, x) \cdot \xi^{2i} \qquad \text{and} \qquad \eta = \sum_i \phi_{2i+1}(t, x) \cdot \xi^{2i+1} .$$

From this we deduce that $\langle\!\langle \partial_t|_{(t,x,\xi)} \| T\phi \rangle\!\rangle \equiv (\partial_t \diamond T\phi)(t, x, \xi)$ is given by

$$\langle\!\langle \partial_t|_{(t,x,\xi)} \| T\phi \rangle\!\rangle = \Big(\sum_i \frac{\partial \phi_{2i}}{\partial t}(t, x) \cdot \xi^{2i}\Big) \cdot \frac{\partial}{\partial y} + \Big(\sum_i \frac{\partial \phi_{2i+1}}{\partial t}(t, x) \cdot \xi^{2i+1}\Big) \cdot \frac{\partial}{\partial \eta} .$$

Using the same shorthand, we write the tangent vector $X(y, \eta)$ as

$$X(y, \eta) = \Big(\sum_i X_{2i}(y) \cdot \eta^{2i}\Big) \cdot \frac{\partial}{\partial y} + \Big(\sum_i X_{2i+1}(y) \cdot \eta^{2i+1}\Big) \cdot \frac{\partial}{\partial \eta} .$$

Substituting $(y, \eta) = \phi(t, x, \xi)$, using $T\phi \circ \partial_t = X \circ \phi$ and expanding as in [III.2.27], we obtain the following two sets of equations:

$$\sum_i \frac{\partial \phi_{2i}}{\partial t}(t, x) \cdot \xi^{2i} =$$

$$\sum_j \Big(\sum_k X_{2j}^{(k)}(\phi_0(t, x)) \cdot \Big(\sum_{i>0} \phi_{2i}(t, x) \cdot \xi^{2i}\Big)^k\Big) \cdot \Big(\sum_\ell \phi_{2\ell+1}(t, x) \cdot \xi^{2\ell+1}\Big)^{2j} ,$$

$$\sum_i \frac{\partial \phi_{2i+1}}{\partial t}(t, x) \cdot \xi^{2i+1} =$$

$$\sum_j \Big(\sum_k X_{2j+1}^{(k)}(\phi_0(t, x)) \cdot \Big(\sum_{i>0} \phi_{2i}(t, x) \cdot \xi^{2i}\Big)^k\Big) \cdot \Big(\sum_\ell \phi_{2\ell+1}(t, x) \cdot \xi^{2\ell+1}\Big)^{2j+1} .$$

Comparing the various powers of ξ gives us the equations

(4.6.a) $$\frac{\partial \phi_0}{\partial t}(t, x) = X_0(\phi_0(t, x))$$

(4.6.b) $$\frac{\partial \phi_1}{\partial t}(t, x) = X_1(\phi_0(t, x)) \cdot \phi_1(t, x)$$

(4.6.c) $$\frac{\partial \phi_{2p}}{\partial t}(t, x) = X_0^{(1)}(\phi_0(t, x)) \cdot \phi_{2p}(t, x) + F_{2p}(t, x) \qquad p \geq 1$$

(4.6.d) $$\frac{\partial \phi_{2p+1}}{\partial t}(t, x) = X_1(\phi_0(t, x)) \cdot \phi_{2p+1}(t, x) + F_{2p+1}(t, x) \qquad p \geq 1 ,$$

where the F_i are expressions in the X_j and ϕ_k with $k < i$.

For arguments sake we now assume that we have at time $t_o \in I$ an initial condition $f : U \to V$, $f(x, \xi) = (y, \eta)$, which we expand into powers of ξ as $y = \sum_i f_{2i}(x) \cdot \xi^{2i}$ and $\eta = \sum_i f_{2i+1}(x) \cdot \xi^{2i+1}$. On $W \cap \{t_o\} \times U$, the requirement $\phi(t_o, m) = f(m)$ translates into the set of initial conditions

$$\phi_i(t_o, x) = f_i(x) \ .$$

Finding a ϕ satisfying (4.4) on W is obviously equivalent to solving the equations (4.6.a-d). These equations are ordinary differential equations in the ordinary real valued smooth functions ϕ_i. These functions are all defined on $\mathbf{B}W$; ϕ_0 takes its values in $\mathbf{B}V$, the others can take their values "anywhere." Since the F_i do not involve the ϕ_j with $j \geq i$, we can solve these equations recursively, starting with ϕ_0.

Solving (4.6.a) might not be possible on the whole of W. However, uniqueness of solutions of ordinary differential equations with initial condition, together with the fact that W is a wave imply that a solution of (4.6.a) on W with initial condition $\phi_0(t_o, m) = f_0(m)$ is unique. Moreover, if such a unique solution for ϕ_0 exists, the form of the equations (4.6.b-d) for ϕ_i with $i > 0$ (they are of the form $y' = ay + b$), the fact that these ϕ_i can take their values anywhere, and the fact that W is a wave together imply that we can solve these equations on W in a unique way, given an initial condition f_i at time t_o.

Applying the body map $\mathbf{B}$ we obtain an even vector field $\mathbf{B}X$ on $\mathbf{B}M$; finding a local flow ψ on $\mathbf{B}W$ for $\mathbf{B}X$ reduces to the single equation (4.6.a) in ψ instead of ϕ_0. Moreover, if ϕ is a local flow for X on W, we automatically have a local flow $\psi = \phi_0 = \mathbf{B}\phi$ for $\mathbf{B}X$ on $\mathbf{B}W$. In [4.7] we summarize the above discussion; in [4.8] we generalize this result to arbitrary open subsets of M.

4.7 Lemma. *Let W be a wave in $I \times U$ around t_o with U a chart and I an interval, let V be another chart for M, and let $f : U \to V$ be a smooth function. There exists a local flow $\phi : W \to V$ for X with initial condition f at time t_o if and only if there exists a local flow $\psi : \mathbf{B}W \to \mathbf{B}V$ for $\mathbf{B}X$ with initial condition $\mathbf{B}f$ at time t_o. If ϕ or ψ exists, it is unique.*

4.8 Lemma. *Let W_o be a wave in $I_o \times U_o$ around t_o with U_o an open set in M and I_o an interval, and let $f : U_o \to M$ be a smooth function. There exists a local flow $\phi_o : W_o \to M$ for X with initial condition f at time t_o if and only if there exists a local flow $\psi : \mathbf{B}W_o \to \mathbf{B}M$ for $\mathbf{B}X$ with initial condition $\mathbf{B}f$ at time t_o. If ϕ_o or ψ exists, it is unique.*

Proof. If the local flow ϕ_o exists, $\psi = \mathbf{B}\phi_o$ is a local flow as desired. Thus suppose we know that ψ exists, and let us show the existence of ϕ_o. Define $\mathcal{W}$ as the set of all waves W in $I_o \times U_o$ around t_o that are contained in W_o, on which there exists a unique local flow ϕ_W with initial condition f at time t_o and such that $\mathbf{B}\phi_W = \psi$.

Our first claim is that $\mathcal{W}$ is not empty. For each $m \in U_o$ we choose local charts $m \in U_m$ and V_m and an open interval $t_o \in I_m$ such that ψ is a local flow on $\mathbf{B}I_m \times \mathbf{B}U_m$ with values in $\mathbf{B}V_m$ and with initial condition $\mathbf{B}f$ at time t_o. According to [4.7] there is a unique local flow ϕ_m on $I_m \times U_m$ with initial condition f at time t_o. This ϕ_m satisfies $\mathbf{B}\phi_m = \psi$. By uniqueness in [4.7] two such flows ϕ_m and $\phi_{m'}$ coincide on the intersection of their domains. We conclude that $W = \cup_{m \in U}(I_m \times U_m)$ is an element of $\mathcal{W}$.

Once we know that $\mathcal{W}$ is not empty, let W_1 be the maximal element of $\mathcal{W}$ with local flow ϕ_1. If W_1 is strictly smaller than W_o, there exists a border point $(t_1, m_1) \in W_o \setminus W_1$ for the wave W_1. Now choose local charts $m_1 \in U$ and V and an open interval $t_1 \in I_1$ such that ψ is a local flow on $\mathbf{B}I_1 \times \mathbf{B}U$ with values in $\mathbf{B}V$. By definition of t_1 there exists a real value $t_2 \in I_1$ such that $(t_2, m_1) \in W_1$. Since W_1 is open, there exists an open interval $I_2 \ni t_2$ such that $I_2 \times U \subset W_1$ (shrinking U if necessary). Now define $W_2 = W_1 \cap I_1 \times U$. It is a wave in $I_1 \times U$ around t_2. Since $\mathbf{B}I_1 \times \mathbf{B}U \subset \mathbf{B}W_o$, there exists a unique local flow ϕ_2 on $I_1 \times U$ with initial condition $\phi_1(t_2, \_)$ at time t_2 [4.7]. But now $\phi_2|_{W_2}$ and $\phi_1|_{W_2}$ are two local flows on the wave W_2 with the same initial condition, hence they coincide. We thus can extend ϕ_1 in a unique way to a local flow on the wave $W_1 \cup I_1 \times U$. Since $\mathbf{B}\phi_2 = \psi$ we deduce that $W_1 \cup I_1 \times U \in \mathcal{W}$. This contradicts maximality of W_1 because $(t_1, m_1) \notin W_1$. We conclude that necessarily $W_1 = W_o$.

We thus have shown that ϕ_o exists if and only if ψ exists. Uniqueness follows with a similar argument, using that in showing that $\mathcal{W}$ is not empty, we did not need that $\mathbf{B}\phi = \psi$ (it was a consequence of the uniqueness in [4.7]). QED

4.9 Proposition. *Any even vector field X admits a unique flow $\phi_X : W_X \to M$.*

Proof. For any point $m \in M$ the existence of solutions of ordinary differential equations tells us that there exists a local flow ψ on some (small) open subset $(0, \mathbf{B}m) \in \mathbf{B}I \times \mathbf{B}U$ with initial condition $id(\mathbf{B}U)$ at time 0. Going from $\mathbf{B}U$ to U and gluing together these local flows, one deduces that there exists a local flow ϕ defined on a wave W in $\mathcal{A}_0 \times M$ around 0 with initial condition $id(M)$ at time 0.

If we have a family of local flows on waves around 0 with initial condition $id(M)$, [4.8] implies that they must coincide on the intersection of their domains. There thus exists a unique local flow defined on the union of their domains. It thus makes sense to talk about the local flow ϕ_X with *maximal* wave W_X in $\mathcal{A}_0 \times M$ with initial condition $id(M)$; uniqueness is guaranteed by [4.8]. QED

4.10 Remark. It follows from [4.8] that (ϕ_X, W_X) is *the* flow of X if and only if $(\mathbf{B}\phi_X, \mathbf{B}W_X)$ is *the* flow $(\psi_{\mathbf{B}X}, \mathbf{B}W_{\mathbf{B}X})$, the latter being the flow of an ordinary vector field on an $\mathbf{R}$-manifold. In particular, the domain of definition W_X of X is completely determined by the domain $\mathbf{B}W_{\mathbf{B}X} \subset \mathbf{R} \times \mathbf{B}M$ of $\mathbf{B}X$.

4.11 Examples. • Let $M = \mathcal{A}_0$ be the $\mathcal{A}$-manifold with (global) coordinate x on which we define the even vector field $X(x) = x^2 \cdot \partial_x$. Since there are no odd coordinates, we only have to solve the equation $\frac{\partial \phi_0}{\partial t}(t,x) = (\phi_0(t,x))^2$. An elementary calculation then shows that the flow ϕ_X on W_X is given by

$$\phi_X(t,x) = \frac{x}{1-tx} \quad , \quad W_X = \{\,(t,x) \mid \mathbf{B}(tx) < 1\,\}\,.$$

• Now let E be an $\mathcal{A}$-vector space of dimension $1|1$. On $M = E_0$ with coordinates (x,ξ) with respect to some basis we define the even vector field $X(x,\xi) = \mathrm{e}^x \cdot \partial_x + x\xi \cdot \partial_\xi$. In this case we have a ϕ_0 and a ϕ_1, as well as a $X_0(y) = \mathrm{e}^y$ and a $X_1(y) = y$. Writing $(y,\eta) = \phi_X(t,x,\xi)$, we have $y = \phi_0(t,x)$ and $\eta = \phi_1(t,x)$. Equation (4.4) translates as $\partial y/\partial t = \mathrm{e}^y$ and $\partial\eta/\partial t = y\eta$; in terms of ϕ_0 and ϕ_1 this becomes system (4.6):

$$\begin{aligned}
\frac{\partial \phi_0}{\partial t}(t,x) &= X_0(\phi_0(x,t)) = \exp(\phi_0(t,x)) \\
\frac{\partial \phi_1}{\partial t}(t,x) &= X_1(\phi_0(t,x)) \cdot \phi_1(t,x) = \phi_0(t,x) \cdot \phi_1(t,x)\,.
\end{aligned}$$

A slightly longer calculation than in the previous example shows that here the flow $(y,\eta) = \phi_X(t,x,\xi)$ on W_X is given by

$$\begin{pmatrix} y \\ \eta \end{pmatrix} = \begin{pmatrix} x - \ln(1 - t\,\mathrm{e}^x) \\ \exp\Big((x+1)\,t + (\mathrm{e}^{-x} - t)\,\ln(1 - t\,\mathrm{e}^x)\Big)\xi \end{pmatrix}$$

$$W_X = \{\,(t,x,\xi) \mid \mathbf{B}(t\,\mathrm{e}^x) < 1\,\}\,.$$

• Finally, let E be an $\mathcal{A}$-vector space of dimension $1|2$. On $M = E_0$ with coordinates (x,ξ_1,ξ_2) we define the vector field $X(x,\xi_1,\xi_2) = (\frac{1}{2}x^3 + x\xi_1\xi_2) \cdot \partial_x + \xi_2 \cdot \partial_{\xi_1} - \xi_1 \cdot \partial_{\xi_2}$, which is even. Since we have two odd coordinates, we have a ϕ_0, a ϕ_1 (a 2×2 matrix), and a ϕ_2, as well as an $X_0(y) = \frac{1}{2}y^3$, an $X_2(y) = y$, and an $X_1(y) = \left(\begin{smallmatrix} 0 & 1 \\ -1 & 0 \end{smallmatrix}\right)$. Writing $(y,\eta_1,\eta_2) = \phi_X(t,x,\xi_1,\xi_2)$ we have the equations $y = \phi_0(t,x) + \phi_2(t,x)\xi_1\xi_2$ and $\left(\begin{smallmatrix} \eta_1 \\ \eta_2 \end{smallmatrix}\right) = \phi_1(t,x)\left(\begin{smallmatrix} \xi_1 \\ \xi_2 \end{smallmatrix}\right)$. Equation (4.4) translates as $\partial_t y = \frac{1}{2}y^3 + y\eta_1\eta_2$, $\partial_t \eta_1 = \eta_2$, and $\partial_t\eta_2 = -\eta_1$; in terms of ϕ_0, ϕ_1, and ϕ_2 this becomes system (4.6):

$$\begin{aligned}
\frac{\partial \phi_0}{\partial t}(t,x) &= \tfrac{1}{2}(\phi_0(t,x))^3 \\
\frac{\partial \phi_1}{\partial t}(t,x) &= X_1(\phi_0(t,x)) \cdot \phi_1(t,x) = \begin{pmatrix} 0 & 1 \\ -1 & 0 \end{pmatrix} \cdot \phi_1(t,x) \\
\frac{\partial \phi_2}{\partial t}(t,x) &= X_0^{(1)}(\phi_0(t,x)) \cdot \phi_2(t,x) + X_2(\phi_0(t,x)) \cdot \mathrm{Det}(\phi_1(t,x)) \\
&= \tfrac{3}{2}(\phi_0(t,x))^2\phi_2(t,x) + \phi_0(t,x) \cdot \mathrm{Det}(\phi_1(t,x)) \\
&= \tfrac{3}{2}(\phi_0(t,x))^2\phi_2(t,x) + \phi_0(t,x) \qquad \text{using the previous equation.}
\end{aligned}$$

Careful analysis, using the initial conditions $\phi_0(0,x) = x$, $\phi_1(0,x) = \mathbf{1}$, and $\phi_2(0,x) = 0$, shows that the flow $(y,\eta_1,\eta_2) = \phi_X(t,x,\xi_1,\xi_2)$ on W_X is given by

$$\begin{pmatrix} y \\ \\ \eta_1 \\ \\ \eta_2 \end{pmatrix} = \begin{pmatrix} \dfrac{x}{\sqrt{1-tx^2}} + \dfrac{xt - \frac{1}{2}x^3t^2}{\sqrt{(1-tx^2)^3}}\,\xi_1\xi_2 \\ \\ \begin{pmatrix} \cos t & \sin t \\ -\sin t & \cos t \end{pmatrix}\begin{pmatrix} \xi_1 \\ \xi_2 \end{pmatrix} \end{pmatrix} \quad , \; W_X = \{\,(t,x,\xi) \mid \mathbf{B}(tx^2) < 1\,\}\,.$$

4.12 Lemma. *Let X be an even vector field on M and $\phi_X : W_X \to M$ its flow. If $\phi_X(t,m)$ is defined, then $\phi_X(-t,\phi_X(t,m))$ is defined; if $\phi_X(t,\phi_X(s,m))$ is defined, then $\phi_X(s+t,m)$ is defined; if s and t have the same sign and if $\phi_X(s+t,m)$ is defined, then $\phi_X(t,\phi_X(s,m))$ is defined.*

Proof. The statements say something about the domain W_X. Since W_X is completely determined by its body $\mathbf{B}W_X \equiv \mathbf{B}W_{\mathbf{B}X} \subset \mathbf{R} \times \mathbf{B}M$, the result follows from the fact that these statements are true for ordinary vector fields on $\mathbf{R}$-manifolds (e.g., [HS, §8.7]).[†] QED

4.13 Lemma. *If both $\phi_X(t_o,\phi_X(s_o,m_o))$ and $\phi_X(s_o+t_o,m_o)$ are defined, they coincide.*

Proof. Continuity of ϕ_X and openness of W_X imply the existence of open intervals $s_o \in I_s$ and $t_o \in I_t$ and an open set $m \in U \subset M$ such that both terms are defined for all $(t,s,m) \in I_t \times I_s \times U$. Moreover, the existence of both terms implies that $\phi_X(s_o+0,m_o)$ and $\phi_X(s_o+t_o,m_o)$ are defined as well as $\phi_X(0,\phi_X(s_o,m_o))$ and $\phi_X(t_o,\phi_X(s_o,m_o))$. Since W_X is a wave, we may assume that $0 \in I_t$.

Now define two maps $\Phi_i : I_t \times I_s \times U \to \mathcal{A}_0 \times M$ by

$$\begin{aligned} \Phi_1 &: (t,s,m) \mapsto (s+t,m) \mapsto (s+t,\phi_X(s+t,m)) \\ \Phi_2 &: (t,s,m) \mapsto (t,s,\phi_X(s,m)) \mapsto (s+t,\phi_X(t,\phi_X(s,m))) \,. \end{aligned}$$

Using the given decomposition and the fact that ϕ_X is the flow of X, it is not hard to show that both Φ_i satisfy the relation $T\Phi_i \circ \partial_t = (\partial_s \times X) \circ \Phi_i$ [2.24]. We conclude that both Φ_i are local flows for the vector field $(\partial_s \times X)$. Since they have the same initial condition at time $t = 0$, they coincide. QED

4.14 Discussion/Definition. In the context of $\mathbf{R}$-manifolds, the above results are usually stated by saying that the family of maps $\{\phi_{X,t}\}$, defined by $\phi_{X,t}(m) = \phi_X(t,m)$, forms a 1-parameter group of local diffeomorphisms of M; local because $\phi_{X,t}$ need not be defined on the whole of M. In the context of $\mathcal{A}$-manifolds such a formulation is not possible because for non-real t the map $\phi_{X,t}$ is in general not smooth.

However, the group property as well as the local diffeomorphism property can be stated in a remarkably simple statement about a *global* diffeomorphism. It follows from [4.12] that the smooth map

$$\Phi_X : W_X \to W_X \quad , \quad (t,m) \mapsto (-t,\phi_X(t,m))$$

[†] For ordinary vector fields, these results follow from the fact that Ω is defined as the collection of all *maximal* integral curves. And then the statement that Ω is *open* needs a proof. In our context we cannot speak of individual integral curves (they are not smooth). Hence the approach in which Ω is automatically open. But then it seems not to be easy to prove the results of [4.12]. I thus "cheat" and rely upon the classical result.

is well defined. It then follows from [4.13] that Φ_X is a diffeomorphism satisfying the equality $\Phi_X \circ \Phi_X = id(W_X)$. In this way the fact that $\phi_{X,-t} = (\phi_{X,t})^{-1}$ is captured in the statement $(\Phi_X)^2 = id(W_X)$ about the global diffeomorphism Φ_X. Since we will use this Φ_X sometimes, we will call it *the global flow of X* to distinguish it from the flow ϕ_X without adjective.

In the same spirit we can capture the group property. We define $W'_X \subset \mathcal{A}_0 \times W_X$ as $W'_X = \{ (s,t,m) \mid (t,m) \in W_X \ \& \ (s, \phi_X(t,m)) \in W_X \}$. It follows from [4.12] that the smooth map

$$\Phi'_X : W'_X \to W'_X \quad , \quad (s,t,m) \mapsto (-s-t, s, \phi_X(t,m))$$

is well defined. It then follows from [4.13] that Φ'_X is a diffeomorphism satisfying the equality $\Phi'_X \circ \Phi'_X \circ \Phi'_X = id(W'_X)$. Taking $s = 0$ in the relation $(\Phi'_X)^3 = id(W'_X)$ and using the initial condition for ϕ_X gives us (again) that $\phi_{X,-t} = (\phi_{X,t})^{-1}$. This plus the general case of $(\Phi'_X)^3 = id(W'_X)$ gives us the group property $\phi_{X,s} \circ \phi_{X,t} = \phi_{X,s+t}$.

4.15 Discussion. So far we have discussed the case of even vector fields. Let us now turn our attention to odd vector fields. To give an odd flavor to this discussion, we will denote the odd time parameter by τ rather than by t, in accordance with our general principle to denote odd coordinates by greek letters.

In order to investigate the existence of a flow ϕ for an odd vector field X, we start looking at condition (4.4) in a local chart. Let U and V be local charts such that $\phi(\mathcal{A}_1 \times U) \subset V$. Because we have an odd time parameter, we will use a completely different approach. We denote by x the (even and odd) coordinates on U, by y the coordinates on V and we expand $y = \phi(\tau, x)$ as

$$y = \phi_0(x) + \tau \cdot \phi_1(x) \ .$$

It follows that $\langle\!\langle \partial_\tau|_{(\tau,x)} \| T\phi \rangle\!\rangle \equiv (\partial_\tau \diamond T\phi)(\tau, x)$ is given by

$$\langle\!\langle \partial_\tau|_{(\tau,x)} \| T\phi \rangle\!\rangle = \phi_1(x) \cdot \frac{\partial}{\partial y} \ .$$

Writing $X(y) = X_c(y) \cdot \partial_y$, substituting $y = \phi(\tau, x)$, and expanding into powers of τ gives us

$$X(y) = \Big(X_c(\phi_0(x)) + \tau \cdot \phi_1(x) \cdot \frac{\partial X_c}{\partial y}(\phi_0(x)) \Big) \cdot \frac{\partial}{\partial y} \ .$$

We conclude that condition (4.4) translates as the equations

$$\text{(4.16)} \qquad \phi_1(x) = X_c(\phi_0(x)) \quad , \quad \phi_1(x) \cdot \frac{\partial X_c}{\partial y}(\phi_0(x)) = 0 \ .$$

It now becomes very important to realize that X is supposed to be an odd vector field. Substituting the first equation of (4.16) in the last one, one should recognize (see [1.23]) the local expression (up to a factor 2) of the coefficients of the commutator $[X, X]$ at the point $\phi_o(x) = \phi(0, x)$. Moreover, since 0 always lies in the odd open interval $\mathcal{A}_1$ (it is *the only* real value in $\mathcal{A}_1$), we can add the initial condition $\phi(0,x) = f(x)$. The whole set of equations, initial condition plus (4.4), thus becomes equivalent to the set of equations

$$\phi_0(x) = f(x) \quad , \quad \phi_1(x) = X_c(f(x)) \quad , \quad [X,X](f(x)) = 0 \; .$$

This discussion is summarized in [4.17].

4.17 Proposition. *Let X be an odd vector field on M and let $f : M \to M$ be smooth. There exists a (unique) local flow $\phi : \mathcal{A}_1 \times M \to M$ for X with initial condition f at time 0 if and only if $[X,X] = 0$ on $\mathrm{im}(f)$.*

In particular, the flow ϕ_X of an odd vector field on M exists if and only if $[X,X] = 0$. If this condition is satisfied, it is unique and $W_X = \mathcal{A}_1 \times M$. It has the following properties.

(i) *For any open set $U \subset M : \phi_X(\mathcal{A}_1 \times U) \subset U$.*

(ii) $\forall \tau, \sigma \in \mathcal{A}_1 \forall m \in M : \phi_X(\tau, \phi_X(\sigma, m)) = \phi_X(\sigma + \tau, m)$.

Proof. Property (i) follows from the facts that open sets are saturated with nilpotent vectors and that the initial condition is the identity $f(x) = x$. One way to prove property (ii) is to use a local chart and to make an explicit calculation, judiciously using $[X,X] = 0$. Another way is to copy the proof of the case for even vector fields. Consider the two smooth maps $\Phi_i : \mathcal{A}_1 \times \mathcal{A}_1 \times M \to \mathcal{A}_1 \times M$:

$$\begin{aligned} \Phi_1 &: (\tau, \sigma, m) \mapsto (\sigma + \tau, \phi_X(\sigma + \tau, m)) \\ \Phi_2 &: (\tau, \sigma, m) \mapsto (\sigma + \tau, \phi_X(\tau, \phi_X(\sigma, m))) \; . \end{aligned}$$

As in the even case, one shows that both maps satisfy $T\Phi_i \circ \partial_\tau = (\partial_\sigma \times X) \circ \Phi_i$ and $\Phi_i(0, \sigma, m) = (\sigma, \phi_X(\sigma, m))$. We deduce from the uniqueness of flows that the Φ_i coincide. QED

4.18 Discussion. If X is an integrable odd vector field on M, the local expression for its flow is essentially given by the simple formula

$$\phi_X(\tau, m) = m + \tau \cdot X_c(m) \; ,$$

where we obviously confuse the coordinates of a point m with the coordinates of a tangent vector $X_m = X_c(m) \cdot \partial_m$.

Unlike the case for even vector fields, for odd vector fields we do not have a domain problem: ϕ_X, if it exists, is defined on the whole of $\mathcal{A}_1 \times M$. As for even vector fields, we can form the global flow $\Phi_X : \mathcal{A}_1 \times M \to \mathcal{A}_1 \times M$, $\Phi_X(\tau, m) = (-\tau, \phi_X(\tau, m))$ with the property $(\Phi_X)^2 = id$. Likewise we can construct the global diffeomorphism $\Phi'_X : \mathcal{A}_1 \times \mathcal{A}_1 \times M \to \mathcal{A}_1 \times \mathcal{A}_1 \times M$ with the property $(\Phi'_X)^3 = id$.

5. Commuting flows

In this section we show that two integrable homogeneous vector fields commute if and only if their flows commute. As a byproduct we show that if X is an integrable homogeneous vector field with flow ϕ_t and Y an arbitrary vector field, then the commutator $[X, Y]$ equals $\partial_t T\phi_{-t} Y$ at $t = 0$.

5.1 Definitions. Let X and Y be two homogeneous integrable vector fields on an $\mathcal{A}$-manifold M with flows ϕ_X and ϕ_Y respectively. We say that *the flows commute* if we have the equality $\phi_X(t, \phi_Y(s, m)) = \phi_Y(s, \phi_X(t, m))$ for all points (t, s, m) for which both sides make sense. It is the purpose of this section to show that the flows commute if and only if the vector fields commute, i.e., $[X, Y] = 0$. The proof of this statement requires some preparatory lemmas, some of which have an interest on their own.

5.2 Lemma. *Let M and N be two $\mathcal{A}$-manifolds. Let I be a homogeneous open interval, W a wave in $I \times M$ around t_o for some real value $t_o \in I$. Let Y be a homogeneous vector field on N of the same parity as I. Finally let $f_i : W \to N$ be two smooth maps that coincide on $\{t_o\} \times M$ such that $Tf_i \circ \partial_t = Y \circ f_i$. Then $f_1 = f_2$.*

Proof. Define $g_i : W \times N \to M \times N$ by $g_i(t, m, n) = (m, f_i(t, m))$. It follows that $Tg_i \circ \partial_t = (\underline{0} \times Y) \circ g_i$ and that the g_i coincide on $\{t_o\} \times M \times N$. The result now follows from uniqueness of (local) flows with initial condition [4.8] and [4.17]. $\boxed{QED}$

5.3 Corollary. *Let $m_o \in \mathbf{B}M$ be a point with real coordinates and X a homogeneous vector field on M with flow ϕ_X such that $X(m_o) = 0$. Then for all t : $\phi_X(t, m_o) = m_o$.*

Proof. Let W_X be the domain of definition of ϕ_X and $I_{m_o} = W_X \cap \mathcal{A}_\alpha \times \{m_o\}$. If $f_1 : I_{m_o} \to M$ is defined as $f_1(t) = \phi_X(t, m_o)$, then by definition of the flow we have $Tf_1 \circ \partial_t = X \circ f_1$. But for $f_2 : I_{m_o} \to M$ defined by $f_2(t) = m_o$ we also have $Tf_2 \circ \partial_t = 0 = X(m_o) = X \circ f_2$. We conclude by [5.2]. $\boxed{QED}$

5.4 Corollary. *Let $f : M \to N$ be a smooth map and let X be an integrable vector field of parity α on M with flow ϕ_X. Then $Tf \circ X = 0$ if and only if $f \circ \phi_X = f \circ \pi_2$ on W_X, where $\pi_2 : \mathcal{A}_\alpha \times M \to M$ denotes the canonical projection.*

Proof. If $f \circ \phi_X = f \circ \pi_2$ then, by applying the tangent map, we obtain the equation $0 = Tf \circ T\pi_2 \circ \partial_t = Tf \circ T\phi_X \circ \partial_t = Tf \circ X \circ \phi_X$. Restriction of this map to $\{0\} \times M$ gives $Tf \circ X = 0$. If on the other hand $Tf \circ X = 0$, we deduce that $T(f \circ \phi_X) \circ \partial_t = \underline{0}$ and $T(f \circ \pi_2) \circ \partial_t = \underline{0}$. Since the two maps $f \circ \phi_X$ and $f \circ \pi_2$ coincide on $\{0\} \times M$, the result follows from [5.2]. $\boxed{QED}$

5.5 Lemma. *Let $f : M \to N$ be smooth, let X be an integrable vector field of parity α on M, and let Y be an integrable vector field of the same parity α on N. If we have $\phi_Y \circ (id \times f) = f \circ \phi_X$ on an open set containing $\{0\} \times M$ then X and Y are related by f. On the other hand, if X and Y are related by f, then $\phi_Y \circ (id \times f) = f \circ \phi_X$ on W_X.*

Proof. Suppose $\phi_Y \circ (id \times f) = f \circ \phi_X$. Applying the tangent map of this relation to the vector field ∂_t on $\mathcal{A}_0 \times M$ (more precisely, on the open set containing $\{0\} \times M$) gives us

$$\begin{aligned} & Tf \circ T\phi_X \circ \partial_t = T\phi_Y \circ T(id \times f) \circ \partial_t \\ \iff \quad & Tf \circ X \circ \phi_X = T\phi_Y \circ (\partial_t \times \underline{0}) \circ (id \times f) \\ \iff \quad & Tf \circ X \circ \phi_X = Y \circ \phi_Y \circ (id \times f) = Y \circ f \circ \phi_X \end{aligned}$$

Restricting this identity to $\{0\} \times M$ proves that $Y \circ f = Tf \circ X$.

To prove the second part we start with a local statement. Since f is smooth, for any point $m \in M$ there exist open sets $m \in U \subset M$ and $f(m) \in V \subset N$, and an open interval I containing 0 such that $f(U) \subset V$, $I \times U \subset W_X$, and $I \times V \subset W_Y$. We compute $T(f \circ \phi_X) \circ \partial_t = Y \circ (f \circ \phi_X)$ and $T(\phi_Y \circ (id \times f)) \circ \partial_t = Y \circ (\phi_Y \circ (id \times f))$. Since the maps $f \circ \phi_X$ and $\phi_Y \circ (id \times f)$ coincide on $\{0\} \times U$, we conclude from [5.2] that the maps coincide on $I \times U$.

By gluing these local subsets $I \times U$ together, we conclude that $f \circ \phi_X$ and $\phi_Y \circ (id \times f)$ are defined and coincide on a wave W in $\mathcal{A}_\alpha \times M$ around 0. Let W_o be the biggest wave with these properties. We claim that $W_o = W_X$, a fact which is obvious in the odd case. In case X and Y are even, suppose W_o is strictly included in W_X. There thus exists a border point $(t_1, m_1) \in W_X \setminus W_o$ for W_o. Since W_o is open and ϕ_X continuous, there exist open $t_1 \in I$ and $m_1 \in U$ such that $(I - I) \times \phi_X(I \times U) \subset W_o$, where $I - I = \{a - b \mid a, b \in I\}$ is an open interval containing 0. Moreover, since W_X is open, we may assume $I \times U \subset W_X$. By definition of (t_1, m_1) there exists a $t_2 \in I$ such that $(t_2, m_1) \in W_o$. Since W_o is open, we may assume $\{t_2\} \times U \subset W_o$ (shrinking U if necessary). For an arbitrary $t \in I$ we compute

$$\begin{aligned} f(\phi_X(t, m)) &= f(\phi_X(t, \phi_X(-t_2, \phi_X(t_2, m)))) \\ &= f(\phi_X(t - t_2, \phi_X(t_2, m))) && \text{by [4.12] and [4.13]} \\ &= \phi_Y(t - t_2, f(\phi_X(t_2, m))) && \text{because } (I - I) \times \phi_X(I \times U) \subset W_o \\ &= \phi_Y(t - t_2, \phi_Y(t_2, f(m))) && \text{because } \{t_2\} \times U \subset W_o \\ &= \phi_Y(t, f(m)) && \text{by [4.12] and [4.13].} \end{aligned}$$

We conclude that $I \times U \subset W_o$ by maximality, contradicting $(t_1, m_1) \in I \times U \setminus W_o$. This proves that $W_X = W_o$. QED

5.6 Corollary. *If $f : M \to N$ is a diffeomorphism and X an integrable homogeneous vector field on X with flow ϕ_X, then $Y = Tf \circ X \circ f^{-1}$ is an integrable vector field on N with flow $\phi_Y = f \circ \phi_X \circ (id \times f^{-1})$.*

5.7 Lemma. *Let X be a homogeneous integrable vector field on M and let P be any open subset of $N \times M$. If $\phi_X : W_X \to M$ is the flow of X on M, then the flow of $\underline{0} \times X$ on P is given by*

$$\phi_{\underline{0}\times X} : (t, n, m) \mapsto (n, \phi_X(t, m)) \; . \tag{5.8}$$

Its domain of definition $W_{\underline{0}\times X}$ is $\mathcal{A}_1 \times P$ for odd X; if X is even it is given by

$$W_{\underline{0}\times X} = \{\, (t, n, m) \in \mathcal{A}_0 \times P \mid (t, m) \in W_X \;\&\; \exists I_{0,t} : \{n\} \times \phi_X(I_{0,t}, m) \subset P \,\} \; ,$$

where $I_{0,t}$ denotes an open interval containing both 0 and t.

Proof. We apply [5.5] to the two canonical projections $\pi_N : P \to N$ and $\pi_M : P \to M$, which satisfy $T\pi_M \circ (\underline{0} \times X) = X \circ \pi_M$ and $T\pi_N \circ (\underline{0} \times X) = \underline{0} \circ \pi_N$. This gives us as result that $\pi_M \circ \phi_{\underline{0}\times X} = \phi_X \circ (id \times \pi_M)$ and $\pi_N \circ \phi_{\underline{0}\times X} = \phi_{\underline{0}} \circ (id \times \pi_N)$. This proves that (5.8) is valid on the domain of definition of $\phi_{\underline{0}\times X}$. This finishes the proof for odd X.

For even X it remains to show that the given $W_{\underline{0}\times X}$ coincides with the domain of definition of $\phi_{\underline{0}\times X}$, which we temporarily denote by W_o. Since ϕ_X is continuous, P open and W_X a wave, it follows that $W_{\underline{0}\times X}$ is a wave in $\mathcal{A}_0 \times P$ around 0 on which the map $\psi : (t, n, m) \mapsto (n, \phi_X(t, m))$ makes sense. An elementary calculation shows that $T\psi \circ \partial_t = \underline{0} \times X \circ \psi$. By maximality of W_o this implies that $W_{\underline{0}\times X} \subset W_o$. On the other hand, if $I, U \subset M$, and $V \subset N$ are open such that $I \times V \times U \subset W_o$, it follows from our first observation that $I \times U \subset W_X$. Moreover, since W_X is a wave, we may assume that $0 \in I$. It follows that $I \times V \times U \subset W_{\underline{0}\times X}$, proving the other inclusion. QED

5.9 Corollary. *Let X and Y be integrable homogeneous vector fields on M. Their flows commute if and only if Y and $\underline{0} \times Y$ are related by ϕ_X (if and only if X and $\underline{0} \times X$ are related by ϕ_Y).*

Proof. Consider the vector field $\underline{0} \times Y$ on W_X. It follows from the explicit expression (5.8) that the flows of X and Y commute if and only if $\phi_Y \circ (id \times \phi_X) = \phi_X \circ \phi_{\underline{0}\times Y}$. According to [5.5] this is true if and only if $T\phi_X \circ (\underline{0} \times Y) = Y \circ \phi_X$. Interchanging the roles of X and Y proves the second part. QED

5.10 Discussion. We continue with the preparations for our characterization of commuting flows and we take a closer look at the tangent map of the flow ϕ_X of a homogeneous vector field X on an $\mathcal{A}$-manifold M. We will use the symbol t for the time parameter, even if the vector field X is odd. In order to simplify this discussion, we will drop (in this discussion) the subscript X in ϕ_X. We start by choosing a point $(s, t, m) \in W'_X$ (i.e., such that $\phi(s, \phi(t, m))$ makes sense) and we choose local coordinate systems x^i_a around m, x^j_b around $\phi(t, m)$, and x^k_c around $\phi(s, \phi(t, m)) \equiv \phi(s + t, m)$. We then note that

$$T\phi : \frac{\partial}{\partial x^i_a}|_{(t,m)} \mapsto \sum_j \frac{\partial \phi^j_b}{\partial x^i_a}(t, m) \cdot \frac{\partial}{\partial x^j_b}|_{\phi(t,m)}$$

$$T\phi : \frac{\partial}{\partial t}|_{(t,m)} \mapsto \sum_j \frac{\partial \phi^j_b}{\partial t}(t, m) \cdot \frac{\partial}{\partial x^j_b}|_{\phi(t,m)} = X^i_b(\phi(t, m)) \frac{\partial}{\partial x^j_b}|_{\phi(t,m)} \; ,$$

where the last equality follows by definition of a flow. Since $\Phi_X^2 = id$, the same is true for the tangent map: $(T\Phi_X)^2 = id$; applied to the vectors $\partial_{x_a^i}|_{(t,m)}$ and $\partial_t|_{(t,m)}$, this identity gives the relations

$$\sum_i \frac{\partial \phi_a^i}{\partial x_b^j}(-t, \phi(t,m)) \cdot \frac{\partial \phi_b^k}{\partial x_a^i}(t,m) = \delta_j^k ,$$

$$\sum_i X_a^i(m) \cdot \frac{\partial \phi_b^j}{\partial x_a^i}(t,m) = X_b^j(\phi(t,m)) .$$

Applying likewise the identity $(T\Phi'_X)^3 = id$ to the vectors $\partial_{x_a^i}|_{(s,t,m)}$ gives the additional relation

(5.11) $$\sum_j \frac{\partial \phi_b^j}{\partial x_a^i}(t,m) \cdot \frac{\partial \phi_c^k}{\partial x_b^j}(s, \phi(t,m)) = \frac{\partial \phi_c^k}{\partial x_a^i}(s+t, m) .$$

Now if s and t are close to zero, we may assume that the coordinate systems are the same, i.e., $a = b = c$. We then obtain in particular the relation

(5.12) $$\frac{\partial \phi_a^i}{\partial x_a^j}(0,m) = \delta_j^i .$$

Applying [III.3.13] then gives the useful identity

(5.13) $$\frac{\partial^2 \phi_a^i}{\partial x_a^k\, \partial x_a^j}(0,m) = 0 .$$

5.14 Discussion. Let us now return to the actual characterization of commuting flows. According to [5.9], the flows of X and Y commute if and only if Y and $\underline{0} \times Y$ are related by ϕ_X, i.e., $Y \circ \phi_X = T\phi_X \circ (\underline{0} \times Y)$. Composing on the right with the diffeomorphism Φ_X shows that this is the case if and only if $Y \circ \pi_2 = T\phi_X \circ (\underline{0} \times Y) \circ \Phi_X$, where $\pi_2 : W_X \to M$ denotes the canonical projection $(t,m) \mapsto m$. Since the left hand side is rather easy to understand, we have to study the right hand side in more detail. We thus define the function $\psi : W_X \to TM$ by

$$\psi = \Phi_X \diamond (\underline{0} \times Y) \diamond T\phi_X .$$

Now Φ_X maps (t,m) to $(-t, \phi_X(t,m))$, then $\underline{0} \times Y$ maps this point to a tangent vector at this point, and finally $T\phi_X$, which maps the base point $(-t, \phi_X(t,m))$ back to m, sends this tangent vector to a tangent vector at m. In short, $\psi(t,m) \in T_m M$. If ψ equals $\pi_2 \diamond Y$, this implies that $\psi(t,m)$ should be equal to $Y_m \in T_m M$, independent of t. It now follows easily from the relation (5.12) that $\psi(0,m) = Y_m$, so it remains to show that $\psi(t,m)$ is independent of t. Applying [5.4] to the vector field ∂_t on W_X shows that this is the case if and only if $\partial_t \diamond T\psi = 0$. Since the vector field ∂_t and the map ψ satisfy the conditions of [3.13] and since ∂_t is homogeneous, we conclude that $\partial_t \diamond T\psi = 0$ if and only if $\partial_t \psi = 0$.

To summarize this discussion, we have shown that the flows of X and Y commute if and only if $\partial_t \psi = 0$. Hence our interest in the quantity $\partial_t \psi$, where we recall that t denotes the time parameter whose parity equals the parity of X.

5.15 Proposition. *Let X be a homogeneous integrable vector field on M and let Y be an arbitrary vector field on M. If $\psi = \Phi_X \diamond (\underline{0} \times Y) \diamond T\phi_X : W_X \to TM$, then $\partial_t\psi : W_X \to TM$ is given by $\partial_t\psi = \Phi_X \diamond (\underline{0} \times [X,Y]) \diamond T\phi_X$. Using the notation of the generalized tangent map [3.19] and abbreviating ϕ_X to ϕ, this can be written as*

$$\partial_t \langle\!\langle Y_{\phi_t(m)} \| T\phi_{-t} \rangle\!\rangle = \langle\!\langle [X,Y]_{\phi_t(m)} \| T\phi_{-t} \rangle\!\rangle \; . \tag{5.16}$$

Proof. To simplify the notation, we will, as in [5.10], write ϕ for ϕ_X throughout this proof. Let us first consider the last statement. Writing the definition of $\psi(t,m)$ explicitly gives

$$\psi(t,m) = \langle\!\langle (\underline{0}_{-t}, Y_{\phi_t(m)}) \| T\phi \rangle\!\rangle = \langle\!\langle Y_{\phi_t(m)} \| T\phi_{-t} \rangle\!\rangle \; .$$

With this formula, (5.16) is immediate.

To prove the real result, we start by noticing that if Y_1 and Y_2 are two vector fields, then the sum of the maps ψ is the same as the map ψ associated to the sum of the Ys. We thus may assume that Y is a homogeneous vector field.

Introducing the function $\chi : W'_X \to TM$ by $\chi(s,t,m) = \psi(s+t,m)$, it follows easily that $(\partial_s\chi)(0,t,m) = (\partial_t\psi)(t,m)$. We now introduce local charts U_a around m (with coordinates x^i_a) and U_b around $\phi(t,m)$. Since we will take s close to zero, we may use the coordinates x^j_b also around $\phi(s,\phi(t,m)) \equiv \phi(s+t,m)$. In the given trivializations the map χ is given by

$$\begin{aligned}\chi(s,t,m) &= \sum_{k,i} Y^k_b(\phi(s+t,m)) \cdot \frac{\partial \phi^i_a}{\partial x^k_b}(-s-t,\phi(s+t,m)) \cdot \frac{\partial}{\partial x^i_a}|_m \\ &= \sum_{i,j,k} Y^k_b(\phi(s,\widehat{m})) \cdot \frac{\partial \phi^j_b}{\partial x^k_b}(-s,\phi(s,\widehat{m})) \cdot \frac{\partial \phi^i_a}{\partial x^j_b}(-t,\widehat{m}) \cdot \frac{\partial}{\partial x^i_a}|_m \; ,\end{aligned}$$

where we introduced $\widehat{m} = \phi(t,m)$ and where we used (5.11). In the same trivialization we now compute $(\partial_s\chi)(0,t,m)$, which, according to (3.12), means that we have to differentiate the fiber function (meaning the coefficients of $\partial_{x^i_a}|_m$) with respect to s and not touching the base point (meaning m, which indeed does not depend upon s).

$$\begin{aligned}\frac{\partial\chi}{\partial s}(0,t,m) = \sum_{i,j}\Bigg(&\sum_{k,\ell} \frac{\partial \phi^\ell_b}{\partial t}(0,\widehat{m}) \cdot \frac{\partial Y^k_b}{\partial x^\ell_b}(\phi(0,\widehat{m})) \cdot \frac{\partial \phi^j_b}{\partial x^k_b}(0,\phi(0,\widehat{m})) \\ &- \sum_k (-1)^{\langle \varepsilon(s) | \varepsilon(Y^k_b)\rangle} \cdot Y^k_b(\phi(0,\widehat{m})) \cdot \frac{\partial^2 \phi^j_b}{\partial t\, \partial x^k_b}(0,\phi(0,\widehat{m})) \\ &+ \sum_{k,\ell} (-1)^{\langle \varepsilon(s) | \varepsilon(Y^k_b)\rangle} \cdot Y^k_b(\phi(0,\widehat{m})) \cdot \frac{\partial \phi^\ell_b}{\partial t}(0,\widehat{m}) \cdot \frac{\partial^2 \phi^j_b}{\partial x^\ell_b\, \partial x^k_b}(0,\phi(0,\widehat{m}))\Bigg) \cdot \\ &\qquad \frac{\partial \phi^i_a}{\partial x^j_b}(-t,\widehat{m}) \cdot \frac{\partial}{\partial x^i_a}|_m \\ = \sum_{i,j,k,\ell} \Big(& X^\ell_b(\widehat{m}) \cdot \frac{\partial Y^k_b}{\partial x^\ell_b}(\widehat{m}) \cdot \delta^j_k - (-1)^{\langle \varepsilon(s) | \varepsilon(Y^k_b) + \varepsilon(x^k_b)\rangle} \cdot Y^k_b(\widehat{m}) \cdot \frac{\partial X^j_b}{\partial x^k_b}(\widehat{m})\Big) \cdot \\ &\qquad \frac{\partial \phi^i_a}{\partial x^j_b}(-t,\widehat{m}) \cdot \frac{\partial}{\partial x^i_a}|_m \; .\end{aligned}$$

To obtain the second equality we have applied in particular (5.12) and (5.13). The additional sign appears because we have to change the order of the partial derivatives ∂_s and $\partial_{x_b^k}$. We now note that $\varepsilon(x_b^k)$ is the parity of the k-th basis vector in E, and thus that $\varepsilon(Y_b^k) + \varepsilon(x_b^k) = \varepsilon(Y)$. We also note that $\varepsilon(s) = \varepsilon(X)$. Now comparing the expression in parentheses with (1.22), we see that it represents the coefficients of $[X,Y](\widehat{m})$ with respect to the basis $\partial_{x_b^j}|_{\widehat{m}}$. We conclude that $(\partial_s\chi)(0,t,m) \equiv (\partial_t\psi)(t,m)$ gives the local expression of $\Phi_X \diamond (\underline{0} \times [X,Y]) \diamond T\phi_X$. Since the point (t,m) is chosen arbitrarily, we have finished the proof. $\boxed{QED}$

5.17 Remark. An analogous result for k-forms is given in [7.27]; a generalization to arbitrary $\mathcal{A}$-Lie group actions is given in [VI.8.3].

5.18 Theorem. *Let X and Y be two integrable homogeneous vector fields on M. Then their flows commute if and only if $[X,Y] = 0$.*

Proof. If $[X,Y] = 0$, then $\partial_t\psi = 0$ by [5.15] and thus by the conclusion of [5.14] the two flows commute. On the other hand, if the two flows commute, we have by the same reasoning that $\partial_t\psi = 0$. In particular we have $0 = (\partial_t\psi)(0,m) = [X,Y]|_m$. $\boxed{QED}$

5.19 Remark. If X is an odd integrable vector field, the proof of [5.18] can be given by a direct computation as follows. If X is odd and integrable, its flow is given by the expression $\phi_X(\tau,m) = m + \tau \cdot X_c(m)$, where we have used the shorthand introduced in [4.18]. Using the same shorthand, we write $Y(m) = Y_c(m) \cdot \partial_m$ and thus we have $(\underline{0} \times Y)(\tau,m) = 0 \cdot \partial_\tau + Y_c(m) \cdot \partial_m$. Computing the explicit expression for the equality $Y \circ \Phi_X = T\phi_X \circ (\underline{0} \times Y)$ at (τ,m) gives us

$$
\begin{aligned}
& & Y_c(m + \tau \cdot X_c(m)) \cdot \partial_m &= \langle\!\langle Y_c(m) \cdot \partial_m \,\|\, T\phi_X \rangle\!\rangle \\
&\Longleftrightarrow & Y_c(m + \tau \cdot X_c(m)) &= Y_c(m) \cdot \partial_m(m + \tau \cdot X_c(m)) \\
&\Longleftrightarrow & Y_c(m) + \tau \cdot X_c(m) \cdot \frac{\partial Y_c}{\partial m}(m) &= Y_c(m) + (-1)^{\varepsilon(Y)} \cdot \tau \cdot Y_c(m) \cdot \frac{\partial X_c}{\partial m}(m) \\
&\Longleftrightarrow & X_c(m) \cdot \frac{\partial Y_c}{\partial m}(m) - (-1)^{\varepsilon(Y)} \cdot Y_c(m) \cdot \frac{\partial X_c}{\partial m}(m) &\equiv [X,Y](m) = 0 \,.
\end{aligned}
$$

5.20 Remark. With [5.18] we have another way to view the integrability condition of an odd vector field X. The property of its flow $\phi_X(t, \phi_X(s,m)) = \phi_X(s+t,m)$ implies that $\phi_X(t,\phi_X(s,m)) = \phi_X(s,\phi_X(t,m))$, i.e., the flow commutes with itself. According to the quoted result, this is the case if and only if $[X,X] = 0$, i.e., exactly the necessary and sufficient condition for X to be integrable.

6. Frobenius' theorem

In this section we first define the notion of an involutive subbundle of the tangent bundle TM. We then show Frobenius' theorem which states that a subbundle is involutive if and only if it is locally generated by $\partial_{x^1}, \ldots, \partial_{x^k}$ with k the rank of the subbundle. Using this we show that through all points with real coordinates there passes a maximal integral manifold (a leaf) of such a subbundle. For points that do not have real coordinates, there is no guarantee that a leaf passes through them.

6.1 Proposition. *Let X_i, $1 \leq i \leq k$ be k homogeneous vector fields on M satisfying $[X_i, X_j] = 0$, $1 \leq i, j \leq k$. If $m \in M$ is a point at which the k vectors $X_i(m)$ are independent, then there exists a local chart $U \ni m$ with coordinates $(x^1, \ldots, x^n)$ such that $X_i|_U = \partial_{x^i}$, $1 \leq i \leq k$.*

Proof. Note first that we have in particular that $[X_i, X_i] = 0$, i.e., all vector fields are integrable. If we define $m_o = \mathbf{B}m \in \mathbf{B}M$, it follows from [II.6.6] and [III.2.7] that the vectors $\mathbf{B}X_i(m) = X_i(m_o)$ are independent. If $(y^1, \ldots, y^n)$ are local coordinates on a chart U_o containing m_o (and thus m), the vectors $X_i(m_o)$ are represented as

$$X_i(m_o) = \sum_{j=1}^{n} r_i^j \cdot \frac{\partial}{\partial y^j} \quad , \quad r_i^j \in \mathbf{R} .$$

By a constant affine transformation with real coefficients (which is smooth!) we may assume that the coordinates of m_o are all zero and that $X_i(m_o) = \partial_{y^i}$. We now define a map ψ by the expression

$$\psi(x^1, \ldots, x^n) = \phi_{X_1}(x^1, \phi_{X_2}(x^2, \cdots \phi_{X_k}(x^k, (0, 0, \ldots, 0, x^{k+1}, \ldots, x^n)) \cdots)) ,$$

i.e., starting from $(0, \ldots, 0, x^{k+1}, \ldots, x^n)$ we first follow the flow of X_k for a time x^k, then we follow the flow of X_{k-1} for a time x^{k-1}, ... , and we finish by following the flow of X_1 for a time x^1. If the $(x^1, \ldots, x^n)$ are taken in a sufficiently small neighborhood $U_1 \subset U_o$ of zero (representing m_o), this expression makes sense and takes its values in the chart U_o. In order to compute the Jacobian $(\partial\psi^j/\partial x^i)(0, \ldots, 0)$ of ψ at zero, it is obvious that it equals δ_i^j for $j > k$, using that a flow is the identity at time zero. With the same argument and the flow property we obtain $(\partial\psi/\partial x^j)(0, \ldots, 0) = X_j(0, \ldots, 0) = \partial_{y^j}$ for $1 \leq j \leq k$. We conclude that $Jac(\psi)(m_o)$ is the identity, and thus, by the inverse function theorem, there exists a neighborhood $m_o \in U$ such that $\psi^{-1} : U \to \psi^{-1}(U)$ is a diffeomorphism.

We claim that, in the coordinates $(x^1, \ldots, x^n) = \psi^{-1}(y^1, \ldots, y^n)$, the vector fields X_i are represented as $X_i = \partial_{x^i}$. For $i = 1$ this is fairly easy to show. Using the group property of a flow it follows that

$$\psi^{-1} \circ \phi_{X_1} \circ (id \times \psi) : (t, x^1, \ldots, x^n) \mapsto (x^1 + t, x^2, \ldots, x^n) .$$

Since this is the flow of the vector field ∂_{x^1}, it follows from [5.6] that X_1 is represented by ∂_{x^1} in the coordinates x. One might think that it is more difficult for $i > 1$, but it is as easy. Since the vector fields X_i commute, their flows commute. For $i = 2$ we thus may write

$$\psi(x^1, \dots, x^n) = \phi_{X_2}(x^2, \phi_{X_1}(x^1, \phi_{X_3}(x^3, \cdots \phi_{X_k}(x^k, 0, 0, \dots, 0, x^{k+1}, \dots, x^n) \cdots) .$$

We thus can apply exactly the same argument to prove that X_i is represented by ∂_{x^i} in the coordinates x for $1 < i \leq k$. QED

6.2 Definitions. Let M be an $\mathcal{A}$-manifold and $\mathcal{D}$ a subbundle of TM. $\mathcal{D}$ is said to be *involutive* if for any two smooth vector fields X and Y that take their values in $\mathcal{D}$, their commutator also takes values in $\mathcal{D}$, i.e., $X, Y \in \Gamma(\mathcal{D}) \implies [X, Y] \in \Gamma(\mathcal{D})$. A subbundle $\mathcal{D}$ is said to be *integrable* if for each $m \in M$ there exists a chart $m \in U$ with coordinates x such that $\mathcal{D}|_U$ is generated by $\partial_{x^1}, \dots, \partial_{x^k}$, where k is the rank of $\mathcal{D}$. An integrable subbundle is also called a *foliation*.

6.3 Theorem (Frobenius). *A subbundle $\mathcal{D} \subset TM$ is involutive if and only if it is integrable.*

Proof. If $\mathcal{D}$ is generated on U by $\partial_{x^1}, \dots, \partial_{x^k}$, it follows immediately from [1.21] that $X, Y \in \Gamma_U(\mathcal{D})$ implies $[X, Y] \in \Gamma_U(\mathcal{D})$. Since this is true all over M, $\mathcal{D}$ is involutive by [1.20].

To prove the implication in the other direction, suppose $\mathcal{D}$ is involutive and let $m \in M$ be arbitrary. Since $\mathcal{D}$ is a subbundle of rank k, there exists a chart $m \in U$ with coordinates $(y^1, \dots, y^n)$ and homogeneous $X_i \in \Gamma_U(TM)$, $1 \leq i \leq n$ generating $\Gamma_U(TM)$ such that the X_i, $1 \leq i \leq k$ generate $\mathcal{D}|_U$ [IV.3.14]. If we define $m_o = \mathbf{B}m$, the coefficients of $X_i(m_o)$ with respect to the basis ∂_{y^j} are real; by a constant linear transformation with real coefficients we may assume that $X_i(m_o) = \partial_{y^i}$, $1 \leq i \leq k$. Using these coordinates, we define a projection $\pi_k : U \to V$, $(y^1, \dots, y^n) \mapsto (y^1, \dots, y^k)$, from U to an open set V in an $\mathcal{A}$-vector space of total dimension k. By construction the map

$$T_{m_o}\pi_k|_{\mathcal{D}_{m_o}} : \mathcal{D}_{m_o} \to T_{\pi_k(m_o)}V$$

is bijective, simply because $X_i(m_o) = \partial_{y^i}$. Writing $X_i(m) = \sum_j X_i^j(m) \cdot \partial_{y^j}$, our assumption implies that $X_i^j(m_o) = \delta_i^j$, $1 \leq i \leq k$, $1 \leq j \leq n$. It follows that the square matrix $X_i^j(m)$, $1 \leq i, j \leq k$ is invertible in a neighborhood U_1 of m_o. And thus $T_m\pi_k|_{\mathcal{D}_m} : \mathcal{D}_m \to T_{\pi_k(m)}V$ is bijective for all $m \in U_1$. Hence there exist vector fields $Y_i \in \Gamma_{U_1}(\mathcal{D})$ generating $\mathcal{D}|_{U_1}$ such that $T\pi_k \circ Y_i = \partial_{y^i} \circ \pi_k$. (The $Y_i(m)$ are given explicitly by multiplying the $X_i(m)$ by the inverse of the matrix $X_i^j(m)$, $1 \leq i, j \leq k$.) We deduce that

$$T\pi_k \circ [Y_i, Y_j] = [\frac{\partial}{\partial y^i}, \frac{\partial}{\partial y^j}] \circ \pi_k = \underline{0} .$$

Since $\mathcal{D}$ is involutive and $T\pi_k$ bijective on $\mathcal{D}|_{U_1}$, it follows that $[Y_i, Y_j] = 0$. Since the Y_i are independent, we can apply [6.1] to conclude. QED

6.4 Definition. Let $\mathcal{D} \subset TM$ be a subbundle of rank k. A smooth map $f : N \to M$ is said to be *tangent to* $\mathcal{D}$ if for all $n \in N$ we have $Tf(T_nN) \subset \mathcal{D}|_{f(n)}$. An *integral manifold of* $\mathcal{D}$ is a pair (i, N) such that (i) $\dim(N) = k$ and (ii) $i : N \to M$ is an injective immersion tangent to $\mathcal{D}$. If $\mathcal{D}$ is a foliation, one defines a *leaf of* $\mathcal{D}$ to be a connected integral manifold $i : L \to M$ such that $i(L)$ is maximal with respect to inclusion.

6.5 Corollary. *Let $\mathcal{D} \subset TM$ be a foliation of rank k and let $m_o \in \mathbf{B}M$ be a point with real coordinates. Then there exists an integral manifold (i, N) of $\mathcal{D}$ such that $i : N \to M$ is an embedding and such that $m_o \in i(N)$.*

Proof. Let U be a chart around m_o as in the definition of an integrable subbundle and let $N \subset U$ be the subset $N = \{m \in U \mid \forall i > k : x^i(m) = x^i(m_o)\}$. Then N is a submanifold of M of dimension k (the $x^i(m_o)$ are real!) and thus the canonical injection $i : N \to M$ is an embedding. Moreover, since $\mathcal{D}$ is spanned by the ∂_{x^i}, $i \leq k$, (i, N) is an integral manifold for $\mathcal{D}$. QED

6.6 (Counter) Examples. One usually says that a subbundle $\mathcal{D} \subset TM$ is integrable if through every point passes an integral manifold. Defined that way, Frobenius' theorem states that $\mathcal{D}$ is involutive if and only if through every point passes an integral manifold. However, in the context of $\mathcal{A}$-manifolds problems arise due to the fact that the image of an immersion has to pass through points with real coordinates. The following examples show what can happen, justifying our definition of integrability.

• Consider first the $\mathcal{A}$-manifold $M = \mathcal{A}_0^2$ of dimension $2|0$ with coordinates (x^1, x^2), on which we define $\mathcal{D}$ as the subbundle of rank 1 generated by the vector field $X_1 = \partial_{x^1}$. This subbundle is involutive: (x^1, x^2) is a coordinate system satisfying [6.2]. If $i : N \to M$ is an integral manifold, it is fairly obvious that $i(N)$ should be contained in a slice x^2 constant (we will show it explicitly in the proof of [6.9]). But a point with real coordinates in N is mapped to a point with real coordinates in M, implying that x^2 should be real. It follows that no integral manifold passes through points (x^1, x^2) with x^2 non-real.

• With this example in mind, one might think that it should be sufficient to demand that through every point with *real* coordinates passes an integral manifold. As the next example will show, this condition is too weak to ensure that a subbundle $\mathcal{D}$ is involutive. Consider the $\mathcal{A}$-vector space E of dimension $1|2$ and define $M = E_0$ with coordinates (x, ξ^1, ξ^2). The subbundle $\mathcal{D}$ of rank 2 is generated by the global vector fields $X_1 = \partial_x$ and $X_2(x, \xi^1, \xi^2) = \partial_{\xi^1} + \xi^1\xi^2\,\partial_{\xi^2}$. Now consider the $\mathcal{A}$-vector space F of dimension $1|1$ and define $N = F_0$ with coordinates (y, η). It is elementary to show that the smooth map $i : N \to M$, $(y, \eta) \mapsto (x, \xi^1, \xi^2) = (y, \eta, 0)$ is an integral manifold of $\mathcal{D}$. Moreover, it passes through every point with real coordinates of M. However, $\mathcal{D}$ is not involutive because $[X_2, X_2] = \xi^2\partial_{\xi^2} \notin \mathcal{D}$.

6.7 Lemma. *Let $\mathcal{D}$ be an involutive subbundle of TM of rank k. Let furthermore U be a chart with coordinates x^i such that $\mathcal{D}|_U$ is generated by ∂_{x^i}, $1 \leq i \leq k$. Finally define* *the slices $S_m \subset U$ as $S_m = \{ m' \in U \mid \forall i > k : x^i(m') = x^i(m) \}$. If $f : N \to M$ is tangent to $\mathcal{D}$, then each connected component of $f(N) \cap U$ is contained in a slice S_m with $m \in \mathbf{B}M$.*

Proof. The main problem of this proof is to show that $f(N)$ cannot "fill up" parts of U; the crucial ingredient is that N is second countable. Let F be the $\mathcal{A}$-vector space of the appropriate dimension such that $x^{k+1}, \ldots, x^n$ ($n = \dim M$) are coordinates on F_0, and define $s : U \to F_0$ by $s(x^1, \ldots; x^n) = (x^{k+1}, \ldots, x^n)$. By definition s is constant on slices and $S_m = s^{-1}(s(m))$. Since f is tangent to $\mathcal{D}$, it follows that $T(s \circ f) = 0$. By [3.21] we conclude that $s \circ f$ is constant on connected components of $f^{-1}(U)$. In particular if B is a connected component of $f^{-1}(U)$ and $b \in B$, then $f(B)$ is contained in the slice $S_{f(b)}$ (because $(s \circ f)(B)$ is constant equal to $s(f(b))$). Since N is locally homeomorphic to the even part of an $\mathcal{A}$-vector space which is locally connected [III.1.3], a connected component B of the open set $f^{-1}(U)$ is open. Hence $\mathbf{B}B \subset B$ and B contains a point $b \in \mathbf{B}B \subset B$ with real coordinates. And thus $s(f(B)) \in \mathbf{B}F$ and $f(B)$ is contained in a slice S_m with $m \in \mathbf{B}M$, namely $m = f(b)$.

Now $f(N) \cap U = f(f^{-1}(U))$ and thus the image $f(B)$ of a connected component B of $f^{-1}(U)$ is contained in a connected component C of $f(N) \cap U$. Since each $f(B)$ is contained in a slice, C is the union of (parts of) slices. It is true that the union of two slices is no longer connected, but an arbitrary union of slices could be connected. For instance, U itself (if it were connected) is the union of all its slices. In order to prove that C is contained in a single slice, we invoke the fact that N is second countable and thus that there are (at most) countably many connected components (open!) B of $f^{-1}(U)$. Since for each B the image $s(f(B))$ is a single point in $\mathbf{B}F$, $s(C)$ is a countable subset of $\mathbf{B}F$. Now C is connected, $\mathbf{B}F$ is homeomorphic to some $\mathbf{R}^d$ and the only countable connected subsets of $\mathbf{R}^d$ are points (the only connected subsets of $\mathbf{R}$ are intervals). Hence $s(C)$ is a single point and thus C is contained in a single slice. QED

6.8 Proposition. *Let $\mathcal{D}$ be an involutive subbundle of TM, let $i : L \to M$ be an integral manifold of $\mathcal{D}$, and let $f : N \to M$ be a smooth map. If $f(N) \subset i(L)$, then there exists a unique smooth map $g : N \to L$ such that $f = i \circ g$.*

Proof. Since i is injective, existence and uniqueness of a set theoretic map g is guaranteed. The only difficulty is in proving that this g is smooth. So let $n_o \in \mathbf{B}N$ be arbitrary, $m_o = f(n_o) \in M$ and $p_o = g(n_o) \in L$. Let furthermore U, N_o, and i_o be as in [6.5]. Finally let U_L be the connected component of $i^{-1}(U)$ containing p_o and let U_N be the connected component of $f^{-1}(U)$ containing n_o. These sets are open because i and f are smooth. Since i is tangent to $\mathcal{D}$, it follows from [6.7] that $i(U_L) \subset i_o(N_o)$. Since i_o is an embedding, there exists a unique smooth $j : U_L \to N_o$ such that $i = i_o \circ j$ [2.18]. Since Ti is injective, Tj is injective; since N_o and L have the same dimension, j is a diffeomorphism onto its image [2.14]. And now: $f(U_N)$ is connected and contained in $i(L) \cap U$, and the connected components of $i(L) \cap U$ are contained in the slices S_m [6.7]. Hence $f(U_N)$ is contained in the slice S_{m_o}, because $f(n_o) = m_o \in S_{m_o}$. Once again because i_o is an embedding, there exists a unique smooth map $f_o : U_N \to N_o$ such that

$f = i_o \circ f_o$. It follows that the set theoretic map g is given on U_N by $g|_{U_N} = j^{-1} \circ f_o$, which is smooth. Since n_o is arbitrary, we conclude that g is smooth. $\boxed{QED}$

6.9 Proposition. *Let $\mathcal{D}$ be an involutive subbundle of TM of rank k. Through every point $m_o \in M$ with real coordinates passes a unique (up to diffeomorphism) leaf (i, L). Moreover, if $f : N \to M$ is tangent to $\mathcal{D}$, if N is connected, and if $f(N) \cap i(L) \neq \phi$, then there exists a unique smooth map $g : N \to L$ such that $f = i \circ g$, and thus in particular $f(N) \subset i(L)$.*

Proof. Let $m \in M$ be arbitrary and choose a chart U as in [6.2]. Shrinking U if necessary, we may assume that all slices $S_{m'}$ with $m' \in U$ are connected. Since M is second countable, there exists a countable set of such charts $\mathcal{U} = \{ U_i \mid i \in \mathbf{N} \}$ covering M. Let $U_o \in \mathcal{U}$ be such that $m_o \in U_o$, and let $S_o = S_{m_o}$ be the slice in U_o containing m_o [6.7].

If S is a slice in $U \in \mathcal{U}$ and S' a slice in $U' \in \mathcal{U}$, we will say that S and S' are related if there exists a sequence $U_i \in \mathcal{U}$, $1 \le i \le \ell$, and slices S_i in U_i such that $S = S_1$ (and thus $U = U_1$), $S_i \cap S_{i+1} \neq \phi$, and $S_\ell = S'$. We now define $\mathcal{S}$ as the set of all slices in any $U \in \mathcal{U}$ that are related to $S_o = S_{m_o}$. Then we define the topological space $X = \coprod_{S \in \mathcal{S}} S$ and the continuous map $j : X \to M$ such that $j|_S$ is just the canonical injection of the slice S in M. We finally define an equivalence relation $\sim$ on X by $x \sim y \iff j(x) = j(y)$, and the topological space $L = X/\!\sim$ with the canonically induced injective continuous map $i : L \to M$. We claim that this (i, L) is the sought for leaf passing through m_o. The proof of this claim breaks down into several steps.

• The first step is to prove that j is an immersion. We will say that a slice S in $U \in \mathcal{U}$ is a real slice if it is of the form $S = S_m$ with $m \in \mathbf{B}M$. It follows that if S is a real slice, the canonical injection $i : S \to M$ is an integral manifold and an embedding. Now let S_a be a real slice in $U_a \in \mathcal{U}$ and $U_b \in \mathcal{U}$ arbitrary. It follows that the connected components of $S_a \cap U_b$ are contained in the slices of U_b. Hence, if S_b is a slice in U_b, the intersection $S_a \cap S_b$ is a union of connected components, and thus open in S_a. In particular, $S_a \cap S_b$ being open in S_a, there is a point with real coordinates in this intersection, i.e., S_b is a real slice. Since S_o is a real slice, we conclude that all $S \in \mathcal{S}$ are real slices, and that $j : X \to M$ is an immersion.

• The next step is to prove that L is a proto $\mathcal{A}$-manifold and that i is an injective immersion. If $S_a, S_b \in \mathcal{S}$ intersect, we have seen that $S_{ab} = S_a \cap S_b$ is open in S_a. We thus can define $\varphi_{ba} : S_{ab} \to S_{ba}$ by $\varphi_{ba} = (j|_{S_b})^{-1} \circ (j|_{S_a})$ (use [2.18] with the embedding $j|_{S_b} : S_b \to M$). It follows immediately that $x \sim y$ if and only if $x \in S_{ab}$, $y \in S_{ba}$ and $y = \varphi_{ba}(x)$ for some indices a, b. We thus have the complete set of ingredients to form a proto $\mathcal{A}$-manifold [III.4.9] (recall that all slices are essentially open sets in a k dimensional $\mathcal{A}$-vector space). We conclude that $L = X/\!\sim$ is a proto $\mathcal{A}$-manifold. The induced map i verifies $j = i \circ \pi$, where π denotes the canonical projection $\pi : X \to L$. Since $\pi|_{S_a}$ is a diffeomorphism, we have $i|_{\pi(S_a)} = (j|_{S_a}) \circ (\pi|_{S_a})$. It follows that $i : L \to M$ is smooth, injective, an immersion, and tangent to $\mathcal{D}$, i.e., an integral manifold of $\mathcal{D}$, except that we do not know that L is an $\mathcal{A}$-manifold.

• Since $i : L \to M$ is injective and smooth, $\mathbf{B}L$ is Hausdorff because $\mathbf{B}M$ is. To prove that L is second countable is harder. We will show that $\mathcal{S}$ is countable, which implies that X, and thus L is second countable. First fix a sequence $U_i \in \mathcal{U}$, $1 \leq i \leq \ell$. If S_i is a slice in U_i, the connected components of the intersection $S_i \cap U_{i+1}$ are contained in slices of U_{i+1}. Since S_i is second countable, there are only countably many slices S_{i+1} in U_{i+1} that intersect S_i. It follows that there are only countably many slices related to $S_o = S_{m_o}$ by a sequence of slices contained in the given sequence of $U_i \in \mathcal{U}$. Since $\mathcal{U}$ is countable, there are only countably many such sequences, proving that $\mathcal{S}$ is countable. We conclude that L is a genuine $\mathcal{A}$-manifold, and that $i : L \to M$ is an integral manifold passing through m_o.

• Two items remain to be proved: that L is connected and that it is maximal. The connectedness follows from the fact that the slices S in the charts $U \in \mathcal{U}$ are all connected. The actual argument is a bit tedious and left to the reader. Maximality will be proved at the end.

• For the second part, let $f : N \to M$ be tangent to $\mathcal{D}$. For any $U_i \in \mathcal{U}$ we define the open sets $V_{ij} \subset N$ as the connected components of $f^{-1}(U_i)$. It follows from [6.7] that $f(V_{ij})$ is contained in a slice S_i of U_i. By construction of L, if a slice S_i in U_i intersects $i(L)$, it must be contained in $i(L)$. Thus if $f(V_{ij}) \cap i(L) \neq \phi$, it must be that $f(V_{ij}) \subset i(L)$. If we define $N_L = \{ n \in N \mid f(n) \in i(L) \}$, this implies that V_{ij} is contained either in N_L or in its complement. Hence N_L is open and closed. By hypothesis N is connected and N_L is not empty, so $N_L = N$ and $f(N) \subset i(L)$. The last conclusion follows from [6.8].

• To finish the proof, suppose $i' : L' \to M$ is an integral manifold passing through m_o. By the previous result, $i'(L') \subset i(L)$, proving that L is maximal. If we have equality $i'(L') = i(L)$, we have induced smooth maps $L \to L'$ and $L' \to L$ by [6.8]. Standard arguments using uniqueness of these factorizations then proves that L and L' are diffeomorphic, proving that (i, L) is unique up to diffeomorphism. QED

6.10 Remark. The natural idea of proving this proposition using Zorn's lemma does not work. It is true that one can construct an upper bound to any chain of integral manifolds (chain with respect to inclusion of their images in M). This upper bound has a canonical structure of a proto $\mathcal{A}$-manifold, and even its body is Hausdorff. However, in this approach it is very hard to prove that it is second countable.

7. The exterior derivative

In this section we define differential forms and the exterior derivative as well as some of its standard properties: it is a derivation of square zero and commutes with pull-backs. For the last property we define the notion of the pull-back of a differential form as well as a generalization using the generalized tangent map. Defining the Lie derivative of *differential forms by the formula of H. Cartan* [H.Ca]*:* $\mathcal{L}(X) = d \circ \iota(X) + \iota(X) \circ d$, *we also show that* $\mathcal{L}(X)\,\alpha$ *equals* $\partial_t \phi_t^* \alpha$ *at* $t = 0$, *where* ϕ_t *denotes the flow of the homogeneous integrable vector field* X.

7.1 Definition/Discussion. Let M be an $\mathcal{A}$-manifold. A *k-form on M* is a section of the bundle $\bigwedge^k {}^*TM$, and thus in particular a *smooth* k-form is an element of $\Gamma(\bigwedge^k {}^*TM)$. In accordance with standard notation, we denote the set of all smooth k-forms on M by $\Omega^k(M)$, i.e.,

$$\Omega^k(M) \equiv \Gamma(\textstyle\bigwedge^k {}^*TM) .$$

A *k-form at $m \in M$* is any point in the fiber of $\bigwedge^k {}^*TM$ above m. It follows that if α is a k-form on M, $\alpha(m) \equiv \alpha_m$ is a k-form at $m \in M$. Since $\bigwedge^0 {}^*TM$ is the trivial bundle $M \times \mathcal{A}$, it follows that a 0-form is just a function on M, and thus in particular $\Omega^0(M) = C^\infty(M)$. For $k = 1$ we find $\Omega^1(M) = \Gamma({}^*TM)$, i.e., a 1-form is a section of the bundle ${}^*TM$, the left dual bundle of the tangent bundle TM. This left dual bundle ${}^*TM$ is usually called the *cotangent bundle of M*.

According to [IV.5.14] and [I.5.5] we have the identifications

$$\begin{aligned}(7.2)\quad \Omega^k(M) &\cong \mathrm{Hom}_L^{sk}(\Gamma(TM)^k; C^\infty(M)) \\ &\cong \mathrm{Hom}_L(\textstyle\bigwedge^k \Gamma(TM); C^\infty(M)) \equiv {}^*(\textstyle\bigwedge^k \Gamma(TM)) ,\end{aligned}$$

which tells us that we may interpret a smooth k-form as a (left) k-linear graded skew-symmetric map of smooth vector fields on M with values in the smooth functions on M. Note that the k-linearity is over $C^\infty(M)$ and not over $\mathcal{A}$ (which does not make sense because $\Gamma(TM)$ is not an $\mathcal{A}$-module). For $k = 1$ the identifications (7.2) reduce to $\Gamma({}^*TM) \cong {}^*\Gamma(TM)$. Even for $k = 0$ (7.2) makes sense: $\Omega^0(M) \equiv \Gamma(M \times \mathcal{A})$ and ${}^*(\bigwedge^0 \Gamma(TM)) \equiv {}^*C^\infty(M)$ because $\bigwedge^0$ gives the basic ring [I.5.3], which here is $C^\infty(M)$. Since the trivial bundle $M \times \mathcal{A}$ comes with its canonical trivialization, we have a canonical identification $\Gamma(M \times \mathcal{A}) \cong C^\infty(M)$. And thus for $k = 0$ (7.2) reduces to the obvious identification $C^\infty(M) \cong {}^*C^\infty(M)$.

For future reference we define $\Omega(M)$ as the direct sum over all k :

$$\Omega(M) = \bigoplus_{k=0}^{\infty} \Omega^k(M) .$$

Obviously $\Omega(M)$ is a $\mathbf{Z} \times \mathbf{Z}_2$-graded $C^\infty(M)$-module, where the $\mathbf{Z}$-grading is given by the k from k-form. Using the wedge product of such sections as defined in [IV.6.1], $\Omega(M)$ becomes a $\mathbf{Z} \times \mathbf{Z}_2$-graded commutative *$C^\infty(M)$-algebra*. This becomes even more explicit when we also use the identification ${}^*(\bigwedge^k \Gamma(TM)) \cong \bigwedge^k {}^*\Gamma(TM)$ [IV.5.20], which tells us that $\Omega(M)$ is (isomorphic to) the exterior algebra $\bigwedge {}^*\Gamma(TM)$. Now if $f \in C^\infty(M) \cong \Omega^0(M)$ is a 0-form and $\alpha \in \Omega^k(M)$ a k-form, we can form the wedge product $f \wedge \alpha$ as well as $f \cdot \alpha$, which uses the $C^\infty(m)$-module structure of $\Omega^k(M)$. Since the wedge product is pointwise, [I.5.8] tells us that these two are equal: $f \wedge \alpha = f \cdot \alpha$ (and similarly $\alpha \cdot f = \alpha \wedge f$).

7.3 Definition. Let α be a k-form on M. The *exterior derivative of* α is the $(k+1)$-form $d\alpha$ on M defined by

$$(7.4)\quad (-1)^k \cdot \iota(X_0,\ldots,X_k)\, d\alpha =$$
$$= \sum_{0\le i\le k} (-1)^{i+\sum_{p<i}\langle\varepsilon(X_p)|\varepsilon(X_i)\rangle} X_i(\iota(X_0,\ldots,X_{i-1},X_{i+1},\ldots,X_k)\,\alpha)$$
$$+ \sum_{0\le i<j\le k} (-1)^{j+\sum_{i<p<j}\langle\varepsilon(X_p)|\varepsilon(X_j)\rangle} \iota(X_0,\ldots,X_{i-1},[X_i,X_j],X_{i+1},\ldots$$
$$\ldots,X_{j-1},X_{j+1},\ldots,X_k)\,\alpha\,.$$

Of course this formula needs a lot of explanation before it can be used as a correct definition. In the first place, the X_i are homogeneous vector fields on M. As such, (7.4) defines a map of $k+1$ homogeneous vector fields to functions on M. Taking the sum over homogeneous parts, it defines a $(k+1)$-additive function of vector fields on M to functions on M.

7.5 Lemma. *The map $d\alpha$ is $(k+1)$-linear over $C^\infty(M)$ and graded skew-symmetric. It thus defines a $(k+1)$-form on M.*

Proof (sketched). The definition of $d\alpha$ breaks up into two parts: a single summation and a double summation. Graded skew-symmetry means that interchanging two neighboring vector fields X_r and X_{r+1} changes the result by a global sign $-(-1)^{\langle\varepsilon(X_r)|\varepsilon(X_{r+1})\rangle}$. Using that α itself is graded skew-symmetric, the only terms in the single summation that do not produce this global sign are the terms with $i = r, r+1$. But these two terms transform into each other, up to this global sign. In the double summation the terms with $i, j \notin \{r, r+1\}$ produce the correct sign change. The summations $i < r$ with $j = r, r+1$ transform into each other up to the needed sign, as do the summations $r+1 < j$ with $i = r, r+1$. The remaining term $i = r$, $j = r+1$ produces the correct sign because of the graded skew-symmetry of the commutator of two vector fields. We thus conclude that $d\alpha$ is graded skew-symmetric.

Knowing that $d\alpha$ is graded skew-symmetric, it suffices to verify that $d\alpha$ is linear over $C^\infty(M)$ in the first vector field X_0, i.e., that

$$\iota(f\cdot X_0, X_1,\ldots,X_k)\, d\alpha = f\cdot(\iota(X_0,X_1,\ldots,X_k)\, d\alpha)\,.$$

The single summation produces, apart from the terms one wants, some terms involving $X_i f$. Using [1.19], it is easy to verify that these unwanted terms are canceled by the unwanted terms appearing in the double summation. QED

7.6 Discussion. Once we know that (7.4) gives a correct definition of the exterior derivative of a k-form, it is useful to write this explicitly in simple cases. So let X be a vector field, and let Y_0, Y_1, Y_2 be homogeneous vector fields. Furthermore, let f be a 0-form, i.e., a smooth function, let θ be a 1-form, and let ω be a 2-form. Their exterior derivatives are given by

$$\iota(X)\, df = Xf$$

$$-\iota(Y_0, Y_1)\, d\theta = Y_0(\iota(Y_1)\,\theta) - (-1)^{\langle\varepsilon(Y_0)|\varepsilon(Y_1)\rangle}\, Y_1(\iota(Y_0)\,\theta) - \iota([Y_0, Y_1])\,\theta$$

$$\begin{aligned}(-1)^{\langle\varepsilon(Y_0)|\varepsilon(Y_2)\rangle}\iota(Y_0,&Y_1, Y_2)\, d\omega = \\ &(-1)^{\langle\varepsilon(Y_0)|\varepsilon(Y_2)\rangle}\big(Y_0(\iota(Y_1, Y_2)\,\omega) + \iota(Y_0, [Y_1, Y_2])\,\omega\big)\\ &+ (-1)^{\langle\varepsilon(Y_1)|\varepsilon(Y_0)\rangle}\big(Y_1(\iota(Y_2, Y_0)\,\omega) + \iota(Y_1, [Y_2, Y_0])\,\omega\big)\\ &+ (-1)^{\langle\varepsilon(Y_2)|\varepsilon(Y_1)\rangle}\big(Y_2(\iota(Y_0, Y_1)\,\omega) + \iota(Y_2, [Y_0, Y_1])\,\omega\big)\,.\end{aligned}$$

In deriving the last (rather symmetric looking) formula, one has to use the graded skew-symmetry of ω and of the commutator. Notice also the minus sign in front of $\iota(Y_0, Y_1)\, d\theta$ (see also [7.14]).

However nice this definition of the exterior derivative might be, it is not one that allows for easy calculation. In order to compute $d\alpha$ in terms of a local expression on a chart, we need a preliminary lemma. If $U \subset M$ is open, it is in particular an $\mathcal{A}$-manifold in its own right. Restricting α and $d\alpha$ to U gives us a k-form $\alpha|_U$ and a $(k+1)$-form $(d\alpha)|_U$ on U. Denoting (temporarily) the exterior derivative on the $\mathcal{A}$-manifold U by d_U, we are thus faced with the question whether $d_U(\alpha|_U)$ equals $(d\alpha)|_U$.

7.7 Lemma. *For any open $U \subset M$: $d_U(\alpha|_U) = (d\alpha)|_U$.*

Proof. Let $m \in U$ be arbitrary, let ρ be a plateau function around m in U, and let $X_0, \ldots, X_k$ be $k+1$ smooth homogeneous vector fields on U. Multiplying the X_i by ρ if necessary, we may assume that the X_i are (smooth!) vector fields on M, zero outside U. According to [IV.6.2] we have $(\iota(X_{i_1}, \ldots, X_{i_k})\,\alpha)|_U = \iota(X_{i_1}|_U, \ldots, X_{i_k}|_U)\,\alpha|_U$. Using [1.5], it follows immediately that we have

$$(\iota(X_0, \ldots, X_k)\, d\alpha)|_U = \iota(X_0|_U, \ldots, X_k|_U)\,(d_U\alpha|_U)\,.$$

Since the values $X_i(m)$ can be chosen "arbitrarily" (see the proof of [IV.5.9]), it follows that $(d_U\alpha|_U)(m) = (d\alpha)|_U(m)$. QED

7.8 Discussion. If $U \subset M$ is a chart with coordinates x^i (even and odd together), we know from [1.8] that $\Gamma(TU) = \Gamma_U(TM)$ is a free graded $C^\infty(U)$-module with the ∂_i as basis. Since the ∂_i are finite in number, it follows in particular that $\Gamma_U(TM)$ is f.g.p. If we denote by dx^i the left dual basis in ${}^*\Gamma(TU) \cong \Gamma({}^*TU) = \Gamma_U({}^*TM)$, i.e., $\iota(\partial_i)\, dx^j = \delta_i^j$, we can apply [I.8.16], [I.8.14], and [I.8.15] to conclude that the space of k-forms on U $\Omega^k(U) = \Gamma_U(\bigwedge^k {}^*TM) \cong \bigwedge^k {}^*\Gamma(TU)$ is a free graded $C^\infty(U)$-module generated by the $dx^{i_1} \wedge \cdots \wedge dx^{i_k}$.

7.9 Proposition. *Let α be a k-form on M, β an ℓ-form, and let f be a smooth function. Furthermore, let $x^1, \ldots, x^n$ be coordinates (even and odd) on a chart $U \subset M$. Then:*

(i) $df|_U = \sum_{i=1}^n dx^i \cdot \partial_i f$;
(ii) *on* U : $d_U\,(dx^{i_1} \wedge \cdots \wedge dx^{i_k} \cdot f|_U) = (-1)^k\, dx^{i_1} \wedge \cdots \wedge dx^{i_k} \wedge df|_U$;
(iii) $d\,(\alpha \wedge \beta) = (d\alpha) \wedge \beta + (-1)^k \alpha \wedge d\beta$;
(iv) $d(d\alpha) = 0$.

Proof. • Using the vector fields ∂_i on U, the equality in (i) is immediate from the definition of df.

• Denote by γ the k-form $dx^{i_1} \wedge \cdots \wedge dx^{i_k}$, and denote by $X_p = \partial_{j_p}$ one of the basis vectors of the tangent space. Since $\iota(X_0, \ldots, X_{i-1}, X_{i+1}, \ldots, X_k)\,\gamma$ is either 0 or ± 1, it follows that

$$X_i(\iota(X_0, \ldots, X_{i-1}, X_{i+1}, \ldots, X_k)\,\gamma \cdot f) = \\ (-1)^{\langle \varepsilon(X_i) | \varepsilon(\gamma) + \sum_{p \neq i} \varepsilon(X_p) \rangle} (\iota(X_0, \ldots, X_{i-1}, X_{i+1}, \ldots, X_k)\,\gamma) \cdot X_i f\ .$$

From this and the fact that the X_i have commutator zero, it follows that

$$(-1)^k \cdot \iota(X_0, \ldots, X_k)\, d(\gamma \cdot f) = \\ \sum_{i=0}^k (-1)^{i + \langle \varepsilon(X_i) | \varepsilon(\gamma) + \sum_{p > i} \varepsilon(X_p) \rangle} \iota(X_0, \ldots, X_{i-1}, X_{i+1}, \ldots, X_k)\,\gamma \cdot X_i f\ .$$

Using that $\iota(X_0, \ldots, X_k)$ equals $\iota(X_0) \circ \cdots \circ \iota(X_k)$ [I.7.16], and that $\iota(X_i)$ is a right derivation of degree $(-1, \varepsilon(X_i))$ [I.6.16], one can show by induction that the right hand side of this formula equals $\iota(X_0, \ldots, X_k)\,(\gamma \wedge df)$. This proves (ii) because the $(X_0, \ldots, X_k)$ generate $(TU)^{k+1}$.

• The direct way to prove (iii) is to use induction as above and the derivation property of vector fields. An easier way is to note that the exterior derivative is additive and commutes with restrictions. It follows that it suffices to prove it in a chart with $\alpha = dx^{i_1} \wedge \cdots \wedge dx^{i_k} \cdot f$ and $\beta = dx^{j_1} \wedge \cdots \wedge dx^{j_\ell} \cdot g$ for some homogeneous functions f and g. Using the definition of d on 0-forms, it is immediate that $d(f \cdot g) = (df) \cdot g + f \cdot dg$. The result then follows from (ii) and the graded skew-symmetry of the wedge product.

• As before, it suffices to prove (iv) on a chart with $\alpha = dx^{i_1} \wedge \cdots \wedge dx^{i_k} \cdot f$. Applying (ii) and (i) twice we obtain

$$d(d\alpha) = (-1)^{k+(k+1)} \sum_{i,j=1}^n dx^{i_1} \wedge \cdots \wedge dx^{i_k} \wedge dx^i \wedge dx^j \cdot \partial_j \partial_i f\ .$$

From $[\partial_i, \partial_j] = 0$ we deduce $\partial_j \partial_i f = (-1)^{\langle \varepsilon(x^i) | \varepsilon(x^j) \rangle} \partial_i \partial_j f$. On the other hand, graded skew-symmetry of the wedge product gives us $dx^i \wedge dx^j = -(-1)^{\langle \varepsilon(x^i) | \varepsilon(x^j) \rangle} dx^j \wedge dx^i$. It follows that $d(d\alpha)$ equals its opposite, and thus is zero. QED

7.10 Discussion. If x^i is a coordinate on a local chart U of an $\mathcal{A}$-manifold M, it is in particular an element of $C^\infty(U)$. As such we can calculate its exterior derivative $d_U(x^i)$, which is a 1-form on U. On the other hand, we have defined the 1-form dx^i on U as being an element of the basis dual to the basis ∂_j of $\Gamma(TU)$. It follows immediately from [7.9-i] that these two 1-forms coincide:

$$d_U(x^i) = dx^i \ .$$

This observation justifies the name dx^i for these 1-forms. In particular it follows that the 1-form dx^i does not depend upon the choice of the other coordinates on U. This in contrast to the vector fields ∂_i, where any one of them depends in general upon the whole set of coordinates.

7.11 Lemma. *Let M be a connected $\mathcal{A}$-manifold and $f : M \to \mathcal{A}$ a smooth function. Then f is constant if and only if $df = 0$.*

Proof. This is a direct consequence of [3.21] (just use all vector fields) and the definition of df [7.3], [7.6]. $\boxed{QED}$

7.12 Discussion. By definition the exterior derivative is a map $d : \Omega^k(M) \to \Omega^{k+1}(M)$, so officially it should be indexed by a k. Taking the direct sum over all k gives us a map (still denoted by d) :

$$d : \Omega(M) \to \Omega(M) \ .$$

Property [7.9-iii] shows that the map d is a right derivation of degree $(1, 0)$. However, there is a pitfall to be avoided: the exterior derivative is *not* linear over $C^\infty(M)$! Luckily $\mathbf{R}$ is a subring of $C^\infty(M)$, and d *is* linear over $\mathbf{R}$. It follows that the exterior derivative is a right derivation of the $\mathbf{Z} \times \mathbf{Z}_2$-graded commutative $\mathbf{R}$-algebra $\Omega(M)$.

Since commutators of derivations are again derivations, we can compute the commutator $[d, d]$. Since the degree of d is $(1, 0)$, it follows that $[d, d] = 2d \circ d \equiv 2d^2$. Property [7.9-iv] then can be stated equivalently as $[d, d] = 0$.

7.13 Summary. *The exterior derivative on differential forms $d : \Omega(M) \to \Omega(M)$ is a right derivation of the $\mathbf{Z} \times \mathbf{Z}_2$-graded $\mathbf{R}$-algebra $\Omega(M)$ of bidegree $(1, 0)$ and of auto commutator zero.*

7.14 Nota Bene. Some readers might wonder about the global factor $(-1)^k$ in the definition of the exterior derivative of a k-form. We will give two explanations. In the first place, this sign can be attributed to our way to identify $\bigwedge^k {}^*E$ with ${}^*(\bigwedge^k E)$, which allowed us to write

$$\iota(X_1, \ldots, X_k)\alpha = \iota(X_1)\big(\iota(X_2)(\ldots(\iota(X_k)\alpha)\ldots)\big) \ ,$$

without an additional sign $(-1)^{k(k-1)/2}$ [I.7.22]. And indeed, introducing this extra sign will give us on the left hand side of (7.4) the factor $(-1)^{k(k+1)/2}$ and on the right hand side the factor $(-1)^{k(k-1)/2}$, and then our global factor $(-1)^k$ disappears.

In the second place, the given d is a *right* derivation of degree $(1,0)$, so applying the inverse transpose $\mathfrak{T}^{-1}$ gives us a *left* derivation $\mathfrak{T}^{-1}d$. Since linearity is over $\mathbf{R}$ for which left and right do not make a difference, the only difference comes from the degree. Now if α is a k-form of parity ($\mathbf{Z}_2$-degree) a, we have $\varepsilon(\alpha) = (k,a)$ and thus

$$\iota(\alpha)\mathfrak{T}^{-1}d = (-1)^{\langle(1,0)|(k,a)\rangle}d\alpha = (-1)^k d\alpha \ .$$

It follows that we could have defined the *left* derivation $\mathfrak{T}^{-1}d$ by (7.4) *without* the additional sign. Looking at [7.9], the conclusions (i) and (iv) remain unchanged when we replace d by $\mathfrak{T}^{-1}d$; in (ii) the sign $(-1)^k$ disappears, and the conclusion in (iii) has to be replaced by $\iota(\alpha\wedge\beta)\mathfrak{T}^{-1}d = \alpha\wedge(\iota(\beta)\mathfrak{T}^{-1}d) + (-1)^\ell(\iota(\alpha)\mathfrak{T}^{-1}d)\wedge\beta$, i.e., by the standard property of a left derivation.

7.15 Definition. If X is a smooth vector field on M, we have defined in [I.6.16] (see also [IV.6.1]) a right derivation $\iota(X)$ of the $\mathbf{Z}\times\mathbf{Z}_2$-graded algebra $\Omega(M)$; if X is homogeneous, $\iota(X)$ has degree $(-1,\varepsilon(X))$. It is a right derivation of the $C^\infty(M)$-algebra structure, and hence a right derivation of the $\mathbf{R}$-algebra structure. Since the exterior derivative d is also a right derivation of the $\mathbf{R}$-algebra structure of $\Omega(M)$, we can define a new right derivation $\mathcal{L}(X)$ by taking the commutator:

$$\mathcal{L}(X) = [d,\iota(X)] \equiv d\circ\iota(X) + \iota(X)\circ d \ .$$

The right derivation $\mathcal{L}(X)$ of $\Omega(M)$ is called *the Lie derivative in the direction* X; if X is homogeneous, $\mathcal{L}(X)$ has bidegree $(0,\varepsilon(X))$. For 0-forms, i.e., for functions, the Lie derivative reduces to $\mathcal{L}(X)f = \iota(X)df = Xf$ because the contraction operator $\iota(X)$ acts as the zero operator on 0-forms.

The definition of the Lie derivative in the direction of a smooth vector field X is extended to include an action on smooth vector fields $Y\in\Gamma(TM)$ by

$$\mathcal{L}(X)\,Y = [X,Y] \ ,$$

i.e., the action of the Lie derivative of Y in the direction of X is just the commutator $[X,Y]$.

7.16 Proposition. *Let X and Y be vector fields on M, then*

(i) $[d,\mathcal{L}(X)] \equiv d\circ\mathcal{L}(X) - \mathcal{L}(X)\circ d = 0$;

(ii) $[\mathcal{L}(X),\iota(Y)] = \iota([X,Y]) \equiv \iota(\mathcal{L}(X)Y) = [\iota(X),\mathcal{L}(Y)]$.

Proof. • The proof of (i) is immediate when using that $d\circ d = 0$.

• To prove the equality $[\mathcal{L}(X),\iota(Y)] = \iota([X,Y])$, we first note that any operator $\iota(Z)$ commutes with restrictions, simply because it is defined pointwise. Since the same is true for d, it follows that it also is true for both sides of the equality $[\mathcal{L}(X), \iota(Y)] = \iota([X,Y])$. It thus suffices to verify this equality on a local chart $U \subset M$. But on a local chart U the $\mathbf{R}$-algebra $\Omega(U)$ is generated by the functions $C^\infty(U)$ and the (local) 1-forms dx^i. Since both sides of the equality $[\mathcal{L}(X), \iota(Y)] = \iota([X,Y])$ are right derivations of this algebra, it suffices to verify it on generators. Now on functions both sides act as the zero operator, hence they are equal. To see what happens on a generator dx^i, write $X|_U = \sum_i X^i \cdot \partial_i$ and $Y|_U = \sum_i Y^i \cdot \partial_i$ with $X^i, Y^i \in C^\infty(U)$. Since the equality is linear in X and Y, we may assume that X and Y are homogeneous. We then compute

$$\begin{aligned}[\mathcal{L}(X), \iota(Y)]\, dx^i &= \mathcal{L}(X)(\iota(Y)\, dx^i) - (-1)^{\langle \varepsilon(X)|\varepsilon(Y)\rangle} \iota(Y)(\mathcal{L}(X)\, dx^i) \\ &= \mathcal{L}(X)\, Y^i - (-1)^{\langle \varepsilon(X)|\varepsilon(Y)\rangle} \iota(Y)(d(\iota(X)\, dx^i)) \\ &= X(Y^i) - (-1)^{\langle \varepsilon(X)|\varepsilon(Y)\rangle} Y(X^i) = \iota([X,Y])\, dx^i\ .\end{aligned}$$

The last equality of (ii) follows by interchanging X and Y and noting that for homogeneous vector fields X and Y we have not only $[X,Y] = -(-1)^{\langle \varepsilon(X)|\varepsilon(Y)\rangle}[Y,X]$ but also $[\mathcal{L}(X), \iota(Y)] = -(-1)^{\langle \varepsilon(X)|\varepsilon(Y)\rangle}[\iota(Y), \mathcal{L}(X)]$. QED

7.17 Corollary. *Let X, Y_1, ..., Y_k be smooth homogeneous vector fields on M, then we have the operator equality*

$$[\mathcal{L}(X)\ ,\ \iota(Y_1)\circ\cdots\circ\iota(Y_k)] = \sum_{i=1}^{k} (-1)^{\sum_{1\le j<i} \langle \varepsilon(X)|\varepsilon(Y_j)\rangle} \iota(Y_1)\circ\cdots\iota(Y_{i-1})\circ\iota([X,Y_i])\circ\iota(Y_{i+1})\circ\cdots\circ\iota(Y_k)\ .$$

7.18 Remarks. • Using the definition of the Lie derivative as well as [7.16-ii] and omitting the k-form α, the defining equations for the exterior derivative can be written as the operator equality

$$\begin{aligned}&(-1)^k\cdot\iota(X_0)\circ\cdots\iota(X_k)\circ d = \\ &= \sum_{0\le i\le k} (-1)^{i+\sum_{p<i}\langle \varepsilon(X_p)|\varepsilon(X_i)\rangle} \mathcal{L}(X_i)\circ\iota(X_0)\circ\cdots\circ\iota(X_{i-1})\circ\iota(X_{i+1})\circ\cdots\circ\iota(X_k) \\ &+ \sum_{0\le i<j\le k} (-1)^{j+\sum_{i<p<j}\langle \varepsilon(X_p)|\varepsilon(X_j)\rangle} \iota(X_0)\circ\cdots\circ\iota(X_{i-1})\circ\iota([X_i,X_j])\circ\iota(X_{i+1})\circ\cdots \\ &\qquad\cdots\circ\iota(X_{j-1})\circ\iota(X_{j+1})\circ\cdots\circ\iota(X_k)\ .\end{aligned}$$

On the other hand, this operator equality could have been derived directly from [7.16-ii]. But then we use a circular argument because the Lie derivative is defined in terms of the exterior derivative.

• For homogeneous vector fields X and Y, and a k-form α, property [7.16-ii] can be written as

$$\mathcal{L}(X)\big(\iota(Y)\alpha\big) = \iota(\mathcal{L}(X)Y)\alpha + (-1)^{\langle\varepsilon(X)|\varepsilon(Y)\rangle}\iota(Y)\mathcal{L}(X)\,\alpha\,.$$

If we interpret the operation $\iota(Y)$ as a kind of multiplication, this equation tells us that $\mathcal{L}(X)$ remains a derivation of bidegree $(0, \varepsilon(X))$.

7.19 Definition. Let M and N be two $\mathcal{A}$-manifolds and let $f : M \to N$ be a smooth map. According to [2.1] we have an induced even vector bundle map $Tf : TM \to TN$. We thus can form the even bundle morphism $\bigwedge^k Tf : \bigwedge^k TM \to \bigwedge^k TN$ and its associated pull-back map [IV.6.3] ${}^*(\bigwedge^k Tf) : \Gamma^{-1}({}^*(\bigwedge^k TN)) \to \Gamma^{-1}({}^*(\bigwedge^k TM))$. Using the identification ${}^*(\bigwedge^k TM) \cong \bigwedge^k {}^*TM$ and [I.7.25], we can interpret this pull-back as going from k-forms on N to k-forms on M; restricting attention to smooth k-forms, it maps $\Omega^k(N)$ to $\Omega^k(M)$. In this interpretation (see [IV.6.9] and [IV.6.10]) it is customary to denote the pull-back map ${}^*(\bigwedge^k Tf)$ between k-forms by f^*. (Note that, although convention dictates that the $*$ is at the right of f, f^* is a left linear pull-back; this does not really pose a problem because f itself does not admit any kind of dual.)

Although f^* is not a vector bundle morphism, its definition is pointwise: it maps a k-form $\alpha_{f(m)}$ at $f(m) \in N$ to a k-form $(f^*\alpha)_m$ at $m \in M$. If $X_1, \dots, X_k \in T_mM$ are tangent vectors at $m \in M$, the definition of the pull-back gives us the formula

$$(7.20) \qquad \iota(X_1, \dots, X_k)\,(f^*\alpha)_m = \iota(\langle\!\langle X_1 \| Tf \rangle\!\rangle, \dots, \langle\!\langle X_k \| Tf \rangle\!\rangle)\,\alpha_{f(m)}\,.$$

In the particular case $k = 0$, this formula tells us that the pull-back of a smooth function $\phi \in C^\infty(M) \equiv \Omega^0(M)$ is given by $f^*\phi = f \diamond \phi = \phi \circ f$. This is the special case alluded to at the end of [IV.6.7], with the map $h = {}^*(\bigwedge^0 Tf) : M \times \mathcal{A} \to N \times \mathcal{A}$, $(m, a) \mapsto (f(m), a)$. For future use we recall that f^* is an algebra morphism [IV.6.10], i.e.,

$$f^*(\alpha \wedge \beta) = (f^*\alpha) \wedge (f^*\beta)\,,$$

and that f^* commutes (more or less) with restrictions [IV.6.6].

7.21 Lemma. *Let $f : M \to N$ be a smooth map and X a vector field on M. Then for any point $m \in M$ we have (with some abuse of notation) the equality*

$$\iota(X_m)\,(f^*\alpha)_m = f^*(\iota(\langle\!\langle X_m \| Tf \rangle\!\rangle)\,\alpha_{f(m)})\,.$$

In particular, if Y is a vector field on N related to X by f, then

$$\iota(X)\,(f^*\alpha) = f^*(\iota(Y)\,\alpha)\,.$$

Proof. The second part follows from the first by noting that $\langle\!\langle X_m \| Tf \rangle\!\rangle = Y_{f(m)}$. The abuse of notation is that the f^* in $f^*(\iota(\langle\!\langle X_m \| Tf \rangle\!\rangle)\,\alpha_{f(m)})$ should be seen as the dual map $\bigwedge^k {}^*Tf$ from $\bigwedge^k {}^*T_{f(m)}N$ to $\bigwedge^k {}^*T_mM$, i.e., as the pointwise expression of the pull-back map, rather then the official definition of the pull-back map on sections. With this explanation, let $Z_1, \ldots, Z_{k-1} \in T_mM$ be $k-1$ arbitrary tangent vectors at a point $m \in M$. With these we obtain the algebraic equalities

$$\begin{aligned}\iota(Z_1, \ldots, Z_{k-1})(\iota(X)\,(f^*\alpha))_m &= \iota(Z_1, \ldots, Z_{k-1}, X_m)(f^*\alpha)_m \\ &= \iota(\langle\!\langle Z_1 \| Tf \rangle\!\rangle, \ldots, \langle\!\langle Z_{k-1} \| Tf \rangle\!\rangle, \langle\!\langle X_m \| Tf \rangle\!\rangle)\, \alpha_{f(m)} \\ &= \iota(\langle\!\langle Z_1 \| Tf \rangle\!\rangle, \ldots, \langle\!\langle Z_{k-1} \| Tf \rangle\!\rangle)(\iota(\langle\!\langle X_m \| Tf \rangle\!\rangle)\, \alpha_{f(m)}) \\ &= \iota(Z_1, \ldots, Z_{k-1})(f^*(\iota(\langle\!\langle X_m \| Tf \rangle\!\rangle)\, \alpha_{f(m)})) \;. \qquad \boxed{QED}\end{aligned}$$

7.22 Proposition. *Let $f : M \to N$ be a smooth map, and let α be a k-form on N. Then $d(f^*\alpha) = f^*(d\alpha)$.*

Proof. To prove this global equality, it suffices to prove it on charts U that cover M. We may even assume that $f(U) \subset V$ for some chart V because such charts U still cover M. Using that d commutes with restrictions to open subsets [7.7], and using the way the pull-back commutes with restrictions [IV.6.6], we thus have to show the equality $d_U((f|_U)^*\alpha) = (f|_U)^*(d_V\alpha|_V)$. Now let x^i be coordinates on U and y^j be coordinates on V, and let us not note the various restrictions. Since f^* is an algebra morphism [IV.6.10] and since d is a right derivation, it suffices to prove the equality on the generators $\phi \in C^\infty(V)$ and dy^j. Using the chain rule [III.3.8] we have the equalities

$$d(f^*\phi) = d(\phi \circ f) = \sum_i dx^i \cdot \frac{\partial}{\partial x^i}(\phi \circ f) = \sum_{i,j} dx^i \cdot \frac{\partial f^j}{\partial x^i} \cdot (\frac{\partial \phi}{\partial y^j} \circ f) \;.$$

On the other hand we can compute $\iota(\partial_i)f^*d\phi$ at any point $m \in U$ by using [7.21]:

$$\begin{aligned}(\iota(\partial_i)\, f^*d\phi)_m &= \iota(\langle\!\langle \partial_i \| Tf \rangle\!\rangle)\,(d\phi)_{f(m)} \\ &= \iota(\sum_j \frac{\partial f^j}{\partial x^i}(m) \cdot \frac{\partial}{\partial y^j})\,(\sum_k (dy^k)_{f(m)} \cdot \frac{\partial \phi}{\partial y^k}(f(m))) \\ &= \sum_j \frac{\partial f^j}{\partial x^i}(m) \cdot \frac{\partial \phi}{\partial y^j}(f(m)) \;.\end{aligned}$$

Since this is the same as $\iota(\partial_i)(d(f^*\phi))_m$, we conclude that $f^*(d\phi) = d(f^*\phi)$. Since we know that dy^j is the exterior derivative of the coordinate function $y^j \in C^\infty(U)$, we immediately obtain $d(f^*(dy^j)) = d^2(f^*y^j) = 0$, as well as $f^*(d\,dy^j) = f^*0 = 0$ (both because $d^2 = 0$). $\boxed{QED}$

7.23 Definition/Construction. Let M, N, and Q be $\mathcal{A}$-manifolds, $O \subset M \times N$ an open subset, $f : O \to Q$ a smooth map, and α a (not necessarily smooth) k-form on Q. With these ingredients we want to define a pull-back of α to N but in an M-dependent way. More precisely, for each $m \in M$ we define the open subset $O_m \subset N$ by $O \cap \{m\} \times N = \{m\} \times O_m$ and the map $f_m : O_m \to Q$ by $f_m(n) = f(m,n)$. Using the definition of the generalized tangent map Tf_m [3.19] we now define for each $m \in M$ the k-form $f_m^*\alpha$ on O_m by its action on tangent vectors:

$$\iota(Y_1|_n, \ldots, Y_k|_n)(f_m^*\alpha)_n = \iota(\langle\!\langle Y_1|_n \| T_n f_m \rangle\!\rangle, \ldots, \langle\!\langle Y_k|_n \| T_n f_m \rangle\!\rangle)\, \alpha_{f_m(n)} \;.$$

In order to discuss smoothness properties of the k-forms $f_m^*\alpha$, we need another way to describe them. We thus define a map $\psi_\alpha : O \to \bigwedge^k {}^*TN$ such that for each $(m,n) \in O$ the image $\psi_\alpha(m,n)$ is a k-form at $n \in N$, i.e., $\psi_\alpha(m,n) \in (\bigwedge^k {}^*TN)_n$. We will call ψ_α an *M-dependent k-form on N* (even though it is only defined for $(m,n) \in O \subset M \times N$). In terms of its action on tangent vectors at a point $n \in N$, ψ_α is defined by

$$\begin{aligned}\iota(Y_1|_n, \ldots, Y_k|_n)\psi_\alpha(m,n) &= \iota((\underline{0}_m, Y_1|_n), \ldots, (\underline{0}_m, Y_k|_n))(f^*\alpha)_{(m,n)} \\ &= \iota(\langle\!\langle (\underline{0}_m, Y_1|_n) \| T_{(m,n)} f \rangle\!\rangle, \ldots, \langle\!\langle (\underline{0}_m, Y_k|_n) \| T_{(m,n)} f \rangle\!\rangle)\, \alpha_{f(m,n)} \;.\end{aligned}$$

For a fixed $m \in M$, the map $\psi_\alpha(m, \_)$ is a k-form on $O_m \subset N$; comparing the definitions immediately shows that $\psi_\alpha(m,n) = (f_m^*\alpha)_n$.

7.24 Lemma. *If the k-form α is smooth, the map ψ_α is also smooth.*

Proof. Being smooth is a local property, so we may restrict attention to (trivializing) charts U for M and $V \subset E_0$ for N (we assume that N is modeled on the $\mathcal{A}$-vector space E). This gives us a local trivialization $V \times \bigwedge^k {}^*E$ for $\bigwedge^k {}^*TN$. In this local trivialization the map ψ_α is given as $(m,n) \mapsto (n, \widetilde{\psi}(m,n))$ for some map $\widetilde{\psi} : U \times V \to \bigwedge^k {}^*E$. The map $\widetilde{\psi}$ is smooth if and only if its coordinates with respect to some basis for $\bigwedge^k {}^*E$ are smooth. Those coordinates (the right version) are given by $\iota(e_{i_1}, \ldots, e_{i_k})\widetilde{\psi}(m,n)$, where (e_i) is a basis for E. According to [1.16] those coordinates are also given by $\iota(\partial_{i_1}, \ldots, \partial_{i_k})\psi_\alpha(m,n)$. This means that we have to prove that the functions

$$(m,n) \mapsto \iota(\langle\!\langle (\underline{0}_m, \partial_{i_1}|_n) \| T_{(m,n)} f \rangle\!\rangle, \ldots, \langle\!\langle (\underline{0}_m, \partial_{i_k}|_n) \| T_{(m,n)} f \rangle\!\rangle)\, \alpha_{f(m,n)}$$

are smooth. But that is obvious because α, f, Tf, the zero section $\underline{0}$, and the vector fields ∂_j are all smooth maps. QED

7.25 Discussion/Definition. Let the ingredients M, N, O, Q, and f be as in [7.23], let α be a k-form on Q and let β be an ℓ-form on Q. We then can form the M-dependent k-form ψ_α and the M-dependent ℓ-form ψ_β. More precisely, for each $(m,n) \in O$ we have $\psi_\alpha(m,n) = (f_m^*\alpha)_n \in (\bigwedge^k {}^*TN)_n$ and $\psi_\beta(m,n) = (f_m^*\beta)_n \in (\bigwedge^\ell {}^*TN)_n$. We thus can form the wedge product $(f_m^*\alpha)_n \wedge (f_m^*\beta)_n \in (\bigwedge^{k+\ell} {}^*TN)_n$. Generalizing the symbol $\wedge$, we define the M-dependent $(k+\ell)$-form $\psi_\alpha \wedge \psi_\beta : O \to \bigwedge^{k+\ell} {}^*TN$ by

$$(\psi_\alpha \wedge \psi_\beta)(m,n) = \psi_\alpha(m,n) \wedge \psi_\beta(m,n) \equiv (f_m^*\alpha)_n \wedge (f_m^*\beta)_n \;.$$

Looking at this formula in local charts shows immediately that $\psi_\alpha \wedge \psi_\beta$ is smooth if ψ_α and ψ_β are smooth. On the other hand, we can form the $(k+\ell)$-form $\alpha \wedge \beta$ on Q and the M-dependent $(k+\ell)$-form $\psi_{\alpha\wedge\beta}$. Nobody will be surprised that this is the same as $\psi_\alpha \wedge \psi_\beta$.

7.26 Lemma. *$\psi_\alpha \wedge \psi_\beta = \psi_{\alpha\wedge\beta}$ or, equivalently, $f_m^*\alpha \wedge f_m^*\beta = f_m^*(\alpha \wedge \beta)$.*

Proof. This is an immediate consequence of the definition of ψ_α in terms of $f^*\alpha$ and the fact that we have $f^*(\alpha \wedge \beta) = f^*\alpha \wedge f^*\beta$ [7.19], [IV.6.10]. QED

7.27 Proposition. *Let X be a smooth homogeneous integrable vector field on an $\mathcal{A}$-manifold M, let $\phi : W_X \to M$ be its flow, and let α be a smooth k-form on M. We then have a smooth map $\psi_\alpha : W_X \to \bigwedge^k {}^*TM$, $(t,m) \mapsto (\phi_t^*\alpha)_m$. The vector field ∂_t and this map satisfy the condition of [3.10] and we have $\partial_t\psi_\alpha = \psi_{\mathcal{L}(X)\alpha}$, i.e.,*

$$\partial_t\,(\phi_t^*\alpha)_m = (\phi_t^*(\mathcal{L}(X)\,\alpha))_m\ .$$

Proof. The property we have to prove is local (equality of maps), so let us choose a local chart V around $\phi(t,m)$ with coordinates y^i, and let $I \subset \mathcal{A}_{\varepsilon(X)}$ be an interval around t and U a coordinate chart around m with coordinates x^j such that $I \times U \subset W_X$ and $\phi(I \times U) \subset V$ (continuity of ϕ assures the existence of such sets). If M is modeled on the $\mathcal{A}$-vector space E, then, as in the proof of [7.24] the map ψ_α is given on $I \times U$ as $(t,m) \mapsto (m, \widetilde{\psi}_\alpha(t,m))$ for some smooth map $\widetilde{\psi}_\alpha : I \times U \to \bigwedge^k {}^*E$. From this it is obvious that indeed the vector field ∂_t and ψ_α verify the conditions of [3.10], so our statement makes sense. Moreover, the map $\partial_t\psi_\alpha$ is, in the same local trivialization, given as $(t,m) \mapsto (m, \partial_t\widetilde{\psi}_\alpha(t,m))$ (3.12). We thus have to show that $\partial_t\widetilde{\psi}_\alpha = \widetilde{\psi}_{\mathcal{L}(X)\alpha}$.

On the local chart V there exist smooth functions $\alpha_{i_1,\dots,i_k} : V \to \mathcal{A}$ such that the k-form α has the form $\alpha = \sum_{i_1,\dots,i_k} \alpha_{i_1\dots i_k} \cdot dy^{i_1} \wedge \cdots \wedge dy^{i_k}$. Using [7.26] we thus have $(\phi_t^*\alpha) = \sum_{i_1,\dots,i_k} (\phi_t^*\alpha_{i_1\dots i_k})(x) \cdot (\phi_t^* dy^{i_1}) \wedge \cdots \wedge (\phi_t^* dy^{i_k})$. Taking a closer look at the separate factors in this expression, we first note that $(\phi_t^*\alpha_{i_1\dots i_k})(x)$ is the function $\alpha_{i_1\dots i_k}(\phi_t(x)) = \alpha_{i_1\dots i_k}(\phi(t,x))$. And then we note that the definition of ϕ_t^* implies that we have $\iota(\partial_{x^j})\phi_t^* dy^i = \iota(\langle\!\langle \partial_{x^j}|_{(t,x)} \,\|\, T\phi \rangle\!\rangle) dy^i = (\partial_{x^j}\phi^i)(t,x)$. Hence all terms of the form $\phi_t^* dy^i$ are of the form

$$\phi_t^* dy^i = \sum_j dx^j \cdot \frac{\partial \phi^i}{\partial x^j}(t,x)\ .$$

We conclude that the local expression $\widetilde{\psi}_\alpha$ of $\phi_t^*\alpha$ is a sum, with signs, of terms of the form $\alpha_{i_1\dots i_k}(\phi(t,x)) \cdot (\partial_{x^{j_1}}\phi^{i_1})(t,x) \cdots (\partial_{x^{j_k}}\phi^{i_k})(t,x) \cdot dx^{j_1} \wedge \cdots \wedge dx^{j_k}$. Since the derivative of an $\mathcal{A}$-vector space-valued function can be computed by deriving each of its components (III.3.9), we thus have to compute the action of ∂_t on the coefficients of $dx^{j_1} \wedge \cdots \wedge dx^{j_k}$. Since ∂_t is a derivation, we thus have to derive each of the factors $\alpha_{i_1\dots i_k}(\phi(t,x))$ and $(\partial_{x^j}\phi^i)(t,x)$ separately.

If on the chart V the vector field X has the local expression $X|_{(y)} = \sum_i X^i(y) \cdot \partial_{y^i}|_{(y)}$, then the definition of the flow gives the explicit expression $(\partial_t \phi^i)(t,x) = X^i(\phi(t,x))$. Instead of computing each $\partial_t \partial_{x^j} \phi^i$ separately, we compute the time derivative for a whole group at a time:

$$\begin{aligned}
\partial_t(\phi_t^* dy^i) &= \sum_j (-1)^{\langle \varepsilon(t) | \varepsilon(x^j) \rangle} dx^j \cdot \frac{\partial^2 \phi^i}{\partial t \partial x^j}(t,x) = \sum_j dx^j \cdot \frac{\partial^2 \phi^i}{\partial x^j \partial t}(t,x) \\
&= \sum_j dx^j \cdot \frac{\partial (X^i \circ \phi)}{\partial x^j}(t,x) = \sum_{j,k} dx^j \cdot \frac{\partial \phi^k}{\partial x^j}(t,x) \cdot \frac{\partial X^i}{\partial y^k}(\phi(t,x)) \\
&= \sum_k \phi_t^* dy^k \cdot \phi_t^* \frac{\partial X^i}{\partial y^k} = \phi_t^* (\sum_k dy^k \cdot \frac{\partial X^i}{\partial y^k}) = \phi_t^*(dX^i) \ .
\end{aligned}$$

Since $X^i = \iota(X) dy^i$, it follows immediately from the definition of the Lie derivative that $dX^i = \mathcal{L}(X) dy^i$. By definition of the flow we have $\partial_t((\alpha_{i_1 \dots i_k} \circ \phi) = (X\alpha_{i_1 \dots i_k}) \circ \phi$. Since, again by the definition of the Lie derivative, we have $X\alpha_{i_1 \dots i_k} = \mathcal{L}(X)\alpha_{i_1 \dots i_k}$, we conclude that

$$\partial_t(\phi_t^* dy^i) = \phi_t^*(\mathcal{L}(X) dy^i) \qquad \text{and} \qquad \partial_t(\phi_t^* \alpha_{i_1 \dots i_k}) = \phi_t^*(\mathcal{L}(X)\alpha_{i_1 \dots i_k}) \ .$$

It now only remains to knit these results correctly together:

$$\begin{aligned}
\partial_t(\phi_t^* \alpha) &= \sum_{i_1, \dots, i_k} \partial_t \Big(\phi_t^* \alpha_{i_1 \dots i_k} \cdot \phi_t^* dy^{i_1} \wedge \dots \wedge \phi_t^* dy^{i_k} \Big) \\
&= \sum_{i_1, \dots, i_k} \Big(\partial_t(\phi_t^* \alpha_{i_1 \dots i_k}) \cdot \phi_t^* dy^{i_1} \wedge \dots \wedge \phi_t^* dy^{i_k} + \\
&\qquad \sum_{p=1}^k (-1)^{\langle \varepsilon(t) | \varepsilon(\alpha_{i_1 \dots i_k}) + \sum_{1 \le q < p} \varepsilon(y^{i_q}) \rangle} \phi_t^* \alpha_{i_1 \dots i_k} \cdot \phi_t^* dy^{i_1} \wedge \dots \\
&\qquad\qquad \dots \wedge \phi_t^* dy^{i_{p-1}} \wedge \partial_t(\phi_t^* dy^{i_p}) \wedge \phi_t^* dy^{i_{p+1}} \wedge \dots \wedge \phi_t^* dy^{i_k} \Big) \\
&= \sum_{i_1, \dots, i_k} \Big(\phi_t^*(\mathcal{L}(X)\alpha_{i_1 \dots i_k}) \cdot \phi_t^* dy^{i_1} \wedge \dots \wedge \phi_t^* dy^{i_k} + \\
&\qquad \sum_{p=1}^k (-1)^{\langle \varepsilon(t) | \varepsilon(\alpha_{i_1 \dots i_k}) + \sum_{1 \le q < p} \varepsilon(y^{i_q}) \rangle} \phi_t^* \alpha_{i_1 \dots i_k} \cdot \phi_t^* dy^{i_1} \wedge \dots \\
&\qquad\qquad \dots \wedge \phi_t^* dy^{i_{p-1}} \wedge \phi_t^*(\mathcal{L}(X) dy^{i_p}) \wedge \phi_t^* dy^{i_{p+1}} \wedge \dots \wedge \phi_t^* dy^{i_k} \Big) \\
&= \phi_t^* \sum_{i_1, \dots, i_k} \Big(\mathcal{L}(X)\alpha_{i_1 \dots i_k} \cdot dy^{i_1} \wedge \dots \wedge dy^{i_k} + \\
&\qquad \sum_{p=1}^k (-1)^{\langle \varepsilon(t) | \varepsilon(\alpha_{i_1 \dots i_k}) + \sum_{1 \le q < p} \varepsilon(y^{i_q}) \rangle} \alpha_{i_1 \dots i_k} \cdot dy^{i_1} \wedge \dots \\
&\qquad\qquad \dots \wedge dy^{i_{p-1}} \wedge \mathcal{L}(X) dy^{i_p} \wedge dy^{i_{p+1}} \wedge \dots \wedge dy^{i_k} \Big)
\end{aligned}$$

$$= \phi_t^*\Big(\mathcal{L}(X) \sum_{i_1,\dots,i_k} \alpha_{i_1\dots i_k} \cdot dy^{i_1} \wedge \dots \wedge dy^{i_k}\Big) .$$

The last equality is a direct consequence of the fact that the Lie derivative $\mathcal{L}(X)$ is a right derivation of parity $(0, \varepsilon(X)) = (0, \varepsilon(t))$. QED

7.28 Remark. An analogous result for vector fields is given in [5.15]; a generalization to arbitrary $\mathcal{A}$-Lie group actions is given in [VI.8.19]

8. DE RHAM COHOMOLOGY

Using Batchelor's theorem we show that the de Rham cohomology of an $\mathcal{A}$-manifold M (closed forms modulo exact forms) equals the de Rham cohomology of the underlying ordinary $\mathbf{R}$-manifold $\mathbf{B}M$.

8.1 Discussion. When we apply the body map to an $\mathcal{A}$-manifold M, we get the $\mathbf{R}$-manifold $\mathbf{B}M$. If we apply $\mathbf{B}$ to a smooth function $f \in C^\infty(M)$, we get a smooth function $\mathbf{B}f$ on $\mathbf{B}M$. Things change a little when we apply $\mathbf{B}$ to vector fields: the vector bundle $\mathbf{B}TM$ splits into its two homogeneous parts $\mathbf{B}TM = \mathbf{B}(TM_0) \oplus \mathbf{B}(TM_1)$. On a local chart of M with coordinates x and ξ, taking the body map means ignoring the coordinates ξ. Since the tangent space is generated by the ∂_x and the ∂_ξ, it follows easily that $\mathbf{B}(TM_0)$ is generated by the ∂_x and that $\mathbf{B}(TM_1)$ is generated by the ∂_ξ. One thus immediately sees that $\mathbf{B}(TM_0)$ can be identified with the tangent bundle of $\mathbf{B}M$. In other words, applying the body map to any vector field on M does in general not yield a vector field on $\mathbf{B}M$, but applying it to an *even* vector field yields a vector field on $\mathbf{B}M$.

If we now turn our attention to k-forms, it becomes even more delicate. Applying $\mathbf{B}$ to a 0-form on M (i.e., a function) yields a function on $\mathbf{B}M$, and applying $\mathbf{B}$ to an *even* 1-form (i.e., a section of ${}^*TM$) yields a 1-form on $\mathbf{B}M$. But already for 2-forms it becomes more complicated. If E is an $\mathcal{A}$-vector space of dimension $p|q$, the dimension of $\bigwedge^2 E$ is $\frac{1}{2}p(p-1) + pq + \frac{1}{2}q(q+1)$, of which pq are odd. In local coordinates this means that the body part of an even 2-form still might contain terms involving $d\xi \wedge d\xi$. We thus are naturally led to ask the question: does there exist a (canonical) way to obtain a k-form on $\mathbf{B}M$ from a k-form on M?

8.2 Construction. Let M be an $\mathcal{A}$-manifold of dimension $p|q$, and let U be a chart with coordinates (x^i, ξ^j). On U we define an even vector field X_E by

$$X_E = \sum_{j=1}^{q} \xi^j \cdot \frac{\partial}{\partial \xi^j} . \tag{8.3}$$

The subscript E stands for *Euler* because X_E is essentially the Euler vector field, although only in the odd coordinates. An elementary calculation shows that

$$\mathcal{L}(X_E)\,(\xi^{j_1}\cdots\xi^{j_\ell}\cdot f_{j_1\ldots j_\ell}(x^1,\ldots,x^p)) = \ell\cdot(\xi^{j_1}\cdots\xi^{j_\ell}\cdot f_{j_1\ldots j_\ell}(x^1,\ldots,x^p))\ .$$

Moreover, $\mathcal{L}(X_E)\,d\xi^j = d\xi^j$ and $\mathcal{L}(X_E)\,dx^i = 0$. Since $\mathcal{L}(X_E)$ is a right derivation of degree $(0,0)$, it follows that any k-form α on U can be written as a sum

$$\alpha = \sum_\ell \alpha^{(\ell)} \qquad , \qquad \mathcal{L}(X_E)\,\alpha^{(\ell)} = \ell\cdot\alpha^{(\ell)}\ .$$

It is easy to describe this decomposition in words: the operator $\mathcal{L}(X_E)$ counts the number of ξ's in a k-form (whether or not preceded by a d) and each term $\alpha^{(\ell)}$ has exactly ℓ ξ's. A simple count shows that there will be no non-zero $\alpha^{(\ell)}$ for $\ell > k+q$.

To show that this local decomposition $\alpha = \sum_{\ell=0}^{k+q}\alpha^{(\ell)}$ is unique, we suppose that $\alpha = 0$ and we apply $\mathcal{L}(X_E)$ s times to get the equations $0 = \sum_{\ell=0}^{k+q}\ell^s\alpha^{(\ell)}$ (writing formally $0^0 = 1$). Taking $s = 0,\ldots,k+q$ we can see these $k+q+1$ equations as linear equations in the $k+q+1$ "unknowns" $\alpha^{(0)},\ldots,\alpha^{(k+q)}$. The entries of the matrix A of these linear equations are given by $A_{s\ell} = \ell^s$. Since the determinant of this matrix is $\prod_{n=1}^{k+q} n!$, the matrix A is invertible. We conclude that all $\alpha^{(\ell)}$ are zero, proving uniqueness of the decomposition.

The obvious candidate for a k-form on $\mathbf{B}M$ associated to a k-form α on M is the form described locally as the zero-th order term $\alpha^{(0)}$ in this decomposition. But we have to be careful to see what happens in another chart. So let V be another chart with coordinates (y^i,η^j), on which we define the Euler vector field $Y_E = \sum_j \eta^j\cdot\partial_{\eta^j}$. In these terms, a k-form α on V can be decomposed according to the number of η's. On the overlap $U\cap V$ we thus have two decompositions: $\alpha = \sum_\ell \alpha_U^{(\ell)} = \sum_\ell \alpha_V^{(\ell)}$, according to X_E and Y_E respectively. In particular the homogeneous terms $\alpha_V^{(\ell_o)}$ decompose according to X_E as $\alpha_V^{(\ell_o)} = \sum_\ell \alpha_{VU}^{(\ell_o\ell)}$. Since in the coordinate change $(x,\xi)\mapsto(y,\eta)$ the η's are at least linear in the ξ's, it follows easily that $\alpha_V^{(\ell_o)} = \sum_{\ell\geq\ell_o}\alpha_{VU}^{(\ell_o\ell)}$, i.e., the number of odd coordinates does never diminish. Uniqueness of the decomposition on U then implies that $\alpha_{VU}^{(00)} = \alpha_U^{(0)}$, i.e., $\alpha_V^{(0)} = \alpha_U^{(0)} + \sum_{\ell>0}\alpha_{VU}^{(0\ell)}$. This formula tells us that the k-form $\alpha_V^{(0)}$, which does not contain any η nor $d\eta$, is given in the (x,ξ) coordinates as $\alpha_U^{(0)}$, which does not contain any ξ nor $d\xi$, plus terms involving ξs and/or $d\xi$s. We now recall that in the coordinate change $(x,\xi)\mapsto(y,\eta)$ the y contain only even powers of ξ. This implies that all terms $\alpha_{VU}^{(0\ell)}$ with $\ell > 0$ must contain ξs and not only $d\xi$s, simply because $d(\xi^i\xi^j) = \xi^i d\xi^j - \xi^j d\xi^i$. We conclude that $\mathbf{B}\alpha_V^{(0)} = \mathbf{B}\alpha_U^{(0)}$. This proves that, if α is a k-form on M, then we can decompose it in a local chart as $\alpha|_U = \sum_\ell \alpha_U^{(\ell)}$, and then the local forms $\mathbf{B}\alpha_U^{(0)}$ coincide on overlaps, i.e., they define a global k-form on $\mathbf{B}M$. But the reader has to be careful: $\mathbf{B}\alpha$ is in general *not* a k-form on $\mathbf{B}M$ (the $d\xi$s do not necessarily disappear), and the local k-forms $\alpha_U^{(0)}$ do in general *not* glue together to form a global k-form on M. The latter can be achieved by making a particular choice for the charts U to be used (see the proof of [8.9]).

8.4 Definition. Let M be an $\mathcal{A}$-manifold, and α a k-form on M. We define the k-form $\mathbf{B}_M\alpha$ on $\mathbf{B}M$ as being the k-form $\mathbf{B}(\alpha^{(0)})$ constructed in [8.2]. If α is a 0-form (a function), this definition coincides with the standard body map: $\mathbf{B}_M f = \mathbf{B}f$. If α is a 1-form, this definition coincides with the standard body map applied to the even part of α : $\mathbf{B}_M\alpha = \mathbf{B}(\alpha_0)$. In the same spirit we define the map $\mathbf{B}_M$ on vector fields by $\mathbf{B}_M X = \mathbf{B}(X_0)$.

8.5 Discussion. Using the Lie derivative $\mathcal{L}(X_E)$ on vector fields ($\mathcal{L}(X_E)(X) = [X_E, X]$) we could have defined an analogous (local) decomposition of a vector field $X = \sum_\ell X^{(\ell)}$ according to the number of times an odd coordinate ξ appears in a (local) description. As for k-forms, one then could associate to any vector field X on M a vector field $\mathbf{B}(X^{(0)})$ on $\mathbf{B}M$. However, as for 1-forms, this coincides with the given definition of $\mathbf{B}_M$: $\mathbf{B}(X^{(0)}) = \mathbf{B}(X_0) = \mathbf{B}_M X$. On the other hand, this approach to $\mathbf{B}_M$ on vector fields can be easily extended to other kinds of tensor fields on M (if the need arises).

8.6 Remark. Our construction of $\mathbf{B}_M$ on k-forms can be stated in a more fancy language. The fact that the number of odd coordinates appearing in a local expression never decreases, implies that we can define the subsets $\mathcal{F}^k_\ell(M) \subset \Omega^k(M)$ as consisting of those k-forms that do contain at least ℓ odd coordinates in any local coordinate system. We thus obtain a filtration $\Omega^k(M) = \mathcal{F}^k_0(M) \supset \mathcal{F}^k_1(M) \supset \mathcal{F}^k_2(M) \supset \ldots$. The existence of our map $\mathbf{B}_M : \Omega^k(M) \to \Omega^k(\mathbf{B}M)$ then can be rephrased as the existence of a (canonical) map $\mathcal{F}^k_0(M)/\mathcal{F}^k_1(M) \to \Omega^k(\mathbf{B}M)$.

8.7 Lemma. *Let M be an $\mathcal{A}$-manifold, α a k-form, β an ℓ-form, and X a vector field on M. Then*

(i) $\mathbf{B}_M : \Omega^k(M) \to \Omega^k(\mathbf{B}M)$ *is* $\mathbf{R}$*-linear and surjective;*
(ii) $\mathbf{B}_M(\alpha \wedge \beta) = (\mathbf{B}_M\alpha) \wedge (\mathbf{B}_M\beta)$;
(iii) $\mathbf{B}_M(\iota(X)\,\alpha) = \iota(\mathbf{B}_M X)\,\mathbf{B}_M\alpha$;
(iv) $\mathbf{B}_M(d\alpha) = d(\mathbf{B}_M\alpha)$.

Proof. The properties (ii), (iii), and (iv) are rather obvious in a local chart, and thus globally when one realizes that the given constructions all commute with restrictions. The only non-trivial part is the surjectivity, which will be proven in [8.9]. $\boxed{QED}$

8.8 Definition. A k-form α on M is said to be *closed* if $d\alpha = 0$; it is said to be *exact* if there exists a $(k-1)$-form β on M such that $\alpha = d\beta$. Since $d^2 = 0$ it follows that any exact k-form is closed. An equivalent way to state these definitions is that α is closed if $\alpha \in \ker(d : \Omega^k(M) \to \Omega^{k+1}(M))$, and it is exact if $\alpha \in \operatorname{im}(d : \Omega^{k-1}(M) \to \Omega^k(M))$. The implication *exact* $\Longrightarrow$ *closed* gets translated into the inclusion

$$\operatorname{im}(d : \Omega^{k-1}(M) \to \Omega^k(M)) \quad \subset \quad \ker(d : \Omega^k(M) \to \Omega^{k+1}(M)) \ .$$

Since d is $\mathbf{R}$-linear and homogeneous, it follows from [I.3.9] and the fact that $\mathbf{R}$ is a field that these subsets of $\Omega^k(M)$ are graded $\mathbf{R}$-vector subspaces of the $\mathbf{R}$-vector space $\Omega^k(M)$. (Of course $\Omega^k(M)$ is also a graded $C^\infty(M)$-module, but that structure is not preserved by these subsets.) We thus can define the (quotient) $\mathbf{R}$-vector space

$$H^k(M) = \ker(d : \Omega^k(M) \to \Omega^{k+1}(M)) / \operatorname{im}(d : \Omega^{k-1}(M) \to \Omega^k(M)) .$$

The $\mathbf{R}$-vector space $H^k(M)$ is called *the kth de Rham cohomology group of M*. If α is a closed k-form on M, its cohomology class in $H^k(M)$ will be denoted by $[\alpha]$. Note that $H^k(M)$ is an $\mathbf{R}$-vector space for $\mathcal{A}$-manifolds M as well as for $\mathbf{R}$-manifolds M.

8.9 Theorem. *If M is an $\mathcal{A}$-manifold, then the map $[\alpha] \mapsto [\mathbf{B}_M\alpha]$, $H^k(M) \to H^k(\mathbf{B}M)$ is an isomorphism of $\mathbf{R}$-vector spaces.*

Proof. From [8.7] we know that this map is a well defined $\mathbf{R}$-linear map, so only injectivity and surjectivity remain to be proven. To do so, let $\mathcal{U}$ be an atlas as in Batchelor's theorem [IV.8.2]. On each chart $U \in \mathcal{U}$ we can define the Euler vector field X_E (8.3), but now the special form of the transition functions for this atlas guarantee that these vector fields coincide on overlaps, i.e., there exists a *globally* defined Euler vector field X_E (Nota Bene. This global vector field X_E depends upon the choice of the special atlas $\mathcal{U}$!). The same argument (the special form of the transition functions for our atlas) shows that the decomposition of a k-form α in homogeneous parts $\alpha^{(\ell)}$ is invariant under a change of coordinates, i.e., each global k-form α decomposes (uniquely) as $\alpha = \sum_\ell \alpha^{(\ell)}$ with $\alpha^{(\ell)}$ a global k-form satisfying $\mathcal{L}(X_E)\,\alpha^{(\ell)} = \ell \cdot \alpha^{(\ell)}$.

If α is closed, the uniqueness of the decomposition implies that all $\alpha^{(\ell)}$ are closed, and thus that $\ell \cdot \alpha^{(\ell)} = \mathcal{L}(X_E)\,\alpha^{(\ell)} = d\,\iota(X_E)\alpha^{(\ell)}$. We thus can write

$$(8.10) \qquad d\alpha = 0 \qquad \Longrightarrow \qquad \alpha = \alpha^{(0)} + d\big(\sum_{l\geq 1} \tfrac{1}{\ell}\,\iota(X_E)\alpha^{(\ell)}\big) .$$

Since $\alpha^{(0)}$ does not contain any ξ nor $d\xi$, we may identify $\alpha^{(0)}$ with its body part $\mathbf{B}\alpha^{(0)}$. In other words, we have a bijection between k-forms on $\mathbf{B}M$ and k-forms α on M satisfying $\mathcal{L}(X_E)\alpha = 0$. This proves en passant the surjectivity of $\mathbf{B}_M$ claimed in [8.7-i].

We now are prepared to prove the isomorphism between $H^k(M)$ and $H^k(\mathbf{B}M)$. If α is a closed k-form on $\mathbf{B}M$, it can be seen as a closed k-form on M satisfying $\mathcal{L}(X_E)\alpha = 0$, proving that the given map $H^k(M) \to H^k(\mathbf{B}M)$ is surjective. Now $[\mathbf{B}_M\alpha] = 0$ is equivalent to saying that $\alpha^{(0)}$ is exact. But then according to (8.10) the whole form α is exact, proving injectivity of the map $H^k(M) \to H^k(\mathbf{B}M)$. QED

8.11 Remark. The essential ingredient in the proof of [8.9] (and thus in the proof that $\mathbf{B}_M$ is surjective) is the existence of a special atlas with transition functions as in Batchelor's theorem [IV.8.2]. Since that theorem depends in an essential way on the existence of partitions of unity, it follows that [8.9] depends in an essential way on the existence of partitions of unity.

Chapter VI

$\mathcal{A}$-Lie groups

Just as differential geometry can be seen as the interplay between analysis and geometry, so can the theory of Lie groups be seen as the interplay between differential geometry and group theory/algebra. An $\mathcal{A}$-Lie group is an $\mathcal{A}$-manifold which happens to be also a group in such a way that the group multiplication is compatible with the $\mathcal{A}$-manifold structure. We have already introduced $\mathcal{A}$-Lie groups in chapter IV, but there they did not play a very important role. On the other hand, this chapter is entirely devoted to the study of $\mathcal{A}$-Lie groups and subjects directly related. Even so, we can only scratch the surface of the theory of $\mathcal{A}$-Lie groups.

Any $\mathcal{A}$-Lie group determines an $\mathcal{A}$-Lie algebra, which is the linearized version of the group in the sense that it is the tangent space at the identity equipped with the linearized version of the group multiplication. Once we have the $\mathcal{A}$-Lie algebra associated to an $\mathcal{A}$-Lie group, we can also go back from the $\mathcal{A}$-Lie algebra to the $\mathcal{A}$-Lie group with the exponential map. This map is defined in terms of the flow of a vector field and generalizes the usual exponential (series) of matrices. The most important aspect of $\mathcal{A}$-Lie groups is that an $\mathcal{A}$-Lie group is completely determined by its $\mathcal{A}$-Lie algebra up to coverings. More precisely, there is, up to isomorphisms, a unique simply connected $\mathcal{A}$-Lie group G associated to each (finite dimensional) $\mathcal{A}$-Lie algebra $\mathfrak{g}$ and any $\mathcal{A}$-Lie group with the same $\mathcal{A}$-Lie algebra is a quotient of G by a countable discrete subgroup. This fact means that a lot of global geometric properties of an $\mathcal{A}$-Lie group can be translated into algebraic properties of its associated $\mathcal{A}$-Lie algebra. For instance, if we define an $\mathcal{A}$-Lie subgroup of an $\mathcal{A}$-Lie group G as an $\mathcal{A}$-Lie group H that admits a smooth injective homomorphism into G, then connected $\mathcal{A}$-Lie subgroups of G are in bijection with the $\mathcal{A}$-Lie subalgebras of its associated $\mathcal{A}$-Lie algebra and normal connected $\mathcal{A}$-Lie subgroups correspond to ideals.

Once we have groups and subgroups, it is natural to consider homogeneous spaces, i.e., the quotient of a group by a subgroup. A homogeneous $\mathcal{A}$-manifold is an $\mathcal{A}$-manifold M with a transitive action of an $\mathcal{A}$-Lie group G. Homogeneous $\mathcal{A}$-manifolds for a given

$\mathcal{A}$-Lie group G are completely determined by the proper $\mathcal{A}$-Lie subgroups of G, where proper $\mathcal{A}$-Lie subgroup means an $\mathcal{A}$-Lie subgroup which is also a submanifold. More precisely, any homogeneous $\mathcal{A}$-manifold is the quotient of G by a proper $\mathcal{A}$-Lie subgroup.

Another example where properties related to an $\mathcal{A}$-Lie group can be translated into properties of its $\mathcal{A}$-Lie algebra is the case of invariant vector fields and invariant differential forms. A vector field on a connected $\mathcal{A}$-manifold is invariant under the group action if and only if it commutes with all fundamental vector fields associated to the $\mathcal{A}$-Lie algebra and a differential form is invariant if and only if the Lie derivative of this form in the direction of the fundamental vector fields is zero.

In a separate section we prove that any action of an $\mathcal{A}$-Lie group on an $\mathcal{A}$-manifold can be transformed into a pseudo effective action of a quotient group. This shows that the restriction to pseudo effective actions for structure groups of a fiber bundle is in reality not a restriction at all.

1. $\mathcal{A}$-Lie groups and their $\mathcal{A}$-Lie algebras

In this section we (re)define an $\mathcal{A}$-Lie group G and we show that topologically it is the direct product of $G^{w.o.d}$ and the even part of an $\mathcal{A}$-vector space with only odd dimension. We then define the associated $\mathcal{A}$-Lie algebra $\mathfrak{g}$ as the set of left-invariant vector fields on G with the usual commutator of vector fields as bracket; as an $\mathcal{A}$-vector space $\mathfrak{g}$ is isomorphic to the tangent space at the identity of G. We give a formula for the structure constants of $\mathfrak{g}$ in terms of the multiplication function in a neighborhood of the identity. We finish with a detailed discussion on how to interpret the $\mathcal{A}$-Lie algebra associated to the $\mathcal{A}$-Lie group $\mathrm{Aut}(E)$. Of course this is $\mathrm{End}_R(E)$, but the identification in terms of matrices is subtle and is not what one would think first.

1.1 Definitions. Some of the definitions we will give here were already given in §IV.1. We recall them because they belong rightfully in this chapter.

• An *$\mathcal{A}$-Lie group* is an $\mathcal{A}$-manifold G that admits at the same time a group structure for which the multiplication $m : G \times G \to G$ is smooth. Depending on context, we will denote the product $m(g, h)$ also by $m(g, h) = g \circ h = g \cdot h = gh$. The identity element will usually be denoted by e, sometimes also by id. We will show in [1.2] that e has necessarily real coordinates, and in [1.6] that for any $\mathcal{A}$-Lie group taking the inverse is automatically a smooth map. The basic example of an $\mathcal{A}$-Lie group is the group $\mathrm{Aut}(E)$ of automorphisms of an $\mathcal{A}$-vector space E (see [IV.1.3]).

• A *smooth left action* of an $\mathcal{A}$-Lie group G on an $\mathcal{A}$-manifold M is a smooth map $\Phi : G \times M \to M$ such that for all $m \in M$: $\Phi(e, m) = m$ and such that for all m, g, h : $\Phi(g, \Phi(h, m)) = \Phi(gh, m)$. For a fixed $g \in G$ we will denote the map $m \mapsto \Phi(g, m)$ by Φ_g. If no confusion is possible, we will denote a left action also as $\Phi(g, m) = \Phi_g(m) = g(m) = g \cdot m = gm$. The evaluation map $\mathrm{Aut}(E) \times E \to E$ is a left action when we equip $\text{Aut}(E) \subset \text{End}_R(E)$ with the usual composition of endomorphisms as group structure (see [IV.1.3]).

• A *smooth right action* of an $\mathcal{A}$-Lie group G on an $\mathcal{A}$-manifold M is a smooth map $\Phi : M \times G \to M$ such that for all $m \in M : \Phi(m, e) = m$ and such that for all $m \in M$, $g, h \in G : \Phi(\Phi(m, g), h) = \Phi(m, gh)$. For a fixed $g \in G$ we will denote the map $m \mapsto \Phi(m, g)$ by Φ_g. If no confusion is possible, we will denote a right action also as $\Phi(m, g) = m \cdot g = mg$.

• A left/right action of an $\mathcal{A}$-Lie group G on an $\mathcal{A}$-manifold M is called *transitive* if $\forall m, m' \in M \, \exists g \in G : \Phi_g(m) = m'$.

• A *(homo)morphism of $\mathcal{A}$-Lie groups* is a smooth map $\rho : G \to H$ between two $\mathcal{A}$-Lie groups G and H that is at the same time a homomorphism of (abstract) groups.

• An *isomorphism of $\mathcal{A}$-Lie groups* is an $\mathcal{A}$-Lie group morphism $\rho : G \to H$ between two $\mathcal{A}$-Lie groups G and H that is at the same time a diffeomorphism of $\mathcal{A}$-manifolds. It follows that ρ^{-1} is also an $\mathcal{A}$-Lie group morphism.

• A *linear representation of G on E*, or just *a representation of G* is an $\mathcal{A}$-Lie group morphism $\rho : G \to \text{Aut}(E)$, E being an $\mathcal{A}$-vector space.

1.2 Discussion. Associated to any $\mathcal{A}$-manifold M we have two subsets: $M^{w.o.d}$, which is a submanifold, and $\mathbf{B}M$, which is an $\mathbf{R}$-manifold. Moreover, restricting a smooth map to one of these subsets maps it to the corresponding subset in the target space. Looking at $\mathbf{B}G$ we thus find that $m : \mathbf{B}G \times \mathbf{B}G \to \mathbf{B}G$. From the commutativity of the diagram in [III.4.22] (i.e., $\mathbf{B} \circ m = m \circ \mathbf{B}$) we deduce $\mathbf{B}(e \cdot e) = (\mathbf{B}e) \cdot (\mathbf{B}e)$ and thus $e = \mathbf{B}e$, i.e., the identity element has real coordinates. From $e = \mathbf{B}(g \cdot g^{-1}) = (\mathbf{B}g) \cdot \mathbf{B}(g^{-1})$ we deduce $(\mathbf{B}g)^{-1} = \mathbf{B}(g^{-1})$, i.e., if g has real coordinates, then so has g^{-1}. It follows immediately that $\mathbf{B}G$ is an $\mathbf{R}$-Lie group ($\mathbf{B}G$ is an $\mathbf{R}$-manifold and restriction of m to $\mathbf{B}G$ remains smooth) and that $\mathbf{B} : G \to \mathbf{B}G$ is a homomorphism of (abstract) groups.

Looking now at $G^{w.o.d}$ we can deduce from the fact that inversion Inv is smooth [1.6] that $G^{w.o.d}$ is an $\mathcal{A}$-Lie group (because Inv then also preserves $G^{w.o.d}$). (Note that the proof of [1.6] uses that e has real coordinates, so we have to be careful in what order we prove our statements.)

1.3 Nota Bene. If Φ is a smooth left action of an $\mathcal{A}$-Lie group G on an $\mathcal{A}$-manifold M, it follows immediately that for $g, h \in G$ we have $\Phi_g \circ \Phi_h = \Phi_{gh}$ (for a right action we obtain $\Phi_g \circ \Phi_h = \Phi_{hg}$), and thus in particular all maps $\Phi_g : M \to M$ are bijective with inverse $\Phi_{g^{-1}}$ (because $\Phi_e = id(M)$). Since the action is smooth, it is in particular continuous. It follows that all maps Φ_g are homeomorphisms of M. Moreover, it follows from [III.1.23-g] that if $g \in G$ has real coordinates (i.e., $g \in \mathbf{B}G$), then $\Phi_g : M \to M$ is smooth. Since $(\mathbf{B}G)^{-1} = \mathbf{B}G$ we conclude that such a Φ_g is a diffeomorphism of M. However, if g does not have real coordinates, there is no reason to suppose that Φ_g is smooth. Consider for example the smooth left action of $\text{Aut}(E)$ on $E = E_0^\sharp$. According to [II.6.22] and [III.1.27] a map $\Phi_g = g \in \text{Aut}(E)$ is smooth if and only if the matrix elements of g (i.e., its coordinates) are real.

Even though Φ_g is in general not smooth, we have defined generalized tangent maps $T\Phi_g$ in [V.3.19] by the formula

$$T\Phi_g : X_m \mapsto \langle\!\langle (\underline{0}_g, X_m) \| T\Phi \rangle\!\rangle \ .$$

The property $\Phi_g \circ \Phi_h = \Phi_{gh}$ then easily gives the property $T\Phi_g \circ T\Phi_h = T\Phi_{gh}$, as if the chain rule were still valid. (For right actions the defining formula for $T\Phi_g$ would be $T\Phi_g(X_m) = T\Phi(X_m, \underline{0}_g)$, and then we get $T\Phi_g \circ T\Phi_h = T\Phi_{hg}$, in accordance with the equality $\Phi_g \circ \Phi_h = \Phi_{hg}$.)

1.4 Definition/Discussion. We can interpret the multiplication $m : G \times G \to G$ as either a left or a right action of the $\mathcal{A}$-Lie group G on the $\mathcal{A}$-manifold G. If we view it as a left action, i.e., $m : G_{\text{gp}} \times G_{\text{mfd}} \to G_{\text{mfd}}$, the maps $m_g = m(g, \_)$ are usually denoted as $m_g = L_g$ and are called *left translations of G over g*. In case we view m as a right action, i.e., $m : G_{\text{mfd}} \times G_{\text{gp}} \to G_{\text{mfd}}$, the maps $m_g = m(\_, g)$ are usually denoted as $m_g = R_g$ and are called *right translations of G over g*. All left and right translations are homeomorphisms of G; they are diffeomorphisms if and only if g has real coordinates (e.g., if L_g is a diffeomorphism, $L_g(e) = g$ must have real coordinates by property (A2) of smooth functions).

1.5 Lemma. *The generalized tangent maps TL_g and TR_h commute, i.e., $\forall g, h \in G$: $TL_g \circ TR_h = TR_h \circ TL_g$.*

Proof. The associativity of the multiplication says $m(g, m(k, h)) = m(m(g, k), h)$. Applying the tangent map to this identity allows us to compute

$$\begin{aligned} TL_g(TR_h(X_k)) &= TL_g(Tm(X_k, \underline{0}_h)) = Tm(\underline{0}_g, Tm(X_k, \underline{0}_h)) \\ &= Tm(Tm(\underline{0}_g, X_k), \underline{0}_h) = TR_h(TL_g(X_k)) \ . \end{aligned}$$ QED

1.6 Lemma. *If G is an $\mathcal{A}$-Lie group, then the map $Inv : G \to G$, $g \mapsto Inv(g) = g^{-1}$ describing the inverse is a smooth map and for an arbitrary $X_g \in T_gG$ we have the equality $TInv(X_g) = -TL_{g^{-1}}(TR_{g^{-1}}(X_g)) = -TR_{g^{-1}}(TL_{g^{-1}}(X_g))$. In particular $TInv(X_e) = -X_e$.*

Proof. Since $h = Inv(g)$ is the unique solution of the equation $m(g, h) = e$, smoothness is given by the implicit function theorem. The details are as follows. For $g_o \in \mathbf{B}G$ we compute the partial derivatives $\partial m/\partial h$ of m with respect to the second variable at the point (g_o, g_o^{-1}). Using [III.3.13] this is exactly the map TL_{g_o}. Since this map is invertible with inverse $TL_{g_o^{-1}}$, we can apply the implicit function theorem [III.3.27] to conclude that $h = Inv(g)$ is the unique smooth solution in a neighborhood of $g_o \in \mathbf{B}G$. Since these neighborhoods cover G [III.4.12], Inv is globally smooth [III.4.18].

To compute $T\mathit{Inv}$, we consider the map $\Phi : G \to G, g \mapsto m(g, \mathit{Inv}(g)) \equiv e$. Applying the chain rule we find

$$\begin{aligned} 0 = T\Phi(X_g) &= Tm(X_g, T\mathit{Inv}(X_g)) = Tm(\underline{0}_g, T\mathit{Inv}(X_g)) + Tm(X_g, \underline{0}_{g^{-1}}) \\ &= TL_g(T\mathit{Inv}(X_g)) + TR_{g^{-1}}(X_g) \,. \end{aligned}$$

Applying $TL_{g^{-1}}$ gives the announced result (also using [1.5]). QED

1.7 Proposition. *Let G be an $\mathcal{A}$-Lie group which is modeled as an $\mathcal{A}$-manifold on the $\mathcal{A}$-vector space E of dimension $p|q$. If we denote by F the $\mathcal{A}$-vector space of dimension $0|q$, then there exists a diffeomorphism $G^{w.o.d} \times F_0 \to G$, i.e., as an $\mathcal{A}$-manifold G is the direct product of $G^{w.o.d}$, which contains only even coordinates, and F_0, which contains only odd coordinates.*

Proof. Let $\varphi : U' \to O' \subset E_0$ be a chart around the identity element $e \in G$. In the $\mathcal{A}$-vector space E we not only have the canonical graded subspace $E^{w.o.d}$ spanned by the even basis vectors (in the equivalence class), but also the graded subspace F spanned by the odd basis vectors. Obviously $\dim F = 0|q$ and $E = E^{w.o.d} \oplus F$. By a translation over a vector with real coordinates (a diffeomorphism) we may assume that $\varphi(e) = (0,0)$, where we now (and for the rest of this proof) use the splitting of E as the direct product $E = E^{w.o.d} \times F$. By definition of the DeWitt topology, the open set $O' \subset E_0$ has the form $O' = P' \times F_0$ for some open set $P' \subset E_0^{w.o.d}$ (odd coordinates do not count for the topology). We now define $\psi : P' \times F_0 \to G$ by $\psi(x,\xi) = \varphi^{-1}(x,0) \cdot \varphi^{-1}(0,\xi)$. Since $T_{(0,0)}\psi = T_{(0,0)}\varphi^{-1}$, we conclude by the inverse function theorem that there exist open sets $P \subset P' \subset E_0^{w.o.d}$ and $U \subset U' \subset G$ such that $\psi : P \times F_0 \to U$ is a diffeomorphism. Applying the body map to the definition of ψ (and using that $\varphi^{-1}(0,0) = e$) gives $\mathbf{B}\psi = \mathbf{B}\varphi^{-1}$ and thus φ^{-1} also maps $P \times F_0$ diffeomorphically to U. Moreover, by definition of $G^{w.o.d}$ we have $G^{w.o.d} \cap U = \varphi^{-1}(P \times \{0\}) = \psi(P \times \{0\})$.

With these preparations we now define the map $\chi : G^{w.o.d} \times F_0 \to G$ by the formula $\chi(g,\xi) = g \cdot \varphi^{-1}(0,\xi)$ and we claim that this is the diffeomorphism we are looking for. To prove this claim, we first note that on $(G^{w.o.d} \cap U) \times F_0$ the map χ is given as $\chi|_{(G^{w.o.d} \cap U) \times F_0} = \psi \circ (\varphi|_{G^{w.o.d} \cap U} \times id)$, just because for $g \in G^{w.o.d} \cap U$ we have $\varphi(g) = (x,0)$. Since both ψ and $\varphi|_{G^{w.o.d} \cap U} \times id(F_0)$ are diffeomorphisms, we conclude that χ is a diffeomorphism (onto its image) when restricted to $(G^{w.o.d} \cap U) \times F_0$.

In the second place, for an arbitrary $g \in G^{w.o.d}$, we have that $L_{\mathbf{B}g}U = V$ is an open neighborhood of g in G and $V \cap G^{w.o.d}$ is an open neighborhood of g in $G^{w.o.d}$ such that $L_{\mathbf{B}g^{-1}}(V \cap G^{w.o.d}) = U \cap G^{w.o.d}$. We then note that $g = \mathbf{B}g \cdot \mathbf{B}g^{-1} \cdot g$ and that $\mathbf{B}g^{-1} \cdot g \in G^{w.o.d} \cap U$, and thus $\chi(g,\xi) = (L_{\mathbf{B}g} \circ \psi \circ (L_{\mathbf{B}g^{-1}}|_{G^{w.o.d} \cap V} \times id(F_0)))(g,\xi)$. Since the open sets $V = L_{\mathbf{B}g}(U)$ cover G, this formula proves at the same time that χ is surjective and that it is a diffeomorphism when restricted to $(G^{w.o.d} \cap V) \times F_0$. It now suffices to prove injectivity to finish the proof. So suppose $\chi(g,\xi) = \chi(h,\eta)$, which is equivalent to $h^{-1}g\varphi^{-1}(0,\xi) = \varphi^{-1}(0,\eta)$. Applying the body map shows immediately that $\mathbf{B}h^{-1}g = e \in U \cap G^{w.o.d}$ and thus $h^{-1}g = \varphi^{-1}(x,0)$. And then $h^{-1}g\varphi^{-1}(0,\xi) = \varphi^{-1}(0,\eta) \iff \psi(x,\xi) = \psi(0,\eta) \Longrightarrow x = 0$ and $\xi = \eta$ (by injectivity of ψ). But $x = 0$ implies $h^{-1}g = e$ and thus χ is injective. QED

1.8 Definition. A (not necessarily smooth) vector field X on an $\mathcal{A}$-Lie group G is called *left-invariant* if it satisfies the condition $Tm \circ (\underline{0} \times X) = X \circ m$, i.e., if we have a commutative diagram

$$\begin{array}{ccc} TG \times TG & \xrightarrow{Tm} & TG \\ {\scriptstyle \underline{0}\times X}\uparrow & & \uparrow{\scriptstyle X} \\ G \times G & \xrightarrow[m]{} & G\,. \end{array}$$

Applying this condition to a point $(g, h) \in G \times G$ and using the generalized tangent maps gives us $TL_g(X_h) = X_{gh}$. An alternative definition of a left-invariant vector field thus is $\forall g \in G : TL_g \circ X = X \circ L_g$, but this involves the generalized tangent map TL_g.

The vector field X is said to be *right-invariant* if it satisfies $Tm \circ (X \times \underline{0}) = X \circ m$ or equivalently $\forall g \in G : TR_g \circ X = X \circ R_g$.

1.9 Lemma. *The map $X \mapsto X_e$ from left-invariant vector fields on G to points in the tangent space at the identity $e \in G$ is a bijection. Moreover, X is smooth if and only if $X_e \in \mathbf{B}T_eG$, i.e., if X_e has real coordinates. If X and Y are two smooth left-invariant vector fields on G, then so is their commutator $[X, Y]$.*

Proof. Since $X_g = TL_gX_e$, the map $X \mapsto X_e$ is injective. And if X is smooth, X_e has real coordinates because of condition (A2) of smooth functions and the fact that e has real coordinates. On the other hand, we can define a vector field X from X_e by $X_g = TL_gX_e$. Since $TL_hX_g = TL_hTL_gX_e = TL_{hg}X_e = X_{hg}$ this vector field is left-invariant. Moreover, if X_e has real coordinates, the map $G \to TG$, $g \mapsto TL_gX_e = Tm(\underline{0}_g, X_e)$ is smooth (essentially by [III.1.23-g]). The last claim follows immediately from [V.2.29] and the fact that $[\underline{0} \times X, \underline{0} \times Y] = \underline{0} \times [X, Y]$. $\boxed{QED}$

1.10 Lemma. *If X is a left-invariant vector field on G, then $TInv \circ X \circ Inv$ is a right-invariant vector field with $(TInv \circ X \circ Inv)(e) = -X_e$.*

Proof. A direct corollary of [1.6]. $\boxed{QED}$

1.11 Definition. In [I.6.1] we defined a ($\mathbf{Z}_2$-) graded $\mathcal{A}$-Lie algebra $(\mathfrak{g}, [\_, \_])$ as a graded $\mathcal{A}$-module $\mathfrak{g}$ equipped with an even bilinear graded skew-symmetric map $[\_, \_]$ (called the bracket) satisfying the graded Jacobi identity. In this chapter we will be more restrictive and we define an *$\mathcal{A}$-Lie algebra* without the adjective graded to be a ($\mathbf{Z}_2$-) graded $\mathcal{A}$-Lie algebra $\mathfrak{g}$ in which $\mathfrak{g}$ is an $\mathcal{A}$-vector space and where the bracket is smooth. In the same spirit we define a *morphism of $\mathcal{A}$-Lie algebras* to be a morphism $\phi : \mathfrak{g} \to \mathfrak{h}$ of ($\mathbf{Z}_2$-) graded $\mathcal{A}$-Lie algebras [I.6.9] which is smooth in the sense of $\mathcal{A}$-vector spaces (and recall that it is already supposed to be even). A morphism ϕ of $\mathcal{A}$-Lie algebras is called an *isomorphism* of $\mathcal{A}$-Lie algebras if it is bijective, in which case ϕ^{-1} is also an $\mathcal{A}$-Lie algebra morphism.

1.12 Construction. To any $\mathcal{A}$-Lie group we will associate an $\mathcal{A}$-Lie algebra which we will usually denote by the same letter as the corresponding $\mathcal{A}$-Lie group, but in gothic (or Fraktur) font. In order to define the *$\mathcal{A}$-Lie algebra $\mathfrak{g}$ associated to an $\mathcal{A}$-Lie group G*, we proceed in two steps. As an $\mathcal{A}$-vector space, $\mathfrak{g}$ is the tangent space at the identity $e \in G$: $\mathfrak{g} = T_eG$. Since e has real coordinates, T_eG is indeed an $\mathcal{A}$-vector space, not only a free graded $\mathcal{A}$-module (see [IV.3.2]). If G, as an $\mathcal{A}$-manifold, is modeled on an $\mathcal{A}$-vector space E, this means that $\mathfrak{g}$ is, as an $\mathcal{A}$-vector space, the space E. We can turn this argument around and say that an $\mathcal{A}$-Lie group G is modeled (as an $\mathcal{A}$-manifold) on its $\mathcal{A}$-Lie algebra $\mathfrak{g}$ (as an $\mathcal{A}$-vector space). According to [1.9] we can and will identify $\mathfrak{g}$ with the space of all left-invariant vector fields on G. In order to distinguish between these two aspects of the $\mathcal{A}$-Lie algebra $\mathfrak{g}$, we will usually denote an element of T_eG by a lower case letter (e.g., x), and the associated left-invariant vector field by putting an arrow over this letter (e.g., $\vec{x}$). In view of [1.9] this means in particular that for $x \in \mathfrak{g}$ we have $\vec{x}_e = x$.

The usual way to define a bracket on $\mathfrak{g}$ is by means of the commutator of left-invariant vector fields. However, in the graded case there are too few smooth left-invariant vector fields to fill $\mathfrak{g} = T_eG$, and for non-smooth vector fields the commutator is not defined. The "obvious" way out (and we will use this trick quite often) is not to consider two fixed left-invariant vector fields at a time, but to consider them all together, i.e., as variables.[†] We thus introduce the $\mathcal{A}$-manifold $\mathfrak{g} \times \mathfrak{g} \times G$, which is modeled on the $\mathcal{A}$-vector space $\mathfrak{g}^\sharp \times \mathfrak{g}^\sharp \times \mathfrak{g}$. On this $\mathcal{A}$-manifold we define two vector fields Z_1 and Z_2 by

$$Z_1(x,y,g) = (\underline{0}_x, \underline{0}_y, Tm(\underline{0}_g, x)) \qquad , \qquad Z_2(x,y,g) = (\underline{0}_x, \underline{0}_y, Tm(\underline{0}_g, y)) \; .$$

These two maps are indeed smooth vector fields because the zero sections are smooth, as is the identity map $\mathfrak{g} \to T_eG$. The vector field Z_1 (and similarly for Z_2) consists of all left-invariant vector fields on G : restricted to $\{x\} \times \{y\} \times G$ it *is* the (not necessarily smooth) left-invariant vector field whose value at $e \in G$ is x. But remember: if either x or y has non-real coordinates, the slice $\{x\} \times \{y\} \times G$ is not a submanifold of $\mathfrak{g} \times \mathfrak{g} \times G$.

Denoting by π_3 the projection $\mathfrak{g} \times \mathfrak{g} \times G \to G$, we finally define the commutator on $\mathfrak{g}$ by

$$[x,y] = T\pi_3([Z_1, Z_2]_{(x,y,e)}) \in T_eG = \mathfrak{g} \; .$$

By construction, this map $\mathfrak{g} \times \mathfrak{g} \to \mathfrak{g}$, $(x,y) \mapsto T\pi_3([Z_1, Z_2]_{(x,y,e)})$ is smooth. Remains to prove that this map gives $\mathfrak{g}$ the structure of an $\mathcal{A}$-Lie algebra.

1.13 Lemma. *The map $(x,y) \mapsto [x,y] \equiv T\pi_3([Z_1, Z_2]_{(x,y,e)})$ is an even graded skew-symmetric bilinear map satisfying the graded Jacobi identity, thus turning $\mathfrak{g}$ into an $\mathcal{A}$-Lie algebra. Moreover, if the coordinates of x and y are real, then $[x,y]$ represents the commutator of the two smooth left-invariant vector fields $\vec{x}$ and $\vec{y}$, i.e., $[x,y] = [\vec{x}, \vec{y}]_e$.*

[†]The same trick can be used to describe the unconventional super Lie groups of DeWitt [DW, §4.1]. For instance, the unconventional super Lie group $G = \mathcal{A}_1$ with group structure $m_\gamma : G \times G \to G$, $m_\gamma(\xi, \eta) = \xi + \eta - \frac{1}{2}\gamma\xi\eta$ with $\gamma \in \mathcal{A}_1$ a fixed parameter is not a $\mathcal{A}$-Lie group in our sense because the multiplication m_γ is not smooth for non-zero γ. However, if we consider γ as a variable, we obtain a family of group structures $m : \mathcal{A}_1 \times G \times G \to G$, $m(\gamma, \xi, \eta) = m_\gamma(\xi, \eta)$ for which the map m is smooth.

Proof. Let (v_i) be a basis of $T_eG = \mathfrak{g}$, which has by definition real coordinates. An elementary calculation shows that the vector field Z_1 is given by

$$Z_1(x,y,g) = (\underline{0}_x, \underline{0}_y, \textstyle\sum_i x^i \cdot \vec{v}_i|_g) ,$$

where the x^i denote the left coordinates of x with respect to the basis $(v_i) : x = \sum_i x^i \cdot v_i$. A similar formula holds for Z_2, and we find for their commutator the expression

$$[Z_1, Z_2](x,y,g) = (\underline{0}_x, \underline{0}_y, \textstyle\sum_{i,j} x^i \cdot \mathfrak{C}^{\varepsilon(v_i)}(y^j) \cdot [\vec{v}_i, \vec{v}_j]_g) .$$

To derive this formula we have used that the vector field $\vec{v}_i$ has parity $\varepsilon(v_i)$, and that the commutator of vector fields is bi-additive. That the coefficients (smooth functions on $\mathfrak{g} \times \mathfrak{g} \times G$!) x^i and y^j come out as they do is because the vector fields Z_i do not contain any derivatives with respect to these *variables*. It thus follows that our bracket is given by

$$(1.14) \qquad [x,y] = [\textstyle\sum_i x^i v_i, \sum_j y^j v_j] = \sum_{i,j} x^i \cdot \mathfrak{C}^{\varepsilon(v_i)}(y^j) \cdot [\vec{v}_i, \vec{v}_j]_e .$$

Since the commutator of vector fields is even, this formula shows immediately that our bracket on $\mathfrak{g}$ is even and bilinear. Since the commutator of vector fields is also graded skew-symmetric and verifies the graded Jacobi identity, it follows easily that our bracket on $\mathfrak{g}$ does so as well. The last part follows immediately from the given formula for $[x,y]$ in terms of the left-invariant vector fields $\vec{v}_i$. QED

1.15 Remarks. • It is not hard to show that the vector field W given by the formula $W_g = T\pi_3([Z_1, Z_2]_{(x,y,g)}) \in T_gG$ is left-invariant; it is the left-invariant vector field whose value at e is $W_e = [x,y] = T\pi_3([Z_1, Z_2]_{(x,y,e)})$. In this way the bracket $[x,y]$ really is the commutator of two vector fields.

• We could have avoided using the $\mathcal{A}$-manifold $\mathfrak{g} \times \mathfrak{g} \times G$ and the vector fields Z_1 and Z_2 by introducing a basis on $\mathfrak{g} = E$ and defining the bracket on $\mathfrak{g}$ directly by (1.14). The advantage of the given way is that it is intrinsic (avoiding the choice of a basis) and that it introduces a technique we will use more often.

• Although it is customary to define the bracket using left-invariant vector fields, nothing prohibits the use of right-invariant vector fields to define a bracket on $\mathfrak{g} = T_eG$. This would mean that one defines the bracket of $x, y \in \mathfrak{g}$ as being the value at $e \in G$ of the commutator of the right-invariant vector fields whose values at e are x and y respectively. It follows immediately from [1.10] and [V.2.29] that the bracket one obtains that way is the opposite of the bracket defined by means of left-invariant vector fields.

1.16 Definition/Discussion. Let $\mathfrak{g}$ be an $\mathcal{A}$-Lie algebra with basis (v_i). Then there exist constants c_{ij}^k such that

$$[v_i, v_j] = \textstyle\sum_k c_{ij}^k v_k .$$

If the bracket is smooth, these constants must be real; they are called *the structure constants of* $\mathfrak{g}$ *with respect to the basis* (v_i). Given that the bracket is bilinear, it is immediate that the bracket is completely determined by these structure constants.

Now if G is an $\mathcal{A}$-Lie group with multiplication map m, it automatically has an associated $\mathcal{A}$-Lie algebra $\mathfrak{g}$ with basis $(\partial_i|_e)$ relative to some coordinate system $(x^1, \ldots, x^n)$ in a neighborhood U of the identity e. By continuity of m there exists a chart $V \subset U$ such that $m(V \times V) \subset U$. Let $y^i = x^i \circ \pi_1$, $y^{n+i} = x^i \circ \pi_2$, $1 \leq i \leq n$ be the $2n$ local coordinates on $V \times V \subset G \times G$. Writing the local coordinates of $m(g, h)$ for $g, h \in V$ as $m^k(g, h)$, $1 \leq k \leq n$, we can compute the structure constants of $\mathfrak{g}$.

1.17 Lemma. *The structure constants of $\mathfrak{g}$ with respect to the basis $(\partial_k|_e)$ are given by*

$$c_{ij}^k = \frac{\partial^2 m^k}{\partial y^i \partial y^{n+j}}(e, e) - \frac{\partial^2 m^k}{\partial y^{n+i} \partial y^j}(e, e) .$$

Proof. Since the $v_i = \partial_i|_e$ have real coordinates, their brackets are the commutators of the associated left-invariant vector fields [1.13]. These are given by the formula

$$\vec{v}_i|_g = TL_g v_i = \langle\!\langle (\underline{0}_g, \partial_i|_e) \| Tm \rangle\!\rangle = \sum_k \frac{\partial m^k}{\partial y^{n+i}}(g, e) \frac{\partial}{\partial x^k}|_g .$$

Using (V.1.22) we thus find for the commutator of $\vec{v}_i$ and $\vec{v}_j$:

$$\begin{aligned}
[\, \textstyle\sum_k \frac{\partial m^k}{\partial y^{n+i}}(x, e) \frac{\partial}{\partial x^k}, \textstyle\sum_\ell \frac{\partial m^\ell}{\partial y^{n+j}}(x, e) \frac{\partial}{\partial x^\ell} \,]|_e = \\
\textstyle\sum_{k,\ell} \frac{\partial m^k}{\partial y^{n+i}}(e, e) \cdot \frac{\partial^2 m^\ell}{\partial y^k \partial y^{n+j}}(e, e) \frac{\partial}{\partial x^\ell}|_e \\
- (-1)^{\langle \varepsilon(x^i) | \varepsilon(x^j) \rangle} \frac{\partial m^k}{\partial y^{n+j}}(e, e) \cdot \frac{\partial^2 m^\ell}{\partial y^k \partial y^{n+i}}(e, e) \frac{\partial}{\partial x^\ell}|_e .
\end{aligned}$$

Since $m(e, g) = g$, we have $(\partial_{n+i} m^k)(e, e) = \delta_i^k$. The conclusion follows from [III.3.6]. QED

1.18 Lemma. *If G is an $\mathcal{A}$-Lie group with $\mathcal{A}$-Lie algebra $\mathfrak{g}$, then its tangent bundle TG is trivial, a trivialization being given by the map $\phi : G \times \mathfrak{g} \to TG$, $(g, x) \mapsto TL_g x = \vec{x}_g$. Moreover, if H is another $\mathcal{A}$-Lie group with $\mathcal{A}$-Lie algebra $\mathfrak{h}$ and if $\rho : G \to H$ is an $\mathcal{A}$-Lie group morphism, then the tangent map $T\rho$ is described in this trivialization by $(g, x) \mapsto (\rho(g), T_e\rho(x))$*

Proof. The map $\phi : G \times \mathfrak{g} \to TG$, $\phi(g, x) = Tm(\underline{0}_g, x)$ is smooth because Tm, the zero section, and the identity $\mathfrak{g} \to T_eG$ are smooth. On the other hand, the map $\psi : TG \to G \times \mathfrak{g}$ given by $\psi(X_g) = (\pi(X_g), Tm(\underline{0}_{\pi(X_g)^{-1}}, X_g))$ is also smooth, just because Tm, the canonical projection $\pi : TG \to G$, and the inverse $g \mapsto g^{-1}$ are smooth. We finish by showing that they are each others inverse: $\psi(\phi(g, x)) = \psi(TL_g x) = (g, TL_{g^{-1}} TL_g x) = (g, x)$ and $\phi(\psi(X_g)) = \phi(g, TL_{g^{-1}} X_g)) = TL_g TL_{g^{-1}} X_g = X_g$.

The second statement follows from the computation: $T\rho(g, x) = T\rho(Tm_G(\underline{0}_g, x)) = Tm_H(T\rho(\underline{0}_g), T\rho(x)) = Tm_H(\underline{0}_{\rho(g)}, T_e\rho(x))$, where we used that ρ is an $\mathcal{A}$-Lie group morphism. QED

1.19 Corollary. *Let G be an $\mathcal{A}$-Lie group with $\mathcal{A}$-Lie algebra $\mathfrak{g}$ with basis (v_i). Then $\Gamma(TG)$ is a free graded $C^\infty(G)$-module with basis $(\vec{v}_i)$. In particular the $(\vec{v}_i|_g)$ generate the tangent space T_gG.*

1.20 Discussion. We now attack the question of how to interpret the $\mathcal{A}$-Lie algebra $\mathfrak{g}$ of the $\mathcal{A}$-Lie group $G = \text{Aut}(E)$ for an $\mathcal{A}$-vector space E. We have interpreted $\text{Aut}(E)$ as consisting of right linear morphisms [IV.1.3], and we already argued that it is an open subset of the even part $\text{End}_R(E)_0$ [III.2.26]. Since for any $U \subset F_0$ open in the even part of a finite dimensional $\mathcal{A}$-vector space F we have a natural identification $TU \cong U \times F$ [V.1.2], it follows that we have a natural identification $T\,\text{Aut}(E) \cong \text{Aut}(E) \times \text{End}_R(E)$, and in particular

$$\mathfrak{g} \equiv T_{id}\,\text{Aut}(E) \cong \text{End}_R(E)\ .$$

In other words, we identify the $\mathcal{A}$-Lie algebra $\mathfrak{g}$ of $\text{Aut}(E)$ with the right linear endomorphism $\text{End}_R(E)$ on E.

But there is more to an $\mathcal{A}$-Lie algebra then just its $\mathcal{A}$-vector space structure: it also has a bracket. Now both $\mathfrak{g}$ and $\text{End}_R(E)$ are naturally equipped with a bracket: $\mathfrak{g}$ by the commutator of left-invariant vector fields, and $\text{End}_R(E)$ by the (right) commutator $[\phi,\psi]_R = \phi\circ\psi - (-1)^{\langle\varepsilon(\phi)|\varepsilon(\psi)\rangle}\psi\circ\phi$ [I.6.3]. We want to show that our identification $\mathfrak{g} \cong \text{End}_R(E)$ respects these brackets.

Our first concern is to find a suitable coordinate system for $\text{Aut}(E)$ in order to describe the left-invariant vector fields. The obvious choice is the matrix representation $\overset{\bullet}{M}_R$ defined in [II.4.1], which nicely respects the group structure. Unfortunately this choice is not really compatible with more important choices already made: coordinates on (the even part of) an $\mathcal{A}$-vector space are supposed to be *left coordinates* with respect to a basis. If $(e_i)_{i=1}^n$ is a basis for E, the natural basis for $\text{End}_R(E)$ is given by $(e_i \otimes e^j)_{i,j=1}^n$ [II.4.1]. Our approach thus *imposes* the coordinate functions $x^i{}_j = ({}^\bullet M_R)^i{}_j$ defined by

$$\phi = \sum_{i,j} x^i{}_j(\phi)\cdot e_i \otimes e^j\ .$$

We know that the composition $\circ$ corresponds to an elementary (inner) contraction [II.5.2], so we obtain the following composition rule in these coordinates:

$$x^i{}_k(\phi\circ\psi) = \textstyle\sum_j x^i{}_j(\phi)\cdot \mathfrak{C}^{\varepsilon(e_i)+\varepsilon(e_j)}(x^j{}_k(\psi))\ . \tag{1.21}$$

The conjugation $\mathfrak{C}^{\varepsilon(e_i)+\varepsilon(e_j)}$ appears because we have to permute the coefficients $x^j{}_k(\psi)$ with the $e_i \otimes e^j$. Writing for an arbitrary element $g \in \text{End}_R(E)$ the coordinates as $g^i{}_j = x^i{}_j(g)$, we obtain as multiplication in $\text{Aut}(E)$:

$$m(g,h)^i{}_k = \sum_j (-1)^{\langle\varepsilon(e_i)+\varepsilon(e_j)|\varepsilon(e_j)+\varepsilon(e_k)\rangle}\cdot g^i{}_j\cdot h^j{}_k = \sum_j h^j{}_k\cdot g^i{}_j\ . \tag{1.22}$$

The conjugation of $h^j{}_k$ gets transformed into the given sign because h is even, and thus the parity of $h^j{}_k$ is $\varepsilon(e_j\otimes e^k) = \varepsilon(e_j)+\varepsilon(e_k)$. Now let X be any element of $\text{End}_R(E)$

with coordinates $X^i{}_j = x^i{}_j(X)$. In the given coordinates on $\mathrm{Aut}(E) \subset \mathrm{End}_R(E)$, the point $(h, X) \in \mathrm{Aut}(E) \times \mathrm{End}_R(E)$ corresponds to the tangent vector

$$\widehat{X}_h = \sum_{i,j} X^i{}_j \cdot \frac{\partial}{\partial h^i{}_j}|_h \in T_h \,\mathrm{Aut}(E) \ .$$

Writing $z = m(g, h)$ we now compute, using (1.21) and (1.22), $\langle\!\langle \widehat{X}_h \| TL_g \rangle\!\rangle$ as

$$\begin{aligned}
\langle\!\langle \widehat{X}_h \| TL_g \rangle\!\rangle = \langle\!\langle (\underline{0}_g, \widehat{X}_h) \| Tm \rangle\!\rangle &= \sum_{i,j,p,r} X^i{}_j \cdot \frac{\partial}{\partial h^i{}_j}\Big(m(g \circ h)^p{}_r\Big) \cdot \frac{\partial}{\partial z^p{}_r}|_{gh} \\
&= \sum_{i,j,p,q,r} X^i{}_j \cdot \frac{\partial}{\partial h^i{}_j}\Big(h^q{}_r \cdot g^p{}_q\Big) \cdot \frac{\partial}{\partial z^p{}_r}|_{gh} = \sum_{i,j,p} X^i{}_j \cdot g^p{}_i \cdot \frac{\partial}{\partial z^p{}_j}|_{gh} \\
&= \sum_{i,j,p} g^p{}_i \cdot \mathfrak{C}^{\varepsilon(e_p)+\varepsilon(e_i)}(X^i{}_j) \cdot \frac{\partial}{\partial z^p{}_j}|_{gh} = \sum_{j,p} (g \circ X)^p{}_j \cdot \frac{\partial}{\partial z^p{}_j}|_{gh} \ .
\end{aligned}$$

Similar computations complete the proof of [1.23].

The above computation shows that for a left-invariant vector field X on $\mathrm{Aut}(E)$ we have, in the natural identification $T\,\mathrm{Aut}(E) \cong \mathrm{Aut}(E) \times \mathrm{End}_R(E)$, $X_g = g \circ X_e$. It follows immediately that the vector fields Z_i on $\mathfrak{g} \times \mathfrak{g} \times \mathrm{Aut}(E)$ [1.12] are given by

$$Z_1(X, Y, g) = \sum_{j,p} (g \circ X)^p{}_j \cdot \frac{\partial}{\partial g^p{}_j} \quad , \quad Z_2(X, Y, g) = \sum_{j,p} (g \circ Y)^p{}_j \cdot \frac{\partial}{\partial g^p{}_j} \ .$$

A simple calculation gives for the commutator

$$[Z_1, Z_2](X, Y, g) = \sum_{j,p} (g \circ ([X, Y]_R))^p{}_j \cdot \frac{\partial}{\partial g^p{}_j} \ ,$$

where $[X, Y]_R$ denotes the commutator on $\mathrm{End}_R(E)$. From this it is clear that the commutator of left-invariant vector fields on $\mathrm{Aut}(E)$ corresponds to the usual commutator on $\mathrm{End}_R(E)$ under our identification $\mathfrak{g} \cong \mathrm{End}_R(E)$.

1.23 Lemma. *In the natural identification* $T\,\mathrm{Aut}(E) \cong \mathrm{Aut}(E) \times \mathrm{End}_R(E)$, *the tangent maps* TL_g, TR_g, *and* $TInv$ *are given by*

$$TL_g(h, X) = (gh, gX) \quad , \quad TR_g(h, X) = (hg, Xg)$$
$$TInv(h, X) = (h^{-1}, -h^{-1}Xh^{-1}) \ .$$

1.24 Remark. Of course, even with $\circ$ as group law, we could have seen $\mathrm{Aut}(E)$ as left linear morphisms. That would have led to a natural identification of $\mathfrak{g}$ with $\mathrm{End}_L(E)$, but an $\mathrm{End}_L(E)$ equipped with $\circ$ as composition law, not with $\diamond$. A careful computation reveals that under this identification the commutator of left-invariant vector fields corresponds, in $\text{End}_L(E)$, to the bracket defined on homogeneous elements by

$$[X,Y] = (-1)^{\langle\varepsilon(X)|\varepsilon(Y)\rangle} X \circ Y - Y \circ X$$
$$= -(X \diamond Y - (-1)^{\langle\varepsilon(X)|\varepsilon(Y)\rangle} Y \diamond X) = -[X,Y]_L \ .$$

It should come as no surprise that this bracket is exactly the bracket obtained from the usual bracket $[X,Y]_R$ on $\text{End}_R(E)$ under the isomorphism $\mathfrak{T}^{-1}$ [I.6.4].

Another variation on the above theme is to consider $\text{Aut}(E)$ equipped with the composition law $\diamond$. In this case the most natural identification for $\mathfrak{g}$ is with $\text{End}_L(E)$ equipped with $\diamond$ as composition. A similar computation as for $\circ$ shows that in this case the commutator of left-invariant vector fields corresponds exactly to the usual bracket $[X,Y]_L$ on $\text{End}_L(E)$.

1.25 Discussion. We have argued that our approach imposes the left coordinates ${}^\bullet M_R$ on $\text{Aut}(E)$ instead of the more natural middle ones given by $\overset{\bullet}{M}_R$ [II.4.1]. A variation upon the given argument is the following. Going from $G = \text{Aut}(E)$ to its Lie algebra $\mathfrak{g} = T_eG = \text{End}_R(E)$ is a kind of derivative (infinitesimal form). Since the map $\overset{\bullet}{M}_R$ on $\text{Aut}(E) \subset \text{End}_R(E)$ is neither left nor right linear, it is not its own derivative (nor its transpose) [III.3.14], and "thus" $\overset{\bullet}{M}_R$ will not give the coordinates on the Lie algebra.

But let us have a look at what would actually happen if we did use the $\overset{\bullet}{M}_R$ coordinates. The identifications $g = \sum_{i,j} x^i{}_j(g) \cdot e_i \otimes e^j = \sum_{i,j} e_i \otimes y^i{}_j(g) \cdot e^j$ immediately give us the relations

$$y^i{}_j(g) = (-1)^{\langle\varepsilon(e_i)|\varepsilon(e_i)+\varepsilon(e_j)\rangle} x^i{}_j(g) \ ,$$

where we have used that the parity of $x^i{}_j(g)$ equals $\varepsilon(e_i)+\varepsilon(e_j)$ (because g is supposed to be an even endomorphism). These two coordinate systems on $\text{Aut}(E)$ thus differ only by some signs. What then could cause such problems that we insisted on the x coordinates? The answer lies in the way one has to identify $\text{End}_R(E)$ with tangent vectors at the identity!

Let $X \in \text{End}_R(E)$ be arbitrary, not necessarily even. We have identified it with the tangent vector $X \simeq \sum_{i,j} x^i{}_j(X) \cdot \partial_{x^i{}_j}$, which in the y coordinates gives us

$$X \simeq \sum_{i,j} (-1)^{\langle\varepsilon(e_i)|\varepsilon(e_i)+\varepsilon(e_j)\rangle} x^i{}_j(X) \cdot \frac{\partial}{\partial y^i{}_j} \ .$$

Unfortunately, the combination $(-1)^{\langle\varepsilon(e_i)|\varepsilon(e_i)+\varepsilon(e_j)\rangle} x^i{}_j(X)$ is *not* the same as $y^i{}_j(X)$, unless X is even. In the general case we get

$$X \simeq \sum_{i,j} (-1)^{\langle\varepsilon(e_i)|\varepsilon(e_i)+\varepsilon(e_j)\rangle} \mathfrak{C}^{\varepsilon(e_i)}(y^i{}_j(X)) \cdot \frac{\partial}{\partial y^i{}_j} \overset{\text{for even } X}{=\!=\!=\!=} \sum_{i,j} y^i{}_j(X) \cdot \frac{\partial}{\partial y^i{}_j} \ .$$

We conclude that there is nothing wrong with using the (more natural) coordinates y on $\text{Aut}(E)$, and that the identification of $\mathfrak{g}$ with $\text{End}_R(E)$ using the coordinates y also poses no problems, *provided we only use even vectors/endomorphisms* (see also [2.8]).

Yet another way to interpret the problems with the $\overset{\bullet}{M}_R$ coordinates is the following observation. The $y^i{}_j(g)$ coordinates are in between the e_i and e^j vectors in the tensor product representation $g = \sum_{i,j} e_i \otimes y^i{}_j(g) \cdot e^j$. It thus would be natural to put the coordinates of a tangent vector also in between these indices. But the basis vectors $\partial_{y^i{}_j}$ can not be written in a natural way as a product of terms with separate indices i and j. Hence there is no natural way to put the coordinates of a tangent vector between the indices i and j in $\partial_{y^i{}_j}$.

2. The exponential map

In this section we start with the definition of the exponential map, which goes from $\mathfrak{g}_0$ to G with $\mathfrak{g}$ being the $\mathcal{A}$-Lie algebra of the $\mathcal{A}$-Lie group G. We then show that the exponential map intertwines an $\mathcal{A}$-Lie group morphism with its associated $\mathcal{A}$-Lie algebra morphism (its tangent map at the identity). We finish with the definition of the Adjoint representation $\mathrm{Ad} : G \to \mathrm{Aut}(\mathfrak{g})$ *and the fact that the derivative of the Adjoint representation is the algebraic adjoint representation:* $T_e \,\mathrm{Ad} = \mathrm{ad}_R : \mathfrak{g} \to \mathrm{End}_R(\mathfrak{g})$.

2.1 Construction. The usual way to define the exponential map is by following the flows of the left-invariant vector fields. However, as we have seen, for $\mathcal{A}$-Lie groups there are far too few smooth left-invariant vector fields. As before we circumvent this difficulty by looking at *all* (even) left-invariant vector fields at the same time. We thus consider the $\mathcal{A}$-manifold $\mathfrak{g}_0 \times G$, on which we define the even smooth vector field Z_L by

$$Z_L(x, g) = (\underline{0}_x, TL_g \vec{x}_e) = (\underline{0}_x, Tm(\underline{0}_g, \vec{x}_e)) \in T_{(x,g)}(\mathfrak{g}_0 \times G) \ .$$

Since the zero sections and the identity map $\mathfrak{g} \to T_eG$ are smooth, this defines indeed a smooth vector field on $\mathfrak{g}_0 \times G$. Obviously this vector field regroups all *even* left-invariant vector fields on G. Since Z_L has no components in the direction of $\mathfrak{g}_0$, its flow has necessarily the form $\phi(t, x, g) = (x, \psi(t, x, g))$.

2.2 Proposition. *Let* $W_{Z_L} \subset \mathcal{A}_0 \times \mathfrak{g}_0 \times G$ *be the domain of definition of the map* $\psi : W_{Z_L} \to G$, *then*

(i) $\psi(t, x, g) = g \cdot \psi(t, x, e) \equiv m(g, \psi(t, x, e))$;
(ii) $W_{Z_L} = \mathcal{A}_0 \times \mathfrak{g}_0 \times G$;
(iii) $\forall \lambda \in \mathcal{A}_0 : \psi(t\lambda, x, g) = \psi(t, \lambda x, g)$.

Proof. • Since ψ is part of the flow of Z_L, we have $T\psi(\partial_t|_{(t_o,x,g)}) = TL_{\psi(t_o,x,g)}\vec{x}_e$. We then define $\widehat{\phi}(t, x, g) = (x, g \cdot \psi(t, x, e))$ wherever it makes sense. This $\widehat{\phi}$ obviously satisfies $\widehat{\phi}(0, x, g) = (x, g)$. Moreover,

$$\begin{aligned} T\widehat{\phi}(\partial_t|_{(t_o,x,g)}) &= (\underline{0}_x, TL_g T\psi(\partial_t|_{(t_o,x,e)}))) \\ &= (\underline{0}_x, TL_g TL_{\psi(t_o,x,e)}\vec{x}_e)) \\ &= (\underline{0}_x, TL_{g\cdot\psi(t_o,x,e)}\vec{x}_e) = Z_L(\widehat{\phi}(t_o, x, g)) \ , \end{aligned}$$

where in the second equality we used the definition of ψ. We thus see that $\widehat{\phi}$ satisfies the conditions of a flow for Z_L; by uniqueness of flows it thus must coincide with ϕ on their common domain of definition.

• Since W_{Z_L} is the maximal domain of definition of ϕ, it must contain the domain of definition $\widehat{W}$ of $\widehat{\phi}$. But if (t, x, e) belongs to W_{Z_L}, it belongs to $\widehat{W}$, hence any (t, x, g) belongs to $\widehat{W}$, and thus (t, x, g) belongs to W_{Z_L}. But then the group law $\phi(t, \phi(s, x, g)) = \phi(s + t, x, g)$, the openness of W_{Z_L}, and the fact that x does not move, these all together imply that (for fixed x) all (t, x, g) belong to W_{Z_L}.

• To prove (iii), consider the $\mathcal{A}$-manifold $\mathcal{A}_0 \times \mathfrak{g}_0 \times G$ on which we define the even smooth vector field Z_L^E by

$$Z_L^E(\lambda, x, g) = (\underline{0}_\lambda, \underline{0}_x, Tm(\underline{0}_g, \lambda\vec{x}_e)) \ .$$

Its flow ϕ^E has necessarily the form $\phi^E(t, \lambda, x, g) = (\lambda, x, \psi^E(t, \lambda, x, g))$. We also define $\widehat{\psi}^E(t, \lambda, x, g) = \psi(t, \lambda x, g)$, and $\widetilde{\psi}^E(t, \lambda, x, g) = \psi(t\lambda, x, g)$. With these we compute

$$\begin{aligned} T\widehat{\psi}^E(\partial_t|_{(t_o,\lambda,x,g)}) &= T\psi(\partial_t|_{(t_o,\lambda x,g)}) = Tm(\underline{0}_{\psi(t_o,\lambda x,g)}, \lambda\vec{x}_e) \\ &= Tm(\underline{0}_{\widehat{\psi}^E(t_o,\lambda,x,g)}, \lambda\vec{x}_e) \end{aligned}$$

and

$$\begin{aligned} T\widetilde{\psi}^E(\partial_t|_{(t_o,g,x,\lambda)}) &= \lambda \cdot T\psi(\partial_t|_{(t_o\lambda,x,g)}) && \text{by the chain rule} \\ &= \lambda \cdot Tm(\underline{0}_{\psi(t_o\lambda,x,g)}, \vec{x}_e) \\ &= Tm(\underline{0}_{\widetilde{\psi}^E(t_o,\lambda,g,x)}, \lambda\vec{x}_e) && \text{by left linearity of } Tm. \end{aligned}$$

From these two computations and the uniqueness of the flow of Z_L^E, it follows immediately that $\psi^E = \widehat{\psi}^E = \widetilde{\psi}^E$. QED

2.3 Definition. The *exponential map* $\exp : \mathfrak{g}_0 \to G$ is defined in terms of the flow of the vector field Z_L on $\mathfrak{g}_0 \times G$ as $\exp(x) = \psi(1, x, e)$.

2.4 Proposition. *The exponential map* $\exp : \mathfrak{g}_0 \to G$ *has the following properties.*

(i) *The flow of* Z_L *is given by* $\phi(t, x, g) = (x, g \cdot \exp(tx))$.
(ii) $\forall s, t \in \mathcal{A}_0\ \forall x \in \mathfrak{g}_0 : \exp(sx) \cdot \exp(tx) = \exp((s + t)x)$.

(iii) *If $\vec{x}$ is any even smooth left-invariant vector field on G, its flow is defined on the whole of $\mathcal{A}_0 \times G$ and is given by $(t, g) \mapsto g \cdot \exp(t\vec{x}_e)$.*

(iv) $T_0 \exp = id(\mathfrak{g})$.

(v) *The flow of the even smooth vector field $Z_R(x, g) = (\underline{0}_x, TR_g x)$ on $\mathfrak{g}_0 \times G$ (the right equivalent of Z_L) is given by $(t, x, g) \mapsto (x, \exp(tx) \cdot g)$.*

(vi) *If X is any even smooth right-invariant vector field on G, its flow is defined on the whole of $\mathcal{A}_0 \times G$ and is given by $(t, g) \mapsto \exp(tX_e) \cdot g$.*

Proof. • (i) is a direct consequence of [2.2-i,iii], and (ii) follows from the group property of the flow of Z_L.

• If $\vec{x}$ is a smooth left-invariant vector field on G, $x = \vec{x}_e$ has real coordinates, and thus the map $\chi : (t, g) \mapsto g \cdot \exp(t\vec{x}_e) = \psi(t, x, g)$ is smooth. We then compute, using the left invariance of $\vec{x}$, $T\chi(\partial_t|_{(t,g)}) = Tm(\underline{0}_{\psi(t,x,g)}, \vec{x}_e) = \vec{x}_{\chi(t,g)}$. Uniqueness of its flow then proves (iii).

• To prove (iv), consider the map $\chi : \mathcal{A}_0 \times \mathfrak{g}_0 \to G$, $(t, x) \mapsto \exp(tx)$. By the chain rule we find that $T\chi(\partial_t|_{(0,x)}) = T_0 \exp(x|_0)$. Note however that there is a change in interpretation of the x in this formula. The tangent map of the map $(t, x) \mapsto tx$ transforms the tangent vector ∂_t at $(0, x)$ into the tangent vector $x \equiv x|_0 \in T_0\mathfrak{g}_0 \cong \mathfrak{g}$ at $0 \in \mathfrak{g}_0$. Since $\chi(t, x) = \psi(t, e, x)$, we have $T\chi(\partial_t|_{(0,x)}) = TL_{\psi(0,x,e)}\vec{x}_e = \vec{x}_e = x$ (use that $\psi(0, x, e) = e$ and $e \cdot g = g$).

• According to (i) we have the equality $\exp(sx) \cdot \exp(tx) = \exp(tx) \cdot \exp(sx)$. When we see this as maps defined on $(s, t, x) \in \mathcal{A}_0 \times \mathcal{A}_0 \times \mathfrak{g}_0$, we can apply the tangent maps to the vector $\partial_s|_{(0,t,x)}$. Using (iv) we obtain the equality $TR_{\exp(tx)}\vec{x}_e = TL_{\exp(tx)}\vec{x}_e$. We now define χ by $\chi(t, x, g) = (x, \exp(tx) \cdot g)$, and we compute

$$\begin{aligned} T\chi(\partial_t|_{(t,x,g)}) &= (\underline{0}_x, TR_g T\psi(\partial_t|_{(t,x,e)})) = (\underline{0}_x, TR_g(TL_{\exp(tx)}\vec{x}_e)) \\ &= (\underline{0}_x, TR_g(TR_{\exp(tx)}\vec{x}_e)) = (\underline{0}_x, TR_{\exp(tx)\cdot g}\vec{x}_e) \\ &= Z_R(\chi(t, x, g)) . \end{aligned}$$

Uniqueness of the flow finishes the proof of (v). The proof of (vi) is a variation of that of (iii). QED

2.5 Nota Bene. The restriction to *even* elements in [2.4-ii] and [2.4-iii] is essential. One might be tempted to think that for a smooth odd $\vec{x}$, i.e., $x \in \mathbf{B}\mathfrak{g}_1$ [1.9], its flow is given by $(\tau, g) \mapsto g \exp(\tau x)$, using that $\tau x \in \mathfrak{g}_0$ (because the time parameter of an odd vector field is odd). In [3.17] we will show that this is the case if x satisfies $[x, x] = 0$, i.e., the standard condition for integrability. We will also show that this is equivalent to the homomorphism property $\exp(\tau x) \cdot \exp(\sigma x) = \exp((\tau + \sigma)x)$. In [2.8] we will give an example in which these conditions are not satisfied.

2.6 Lemma. *Let G be a connected $\mathcal{A}$-Lie group and U an open neighborhood of $e \in G$. Then G is generated by U, i.e., any element of G is a finite product of elements of U.*

Proof. Define $V = U \cap U^{-1} = \{g \in U \mid g^{-1} \in U\}$, which is an open neighborhood of e because Inv is a diffeomorphism ($\Rightarrow$ homeomorphism), and denote by G^1 the (abstract) subgroup generated by V. For any $g \in G^1$ it follows that L_gV is an open neighborhood of g (because L_g is a homeomorphism) which is contained in G^1. Hence G^1 is open.

On the other hand, suppose $g \in G \setminus G^1$ and $h \in L_gV \cap G^1$, then $\exists v \in V : g = hv^{-1}$, i.e., $g \in G^1$ (because $V = V^{-1}$). Since this contradicts $g \notin G^1$, we conclude that $L_gV \subset G \setminus G^1$, i.e., that $G \setminus G^1$ is open. We conclude that G^1 is open, closed, and non-empty. Since G is connected, we conclude $G = G^1$. QED

2.7 Corollary. *There exists an open set $U \subset \mathfrak{g}_0$ containing 0 and an open set $V \subset G$ containing e such that* $\exp : U \to V$ *is a diffeomorphism. In particular, if G is connected, it is generated by elements of the form* $\exp(x)$ *with* $x \in \mathfrak{g}_0$.

Proof. The first part is a direct consequence of [2.4-iv] and the inverse function theorem [III.3.23]. The second part follows from [2.6]. QED

2.8 Example. Let G be the multiplicative $\mathcal{A}$-Lie group of invertible elements in $\mathcal{A}$, i.e., $G = \{\, x + \xi \in \mathcal{A} \mid \mathbf{B}x \neq 0 \,\}$. It is modeled on an $\mathcal{A}$-vector space of dimension $1|1$ and its multiplication is given by

$$(x, \xi) \cdot (y, \eta) = (xy + \xi\eta, x\eta + y\xi) \ . \tag{2.9}$$

A basis of $\mathfrak{g} = T_eG$ at $e = (1, 0)$ is given by the vectors $v_1 = \partial_x|_e$ and $v_2 = \partial_\xi|_e$. The associated left-invariant vector fields are given by

$$\vec{v}_1 = x \cdot \partial_x + \xi \cdot \partial_\xi \qquad , \qquad \vec{v}_2 = x \cdot \partial_\xi - \xi \cdot \partial_x \ .$$

An elementary computation reveals $[\vec{v}_1, \vec{v}_1] = [\vec{v}_1, \vec{v}_2] = 0$ and $[\vec{v}_2, \vec{v}_2] = -2\vec{v}_1$. We could also have used [1.17] to obtain these commutators (structure constants): using (2.9) one obtains

$$\begin{aligned} \partial_1\partial_{2+1}m &= \partial_x\partial_y m = (1, 0) \qquad , \qquad & \partial_1\partial_{2+2}m &= \partial_x\partial_\eta m = (0, 1) \ , \\ \partial_2\partial_{2+1}m &= \partial_\xi\partial_y m = (0, 1) \qquad , \qquad & \partial_2\partial_{2+2}m &= \partial_\xi\partial_\eta m = (-1, 0) \ . \end{aligned}$$

Inserting the appropriate signs immediately gives $c_{11}^* = c_{12}^* = c_{21}^* = c_{22}^2 = 0$ and $c_{22}^1 = -2$, in accordance with the previously calculated commutators of the v_i. If we denote by (f^1, f^2) the left coordinates of an element $f \in \mathfrak{g}$ with respect tot the basis (v_i), we obtain the full bracket in $\mathfrak{g}$ by (1.14) as

$$[f, g] = [(f^1, f^2), (g^1, g^2)] = (-2f^2 \cdot \mathfrak{C}(g^2), 0) \ .$$

Integrating the vector field $Z_L(f^1, f^2, x, \xi) = f^1 \cdot \vec{v}_1 + f^2 \cdot \vec{v}_2$ (but now $f \in \mathfrak{g}_0$, i.e., f^1 is even and f^2 is odd), one finds the flow

$$\phi(t, (f^1, f^2, x, \xi)) = (f^1, f^2, (x + t\xi f^2)\,\mathrm{e}^{tf^1}, (\xi + txf^2)\,\mathrm{e}^{tf^1}) \ ,$$

which gives for the exponential map $(av_1 + \alpha v_2) \mapsto \exp(a, \alpha)$:

$$\exp(a, \alpha) = (\mathrm{e}^a, \mathrm{e}^a \cdot \alpha) \ .$$

Combining the even and odd coordinates in a single $\mathcal{A}$-valued "coordinate" on both sides, we can write this expression as $\exp(a + \alpha) = \mathrm{e}^a \cdot (1 + \alpha)$. If we realize that α is odd and thus $\alpha^2 = 0$, we can see the term $1 + \alpha$ as the Taylor expansion of e^α, i.e., we can write

$$\exp(a + \alpha) = \mathrm{e}^{a+\alpha} \ .$$

In this visualization the exponential map of G thus becomes the ordinary exponential map extended to $\mathcal{A}$ (see also [3.11]).

Once we know the exponential map, it is easy to compute for odd $\sigma, \tau \in \mathcal{A}_1$ the product $\exp(\sigma v_2) \cdot \exp(\tau v_2) = \exp(0, \sigma) \cdot \exp(0, \tau) = (1, \sigma) \cdot (1, \tau) = (1 + \sigma\tau, \sigma + \tau)$. Since this is not equal to $\exp((\sigma + \tau)v_2) = (1, \sigma + \tau)$, we here have an example in which [2.4-ii] is not true for odd vector fields (and odd coefficients). We can also consider the map $\phi : \mathcal{A}_1 \times G \to G$ defined by $\phi(\tau, (x, \xi)) = (x, \xi) \cdot \exp(\tau v_2) = (x + \xi\tau, \xi + x\tau)$. A direct calculation gives $T\phi(\partial_\tau|_{(\tau,x,\xi)}) = x\partial_\xi|_{(x+\xi\tau,\xi+x\tau)} - \xi\partial_x|_{(x+\xi\tau,\xi+x\tau)}$, which is *not* equal to $\vec{v}_2|_{(x+\xi\tau,\xi+x\tau)} = (x + \xi\tau)\partial_\xi|_{(x+\xi\tau,\xi+x\tau)} - (\xi + x\tau)\partial_x|_{(x+\xi\tau,\xi+x\tau)}$. This shows that [2.4-iii] need not be true in the odd case.

This example is also perfectly suited to illustrate the truth of [1.25]. The group G can be realized as a group of 2×2 matrices equipped with the usual matrix multiplication:

$$(x, \xi) \simeq \begin{pmatrix} x & \xi \\ \xi & x \end{pmatrix} .$$

It seems reasonable that the corresponding matrix representation of $\mathfrak{g}$ is given by

$$f = f^1 \cdot v_1 + f^2 \cdot v_2 \simeq \begin{pmatrix} f^1 & f^2 \\ f^2 & f^1 \end{pmatrix} .$$

And indeed, if $f, g \in \mathfrak{g}$ are even (meaning $f^1, g^1 \in \mathcal{A}_0$, $f^2, g^2 \in \mathcal{A}_1$), a direct computation shows

$$\begin{aligned} [f, g] \simeq \left[\begin{pmatrix} f^1 & f^2 \\ f^2 & f^1 \end{pmatrix}, \begin{pmatrix} g^1 & g^2 \\ g^2 & g^1 \end{pmatrix} \right] &= \begin{pmatrix} 2f^2 \cdot g^2 & 0 \\ 0 & 2f^2 \cdot g^2 \end{pmatrix} \\ &\simeq (2f^2 \cdot g^2, 0) = (-2f^2 \cdot \mathfrak{C}(g^2), 0) \ , \end{aligned}$$

in complete agreement with the official bracket. But ... when we take for f and g the *odd* element v_2, i.e., $f = g \cong (0, 1)$, the official bracket equals $-2v_1 \cong (-2, 0)$, while the bracket of the corresponding (odd!) matrices gives $(2, 0)$. As said in [1.25] the origin of this problem lies in the identification between tangent vectors and matrices. Since we use the standard matrix multiplication, this means that we use the coordinates $\overset{\bullet}{M}_R$. According to the formulas in [II.4.1] we thus have

$$\begin{pmatrix} x & \xi \\ \xi & x \end{pmatrix} = x \cdot \begin{pmatrix} 1 & 0 \\ 0 & 1 \end{pmatrix} + \xi \cdot \begin{pmatrix} 0 & 1 \\ -1 & 0 \end{pmatrix} ,$$

i.e., the matrices $\left(\begin{smallmatrix}1&0\\0&1\end{smallmatrix}\right)$ and $\left(\begin{smallmatrix}0&1\\-1&0\end{smallmatrix}\right)$ form a basis. For the Lie algebra we thus find, using the same formulas,

$$f = f^1 \cdot v_1 + f^2 \cdot v_2 \simeq f^1 \cdot \begin{pmatrix}1 & 0\\ 0 & 1\end{pmatrix} + f^2 \cdot \begin{pmatrix}0 & 1\\ -1 & 0\end{pmatrix} = \begin{pmatrix} f^1 & f^2 \\ -\mathfrak{C}(f^2) & \mathfrak{C}(f^1)\end{pmatrix} .$$

A careful calculation reveals that *this* matrix representation indeed effectively represents the bracket in $\mathfrak{g}$; and it corresponds to the previous identification for even elements.

2.10 Definitions. Let Q be an $\mathcal{A}$-manifold, and let G and H be $\mathcal{A}$-Lie groups. A smooth map $\Phi : Q \times G \to H$ is called *a family of ($\mathcal{A}$-Lie group) homomorphisms from G to H* if for all $q \in Q$ the map $\Phi_q : G \to H, g \mapsto \Phi(q, g)$ is a homomorphism of (abstract) groups.

If $\mathfrak{g}$ and $\mathfrak{h}$ are $\mathcal{A}$-Lie algebras, then a smooth map $\phi : Q \times \mathfrak{g} \to \mathfrak{h}$ is called *a family of ($\mathcal{A}$-Lie algebra) morphisms from $\mathfrak{g}$ to $\mathfrak{h}$* if for all $q \in Q$ the map $\phi_q : \mathfrak{g} \to \mathfrak{h}, X \mapsto \phi(q, X)$ is a morphism of $\mathcal{A}$-Lie algebras [I.6.9] (and thus in particular even). According to [IV.3.17], such a family is equivalent to a smooth map $\phi : Q \to \mathrm{Hom}_R(\mathfrak{g}; \mathfrak{h})_0$ such that all $\phi(q) \equiv \phi_q$ are $\mathcal{A}$-Lie algebra morphisms.

2.11 Proposition. *Let G and H be $\mathcal{A}$-Lie groups, let $\mathfrak{g}$ and $\mathfrak{h}$ be their $\mathcal{A}$-Lie algebras, and let $\Phi : Q \times G \to H$ be a family of homomorphisms. Then the map $T'\Phi : Q \times \mathfrak{g} \to \mathfrak{h}$ defined by $T'\Phi(q, x) = T\Phi_q(\vec{x}_e) = T\Phi(\underline{0}_q, \vec{x}_e)$ [V.3.19] is a family of morphisms from $\mathfrak{g}$ to $\mathfrak{h}$.*

In case Q contains a single point, this reduces to the fact that the tangent map at e of a homomorphism between $\mathcal{A}$-Lie groups is a morphism between their $\mathcal{A}$-Lie algebras.

Proof. First of all note that $T'\Phi$ is indeed a smooth map, and that it is even and linear in x (because $T\Phi$ is a smooth even vector bundle map). Since Φ is a family of homomorphisms, the map Φ_q sends the identity of G to the identity of H, proving that $T'\Phi$ has indeed $\mathfrak{h} = T_eH$ as target space.

In order to prove that $T'\Phi$ preserves brackets, we recall that the bracket on $\mathfrak{g}$ is defined by the commutator of the vector fields Z_i^G on $\mathfrak{g} \times \mathfrak{g} \times G$ (and similarly for $\mathfrak{h}$). We now extend these vector fields to vector fields $\widehat{Z}_i^G$ on $Q \times \mathfrak{g} \times \mathfrak{g} \times G$ by $\widehat{Z}_i^G = \underline{0} \times Z_i^G$. We also extend the map Φ to a smooth map $\widehat{\Phi} : Q \times \mathfrak{g} \times \mathfrak{g} \times G \to \mathfrak{h} \times \mathfrak{h} \times H$ by

$$\widehat{\Phi}(q, x, y, g) = (T'\Phi(q, x), T'\Phi(q, y), \Phi(q, g)) .$$

With these ingredients we compute $(T\widehat{\Phi} \circ \widehat{Z}_1^G)(q, x, y, g)$ as

$$\begin{aligned}(T\widehat{\Phi} \circ \widehat{Z}_1^G)(q, x, y, g) &= T\widehat{\Phi}(\underline{0}_q, \underline{0}_x, \underline{0}_y, TL_g(\vec{x}_e)) \\ &= (\underline{0}_{T'\Phi(q,x)}, \underline{0}_{T'\Phi(q,y)}, T\Phi_q(TL_g(\vec{x}_e))) \\ &= (\underline{0}_{T'\Phi(q,x)}, \underline{0}_{T'\Phi(q,y)}, TL_{\Phi(q,g)}(T'\Phi(q, x)) \\ &= Z_1^H(T'\Phi(q, x), T'\Phi(q, y), \Phi(q, g))\end{aligned}$$

$$= (Z_1^H \circ \widehat{\Phi})(q, x, y, g) ,$$

where, in going from the second to the third line, we have used that Φ is a family of homomorphisms from G to H. We thus have proved that the vector fields $\widehat{Z}_i^G$ and Z_i^H are related by the map $\widehat{\Phi}$ (for $\widehat{Z}_2^G$ the computations are similar). By [V.2.29] their commutators are also related by $\widehat{\Phi}$, in particular at the point (q, x, y, e) and its image $(T'\Phi(q, x), T'\Phi(q, y), e)$ where we have

$$\begin{aligned}(\underline{0}, \underline{0}, T'\Phi(q, [x, y])) &= T\widehat{\Phi}(\underline{0}, \underline{0}, \underline{0}, \overrightarrow{[x, y]}|_e) = T\widehat{\Phi}([\widehat{Z}_1^G, \widehat{Z}_2^G]) \\ &= [Z_1^H, Z_2^H] = (\underline{0}, \underline{0}, [\, T'\Phi(q, x), T'\Phi(q, y)\,]) \, .\end{aligned}$$ QED

2.12 Definition. Let G be an $\mathcal{A}$-Lie group and $\mathfrak{g}$ its $\mathcal{A}$-Lie algebra. We define the map $I : G \times G \to G$ by $I(g, h) = ghg^{-1}$. It is elementary to verify that I is a left action of G on itself and at the same time a family of homomorphisms from G to G. We thus can define the associated family $T'I$ of morphisms from $\mathfrak{g}$ to $\mathfrak{g}$ by $T'I(g, x) = TI_g\vec{x}_e$. Formally the tangent map is left linear, but since it is also even, it is right linear too. We thus can apply [IV.3.17] to obtain a smooth map $\mathrm{Ad} : G \to \mathrm{End}_R(\mathfrak{g})_0$, i.e., $\mathrm{Ad}(g)(x) = T'I(g, x) = TI_g\vec{x}_e$. According to the definition of $I(g, h)$, this can also be written as

$$\mathrm{Ad}(g)(x) = TL_gTR_{g^{-1}}\vec{x}_e \, .$$

According to [2.11], each $\mathrm{Ad}(g)$ is an $\mathcal{A}$-Lie algebra morphism, i.e., we have the equality $\mathrm{Ad}(g)([x, y]) = [\mathrm{Ad}(g)(x), \mathrm{Ad}(g)(y)]$. And even more: the map $\mathrm{Ad} : G \to \mathrm{End}_R(\mathfrak{g})$ is an $\mathcal{A}$-Lie group homomorphism. To prove this, we compute

$$\mathrm{Ad}(h)(\mathrm{Ad}(g)(x)) = TI_h(TI_g(\vec{x}_e)) = TI_{hg}(\vec{x}_e) = \mathrm{Ad}(hg)(x) \, .$$

Since obviously $\mathrm{Ad}(e) = id(\mathfrak{g})$, this shows that Ad takes its values in $\mathrm{Aut}(\mathfrak{g})$ and that it is a linear representation of G on $\mathfrak{g}$. This representation is called *the Adjoint representation of G*.

2.13 Example. Let G be the group $\mathrm{Aut}(E)$ of automorphisms of some finite dimensional $\mathcal{A}$-vector space E, and recall that we have identified its $\mathcal{A}$-Lie algebra $\mathfrak{g}$ with $\mathrm{End}_R(E)$. We now want to compute the Adjoint representation explicitly. Using [1.23], we find for $\mathrm{Ad}(g)x = TL_g\,TR_{g^{-1}}\,\vec{x}_e$ in the trivialization $T\,\mathrm{Aut}(E) \cong \mathrm{Aut}(E) \times \mathrm{End}_R(E)$ the expression $TL_g\,TR_{g^{-1}}\,(e, x) = (e, gxg^{-1})$. We thus obtain for the Adjoint representation of $\mathrm{Aut}(E)$ on $\mathrm{End}_R(E)$ the following formula:

$$\forall g \in \mathrm{Aut}(E),\ \forall X \in \mathrm{End}_R(E) \cong \mathfrak{g} \ :\ \mathrm{Ad}(g)(X) = g \circ X \circ g^{-1} \, .$$

2.14 Discussion/Definition. We can apply [2.11] to the morphism $\mathrm{Ad} : G \to \mathrm{Aut}(\mathfrak{g})$ of $\mathcal{A}$-Lie groups to obtain a morphism $T_e \mathrm{Ad} : \mathfrak{g} \to \mathrm{End}_R(\mathfrak{g})$ of $\mathcal{A}$-Lie algebras. In [2.15] we will show that this map is the right adjoint representation of the $\mathcal{A}$-Lie algebra $\mathfrak{g}$ defined by $\mathrm{ad}_R(x)(y) = [x, y]$, i.e., $T_e \mathrm{Ad} = \mathrm{ad}_R$. Once we know this, the fact that it is an $\mathcal{A}$-Lie algebra morphism, i.e.,

$$\forall x, y \in \mathfrak{g} : \mathrm{ad}([x, y]) = [\mathrm{ad}(x), \mathrm{ad}(y)]_R ,$$

is just a reformulation of the graded Jacobi identity.

2.15 Proposition. *Let G be an $\mathcal{A}$-Lie group, and let $\mathrm{Ad} : G \to \mathrm{Aut}(\mathfrak{g})$ be its Adjoint representation. Then the $\mathcal{A}$-Lie algebra morphism $T_e \mathrm{Ad} : \mathfrak{g} \to \mathrm{End}_R(\mathfrak{g})$ is the algebraic adjoint representation ad_R : $(T_e \mathrm{Ad}(\vec{x}_e))(y) = [x, y]$.*

Proof. Since $T_e \mathrm{Ad}$ is an even linear map, it is sufficient to know its values on homogeneous vectors with real coordinates (these contain at least a basis). Similarly, to know the linear map $T_e \mathrm{Ad}(\vec{x}_e) \in \mathrm{End}_R(\mathfrak{g})$ it is sufficient to know its values on homogeneous vectors with real coordinates. Since $\mathrm{Aut}(\mathfrak{g}) \subset \mathrm{End}_R(\mathfrak{g})_0$, we can see Ad as a smooth function with values in the even part of the $\mathcal{A}$-vector space $\mathrm{End}_R(\mathfrak{g})$. We know from [V.3.2] that $\langle\!\langle \vec{x}_e \| T \mathrm{Ad} \rangle\!\rangle = \vec{x}_e \mathrm{Ad} \in \mathrm{End}_R(\mathfrak{g})$. Using [1.9] it thus is sufficient to prove the equality $(\vec{x}_e \mathrm{Ad})(y) = [\vec{x}, \vec{y}\,]_e$ for *smooth* homogeneous left-invariant vector fields.

So let $\vec{x}$ and $\vec{y}$ be smooth homogeneous left-invariant vector fields on G and let $\alpha = \varepsilon(x)$ be the parity of $\vec{x}$. We define the maps $\phi : \mathcal{A}_\alpha \times G \to G$ and $\Phi : \mathcal{A}_\alpha \times G \to \mathcal{A}_\alpha \times G$ by $\phi(t, g) = g \cdot \exp(tx)$ and $\Phi(t, g) = (-t, \phi(t, g))$. This looks like the flow of the left-invariant vector field $\vec{x}$, but for a general x this needs not be the case (see [2.5] and [2.8] for more details). What we do have is the property $\langle\!\langle \partial_t|_{(0,g)} \| T\phi \rangle\!\rangle = \vec{x}_g$ (but not necessarily for values of t different from zero). Moreover, since $\phi(0, g) = g$ we also have the equalities (V.5.12) and (V.5.13). As in [V.5.14] and [V.5.15] we define $\psi : \mathcal{A}_\alpha \times G \to TG$ by $\psi = T\phi \circ (\underline{0} \times \vec{y}\,) \circ \Phi$. Unlike [V.5.15] we will not compute $\partial_t \psi$ for all values of t, but only at $t = 0$. The computations are completely similar to those of the proof of [V.5.15] (without the additional s) and we find $(\partial_t \psi)(0, g) = [\vec{x}, \vec{y}\,]_g$.

On the other hand, we can use the explicit form $\phi(t, g) = R_{\exp(tx)} g$ to compute ψ directly

$$\begin{aligned} \psi(t, g) &= T\phi(\underline{0}_{-t}, \vec{y}_{g \exp(tx)}) = TR_{\exp(-tx)} \vec{y}_{g \exp(tx)} \\ &= TL_g TL_{\exp(tx)} TR_{\exp(tx)^{-1}} \vec{y}_e = TL_g TI_{\exp(tx)} \vec{y}_e \\ &= TL_g(\mathrm{Ad}(\exp(tx))(y)) , \end{aligned}$$

where for the third equality we used that the vector field $\vec{y}$ is left-invariant. We thus find that $\psi(t, e) = \mathrm{Ad}(\exp(tX))(Y)$ and thus $\partial_t \psi(0, e) = (\vec{x}_e \mathrm{Ad})(y)$, where we used that $T_0 \exp = id$. But restriction to real values and differentiating other coordinates commute [III.3.13], so we find $(\vec{x}_e \mathrm{Ad})(y) = [\vec{x}, \vec{y}\,]_e$. QED

2.16 Proposition. *Let $\Phi : Q \times G \to H$ be a family of homomorphisms from G to H, and let $T'\Phi$ be the associated family of $\mathcal{A}$-Lie algebra morphisms from $\mathfrak{g}$ to $\mathfrak{h}$. Then for all $q \in Q$ and all $x \in \mathfrak{g}_0$:*

$$\Phi(q, \exp_G(x)) = \exp_H(T'\Phi(q,x)) \ .$$

In case Q contains a single point, this reduces to $\Phi(\exp_G(x)) = \exp_H(T_e\Phi(x))$.

Proof. The exponential maps are defined in terms of the flows of the vector fields Z_L. We extend the vector field Z_L^G on $\mathfrak{g} \times G$ to a vector field $\widehat{Z}_L^G$ on $Q \times \mathfrak{g} \times G$ by the formula $\widehat{Z}_L^G = \underline{0} \times Z_L^G$. We also extend the map Φ to a map $\widehat{\Phi} : Q \times \mathfrak{g} \times G \to \mathfrak{h} \times H$ by $\widehat{\Phi}(q,x,g) = (T'\Phi(q,x), \Phi(q,g))$. With these definition we compute

$$\begin{aligned}(T\widehat{\Phi} \circ \widehat{Z}_L^G)(q,x,g) &= T\widehat{\Phi}(\underline{0}_q, \underline{0}_x, TL_g(\vec{x}_e)) = (\underline{0}_{T'\Phi(q,x)}, T\Phi_q(TL_g(\vec{x}_e))) \\ &= (\underline{0}_{T'\Phi(q,x)}, TL_{\Phi(q,g)}(T\Phi'(q,x))) = (Z_L^H \circ \widehat{\Phi})(q,x,g) \ .\end{aligned}$$

We conclude that the vector fields $\widehat{Z}_L^G$ and Z_L^H are related by the map $\widehat{\Phi}$. Combining [V.5.7] and [V.5.5], the result follows. $\boxed{QED}$

2.17 Corollary. *Let $\phi_i : G \to H$ be two $\mathcal{A}$-Lie group homomorphisms. If $T_e\phi_1$ and $T_e\phi_2$ are the same as linear maps $\mathfrak{g} \to \mathfrak{h}$, then ϕ_1 and ϕ_2 coincide on the connected component of G containing the identity e.*

Proof. According to [2.16] we have $\phi_1(\exp(x)) = \exp(T_e\phi_1(x)) = \exp(T_e\phi_2(x)) = \phi_2(\exp(x))$. The result now follows from [2.7]. $\boxed{QED}$

2.18 Example. In [III.3.14] we essentially proved that $T_e\,\mathrm{gDet} = \mathrm{gtr}_R$ when we view $\mathrm{Aut}(E)$ as a subset of $\mathrm{End}_R(E)_0$. From [2.17] we deduce that gDet is the unique (on the connected component) $\mathcal{A}$-Lie group homomorphism $\mathrm{Aut}(E) \to \mathcal{A}$ whose associated algebra morphism is gtr_R.

2.19 Corollary. *Let G be an $\mathcal{A}$-Lie group and let $\mathfrak{g}$ be its $\mathcal{A}$-Lie algebra. For any $g \in G$ and any $x \in \mathfrak{g}_0$ we have*

$$g \cdot \exp_G(x) \cdot g^{-1} = \exp_G(\mathrm{Ad}(g)x) \qquad \textit{and} \qquad \mathrm{Ad}(\exp_G(x)) = \exp_{\mathrm{Aut}(\mathfrak{g})}(\mathrm{ad}_R(x)) \ .$$

In particular if $G = \mathrm{Aut}(E)$ we have $g \circ \exp(X) \circ g^{-1} = \exp(g \circ X \circ g^{-1})$.

Proof. This is a direct consequence of [2.15] and [2.16]. The particular case $G = \mathrm{Aut}(E)$ follows from [2.13]. $\boxed{QED}$

3. Convergence and the exponential of matrices

In this section we show that the derivative of the exponential map is given by the formula $T_x \exp = TL_{\exp(x)} \frac{1-\mathrm{e}^{-\operatorname{ad}(x)}}{\operatorname{ad}(x)}$. *In order to prove this we have to introduce a notion of convergence of a sequence of smooth functions. We use this notion also to define the exponential of a matrix, and we show that it corresponds to the exponential map defined previously for abstract* $\mathcal{A}$*-Lie groups.*

3.1 Construction/Definition. Let E be an $\mathcal{A}$-vector space of dimension $p|q$, let F be a finite dimensional $\mathcal{A}$-vector space, and let U be open in E_0. If we want to define a suitable notion of convergence of functions in $C^\infty(U; F)$, the usual pointwise convergence will not do, because the non-Hausdorff topology of F prohibits uniqueness. In order to obtain a satisfactory notion of convergence, we decompose a function $f \in C^\infty(U; F)$ according to [III.2.23] as

$$f(x, \xi) = \sum_{k=0}^{q} \sum_{1 \le i_1 < \cdots < i_k \le q} \xi^{i_1} \cdots \xi^{i_k} \cdot f_{i_1 \ldots i_k}(x) ,$$

where the $f_{i_1 \ldots i_k}(x)$ are ordinary smooth functions on $\mathbf{B}U$ with values in $\mathbf{B}F$. Since these ordinary vector valued functions $f_{i_1 \ldots i_k}$ are uniquely determined by f, it seems natural to define convergence in terms of these components $f_{i_1 \ldots i_k}$. We thus will say that a sequence $f_n \in C^\infty(U; F)$ *converges (pointwise/uniformly on compacta)* to $f \in C^\infty(U; F)$ if and only if all components $(f_n)_{i_1 \ldots i_k} \in C^\infty(\mathbf{B}U; \mathbf{B}F)$ converge (pointwise/uniformly on compacta) to the corresponding component $f_{i_1 \ldots i_k} \in C^\infty(\mathbf{B}U; \mathbf{B}F)$, using any suitable norm on $\mathbf{B}F$ to define these notions of convergence in $C^\infty(\mathbf{B}U; \mathbf{B}F)$. Note that with this definition we have uniqueness of convergence: if f_n converges to f and to g, then $f = g$. This is an immediate consequence of the bijection between a function f and the set of its components $f_{i_1 \ldots i_k}$.

3.2 Discussion. Our main application of the notion of convergence will be in the construction of functions on $\operatorname{End}_R(E)_0$, with E an $\mathcal{A}$-vector space of dimension $p|q$. On $\operatorname{End}_R(E)$ we will use left coordinates $X = \sum_{i,j} X^i{}_j \cdot e_i \otimes e^j$. However, in order to simplify notation, we denote the $p^2 + q^2$ even ones among the $X^i{}_j$ by x^i, and the $2pq$ odd ones by ξ^i. Finally we define the smooth functions $g_n : \operatorname{End}_R(E)_0 \to \operatorname{End}_R(E)_0$ by

$$g_n(X) = X^n = \underbrace{X \circ \cdots \circ X}_{n \text{ times}} .$$

In terms of coordinates, these functions decompose into components

$$g_n(x, \xi) = \sum_k \sum_{i_1 < \cdots < i_k} \xi^{i_1} \cdots \xi^{i_k} \cdot (g_n)_{i_1 \ldots i_k}(x) .$$

Each $(g_n)_{i_1\dots i_k}$ takes it values in $\mathbf{B}\,\mathrm{End}_R(E)$, i.e., in the space of $(p+q)\times(p+q)$ matrices with real entries. It is immediate from matrix multiplication that each matrix entry of $(g_n)_{i_1\dots i_k}$ is a homogeneous polynomial of degree $n-k$ in the x^i (provided $n\geq k$, else it is zero). Nearly as immediate is the estimate

$$(3.3)\quad |\text{entry of } (g_n)_{i_1\dots i_k}(x)| \leq (p+q)^{n-1}\cdot(\max_i |x^i|)^{n-k} \leq \Big((p+q)\cdot\max_i |x^i|\Big)^n ,$$

This estimate will permit us to define functions on $\mathrm{End}_R(E)_0$ by means of converging power series.

3.4 Lemma. *Let $f(z)=\sum_{n=0}^{\infty} a_n z^n$ be a convergent power series on the whole of* $\mathbf{C}$*, and let E be a finite dimensional $\mathcal{A}$-vector space. Then the sequence of functions $f_i:\mathrm{End}_R(E)_0\to\mathrm{End}_R(E)_0$ defined by $f_i(X)=\sum_{n=0}^{i} a_n X^n$ converges uniformly on compacta to a smooth function $f:\mathrm{End}_R(E)_0\to\mathrm{End}_R(E)_0$ (slight abuse of notation).*

Proof. In terms of the coordinates x and ξ, the functions f_i obviously decompose as

$$f_i(x,\xi)=\sum_k\sum_{i_1<\dots<i_k}\xi^{i_1}\cdots\xi^{i_k}\cdot\Big(\sum_{n=0}^{i} a_n\cdot(g_n)_{i_1\dots i_k}(x)\Big).$$

Using the estimate (3.3) and the fact that $f(z)$ is convergent on the whole of $\mathbf{C}$ shows that the series $\sum_{n=0}^{i} a_n\cdot(g_n)_{i_1\dots i_k}(x)$ converges to a function $f_{i_1\dots i_k}(x)$. Standard arguments of real analysis show that the convergence is uniform on compacta and that the resulting function is smooth. We thus can define the function $f(X)$ by

$$\begin{aligned} f(x,\xi) &= \sum_k\sum_{i_1<\dots<i_k}\xi^{i_1}\cdots\xi^{i_k}\cdot f_{i_1\dots i_k}(x)\\ &= \sum_k\sum_{i_1<\dots<i_k}\xi^{i_1}\cdots\xi^{i_k}\cdot\Big(\sum_{n=0}^{\infty} a_n\cdot(g_n)_{i_1\dots i_k}(x)\Big). \end{aligned}$$

QED

3.5 Definition. If E is a finite dimensional $\mathcal{A}$-vector space, we define the *exponential map* $\mathrm{e}:\mathrm{End}_R(E)_0\to\mathrm{End}_R(E)_0$, $X\mapsto \mathrm{e}^X$ as the smooth function associated to the power series $\mathrm{e}^z=\sum_{n=0}^{\infty} z^n/n!$.

3.6 Example. Let E be an $\mathcal{A}$-vector space of total dimension 2. If its graded dimension is either $2|0$ or $0|2$, the space $\mathrm{End}_R(E)_0$ is described by 4 even coordinates and the exponential map is just the $\mathbf{G}$-extension of the ordinary exponential of a 2×2 matrix. However, if the dimension of E is $1|1$ it gets more interesting. Using the matrix representation $\overset{\bullet}{M}_R$ [II.4.1] we describe an even endomorphism X with two even coordinates a and d and two odd coordinates β and γ by $\overset{\bullet}{M}_R(X) = \left(\begin{smallmatrix} a & \beta \\ \gamma & d \end{smallmatrix}\right)$. An elementary calculation then gives

$$\begin{pmatrix} a & \beta \\ \gamma & d \end{pmatrix}^n = \begin{pmatrix} a^n & 0 \\ 0 & d^n \end{pmatrix} + \frac{a^n - d^n}{a - d} \begin{pmatrix} 0 & \beta \\ \gamma & 0 \end{pmatrix}$$
$$+ \frac{\partial}{\partial a}\left(\frac{a^n - d^n}{a - d}\right) \begin{pmatrix} \beta\gamma & 0 \\ 0 & 0 \end{pmatrix} - \frac{\partial}{\partial d}\left(\frac{a^n - d^n}{a - d}\right) \begin{pmatrix} 0 & 0 \\ 0 & \beta\gamma \end{pmatrix} .$$

As a final we result one finds

$$\exp(\begin{pmatrix} a & \beta \\ \gamma & d \end{pmatrix}) = \begin{pmatrix} \mathrm{e}^a + \mathrm{e}^d\, B'(a-d)\beta\gamma & \mathrm{e}^d\, B(a-d)\beta \\ \mathrm{e}^a\, B(d-a)\gamma & \mathrm{e}^d - \mathrm{e}^a\, B'(d-a)\beta\gamma \end{pmatrix} ,$$

with $B(z) = (\mathrm{e}^z - 1)/z$, and $B'(z)$ its derivative.

We can apply this result to the matrix representation of [2.8], in which case we have $a = d = f^1 \in \mathcal{A}_0$ and $\beta = \gamma = f^2 \in \mathcal{A}_1$. Our exponential of matrices then yields the result $\left(\begin{smallmatrix} \mathrm{e}^{f^1} & \mathrm{e}^{f^1} \cdot f^2 \\ \mathrm{e}^{f^1} \cdot f^2 & \mathrm{e}^{f^1} \end{smallmatrix}\right)$, in agreement with the exponential found in [2.8].

3.7 Lemma. *Let $g, f, f_n \in C^\infty(U; F)$ be smooth functions, and let x^i denote any (even or odd) coordinate on U. If the sequence f_n converges pointwise to f and if the sequence $\partial_{x^i} f_n$ converges uniformly on compacta to g, then $g = \partial_{x^i} f$.*

Proof. From (III.3.5) we see that for an odd coordinate x^i, the set of components of $\partial_{x^i} f_n$ is essentially the same as that of f_n. The announced result then follows immediately. From (III.3.5) we also see that if x^i is an even coordinate, we have $(\partial_{x^i} f)_{i_1 \dots i_k} = \partial_{x^i}(f_{i_1 \dots i_k})$. The result now follows from the corresponding result for ordinary vector valued functions in $C^\infty(\mathbf{B}U; \mathbf{B}F)$. QED

3.8 Lemma. *Let $f, f_n \in C^\infty(U; F)$ be smooth functions, let $V \subset F_0$ be open, and let $g \in C^\infty(V; G)$ be arbitrary. Suppose that the sequence f_n converges (pointwise/uniformly on compacta) to f. If $g \circ f_n$ and $g \circ f$ are defined, then $g \circ f_n$ converges (pointwise/uniformly on compacta) to $g \circ f$*

Proof. Let $\widehat{f} \in C^\infty(U; F)$ be an arbitrary function, but such that $h = g \circ \widehat{f}$ is defined. If we denote the coordinates on U by (x, ξ) and those on V by (y, η), we can write $\widehat{f}(x, \xi) = (y(x, \xi), \eta(x, \xi))$ and thus:

$$(g \circ \widehat{f})(x, \xi) = g(y(x, \xi), \eta(x, \xi)) = \sum \sum \eta^{j_1}(x, \xi) \cdots \eta^{j_\ell}(x, \xi) \cdot g_{j_1 \dots j_\ell}(y(x, \xi)) .$$

From this expression, it follows that the components $h_{i_1 \dots i_k}$ of the composite function $h = g \circ \widehat{f}$ are of the form $h_{i_1 \dots i_k}(x) = \phi_{i_1 \dots i_k}(f_{j_1 \dots j_\ell}(x))$, where each $\phi_{i_1 \dots i_k}(z)$ is a smooth function in z whose structure depends only on g, not on $\widehat{f}$.

Since all components of f_n converge (pointwise/uniformly on compacta) to the corresponding components of f, it follows that the functions $\phi_{i_1 \dots i_k}((f_n)_{j_1 \dots j_\ell})$ converge (pointwise/uniformly on compacta) to $\phi_{i_1\dots i_k}(f_{j_1\dots j_\ell})$. For pointwise convergence this is immediate from the continuity of $\phi_{i_1\dots i_k}$; for uniform convergence on compacta this is a more delicate reasoning using that a continuous function is uniformly continuous on compacta and using that a compact set in $\mathbf{B}V$ can always be enlarged a small amount while remaining inside $\mathbf{B}V$. $\boxed{QED}$

3.9 Corollary. *Let $f, g, f_n, g_n \in C^\infty(U; \mathrm{End}_R(F))$ be smooth functions such that f_n converges to f and g_n converges to g.*

(i) *The functions $\phi_n(m, v) = f_n(m)(v)$ defined on $U \times F$ converge to the function $\phi(m, v) = f(m)(v)$.*

(ii) *The functions $h_n(m) = f_n(m) \circ g_n(m)$ converge to $h(m) = f(m) \circ g(m)$.*

Proof. If the functions f_n converge to f, then the functions $\widehat{f_n} : U \times F \to \mathrm{End}_R(F) \times F$ defined by $\widehat{f_n}(m, v) = (f_n(m), v)$ converge to $\widehat{f}(m, v) = (f(m), v)$. This is immediate when we look at the coordinates in $\mathrm{End}_R(F) \times F$ separately. Since the evaluation map $\mathrm{End}_R(F) \times F \to F$ is smooth, the result follows from [3.8].

A similar argument applies in (ii): if the functions f_n converge to f and the g_n to g, then the functions $\widehat{f_n}(m) = (f_n(m), g_n(m))$ converge to (f, g). Since the composition map $\mathrm{End}_R(F) \times \mathrm{End}_R(F) \to \mathrm{End}_R(F)$ is smooth the result follows again from [3.8]. $\boxed{QED}$

3.10 Discussion. If E is a finite dimensional $\mathcal{A}$-vector space we now have two exponential maps on $\mathrm{End}_R(E)_0$: $\exp(X)$ defined by means of the flow of left-invariant vector fields on $\mathrm{Aut}(E)$, and e^X defined by means of a converging power series. Let us show that these two are actually the same.

Consider the smooth map $\widehat{\psi} : \mathcal{A}_0 \times \mathrm{End}_R(E)_0 \times \mathrm{Aut}(E) \to \mathrm{End}_R(E)_0$ defined by

$$\widehat{\psi}(t, X, g) = g \circ \mathrm{e}^{tX} ,$$

and use it to define the map $\widehat{\phi} : \mathcal{A}_0 \times \mathrm{End}_R(E)_0 \times \mathrm{Aut}(E) \to \mathrm{End}_R(E)_0 \times \mathrm{End}_R(E)_0$ by

$$\widehat{\phi}(t, X, g) = (X, \widehat{\psi}(t, X, g)) .$$

If we can show that this $\widehat{\phi}$ satisfies the equation $T\widehat{\phi} \circ \partial_t = Z_L \circ \widehat{\phi}$, then uniqueness of flows proves our claim.

In [1.20] we already proved that we have $Tm(\underline{0}_g, X_e) = g \circ X$, giving us the equality $Z_L(\widehat{\phi}(t, X, g)) = (\underline{0}_X, \widehat{\psi}(t, X, g) \circ X) = (\underline{0}_X, g \circ \mathrm{e}^{tX} \circ X)$. On the other hand we have the equality $T\widehat{\phi}(\partial_t|_{\widehat{\phi}(t,X,g)}) \equiv \partial_t|_{\widehat{\phi}(t,X,g)}(\widehat{\phi}) = (\underline{0}_X, \partial_t \widehat{\psi}(t, X, g)) = (\underline{0}_X, g \circ \partial_t(\mathrm{e}^{tX}))$. But then we are in position to apply our lemmas. The function $(t, X) \mapsto \mathrm{e}^{tX}$ is the limit of the partial sums $(t, X) \mapsto \sum_{n=0}^{i} t^n \cdot X^n/n!$. The partial derivative with respect to t of these sums is given by $(t, X) \mapsto \sum_{n=0}^{i-1} t^n \cdot X^{n+1}/n! = (\sum_{n=0}^{i-1} t^n \cdot X^n/n!) \circ X$. Combining [3.7] and [3.9-ii] we obtain that $\partial_t(\mathrm{e}^{tX}) = \mathrm{e}^{tX} \circ X$. We conclude that $\widehat{\phi}$ must be the flow of Z_L and thus both definitions of the exponential map coincide. In particular we deduce that the power series takes its values in $\mathrm{Aut}(E)$.

3.11 Remark. Some readers might notice that there is actually nothing that prevents us from defining the exponential in terms of the power series on the whole of $\mathrm{End}_R(E)$. The obvious question is whether this gives us something new. The answer is negative. The ring $\mathrm{End}_R(E)$ can be identified in a natural way with a subring of $\mathrm{End}_R(E^\sharp)_0$, in the same spirit we used to identify the invertible elements of $\mathcal{A}$ with 2×2 matrices [1.25], [2.8]. Taking the exponential on $\mathrm{End}_R(E^\sharp)_0$ preserves this subring, and thus we find that the map $X \mapsto \mathrm{e}^X$ takes its values in the $\mathcal{A}$-Lie group G of invertible elements in $\mathrm{End}_R(E)$, i.e., $G = \{\, A \in \mathrm{End}_R(E) \mid \exists B \in \mathrm{End}_R(E) : A \circ B = id \,\}$.

Another way to interpret the exponential on the whole of $\mathrm{End}_R(E)$ is in terms of the $\mathcal{A}$-Lie group G introduced above. The space $\mathrm{End}_R(E)$ can be identified with *the even part* of the $\mathcal{A}$-Lie algebra $\mathfrak{g}$ of G. As such, the exponential as power series is exactly the exponential map $\exp : \mathfrak{g}_0 \to G$.

3.12 Proposition. *Let E be a finite dimensional $\mathcal{A}$-vector space, and let the convergent power series f and f_n be defined on $\mathbf{C}$ as*

$$f(z) = \frac{\mathrm{e}^z - 1}{z} = \sum_{i=0}^{\infty} \frac{z^i}{(i+1)!}$$

$$f_n(z) = \frac{\mathrm{e}^z - 1}{n(\mathrm{e}^{z/n} - 1)} = \sum_{\ell=0}^{n-1} \tfrac{1}{n}\, \mathrm{e}^{\ell z/n} = \sum_{i=0}^{\infty} \Big(\sum_{\ell=0}^{n-1} \frac{\ell^i}{i!\, n^{i+1}} \Big) \cdot z^i \,.$$

Then f_n converges uniformly on compacta to f, seen as functions on $\mathrm{End}_R(E)_0$.

Proof. It is an elementary computation in real analysis to show that the power series $f_n(z)$ converge uniformly on compacta to f. This plus the estimate (3.3) and the usual arguments on interchanging the order of limits show that the functions $\sum_{i=0}^{\infty} \big(\sum_{\ell=0}^{n-1} \frac{\ell^i}{i!\, n^{i+1}} \big) \cdot (g_i)_{i_1 \dots i_k}$ converge uniformly on compacta to the function $\sum_{i=0}^{\infty} \frac{1}{(i+1)!} \cdot (g_i)_{i_1 \dots i_k}$. QED

3.13 Discussion. We are now finally in position to give an explicit expression for the tangent map of the exponential map. Let G be an $\mathcal{A}$-Lie group, $\mathfrak{g}$ its $\mathcal{A}$-Lie algebra, and let $\exp : \mathfrak{g}_0 \to G$ be the associated exponential map. If y is a tangent vector at $x \in \mathfrak{g}_0$, the map $T_x \exp$ maps it to a tangent vector at $\exp(x) \in G$. Since it is easier to deal with the tangent space at $e \in G$ than with the tangent space at a general point of G, we left-translate this vector back to $e \in G$. Since $T_x(\mathfrak{g}_0)$ is in a natural way isomorphic to $\mathfrak{g}$, we thus are led to the definition of the smooth map $M : \mathfrak{g}_0 \times \mathfrak{g} \to \mathfrak{g}$ by

(3.14)
$$M(x,y) = TL_{\exp(-x)}((T_x \exp)(y)) = Tm(\underline{0}_{\exp(-x)}, (T_x \exp)(y)) \in T_e G = \mathfrak{g} \,.$$

Since this map is obviously even and linear in y, we can interpret this as a smooth map $M : \mathfrak{g}_0 \to \mathrm{End}_R(E)_0$ [IV.3.17].

3.15 Proposition. *For all $x \in \mathfrak{g}_0$ the map $M(x)$ [3.14] is given by*

$$M(x) = \frac{\mathbf{1} - \mathrm{e}^{-\operatorname{ad}(x)}}{\operatorname{ad}(x)} ,$$

i.e., $M(x) = f(-\operatorname{ad}(x))$ with f as in [3.12].

Proof. Let us denote by m_k the map $m_k : G^k \to G$, $(g_1, \ldots, g_k) \mapsto g_1 \cdots g_k$ of k-fold multiplication. For any $n \in \mathbf{N}^*$ it follows easily from [2.4-ii] that we have the equality $\exp(x) = m_n(\exp(\frac{1}{n}x), \ldots, \exp(\frac{1}{n}x))$. In the canonical identification $T\mathfrak{g}_0 = \mathfrak{g}_0 \times \mathfrak{g}$ [V.1.2], we obtain by the chain rule that for $\psi(x) = \exp(\frac{1}{n}x)$ and $y \in \mathfrak{g}$ we have $(T_x\psi)(y) = \frac{1}{n}(T_{x/n}\exp)(y) = (T_{x/n}\exp)(\frac{1}{n}y)$. Using extensively the associativity of the multiplication we then compute:

$$\begin{aligned}
M(x,y) &= Tm_2(\underline{0}_{\exp(-x)}, (T_x\exp)(y)) \\
&= Tm_{n+1}(\underline{0}_{\exp(-x)}, \underbrace{(T_{x/n}\exp)(\tfrac{1}{n}y), \ldots, (T_{x/n}\exp)(\tfrac{1}{n}y)}_{n\text{ terms}}) \\
&= \sum_{k=0}^{n-1} Tm_{n+1}(\underline{0}_{\exp(-x)}, \underbrace{\underline{0}_{\exp(x/n)}, \ldots, \underline{0}_{\exp(x/n)}}_{n-1-k\text{ terms}}, (T_{x/n}\exp)(\tfrac{1}{n}y), \\
&\qquad\qquad \underbrace{\underline{0}_{\exp(x/n)}, \ldots, \underline{0}_{\exp(x/n)}}_{k\text{ terms}}) \\
&= \sum_{k=0}^{n-1} Tm_4(\underline{0}_{\exp(-kx/n)}, \underline{0}_{\exp(-x/n)}, (T_{x/n}\exp)(\tfrac{1}{n}y), \underline{0}_{\exp(kx/n)}) \\
&= \sum_{k=0}^{n-1} Tm_3(\underline{0}_{\exp(-kx/n)}, M(\tfrac{1}{n}x, \tfrac{1}{n}y), \underline{0}_{\exp(kx/n)}) \\
&= \sum_{k=0}^{n-1} \operatorname{Ad}(\exp(-\tfrac{k}{n}x), M(\tfrac{1}{n}x, \tfrac{1}{n}y)) ,
\end{aligned}$$

where we used that $\operatorname{Ad}(g,x) = Tm_3(\underline{0}_g, \vec{x}_e, \underline{0}_{g^{-1}})$. Using [IV.3.17], [2.19], and the linearity of M in its second variable, we thus have established, for all $n \in \mathbf{N}^*$, the equality

$$M(x) = \Big(\sum_{k=0}^{n-1} \tfrac{1}{n} \cdot \exp(-\tfrac{k}{n}\operatorname{ad}(x))\Big) \circ M(\tfrac{1}{n}x) .$$

But in the limit $n \to \infty$ the sequence $M(\frac{1}{n}x)$ converges uniformly on compacta to the function $M(0)$, which is the identity by [2.4-iv]. The result then follows from [3.9-ii] and [3.12]. QED

3.16 Corollary. *For $x \in \mathfrak{g}_0$ and $y \in \mathfrak{g}$ we have $(T_x\exp)(y) = TL_{\exp(x)}\, M(x)(y) = TR_{\exp(x)}\, M(-x)(y)$.*

Proof. The first equality follows immediately from the definition (3.14) of $M(x)$. The second one follows from the observation that $TR_{\exp(-x)}(T_x\exp)(y) = \operatorname{Ad}(\exp(x))M(x,y)$ and that $\operatorname{Ad}(\exp(x)) \circ M(x) = \exp(\operatorname{ad}(x)) \circ M(x) = M(-x)$. QED

3.17 Proposition. *Let $\vec{x}$ be an odd smooth left-invariant vector field on an $\mathcal{A}$-Lie group G. Then $[\vec{x}, \vec{x}\,] = 0$ if and only if $\forall \sigma, \tau \in \mathcal{A}_1 : \exp(\sigma x) \cdot \exp(\tau x) = \exp((\sigma + \tau)x)$. In that case its flow is given by $\phi(\tau, g) = g \cdot \exp(\tau x)$.*

Proof. Suppose first that $[\vec{x}, \vec{x}\,] = 0$, which means that $\vec{x}$ is integrable. Since obviously $\phi(0, g) = g$, it follows from the uniqueness of flows that it is sufficient to prove that $T\phi(\partial_\tau|_{(\tau,g)}) = \vec{x}_{\phi(\tau,g)}$. To do so, we first note that $T\phi(\partial_\tau|_{(\tau,g)}) = TL_g T\exp(x|_{\tau x})$. According to (3.14) and [3.15] we find for $T\exp(x_{\tau x})$ the value $TL_{\exp(\tau x)} M(\tau x)(x)$. Since $\mathrm{ad}(\tau x)(x) = [\tau x, x] = \tau[x, x] = 0$, we conclude from the explicit expression for M that $M(\tau x)(x) = \vec{x}_e$ and thus $T\phi(\partial_\tau|_{(\tau,g)}) = TL_g TL_{\exp(\tau x)}(\vec{x}_e) = \vec{x}_{g\cdot\exp(\tau x)}$. That we have $\exp(\sigma x) \cdot \exp(\tau x) = \exp((\sigma + \tau)x)$ follows immediately from the group property of a flow [V.4.13].

Now suppose that for all odd $\sigma, \tau \in \mathcal{A}_1$ we have $\exp(\sigma x) \cdot \exp(\tau x) = \exp((\sigma + \tau)x)$. Since $\vec{x}$ is smooth, x has real coordinates, and thus the map $\psi : (\sigma, \tau) \mapsto \exp(\sigma x) \cdot \exp(\tau x)$ is smooth. Since σ and τ are odd, $\mathbf{B}\psi$ is constant the identity $e \in G$. This implies that $\psi(\sigma, \tau)$ remains in a chart U around e. Working in local coordinates, we can expand ψ as

$$\begin{aligned}\psi(\sigma, \tau) &= \psi(0, 0) + \sigma\partial_\sigma\psi(0, 0) + \tau\partial_\tau\psi(0, 0) + \sigma\tau\partial_\tau\partial_\sigma\psi(0, 0) \\ &= e + (\sigma + \tau)\vec{x}_e + \sigma\tau\partial_\tau\partial_\sigma\psi(0, 0)\ .\end{aligned}$$

Since σ and τ are both odd, we find as in the proof of [1.17] that in the chart U $\partial_\tau\partial_\sigma\psi(0, 0)$ is the vector $[\vec{x}, \vec{x}\,]|_e$. On the other hand, we know that $\psi(\sigma, \tau)$ equals $\exp((\sigma + \tau)x)$, which develops as $e + (\sigma + \tau)\vec{x}_e$. We conclude that $[\vec{x}, \vec{x}\,]_e = 0$. But the value of the commutator of two left-invariant vector fields is determined by its value at the identity, hence $[\vec{x}, \vec{x}\,] = 0$. QED

3.18 Remark. If in [3.17] we do not have $[\vec{x}, \vec{x}\,] = 0$, then the fact that $\tau^2 = 0$ immediately gives that $M(\tau x)(x) = \vec{x}_e - \frac{1}{2}\tau\overrightarrow{[x, x]}|_e$, which allows one to compute $T\phi(\partial_\tau|_{(\tau,g)})$ with $\phi(\tau, g) = g\exp(\tau x)$ for non integrable smooth odd left-invariant vector fields on G.

4. Subgroups and subalgebras

In this section we define an $\mathcal{A}$-Lie subgroup to be an injective $\mathcal{A}$-Lie group morphism. We then show that $\mathcal{A}$-Lie subgroups of an $\mathcal{A}$-Lie group G correspond to subalgebras of $\mathfrak{g}$ and that normal $\mathcal{A}$-Lie subgroups of G correspond to ideals of $\mathfrak{g}$.

4.1 Definition. Let G be an $\mathcal{A}$-Lie group. An *$\mathcal{A}$-Lie subgroup of G* is a pair (i, H) such that H is an $\mathcal{A}$-Lie group and such that $i : H \to G$ is an injective $\mathcal{A}$-Lie group homomorphism. If no confusion is possible we will denote an $\mathcal{A}$-Lie subgroup just by H.

4.2 Lemma. *If $i : H \to G$ is an $\mathcal{A}$-Lie subgroup of G, then the map i is an injective immersion.*

Proof. We know that i is injective, and we want to prove that Ti is injective. According to [1.18] it suffices to prove that $T_e i$ is injective. So suppose $T_e i(x) = T_e i(y)$ for some $x \neq y \in \mathfrak{h}$. We may assume, multiplying them by a small real number if necessary, that both belong to the open neighborhood of $0 \in \mathfrak{h}$ on which the exponential is bijective [2.7]. We thus have $\exp(x) \neq \exp(y)$. But by [2.16] we have $i(\exp(x)) = \exp(T_e i(x)) = \exp(T_e i(y)) = i(\exp(y))$, contradicting injectivity of i. QED

4.3 Definition. Let $\mathfrak{g}$ be an $\mathcal{A}$-Lie algebra. A *subalgebra* of $\mathfrak{g}$ is a graded subspace (in the sense of $\mathcal{A}$-vector spaces [II.6.23]) $\mathfrak{h} \subset \mathfrak{g}$ that is invariant under the bracket, i.e., $\forall x, y \in \mathfrak{h} : [x, y] \in \mathfrak{h}$. It follows that $\mathfrak{h}$ itself, equipped with the bracket inherited from $\mathfrak{g}$, is an $\mathcal{A}$-Lie algebra.

4.4 Proposition. *Let G be an $\mathcal{A}$-Lie group, and let $i : H \to G$ be an $\mathcal{A}$-Lie subgroup. Then $\mathring{\mathfrak{h}} = T_e i(\mathfrak{h})$ is a subalgebra of $\mathfrak{g}$ isomorphic to $\mathfrak{h}$. Moreover, for all $g \in i(H)$: $\mathrm{Ad}_G(g)(\mathring{\mathfrak{h}}) \subset \mathring{\mathfrak{h}}$.*

Proof. Since $e \in H$ has real coordinates, $T_e i$ is smooth; since Ti is even, $T_e i(\mathfrak{h})$ is a graded subspace of $\mathfrak{g}$ in the sense of $\mathcal{A}$-vector spaces by [II.6.24]. Since $T_e i$ is injective [4.2] and an $\mathcal{A}$-Lie algebra morphism [2.11], the first result follows. For the second part, we compute for $h \in H$ and $x \in \mathfrak{h}$, using that i is a homomorphism, $Ti\,\mathrm{Ad}_H(h)x = Ti(TL_h(TR_{h^{-1}}(x))) = TL_{i(h)}(TR_{i(h)^{-1}}(Ti(x))) = \mathrm{Ad}_G(i(h))Ti(x)$. QED

4.5 Definition. Let G be an $\mathcal{A}$-Lie group. Two $\mathcal{A}$-Lie subgroups $i : H \to G$ and $j : K \to G$ of G are said to be *equivalent* if there exists an $\mathcal{A}$-Lie group isomorphism $\rho : K \to H$ such that for all $k \in K : j(k) = i(\rho(k))$ (and thus in particular $i(H) = j(K)$).

4.6 Corollary. *Let $i : H \to G$ and $j : K \to G$ be two equivalent $\mathcal{A}$-Lie subgroups of an $\mathcal{A}$-Lie group G with $\mathcal{A}$-Lie algebras $\mathfrak{h}$ and $\mathfrak{k}$ respectively. Then $\mathfrak{h}$ and $\mathfrak{k}$ are isomorphic and $T_e i(\mathfrak{h}) = T_e j(\mathfrak{k})$.*

4.7 Theorem. *Let G be an $\mathcal{A}$-Lie group with $\mathcal{A}$-Lie algebra $\mathfrak{g}$. For each subalgebra $\mathring{\mathfrak{h}}$ of $\mathfrak{g}$ there exists a unique (up to equivalence) connected $\mathcal{A}$-Lie subgroup (i, H) of G with $\mathcal{A}$-Lie algebra $\mathfrak{h}$ such that $T_e i(\mathfrak{h}) = \mathring{\mathfrak{h}}$.*

Proof. If $\mathring{\mathfrak{h}}$ is a subalgebra of total dimension k, then in particular there exists a basis $(v_i)_{i=1}^n$ of $\mathfrak{g}$ such that $(v_i)_{i=1}^k$ is a basis of $\mathring{\mathfrak{h}}$. We denote by $\mathcal{F}$ the subbundle of total dimension k of TG generated by the vector fields $(\vec{v}_i)_{i=1}^k$ (see also [IV.3.14]). (In terms of the trivialization [1.18] of TG this subbundle is given as $\mathring{\mathfrak{h}} \times G$.) Since $\mathring{\mathfrak{h}}$ is a subalgebra, the commutator $[\vec{v}_i, \vec{v}_j]$, $1 \le i, j \le k$ belongs to $\mathfrak{h}^\bullet$, in other words, it is a linear combination of the $\vec{v}_\ell$, $1 \le \ell \le k$. From this and [V.1.19] it follows that $\mathcal{F}$ is an involutive subbundle of TG of total dimension k. Using [V.6.9], we define (i, H) as the (unique) leaf passing through $e \in G$. We will denote by $e_o \in H$ the (unique) element such that $i(e_o) = e$.

The proof now proceeds in several steps. The first step is to prove that the abstract subgroup H_o of G generated by $\exp(\mathfrak{h}^\bullet_0)$ is contained in $i(H)$. The next step is to prove that H_o is the whole of $i(H)$. This proves that $i(H)$ is an abstract group. The third step is to transport this group structure to H such that (i, H) becomes an $\mathcal{A}$-Lie subgroup with $\mathcal{A}$-Lie algebra $\mathfrak{h}$ and $T_e i(\mathfrak{h}) = \mathfrak{h}^\bullet$. The last step is to prove uniqueness of this $\mathcal{A}$-Lie subgroup up to equivalence.

• We start with the first step. We define the subset $H_o \subset G$ as the set of all finite products of elements of the form $\exp(x)$ with $x \in \mathfrak{h}^\bullet_0$. Since $\exp(x)^{-1} = \exp(-x)$, this is an abstract subgroup of G. Next we consider the map $\phi : H \times \mathfrak{h}^\bullet_0 \to G$ defined by $\phi(h, x) = i(h) \cdot \exp(x)$. In [4.8] we prove that it is tangent to $\mathcal{F}$. But H and $\mathfrak{h}^\bullet_0$ are connected, and $\phi(e_o, 0) = i(e_o) \in i(H)$. By [V.6.9] we conclude that $\phi(H \times \mathfrak{h}^\bullet_0) \subset i(H)$. We now start induction with $h = e_o \in H$ to prove that $H_o \subset i(H)$.

• The second step is more or less a topological argument to prove that $H_o = i(H)$. In [4.9] we prove that $\widehat{H}_o = i^{-1}(H_o) \subset H$ is open and closed in H. Now H is connected and hence $\widehat{H}_o = H$ and thus $i(H) = i(\widehat{H}_o) = H_o \cap i(H) = H_o$ as wanted.

• Once we know that $i(H) = H_o$, life becomes easy. Consider the map $\phi : H \times H \to G$ defined by $\phi(g, h) = i(g) \cdot i(h)$. Since $i(H) = H_o$ is an abstract subgroup of G, we have $\phi(H \times H) \subset i(H)$, and thus by [V.6.8] there exists a smooth map $m_H : H \times H \to H$ such that $\phi = i \circ m_H$. It is immediate that this m_H makes H into an $\mathcal{A}$-Lie group, i into an injective $\mathcal{A}$-Lie group homomorphism, and that $T_e i(\mathfrak{h}) = \mathfrak{h}^\bullet$.

• The fourth step is to prove uniqueness up to equivalence. Let (j_1, K_1) and (j_2, K_2) be two connected $\mathcal{A}$-Lie subgroups of G with $\mathcal{A}$-Lie algebras $\mathfrak{k}_1$ and $\mathfrak{k}_2$ respectively such that $T_e j_i(\mathfrak{k}_i) = \mathfrak{h}^\bullet$. In [4.10] we prove that both must be equivalent to the $\mathcal{A}$-Lie subgroup (i, H) constructed above. Hence (j_1, K_1) and (j_2, K_2) are equivalent. QED

4.8 Technical lemma. *The map $\phi : H \times (\mathfrak{h}^\bullet)_0 \to G$ defined by $\phi(h, x) = i(h) \cdot \exp(x)$ is tangent to $\mathcal{F}$.*

Proof. We have to show that for any $(A, B) \in T_h H \times T_x(\mathfrak{h}^\bullet)_0 \cong T_h H \times \mathfrak{h}^\bullet$ we have $T_{(h,x)}\phi(A, B) \in \mathcal{F}|_{\phi(h,x)}$. By definition of H, $T_h i(A)$ belongs to $\mathcal{F}|_{i(h)}$, and thus there exists $y \in \mathfrak{h}^\bullet$ such that $T_h i(A) = TL_{i(h)}(\vec{y}_e)$. We now compute

$$\begin{aligned} T_{(h,x)}\phi(A, B) &= Tm(\,(T_h i(A), \underline{0}_{\exp(x)}) + (\underline{0}_{i(h)}, T_x \exp(B))\,) \\ &= TR_{\exp(x)}(TL_{i(h)}(\vec{y}_e)) + TL_{i(h)}(T_x \exp(B)) \\ &= TL_{h\cdot\exp(x)}(\mathrm{Ad}(\exp(-x)(y)) + TL_{h\cdot\exp(x)}(M(x)(B))\ , \end{aligned}$$

where we used (3.14). Since x, y, and B belong to $\mathfrak{h}^\bullet$ which is a subalgebra, the elements $\mathrm{ad}(x)(y)$ and $\mathrm{ad}(x)(B)$ belong to $\mathfrak{h}^\bullet$. But now $\mathrm{Ad}(\exp(-x)) = \mathrm{e}^{-\mathrm{ad}(x)}$ and $M(x)$ are convergent power series in $-\mathrm{ad}(x)$ [3.15], hence by [3.9-i] $\mathrm{Ad}(\exp(-x))(y)$ and

$M(x)(B)$ belong to $\mathfrak{h}^\bullet$. Substituting this back in the formula for $T_{(h,x)}\phi(A,B)$, we see that it belongs to $\mathcal{F}$. QED

4.9 Technical lemma. *The subset $\widehat{H}_o = i^{-1}(H_o) \subset H$ is open and closed in H.*

Proof. Let $0 \in U \subset \mathfrak{g}_0$ be an open subset such that $\exp : U \to V = \exp(U) \subset G$ is a diffeomorphism. By [4.8] the map $\exp : \mathfrak{h}^\bullet \cap U \to V$ is integral for $\mathcal{F}$ and thus there exists a smooth map $\widehat{\exp} : \mathfrak{h}^\bullet \cap U \to H$ such that $i \circ \widehat{\exp} = \exp : \mathfrak{h}^\bullet \cap U \to G$ [V.6.9]. Since $T\exp$ is injective on $T(\mathfrak{h}^\bullet \cap U)$, $T\widehat{\exp}$ must be too. Since the graded dimensions of $\mathfrak{h}^\bullet \cap U$ and H are the same, it follows that $\widehat{\exp}$ is a diffeomorphism [V.2.14]. We denote by $U_o \subset H$ the open set $U_o = \widehat{\exp}(\mathfrak{h}^\bullet \cap U)$; by construction $U_o \subset \widehat{H}_o$.

Now let $h_o \in \mathbf{B}H$, for which we define $\phi_{h_o} : H \to G$ by $\phi_{h_o}(h) = i(h_o)^{-1} \cdot i(h)$. This is a smooth map because h_o has real coordinates and because i and m are smooth. Moreover, ϕ_{h_o} is integral for $\mathcal{F}$ because $\mathcal{F}$ is generated by left-invariant vector fields. Since $\phi_{h_o}(h_o) = i(e_o)$ it follows that there exists a smooth map $\widehat{\phi}_{h_o} : H \to H$ such that $\phi_{h_o} = i \circ \widehat{\phi}_{h_o}$ [V.6.9]. Denoting $h_o' = \widehat{\phi}_{h_o}(e_o) \in \mathbf{B}H$, we find $i(h_o') = i(h_o)^{-1}$ (and remember, we do not yet know that H has a group structure). Since obviously $\widehat{\phi}_{h_o}$ and $\widehat{\phi}_{h_o'}$ are each others inverses, we find an open neighborhood $U_{h_o} = \widehat{\phi}_{h_o'}(U_o)$ of any $h_o \in \mathbf{B}H$.

Since $\mathbf{B}(\exp(x_1)\cdots\exp(x_\ell)) = \exp(\mathbf{B}x_1)\cdots\exp(\mathbf{B}x_\ell))$, it follows that if $h_1 \in \widehat{H}_o$, then also $h_o = \mathbf{B}h_1 \in \widehat{H}_o$. We thus have an open neighborhood $U_{h_o} = \widehat{\phi}_{h_o'}(U_o)$ of h_o satisfying

$$i(U_{h_o}) = i(\widehat{\phi}_{h_o'}(U_o)) = i(h_o) \cdot i(U_o) .$$

Hence $U_{h_o} \subset \widehat{H}_o$ and thus $\widehat{H}_o$ is open in H, just because any neighborhood of h_o is also a neighborhood of h_1. On the other hand, suppose $h_1 \in H \setminus \widehat{H}_o$, then $h_o \in H \setminus \widehat{H}_o$ because $\widehat{H}_o$ is open. But then $U_{h_o} \subset H \setminus \widehat{H}_o$. For suppose $g \in U_{h_o} \cap \widehat{H}_o = \widehat{\phi}_{h_o'}(U_o) \cap \widehat{H}_o$, then $i(g) = i(h_o) \cdot i(h) \in H_o$ for some $h \in U_o$. But then $i(h_o) = i(g) \cdot i(h)^{-1} \in H_o$, contradicting the hypothesis. It follows that $\widehat{H}_o$ is also closed in H. QED

4.10 Technical lemma. *Let (j, K) be a connected $\mathcal{A}$-Lie subgroup of G with $\mathcal{A}$-Lie algebra $\mathfrak{k}$ such that $T_e j(\mathfrak{k}) = \mathfrak{h}^\bullet$. Then (j, K) and (i, H) are equivalent $\mathcal{A}$-Lie subgroups.*

Proof. Since $T_e i$ and $T_e j$ are smooth and injective, the equality $T_e i(\mathfrak{h}) = \mathfrak{h}^\bullet = T_e j(\mathfrak{k})$ implies the existence of a smooth isomorphism $\phi : \mathfrak{k} \to \mathfrak{h}$ such that $T_e j = T_e i \circ \phi$. By [2.16] we obtain for $x \in \mathfrak{k}_0$:

$$j(\exp_K(x)) = \exp_G(T_e j(x)) = \exp_G(T_e i(\phi(x))) = i(\exp_H(\phi(x))) .$$

Since ϕ is an isomorphism, it follows that $j(K) = i(H)$ (use [2.7]). By [V.6.8] there thus exists a smooth map $\widehat{j} : K \to H$ such that $j = i \circ \widehat{j}$. Since Tj is injective, the same holds for $T\widehat{j}$, but then the fact that H and K both have total dimension k implies that $\widehat{j}$ is a diffeomorphism onto its image [V.2.14]. From $j(K) = i(H)$ we deduce that this image equals H and thus $\widehat{j} : K \to H$ is a diffeomorphism. It is elementary to show that it is an $\mathcal{A}$-Lie group homomorphism, proving that both $\mathcal{A}$-Lie subgroups are equivalent. QED

4.11 Lemma. *Let G be an $\mathcal{A}$-Lie group, (i, H) an $\mathcal{A}$-Lie subgroup of G, and $f : M \to G$ a smooth map. If $f(M) \subset i(H)$, then $i^{-1} \circ f : M \to H$ is smooth.*

Proof. Using the trivialization of [1.18], the map $Ti : TH \to TG$ is described as $Ti(h, x) = (i(h), T_e i(x)) \in G \times T_e i(\mathfrak{h})$. Hence $Ti(TH)$ is contained in the involutive subbundle $\mathcal{F}$ as defined in the proof of [4.7]. It follows that (i, H) is an integral manifold of $\mathcal{F}$. The conclusion now follows from [V.6.8]. QED

4.12 Definitions. Let G be an $\mathcal{A}$-Lie group and (i, H) an $\mathcal{A}$-Lie subgroup. H is called a *normal $\mathcal{A}$-Lie subgroup* if

$$\forall g \in G \quad \forall h \in H \quad \exists \widehat{h} \in H : g \cdot i(h) \cdot g^{-1} = i(\widehat{h}) \ .$$

If $\mathfrak{g}$ is an $\mathcal{A}$-Lie algebra and $\mathfrak{h}$ a subalgebra, $\mathfrak{h}$ is said to be an *ideal in* $\mathfrak{g}$ if

$$\forall x \in \mathfrak{g} \quad \forall y \in \mathfrak{h} : [x, y] \in \mathfrak{h} \ .$$

4.13 Lemma. *Let G be an $\mathcal{A}$-Lie group, $\mathfrak{g}$ its $\mathcal{A}$-Lie algebra, (i, H) an $\mathcal{A}$-Lie subgroup with $\mathcal{A}$-Lie algebra $\mathfrak{h}$, and $\mathring{\mathfrak{h}} = T_e i(\mathfrak{h})$.*

(i) *If H is a normal $\mathcal{A}$-Lie subgroup of G, then $\mathring{\mathfrak{h}}$ is an ideal in $\mathfrak{g}$ and $\forall g \in G$: $\mathrm{Ad}(g)(\mathring{\mathfrak{h}}) \subset \mathring{\mathfrak{h}}$.*

(ii) *If G and H are connected and if $\mathring{\mathfrak{h}}$ is an ideal in $\mathfrak{g}$, then H is a normal $\mathcal{A}$-Lie subgroup of G.*

Proof. To prove (i), consider the family of homomorphism $\Phi : G \times H \to G$ from H to G defined by $\Phi(g, h) = g \cdot i(h) \cdot g^{-1}$. Since H is normal, $\Phi(g, h)$ lies in $i(H)$ and thus there exists a smooth map $\widehat{\Phi} : G \times H \to H$ such that $\Phi = i \circ \widehat{\Phi}$ [4.11]. Uniqueness of $\widehat{\Phi}$ proves that it is a family of homomorphisms from H to H. Applying [2.16] we find $\exp_H(T'\widehat{\Phi}(g, y)) = \widehat{\Phi}(g, \exp_H(y))$, valid for all $y \in \mathfrak{h}_0$. Composing with i, using [2.16] and [2.19], gives us

$$\exp_G(T_e i(T'\widehat{\Phi}(g, y))) = \Phi(g, \exp_H(y)) = \exp_G(\mathrm{Ad}(g) T_e i(y)) \ .$$

Since $T_e i(T'\widehat{\Phi}(g, y))$ and $\mathrm{Ad}(g) T_e i(y)$ are smooth in (g, y), it follows that if y is sufficiently close to $0 \in \mathfrak{h}_0$, both arguments of $\exp_G$ lie in the neighborhood on which $\exp_G$ is bijective. We deduce (by linearity) that $\mathrm{Ad}(g) T_e i(y) = T_e i(T'\widehat{\Phi}(g, y))$ for all $y \in \mathfrak{h}_0$, i.e., $\mathrm{Ad}(g)(y) \in \mathring{\mathfrak{h}}$ for all $y \in \mathring{\mathfrak{h}}_0$. In order to prove that this is also true for all $y \in \mathring{\mathfrak{h}}$, we denote by π the canonical projection $\pi : \mathring{\mathfrak{h}} \subset \mathfrak{g} \to \mathfrak{g}/\mathring{\mathfrak{h}}$ and we define the smooth map $\mathrm{Ad}' : G \to \mathrm{Hom}_R(\mathring{\mathfrak{h}}, \mathfrak{g}/\mathring{\mathfrak{h}})$ by $\mathrm{Ad}'(g) = \pi \circ \mathrm{Ad}(g)$. We thus have shown that $\forall g \in G$: $\mathrm{Ad}'(g)|_{\mathring{\mathfrak{h}}_0} = 0$. By [III.2.31] we conclude that $\forall g \in G : \mathrm{Ad}'(g) = 0$, i.e., $\mathrm{Ad}(g)(\mathring{\mathfrak{h}}) \subset \mathring{\mathfrak{h}}$ as wanted. To finish the proof of (i), consider the *smooth* map $\phi : G \times \mathring{\mathfrak{h}} \to \mathring{\mathfrak{h}} \subset \mathfrak{g}$ defined by $\phi(g, y) = \mathrm{Ad}(g) y$. Computing the derivative $T\phi(\vec{x}_e, \underline{0}_y)$ for $x \in \mathfrak{g}$, $y \in \mathring{\mathfrak{h}}$, we find $\mathrm{ad}(x) y$. Since ϕ takes values in $\mathring{\mathfrak{h}}$, this belongs to $\mathring{\mathfrak{h}} \subset \mathfrak{g}$, i.e., $\mathring{\mathfrak{h}}$ is an ideal of $\mathfrak{g}$.

To prove (ii), let $X \in \mathfrak{g}_0$ and $Y \in \mathfrak{h}_0$ be arbitrary. Using [2.16], [2.19], and [2.15] we find

$$\exp(x) \cdot i(\exp(y)) \cdot \exp(-x) = \exp(\mathrm{Ad}(\exp(x))T_e i(y)) = \exp(\mathrm{e}^{\mathrm{ad}(x)}\, T_e i(y)) \,.$$

Since $T_e i(\mathfrak{h})$ is an ideal, $\mathrm{ad}(x)T_e i(y) = [x, T_e i(y)] \in T_e i(\mathfrak{h})$. From this and [3.9-i] we deduce that $\mathrm{e}^{\mathrm{ad}(x)}\, T_e i(y)$ belongs to $T_e i(\mathfrak{h})$, say $\mathrm{e}^{\mathrm{ad}(x)}\, T_e i(y) = T_e i(\widehat{y})$. It follows that

$$\exp(x) \cdot i(\exp(y)) \cdot \exp(-x) = \exp(T_e i(\widehat{y})) = i(\exp(\widehat{y})) \,.$$

Since elements of the form $\exp(x)$ generate G and elements of the form $\exp(y)$ generate H [2.7], it follows immediately that H is a normal $\mathcal{A}$-Lie subgroup of G. $\boxed{QED}$

4.14 Example. Consider the set $G = \{\, (x, y, \xi, \eta) \in \mathcal{A}_0^2 \times \mathcal{A}_1^2 \mid \mathbf{B}(x^2 + y^2) \neq 0 \,\}$ with its obvious structure of an$\mathcal{A}$-manifold of dimension 2|2. We endow G with the structure of an $\mathcal{A}$-Lie group by giving its multiplication law:

$$\begin{aligned}(x^1, y^1, \xi^1, \eta^1) \cdot (x^2, y^2, \xi^2, \eta^2) = \\ &(x^1x^2 - y^1y^2 + \xi^1\xi^2 - \eta^1\eta^2 \,,\ x^1y^2 + y^1x^2 + \xi^1\eta^2 + \eta^1\xi^2 \,, \\ &\quad x^1\xi^2 - y^1\eta^2 + \xi^1x^2 - \eta^1y^2 \,,\ x^1\eta^2 + y^1\xi^2 + \xi^1y^2 + \eta^1x^2) \,.\end{aligned}$$

Attentive readers will recognize this $\mathcal{A}$-Lie group as $\mathcal{A}^{\mathbf{C}^*}$, the multiplicative group of invertible elements $(x + \xi) + i(y + \eta) \in \mathcal{A} \oplus i\mathcal{A}$. An elementary calculation shows that a basis for the left-invariant vector fields is given by

$$\begin{aligned} \vec{v}_x &= x\partial_x + y\partial_y + \xi\partial_\xi + \eta\partial_\eta & \vec{v}_y &= -y\partial_x + x\partial_y - \eta\partial_\xi + \xi\partial_\eta \\ \vec{v}_\xi &= -\xi\partial_x - \eta\partial_y + x\partial_\xi + y\partial_\eta & \vec{v}_\eta &= \eta\partial_x - \xi\partial_y - y\partial_\xi + x\partial_\eta \,. \end{aligned}$$

Another elementary calculation shows that the only non-zero commutators among these basis vectors are

$$(4.15) \qquad [\vec{v}_\xi, \vec{v}_\xi] = -2 \cdot \vec{v}_x \quad , \quad [\vec{v}_\xi, \vec{v}_\eta] = -2 \cdot \vec{v}_y \quad , \quad [\vec{v}_\eta, \vec{v}_\eta] = 2 \cdot \vec{v}_x \,.$$

Of course these commutators could also have calculated by means of [1.17]. For instance, the structure constants $c^*_{\xi\eta}$ are given by $\partial_{\xi^1}\partial_{\eta^2}m + \partial_{\eta^1}\partial_{\xi^2}m = (0, -1, 0, 0) + (0, -1, 0, 0)$, giving as above $[v_\xi, v_\eta] = -2v_y$.

Yet another elementary calculation, but this time a rather tedious one, shows that the exponential map $\exp : \mathfrak{g}_0 \to G$ is given by

$$\exp(a, b, \alpha, \beta) = (\mathrm{e}^a \cos(b), \mathrm{e}^a \sin(b), \mathrm{e}^a \cos(b)\alpha - \mathrm{e}^a \sin(b)\beta, \mathrm{e}^a \cos(b)\beta + \mathrm{e}^a \sin(b)\alpha) \,,$$

where (a, b, α, β) denote the coordinates in $\mathfrak{g}_0$ of dimension 2|2 with respect to the basis v_x, v_y, v_ξ, v_η. Given this formula, the reader can ascertain that the expression [2.4-i] verifies the conditions of the flow of Z_L. If we identify G with the multiplicative group of $\mathcal{A} \oplus i\mathcal{A}$ and $\mathfrak{g}_0$ with $\mathcal{A} \oplus i\mathcal{A}$, this exponential gets the much nicer form

$$\mathrm{e}^{(a+\alpha)+i(b+\beta)} = \mathrm{e}^{(a+ib)+(\alpha+i\beta)} = \mathrm{e}^{a+ib} \cdot (1 + \alpha + i\beta) .$$

Looking at the commutators (4.15) we see that the graded subspace $\mathfrak{h}^{\bullet}$ generated by the two vectors $w_0 = v_y$ and $w_1 = v_\xi + v_\eta$ is a subalgebra of dimension $1|1$. In order to find the $\mathcal{A}$-Lie subgroup (i, H) that corresponds to this subalgebra, we compute the exponentials $\exp(x)$ with x belonging to $\mathfrak{h}^{\bullet}_0$. Our formula gives us

$$\exp(a \cdot w_0 + \alpha \cdot w_1) = \mathrm{e}^{ia} \cdot (1 + (1+i) \cdot \alpha) .$$

The product of two of such elements is given by

$$(\mathrm{e}^{ia} \cdot (1 + (1+i) \cdot \alpha)) \cdot (\mathrm{e}^{ib} \cdot (1 + (1+i) \cdot \beta)) = \mathrm{e}^{i(a+b+2\alpha\beta)} \cdot (1 + (1+i) \cdot (\alpha + \beta)) .$$

From this we "deduce" that H is the $\mathcal{A}$-manifold $(\mathbf{G}S^1) \times \mathcal{A}_1$ with multiplication

$$(\mathrm{e}^{ia}, \alpha) \cdot (\mathrm{e}^{ib}, \beta) = (\mathrm{e}^{ia}\, \mathrm{e}^{ib}\, \mathrm{e}^{2i\alpha\beta}, \alpha + \beta) ,$$

and with embedding $i : H \to G$ given by $i(\mathrm{e}^{ia}, \alpha) = \mathrm{e}^{ia}(1 + (1+i)\alpha)$. It is elementary to verify that this is indeed an embedding and that $i(H) \subset G$ is the submanifold defined by the equations $x^2 + y^2 = 1$ and $y \cdot (\xi + \eta) + x \cdot (\xi - \eta) = 0$.

The most general subalgebra $\mathfrak{h}^{\bullet} \subset \mathfrak{g}$ of dimension $1|1$ is generated by the vectors $w_0 = av_\xi + bv_\eta$ and $w_1 = (a^2 - b^2)v_x + 2abv_y$, $(a, b) \in \mathbf{R}^2 \setminus \{(0,0)\}$. Except for the cases $a = \pm b$, all the corresponding $\mathcal{A}$-Lie subgroups are isomorphic to $\mathcal{A}$ as $\mathcal{A}$-manifolds. The case $a = -b$ is the complex conjugate of the example treated above; the corresponding $\mathcal{A}$-Lie subgroup thus is the complex conjugate of H, i.e., as $\mathcal{A}$-Lie group it is H, but with embedding $(a, \alpha) \mapsto \mathrm{e}^{-ia}(1 + (1-i)\alpha)$.

5. Homogeneous $\mathcal{A}$-manifolds

In this section we are concerned with actions of $\mathcal{A}$-Lie groups on $\mathcal{A}$-manifolds. We therefore introduce the notion of fundamental vector field on an $\mathcal{A}$-manifold associated to an element of the $\mathcal{A}$-Lie algebra. We also introduce the notion of a proper $\mathcal{A}$-Lie subgroup which replaces the notion of closed subgroup in the non-graded situation. With these definitions we prove that if H is a proper $\mathcal{A}$-Lie subgroup of G, then G/H is an $\mathcal{A}$-manifold. Moreover, if $m_o \in M$ has real coordinates and if H is the isotropy group at m_o, then H is a proper $\mathcal{A}$-Lie subgroup and the map $G/H \to M$ is an injective immersion.

5.1 Definition. Let $\Phi : G \times M \to M$ be a smooth left action of an $\mathcal{A}$-Lie group G on an $\mathcal{A}$-manifold M. For each $x \in \mathfrak{g}$ we define a vector field x^M on M by the formula

$$x^M|_m = -T\Phi_m(\vec{x}_e) = -T_{(e,m)}\Phi(x, \underline{0}_m) ,$$

where $T\Phi_m$ denotes the generalized tangent map [V.3.19]. This vector field is called *the fundamental vector field on M associated to $x \in \mathfrak{g}$*. If x has real coordinates, i.e., $x \in \mathbf{B}\mathfrak{g}$, the vector field x^M is smooth (because $T\Phi$ is smooth and e has real coordinates). A particular case of this definition is when G acts on itself by multiplication, i.e., $\Phi = m_G : G \times G \to G$. Comparing definitions shows that x^G is the right-invariant vector field on G whose value at $e \in G$ is $-x$ (note the minus sign!). Comparing this with [1.10] shows that x^G is exactly the vector field $TInv \circ \vec{x} \circ Inv$.

5.2 Lemma. *The notion of fundamental vector field enjoys the following properties.*

(i) $T\Phi_g x^M|_m = (\mathrm{Ad}(g)x)^M|_{\Phi(g,m)}$ ($\iff T\Phi_g \circ x^M = (\mathrm{Ad}(g)x)^M \circ \Phi_g$).

(ii) $\forall x \in \mathfrak{g} : T\Phi \circ (x^G \times \underline{0}) = x^M \circ \Phi$, *or roughly in words, the right-invariant vector field x^G on G and the fundamental vector field x^M on M are related by Φ.*

(iii) *For $x, y \in \mathbf{B}\mathfrak{g}$ we have $[x^M, y^M] = [x, y]^M$.*

(iv) *The flow ϕ of the even vector field Z_M on $\mathfrak{g}_0 \times M$ defined by the formula $Z_M(x, m) = (\underline{0}_x, x^M|_m)$ is given by $\phi(t, x, m) = (x, \Phi(\exp(tx), m))$; it is defined on the whole of $\mathcal{A}_0 \times \mathfrak{g}_0 \times M$.*

Proof. For (i) we use that Φ is a left action of G on M, i.e., $\Phi_g \circ \Phi_m = \Phi_m \circ L_g$ and $\Phi_m \circ R_g = \Phi_{\Phi(g,m)}$. This allows us to make the following computation: $T\Phi_g(T\Phi_m\vec{x}_e) = T\Phi_m(TL_g\vec{x}_e) = T\Phi_m(TR_g(\mathrm{Ad}(g)x)) = T\Phi_{\Phi(g,m)}(\mathrm{Ad}(g)x)$. For (ii) the computations are similar. Property (iii) is a direct consequence of (ii), [1.10], and [V.2.29]. For (iv) we note that, by (ii), the vector field Z_M on $\mathfrak{g}_0 \times M$ is related to the vector field Z_G on $\mathfrak{g}_0 \times G \times M$ defined as $Z_G(x, g, m) = (\underline{0}_x, x^G|_g, \underline{0}_m)$ by the map $id(\mathfrak{g}_0) \times \Phi$. Since the flow of Z_G is given by $(x, g, m) \mapsto (x, \exp(-tx)g, m)$ [2.4-v], the result follows from [V.5.5], taking $g = e$. QED

5.3 Corollary. *For $x \in \mathbf{B}\mathfrak{g}_0$ the flow of x^M on M is given by $(t, m) \mapsto \Phi(\exp(tx), m)$.*

5.4 Discussion. If $\Phi : M \times G \to M$ is a right action, the definition of a fundamental vector field becomes

$$x^M|_m = T\Phi_m(\vec{x}_e) = T_{(m,e)}\Phi(\underline{0}_m, x) ,$$

without the minus sign. For the natural right action $m : G \times G \to G$ of G on itself by multiplication, the fundamental vector field x^G is exactly the left-invariant vector field $\vec{x}$. With this notion of fundamental vector field, the results [5.2-ii] and [5.2-iii] remain unchanged; [5.2-i] changes to $T\Phi_g x^M|_m = (\mathrm{Ad}(g^{-1})x)^M|_{\Phi(m,g)}$ and [5.2-iv] changes to $\phi(t, x, m) = (x, \Phi(m, \exp(tx)))$.

The reason to introduce the minus sign for fundamental vector field associated to left actions is twofold. In the first place, it makes that [5.2-iii] comes out without a minus sign, making the map $x \mapsto x^M$ from the $\mathcal{A}$-Lie algebra $\mathfrak{g}$ to the set of vector fields on M a morphism of $\mathcal{A}$-Lie algebras. In the second place, if $\Phi : G \times M \to M$ is a left action of G on M, the map $\Psi : M \times G \to M$ defined by $\Psi(m, g) = \Phi(g^{-1}, m)$ is a right action of G on M. The use of the minus sign for fundamental vector fields associated to left actions (and its absence for right actions) makes that the two fundamental vector fields corresponding to these left and right actions are the same (this follows from [1.10]).

5.5 Definition/Discussion. Let G be an $\mathcal{A}$-Lie group and (i, H) an $\mathcal{A}$-Lie subgroup. H will be called a *proper $\mathcal{A}$-Lie subgroup* if $i : H \to G$ is an embedding, i.e., if $i(H)$ is a submanifold of G. According to [V.2.12], an $\mathcal{A}$-Lie subgroup (i, H) is proper if and only if $i : \mathbf{B}H \to \mathbf{B}G$ is an embedding of $\mathbf{R}$-manifolds. Moreover, for $\mathbf{R}$-Lie groups an $\mathbf{R}$-Lie subgroup is embedded if and only if its image is closed [Wa, Thm 3.21]. We conclude that (i, H) is a proper $\mathcal{A}$-Lie subgroup of G if and only if $i(\mathbf{B}H)$ is closed in $\mathbf{B}G$. On the other hand, for $\mathbf{R}$-Lie groups any closed abstract subgroup is a Lie subgroup [Wa, Thm 3.42]. A similar statement does not hold for $\mathcal{A}$-Lie groups because the topology ignores the odd coordinates completely. [5.8] gives a characterization of proper $\mathcal{A}$-Lie subgroups in the context of $\mathcal{A}$-Lie groups.

5.6 Example. Let $G = (\mathcal{A}_1)^2$ be the abelian additive group of dimension $0|2$. The $\mathcal{A}$-Lie subgroup H defined as $\{(\xi, 0) \mid \xi \in \mathcal{A}_1\}$ is a proper $\mathcal{A}$-Lie subgroup, but it is neither closed nor open in G.

5.7 Discussion. In the sequel of this section we will often introduce a variant of the exponential function associated to a graded subspace of the $\mathcal{A}$-Lie algebra. The idea is the following. Let G be an $\mathcal{A}$-Lie group, $\mathfrak{g}$ its $\mathcal{A}$-Lie algebra, and let $\mathfrak{h} \subset \mathfrak{g}$ be a graded subspace. Since $\mathfrak{h}$ is a graded subspace, there exists a supplement $\mathfrak{s} \subset \mathfrak{g}$ for $\mathfrak{h}$ (see [II.6.23]). Using the decomposition $\mathfrak{g} = \mathfrak{s} \oplus \mathfrak{h}$ we define the map $\widehat{\exp} : \mathfrak{g}_0 \to G$ by $\widehat{\exp}(s, z) = \exp(s)\exp(z)$ for $s \in \mathfrak{s}_0$, $z \in \mathfrak{h}_0$. By [2.4-iv] $T_{(0,0)}\,\widehat{\exp} = id$ and hence $\widehat{\exp}$ is a diffeomorphism from a neighborhood of $(0, 0) \in \mathfrak{s}_0 \oplus \mathfrak{h}_0 = \mathfrak{g}_0$ to a neighborhood of $e \in G$. Each time we need this variant of the exponential map, we will give the definition adapted to the circumstances, but we will no longer justify the existence of the supplement, nor will we justify the fact that it is a diffeomorphism in a neighborhood of $(0, 0)$. And obviously we will never say explicitly that $\widehat{\exp}|_{\mathfrak{s}_0} = \exp|_{\mathfrak{s}_0}$ or that $\widehat{\exp}|_{\mathfrak{h}_0} = \exp|_{\mathfrak{h}_0}$

5.8 Lemma. *Let G be an $\mathcal{A}$-Lie group.*

(i) *Let $H \subset G$ be an abstract subgroup of G, let $\mathfrak{h} \subset \mathfrak{g}$ be a graded subspace, let $\mathfrak{s}$ be a supplement to $\mathfrak{h}$, and let $\widehat{\exp}$ be as in [5.7]. Suppose there exists an open neighborhood U of $0 \in \mathfrak{g}_0$ such that $\widehat{\exp} : U \to V = \widehat{\exp}(U)$ is a diffeomorphism and such that* $\exp(U \cap \mathfrak{h}_0) = V \cap H$. *If in addition $\mathbf{B}H \subset H$,* *then H is a submanifold of G and the canonical embedding $i : H \to G$ turns (i, H) into a proper $\mathcal{A}$-Lie subgroup of G with $\mathcal{A}$-Lie algebra (isomorphic to) $\mathfrak{h}$.*

(ii) *Let (i, H) be a proper $\mathcal{A}$-Lie subgroup with $\mathcal{A}$-Lie algebra $\mathfrak{h}$, let $\mathfrak{s}$ be a supplement to $\mathring{\mathfrak{h}} = T_e i(\mathfrak{h}) \subset \mathfrak{g}$, and let the map $\widehat{\exp} : \mathring{\mathfrak{h}}_0 \oplus \mathfrak{s}_0 \to G$ be defined as in [5.7]: $\widehat{\exp}(s, z) = \exp(s)\exp(z)$. Then there exists an open neighborhood U of $0 \in \mathfrak{g}_0$ such that $\widehat{\exp} : U \to V = \exp(U)$ is a diffeomorphism and such that $\exp(U \cap \mathring{\mathfrak{h}}_0)) = V \cap i(H)$.*

Proof. • Let $h_1 \in H$ be arbitrary, and denote $h_o = \mathbf{B}h_1 \in H$. Since h_o has real coordinates, L_{h_o} is a diffeomorphism, and thus $W = L_{h_o}(V)$ is an open neighborhood of $h_o = \mathbf{B}h_1$, and thus of h_1. We claim that $\phi : W \to U \subset \mathfrak{g}_0$ defined by $\phi = \widehat{\exp}^{-1} \circ L_{h_o}^{-1}$ is a chart satisfying $\phi(W \cap H) = U \cap \mathfrak{h}_0$. Suppose $x \in \phi(W \cap H)$, then $\widehat{\exp}(x) = h_o^{-1} \cdot h$ for some $h \in W \cap H$. But then $\widehat{\exp}(x) \in V \cap H = \widehat{\exp}(U \cap \mathfrak{h}_0)$ and thus $x \in \mathfrak{h}_0$. This proves $\phi(W \cap H) \subset U \cap \mathfrak{h}_0$; the other inclusion being obvious, it follows from [III.5.1] that H is a submanifold of G. Hence the restriction of the multiplication m_G to H is smooth, making H into an $\mathcal{A}$-Lie group.

• Since $i : H \to G$ is an embedding, $i(H)$ is a submanifold of G. In particular there exists a chart $\phi : V' \to W \subset \mathfrak{g}_0$ around $e \in G$ (G is modeled on $\mathfrak{g}$) and a graded subspace F of $\mathfrak{g}$ such that $\phi(V' \cap i(H)) = W \cap F_0$ [III.5.1]. According to [V.2.16] the graded subspace F must be isomorphic to $\mathfrak{h}$. By taking a smaller V' and W if necessary we may assume that there exists an open neighborhood U' of $0 \in \mathfrak{g}_0$ such that $\widehat{\exp} : U' \to V'$ is a diffeomorphism. Since i is a homomorphism, we deduce that $\exp(U' \cap \mathring{\mathfrak{h}}_0) \subset V' \cap i(H)$ [2.16], and thus $(\phi \circ \widehat{\exp})(U' \cap \mathring{\mathfrak{h}}_0) \subset W \cap F_0$. Since $\phi \circ \widehat{\exp}$ is a smooth injective map and since F and $\mathring{\mathfrak{h}}$ are isomorphic, it follows from [V.2.14] that $\phi(\widehat{\exp}(U' \cap \mathring{\mathfrak{h}}_0))$ is open in $W \cap F_0$. Since $W \cap F_0$ has the induced topology from W and since ϕ is a diffeomorphism, there exists an open $V \subset V'$ such that $\phi(\exp(U' \cap \mathring{\mathfrak{h}}_0)) = \phi(V) \cap F_0 = \phi(V \cap i(H))$. Taking $U = \widehat{\exp}^{-1}(V) \subset U'$ we find $\phi(\exp(U' \cap \mathring{\mathfrak{h}}_0)) = \phi(\widehat{\exp}(U) \cap i(H))$ and in particular $U' \cap \mathring{\mathfrak{h}}_0 \subset U$, i.e., $U' \cap \mathring{\mathfrak{h}}_0 = U \cap \mathring{\mathfrak{h}}_0$. Hence $\exp(U \cap \mathring{\mathfrak{h}}_0) = V \cap i(H)$. QED

5.9 Theorem. *Let G be an $\mathcal{A}$-Lie group and (i, H) a proper $\mathcal{A}$-Lie subgroup. Then:*

(i) *the coset space G/H admits the structure of an $\mathcal{A}$-manifold modeled on $\mathfrak{g}/\mathring{\mathfrak{h}}$ such that the canonical left action $\Phi : G \times G/H \to G/H$ is smooth;*

(ii) *$\pi : G \to G/H$ with the natural right action of H on G becomes a principal fiber bundle*

with structure group H;

(iii) *if H is also normal, then G/H is an $\mathcal{A}$-Lie group with $\mathcal{A}$-Lie algebra $\mathfrak{g}/\mathring{\mathfrak{h}}$ and $\pi : G \to G/H$ is a morphism of $\mathcal{A}$-Lie groups.*

Proof. • As a topological space we equip G/H with the quotient topology. By surjectivity of π, any subset O of G/H is of the form $\pi(V)$ for some subset V of G. According to the quotient topology, a set $\pi(V)$ is open if and only if $\pi^{-1}(\pi(V)) = V \cdot H$ is open in G. Since right translations are homeomorphisms [1.3], $V \cdot H = \cup_{h \in H} R_h(V)$ is open whenever V is open in G, i.e., we have proven that the canonical projection $\pi : G \to G/H$ is an open map.

Let $\mathfrak{s}$ be a supplement to $\mathfrak{h}^\bullet$ and let $\widehat{\exp} : \mathfrak{s}_0 \oplus \mathfrak{h}^\bullet_0 \to G$, $\widehat{\exp}(x,y) = \exp(x)\exp(y)$ be as in [5.7]. Now let U and V be as in [5.8-ii] and denote by $\pi_{\mathfrak{s}} : \mathfrak{g} \to \mathfrak{s}$ the associated projection. Our first goal is to find an open neighborhood $0 \in U_o \subset \mathfrak{s}_0$ such that the map $\chi : U_o \times H \to G$, $\chi(s,h) = \exp(s) \cdot i(h)$ is a diffeomorphism onto its (open) image. Therefore we consider the smooth map $m_{\mathfrak{s}} : \mathfrak{s}_0 \times \mathfrak{s}_0 \to G$ defined by $m_{\mathfrak{s}}(s,z) = \exp(-s)\exp(z)$. Let W be an open neighborhood of $(0,0) \in \mathfrak{s}_0 \times \mathfrak{s}_0$ such that $m_{\mathfrak{s}}(W) \subset V$, and define $\omega : W \to \mathfrak{s}_0$ by $\omega = \pi_{\mathfrak{s}} \circ \widehat{\exp}^{-1} \circ m_{\mathfrak{s}}$. Since $\omega(s=0,z) = z$, $(\partial\omega/\partial y)(s=0, z=0) = id$ and hence by the implicit function theorem [III.3.27] there exists a (local) function $f : \mathfrak{s}_0 \to \mathfrak{s}_0$ such that $\omega(s,z) = 0 = \omega(0,0)$ is equivalent to $z = f(s)$. But since we know that $z = s$ is such a local function, we conclude that there exists an open neighborhood U_o of $0 \in \mathfrak{s}_0$ with $U_o \times U_o \subset W$ such that $\forall s, z \in U_o \subset \mathfrak{s}_0 :$ $\omega(s,z) = 0 \iff s = z$. Since $m_{\mathfrak{s}}(W) \subset V$, $\widehat{\exp}^{-1}(m_{\mathfrak{s}}(W)) \subset U$. Hence on $U_o \times U_o$ the equation $\omega(s,z) = 0$ is equivalent to $\widehat{\exp}^{-1}(m_{\mathfrak{s}}(s,z)) \in U \cap \mathfrak{h}^\bullet_0$. By definition of $\widehat{\exp}$ this is equivalent to $\exp(-s)\exp(z) \in \exp(U \cap \mathfrak{h}^\bullet_0) = V \cap i(H)$. Hence we have proven that

$$\forall s, s' \in U_o \subset \mathfrak{s}_0, \forall h, h' \in H : \exp(s) \cdot i(h) = \exp(s') \cdot i(h') \iff s = s' . \tag{5.10}$$

By taking a smaller U_o if necessary, we may assume that $U_o \subset U \cap \mathfrak{s}_0$. It is this U_o that we will use.

In order to prove that χ is a diffeomorphism from $U_o \times H$ onto its (open) image, we will use [V.2.14]. We first compute $T_{(s,x)}\,\widehat{\exp}$ for $(s,x) \in U \subset \mathfrak{s}_0 \oplus \mathfrak{h}^\bullet_0$:

$$\begin{aligned} T_{(s,x)}\,\widehat{\exp}(y,z) &= TR_{\exp x} T_s \exp y + TL_{\exp s} T_x \exp z \\ &= TR_{\exp x}(T_s \exp y + TL_{\exp s}\,\mathrm{Ad}(\exp x) M(x,z)) . \end{aligned}$$

Since (i,H) is a normal $\mathcal{A}$-Lie subgroup and since $x \in \mathfrak{h}^\bullet_0$, $z \in \mathfrak{h}^\bullet$, it follows from [4.13] and [3.15] that $\mathrm{Ad}(\exp x)M(x,z)$ remains in $\mathfrak{h}^\bullet$. Bijectivity of $T_{(s,x)}\,\widehat{\exp}$ then shows that $(T_s \exp)(\mathfrak{s})$ is a supplement for $(TL_{\exp(s)})(\mathfrak{h}^\bullet)$. We next compute $T_{(s,h)}\chi$ for $(s,h) \in U_o \times H$:

$$T_{(s,h)}\chi(y,z) = TR_{i(h)}(T_s \exp y + TL_{\exp s}\,\mathrm{Ad}(i(h)) T_e i(TL_{h^{-1}} z)) .$$

Since $\mathrm{Ad}(i(h)) T_e i(TL_{h^{-1}} z)$ is in $\mathfrak{h}^\bullet$, we conclude from the bijectivity of $T_{(s,x)}\,\widehat{\exp}$ that $T_{(s,h)}\chi$ is bijective. Since we have proven injectivity of χ in (5.10), we conclude by [V.2.14] that χ is a diffeomorphism from $U_o \times H$ onto its open image.

To finish our preparations, we define for an arbitrary $g_o \in \mathbf{B}G$ the smooth map $\chi_{g_o} : U_o \times H \to G$ by $\chi_{g_o} = L_{g_o} \circ \chi$ (and thus $\chi \equiv \chi_e$). Since for such g_o the map L_{g_o} is a diffeomorphism of G, it follows that χ_{g_o} is a diffeomorphism from $U_o \times H$ onto $V_{g_o} = \chi_{g_o}(U_o \times H)$, which is open in G.

We have now all ingredients needed to start the construction of the $\mathcal{A}$-manifold structure on G/H. Since π is an open map, $U_{g_o} = \pi(V_{g_o}) \subset G/H$ is open. Moreover, the map $\widehat{\chi}_{g_o} : U_o \to U_{g_o}$ defined as $\widehat{\chi}_{g_o}(x) = \pi(\chi_{g_o}(x,e))$ is a homeomorphism. Bijectivity is immediate, $U' \subset U_o$ is open if and only if $U' \times H$ is open in $U_o \times H$, and (to finish) $\chi_{g_o}(U' \times H) = \pi^{-1}(\widehat{\chi}_{g_o}(U'))$. We conclude that the map $\phi_{g_o} = \widehat{\chi}_{g_o}^{-1} : U_{g_o} \to U_o$ is a chart for G/H. Moreover, denoting by $\pi_1 : U_o \times H \to U_o$ the canonical projection, one verifies immediately that

$$\text{(5.11)}\qquad \begin{array}{ccc} U_o \times H & \underset{\chi_{g_o}}{\overset{\chi_{g_o}^{-1}}{\rightleftarrows}} & V_{g_o} \\ {\scriptstyle \pi_1}\big\downarrow & & \big\downarrow{\scriptstyle \pi} \\ U_o & \underset{\phi_{g_o}}{\overset{\widehat{\chi}_{g_o}}{\rightleftarrows}} & U_{g_o} \end{array}$$

is a commutative diagram, proving that π is a smooth map on these charts. To compute a change of charts, let (U_{g_o}, ϕ_{g_o}) and (U_{g_1}, ϕ_{g_1}) be two charts. Since ϕ_{g_o} is a homeomorphism, there exists an open $U_{g_o g_1} \subset U_{g_o}$ such that $\phi_{g_o}(U_{g_o g_1}) = U_{g_o} \cap U_{g_1}$. Since $(\phi_{g_1} \circ \phi_{g_o}^{-1})(x) = (\pi_1 \circ \chi_{g_1}^{-1} \circ \chi_{g_o})(x, e)$ is clearly a smooth map, we find that the charts (U_{g_o}, ϕ_{g_o}) and (U_{g_1}, ϕ_{g_1}) are compatible. Since they cover G/H we conclude that G/H is a proto $\mathcal{A}$-manifold. Since U_o is an open subset of $\mathfrak{s}_0$, which is isomorphic to $\mathfrak{g}/\mathfrak{h}^\bullet$ as an $\mathcal{A}$-vector space, this proto $\mathcal{A}$-manifold is modeled on $\mathfrak{g}/\mathfrak{h}^\bullet$.

Remains to prove that $\mathbf{B}(G/H)$ is a second countable Hausdorff space. Since all maps are smooth, they commute with the body map, hence $\mathbf{B}(G/H) = (\mathbf{B}G)/(\mathbf{B}H)$. Since G and H are $\mathcal{A}$-manifolds, $\mathbf{B}G$ and $\mathbf{B}H$ are second countable Hausdorff spaces. By definition of the quotient topology $(\mathbf{B}G/\mathbf{B}H)$ is also second countable. Consider $\mathbf{B}G \times \mathbf{B}G$ with the subset $R = \widehat{m}^{-1}(\mathbf{B}H)$ where $\widehat{m} : \mathbf{B}G \times \mathbf{B}G \to \mathbf{B}G$ is the smooth map $(g, h) \mapsto g^{-1} \cdot h$. Since $\mathbf{B}H$ is closed in $\mathbf{B}G$ [5.5], R is closed in $G \times G$. Moreover, by definition of cosets, $\pi(g) \neq \pi(h)$ if and only if $(g, h) \notin R$. Hence if $\pi(g)$ and $\pi(h)$ are distinct points in $\mathbf{B}G/\mathbf{B}H$, there exist open sets U, V in $\mathbf{B}G$ such that $(g, h) \in U \times V \subset (G \times G) \setminus R$. It follows that $\pi(U) \cap \pi(V) = \phi$. Since π is an open map, it follows that $\mathbf{B}G/\mathbf{B}H$ is Hausdorff.

The last item of (i) that remains to be proven is that the canonical left action of G on G/H is smooth. Set theoretically this action $\Phi : G \times G/H \to G/H$ is defined as $\Phi(g, m) = \pi(g \cdot \pi^{-1}(m))$. It follows that on the local chart U_{g_o} we can write $\Phi(g, m) = \pi(g \cdot \chi_{g_o}(\phi_{g_o}(m), e))$. Since the right hand side is composed of smooth functions, we deduce that Φ is smooth.

• To prove (ii), we first note that $\pi^{-1}(U_{g_o}) = V_{g_o}$ by definition of χ_{g_o}. We then define the map $\psi_{g_o} : \pi^{-1}(U_{g_o}) = V_{g_o} \to U_{g_o} \times H$ by the equation $\psi_{g_o}(g) = (m, h) \iff g = \chi_{g_o}(\phi_{g_o}(m), h)$. It follows from (5.11) that ψ_{g_o} is a trivializing chart for π. Moreover, an elementary computation shows that it is compatible with the right actions of H on G and $U_{g_o} \times H$, i.e., $\psi_{g_o}(g \cdot h) = \psi_{g_o}(g) \cdot h$, with $(m, h') \cdot h = (m, h' \cdot h)$. Finally, for two trivializing charts one easily finds, using (5.11), that we have the equality $(\psi_{g_1} \circ \psi_{g_o}^{-1})(m, h) = (m, \widetilde{\psi}_{g_1 g_o}(m) \cdot h)$, where the smooth map $\widetilde{\psi}_{g_1 g_o} : U_{g_1} \cap U_{g_o} \to H$ is defined by the equation $(\psi_{g_1} \circ \psi_{g_o}^{-1})(m, e) = (m, \widetilde{\psi}_{g_1 g_o}(m))$. We conclude that the map $\pi : G \to G/H$ is a principal fiber bundle with structure group H acting in the natural way on the right on G.

• To prove that G/H is an $\mathcal{A}$-Lie group it suffices to show that the multiplication is smooth. To that end, let $s_i : U_i \to G$, $i = 1, 2$ be two local smooth sections of the principal bundle $\pi : G \to G/H$. The definition of the group structure on G/H shows that for $x_i \in U_i$ the multiplication is defined as

$$m_{G/H}(x_1, x_2) = \pi(s_1(x_1), s_2(x_2)) .$$

It follows immediately that the multiplication $m_{G/H}$ is smooth on $U_1 \times U_2$. Since the domains of such local sections cover G/H, the multiplication is globally smooth.

To prove that the $\mathcal{A}$-Lie algebra of G/H is isomorphic to $\mathfrak{g}/\mathfrak{h}^\bullet$, we first note that π is a morphism of $\mathcal{A}$-Lie groups and hence that $T_e\pi$ is a morphism of $\mathcal{A}$-Lie algebras. Since $\pi : G \to G/H$ is a fiber bundle, $T_e\pi$ is surjective. Since $\pi(\exp_G(x)) = \exp_{G/H}(T_e\pi(x))$, it follows that $\mathfrak{h}^\bullet \subset \ker T_e\pi$. A simple dimension argument then shows that $\mathfrak{h}^\bullet = \ker T_e\pi$. It then follows from [II.3.11] and [II.6.24] that the $\mathcal{A}$-Lie algebra of G/H is isomorphic to $\mathfrak{g}/\mathfrak{h}^\bullet$ as an $\mathcal{A}$-vector space (which we already knew), but because $T_e\pi$ is a morphism of $\mathcal{A}$-Lie algebras, this isomorphism is also an isomorphism of $\mathcal{A}$-Lie algebras. QED

5.12 Definition. Let $\Phi : G \times M \to M$ be a smooth left action of an $\mathcal{A}$-Lie group G on an $\mathcal{A}$-manifold M, and let $m_o \in M$ be arbitrary. Then $H = \{\, g \in G \mid \Phi(g, m_o) = m_o \,\}$ is called *the isotropy subgroup at* m_o.

5.13 Proposition. *Let $\Phi : G \times M \to M$ be a smooth left action of an $\mathcal{A}$-Lie group G on an $\mathcal{A}$-manifold M, and let $m_o \in \mathbf{B}M$ be arbitrary. Then the isotropy subgroup H at m_o is a submanifold of G and the canonical embedding $i : H \to G$ is a proper $\mathcal{A}$-Lie subgroup of G with $\mathcal{A}$-Lie algebra $\mathfrak{h} = \{\, x \in \mathfrak{g} \mid x^M|_{m_o} = 0 \,\}$.*

Proof. Note first that H is an abstract subgroup of G (because Φ is a left action) and that $\mathfrak{h} = \ker(T_e\psi)$, where $\psi : G \to M$ is defined as $\psi(g) = \Phi(g, m_o)$. Since ψ and $T_e\psi$ are smooth it follows that $\mathfrak{h}$ is a graded subspace of $\mathfrak{g}$. We thus can choose a supplement $\mathfrak{s} \subset \mathfrak{g}$ for $\mathfrak{h}$. The proof now proceeds in two steps. We first show that $\exp(\mathfrak{h}_0) \subset H$, and then that there exists an open neighborhood U of $0 \in \mathfrak{s}_0 \oplus \mathfrak{h}_0 = \mathfrak{g}_0$ such that $\widehat{\exp} : U \to V = \exp(U)$, $\widehat{\exp}(s, x) = \exp(s)\exp(x)$ [5.7] is a diffeomorphism and such that $\exp(U \cap \mathfrak{h}_0) = V \cap H$. Applying [5.8-i] finishes the proof.

For the first step we define the vector field Z_M on $\mathfrak{h}_0 \times M$ by $Z_M(x, m) = (\underline{0}_x, x^M|_m)$, i.e., Z_M is the collection of all fundamental vector fields on M with $x \in \mathfrak{h}_0$. We then consider the two smooth maps $f_i : \mathcal{A}_0 \times \mathfrak{h}_0 \to \mathfrak{h}_0 \times M$ defined by $f_1(t, x) = (x, m_o)$ and $f_2(t, x) = (x, \psi(\exp(-tx)))$. Obviously $Tf_1 \circ \partial_t = 0 = Z_M(x, m_o) = Z_M \circ f_1$, but also $Tf_2 \circ \partial_t = Z_M \circ f_2$ because f_2 is essentially the flow of Z_M [5.2-iv]. Since we also have $f_1(0, x) = (x, m_o) = f_2(0, x)$ we conclude by [V.5.2] that $f_1 = f_2$, i.e., $\psi(\exp(-tx)) = m_o$. We thus have proven that $\exp(\mathfrak{h}_0) \subset H$.

To prove that there exists an open neighborhood U with the desired properties, we first choose any U such that $\widehat{\exp} : U \to V = \widehat{\exp}(U)$ is a diffeomorphism. We now focus our attention on the map $\Psi : \mathfrak{s}_0 \to M$ defined by $\Psi(y) = \psi(\exp(y))$. By definition of $\mathfrak{h}$ the map $T_0\Psi$ is injective. It then follows from [III.3.30] that there exists a neighborhood $0 \in U' \subset \mathfrak{s}_0$ such that $\Psi|_{U'}$ is injective. By taking a smaller $U \subset \mathfrak{s}_0 \oplus \mathfrak{h}_0$

we may suppose that Ψ is injective on $U \cap \mathfrak{s}_0$. We claim that such an U satisfies our conditions. Obviously $\exp(U \cap \mathfrak{h}_0) \subset \widehat{\exp}(U) \cap H$ because $\exp(\mathfrak{h}_0) \subset H$. So suppose $h \in \widehat{\exp}(U) \cap H$, i.e., $h = \exp(s)\exp(x) \in H$ for some $s \in \mathfrak{s}_0$ and $x \in \mathfrak{h}_0$. But then $\exp(s) = h \cdot \exp(-x) \in H$ and thus $\Psi(s) = \Psi(0)$. Injectivity of Ψ on $U \cap \mathfrak{s}_0$ then proves $s = 0$. And thus $h = \exp(x) \in \exp(U \cap \mathfrak{h}_0)$. QED

5.14 Proposition. *Let $\Phi : G \times M$ be a smooth left action of an $\mathcal{A}$-Lie group G on an $\mathcal{A}$-manifold M, let $m_o \in \mathbf{B}M$, and let $H \subset G$ be the isotropy subgroup at m_o. Then the canonical injection $\Psi : G/H \to M$ defined by $\Psi(gH) = \Phi(g, m_o)$ is an injective immersion, equivariant with respect to the G-actions. In particular if the action of G is transitive, then $\Psi : G/H \to M$ is a diffeomorphism.*

Proof. First note that H is a proper $\mathcal{A}$-Lie subgroup of G [5.13], and thus G/H is an $\mathcal{A}$-manifold [5.9]. Since Ψ is injective by definition of H, it remains to show that Ψ is smooth and that $T\Psi$ is injective at all points. Let $\phi_{g_o} : U_{g_o} \to U_o \subset \mathfrak{s}_0$, $g_o \in \mathbf{B}G$ be a local chart as in the proof of [5.9] with $\mathfrak{s}$ a supplement to $\mathfrak{h}$ in $\mathfrak{g}$. In terms of this chart, we have $(\Psi \circ \phi_{g_o}^{-1})(x) = \Phi(g_o \cdot \exp(x), m_o) = \Phi(g_o, \Phi(\exp(x), m_o))$. This shows that Ψ is smooth. Moreover, since $\mathfrak{h} = \ker(T_{(e,m_o)}\Phi)$, since $\mathfrak{s}$ is a supplement to $\mathfrak{h}$, and since $\Phi(g_o, \_)$ is a diffeomorphism, it follows that $T_x\Psi$ is injective. Being equivariant with respect to the G-actions means that $\Psi(\widehat{g} \cdot (gH)) = \Phi(\widehat{g}, \Psi(gH))$, which is an immediate consequence of the definition of a left action. The last part of the statement follows from [V.2.14]. QED

5.15 Corollary. *The structure of G/H given in [5.9] is uniquely determined (up to diffeomorphisms) by the stated properties.*

Proof. Denote by M the set G/H equipped with some structure of an $\mathcal{A}$-manifold such that the canonical left action $\Phi : G \times M \to M$ is smooth. Then by [5.14] we obtain a diffeomorphism between G/H with the structure given in [5.9] and M. QED

5.16 Corollary. *Let $\rho : G \to H$ be a morphism of $\mathcal{A}$-Lie groups, then $\ker(\rho)$ is a proper $\mathcal{A}$-Lie subgroup of G, and the induced morphism $\hat{\rho} : G/\ker(\rho) \to H$ is an $\mathcal{A}$-Lie subgroup of H.*

Proof. If we consider the left action of G on H defined as $\Phi(g, h) = \rho(g) \cdot h$, then $\ker(\rho)$ is the isotropy subgroup at $e \in H$. Hence it is a proper $\mathcal{A}$-Lie subgroup by [5.13]. Since it is also a normal $\mathcal{A}$-Lie subgroup, we conclude by [5.9] that $G/\ker(\rho)$ is an $\mathcal{A}$-Lie group. To prove that the induced map $\hat{\rho} : G/\ker(\rho) \to H$ is an $\mathcal{A}$-Lie subgroup, it suffices to prove that $\hat{\rho}$ is smooth, because it is an injective homomorphism by construction. On a neighborhood $U \subset G/\ker(\rho)$ with a smooth section $s : U \to G$ the map $\hat{\rho}$ is defined as $\hat{\rho}(x) = \rho(s(x))$. Hence $\hat{\rho}$ is smooth on U. Since such U cover $G/\ker(\rho)$, the conclusion follows. QED

6. Pseudo effective actions

In this section we prove that every action can be transformed into a pseudo effective action: if G acts on M, there exists a proper $\mathcal{A}$-Lie subgroup $G^{\psi o}$ of G such that $G^{\psi o}$ acts as the identity on M and such that the induced action of $G/G^{\psi o}$ on M is pseudo effective.

6.1 Discussion. In chapter IV we defined fiber bundles with a structure group and we required the action of the structure group G on the typical fiber F to be pseudo effective. In the remaining part of this section we will show that we can transform any smooth (left) action into a pseudo effective action. More precisely, we will show that if $\Phi : G \times M \to M$ is a smooth left action of an $\mathcal{A}$-Lie group G on an $\mathcal{A}$-manifold M, then there exists a proper normal $\mathcal{A}$-Lie subgroup $G^{\psi o} \subset G$ such that (i) all elements of $G^{\psi o}$ act as the identity on M, and (ii) the induced action of $G/G^{\psi o}$ on M is smooth and pseudo effective.

Forgetting for the moment the smoothness conditions, the natural approach to obtain an effective action would be the following. One would first define $G^o \subset G$ as the set of all elements of G that act as the identity on M, i.e., $g \in G^o$ if and only if $\forall m \in M : \Phi(g, m) = m$. Obviously G^o is a normal abstract subgroup of G and G/G^o acts effectively on M. When one tries to prove that G^o is an $\mathcal{A}$-Lie subgroup, it is natural to think that its $\mathcal{A}$-Lie algebra $\mathfrak{g}^o$ consists of those $x \in \mathfrak{g}$ whose associated fundamental vector field x^M is identically zero. The next logical step would be to apply [5.8] to really prove what one wants. The problem with this approach is that it is hard (if possible at all) to prove that $\mathfrak{g}^o$ is a graded subspace of $\mathfrak{g}$. The definition of $\mathfrak{g}^o$ is in terms of equations, but these equations depend upon even and odd parameters (the local coordinates of M), and we do not have much control over them. Our approach will be to define subsets $G^{\psi o} \subset G^o$ and $\mathfrak{g}^{\psi o} \subset \mathfrak{g}^o$ described by smooth families in G^o and $\mathfrak{g}^o$ respectively. Then our equations are "parameterized" by smooth functions over which we have complete control.

6.2 Definitions. • Let K be an $\mathcal{A}$-manifold and $X \subset K$ a subset. We will say that $k \in K$ is *part of a smooth family in* X if there exists an $\mathcal{A}$-manifold N and a smooth map $\psi : N \to K$ such that $k \in \psi(N) \subset X$.

• Let $\Phi : G \times M \to M$ be a smooth (left) action of an $\mathcal{A}$-Lie group G on an $\mathcal{A}$-manifold M and let $\mathfrak{g}$ be the $\mathcal{A}$-Lie algebra of G. The subsets $G^{\psi o} \subset G^o \subset G$ and $\mathfrak{g}^{\psi o} \subset \mathfrak{g}^o \subset \mathfrak{g}$ are defined as :

$$\begin{aligned} G^o &= \{\, g \in G \mid g \text{ acts as the identity on } M \,\} \\ G^{\psi o} &= \{\, g \in G \mid g \text{ is part of a smooth family in } G^o \,\} \\ \mathfrak{g}^o &= \{\, x \in \mathfrak{g} \mid x^M \text{ is identically zero} \,\} \\ \mathfrak{g}^{\psi o} &= \{\, x \in \mathfrak{g} \mid x \text{ is part of a smooth family in } \mathfrak{g}^o \,\}\ . \end{aligned}$$

6.3 Remark. With hindsight we now can say that the action of G on M is pseudo effective if and only if $G^{\psi o}$ reduces to the identity element of G.

6.4 Discussion. If $g \in G$ is part of a smooth family $\psi : N \to G$ in G^o, then obviously each element $g' \in \psi(N)$ is part of a (the same!) smooth family in G^o. In other words, $\psi(N) \subset G^{\psi o}$. It follows that $G^{\psi o}$ is the union of images of smooth maps into G^o. Similarly, if $x \in \mathfrak{g}$ is part of a smooth family $\psi : N \to \mathfrak{g}$ in $\mathfrak{g}^o$, then $\psi(N) \subset \mathfrak{g}^{\psi o}$.

Since Φ is a left action, it is immediate that G^o is a normal (abstract) subgroup of G. Moreover, since multiplication and inverse are smooth operations in G, $G^{\psi o}$ is also a normal abstract subgroup. On the other hand, it is not clear whether G^o or $G^{\psi o}$ are $\mathcal{A}$-Lie subgroups of G. Similarly, linearity of the tangent map proves that $\mathfrak{g}^o$ is a graded submodule of $\mathfrak{g}$. Continuity of the module operations then shows that $\mathfrak{g}^{\psi o}$ also is a graded submodule. But again, it is not obvious whether they are subspaces, nor is it completely obvious that they are stable under the bracket operations.

6.5 Lemma. *Let E be an $\mathcal{A}$-vector space and let $\phi_1, \dots, \phi_k \in \mathbf{B}^*E$ be a finite number of smooth homogeneous left linear maps. Then $E^\phi = \{e \in E \mid \forall 1 \leq i \leq k : \langle\!\langle e \| \phi_i \rangle\!\rangle = 0\}$ is a graded subspace of E in the sense of [II.6.23].*

Proof. Let $G = \{g_1, \dots, g_k\}$ be a set of k elements with parity map $\varepsilon : G \to \mathbf{Z}_2$ given by $\varepsilon(g_i) = \varepsilon(\phi_i)$, and let $F = F(G, \varepsilon)$ be the free $\mathcal{A}$-module (i.e., an $\mathcal{A}$-vector space) on these generators. With these we define the even (left) linear map $\Phi : E \to F$ as $\langle\!\langle e \| \Phi \rangle\!\rangle = \sum_i \langle\!\langle e \| \phi_i \rangle\!\rangle g_i$. Since the ϕ_i are smooth, Φ is smooth. Moreover, $E^\phi = \ker \Phi$. The conclusion then follows from [II.6.24]. QED

6.6 Lemma. *$\mathfrak{g}^{\psi o}$ is a subalgebra of $\mathfrak{g}$.*

Proof. Let M be modeled on the $\mathcal{A}$-vector space of dimension $p|q$ and let $\varphi : W \to O$ be a local chart for M with coordinates $(x^1, \dots, x^p, \xi^1, \dots, \xi^q)$. Let $e_1, \dots, e_d$ be a homogeneous basis of $\mathfrak{g}$ and let $\psi : N \to \mathfrak{g}$ be a smooth family in $\mathfrak{g}^o$. This implies in particular that $\psi(N) \subset \mathfrak{g}^{\psi o} \subset \mathfrak{g}^o$. On the local chart $W \subset M$, the fundamental vector fields e_i^M can be written as $e_i^M|_m = \sum_{j=1}^{p+q} E_i^j(m)\partial_j$ with smooth functions $E_i^j : W \to \mathcal{A}$. Since ψ is smooth, there exist smooth functions $\psi^i : N \to \mathcal{A}$ such that $\psi(n) = \sum_{i=1}^d \psi^i(n) e_i$. In the local chart W, the fundamental vector field $\psi(n)^M$ thus can be written as

$$\psi(n)^M|_m = \sum_{i,j} \psi^i(n) E_i^j(m) \partial_j \ .$$

The fact that $\psi(n)$ belongs to $\mathfrak{g}^o$ is equivalent to the fact that for all $j = 1, \dots, p+q$ we must have $\sum_i \psi^i(n) E_i^j(m) = 0$ as smooth functions of $(m, n) \in W \times N$. Using local coordinates (y, η) on N and using the notational shorthand of [III.3.17], these equations can be written as $\sum_{i,I,J} \psi_I^i(y)\eta^I E_{i,J}^j(x)\xi^J = 0$. Since we require $\mathcal{A}$ to satisfy (C[∞]), we can apply [III.2.21] or [III.2.22] to conclude that these equations are equivalent to the system of equations

$$\textstyle\sum_i \psi_I^i(y) E_{i,J}^j(x) = 0 \ ,$$

with x and y taking real values.

This suggests that we introduce the functions $\phi^j_J(x) = \sum_i {}^ie \cdot E^j_{i,J}(x) \in \mathbf{B}^*\mathfrak{g}$, where $({}^ie)$ denotes the left dual basis. Since the parity of the fundamental vector field e_i^M is the parity of e_i, it follows that the parity of the function $E^j_{i,J}$ equals $\varepsilon(e_i) + \varepsilon(\partial_j) + \varepsilon(\xi^J)$. Hence ϕ^j_J is a homogeneous function of parity $\varepsilon(\partial_j) + \varepsilon(\xi^J)$. Moreover, if $(\widehat{e}_i)$ is another basis of $\mathfrak{g}$ related to (e_i) by $\widehat{e}_i = \sum_j a_i^j e_j$, then the left dual basis is given by $\sum_i {}^i\widehat{e}\, a_i^j = {}^je$. It follows that the functions $\phi^j_J(x)$ are independent of the choice of a basis for $\mathfrak{g}$.

Our analysis shows that if $\psi(N) \subset \mathfrak{g}^{\psi o}$, then $\langle\!\langle \psi(n) \,\|\, \phi^j_J(x) \rangle\!\rangle = 0$ for all (j, J, x) (with of course the coordinates x only taking real values). On the other hand, if $\psi : N \to \mathfrak{g}$ is a smooth map such that for all (j, J, x) we have $\langle\!\langle \psi(n) \,\|\, \phi^j_J(x) \rangle\!\rangle = 0$, then the same analysis shows that the fundamental vector field $\psi(n)^M$ is identically zero on the chart W. It follows that $\mathfrak{g}^{\psi o}$ is described by the system of equations $\langle\!\langle z \,\|\, \phi^j_J(x) \rangle\!\rangle = 0$, where we also have to vary the chart W in order to cover the whole of M. The set of all $\phi^j_J(x)$ forms a subset of the d-dimensional vector space $\mathbf{B}^*\mathfrak{g}$ over $\mathbf{R}$. We thus can find $k \le d$ independent elements $\phi^{j_1}_{J_1}(x_1), \ldots, \phi^{j_k}_{J_k}(x_k)$, with x_r the (real) coordinates of a point $m_r \in \mathbf{B}W_r$ for local charts W_r of M, such that all other elements $\phi^j_J(x)$ are linear combinations with real coefficients of these k elements. It follows that $\mathfrak{g}^{\psi o}$ is described by the k equations $\langle\!\langle z \,\|\, \phi^{j_r}_{J_r}(x_r) \rangle\!\rangle = 0$, $1 \le r \le k$. We conclude by [6.5] that $\mathfrak{g}^{\psi o}$ is a graded subspace of $\mathfrak{g}$.

To show that $\mathfrak{g}^{\psi o}$ is a subalgebra, we argue as follows. Since $\mathfrak{g}^{\psi o}$ is a subspace, it has a basis $f_1, \ldots, f_k \in \mathbf{B}\mathfrak{g}$. Applying [5.2-iii] we conclude that $[f_i, f_j]^M \in \mathfrak{g}^o$. We now consider the map $\psi : \mathfrak{g}^{\psi o} \times \mathfrak{g}^{\psi o} \to \mathfrak{g}$ defined by $\psi(z_1, z_2) = [z_1, z_2]$. Since the bracket is bilinear, ψ takes values in $\mathfrak{g}^o$. Since the bracket is smooth, ψ is smooth. Hence ψ takes values in $\mathfrak{g}^{\psi o}$, i.e., $\mathfrak{g}^{\psi o}$ is a subalgebra. QED

6.7 Proposition. *$G^{\psi o}$ is a proper normal $\mathcal{A}$-Lie subgroup of G with $\mathcal{A}$-Lie algebra $\mathfrak{g}^{\psi o}$. Furthermore, the induced left action of $G/G^{\psi o}$ on M is smooth and pseudo effective.*

Proof. • Let us start with the hardest part: the proof that $G^{\psi o}$ is a proper $\mathcal{A}$-Lie subgroup with $\mathcal{A}$-Lie algebra $\mathfrak{g}^{\psi o}$, for which we will use [5.8]. We first introduce the vector fields Z_M and Z_G on $\mathfrak{g}_0^{\psi o} \times M$ and $\mathfrak{g}_0^{\psi o} \times G \times M$ respectively, defined as $Z_M(x, m) = (\underline{0}_x, x^M|_m)$ and $Z_G(x, g, m) = (\underline{0}_x, x^G|_g, \underline{0}_m)$ (see the proof of [5.2-iv]). These two vector fields are related by the map $id(\mathfrak{g}_0^{\psi o}) \times \Phi$. But the flow of Z_G is given by the expression $(t, x, g, m) \mapsto (x, \exp(-tx)g, m)$ and the flow of Z_M by $(t, x, m) \mapsto (x, m)$ because for $x \in \mathfrak{g}_0^{\psi o}$ we have $x^M \equiv 0$. We conclude by [V.5.5] that $\exp(\mathfrak{g}_0^{\psi o})$ is contained in G^o. Since exp is smooth, this means that $\exp(\mathfrak{g}_0^{\psi o})$ is contained in $G^{\psi o}$.

If $g \in G^{\psi o}$, there exists a smooth $\psi : N \to G^{\psi o}$ with $g \in \psi(N)$. Applying the body map we find $\mathbf{B}g \in \psi(\mathbf{B}N)$, i.e., $\mathbf{B}g$ also belongs to $G^{\psi o}$. In other words, $\mathbf{B}G^{\psi o} \subset G^{\psi o}$.

Now let $\mathfrak{s}$ be a supplement to $\mathfrak{g}^{\psi o}$. Using the decomposition $\mathfrak{g} = \mathfrak{s} \oplus \mathfrak{g}^{\psi o}$, we define as usual the map $\widehat{\exp} : \mathfrak{g}_0 \to G$ by $\widehat{\exp}(s, z) = \exp(s)\exp(z)$ for $s \in \mathfrak{s}_0$ and $z \in \mathfrak{g}_0^{\psi o}$ [5.7]. To find a neighborhood V of $e \in G$ and a neighborhood U of $0 \in \mathfrak{g}_0$ as required by [5.8], we proceed as follows. In the proof of [6.6] we found k independent elements $\phi^{j_r}_{J_r}(x_r)$, $1 \le r \le k$ in $\mathbf{B}^*\mathfrak{g}$ associated to coordinate charts W_r of M, points $m_r \in \mathbf{B}W_r$ with (real) coordinates x_r, and indices j_r (for a generic coordinate on W_r) and J_r (a multi-index for a product of odd coordinates on W_r). Continuity of Φ implies that there exists neighborhoods $W_r' \subset W_r$ of m_r and a coordinate neighborhood V of $e \in G$ such that $\forall r = 1, \ldots, k : \Phi(V, W_r') \subset W_r$. We thus can use the coordinates on W_r also for points in the image $\Phi(V, W_r')$. Let $(g^i)_{i=1}^d$ denote coordinates on V and let $(m^i)_{i=1}^{p+q}$ denote coordinates on W_r. If we wish to distinguish even and odd coordinates of a point $m \in W_r$, we will denote them as (x^i, ξ^j). It follows that the fundamental vector field associated to $\partial_{g^i}|_e \in \mathfrak{g}$ is given on W_r' as

$$(\partial_{g^i}|_e)^M|_m = -\sum\nolimits_j \frac{\partial \Phi^j}{\partial g^i}(g = e, m)\partial_{m^j} \ .$$

This means that the function $E_i^j(m)$ of the proof of [6.6] is the function $-\frac{\partial \Phi^j}{\partial g^i}(g = e, m)$. Using that $(dg^i|_e)$ is the left dual basis of $(\partial_{g^i}|_e)$ and using [III.3.7] to express the function $E_{i,J}^j$ in terms of partial derivatives of E_i^j, we can write the functions $\phi_J^j(x)$ (defined in the proof of [6.6]) as

$$\phi_J^j(x) = -\sum\nolimits_i (dg^i|_e) \cdot \Big(\frac{\partial}{\partial \xi^J}\frac{\partial}{\partial g^i}\Phi^j\Big)(g = e, x, \xi = 0) \ .$$

We then note that the functions Φ^j are defined on the whole of $V \times W_r'$ and hence that we can consider the functions $V \to \mathcal{A}$, $g \mapsto -\frac{\partial \Phi^{j_r}}{\partial \xi^{J_r}}(g, x = x_r, \xi = 0)$, $1 \le r \le k$, whose parity is given as $\varepsilon(m^{j_r}) + \varepsilon(\xi^{J_r})$. In order to combine these functions in a single even map, we introduce an $\mathcal{A}$-vector space F of total dimension k whose basis vectors f_r have parity $\varepsilon(f_r) = \varepsilon(m^{j_r}) + \varepsilon(\xi^{J_r})$. With this $\mathcal{A}$-vector space we define an even map $\chi : V \to F_0$ by

$$\chi(g) = -\sum_{r=1}^{k} \frac{\partial \Phi^{j_r}}{\partial \xi^{J_r}}(g, x = x_r, \xi = 0) \cdot f_r \ .$$

The fact that the functions $\phi_{J_r}^{j_r}(x_r)$ form k independent elements of ${}^*\mathfrak{g}$ is equivalent to saying that the Jacobian of χ at e has maximal rank k (note that $\partial_{\xi^J}\partial_{g^i}\Phi$ and $\partial_{g^i}\partial_{\xi^J}\Phi$ only differ by a sign $(-1)^{\langle \varepsilon(g^i)|\varepsilon(\xi^J)\rangle}$). Said in yet another way, using that $\mathfrak{g}^{\psi o}$ is the null space of the equations $\phi_{J_r}^{j_r}(x_r) = 0$, this is equivalent to the fact that $\mathfrak{g}^{\psi o}$ is the kernel of $T_e\chi$. It follows that the Jacobian at $0 \in \mathfrak{s}_0$ of the composite map $\mathfrak{s}_0 \to G \to F_0$, $s \mapsto \widehat{\exp}(s, 0) = \exp(s) \mapsto \chi(\exp(s))$ is invertible. Hence if we take the neighborhood V of $e \in G$ sufficiently small, we may assume that $\widehat{\exp}$ is a diffeomorphism between $U = \widehat{\exp}^{-1}(V) \subset \mathfrak{g}_0$ and V and at the same time that $\chi \circ \widehat{\exp}$ is a diffeomorphism between $U \cap \mathfrak{s}_0$ and its image in F_0.

Now let $g \in G^{\psi o} \cap V$ be arbitrary. By definition of $G^{\psi o}$ there exists a smooth map $\psi : N \to G^{\psi o}$ such that $g \in \mathrm{im}(\psi)$. By taking a smaller N we also may assume that $\mathrm{im}(\psi) \subset V$, i.e., $\mathrm{im}(\psi) \subset G^{\psi o} \cap V$. Composing ψ with χ, using that taking derivatives with respect to the ξ variables does not interfere with what happens with the g coordinates [III.3.13], and using the definition of $G^{\psi o}$ as acting as the identity on M, we deduce that $\chi(\psi(n))$ is constant equal to $\chi(e)$. Since $\psi(n)$ lies in V, there exists $(s, z) \in U \subset \mathfrak{s}_0 \times \mathfrak{g}_0^{\psi o}$ such that $\psi(n) = \widehat{\exp}(s, z)$. Since Φ is a left action and since $\exp(\mathfrak{g}^{\psi o}) \subset G^{\psi o}$ acts as the identity on M, it follows as above that $\chi(\psi(n)) = \chi(\exp(s)\exp(z)) = \chi(\exp(s))$. Hence $\chi(\exp(s)) = \chi(e)$, so $s = 0$ because $\chi \circ \exp$ is a diffeomorphism between $U \cap \mathfrak{s}_0$ and its image in F_0. This proves that $G^{\psi o} \cap V$ is included in $\exp(\mathfrak{g}_0^{\psi o} \cap U)$. Since we already know that $\exp(\mathfrak{g}_0^{\psi o} \cap U)$ is contained in $G^{\psi o} \cap V$, we can apply [5.8] to conclude that $G^{\psi o}$ is a proper $\mathcal{A}$-Lie subgroup of G with $\mathcal{A}$-Lie algebra $\mathfrak{g}^{\psi o}$.

• To show that $G^{\psi o}$ is normal, we first note that G^o is normal. We then consider the smooth map $\psi : G \times G^{\psi o} \to G^o$ defined by $\psi(g,h) = ghg^{-1}$. Since this is smooth, we conclude by definition of $G^{\psi o}$ that the image of ψ lies in $G^{\psi o}$, i.e., $G^{\psi o}$ is normal.

• The induced action Φ_{ind} of $G/G^{\psi o}$ on M is defined by $\Phi_{ind}(gG^{\psi o}, m) = \Phi(g, m)$. This is well defined because the elements of $G^{\psi o}$ all act as the identity on M. Since $\pi : G \to G/G^{\psi o}$ is a locally trivial (principal) fiber bundle [5.9], we choose a local smooth section $s : V \to G$ for some trivializing chart $V \subset G/G^{\psi o}$. It follows that the restriction $\Phi_{ind} : V \times M \to M$ is given by $\Phi_{ind}(z, m) = \Phi(s(z), m)$, which is smooth. Since being smooth is a local property, we conclude that Φ_{ind} is globally smooth.

• To prove that Φ_{ind} acts pseudo effectively, we consider an arbitrary smooth map $\psi : N \to G/G^{\psi o}$ such that all $\psi(n)$ act as the identity on M. As above we choose a local trivializing chart V and a smooth section $s : V \to G$. We then consider the smooth map $\chi : \psi^{-1}(V) \to G$ defined by $\chi(n) = s(\psi(n))$. By definition of ψ, all $\chi(n)$ act as the identity on M. By definition of $G^{\psi o}$ this means that all $\chi(n)$ lie in $G^{\psi o}$. But that means that all $\psi(n)$ are the identity element in $G/G^{\psi o}$ for $n \in \psi^{-1}(V)$. Since the local trivializing charts cover $G/G^{\psi o}$, we conclude that ψ is constant the identity element. This means that Φ_{ind} acts pseudo effectively. QED

6.8 Lemma. *Let G be an $\mathcal{A}$-Lie group and (i, H) an $\mathcal{A}$-Lie subgroup. If there exists a neighborhood V of the identity $e \in G$ such that $V \cap i(H) = \{e\}$, then $i(H)$ is a closed discrete subgroup of $\mathbf{B}G \subset G$. If in addition G is connected and H is normal, then $i(H)$ is contained in the center of G (and in particular H is abelian).*

Proof. • Since H is an $\mathcal{A}$-Lie subgroup we have $g \in i(H) \Rightarrow \mathbf{B}g \in i(H)$. Hence for any $g \in i(H)$ we have $(\mathbf{B}g)^{-1} \cdot g \in V$, and thus $g = \mathbf{B}g$, i.e., $i(H) \subset \mathbf{B}G$. Now if $g \in i(H)$, then L_g is a diffeomorphism and $V_g = L_g V$ is a neighborhood of g such that $V_g \cap i(H) = \{g\}$, i.e., $i(H)$ is a discrete subgroup.

Since inversion and multiplication are smooth operations, there exists a neighborhood $W \subset V$ of the identity such that $W^{-1} \cdot W \subset V$. If $g \in \mathbf{B}G \cap \overline{i(H)} \setminus i(H)$, then $L_g W \cap i(H) \neq \phi$, i.e., $\exists x_1 \in \mathbf{B}W \ \exists h_1 \in i(H) : gx_1 = h_1$. Since $g \notin i(H)$, $x_1 \neq e$ and thus there exists a neighborhood $U \subset W$ of the identity such that $x_1 \notin U$ (because $\mathbf{B}G$ is Hausdorff!). But then again $L_g U \cap i(H) \neq \phi$, and thus $\exists x_2 \in \mathbf{B}U \ \exists h_2 \in i(H) : gx_2 = h_2$. But then $h_1^{-1}h_2 = x_1^{-1}x_2 \in W^{-1} \cdot U \cap i(H) \subset V \cap i(H) = \{e\}$. But this contradicts $x_1 \neq x_2$ (because $x_2 \in U$ and $x_1 \notin U$). This proves that $i(H)$ is closed in $\mathbf{B}G$.

• If H is normal, we fix $h \in H$ and we consider the smooth map $f : \mathfrak{g}_0 \to G$ defined by $f(x) = \exp(x) \cdot i(h) \cdot \exp(-x)$. Since H is normal, f takes values in $i(H)$. Since f is smooth ($i(h) \in \mathbf{B}G$), there exists a neighborhood U of $0 \in \mathfrak{g}_0$ such that $f(U) \subset L_{i(h)}V$. But $i(H) \cap L_{i(h)}V = \{i(h)\}$ ($i(H)$ is discrete), hence f is constant $i(h)$ on U. If G

is connected, $\exp(U)$ generates G [2.7], and thus we have proven that for all $g \in G$ necessarily $g \cdot i(h) \cdot g^{-1} = i(h)$, i.e., that $i(H)$ is contained in the center of G QED

6.9 Remark. Unless $\mathcal{A}$ does not contain nilpotent elements, a closed subgroup of $\mathbf{B}G \subset G$ is never closed in G, simply because, by definition of the DeWitt topology, each closed set of G containing e also contains $\mathbf{B}^{-1}\{e\}$.

6.10 Corollary. *Let $\Phi : G \times M \to M$ be a smooth left action of an $\mathcal{A}$-Lie group G on an $\mathcal{A}$-manifold M and let $G^{\psi o}$ and $\mathfrak{g}^{\psi o}$ be as in [6.2].*

(i) *The action Φ is pseudo effective $\iff G^{\psi o} = \{e\} \implies \mathfrak{g}^{\psi o} = \{0\}$.*
(ii) *If $\mathfrak{g}^{\psi o} = \{0\}$, then $G^{\psi o}$ is a closed discrete subgroup of $\mathbf{B}G \subset G$ contained in the center of G.*

Proof. • If $\mathfrak{g}^{\psi o}$ were not $\{0\}$, $\exp(\mathfrak{g}_0^{\psi o})$ would contain elements different from the identity acting as the identity on M [6.7], which contradicts that the action is pseudo effective.

• Assume $\mathfrak{g}^{\psi o} = \{0\}$. According to [6.7] $G^{\psi o}$ is a proper normal $\mathcal{A}$-Lie subgroup, and thus by [5.8] there exists a neighborhood V of $e \in G$ such that $V \cap G^{\psi o} = \{\exp(0)\}$. The conclusion then follows from [6.8]. QED

7. Covering spaces and simply connected $\mathcal{A}$-Lie groups

In this short section we prove that a morphism of $\mathcal{A}$-Lie algebras determines a unique morphism of the associated $\mathcal{A}$-Lie groups, provided the source group is simply connected. We thus start with a brief review of covering spaces, universal coverings, and simply connected spaces, and we prove that the simply connected cover of an $\mathcal{A}$-Lie group has a unique structure of an $\mathcal{A}$-Lie group such that the projection is a morphism of $\mathcal{A}$-Lie groups.

7.1 Definitions. Let X and Y be a topological spaces. A continuous surjection $p : Y \to X$ is called *a (topological) covering of X* if every $x \in X$ admits an open neighborhood U such that $p^{-1}(U) = \cup_{i\in I} U_i$ such that

(i) each U_i is open in Y,
(ii) the $(U_i)_{i\in I}$ are pairwise disjoint, and
(iii) $p : U_i \to U$ is a homeomorphism.

It is easy to show that if $p : Y \to X$ is a covering of X and if $q : Z \to Y$ is a covering of Y, then $p \circ q : Z \to X$ is a covering of X.

A covering of X is called *connected* if Y is connected (and thus X has to be connected too). A connected covering $p : Y \to X$ is called *universal* if Y is connected and if for any connected covering $f : Z \to X$ there exists a covering $g : Y \to Z$ such that $p = f \circ g$. A connected space is called *simply connected* if the identity map is a universal covering. It follows immediately from the definition of a universal covering that it is unique up to homeomorphisms and that it is simply connected.

Theorem [7.2] collects the main results concerning covering spaces that we will need; its proof can be found in most textbooks on algebraic topology (e.g., [Spa]).

7.2 Theorem. *Let X be a connected, locally path connected, second countable space admitting an open cover consisting of contractible sets.*

(i) *If X and X' are simply connected, then $X \times X'$ is simply connected.*
(ii) *There exists a universal covering $p : \widetilde{X} \to X$.*
(iii) *If $p : Y \to X$ is a connected covering of X, then Y is second countable.*
(iv) *Let $p : Y \to X$ be a connected covering of X, and let $g_i : Z \to Y$ be two continuous maps such that $p \circ g_1 = p \circ g_2 : Z \to X$. If Z is connected and if there exists a point $z \in Z$ such that $g_1(z) = g_2(z)$, then $g_1 = g_2$.*
(v) *Let $p : Y \to X$ be a covering, $f : Z \to X$ a continuous map, and $(z_o, y_o) \in Z \times Y$ such that $f(z_o) = p(y_o)$. If Z is simply connected, then there exists a (unique) continuous* lift *$g : Z \to Y$ such that $f = p \circ g$ and $g(z_o) = x_o$.*

7.3 Definitions. In the context of $\mathcal{A}$-manifolds we will call a map $p : N \to M$ a *covering* if it is a topological covering [7.1] such that p is smooth and such that the restrictions $p : U_i \to U$ in condition [7.1-iii] are diffeomorphisms. These additional conditions on a covering exclude maps such as $p : \mathcal{A}_0 \to \mathcal{A}_0$, $x \mapsto x^3$, which is a smooth bijection, but not a diffeomorphism in any neighborhood of 0. In view of the inverse function theorem [III.3.23], the additional conditions on a topological covering can be rephrased as saying that p should be smooth and $T_n p$ should be a bijection for every $n \in N$, i.e., p is everywhere a local diffeomorphism. The definitions of a universal covering and of simply connected remain the same (but with the changed notion of covering).

7.4 Lemma. *Let M be a connected $\mathcal{A}$-manifold.*

(i) *If $p : N \to M$ is a connected topological covering, then N has a unique structure of an $\mathcal{A}$-manifold such that p becomes an $\mathcal{A}$-manifold covering.*
(ii) *There exists a universal $\mathcal{A}$-manifold covering $p : \widetilde{M} \to M$.*
(iii) *Let $p : N \to M$ be a covering, $f : L \to M$ a smooth map, and $(\ell_o, n_o) \in L \times N$ such that $f(\ell_o) = p(n_o)$. If L is simply connected, then there exists a unique smooth lift $g : L \to N$ such that $f = p \circ g$ and $g(\ell_o) = n_o$.*

Proof. First note that if M is a connected $\mathcal{A}$-manifold, then it is in particular a connected, locally path connected, second countable topological space with an open cover consisting of contractible sets, just by choosing contractible charts. We thus can apply [7.2].

• To prove (i), let $p : N \to M$ be a topological covering. For any $n \in N$ we choose a chart $\varphi : U \to O$ around $m = p(n)$ in M satisfying condition [7.1-ii] (this can always be done by shrinking U if necessary). Since n lies in a unique U_i, we define $\varphi \circ p : U_i \to O$ to be a chart for N. We leave it as an exercise for the reader to prove that two charts for N defined in this way are compatible. We thus have constructed an atlas for N, i.e., we have made N into a proto $\mathcal{A}$-manifold for which p satisfies the conditions of an $\mathcal{A}$-manifold covering. With this structure it is elementary to show that $\mathbf{B}p : \mathbf{B}N \to \mathbf{B}M$ is a connected covering of $\mathbf{B}M$. Since $\mathbf{B}M$ is second countable and Hausdorff, it follows easily from the definition that any covering must also be Hausdorff. We conclude from [7.2-ii] that N is an $\mathcal{A}$-manifold.

To prove uniqueness of this structure, it suffices to note that, if p is a smooth local diffeomorphism, then $\varphi \circ p : U_i \to O$ is a diffeomorphism, and hence a chart for N [III.4.20].

• Let $p : \widetilde{M} \to M$ be the topological universal covering of M. According to (i) we may assume that it is an $\mathcal{A}$-manifold covering, so it remains to prove that it has the right properties. Therefore, let $f : N \to M$ be a connected $\mathcal{A}$-manifold covering. It then is in particular a topological covering, and thus there exists a topological covering $g : \widetilde{M} \to N$ such that $p = f \circ g$. It now suffices to prove that this g is smooth. Since p and f are local diffeomorphisms and g a local homeomorphism, we can locally say that $g = f^{-1} \circ p$. But this proves that g is locally a diffeomorphism, and in particular globally smooth.

• In (ii) we have seen that the universal $\mathcal{A}$-manifold covering is the same as the universal topological covering. This means in particular that the notion of simply connected does not depend upon whether we use $\mathcal{A}$-manifold coverings or topological coverings. Applying [7.2-v] we thus find a continuous lift $g : L \to N$ with the desired properties. To prove that g is smooth, we choose $\ell \in L$ and a chart $U \ni f(\ell)$ of M satisfying the conditions of a covering. Since the U_i are disjoint, there is a unique U_i such that $g(\ell) \in U_i$. Since $p : U_i \to U$ is a diffeomorphism, it follows that g restricted to $f^{-1}(U)$ coincides with $p^{-1} \circ f$ (use [7.2-iii] if needed). Hence g is smooth in an open neighborhood of ℓ. *QED*

7.5 Discussion. It is elementary to show that if $p : N \to M$ is a covering (in the sense of $\mathcal{A}$-manifolds), then $\mathbf{B}p : \mathbf{B}N \to \mathbf{B}M$ is a covering (in the sense of $\mathbf{R}$-manifolds). On the other hand, if $q : Z \to \mathbf{B}M$ is a covering in the sense of $\mathbf{R}$-manifolds, then one can show (it is elementary but not immediate) that there exists a unique covering $p : N \to M$ such that $Z = \mathbf{B}N$ and $q = \mathbf{B}p$. In this way one obtains a bijection between coverings of M and coverings of $\mathbf{B}M$. This implies in particular that M is simply connected if and only if $\mathbf{B}M$ is simply connected.

7.6 Lemma. *Let G be a connected $\mathcal{A}$-Lie group, $p : \widetilde{G} \to G$ its universal covering, and let $\widetilde{e} \in \widetilde{G}$ be such that $p(\widetilde{e}) = e \in G$. Then $\widetilde{G}$ admits a unique structure of an $\mathcal{A}$-Lie group such that $\widetilde{e}$ is the identity element and $p : \widetilde{G} \to G$ a morphism of $\mathcal{A}$-Lie groups.*

Proof. If $f : \widetilde{G} \times \widetilde{G} \to G$ is the smooth map defined by $f(x, y) = m(p(x), p(y))$, then by [7.4] there exists a unique smooth map $\widetilde{m} : \widetilde{G} \times \widetilde{G} \to \widetilde{G}$ such that $p \circ \widetilde{m} = f$ and such that $\widetilde{m}(\widetilde{e},\widetilde{e}) = \widetilde{e}$. By construction $p(\widetilde{m}(x,y)) = m(p(x),p(y))$, i.e., p is a morphism of $\mathcal{A}$-Lie groups once we know that $\widetilde{m}$ defines a group structure. To construct the inverse, we consider the map $Inv \circ p : \widetilde{G} \to G$. By [7.4] there exists a unique smooth map $\widetilde{Inv} : \widetilde{G} \to \widetilde{G}$ such that $p \circ \widetilde{Inv} = Inv \circ p$. The map $h : \widetilde{G} \to \widetilde{G}$, $h(x) = \widetilde{m}(x, \widetilde{Inv}(x))$ is such that $h(\widetilde{e}) = \widetilde{e}$ and $p \circ h$ is constant $e \in G$. Since the constant map $\widetilde{e}$ is a lift of this map, by [7.2] we have $h(x) = \widetilde{e}$, i.e., $\widetilde{Inv}$ is indeed the inverse and $\widetilde{e}$ the identity in $(\widetilde{G}, \widetilde{m})$. To prove associativity, we consider the smooth map $a : \widetilde{G} \times \widetilde{G} \times \widetilde{G} \to G$ defined by $a(x,y,z) = p(x) \cdot p(y) \cdot p(z)$. We have two lifts of this map to $\widetilde{G}$: $\widetilde{m}(x, \widetilde{m}(y,z))$ and $\widetilde{m}(\widetilde{m}(x,y),z)$. Since both send $(\widetilde{e},\widetilde{e},\widetilde{e})$ to $\widetilde{e}$, they must be the same by [7.2-iii]. This proves that $\widetilde{m}$ is associative. The other properties of an abstract group are proved in the same way. QED

7.7 Discussion. If $p : \widetilde{G} \to G$ is the universal covering of a connected $\mathcal{A}$-Lie group G, then $\pi_1(G) \equiv \ker(p)$ is a normal (abstract) subgroup of $\widetilde{G}$. By definition of a covering, there exists a neighborhood V of $\widetilde{e} \in \widetilde{G}$ such that $\pi_1(G) \cap V = \{\widetilde{e}\}$. By [6.8] it follows that $\pi_1(G)$ is a discrete central subgroup of $\widetilde{G}$. Since $\pi_1(G)$ is the first homotopy group of G, this proves that the first homotopy group of an $\mathcal{A}$-Lie group is abelian.

7.8 Lemma. *Let $\rho : G \to H$ be an $\mathcal{A}$-Lie group homomorphism and let H be connected. Then ρ is a covering if and only if $T_e\rho : \mathfrak{g} \to \mathfrak{h}$ is a bijection.*

Proof. If ρ is a covering, there exist neighborhoods U of $e_H \in H$ and U_i of $e_G \in G$ such that $\rho : U_i \to U$ is a diffeomorphism. Hence by the inverse function theorem [III.3.23] $T_e\rho$ is a bijection.

If we assume that $T_e\rho$ is a bijection, it follows from [1.18] and [V.2.14] that ρ is everywhere a local diffeomorphism; it follows from [III.3.23] that there exist neighborhoods $\widehat{U} \ni e_H$ and $\widehat{V} \ni e_G$ such that $\rho : \widehat{V} \to \widehat{U}$ is a diffeomorphism. Since ρ is a homomorphism and H connected, it follows from [2.6] that ρ must be surjective. We now consider the (smooth) map $f : G \times G \to G$, $(g,h) \mapsto gh^{-1}$. By continuity of f there exists an open neighborhood $V \subset \widehat{V}$ of e_G such that $f(V \times V) \subset \widehat{V}$. We finally define $D = \ker(\rho)$ and $U = \rho(V) \subset \widehat{U}$. With these ingredients we can prove that ρ is a covering.

Let $h \in H$ and $g \in \rho^{-1}(h)$ be arbitrary, then the set $L_g(V)$ is an open neighborhood of g and $L_h(U)$ is an open neighborhood of h. Moreover, since ρ is a homomorphism, we have the equality $\rho = L_h \circ \rho \circ L_{g^{-1}}$. Since ρ is a homeomorphism from V to U and because $L_{g^{-1}}$ and L_h are (global) homeomorphisms, $\rho = L_h \circ \rho \circ L_{g^{-1}}$ is also a homeomorphism from $L_g(V)$ to $L_h(U)$. Since ρ is everywhere a local diffeomorphism, it is a diffeomorphism from $L_g(V)$ to $L_h(U)$. Fixing $g_o \in \rho^{-1}(h)$ we claim that the decomposition $\rho^{-1}(L_hU) = \cup_{d\in D} L_{g_od}V$ satisfies the conditions of a covering. To prove equality, we choose $g \in \rho^{-1}(L_hU)$ and then $\exists \widehat{g} \in V : \rho(g) = h \cdot \rho(\widehat{g})$, which is equivalent to $g_o^{-1} g \widehat{g}^{-1} = d \in D$. Hence $g = g_o d \widehat{g}$, i.e., $g \in L_{g_od}V$. Since obviously $\cup_{d\in D} L_{g_od}V \subset \rho^{-1}(L_hU)$, we thus have equality. To show that they are mutually disjoint, suppose $g \in L_{g_od_1}V \cap L_{g_od_2}V \iff \exists g_1, g_2 \in V : g_od_1g_1 = g = g_od_2g_2$, hence

$d_2^{-1}d_1 = g_2 g_1^{-1}$. But $f(V \times V) \subset \widehat{V}$ and thus $d_2^{-1}d_1 \in \widehat{V}$. Since $\rho(d_2^{-1}d_1) = e_H$, we must have $d_2^{-1}d_1 = e_G$ and thus $d_1 = d_2$. QED

7.9 Proposition. *Let G and H be $\mathcal{A}$-Lie groups with $\mathcal{A}$-Lie algebras $\mathfrak{g}$ and $\mathfrak{h}$ respectively and G simply connected. If $r : \mathfrak{g} \to \mathfrak{h}$ is an $\mathcal{A}$-Lie algebra morphism, there exists a unique $\mathcal{A}$-Lie group morphism $\rho : G \to H$ such that $T_e\rho = r$.*

Proof. Uniqueness follows from [2.17], so we only have to show existence. Therefore we consider the $\mathcal{A}$-Lie group $G \times H$ with its $\mathcal{A}$-Lie algebra $\mathfrak{g} \times \mathfrak{h}$ and the canonical projections $\pi_1 : G \times H \to G$ and $\pi_2 : G \times H \to H$. Inside $\mathfrak{g} \times \mathfrak{h}$ we have the subalgebra $\mathfrak{s} = \{ (x, r(x)) \mid x \in \mathfrak{g} \}$. According to [4.7] we thus have an associated connected $\mathcal{A}$-Lie subgroup $j : S \to G \times H$. Now $T_e j$ is an isomorphism from $T_e S$ to the subalgebra $\mathfrak{s}$, hence $T_e(\pi_1 \circ j) : T_e S \to T_e G$ is a bijection. Since $\pi_1 \circ j$ is an $\mathcal{A}$-Lie group morphism, it is a covering [7.8]. Since S is connected and G simply connected, $\pi_1 \circ j$ must be a diffeomorphism, i.e., an isomorphism of $\mathcal{A}$-Lie groups. We now define $\rho = (\pi_2 \circ j) \circ (\pi_1 \circ j)^{-1} : G \to H$, which obviously is a homomorphism satisfying $T_e\rho = r$. QED

8. Invariant vector fields and forms

In this section we define the notions of invariant vector field and invariant differential form on an $\mathcal{A}$-manifold on which an $\mathcal{A}$-Lie group acts smoothly. This generalizes the notion of left/right-invariant vector field on an $\mathcal{A}$-Lie group. The main results of this section are that on a connected $\mathcal{A}$-manifold a vector field is invariant if and only if it commutes with the fundamental vector fields and that a differential form is invariant if and only if the Lie derivative in the direction of the fundamental vector fields is zero. To prove these results we generalize [V.5.15] and [V.7.27], which are essentially the case of the action of a 1-dimensional $\mathcal{A}$-Lie group.

8.1 Definition. Let $\Phi : G \times M \to M$ be a smooth left action of an $\mathcal{A}$-Lie group G on an $\mathcal{A}$-manifold M and let Y be a (not necessarily smooth) vector field on M. Extending the notion of a (left/right) invariant vector field on an $\mathcal{A}$-Lie group [1.8], we will say that Y *is invariant under the G-action* if it satisfies the condition $T\Phi \circ (\underline{0} \times Y) = Y \circ \Phi$. Using the generalized tangent map $T\Phi_g$, $g \in G$ [V.3.19], we can reformulate this definition as $\forall g \in G : T\Phi_g \circ Y = Y \circ \Phi_g$, which means that for $g \in G$ and $m \in M$ we have $T\Phi_g(Y_m) = Y_{\Phi_g(m)}$. If $\Phi : M \times G \to M$ is a smooth right action, then we will say that Y is invariant under the G-action if it satisfies the condition $T\Phi \circ (Y \times \underline{0}) = Y \circ \Phi$. In terms of the generalized tangent map $T\Phi_g$ this also reads as $T\Phi_g(Y_m) = Y_{\Phi_g(m)}$, but here the map Φ_g is different from the one in the case of a left action. And of course the left/right invariant vector fields on an $\mathcal{A}$-Lie group G are special cases of these definitions when viewing the multiplication $m : G \times G \to G$ as left/right action of G on itself.

8.2 Discussion/Notation. We intend to show that for a connected $\mathcal{A}$-Lie group G with its $\mathcal{A}$-Lie algebra $\mathfrak{g}$, the vector field Y is invariant under the G-action if and only if Y commutes with the fundamental vector fields. The actual proof is a bit long, but the idea behind it can easily be explained. For $x \in \mathfrak{g}$ we have the fundamental vector field x^M, whose flow is given by $\Phi_{\exp(tx)}$. Using [V.5.15] we deduce that Y is invariant under the action of the subgroup formed by the $\exp(tx)$ if and only if Y and x^M commute. Since the elements of the form $\exp(x)$ generate G the result follows. A first problem with this idea is that vector fields must be smooth to be integrable, which restricts attention to $x \in \mathbf{B}\mathfrak{g}$. However, the main problem is that for odd elements x there is no guarantee that the odd vector field x^M satisfies $[x^M, x^M] = 0$, a condition necessary for x^M to be integrable. Since the even elements in $\mathbf{B}\mathfrak{g}$ do not generate $\mathfrak{g}$, we can not reach the whole group G.

To overcome this problem, we note that [V.5.15] concerns the flow of a vector field. And the flow ϕ_X of a vector field X can be seen as the action of the 1-dimensional $\mathcal{A}$-Lie group $\mathcal{A}_{\varepsilon(X)}$ on M (apart from the fact that the domain W_X need not be the whole of $\mathcal{A}_{\varepsilon(X)} \times M$). We thus generalize this result to the setting of general group actions: the flow ϕ_X will be replaced by the group action Φ, the time parameter t will be replaced by a group element $g \in G$ and the vector field X will be replaced by a fundamental vector field x^M.

In order to prepare the actual statement, we use the generalized tangent map to form the G-dependent vector field ψ_Y on M defined by

$$\psi_Y(g, m) = T\Phi_{g^{-1}} Y_{\Phi_g(m)} \ .$$

If we introduce the function $\Phi^{(1)} : G \times M \to G \times M$, $(g, m) \mapsto (g^{-1}, \Phi_g(m))$ (note the analogy with the flow of a vector field), then we can write the definition of ψ_Y as

$$\psi_Y = T\Phi \circ (\underline{0} \times Y) \circ \Phi^{(1)} : G \times M \to TM \ .$$

It is immediate that Y is invariant under the G-action if and only if ψ_Y is independent of g, i.e., $\psi_Y(g, m) = Y_m$.

If Z is a smooth vector field on G, we can form the vector field $Z \times \underline{0}$ on $G \times M$. By abuse of notation we will denote this vector field also as Z. It is immediate from the definition that $\psi_Y(g, m) \in T_m M$. And thus the map ψ_Y and the vector field Z satisfy the requirements of [V.3.10]. Hence it makes sense to talk about the derivative of ψ_Y in the direction of Z, i.e., about the map $Z\psi_Y : G \times M \to TM$. Now recall that for $x \in \mathfrak{g}$ we have defined a corresponding fundamental vector field x^M on M and that in the same vein we have the right-invariant vector field x^G on G [5.1]. Moreover, the right-invariant vector field x^G and the fundamental vector field x^M are related by Φ: $T\Phi_m\, x^G|_g = x^M|_{\Phi_g(m)}$ [5.2]. Since x^M and x^G are smooth if $x \in \mathbf{B}\mathfrak{g}$, the following statement makes sense.

8.3 Proposition. *Let $\Phi : G \times M \to M$ be a smooth left action of an $\mathcal{A}$-Lie group G with associated $\mathcal{A}$-Lie algebra $\mathfrak{g}$ on an $\mathcal{A}$-manifold M, let $x \in \mathbf{B}\mathfrak{g}$ and let Y be a smooth vector field on M. Then*

$$x^G \psi_Y = \psi_{[x^M, Y]} \ .$$

Proof. This proof is a close copy of the proof of [V.5.15]. We first note that the result is additive in the vector field Y and the $\mathcal{A}$-Lie algebra element x, so we may assume that Y and x are homogeneous.

Copying the approach of [V.5.10], we choose a point $(g, h, m) \in G \times G \times M$ and we imagine that g is close to the identity $e \in G$. If g^r are coordinates around $e \in G$, there exist λ^r such that $x^G|_e = \sum_r \lambda^r \partial_{g^r}|_e = -x$. Since $x \in \mathbf{B}\mathfrak{g}$, the λ are real (and thus even); since x is supposed to be homogeneous we have $\varepsilon(\partial_{g^r}) = \varepsilon(x)$ for all ∂_{g^r} contributing to the sum. Finally we choose local coordinates systems x_a^i around m and x_b^j around $\Phi(h, m)$. Since we assume that g is close to e, we may assume that we can use the coordinates x_b^j also around $\Phi(g, \Phi(h, m)) \equiv \Phi(gh, m)$. With these coordinate systems we are going to establish a certain number of identities that will be needed later.

Since the x_b^j are coordinates around $\widehat{m} = \Phi(h, m)$, there exist local functions $X_b^j(\widehat{m})$ such that $x^M|_{\widehat{m}} = \sum_j X_b^j(\widehat{m}) \partial_{x_b^j}|_{\widehat{m}}$. The definition of the fundamental vector field immediately gives

$$X_b^j(\widehat{m}) = \sum_r \lambda^r \frac{\partial \Phi_b^j}{\partial g^r}(e, \widehat{m}) \ . \tag{8.4}$$

Using the generalized tangent map for the equality $\Phi_{g^{-1}} \circ \Phi_{h^{-1}} = \Phi_{(gh)^{-1}}$, applying it to the tangent vector $\partial_{x_b^k}|_{\widetilde{m}}$ with $\widetilde{m} = \Phi(g, \widehat{m}) = \Phi(gh, m)$ and comparing the coefficients of $\partial_{x_a^i}|_m$ gives us the identity

$$\frac{\partial \Phi_a^i}{\partial x_b^k}((gh)^{-1}, \widetilde{m}) = \sum_j \frac{\partial \Phi_b^j}{\partial x_b^k}(g^{-1}, \widetilde{m}) \cdot \frac{\partial \Phi_a^i}{\partial x_b^j}(h^{-1}, \widehat{m}) \ . \tag{8.5}$$

If we assume for the moment that h is also close to e so that we can use the coordinate system x_b^j also around m, and if we take $g = h = e$ in (8.5) (implying $m = \widehat{m} = \widetilde{m}$), we obtain immediately

$$\frac{\partial \Phi_b^j}{\partial x_b^k}(e, \widehat{m}) = \delta_k^j \ . \tag{8.6}$$

Since this is valid for all $\widehat{m}$ in the coordinate neighborhood, we can apply [III.3.13] to obtain

$$\frac{\partial^2 \Phi_b^j}{\partial x_b^\ell \partial x_b^k}(e, \widehat{m}) = 0 \ . \tag{8.7}$$

We now introduce the function $\chi : G \times G \times M \to TM$ by $\chi(g, h, m) = \psi_Y(gh, m)$. It follows easily that $(x^G \times \underline{0} \times \underline{0})|_{(e,h,m)} \chi = (x^G \times \underline{0})|_{(h,m)} \psi_Y \equiv (x^G \psi_Y)(h, m)$, simply because $\langle\!\langle x^G|_g \| TR_h \rangle\!\rangle = x^G|_{gh}$ by right invariance. In the given trivializations the map χ is given by

$$\psi_Y(gh,m) \equiv \chi(g,h,m) =$$

$$= \sum_{k,i} Y_b^k(\Phi(gh,m)) \cdot \frac{\partial \Phi_a^i}{\partial x_b^k}((gh)^{-1}, \Phi(gh,m)) \cdot \frac{\partial}{\partial x_a^i}|_m \tag{8.8}$$

$$= \sum_{i,j,k} Y_b^k(\Phi(g,\widehat{m})) \cdot \frac{\partial \Phi_b^j}{\partial x_b^k}(g^{-1}, \Phi(g,\widehat{m})) \cdot \frac{\partial \Phi_a^i}{\partial x_b^j}(h^{-1},\widehat{m}) \cdot \frac{\partial}{\partial x_a^i}|_m \,, \tag{8.9}$$

where we used (8.5) to obtain the last equality. In the same trivialization we now compute $(x^G \times \underline{0} \times \underline{0})|_{(e,h,m)}\chi$ using (V.3.12) and the fact that $\widehat{m}$ does not depend upon g :

$$(x^G\psi_Y)(h,m) \equiv (x^G \times \underline{0} \times \underline{0})|_{(e,h,m)}\chi =$$

$$\begin{aligned}
= \sum_{i,j}\Bigg(&\sum_{k,\ell,r} \lambda^r \frac{\partial \Phi_b^\ell}{\partial g^r}(e,\widehat{m}) \cdot \frac{\partial Y_b^k}{\partial x_b^\ell}(\Phi(e,\widehat{m})) \cdot \frac{\partial \Phi_b^j}{\partial x_b^k}(e, \Phi(e,\widehat{m})) \\
&- \sum_{k,r} (-1)^{\langle \varepsilon(x)|\varepsilon(Y_b^k)\rangle} \cdot Y_b^k(\Phi(e,\widehat{m})) \cdot \lambda^r \cdot \frac{\partial^2 \Phi_b^j}{\partial g^r \partial x_b^k}(e, \Phi(e,\widehat{m})) \\
&+ \sum_{k,\ell,r} (-1)^{\langle \varepsilon(x)|\varepsilon(Y_b^k)\rangle} \cdot Y_b^k(\Phi(e,\widehat{m})) \cdot \lambda^r \cdot \frac{\partial \Phi_b^\ell}{\partial g^r}(e,\widehat{m}) \cdot \frac{\partial^2 \Phi_b^j}{\partial x_b^\ell \partial x_b^k}(e, \Phi(e,\widehat{m}))\Bigg) \cdot \\
&\qquad\qquad \frac{\partial \Phi_a^i}{\partial x_b^j}(h^{-1},\widehat{m}) \cdot \frac{\partial}{\partial x_a^i}|_m \\
= \sum_{i,j,k,\ell} \Bigg(& X_b^\ell(\widehat{m}) \cdot \frac{\partial Y_b^k}{\partial x_b^\ell}(\widehat{m}) \cdot \delta_k^j - (-1)^{\langle \varepsilon(x)|\varepsilon(Y_b^k)+\varepsilon(x_b^k)\rangle} \cdot Y_b^k(\widehat{m}) \cdot \frac{\partial X_b^j}{\partial x_b^k}(\widehat{m})\Bigg) \cdot \\
&\qquad\qquad \frac{\partial \Phi_a^i}{\partial x_b^j}(h^{-1},\widehat{m}) \cdot \frac{\partial}{\partial x_a^i}|_m \,.
\end{aligned}$$

To obtain the last equality we applied (8.4), (8.6) and (8.7). The additional sign appears because we have to change the order of the partial derivatives ∂_{g^r} and $\partial_{x_b^k}$. Since we have $\varepsilon(Y_b^k) + \varepsilon(x_b^k) = \varepsilon(Y)$, we see that the expression in parentheses represents the coefficients of $[X,Y](\widehat{m})$ with respect to the basis $\partial_{x_b^j}$. Comparing this result with the local expression (8.8) for χ, which is essentially the local expression for ψ_Y, we see that this is the local expression of $\psi_{[X,Y]}(h,m)$. QED

8.10 Corollary. *Let $\Phi : M \times G \to M$ be a smooth right action of an $\mathcal{A}$-Lie group G with associated $\mathcal{A}$-Lie algebra $\mathfrak{g}$ on an $\mathcal{A}$-manifold M, let $x \in \mathbf{B}\mathfrak{g}$ and let Y be a smooth vector field on M. Then*

$$\vec{x}\,\psi_Y = \psi_{[x^M,Y]} \,.$$

Proof. For a right action ψ_Y is defined as for a left action: $\psi_Y(g,m) = T\Phi_{g^{-1}}Y_{\Phi_g(m)}$, but now $\Phi_g : M \to M$ is defined as $\Phi_g(m) = \Phi(m,g)$. If Φ is a right action, the map $\Psi : G \times M \to M$ defined by $\Psi(g,m) = \Phi(m,g^{-1})$ is a left action of G on M. Looking at the definition of a fundamental vector field, it is immediate that the fundamental vector field $x^{M,\Phi}$ associated to $x \in \mathfrak{g}$ for the (right) Φ-action is the same as the fundamental vector field $x^{M,\Psi}$ for the (left) Ψ-action [5.4]. Defining $\widehat{\psi}_Y(g,m) = T\Psi_{g^{-1}}Y_{\Psi_g(m)}$ we have by [8.3] $x^G\widehat{\psi}_Y = \widehat{\psi}_{[x^M,Y]}$ but also $\widehat{\psi}_Y(g,m) = \psi_Y(g^{-1},m)$, i.e., $\psi_Y = \widehat{\psi}_Y \circ (\mathit{Inv} \times \mathit{id})$. Leaving it to the reader to prove that $\vec{x}\,\psi_Y$ makes sense, i.e., that the condition of [V.3.10] is satisfied, we compute it according to the official definition:

$$(\vec{x}\,\psi_Y)(g,m) = \pi_\partial(T_{(g,m)}\psi_Y(\vec{x}_g)) = \pi_\partial(T_{(g^{-1},m)}\widehat{\psi}_Y(T_{(g,m)}(\mathit{Inv} \times \mathit{id})(\vec{x}_g, \underline{0}_m)))$$
$$= \pi_\partial(T_{(g^{-1},m)}\widehat{\psi}_Y(x^G_{g^{-1}}, \underline{0}_m)) = \widehat{\psi}_{[x^M,Y]}(g^{-1},m) = \psi_{[x^M,Y]}(g,m)\ ,$$

where for the third equality we used [1.10] and the fact that x^G is the right-invariant vector field on G satisfying $x^G_e = -x$ [5.1]. QED

8.11 Corollary. *Let $\Phi : G \times M \to M$ be a smooth left action of an $\mathcal{A}$-Lie group G on an $\mathcal{A}$-manifold M and let Y be a smooth vector field on M.*

(i) *If Y is invariant under the G-action, then $\forall x \in \mathbf{B}\mathfrak{g} : [x^M, Y] = 0$.*

(ii) *If G is connected and if $\forall x \in \mathbf{B}\mathfrak{g} : [x^M, Y] = 0$, then Y is invariant under the G-action.*

Proof. It is obvious from the definition that ψ_Y is zero if and only if Y is zero. Now if Y is invariant, then ψ_Y is independent of the G-coordinates, and thus $x^G\psi_Y = 0$ for all $x \in \mathbf{B}\mathfrak{g}$. By [8.3] we conclude that $[x^M, Y] = 0$, proving the first part.

For the second part we first invoke [8.3] to conclude that $x^G\psi_Y = 0$ for all $x \in \mathbf{B}\mathfrak{g}$. Since the vector fields x^G with $x \in \mathbf{B}\mathfrak{g}$ span the tangent space T_gG at each point $g \in G$, we conclude by [V.3.20] that there exists a function $f : M \to TM$ such that $\psi_Y = f \circ \pi_M$. Since $\psi_Y(e,m) = Y_m$, it follows that f equals Y. In other words, Y is invariant under the G-action. QED

8.12 Definition. Let $\Phi : G \times M \to M$ be a smooth left action of an $\mathcal{A}$-Lie group G on an $\mathcal{A}$-manifold M and let ω be a (not necessarily smooth) k-form on M, i.e., a section of $\bigwedge^k {}^*TM$. Using the generalized tangent map $T\Phi_g$, $g \in G$ [V.3.19], we define the generalized pull-back $\Phi_g^*\omega$ by the same formula (V.7.20) as for smooth k-forms and smooth maps. For $X_i, \ldots, X_k \in T_mM$ the k-form $(\Phi_g^*\omega)_m$ is defined by

$$\iota(X_1, \ldots, X_k)(\Phi_g^*\omega)_m = \iota(\langle\!\langle X_1 \,\|\, T_m\Phi_g \rangle\!\rangle, \ldots, \langle\!\langle X_k \,\|\, T_m\Phi_g \rangle\!\rangle)\omega_{\Phi_g(m)}\ .$$

By definition of an action we have $\Phi_e(m) = m$. It then follows directly from [III.3.13] that $T_m\Phi_e = \mathit{id}$. Hence we always have $\Phi_e^*\omega = \omega$. If $\Phi : M \times G \to M$ is a smooth right action, we define the generalized pull-back $\Phi_g^*\omega$ by the same formula, but, as for invariant vector fields, it is the definition of the map Φ_g that changes.

The k-form ω is said to be *invariant under the action of* G if for all $g \in G$ we have $\Phi_g^* \omega = \omega$. As a particular case we mention that a k-form ω on G itself is said to be *left/right-invariant* if it is invariant under the natural left/right action $m : G \times G \to G$ of G on itself.

Recalling that (for left actions) $T_m \Phi_g(X) = T_{(g,m)} \Phi(\underline{0}_g, X)$, it is elementary to see that ω is invariant under the G-action if and only if for all (not necessarily smooth) vector fields $X_1, \ldots, X_k$ on M we have

$$\iota(\underline{0} \times X_1, \ldots, \underline{0} \times X_k)\Phi^* \omega = (\iota(X_1, \ldots, X_k)\omega) \circ \pi_M \ , \tag{8.13}$$

where $\pi_M : G \times M \to M$ denotes the canonical projection. In fact, in order to be invariant, (8.13) need only be verified for a set of vector fields that span the tangent space at each point $m \in M$. Multilinearity then does the rest. For example it is sufficient to verify (8.13) for smooth vector fields (even if ω is not smooth).

Yet another way to say the same is to note that the above definition is a particular case of the generalized pull-back given in [V.7.23]. This means that we look at the map $\psi_\omega : G \times M \to \bigwedge^k {}^*TM$ by $\psi_\omega(g, m) = (\Phi_g^* \omega)_m$. And then the definition of ω being invariant becomes the condition $\psi_\omega = \omega \circ \pi_M$.

For right actions (8.13) changes in the obvious way; the defining formula of ψ_ω does not change, but as before it is Φ_g that changes.

8.14 Lemma. *Let ω be a smooth k-form. If ω is invariant under the G-action, then so is $d\omega$.*

Proof. Using [V.7.22], the equality $(\underline{0} \times X)(f \circ \pi_M) = (Xf) \circ \pi_M$ and the equality $[\underline{0} \times X, \underline{0} \times Y] = \underline{0} \times [X, Y]$, it follows immediately from (8.13) and the definition of the exterior derivative that $d\omega$ is invariant if ω is. $\boxed{QED}$

8.15 Lemma. *The map $\omega \mapsto \omega_e$ from left-invariant k-forms on an $\mathcal{A}$-Lie group G to $\bigwedge^k {}^*\mathfrak{g} = \bigwedge^k {}^*T_eG$ is a bijection. Moreover, ω is smooth if and only if ω_e has real coordinates, i.e., $\omega_e \in \mathbf{B}\bigwedge^k {}^*\mathfrak{g}$.*

Proof. • Suppose ω is left-invariant and $\omega_e = 0$, then for $X_1, \ldots, X_k \in T_gG$ we have by left-invariance:

$$\begin{aligned}\iota(X_1, \ldots, X_k)\omega_g &= \iota(X_1, \ldots, X_k)(L_{g^{-1}}^* \omega)_g \\ &= \iota(\langle\!\langle X_1 \| TL_{g^{-1}} \rangle\!\rangle, \ldots, \langle\!\langle X_k \| TL_{g^{-1}} \rangle\!\rangle)\omega_e = 0 \ .\end{aligned}$$

Hence $\omega = 0$, proving injectivity.

• To prove surjectivity, we construct the global section ω from the given value of ω_e by specifying all k-linear graded skew-symmetric maps $\omega_g : (T_gG)^k \to \mathcal{A}$. For $X_1, \ldots, X_k \in T_gG$ we write

$$\iota(X_1, \ldots, X_k)\omega_g = \iota(\langle\!\langle X_1 \| TL_{g^{-1}} \rangle\!\rangle, \ldots, \langle\!\langle X_k \| TL_{g^{-1}} \rangle\!\rangle)\omega_e \ . \tag{8.16}$$

The global k-form ω defined by this formula is left-invariant because

$$\begin{aligned}\iota(X_1,\ldots,X_k)(L_h^*\omega)_g &= \iota(\langle\!\langle X_1 \| TL_h \rangle\!\rangle,\ldots,\langle\!\langle X_k \| TL_h \rangle\!\rangle)\omega_{hg} \\ &= \iota(\langle\!\langle X_1 \| TL_h \diamond TL_{(hg)^{-1}} \rangle\!\rangle,\ldots,\langle\!\langle X_k \| TL_h \diamond TL_{(hg)^{-1}} \rangle\!\rangle)\omega_e \\ &\qquad \text{by (8.16) and (I.2.14)} \\ &= \iota(\langle\!\langle X_1 \| TL_{g^{-1}} \rangle\!\rangle,\ldots,\langle\!\langle X_k \| TL_{g^{-1}} \rangle\!\rangle)\omega_e = \iota(X_1,\ldots,X_k)\omega_g \ .\end{aligned}$$

- If $X_1,\ldots,X_k$ are smooth vector fields, the function $\iota(X_1,\ldots,X_k)\omega$ is given by

$$g \mapsto \iota(Tm(\underline{0}_{g^{-1}}, X_1|_g),\ldots,Tm(\underline{0}_{g^{-1}}, X_k|_g))\omega_e \ .$$

Since $g \mapsto g^{-1} \mapsto \underline{0}_{g^{-1}}$ is smooth, the map $g \mapsto Tm(\underline{0}_{g^{-1}}, X_i|_g)$ is smooth. Since ω_e is a smooth multi-linear map if and only if ω_e has real coordinates [II.6.22], [III.1.27], the conclusion follows from [IV.5.9]. QED

8.17 Corollary. *A k-form ω on G is left-invariant if and only if for all left-invariant vector fields $X_1,\ldots,X_k$ on G the function $\iota(X_1,\ldots,X_k)\omega$ is constant.*

Proof. A direct consequence of (8.16) and the equality $\langle\!\langle X_g \| TL_{g^{-1}} \rangle\!\rangle = X_e$ for left-invariant vector fields. QED

8.18 Discussion/Notation. Let $\Phi : G \times M \to M$ be a smooth left action of an $\mathcal{A}$-Lie group G with associated $\mathcal{A}$-Lie algebra $\mathfrak{g}$ on an $\mathcal{A}$-manifold M and let ω be a smooth k-form on M. We want to show a statement analogous to the statement for invariant vector fields, i.e., that ω is invariant if and only if the Lie derivative of ω in the direction of the fundamental vector fields is zero. Here the basic idea is to use [V.7.27] to prove this, but the objections are the same as in the vector field case. Luckily the solution to this problem is also the same: the flow ϕ_X will be replaced by the group action Φ, the time parameter t will be replaced by a group element $g \in G$ and the vector field X will be replaced by a fundamental vector field z^M.

The preparations for the actual statement are quite analogous to those for [V.7.27]. In [8.12] we have seen that the invariance of ω can be stated in terms of the G-dependent k-form $\psi_\omega : G \times M \to \bigwedge^k {}^*TM$, $\psi_\omega(g,m) = (\Phi_g^*\alpha)_m$ [V.7.23]. Since we have $\psi_\omega(g,m) \in (\bigwedge^k {}^*TM)_m$, it follows that any vector field Z on G extended by $\underline{0}$ to $G \times M$ and the map ψ_ω satisfy the conditions of [V.3.10]. We thus can speak about the derivative of ψ_ω in the direction of Z. If one now recalls that for $z \in \mathfrak{g}$, z^G is a right-invariant vector field on G and z^M the associated fundamental vector field on M, then [8.19] makes sense; [8.20] and [8.21] are proved as [8.10] and [8.11].

8.19 Proposition. *Let $\Phi : G \times M \to M$ be a smooth left action of an $\mathcal{A}$-Lie group G with associated $\mathcal{A}$-Lie algebra $\mathfrak{g}$ on an $\mathcal{A}$-manifold M, let $z \in \mathbf{B}\mathfrak{g}$ and let ω be a smooth k-form on M. Then*

$$z^G \psi_\omega = \psi_{\mathcal{L}(z^M)\omega} \ .$$

Proof. This proof is a very close copy of the proof of [V.7.27], so we will skip some of the motivations for the computations. Since the result is clearly additive in both z and ω we may assume that z and ω are homogeneous. We choose a coordinate neighborhood V around $\Phi(g,m)$ with coordinates y^i, and let $W \subset G$ be a neighborhood of g with coordinates g^ℓ and U a coordinate neighborhood around m with coordinates x^j such that $\Phi(W \times U) \subset V$. On the local chart V there exist smooth functions $\omega_{i_1,\ldots,i_k} : V \to \mathcal{A}$ such that the k-form ω has the form $\omega = \sum_{i_1,\ldots,i_k} \omega_{i_1\ldots i_k} \cdot dy^{i_1} \wedge \cdots \wedge dy^{i_k}$. There also exist smooth functions Z^ℓ on W and X^i on V such that the vector fields z^G and z^M have the local form $z^G_g = \sum_\ell Z^\ell(g)\partial_{g^\ell}|_g$ and $z^M_{(y)} = \sum_i X^i(y)\partial_{y^i}|_{(y)}$. The relation $T\Phi_m z^G_g = z^M_{\Phi(g,m)}$ between z^G and z^M [5.2] then translates as the equation

$$X^i(\Phi(g,x)) = \sum_\ell Z^\ell(g) \cdot \frac{\partial \Phi^i}{\partial g^\ell}(g,x) \ .$$

Moreover, it follows directly from the definition of the generalized pull-back Φ^*_g that we have $\iota(\partial_{x^j})\Phi^*_g dy^i = \iota(\langle\!\langle \partial_{x^j}|_{(g,x)} \| T\Phi \rangle\!\rangle)dy^i = (\partial_{x^j}\Phi^i)(g,x)$. Hence all terms of the form $\Phi^*_g\, dy^i$ are of the form

$$\Phi^*_g\, dy^i = \sum_j dx^j \cdot \frac{\partial \Phi^i}{\partial x^j}(g,x) \ .$$

With these preparations we compute:

$$\begin{aligned} z^G(\Phi^*_g\, dy^i) &= \sum_{j,\ell} (-1)^{\langle \varepsilon(z)|\varepsilon(x^j)\rangle} dz^j \cdot Z^\ell(g) \cdot \frac{\partial^2 \Phi^i}{\partial g^\ell \partial x^j}(g,x) \\ &= \sum_j dx^j \cdot \frac{\partial}{\partial x^j}\Big(\sum_\ell Z^\ell \cdot \frac{\partial \Phi^i}{\partial g^\ell}\Big)(g,x) = \sum_j dx^j \cdot \frac{\partial (X^i \circ \Phi)}{\partial x^j}(g,x) \\ &= \sum_{j,k} dx^j \cdot \frac{\partial \Phi^k}{\partial x^j}(g,x) \cdot \frac{\partial X^i}{\partial y^k}(\Phi(g,x)) = \sum_k \Phi^*_g dy^k \cdot \Phi^*_g \frac{\partial X^i}{\partial y^k} \\ &= \Phi^*_g(\sum_k dy^k \cdot \frac{\partial X^i}{\partial y^k}) = \Phi^*_g(dX^i) = \Phi^*_g(\mathcal{L}(z^M)dy^i) \ . \end{aligned}$$

For the second equality we used that the functions Z^ℓ do not depend upon the coordinates x^j and for the last equality we used that $X^i = \iota(z^M)dy^i$.

For the coefficient function $\omega_{i_1\ldots i_k}$ we note that $\Phi^*_g\omega_{i_1\ldots i_k}(x) = \omega_{i_1\ldots i_k}(\Phi(g,x))$. Hence $z^G(\Phi^*_g\omega_{i_1\ldots i_k}) = (z^M\omega_{i_1\ldots i_k}) \circ \Phi = \Phi^*_g(z^M\omega_{i_1\ldots i_k})$. The final computation to knit these results together is an exact copy of the analogous computation in the proof of [V.7.27] QED

8.20 Corollary. *Let $\Phi : M \times G \to M$ be a smooth right action of an $\mathcal{A}$-Lie group G with associated $\mathcal{A}$-Lie algebra $\mathfrak{g}$ on an $\mathcal{A}$-manifold M, let $z \in \mathbf{B}\mathfrak{g}$ and let ω be a smooth k-form on M. Then*

$$\vec{z}\,\psi_\omega = \psi_{\mathcal{L}(z^M)\omega} \ .$$

8.21 Corollary. *Let $\Phi : G \times M \to M$ be a smooth left action of an $\mathcal{A}$-Lie group G on an $\mathcal{A}$-manifold M and let ω be a smooth k-form on M.*

(i) *If ω is invariant under the G-action, then $\forall x \in \mathbf{B}\mathfrak{g} : \mathcal{L}(x^M)\omega = 0$.*

(ii) *If G is connected and if $\forall x \in \mathbf{B}\mathfrak{g} : \mathcal{L}(x^M)\omega = 0$, then ω is invariant under the G-action.*

9. Lie's third theorem

In this section we prove that for each finite dimensional $\mathcal{A}$-Lie algebra $\mathfrak{g}$ there exists a unique (up to isomorphism) simply connected $\mathcal{A}$-Lie group G with $\mathfrak{g}$ as its $\mathcal{A}$-Lie algebra. For this proof we need to introduce the notion differential forms with values in an $\mathcal{A}$-vector space, a notion that will be generalized and studied in more detail in chapter VII.

9.1 Definitions (forms with values in an $\mathcal{A}$-vector space). Let M be an $\mathcal{A}$-manifold and E an $\mathcal{A}$-vector space. We have seen that a smooth k-form on M can be interpreted as a left k-linear (over $C^\infty(M)$) skew-symmetric map from vector fields on M to (ordinary, smooth) functions on M, i.e., an element of $\mathrm{Hom}_L^{sk}(\Gamma(TM)^k; C^\infty(M))$. In analogy, we define a *smooth k-form with values in E* or *smooth E-valued k-form* as being a left k-linear (over $C^\infty(M)$) skew-symmetric map from vector fields on M to smooth functions on M with values in E, i.e., an element of $\mathrm{Hom}_L^{sk}(\Gamma(TM)^k; C^\infty(M; E))$. Playing around with the various identifications [I.5.5], [I.8.8], [III.1.24], [V.7.1], this space is isomorphic to ${}^*(\bigwedge^k \Gamma(TM)) \otimes C^\infty(M; E) \cong \Omega^k(M) \otimes C^\infty(M; E)$, where the tensor product is taken over the graded ring $C^\infty(M)$. In analogy with ordinary k-forms, the set of all smooth E-valued k-forms will be denoted as $\Omega^k(M; E)$, i.e.,

$$\Omega^k(M; E) \equiv \mathrm{Hom}_L^{sk}(\Gamma(TM)^k; C^\infty(M; E)) \cong \Omega^k(M) \otimes C^\infty(M; E) \ .$$

If ω is a k-form on M and if $v : M \to E$ is an element of $C^\infty(M; E)$, then $\omega \otimes v$ is (isomorphic to) the E-valued k-form

$$X_1, \ldots, X_k \mapsto \iota(X_1, \ldots, X_k)\omega \cdot v \tag{9.2}$$

from k vector fields $X_1, \ldots, X_k$ to a smooth E-valued function. Conversely, let ω be a smooth E-valued k-form, let (e_i) be a basis for F, and let $({}^j e)$ be the associated left dual basis of ${}^*E$. We then can form the map

$$\omega^j : (X_1, \ldots, X_k) \mapsto \langle\!\langle \iota(X_1, \ldots, X_k)\omega \,\|\, {}^j e \rangle\!\rangle$$

from k vector fields on M to a smooth function on M. Since this map is obviously skew-symmetric and k-linear over $C^\infty(M)$, it represents an ordinary k-form on M. Since $v = \sum_j \langle\!\langle v \,\|\, {}^j e \rangle\!\rangle \, e_j$ for all $v \in E$, we can write

$$\omega = \sum_j \omega^j \otimes e_j \ , \tag{9.3}$$

where we see each e_j as a constant (smooth because $e_j \in \mathbf{B}E$) E-valued function on M.

In the particular case $k = 0$ the ω^j are smooth functions on M. It then is customary to omit the tensor product symbol $\otimes$ and to write $\omega = \sum_j \omega^j \, e_j$. Formally, a 0-form is an element of $\mathrm{Hom}_L(C^\infty(M); C^\infty(M; E))$ [I.5.3]. Omitting the tensor product symbol amounts to applying the identification $\mathrm{Hom}_L(C^\infty(M); C^\infty(M; E)) \cong C^\infty(M; E)$ [I.2.19], [I.8.8], [III.4.21], where one should not forget that in this context the basic graded ring is $C^\infty(M)$. In short, a 0-form with values in E is just an E-valued function.

We thus have shown that for a fixed basis of E, an E-valued k-form is uniquely represented by $\dim E$ ordinary k-forms. However, these k-forms depend upon the choice of the basis for E. More precisely, if $(\widehat{e}_i)$ is another basis (in its equivalence class) for E, it is related to the old basis by (real) matrix elements $A_i{}^j$ by $\widehat{e}_i = \sum_j A_i{}^j e_j$. In the new basis the E-valued k-form defines ordinary k-forms $\widehat{\omega}^j$. Since ${}^je = \sum_i {}^i\widehat{e} A_i{}^j$, we also have

$$\omega^j = \sum_i \widehat{\omega}^i A_i{}^j \ .$$

One usually fixes a basis for E and then a smooth E-valued k-form is just a useful abbreviation for a system of $\dim E$ ordinary k-forms. However, one should be careful with parities: ω is even if $\varepsilon(\omega^i) = \varepsilon(e_i)$ for all i and ω is odd if $\varepsilon(\omega^i) = 1 - \varepsilon(e_i)$.

Once we know that a *smooth* k-form with values in E is completely determined by $\dim E$ ordinary *smooth* k-forms, we can define a not necessarily smooth k-form with values in E as being determined by $\dim E$ ordinary (not necessarily smooth) k-forms for which (9.3) is valid (by definition). Since the arguments of an ordinary k-form can be any vector field, smooth or not, we can use (9.3) and (9.2) to extend an E-valued k-form to act also on arbitrary vector fields, yielding an E-valued function (not necessarily smooth) on M. Seen this way, a smooth E-valued k-form is one that maps smooth vector fields to smooth f-valued functions (see also [IV.5.11]).

As for ordinary k-forms, we define the notions of contraction with a vector field, exterior derivative, and pull-back by the natural formulæ. More precisely, if $\omega = \sum_i \omega^i \otimes e_i$ is an E-valued k-form and X a vector field on M, we define $\iota(X)\omega$ by

$$\iota(X)\omega = \sum_i \iota(X)\omega^i \otimes e_i \ .$$

Using the natural action of vector fields on smooth E-valued functions [V.1.24], the exterior derivative of an E-valued k-form is defined by (V.7.4) (and thus in particular by [V.7.6]). Since the e_i are seen as *constant* functions on M, it is immediate that we have the formula

$$d\omega = \sum_i d\omega^i \otimes e_i \ . \tag{9.4}$$

If $g : N \to M$ is a smooth map, the pull-back $g^*\omega$ is the E-valued k-form on N defined as

$$g^*\omega = \sum_i g^*\omega^i \otimes e_i \ .$$

Finally, if an $\mathcal{A}$-Lie group G acts on M, we say that ω is invariant under the G-action if all ω^i are invariant under the G-action. We leave it to the reader to verify that these definitions are independent of the choice of the basis for E. It is a simple consequence of the fact that the $A_i{}^j$ are real constants.

9.5 Discussion. In [Lie, Ch 25] Sophus Lie stated three fundamental properties of Lie groups, of which the third (in §115) says that to every finite dimensional Lie group is associated a finite dimensional Lie algebra and to each finite dimensional Lie algebra a finite dimensional Lie group. For the second part of this statement, several essentially different proofs by various authors have been given. Here we will present, in the context of $\mathcal{A}$-Lie groups, a geometrical proof based upon a proof of Élie Cartan [É.Ca] (see also [vE], [Gor], [Tu1]).

The idea of this proof is as follows. Let $\mathfrak{g}$ be a finite dimensional $\mathcal{A}$-Lie algebra. If the adjoint representation $\mathrm{ad}_R : \mathfrak{g} \to \mathrm{End}_R(\mathfrak{g})$ is injective, we can interpret $\mathfrak{g}$ as a subalgebra of $\mathrm{End}_R(\mathfrak{g})$. Since we know that $\mathrm{Aut}(\mathfrak{g})$ is an $\mathcal{A}$-Lie group, we can apply [4.7] to conclude that there exists an $\mathcal{A}$-Lie group with associated algebra $\mathfrak{g}$.

If the adjoint representation is not injective, it has a kernel $\mathfrak{c} = \ker(\mathrm{ad}_R)$, called *the center of* $\mathfrak{g}$. Since ad_R is smooth, $\mathfrak{c}$ is a graded subspace of $\mathfrak{g}$ in the sense of $\mathcal{A}$-vector spaces. Moreover, for any two $x, y \in \mathfrak{c}$ we have $0 = \mathrm{ad}_R(x)(y) = [x, y]$, i.e., $\mathfrak{c}$ is a subalgebra of $\mathfrak{g}$ with trivial bracket. Standard arguments then show that the quotient $\mathfrak{h} = \mathfrak{g}/\mathfrak{c}$ inherits the structure of an $\mathcal{A}$-Lie algebra from $\mathfrak{g}$ and that the image $\mathfrak{h}^\bullet = \mathrm{ad}_R(\mathfrak{g}) \subset \mathrm{End}_R(\mathfrak{g})$ is a subalgebra of $\mathrm{End}_R(\mathfrak{g})$ isomorphic to $\mathfrak{h}$. It follows that $\mathfrak{g}$ is a central extension of $\mathfrak{h}$ (extended by $\mathfrak{c}$), and that a simple algebraic ingredient, a Lie algebra 2-cocycle, allows us to reconstruct $\mathfrak{g}$ from $\mathfrak{h}$ and $\mathfrak{c}$. Since $\mathfrak{h}$ is (isomorphic to) a subalgebra of $\mathrm{End}_R(\mathfrak{g})$, we can apply as before [4.7] to obtain an $\mathcal{A}$-Lie group H whose $\mathcal{A}$-Lie algebra is $\mathfrak{h}$. Cartan's construction then proceeds to transform the Lie algebra 2-cocycle into a group 2-cocycle, which allows us to construct a new $\mathcal{A}$-Lie group G out of H and $\mathfrak{c}$, and whose $\mathcal{A}$-Lie algebra is (isomorphic to) $\mathfrak{g}$.

We will not develop the theory of Lie algebra and Lie group cohomology, nor will we go into the details of central extensions and their relation to cohomology in dimension 2. The interested reader is referred to [Br], [Sta], [TW] and references therein for more details. Of course these references do not treat $\mathcal{A}$-Lie algebras and $\mathcal{A}$-Lie groups, but the generalization is straightforward. In the actual proof we will introduce only the strict minimum needed to construct the $\mathcal{A}$-Lie group G. In [9.18] we will give some comments on how the various parts fit into a larger context.

9.6 Theorem. *Let $\mathfrak{g}$ be a finite dimensional $\mathcal{A}$-Lie algebra. Then there exists a simply connected $\mathcal{A}$-Lie group G, unique up to isomorphisms, such that $\mathfrak{g}$ is (isomorphic to) the $\mathcal{A}$-Lie algebra associated to G.*

Proof. We start as indicated in [9.5] by defining $\mathfrak{c} = \ker(\mathrm{ad}_R : \mathfrak{g} \to \mathrm{End}_R(\mathfrak{g}))$, $\mathfrak{h} = \mathfrak{g}/\mathfrak{c}$, and $\mathfrak{h}^\bullet = \mathrm{ad}_R(\mathfrak{g}) \subset \mathrm{End}_R(\mathfrak{g})$. Since $\mathrm{End}_R(\mathfrak{g})$ is the $\mathcal{A}$-Lie algebra of the $\mathcal{A}$-Lie group $\mathrm{Aut}(\mathfrak{g})$, there exists a connected $\mathcal{A}$-Lie subgroup $(i, \widetilde{H})$ of $\mathrm{Aut}(\mathfrak{g})$ whose $\mathcal{A}$-Lie algebra

$\mathfrak{h}$ is isomorphic (by Ti) to $\mathfrak{h}^\bullet$ [4.7]. We denote by H the simply connected cover of $\widetilde{H}$. By [7.6] and the definition of a covering, H is an $\mathcal{A}$-Lie group with $\mathcal{A}$-Lie algebra $\mathfrak{h}$.

If $\mathfrak{c}$ were $\{0\}$, the $\mathcal{A}$-Lie algebra $\mathfrak{h}$ would be isomorphic to $\mathfrak{g}$ and $G = H$ would be a simply connected $\mathcal{A}$-Lie group with $\mathcal{A}$-Lie algebra $\mathfrak{g}$. If $\mathfrak{c}$ is not $\{0\}$, we choose a smooth injective even linear map $\sigma : \mathfrak{h} \to \mathfrak{g}$ such that $\pi \circ \sigma = id(\mathfrak{h})$, where $\pi : \mathfrak{g} \to \mathfrak{g}/\mathfrak{c} = \mathfrak{h}$ is the canonical projection. Such maps certainly exist as can be seen by using a basis (see [II.6.24] and [II.6.23]). Associated to this σ is a left bilinear map $\Phi : \mathfrak{h} \times \mathfrak{h} \to \mathfrak{c}$ defined by

$$\iota(x, y)\Phi = \sigma([x, y]_\mathfrak{h}) - [\sigma(x), \sigma(y)]_\mathfrak{g} \;.$$

Φ indeed takes its values in $\mathfrak{c}$ because π is an $\mathcal{A}$-Lie algebra morphism. By definition this Φ is smooth, bilinear, even, and graded skew-symmetric (because the brackets in $\mathfrak{h}$ and $\mathfrak{g}$ are). Given this Φ and the $\mathcal{A}$-Lie algebra $\mathfrak{h}$, we can reconstruct the $\mathcal{A}$-Lie algebra $\mathfrak{g}$ in the following way. On the $\mathcal{A}$-vector space $\mathfrak{h} \oplus \mathfrak{c}$ we define the bracket

$$[(x, a), (y, b)] = ([x, y]_\mathfrak{h}, -\iota(x, y)\Phi) \;, \tag{9.7}$$

and we define the map $\phi : \mathfrak{h} \oplus \mathfrak{c} \to \mathfrak{g}$ by $\phi(x, a) = \sigma(x) + a$. It is easy to show that this ϕ is an isomorphism of $\mathcal{A}$-vector spaces, but also that ϕ preserves brackets:

$$\begin{aligned}\phi([(x, a), (y, b)]) &= \phi([x, y]_\mathfrak{h}, -\iota(x, y)\Phi) = \sigma([x, y]) - \iota(x, y)\Phi \\ &= [\sigma(x), \sigma(y)] = [\phi(x, a), \phi(y, b)] \qquad \text{because } \mathfrak{c} \text{ is the center.}\end{aligned}$$

It follows that $\mathfrak{g}$ and $\mathfrak{h} \oplus \mathfrak{c}$ with the bracket (9.7) are isomorphic $\mathcal{A}$-Lie algebras.

We now define Ω as being the smooth left-invariant $\mathfrak{c}$-valued 2-form on H whose value at the identity is Φ [8.15], [9.1]. This 2-form is explicitly given by the formula $\iota(X_g, Y_g)\Omega = \iota(TL_{g^{-1}}X_g, TL_{g^{-1}}Y_g)\Phi$, and more in particular for left-invariant vector fields by

$$\iota(\vec{x}, \vec{y})\Omega \; : \; g \mapsto \iota(x, y)\Phi \;. \tag{9.8}$$

But there is more. The graded Jacobi identity for the brackets on $\mathfrak{h}$ and $\mathfrak{g}$ imply that for homogeneous $x, y, z \in \mathfrak{h}$ we have

$$\iota(x, [y, z])\Phi = \iota([x, y], z)\Phi + (-1)^{\langle\varepsilon(x)|\varepsilon(y)\rangle}\iota(y, [x, z])\Phi \;. \tag{9.9}$$

We will show that this implies (actually it is if and only if, but that is of no importance here) that Ω is closed. Using the definition of the exterior derivative [V.7.3] or [V.7.6], and using that a vector field applied on a constant function yields zero, we find for three left-invariant homogeneous vector fields $\vec{x}$, $\vec{y}$, and $\vec{z}$ on H :

$$\begin{aligned}\iota(\vec{x}, \vec{y}, \vec{z})d\Omega &= \iota(\vec{x}, [\vec{y}, \vec{z}])\Omega - \iota([\vec{x}, \vec{y}], \vec{z})\Omega - (-1)^{\langle\varepsilon(x)|\varepsilon(y)\rangle}\iota(\vec{y}, [\vec{x}, \vec{z}])\Omega \\ &= \iota(x, [y, z])\Phi - \iota([x, y], z])\Phi - (-1)^{\langle\varepsilon(x)|\varepsilon(y)\rangle}\iota(y, [x, z])\Phi \quad \text{by [9.8]} \\ &= 0 \quad \text{by (9.9).}\end{aligned}$$

Since homogeneous left-invariant vector fields generate $\Gamma(TH)$ as a graded $C^\infty(H)$-module [1.19], we conclude $d\Omega = 0$ as claimed.

Let (v_i) be a basis of $\mathfrak{h}$ and $({}^i v)$ the associated left-dual basis of ${}^*\mathfrak{h}$. As in [VII.1.5] we define ω^i to be the left-invariant 1-form on H satisfying $\omega^i|_e = {}^i v$. We then have in particular for two left-invariant vector fields $\vec{x}$ and $\vec{y}$ the identity $\iota(\vec{x}, \vec{y}\,)d\omega^i = \langle\!\langle [x,y] \,\|\, {}^i v \rangle\!\rangle$. Now recall that for $x \in \mathfrak{h}$ we have introduced the fundamental vector field x^H on H associated to the natural left action of H on itself. This x^H is the right-invariant vector field on H such that $x^H|_e = -x = -\vec{x}_e$ [5.1]. In order to avoid this minus sign, we define x^R as $x^R = -x^H$, i.e., x^R is the right-invariant vector field on H such that $x^R|_e = x$. Said differently, $x^R|_g = TR_g(x)$.

We now invoke a theorem about ordinary Lie groups that is far from trivial: if G is an ordinary simply connected Lie group, then the de Rham cohomology groups $H^1(G)$ and $H^2(G)$ are zero: $H^1(G) = H^2(G) = \{0\}$. A proof can be found in [God, Ch XIII, 6.3]; it relies on the Künneth formula and a theorem of Iwasawa [Iw] that any Lie group is topologically equivalent to the direct product of a maximal compact subgroup and a vector space (over $\mathbf{R}$). Since H is (by definition) simply connected, the same is true for $\mathbf{B}H$. Using [V.8.9] we conclude that $H^1(H) = H^2(H) = \{0\}$. This has several immediate consequences. We will prove in [9.13] that the 1-forms $\iota(v_i^R)\Omega$ are closed. Our first consequence is that these 1-forms must be exact, i.e., there exist smooth functions f_i on H with values in $\mathfrak{c}$ such that $\iota(v_i^R)\Omega = df_i$. In the second place, the 2-form Ω is closed, and thus must be exact, i.e., there exists a 1-form α on H, not necessarily left-invariant, with values in $\mathfrak{c}$ such that $d\alpha = \Omega$. Now suppose that $\alpha_e = \sum_i a_i\, dx^i$ with respect to some coordinate system $(x^1, \ldots, x^n)$ in a neighborhood U of e. Since α is smooth, the coefficients a_i are real. We thus can define the constant function $A(x) = \sum_i a_i x^i$ on the neighborhood U. If we multiply this function A on U by a plateau function ρ around e in U, we get a global smooth function ρA on H. Since ρ is constant 1 on a (smaller) neighborhood of e, $(d\rho)_e = 0$ and hence $(d(\rho A))_e = \sum_i a_i\, dx^i = \alpha_e$. If we define $\widehat{\alpha} = \alpha - d(\rho A)$, it satisfies $d\widehat{\alpha} = \Omega$, but it also satisfies $\widehat{\alpha}_e = 0$. We conclude that we may assume without loss of generality that $\alpha_e = 0$.

We now reach the final stage of our preparations. Denoting by $\pi_i : H \times H \to H$, $i = 1, 2$ the canonical projections on the first and second factor, we define the 1-form Ψ on $H \times H$ with values in $\mathfrak{c}$ by the formula

$$\Psi = m^*\alpha - \pi_1^*\alpha - \pi_2^*\alpha + \textstyle\sum_i (\pi_1^*\omega^i) \cdot (\pi_2^* f_i - f_i(e)) \ . \tag{9.10}$$

In [9.14] we prove that Ψ is closed. Since H and thus $H \times H$ is simply connected, we have in particular $H^1(H \times H) = \{0\}$. This implies that there exists a smooth function $F : H \times H \to \mathfrak{c}$ such that $\Psi = dF$. In [9.15] we prove that the function F satisfies the relation

$$\forall g, h, k \in H : F(g,h) + F(gh,k) = F(g,hk) + F(h,k) \ . \tag{9.11}$$

We finally define the simply connected $\mathcal{A}$-Lie group G as $G = H \times \mathfrak{c}_0$ [7.2-i] with multiplication m_G defined by

$$m_G((g,a),(h,b)) = (gh, a + b + F(g,h)) \ .$$

Since the multiplication in H is smooth, as is F, the map m_G is smooth. From (9.11) we deduce first (putting $h = e$) that $F(g, e) = F(e, k) = F(e, e)$, and then that m_G is a group structure with $(e, -F(e, e))$ as the identity. We claim that the $\mathcal{A}$-Lie algebra of this G is (isomorphic to) $\mathfrak{g}$.

By construction, the tangent space T_eG is naturally isomorphic (as $\mathcal{A}$-vector space) with $T_eH \times \mathfrak{c} = \mathfrak{h} \times \mathfrak{c}$. Using [1.17] we compute the brackets of the vectors $\partial_{x^i}|_e$, $\partial_{a^j}|_e$ where $(x^1, \ldots, x^n)$ are (local) coordinates on H and (a^j) (global) coordinates on $\mathfrak{c}_0$. Since m_G depends only linearly on the coordinates a, any structure constant c_{ij}^k with either i or j referring to a coordinate a is zero. On the other hand, whenever $k \leq n$, the structure constant c_{ij}^k involves only m_H, and thus for $1 \leq i, j, k \leq n$ we recover the structure constants of $\mathfrak{h}$. The only remaining unknown structure constants are thus those with $i, j \leq n$ and $k > n$, i.e., those involving the term $a + b + F(g, h)$. Since we only derive with respect to the coordinates of g and h, we can forget about the $a + b$ and concentrate on $F(g, h)$. We thus have to compute $\partial_i\partial_{n+j}F - \partial_{n+i}\partial_jF$ at (e, e). We therefore compute

$$\begin{aligned}(\partial_{n+i}F)(g, e) &= \iota(\partial_{n+i}|_{(g,e)})dF = \iota((\underline{0}_g, \partial_i|_e))\Psi \\ &= \iota((\underline{0}_g, \partial_i|_e))m^*\alpha - \iota((\underline{0}_g, \partial_i|_e))\pi_2^*\alpha \\ &= \iota(TL_g\partial_i|_e)\alpha - \iota(\partial_i|_e)\alpha = \iota(\vec{\partial}_i|_g)\alpha \qquad \text{(because } \alpha_e = 0\text{)},\end{aligned}$$

where (as usual) $\vec{\partial}_i$ denotes the left-invariant vector field whose value at the identity is ∂_i. It follows that $(\partial_j\partial_{n+i}F)(e, e) = \partial_j|_e(\iota(\vec{\partial}_i)\alpha)$. We thus find

$$\begin{aligned}\big(\partial_i\partial_{n+j}F - \partial_{n+i}\partial_jF\big)(e, e) &= \big(\partial_i\partial_{n+j}F - (-1)^{\langle\varepsilon(x^i)|\varepsilon(x^j)\rangle}\partial_j\partial_{n+i}F\big)(e, e) \\ &= \partial_i|_e(\iota(\vec{\partial}_j)\alpha) - (-1)^{\langle\varepsilon(x^i)|\varepsilon(x^j)\rangle}\partial_j|_e(\iota(\vec{\partial}_i)\alpha) \\ &= -\iota(\vec{\partial}_i|_e, \vec{\partial}_j|_e)d\alpha + \iota([\vec{\partial}_i, \vec{\partial}_j]|_e)\alpha = -\iota(v_i, v_j)\Phi\ .\end{aligned}$$

Comparing these results with the bracket on $\mathfrak{h} \oplus \mathfrak{c}$ given in (9.7), we see that they are the same. The conclusion is that the $\mathcal{A}$-Lie algebra associated to the $\mathcal{A}$-Lie group G is (isomorphic to) $\mathfrak{g}$. To finish the proof, we note that uniqueness up to isomorphisms follows from [7.9]. $\boxed{QED}$

9.12 Lemma. *Let G be an $\mathcal{A}$-Lie group and let $\phi : G \to \mathrm{Aut}(\mathfrak{g}) \subset \mathrm{End}_R(\mathfrak{g})_0$ be defined by $\phi(g) = \mathrm{Ad}(g^{-1})$. Then for $x \in \mathfrak{g}$: $\vec{x}_g\phi = -\,\mathrm{ad}_R(x) \circ \mathrm{Ad}(g^{-1}) \in \mathrm{End}_R(\mathfrak{g})$.*

Proof. Consider the map $\Phi : G \times G \to \mathrm{Aut}(\mathfrak{g})$ defined as $\Phi = \phi \circ m$. It is immediate that $\langle\!\langle \vec{x}_g \,\|\, T\phi \rangle\!\rangle = \langle\!\langle (\underline{0}_g, x) \,\|\, T\Phi \rangle\!\rangle$. But we can define Φ in a different way, using that Ad is a homomorphism: $\Phi = \mathit{Inv} \circ m \circ (\mathrm{Ad} \times \mathrm{Ad})$, where here the multiplication is in $\mathrm{Aut}(\mathfrak{g})$. Using this interpretation of Φ we can compute $\langle\!\langle (\underline{0}_g, x) \,\|\, T\Phi \rangle\!\rangle$ as

$$\begin{aligned}(\underline{0}_g, \vec{x}_e) &\mapsto (\underline{0}_{\mathrm{Ad}(g)}, (T_e\,\mathrm{Ad})(x)) = (\underline{0}_{\mathrm{Ad}(g)}, \mathrm{ad}_R(x)) \mapsto TL_{\mathrm{Ad}(g)}\,\mathrm{ad}_R(x) \\ &\mapsto -TR_{\mathrm{Ad}(g^{-1})}\,\mathrm{ad}_R(x)\ ,\end{aligned}$$

where we used [2.15] and [1.6]. In the identification $T\,\mathrm{Aut}(\mathfrak{g}) = \mathrm{Aut}(\mathfrak{g}) \times \mathrm{End}_R(\mathfrak{g})$ [1.20] we have, according to [V.3.2], $\langle\!\langle \vec{x}_g \,\|\, T\phi \rangle\!\rangle = (\phi(g), \vec{x}_g\phi)$. Using [1.23] we find in the same identification $\langle\!\langle (\underline{0}_g, x) \,\|\, T\Phi \rangle\!\rangle = (\mathrm{Ad}(g^{-1}), -\,\mathrm{ad}_R(x) \circ \mathrm{Ad}(g^{-1}))$. Comparing these two gives the desired result. QED

9.13 Technical lemma. *The 1-forms $\iota(v_i^R)\Omega$ on H are closed.*

Proof. Using the definition of Ω, we find for an arbitrary left-invariant vector field $\vec{z}$:

$$\iota(\vec{z}, v_i^R)\Omega|_g = \iota(z, TL_{g^{-1}}v_i^R|_g)\Phi = \iota(z, \mathrm{Ad}(g^{-1})v_i)\Phi \ .$$

Using the definition [V.7.3] or [V.7.6], we compute for two smooth homogeneous left-invariant vector fields $\vec{x}$ and $\vec{y}$:

$$\begin{aligned}
-\,\iota(\vec{x}_g, \vec{y}_g)\ & d(\iota(v_i^R)\Omega) \\
&= \vec{x}_g(\iota(\vec{y}, v_i^R)\Omega) - (-1)^{\langle \varepsilon(x)|\varepsilon(y)\rangle}\vec{y}_g(\iota(\vec{x}, v_i^R)\Omega) - \iota([\vec{x}, \vec{y}]_g, v_i^R|_g)\Omega_g \\
&= \vec{x}_g(\iota(y, \mathrm{Ad}(g^{-1})v_i)\Phi) - (-1)^{\langle \varepsilon(x)|\varepsilon(y)\rangle}\vec{y}_g(\iota(x, \mathrm{Ad}(g^{-1})v_i)\Phi) \\
&\qquad - \iota([x,y], \mathrm{Ad}(g^{-1})v_i)\Phi \\
&= (-1)^{\langle \varepsilon(x)|\varepsilon(y)\rangle}\iota(y, \vec{x}_g(\mathrm{Ad}(g^{-1})v_i))\Phi - \iota(x, \vec{y}_g(\mathrm{Ad}(g^{-1})v_i))\Phi \\
&\qquad - \iota([x,y], \mathrm{Ad}(g^{-1})v_i)\Phi \\
&= -(-1)^{\langle \varepsilon(x)|\varepsilon(y)\rangle}\iota(y, [x, \mathrm{Ad}(g^{-1})v_i])\Phi + \iota(x, [y, \mathrm{Ad}(g^{-1})v_i])\Phi \\
&\qquad - \iota([x,y], \mathrm{Ad}(g^{-1})v_i)\Phi \\
&= 0\ ,
\end{aligned}$$

where we used [9.12] to obtain the fifth line and (9.9) for the final result. Since smooth homogeneous left-invariant vector fields generate $\Gamma(TH)$ [1.19], the result follows.

A faster proof of this result, which is less direct because it invokes [8.21], goes as follows. Since Ω is left-invariant, $\mathcal{L}(x^H)\Omega = 0$ for all $x \in \mathbf{B}\mathfrak{h}$, and thus in particular $\mathcal{L}(v_i^H)\Omega = 0$. But $\mathcal{L}(X)\Omega = d\iota(X)\Omega + \iota(X)d\Omega = d\iota(X)\Omega$ because Ω is closed. Since $v_i^H = -v_i^R$ we obtain $d\iota(v_i^R)\Omega = 0$ as wanted. QED

9.14 Technical lemma. *The 1-form Ψ (9.10) on $H \times H$ is closed.*

Proof. Using [V.7.9-iii] and [V.7.22] we find

$$d\Psi = m^*\Omega - \pi_1^*\Omega - \pi_2^*\Omega + \textstyle\sum_i(\pi_1^*\, d\omega^i)\cdot(\pi_2^* f_i - f_i(e)) - \sum_i(\pi_1^*\, \omega^i) \wedge (\pi_2^*(\iota(v_i^R)\Omega))\ .$$

The tangent space at each point in $H \times H$ is generated by vectors of the form $(\vec{x}_1, \vec{x}_2)$ where the $\vec{x}_i$ are (smooth) left-invariant vector fields on H. In order to prove $d\Psi = 0$, it is thus sufficient to prove $\iota((\vec{x}_1, \vec{x}_2), (\vec{y}_1, \vec{y}_2))d\Psi = 0$ for all $x_i, y_i \in \mathfrak{h}$. By graded skew-symmetry and bilinearity it is sufficient to verify the three cases $\iota((\underline{0}, \vec{x}), (\underline{0}, \vec{y}))d\Psi = 0$, $\iota((\vec{x}, \underline{0}), (\underline{0}, \vec{y}))d\Psi = 0$, and $\iota((\vec{x}, \underline{0}), (\vec{y}, \underline{0}))d\Psi = 0$ for $x, y \in \mathfrak{h}$.

Before we start these three verifications, we recall that for $x \in \mathfrak{h}$ we have $TL_g\vec{x}_h = \vec{x}_{gh}$ and $TR_h\vec{x}_g = TL_gTR_hx = TL_{gh}(\mathrm{Ad}(h^{-1})x) = \overrightarrow{\mathrm{Ad}(h^{-1})x}|_{gh}$. In the same spirit we find for x^R the equality $x^R|_h = TR_hx = \overrightarrow{\mathrm{Ad}(h^{-1})x}|_h$. Using [V.7.20] we now turn our attention to the three equalities we have to prove.

• The first equation yields

$$\iota((\underline{0}_g, \vec{x}_h), (\underline{0}_g, \vec{y}_h))\, d\Psi = \iota(x,y)\Phi - 0 - \iota(x,y)\Phi + \textstyle\sum_i 0 - \sum_i 0 = 0\ .$$

• The second one yields

$$\begin{aligned}
&\iota((\vec{x}_g, \underline{0}_h), (\underline{0}_g, \vec{y}_h))\, d\Psi = \\
&\quad = \iota(\mathrm{Ad}(h^{-1})x, y)\Phi - 0 - 0 + \textstyle\sum_i 0 + \sum_i (-1)^{\langle \varepsilon(y)|\varepsilon(\omega^i)\rangle}(\langle\!\langle x \,\|\, {}^i v\rangle\!\rangle \cdot \iota(\vec{y}_h, v_i^R|_h)\Omega \\
&\quad = \iota(\mathrm{Ad}(h^{-1})x, y)\Phi + \textstyle\sum_i (-1)^{\langle \varepsilon(y)|\varepsilon(v_i)\rangle}(\langle\!\langle x \,\|\, {}^i v\rangle\!\rangle \iota(y, \mathrm{Ad}(h^{-1})v_i)\Phi \\
&\quad = \iota(\mathrm{Ad}(h^{-1})x, y)\Phi + (-1)^{\langle \varepsilon(x)|\varepsilon(y)\rangle}\iota(y, \mathrm{Ad}(h^{-1})\Big(\textstyle\sum_i (\langle\!\langle x \,\|\, {}^i v\rangle\!\rangle \cdot v_i\Big))\Phi \\
&\quad = \iota(\mathrm{Ad}(h^{-1})x, y)\Phi + (-1)^{\langle \varepsilon(x)|\varepsilon(y)\rangle}\iota(y, \mathrm{Ad}(h^{-1})x)\Phi = 0\ .
\end{aligned}$$

• Finally for the third equality we find

$$\begin{aligned}
&\iota((\vec{x}_g, \underline{0}_h), (\vec{y}_g, \underline{0}_h))\, d\Psi \\
&\quad = \iota(\mathrm{Ad}(h^{-1})x, \mathrm{Ad}(h^{-1})y)\Phi - \Phi(x,y) + \textstyle\sum_i (\iota(\vec{x}_g, \vec{y}_g)d\omega^i) \cdot (f_i(h) - f_i(e)) \\
&\quad = \iota(\mathrm{Ad}(h^{-1})x, \mathrm{Ad}(h^{-1})y)\Phi - \Phi(x,y) + \textstyle\sum_i \langle\!\langle [x,y] \,\|\, {}^i v\rangle\!\rangle \cdot (f_i(h) - f_i(e))\ .
\end{aligned}$$

To prove that this is identically zero, we introduce the functions $\chi_1, \chi_2 : H \to \mathfrak{c}$ by

$$\begin{aligned}
\chi_1(h) &= \Phi(\mathrm{Ad}(h^{-1})x, \mathrm{Ad}(h^{-1})y) - \Phi(x,y) \\
\chi_2(h) &= \textstyle\sum_i \langle\!\langle [x,y] \,\|\, {}^i v\rangle\!\rangle \cdot (f_i(h) - f_i(e))\ .
\end{aligned}$$

We then compute for an arbitrary homogeneous left-invariant vector field $\vec{z}$ on H the functions $\vec{z}\chi_i$ (using [9.12]) :

$$\begin{aligned}
(\vec{z}\chi_1)(h) = -\iota([z, \mathrm{Ad}(h^{-1})x], \mathrm{Ad}(h^{-1})y)\Phi& \\
- (-1)^{\langle \varepsilon(z)|\varepsilon(x)\rangle}\iota(\mathrm{Ad}(h^{-1})x, [z, \mathrm{Ad}(h^{-1})y])\Phi&
\end{aligned}$$

$$\begin{aligned}
(\vec{z}\chi_2)(h) &= \textstyle\sum_i (-1)^{\langle \varepsilon(z)|\varepsilon(x)+\varepsilon(y)+\varepsilon(v_i)\rangle}\langle\!\langle [x,y] \,\|\, {}^i v\rangle\!\rangle \cdot \vec{z}_h f_i \\
&= \textstyle\sum_i (-1)^{\langle \varepsilon(z)|\varepsilon(x)+\varepsilon(y)+\varepsilon(v_i)\rangle}\langle\!\langle [x,y] \,\|\, {}^i v\rangle\!\rangle \cdot \iota(\vec{z}_h)\iota(v_i^R)\Omega \\
&= \textstyle\sum_i (-1)^{\langle \varepsilon(z)|\varepsilon(x)+\varepsilon(y)+\varepsilon(v_i)\rangle}\langle\!\langle [x,y] \,\|\, {}^i v\rangle\!\rangle \cdot \iota(z, \mathrm{Ad}(h^{-1})v_i)\Phi \\
&= \iota(z, \mathrm{Ad}(h^{-1})\textstyle\sum_i \langle\!\langle [x,y] \,\|\, {}^i v\rangle\!\rangle \cdot v_i)\Phi \\
&= \iota(z, \mathrm{Ad}(h^{-1})([x,y]))\Phi\ .
\end{aligned}$$

It follows from [2.19] and (9.9) that $\vec{z}(\chi_1 + \chi_2) = 0$, and then it follows from [V.3.21] and [1.19] that $\chi_1 + \chi_2$ is a constant function. Since $\chi_1(e) + \chi_2(e) = 0$, we conclude that $\chi_1 + \chi_2$ is identically zero. $\boxed{QED}$

9.15 Technical lemma. *The function F satisfies the relation (9.11).*

Proof. We introduce the function $K : H \times H \times H \to \mathfrak{c}$ by

$$K(g,h,k) = F(g,h) + F(gh,k) - F(h,k) - F(g,hk) ,$$

which means that we have to prove that K is identically zero. We do this by noting that $K(e,e,e) = 0$ and proving that $dK = 0$ as a 1-form. The result then follows from [V.7.11]. If we denote by $\pi_{12} : H \times H \times H \to H \times H$ the map $(g,h,k) \mapsto (g,h)$, by $m_{12} : H \times H \times H \to H \times H$ the map $(g,h,k) \mapsto (gh,k)$, and similarly for π_{23} and m_{23}, the function K can be written as

$$K = F \circ \pi_{12} + F \circ m_{12} - F \circ \pi_{23} - F \circ m_{23} .$$

We thus find for dK, using $dF = \Psi$ and [V.7.19],

$$dK = \pi_{12}^* \Psi + m_{12}^* \Psi - \pi_{23}^* \Psi - m_{23}^* \Psi .$$

Now Ψ contains terms with α and terms with a sum over f_is. Using the easily proved relations for functions from $H^3 \to H$: $m \circ m_{12} = m \circ m_{23}$ (associativity of the multiplication), $m \circ \pi_{12} = \pi_1 \circ m_{12}$, $m \circ \pi_{23} = \pi_2 \circ m_{23}$, $\pi_1 \circ \pi_{12} = \pi_1 \circ m_{23}$, $\pi_2 \circ m_{12} = \pi_2 \circ \pi_{23}$, and $\pi_1 \circ \pi_{23} = \pi_2 \circ \pi_{12}$, the reader can easily show that the parts involving α in dK cancel. We thus have to prove that the remaining terms given by

$$(9.16) \quad \sum_i \pi_1^* \omega^i \cdot (\pi_2^* f_i - f_i(e)) + (\pi_1 \circ m_{12})^* \omega^i \cdot (\pi_3^* f_i - f_i(e)) \\ - \pi_2^* \omega^i \cdot (\pi_3^* f_i - f_i(e)) - \pi_1^* \omega^i \cdot ((\pi_2 \circ m_{23})^* f_i - f_i(e))$$

add up to zero. If we introduce the matrix elements $\mathrm{Ad}(k)_j{}^i$ by $\mathrm{Ad}(k) v_j = \sum_i \mathrm{Ad}(k)_j{}^i v_i$ (the $\overset{\bullet}{M}_L$ matrix elements [II.4.1] of the even automorphism $\mathrm{Ad}(k)$, seen for the occasion as left linear), then we can compute for three left-invariant vector fields $\vec{x}$, $\vec{y}$, and $\vec{z}$ on H (using that $TR_h \vec{x}_g = \overrightarrow{\mathrm{Ad}(h^{-1})x}|_{gh}$)

$$\begin{aligned} \iota((\vec{x}_g, \vec{y}_h, \vec{z}_k))(\pi_1 \circ m_{12})^* \omega^i &= \iota((\vec{x}_g, \vec{y}_h)) m^* \omega^i \\ &= \iota(\langle\!\langle (\underline{0}_g, \vec{y}_h) \| Tm \rangle\!\rangle + \langle\!\langle (\vec{x}_g, \underline{0}_h) \| Tm \rangle\!\rangle) \omega^i \\ &= \iota(\vec{y}_{gh}) \omega^i + \iota(\overrightarrow{\mathrm{Ad}(h^{-1})x}|_{gh}) \omega^i \\ &= \langle\!\langle y \|^i v \rangle\!\rangle + \textstyle\sum_j \langle\!\langle x \|^j v \rangle\!\rangle \cdot \mathrm{Ad}(h^{-1})_j{}^i \\ &= \iota(\vec{x}_g, \vec{y}_h, \vec{z}_k) \pi_2^* \omega^i + \textstyle\sum_j \iota(\vec{x}_g, \vec{y}_h, \vec{z}_k) \pi_1^* \omega^j \cdot \mathrm{Ad}(h^{-1})_j{}^i . \end{aligned}$$

We conclude that

$$(\pi_1 \circ m_{12})^* \omega^i|_{(g,h,k)} = \pi_2^* \omega^i|_{(g,h,k)} + \textstyle\sum_j \pi_1^* \omega^j|_{(g,h,k)} \cdot \mathrm{Ad}(h^{-1})_j{}^i .$$

Substituting this in (9.16) and evaluating this 1-form in the point $(g,h,k) \in H^3$ we find

$$\sum_i (\pi_1^* \omega^i)_{(g,h,k)} \cdot \Big(f_i(h) - f_i(hk) + \textstyle\sum_j \mathrm{Ad}(h^{-1})_i{}^j \cdot (f_j(k) - f_j(e)) \Big) .$$

Since the ω^i are independent, we concentrate on the functions $\chi_i : H \times H \to \mathfrak{c}$ defined as

$$\chi_i(h,k) = f_i(h) - f_i(hk) + \sum_j \operatorname{Ad}(h^{-1})_i{}^j \cdot (f_j(k) - f_j(e)) \,. \tag{9.17}$$

In order to prove that these are identically zero, we want to apply [V.3.20]. Therefore we compute $(\underline{0}_h, \vec{x}_k)\chi_i = \iota((\underline{0}_h, \vec{x}_k))d\chi_i$ for an arbitrary left-invariant vector field $\vec{x}$.

$$\begin{aligned}
(\underline{0}_h, \vec{x}_k)\chi_i &= -\vec{x}_{hk} f_i + \textstyle\sum_j (-1)^{\langle \varepsilon(x)|\varepsilon(v_i)+\varepsilon(v_j)\rangle} \operatorname{Ad}(h^{-1})_i{}^j \cdot \vec{x}_k f_j \\
&= -\iota(\vec{x}_{hk})\iota(v_i^R)\Omega + \textstyle\sum_j (-1)^{\langle \varepsilon(x)|\varepsilon(v_i)+\varepsilon(v_j)\rangle} \operatorname{Ad}(h^{-1})_i{}^j \cdot \iota(\vec{x}_k)\iota(v_j^R)\Omega \\
&= -\iota(x, \operatorname{Ad}((hk)^{-1})v_i)\Phi + \iota(\vec{x}_k)\iota\Big(\big(\textstyle\sum_j \operatorname{Ad}(h^{-1})_i{}^j v_j^R\big)\Big)\Omega \\
&= -\iota((x, \operatorname{Ad}((hk)^{-1})v_i))\Phi + \iota(x, \operatorname{Ad}(k^{-1})(\operatorname{Ad}(h^{-1})v_i))\Phi \\
&= 0 \qquad \text{because Ad is a homomorphism.}
\end{aligned}$$

We conclude by [1.19] and [V.3.20] that $\chi_i(h,k)$ is independent of k. But if we take $k = e$, we find $\chi_i(h,e) = 0$, and thus χ_i is identically zero. This proves that the terms in (9.16) add up to zero, which proves that $dK = 0$, and thus that K is a constant function. Since $K(e,e,e) = 0$, we have shown that K is identically zero, i.e., that F satisfies the relation (9.11). QED

9.18 Comments on the proof of [9.6]. In order to put some of the items of the proof of [9.6] in a wider perspective, we give some remarks for the interested reader.

- The graded skew-symmetric bilinear function Φ is a 2-cocycle in $\mathcal{A}$-Lie algebra cohomology, the cocycle condition being given by (9.9). Changing the section σ changes this cocycle with the coboundary of a 1-cochain. The reconstruction of $\mathfrak{g}$ as the $\mathcal{A}$-Lie algebra $\mathfrak{h} \oplus \mathfrak{c}$ is part of the standard isomorphism between cohomology classes in dimension 2 of $\mathcal{A}$-Lie algebra cohomology and equivalence classes of central extensions of $\mathfrak{h}$ by $\mathfrak{c}$.
- The function F is a 2-cocycle in $\mathcal{A}$-Lie group cohomology, the cocycle condition for this cohomology being given by (9.11).
- The functions f_i can be seen as a generalization of a momentum map known from symplectic geometry; here Ω plays the role of the symplectic form and the v_i^R the role of the fundamental vector fields of the group action on the symplectic manifold.
- The functions $f_i - f_i(e)$ can be put together to form a function on H with values in ${}^*\mathfrak{h} \otimes \mathfrak{c}$. The fact that the functions χ_i in (9.17) are identically zero then says that this new function can be seen as a 1-cocycle on H with values in the H-module ${}^*\mathfrak{h} \otimes \mathfrak{c}$. More details can be found in [So, Thm 11.17].

9.19 Examples. An $\mathcal{A}$-Lie algebra of dimension $1|1$ has a basis v_0, v_1 in which v_ε has parity ε. Since the bracket is even, we have $[v_0, v_0] = 0$, $[v_0, v_1] = \lambda v_1$, and $[v_1, v_1] = \mu v_0$, where λ, μ are real numbers because the bracket is supposed to be smooth. The graded Jacobi identity applied to v_0, v_1, v_1 tells us that $\lambda\mu = 0$. We conclude that, up to rescaling, there exists three $\mathcal{A}$-Lie algebras of dimension $1|1$: an abelian one ($\lambda = \mu = 0$), one with $[v_0, v_0] = [v_1, v_1] = 0$ and $[v_0, v_1] = v_1$ ($\lambda = 1$, $\mu = 0$), and a third with $[v_0, v_0] = [v_0, v_1] = 0$ and $[v_1, v_1] = v_0$ ($\lambda = 0$, $\mu = 1$). We intend to apply the construction of the proof of [9.6] to find the corresponding $\mathcal{A}$-Lie groups.

• In the abelian case we find $\mathfrak{c} = \mathfrak{g}$, and thus $\mathfrak{h} = \{0\}$. The corresponding simply connected $\mathcal{A}$-Lie group H is obviously $\{e\}$ (of dimension 0). Since on an $\mathcal{A}$-manifold of dimension 0 there are no non-zero k-forms with $k > 0$, we have $\alpha = 0$, $f_i = 0$, $\Psi = 0$, and hence $F = 0$. We conclude that $G = \mathfrak{c}_0$ with group law

$$m_G(a, b) = a + b \; .$$

In other words, G is the additive abelian group $\mathfrak{c}_0 \cong \mathcal{A}$.

• For the second case with $[v_0, v_0] = [v_1, v_1] = 0$ and $[v_0, v_1] = v_1$ we find $\mathrm{ad}_R(av_0 + bv_1) : v_0 \mapsto -bv_1,\ v_1 \mapsto av_1$. It follows that $\mathfrak{c} = \{0\}$, and that the image $\mathrm{ad}_R(\mathfrak{g}) \subset \mathrm{End}_R(\mathfrak{g})$ is given by

$$\mathrm{ad}_R(\mathfrak{g}) = \{ \begin{pmatrix} 0 & 0 \\ -b & a \end{pmatrix} \mid a, b \in \mathcal{A} \} \; .$$

Using [3.6] and the proof of [4.7], the corresponding $\mathcal{A}$-Lie subgroup G of $\mathrm{Aut}(\mathfrak{g})$ can be found to be

$$G = \{ \begin{pmatrix} 1 & 0 \\ \alpha & a \end{pmatrix} \mid a \in \mathcal{A}_0, \alpha \in \mathcal{A}_0, \mathbf{B}a > 0 \} \; .$$

As an $\mathcal{A}$-manifold this is an open subset of $\mathfrak{g}_0 \cong \mathcal{A}$, but whose group law is given by $m_G((a, \alpha), (b, \beta)) = (ab, \alpha + a\beta)$. This can be interpreted as the $a\xi + \alpha$ group of affine transformations of the odd affine line $\mathcal{A}_1$.

• The third case with $[v_0, v_0] = [v_0, v_1] = 0$ and $[v_1, v_1] = v_0$ presents the most interesting application of [9.6]. It is easily seen that $\mathrm{ad}_R(av_0 + bv_1) : v_0 \mapsto 0, v_1 \mapsto bv_0$. It follows that $\mathfrak{c}$ is the graded subspace generated by v_0 of dimension $1|0$, that $\mathfrak{h} = \mathfrak{g}/\mathfrak{c}$ is the abelian $\mathcal{A}$-Lie algebra of dimension $0|1$ with single basis vector $\hat{v}_1$, and that the image $\mathrm{ad}_R(\mathfrak{g}) \subset \mathrm{End}_R(\mathfrak{g})$ is given by

$$\mathrm{ad}_R(\mathfrak{g}) = \{ \begin{pmatrix} 0 & b \\ 0 & 0 \end{pmatrix} \mid b \in \mathcal{A} \} \; .$$

Again using (the proof of) [4.7] and [3.6], the corresponding $\mathcal{A}$-Lie subgroup H of $\mathrm{Aut}(\mathfrak{g})$ can be found to be

$$H = \{ \begin{pmatrix} 1 & \alpha \\ 0 & 1 \end{pmatrix} \mid \alpha \in \mathcal{A}_1 \} \; .$$

In other words, $H = \mathcal{A}_1$ with the usual addition as group operation. Using the section $\sigma : \mathfrak{h} \to \mathfrak{g}$ defined by $\sigma(\hat{v}_1) = v_1$, we find for the map Φ :

$$\iota(\hat{v}_1, \hat{v}_1)\Phi = \sigma(0) - [v_1, v_1] = -v_0 \; .$$

Denoting by ξ the odd coordinate on H, we find that the left-invariant 2-form Ω is given by

$$\Omega = -\tfrac{1}{2} d\xi \wedge d\xi = d(-\tfrac{1}{2}\xi \, d\xi) \; .$$

Hence $\alpha = -\frac{1}{2}\xi \, d\xi$, which satisfies indeed the condition $\alpha_e = 0$ (because $e = 0$ in this group). The right-invariant vector field associated to $\hat{v}_1 = \partial_\xi|_e$ is the vector field ∂_ξ, hence

$\iota(\partial_\xi)\Omega = -d\xi$, and thus $f(\xi) = -\xi$ is a solution. Together with the left-invariant 1-form $d\xi$ on H we find, using coordinates (ξ, η) on $H \times H$, for Ψ the 1-form

$$\Psi = -\tfrac{1}{2}(\xi+\eta)\, d(\xi+\eta) + \tfrac{1}{2}\xi\, d\xi + \tfrac{1}{2}\eta\, d\eta - (d\xi)\cdot\eta = d(-\tfrac{1}{2}\xi\eta)\ .$$

For F we thus find the function $F(\xi,\eta) = -\frac{1}{2}\xi\eta$; for the group $G = \mathcal{A}_1 \times \mathcal{A}_0 \cong \mathcal{A}$ this gives us the multiplication

$$m_G((\alpha, a), (\beta, b)) = (\alpha+\beta, a+b-\tfrac{1}{2}\alpha\beta)\ .$$

This group is the simply connected covering of the $\mathcal{A}$-Lie subgroup $H = \mathbf{G}S^1 \times \mathcal{A}_1$ discussed in [4.14]. The difference in constants is explained by the fact that there the scaling is such that $[w_1, w_1] = -4w_0$. In fact, the covering map is given by the morphism $\rho : G \to \mathbf{G}S^1 \times \mathcal{A}_1$, $(\alpha, a) \mapsto (\mathrm{e}^{-4ia}, \alpha)$.

We have also encountered this $\mathcal{A}$-Lie algebra as the $\mathcal{A}$-Lie algebra of the multiplicative group $\mathcal{A}^* = \{a \in \mathcal{A} \mid \mathbf{B}a \neq 0\}$ discussed in [2.8]. This is a non-connected $\mathcal{A}$-Lie group, whose connected components are simply connected. The covering map from G to the connected component containing the identity of $\mathcal{A}^*$ is given by $\rho : G \to \mathcal{A}^*$, $(\alpha, a) \mapsto \mathrm{e}^{-2a+\alpha} = \mathrm{e}^{-2a} + \mathrm{e}^{-2a}\,\alpha$, which is actually an isomorphism because both are simply connected.

9.20 Remark. The three groups of dimensions $1|1$ are exactly the three special cases considered in [MS-V] in a more general approach to integration of (non-homogeneous) vector fields.

Chapter VII

Connections

In a direct product with the two projections on the separate factors, we know what horizontal and vertical directions are: those that project to zero under the tangent map of one of these two projections. A fiber bundle $\pi : B \to M$ with typical fiber F and structure group G is locally a direct product, but only one of the two projections is independent of such a local trivialization: the one corresponding to the bundle projection π. By convention the directions in B that project to zero under the tangent map $T\pi$ are called vertical. It follows that on a fiber bundle we do not have a natural definition of what horizontal directions are; the local idea of horizontal directions is not independent of the choice of the local trivialization. A connection on a fiber bundle is an additional structure which provides the notion of horizontal directions. This additional structure can take various forms. The most natural one is to define exactly the horizontal directions, i.e., a subbundle $\mathcal{H} \subset TB$ which is a supplement to the subbundle of vertical directions $V = \ker T\pi \subset TB$. In this form it is called an Ehresmann connection. But other forms for the additional structure are sometimes useful: a connection 1-form on a principal fiber bundle, a covariant derivative on a vector bundle, or parallel transport.

The notion of an Ehresmann connection is too general for most purposes. A much more interesting subclass of connections is formed by FVF connections, whose form is determined, in a sense to be made precise, by the fundamental vector fields of the structure group on the typical fiber. The connection 1-form, the covariant derivative and linear connections all fall in this subclass. Moreover, for the subclass of fiber bundles concerned (principal/vector), they are equivalent to FVF connections.

In this chapter we define the above mentioned notions of a connection and we show how they are related. On principal fiber bundles the FVF connection is also described as the kernel of the connection 1-form, whereas on vector bundles the covariant derivative of a section describes how far the section is from being horizontal. Moreover, a (vector) bundle B can be seen as associated to a principal fiber bundle P: the structure bundle. Sections of B then can be seen as a special kind of functions on P and the covariant derivative gets *transformed into the exterior covariant derivative on P associated to its FVF connection. This correspondence can be generalized to differential forms with values in an $\mathcal{A}$-vector space or in a vector bundle. And then a covariant derivative and the exterior covariant derivative can be seen as generalizations of the usual exterior derivative of (ordinary) differential forms.*

The last aspect of connections that is treated here is the notion of curvature. An Ehresmann connection $\mathcal{H}$ on a bundle B is in particular a subbundle of TB. As such one can ask whether $\mathcal{H}$ is a foliation, i.e., is involutive. In general the answer will be negative, but there are several cases in which one can measure to what extent it is not involutive. For principal fiber bundles with a connection 1-form ω this is done by the curvature 2-form $\Omega = D\omega$, the exterior covariant derivative of the connection 1-form. For vector bundles with a covariant derivative ∇ this is done by the curvature tensor R. In these cases the statement is that the FVF connection is involutive if and only if the curvature is zero. Moreover, we show that Ω and R correspond under the identifications which link connections on principal fiber bundles with those on associated vector bundles.

1. More about vector valued forms

In this technical section we generalize operations concerning $\mathcal{A}$-vector spaces (composition, evaluation, bracket, etcetera) to vector valued differential forms. We prove some elementary but useful formulæ and we introduce the all important Maurer-Cartan 1-form Θ_{MC} on an $\mathcal{A}$-Lie group.

1.1 Definition. Let E, F, and G be three $\mathcal{A}$-vector spaces with homogeneous bases (e_i), (f_j), and (g_k) respectively, and let $\Phi : E \times F \to G$ be an even smooth bilinear map. With these ingredients we define the *Φ-wedge product* $\wedge_\Phi$, which associates to an E-valued p-form α and an F-valued q-form β, a G-valued $(p+q)$-form $\alpha \wedge_\Phi \beta$, all on an $\mathcal{A}$-manifold M. The construction is as follows. The forms α and β are uniquely determined by ordinary differential forms α^i and β^j according to $\alpha = \sum_i \alpha^i \otimes e_i$, $\beta = \sum_j \beta^j \otimes f_j$ [VI.9.1]. And then $\alpha \wedge_\Phi \beta$ is defined by

$$\Big(\sum_i \alpha^i \otimes e_i\Big) \wedge_\Phi \Big(\sum_j \beta^j \otimes f_j\Big) = \sum_{i,j} \alpha^i \wedge \mathfrak{C}^{\varepsilon(e_i)}(\beta^j) \otimes \Phi(e_i, f_j)\ . \tag{1.2}$$

Introducing matrix elements for Φ by $\Phi(e_i, f_j) = \sum_k \Phi^k_{ij}\, g_k$, the G-valued $(p+q)$-form $\alpha \wedge_\Phi \beta$ is defined by the ordinary $(p+q)$-forms $(\alpha \wedge_\Phi \beta)^k$ given by

$$(\alpha \wedge_\Phi \beta)^k = \sum_{i,j} \alpha^i \wedge \mathfrak{C}^{\varepsilon(e_i)}(\beta^j) \cdot \Phi^k_{ij}\ .$$

It is elementary to check that the definition of $\alpha \wedge_\Phi \beta$ is independent of the choice of the bases for E and F, thus guaranteeing a correct definition of the Φ-wedge product.

1.3 Notation. Each map Φ has its associated Φ-wedge product which we denoted as $\wedge_\Phi$. However, specific maps Φ have their own notation for the associated wedge product. We will need the following four specific maps with the associated notation.

• *Multiplication by scalars:* in this application the $\mathcal{A}$-vector space E is $\mathcal{A}$, $F = G$, and Φ is (left) multiplication: $\Phi(a, v) = a \cdot v$ In this case the Φ-wedge product $\wedge_\Phi$ is simply denoted as $\wedge$.

• *Applying a linear map to a vector:* here F and G are arbitrary $\mathcal{A}$-vector spaces, $E = \mathrm{Hom}_R(F; G)$, and Φ is the evaluation map: $\Phi(A, v) = A(v)$. In this case the Φ-wedge product $\wedge_\Phi$ is denoted as $\mathbin{\dot\wedge}$.

• *Composition of linear maps:* here E, F and G are all equal to $\mathrm{End}_R(C)$, the set of (right linear) endomorphisms of an $\mathcal{A}$-vector space C, and Φ is composition: $\Phi(A, B) = A \circ B$. In this case the Φ-wedge product $\wedge_\Phi$ is denoted as $\mathbin{\mathring\wedge}$.

• *The bracket in $\mathcal{A}$-Lie algebras:* here $E = F = G = \mathfrak{g}$ are all equal to an $\mathcal{A}$-Lie algebra $\mathfrak{g}$ and Φ is the bracket: $\Phi(x, y) = [\, x, y\,]$. In this case the Φ-wedge product of α and β is denoted by $[\, \alpha \mathbin{\overset{,}{\wedge}} \beta\,]$.

As is usual with notation, there is sometimes more than one way to write things. Here the exceptions all occur when either α or β is a 0-form. The most obvious case is in the first case when g is a 0-form on M, i.e., an ordinary function, and β an F-valued k-form. In that case the F-valued k-form $g \wedge \beta$ is the same as $g \cdot \beta$. This is a direct consequence of the similar fact for ordinary k-forms [V.7.1]. Less obvious is the similar situation in the second case when A is a $\mathrm{Hom}_R(F; G)$-valued 0-form, i.e., a smooth function $A : M \to \mathrm{Hom}_R(F; G)$, and β an F-valued k-form. Then it is customary to write $A \circ \beta$ instead of $A \mathbin{\dot\wedge} \beta$. The idea is that at each point $m \in M$ it is the composition of the map $\beta|_m$ from $(T_m M)^k$ to F with the map A_m from F to G. Similarly in the third case: if A is a $\mathrm{Hom}_R(F; G)$-valued 0-form and β a $\mathrm{Hom}_R(F; G)$-valued k-form, then it is customary to write $A \circ \beta$ instead of $A \mathbin{\mathring\wedge} \beta$. Coming back to the second case, if α is a $\mathrm{Hom}_R(F; G)$-valued k-form and g an F-valued 0-form, i.e., a smooth function $s : M \to F$, then it is customary to write $\alpha(g)$ or $\alpha \cdot g$ instead of $\alpha \mathbin{\dot\wedge} g$, the idea being that for fixed $m \in M$ and $X_i \in T_m M$ it is the action of the homomorphism $\iota(X_1, \ldots, X_k)\alpha_m$ on the vector $g(m)$.

1.4 Lemma. *Let $\mathfrak{g}$ be an $\mathcal{A}$-Lie algebra, let α be an even $\mathfrak{g}$-valued 1-form, β an even $\mathfrak{g}$-valued 2-form, and let X, Y, and Z be homogeneous vector fields on M. Then:*

$$\iota(X, Y)[\, \alpha \mathbin{\overset{,}{\wedge}} \alpha\,] = -2[\, \iota(X)\alpha, \iota(Y)\alpha\,] \qquad \textit{and}$$

$$\begin{aligned}\iota(X, Y, Z)[\, \beta \mathbin{\overset{,}{\wedge}} \alpha\,] = {}& \\ & [\, \iota(X, Y)\beta, \iota(Z)\alpha\,] + (-1)^{\langle \varepsilon(X) | \varepsilon(Y) + \varepsilon(Z)\rangle}[\, \iota(Y, Z)\beta, \iota(X)\alpha\,] \\ & \qquad + (-1)^{\langle \varepsilon(Z) | \varepsilon(X) + \varepsilon(Y)\rangle}[\, \iota(Z, X)\beta, \iota(Y)\alpha\,]\,.\end{aligned}$$

In case $\mathfrak{g} = \mathrm{End}_R(E)$ we also have the equality $[\, \alpha \mathbin{\overset{,}{\wedge}} \alpha\,] = 2\alpha \mathbin{\mathring\wedge} \alpha$.

Proof. For the first equality we compute for homogeneous 1-forms α and γ

$$
\begin{aligned}
\iota(X,Y)[\,\alpha \wedge\!\!\!\!,\; \gamma\,] &= \sum_{ij} \iota(X,Y)(\alpha^i \wedge \gamma^j) \otimes (-1)^{\langle \varepsilon(v_i)|\varepsilon(v_j)+\varepsilon(\gamma)\rangle} [\, v_i, v_j \,] \\
&= \sum_{ij} \Big((-1)^{\langle \varepsilon(X)|\varepsilon(Y)+\varepsilon(\alpha)+\varepsilon(v_i)\rangle} \iota(Y)\alpha^i \, \iota(X)\gamma^j \\
&\qquad\qquad - (-1)^{\langle \varepsilon(Y)|\varepsilon(\alpha)+\varepsilon(v_i)\rangle} \iota(X)\alpha^i \, \iota(Y)\gamma^j \Big) \\
&\qquad \otimes (-1)^{\langle \varepsilon(v_i)|\varepsilon(v_j)+\varepsilon(\gamma)\rangle} [\, v_i, v_j \,] \\
&= (-1)^{\langle \varepsilon(X)|\varepsilon(Y)+\varepsilon(\alpha)\rangle} [\, \iota(Y)\alpha, \iota(X)\gamma \,] - (-1)^{\langle \varepsilon(Y)|\varepsilon(\alpha)\rangle} [\, \iota(X)\alpha, \iota(Y)\gamma \,] \,.
\end{aligned}
$$

The special case follows immediately from this result because for even α we have the equality $[\, \iota(Y)\alpha, \iota(X)\alpha \,] = -(-1)^{\langle \varepsilon(X)|\varepsilon(Y)\rangle} [\, \iota(X)\alpha, \iota(Y)\alpha \,]$. For the second equality we compute for homogeneous α and β:

$$
\begin{aligned}
&\iota(X,Y,Z)[\,\beta \wedge\!\!\!\!,\; \alpha\,] = \\
&= \sum_{ij} \iota(X,Y,Z)(\beta^i \wedge \alpha^j) \otimes (-1)^{\langle \varepsilon(v_i)|\varepsilon(v_j)+\varepsilon(\alpha)\rangle} [\, v_i, v_j \,] \\
&= \sum_{ij} \Big((-1)^{\langle \varepsilon(X)|\varepsilon(Y)+\varepsilon(Z)+\varepsilon(\beta)+\varepsilon(v_i)\rangle} \iota(Y,Z)\beta^i \, \iota(X)\alpha^j \\
&\qquad + (-1)^{\langle \varepsilon(Z)|\varepsilon(X)+\varepsilon(Y)\rangle} (-1)^{\langle \varepsilon(Y)|\varepsilon(v_i)+\varepsilon(\beta)\rangle} \iota(Z,X)\beta^i \, \iota(Y)\alpha^j \\
&\qquad + (-1)^{\langle \varepsilon(Z)|\varepsilon(v_i)+\varepsilon(\beta)\rangle} \iota(X,Y)\beta^i \, \iota(Z)\alpha^j \Big) \otimes (-1)^{\langle \varepsilon(v_i)|\varepsilon(v_j)+\varepsilon(\alpha)\rangle} [\, v_i, v_j \,] \\
&= (-1)^{\langle \varepsilon(Z)|\varepsilon(\beta)\rangle} [\, \iota(X,Y)\beta, \iota(Z)\alpha \,] + (-1)^{\langle \varepsilon(X)|\varepsilon(Y)+\varepsilon(Z)+\varepsilon(\beta)\rangle} [\, \iota(Y,Z)\beta, \iota(X)\alpha \,] \\
&\qquad + (-1)^{\langle \varepsilon(Y)|\varepsilon(\beta)\rangle} (-1)^{\langle \varepsilon(Z)|\varepsilon(X)+\varepsilon(Y)\rangle} [\, \iota(Z,X)\beta, \iota(Y)\alpha \,] \,.
\end{aligned}
$$

For $\mathfrak{g} = \mathrm{End}_R(E)$ we have the canonical basis $e_i \otimes e^j$ and we compute:

$$
\begin{aligned}
\iota(X,Y)(\alpha \wedge\!\!\!\!\circ\; \alpha) &= \sum_{ijpq} \iota(X,Y)({\alpha^i}_j \wedge {\alpha^p}_q) \\
&\qquad\qquad \otimes (-1)^{\langle \varepsilon(e_i)+\varepsilon(e_j)|\varepsilon(e_p)+\varepsilon(e_q))\rangle} e_i \otimes e^j \circ e_p \otimes e^q \\
&= \sum_{ijpq} \Big((-1)^{\langle \varepsilon(Y)|\varepsilon(e_p)+\varepsilon(e_q))\rangle} \iota(X){\alpha^p}_q \, \iota(Y){\alpha^i}_j \\
&\qquad - (-1)^{\langle \varepsilon(e_i)+\varepsilon(e_j)|\varepsilon(Y)+\varepsilon(e_p)+\varepsilon(e_q))\rangle} \iota(X){\alpha^i}_j \, \iota(Y){\alpha^p}_q \Big) e_i \otimes e^j \circ e_p \otimes e^q \\
&= (-1)^{\langle \varepsilon(X)|\varepsilon(Y)\rangle} \iota(Y)\alpha \circ \iota(X)\alpha - \iota(X)\alpha \circ \iota(Y)\alpha = -[\, \iota(X)\alpha, \iota(Y)\alpha \,] \,.
\end{aligned}
$$

We thus have shown that $\iota(X,Y)(\alpha \wedge\!\!\!\!\circ\; \alpha) = -[\, \iota(X)\alpha, \iota(Y)\alpha \,]$ for all homogeneous X, Y. Combining this with the first result finishes the proof. QED

1.5 Example (the Maurer-Cartan 1-form). Let G be an $\mathcal{A}$-Lie group, $\mathfrak{g}$ its $\mathcal{A}$-Lie algebra, (v_i) a basis of $\mathfrak{g}$, and $({}^iv)$ the associated left-dual basis of ${}^*\mathfrak{g} = {}^*T_eG$. Using [VI.8.15] we define ω^i to be the left-invariant 1-form on G satisfying $\omega^i|_e = {}^iv$. We then define the $\mathfrak{g}$-valued 1-form Θ_{MC} on G as

$$\Theta_{MC} = \sum_i \omega^i \otimes v_i \ ,$$

which is called the *Maurer-Cartan 1-form of* G. Since all ω^i are left-invariant, Θ_{MC} is a left-invariant $\mathfrak{g}$-valued even 1-form on G (that Θ_{MC} is even follows from the fact that the parity of ω^i is the same as that of v_i). Moreover, if $\vec{x} = \sum x^i \cdot \vec{v}_i$ ($x^i \in \mathcal{A}$) is a left-invariant vector field on G, the contraction $\iota(\vec{x})\Theta_{MC}$ yields

$$\iota(\vec{x})\Theta_{MC} = \sum_{i,j} x^j \iota(\vec{v}_j)\omega^i \otimes v_i = \sum_i x^i \cdot v_i = x \ ,$$

i.e., Θ_{MC} is the tautological $\mathfrak{g}$-valued 1-form on G. Another way to state the tautological nature of Θ_{MC} is the following. Let $X_g \in T_gG$ be an arbitrary tangent vector, then $x = TL_{g^{-1}}X_g \in T_eG \equiv \mathfrak{g}$ satisfies by definition $\vec{x}_g = X_g$. It follows immediately that $\iota(X_g)\Theta_{MC} = x$. Identifying $\mathfrak{g}$ with the set of left-invariant vector fields on G, we conclude that $\iota(X_g)\Theta_{MC}$ is the left-invariant vector field on G whose value at g is the given tangent vector $X_g \in T_gG$. We also deduce that Θ_{MC} can be defined by $\iota(X_g)\Theta_{MC} = TL_{g^{-1}}X_g$.

We know that Θ_{MC} is left-invariant, i.e., using the generalized pull-back we have $L_g^*\Theta_{MC} = \Theta_{MC}$ for all $g \in G$. To see its behavior under right translations, we first note that by definition we have $R_g^*\Theta_{MC} = \sum_i R_g^*\omega^i \otimes v_i$. And then we compute:

$$\iota(\vec{x}_h)R_g^*\omega^i = \iota(TR_g\,TL_hx)\omega^i|_{hg} = \iota(TL_{hg}\,TL_{g^{-1}}\,TR_gx)\omega^i|_{hg} = \iota(\overrightarrow{\mathrm{Ad}(g^{-1})x})\omega^i \ .$$

It follows that $\iota(X)\big(R_g^*\Theta_{MC}\big) = \mathrm{Ad}(g^{-1})\big(\iota(X)\Theta_{MC}\big)$. In other words, using the notation of [1.3], we can write $R_g^*\Theta_{MC} = \mathrm{Ad}(g^{-1}) \circ \Theta_{MC}$.

Let us now consider the special case $G = \mathrm{Aut}(E)$ with $\mathfrak{g} = \mathrm{End}_R(E)$ for some $\mathcal{A}$-vector space E. As explained in [VI.1.20], we use the basis $e_i \otimes e^j$ for $\mathrm{End}_R(E)$. Using the left coordinates $g^i{}_j = {}^\bullet M_R(g)^i{}_j$, i.e., $g = \sum_{i,j} g^i{}_j\, e_i \otimes e^j$, the Maurer-Cartan form can be written as

$$\Theta_{MC} = \sum_{p,q,r,s} dg^p{}_q\, \lambda^{qr}_{ps} \otimes e_r \otimes e^s$$

with coefficients λ that have to be determined. For $X \in \mathfrak{g}$ the corresponding left-invariant vector field $\vec{X}$ is given by $\vec{X}_g = \sum_{i,j,k} X^i{}_j\, g^k{}_i\, \partial_{g^k{}_j}$ [VI.1.20]. The condition $\langle\!\langle \vec{X} \,\|\, \Theta_{MC} \rangle\!\rangle = X$ thus leads to the equations

$$\sum_{i,j,k,r,s} X^i{}_j\, g^k{}_i\, \lambda^{jr}_{ks}\, e_r \otimes e^s = \sum_{r,s} X^r{}_s\, e_r \otimes e^s \ ,$$

i.e., $\sum_{i,j,k} X^i{}_j\, g^k{}_i\, \lambda^{jr}_{ks} = X^r{}_s$ for all j, r. Since this must be true for all X, we deduce that $\sum_k g^k{}_i\, \lambda^{jr}_{ks} = \delta^r_i\, \delta^j_s$, i.e., $\lambda^{jr}_{ks} = (g^{-1})^r{}_k\, \delta^j_s$ (use (VI.1.22)). We find for the Maurer-Cartan form Θ_{MC} :

$$\Theta_{MC} = \sum_{p,q,r} dg^p{}_q\, (g^{-1})^r{}_p \otimes e_r \otimes e^q = \sum_{q,r} (g^{-1} \circ dg)^r{}_q \otimes e_r \otimes e^q = g^{-1} \circ dg\ ,$$

where for the second equality we used (VI.1.22) and the fact that $g^i{}_j$ (and thus $dg^i{}_j$) has parity $\varepsilon(e_i) + \varepsilon(e_j)$. To interpret the last equality, we note that the canonical inclusion $\mathrm{Aut}(E) \to \mathrm{End}_R(E)$, $g \mapsto g$ can be seen as an $\mathrm{End}_R(E)$-valued 0-form (function) on $\mathrm{Aut}(E)$, as can be the map $g \mapsto g^{-1}$, $\mathrm{Aut}(E) \to \mathrm{End}_R(E)$. The 1-form dg thus is an $\mathrm{End}_R(E)$-valued 1-form, exterior derivative of the 0-form g. The composition $g^{-1} \circ dg$ is the wedge composition of the 0-form g^{-1} with the 1-form dg, where as usual we have omitted the wedge symbol because the first factor is a 0-form [V.7.1], [1.3].

We now go back to the general case and we look at the exterior derivative of the Maurer-Cartan form $d\Theta_{MC} = \sum_i d\omega^i \otimes v_i$, which is a left-invariant $\mathfrak{g}$-valued 2-form on G. Since the contraction of a left-invariant 1-form with a left-invariant vector field is a constant [VI.8.17], the formula for the exterior derivative [V.7.6] gives us

$$-\iota(\vec{v}_i, \vec{v}_j) d\omega^k = -\iota([\vec{v}_i, \vec{v}_j])\omega^k = -c^k_{ij}\ ,$$

where we have used the (real) structure constants c^k_{ij} of $\mathfrak{g}$ [VI.1.16]. If we now consider the left-invariant 2-form $\theta^k = \frac{1}{2} \sum_{pq} c^k_{pq} \omega^q \wedge \omega^p$ (beware of the order of the indices), we can compute

$$\iota(\vec{v}_i, \vec{v}_j)\theta^k = \iota(\vec{v}_i)\big(\iota(\vec{v}_j)\theta^k\big) = \tfrac{1}{2} \sum_{pq} c^k_{pq}\big(\delta^q_j \delta^p_i - (-1)^{\langle \varepsilon(v_i)|\varepsilon(v_j)\rangle} \delta^q_i \delta^p_j\big) = c^k_{ij}\ ,$$

where we used that $c^k_{ji} = -(-1)^{\langle \varepsilon(v_i)|\varepsilon(v_j)\rangle} c^k_{ij}$ due to graded skew-symmetry of the bracket on $\mathfrak{g}$. Since the values of a basis of left-invariant vector fields generate the tangent space at each point [VI.1.18], we deduce that $d\omega^k = \frac{1}{2} \sum_{ij} c^k_{ji} \omega^i \wedge \omega^j$. In terms of $d\Theta_{MC}$ this gives us the formula

$$d\Theta_{MC} = \tfrac{1}{2} \sum_{ijk} \omega^i \wedge \omega^j \otimes c^k_{ji} v_k = -\tfrac{1}{2} \sum_{ij} \omega^i \wedge \omega^j \otimes (-1)^{\langle \varepsilon(v_i)|\varepsilon(v_j)\rangle} [v_i, v_j]\ .$$

Comparing this expression with the definition of the wedge Lie bracket shows that we can write this equality as $d\Theta_{MC} = -\frac{1}{2}[\,\Theta_{MC} \wedge \Theta_{MC}\,]$ (remember that $\varepsilon(v_j) = \varepsilon(\omega^j)$); it is called *the structure equations of* G.

1.6 Lemma. *The Maurer-Cartan 1-form Θ_{MC} on an $\mathcal{A}$-Lie group G satisfies the equation $d\Theta_{MC} = -\frac{1}{2}[\,\Theta_{MC} \wedge \Theta_{MC}\,]$.*

Proof. The proof of this result has already been given in [1.5]. Here we give another proof using [1.4] and (V.7.6). For homogeneous $x, y \in \mathfrak{g}$ we have the equalities

$$\begin{aligned} -\iota(\vec{x}, \vec{y}) d\Theta_{MC} &= \vec{x}(\iota(\vec{y})\Theta_{MC}) - (-1)^{\langle \varepsilon(x)|\varepsilon(y)\rangle} \vec{y}(\iota(\vec{x})\Theta_{MC}) \\ &\quad - \iota([\,\vec{x}, \vec{y}\,])\Theta_{MC} = -\iota(\overrightarrow{[\,x, y\,]})\Theta_{MC} = -[\,x, y\,] \end{aligned}$$

and $\iota(\vec{x}, \vec{y})[\,\Theta_{MC} \wedge \Theta_{MC}\,] = -2[\,\iota(\vec{x})\Theta_{MC}, \iota(\vec{y})\Theta_{MC}\,] = -2[\,x, y\,]$. QED

2. Ehresmann connections and FVF connections

In this section we introduce the notion of an Ehresmann connection on an arbitrary fiber bundle and we show that there is a natural way to transport an Ehresmann connection to a pull-back bundle. We then introduce the more restrictive notion of an FVF connection, which can be described by local 1-forms Γ^a with values in the $\mathcal{A}$-Lie algebra of the structure group of the fiber bundle. We show that transporting an FVF connection to a pull-back bundle still gives an FVF connection and that it is described by the pull-backs of the local 1-forms Γ^a. We finish by showing that an FVF connection is integrable if and only if the local 2-forms $d\Gamma^a + \frac{1}{2}[\Gamma^a \wedge \Gamma^a]$ are all identically zero.

2.1 Discussion. If A and B are two sets, a function $f : A \to B$ is constant if and only if the image $f(A)$ consists of a single point: $f(A) = \{b\}$. If we have a differentiable structure, f is locally constant if and only if its tangent map Tf is zero [V.3.21]. Thinking in terms of bundles, these elementary facts obtain a new formulation. If $\pi : A \times B \to A$ denotes the projection on the first factor, there is a bijection between functions $f : A \to B$ and sections $s : A \to A \times B$ of the (trivial) bundle $\pi : A \times B \to A$, the identification given by $s(a) = (a, f(a))$. In the direct product $A \times B$ it is customary to call the subsets $\{a\} \times B$ vertical and the subsets $A \times \{b\}$ horizontal. The reason for this choice is that it is customary to draw the target space A as a horizontal line and the source space $A \times B$ as a rectangle above it. Given the map π, the vertical subspaces $\{a\} \times B$ can also be described as $\pi^{-1}(a)$. If A and B are $\mathcal{A}$-manifolds, we can also talk about vertical and horizontal directions: a tangent vector $(X, Y) \in T_aA \times T_bB \cong T_{(a,b)}(A \times B)$ [V.2.21] is vertical if X is zero, it is horizontal if Y is zero. Again using the map π, the vertical directions can be described as those tangent vectors that map to zero under $T\pi$. The set of all horizontal/vertical directions forms a foliation whose leaves are the horizontal/vertical subsets (as long as they are submanifolds). In terms of these definitions, a section $s : A \to A \times B$ is constant if and only if its image is a horizontal subset; it is locally constant if and only if its tangent map Ts maps vectors $X \in T_aA$ to horizontal vectors. Under the identification $s(a) = (a, f(a))$ this corresponds exactly to constant and locally constant functions $f : A \to B$.

We now generalize the above picture to a fiber bundle $\pi : B \to M$ with typical fiber F. Above a local trivializing chart $U \subset M$ the bundle is isomorphic to the direct product $U \times F$ with projection on the first factor. As such we can speak about horizontal and vertical subsets and about horizontal and vertical directions. And as before, the vertical subsets can be described as $\pi^{-1}(m)$ and the vertical directions as those tangent vectors that map under $T\pi$ to zero. Obviously the notion of a vertical direction does not change when we change the local trivializing chart; it can be described intrinsically by the projection map π. On the other hand, there is no reason to think that what is horizontal in terms of one trivialization remains horizontal in another trivialization. Said differently, a local section $s \in \Gamma_U(B)$ can be constant in one trivialization and non-constant in another, i.e., the notion of a (locally) constant section is not well defined. The purpose of a connection is to give a definition of what directions will be called horizontal, and thus what sections will be called (locally) constant. The fact that one concentrates on horizontal *directions* instead of horizontal *subsets* (submanifolds) is because the latter is too restrictive a notion. If the set of horizontal directions is involutive, we can recover the horizontal submanifolds by means of Frobenius' theorem. However, the set of horizontal directions need not be involutive at all. And indeed, the concept of curvature, which measures more or less the lack of involutivity of the horizontal directions, plays an important rôle in differential geometry and physics.

2.2 Definitions. On any fiber bundle $\pi : B \to M$ (locally trivial, with typical fiber F) we have the vertical subbundle $V \subset TB$, which is defined as the kernel of $T\pi$: $V = \ker(T\pi)$. Its elements are called *vertical (tangent) vectors*. An *Ehresmann connection* $\mathcal{H}$ *on* B is a subbundle of $\mathcal{H} \subset TB$ which is a supplement to V, i.e., $\mathcal{H} \oplus V \cong TB$ [IV.4.6]. A different way to characterize an Ehresmann connection is to require that for all $b \in B$ the map $T\pi : \mathcal{H}_b \to T_{\pi(b)}M$ is a bijection. Since $T\pi$ is an even linear map, this implies that $\mathcal{H}_b$ is isomorphic to $T_{\pi(b)}M$. Elements of $\mathcal{H}$ are usually called *horizontal (tangent) vectors*. An Ehresmann connection $\mathcal{H}$ on a bundle B automatically defines a projection $h : T_bB \to \mathcal{H}_b$ as the $\mathcal{H}$-part in the direct sum $T_bB = V_b \oplus \mathcal{H}_b$; $h(Y)$ is called *the horizontal part* of the tangent vector $Y \in T_bB$.

Since the map $T_b\pi : \mathcal{H}_b \to T_mM$ with $m = \pi(b)$ is an isomorphism, the inverse map $(T_b\pi)^{-1} : T_mM \to H_b$ exists. For any $X \in T_mM$ the image $\widetilde{X}^h = (T_b\pi)^{-1}(X) \in H_b$ is called the *horizontal lift of* X *at* b. Similarly, if X is a vector field on M, its *horizontal lift* $\widetilde{X}^h$ is the (unique) vector field on B such that $\widetilde{X}^h_b$ is the horizontal lift of $X_{\pi(b)}$ at b.

In the context of connections some terminology changes: a (smooth) map $f : N \to B$ is said to be *horizontal* if it is tangent to $\mathcal{H}$ [V.6.4]. In particular a (local) section s of the bundle B is horizontal if $Ts(T_mM) = \mathcal{H}_{s(m)}$ for all m in the domain of definition of s, and a submanifold $C \subset B$ (with its canonical injection) is called horizontal if $TC \subset \mathcal{H}$. The connection $\mathcal{H}$ is said to be *integrable* or *flat* if the subbundle $\mathcal{H} \subset TB$ is integrable [V.6.2].

2.3 Proposition. *Let* (U, ψ) *be a local trivializing coordinate chart for the fiber bundle* $\pi : B \to M$ *with coordinates* x^i*, i.e., in particular,* $\psi : \pi^{-1}(U) \to U \times F$ *is a diffeomorphism. If* $\mathcal{H}$ *is an Ehresmann connection, then there exist unique smooth functions* $\gamma_i : U \times F \to TF$, $\gamma_i(m, f) \in T_fF$ *with* $\varepsilon(\gamma_i) = \varepsilon(x^i)$ *such that the restriction* $\mathcal{H}|_{\pi^{-1}(U)}$ *in terms of the trivialization* ψ *is spanned by the* dim M *tangent vectors*

$$\partial_{x^i}|_m + \gamma_i(m, f) \in T_mM \times T_fF \ . \tag{2.4}$$

Conversely, if a subbundle $\mathcal{H} \subset TB$ *is spanned on local trivializing charts by* dim M *vectors of the form (2.4), then* $\mathcal{H}$ *is an Ehresmann connection on* B.

Proof. In terms of the trivialization, the projection map $\pi : B \to M$ is given as projection on the first factor $\pi_1 : U \times F \to U$. Since $T\pi_1 : H_{(m,f)} \to T_mM$ is a bijection by definition of an Ehresmann connection, it follows that $H_{(m,f)}$ is spanned by vectors of the form $\partial_{x^i}|_m - \gamma_i(m,f)$ for uniquely determined functions $\gamma_i : U \times F \to TF$, $\gamma_i(m,f) \in T_f F$. To prove that these γ are smooth, we argue as follows. By definition of a subbundle, $\mathcal{H}$ is locally spanned by smooth vector fields X_i in a neighborhood of (m,f). Using the decomposition $T(U \times F) = TU \times TF$, each X_i can be written as $X_i = \sum_j X_i^j \, \partial_{x^j} - \widehat{X}_i$ for smooth functions X_i^j and smooth maps $\widehat{X}_i : U \times F \to TF$ with $\widehat{X}_i = \sum_j X_i^j \, \gamma_j$. By definition of an Ehresmann connection the matrix X_i^j must be invertible. Hence, at least in a (small) neighborhood of (m,f), there exists a smooth inverse to it. This shows that γ_i is smooth in such a (small) neighborhood of (m,f). But smoothness is a local property and thus the γ_i are smooth on the whole of U.

To prove the converse, it suffices to note that the conditions guarantee that it is locally generated by smooth independent vector fields, showing that $\mathcal{H}$ is a well defined subbundle, and that $T\pi_1$ is a bijection between $\mathcal{H}|_{(m,f)}$ and $T_m M$, proving that it satisfies the condition for an Ehresmann connection. QED

2.5 Proposition. *Let $\pi : B \to M$ be a fiber bundle, let $\mathcal{H}$ be an Ehresmann connection on B, and let $g : N \to M$ be a smooth map. Using notation as in [IV.2.2], there exists a unique Ehresmann connection $g^*\mathcal{H}$ on the pull-back bundle g^*B such that for all $c \in g^*B$ we have $T\breve{g}\,(g^*\mathcal{H})_c \subset \mathcal{H}_{\breve{g}(c)}$.*

Proof. In [IV.2.2] we have seen that for any trivializing atlas $\mathcal{U}$ for B, there exists a trivializing atlas $\mathcal{V}$ for g^*B and an induced map (also denoted by g) from $\mathcal{V}$ to $\mathcal{U}$ such that the transition functions of g^*B are given by $\psi_{g(a)g(b)} \circ g$ when the $\psi_{\alpha\beta}$ are the transition functions of B associated to the atlas $\mathcal{U}$. Moreover, in these trivializations, the induced fiber bundle map $\breve{g} : g^*B \to B$ takes the form $(n,f) \mapsto (g(n), f)$. Now let y^j be local coordinates on $V_a \in \mathcal{V}$ and let x^i be local coordinates on $U_{g(a)} \in \mathcal{U}$. For any connection $\breve{\mathcal{H}}$ on g^*B there exist local functions $\breve{\gamma}_j : V_a \times F \to TF$ such that $\breve{\mathcal{H}}$ is generated in the trivialization of g^*B determined by V_a by the vectors

$$\partial_{y^j}|_n + \breve{\gamma}_j(n,f) \in T_n V_a \times T_f F \ .$$

The map $T\breve{g}$ maps these vectors to the vectors

$$\breve{\gamma}_j(n,f) + \sum_i \frac{\partial g^i}{\partial y^j}(n) \cdot \frac{\partial}{\partial x^i}|_{g(n)}$$

in the trivialization of B determined by $U_{g(a)}$. If we require that this image lies in $\mathcal{H}_{(g(n),f)}$, which is generated by the vectors $\partial_{x^i}|_{g(n)} + \gamma_i(g(n),f)$, then we must have

$$\breve{\gamma}_j(n,f) = \sum_i \frac{\partial g^i}{\partial y^j}(n) \cdot \gamma_i(g(n),f) \ .$$

We conclude that the $\breve{\gamma}_j$ are uniquely determined by the condition $T\breve{g}\,(g^*\mathcal{H})_c \subset \mathcal{H}_{\breve{g}(c)}$ and that they are indeed smooth. Since the condition is independent of the local trivialization, we conclude that this condition determines a unique Ehresmann connection $g^*\mathcal{H}$ on g^*B. QED

2.6 Discussion. A special case of a pull-back bundle is the restriction of B to a submanifold N of M [IV.2.3]. We thus see that [2.5] tells us in particular that an Ehresmann connection $\mathcal{H}$ on B induces a unique Ehresmann connection on the restriction $B|_N = \pi^{-1}(N)$ of B to the submanifold N. It is not hard to see that this induced connection is just the restriction $\mathcal{H}|_{\pi^{-1}(N)}$ of $\mathcal{H}$ to $B|_N$.

Slightly more general is the case of an immersion $g : N \to M$, in which case again we have an induced Ehresmann connection on g^*B. This application of [2.5] plays a fundamental rôle in the concept of parallel transport along a curve in M. Without going into details because that is outside the scope of this book, we briefly sketch the idea of parallel transport.

We first note that a curve in M is an immersion $g : N \to M$ of a 1-dimensional connected $\mathcal{A}$-manifold N in M. Given a fiber bundle $\pi : B \to M$ we have a pull-back fiber bundle $g^*\pi : g^*B \to N$, and if $\mathcal{H}$ is an Ehresmann connection on B, we have an induced Ehresmann connection $g^*\mathcal{H}$ on g^*B. We now fix $n_o \in \mathbf{B}N$ and $b_o \in \mathbf{B}B_{g(n_o)}$. Parallel transport of b_o along the curve is a smooth horizontal map $pt : N \to B$ with $pt(n) \in B_{g(n)}$ and $pt(n_o) = b_o$, and thus in particular Tpt maps T_nN into $\mathcal{H}_{pt(n)}$, i.e., the vectors tangent to $pt(N)$ are horizontal. A sufficient condition for such a map to exist (and then it is unique) is that $g^*\mathcal{H}$ is integrable *and* that its leaves are diffeomorphic to N via the projection map $g^*\pi : g^*B \to N$. Since N is 1-dimensional, the induced Ehresmann connection $g^*\mathcal{H}$ on g^*B is a 1-dimensional subbundle of $T(g^*B)$, and thus the integrability condition is automatically satisfied when N is even, i.e., of dimension $1|0$. Given these conditions, the map pt is constructed as follows: since $g^*\mathcal{H}$ is integrable, there exists a leaf L passing through $\breve{g}^{-1}(b) \in (g^*B)_n$ (the map $\breve{g}$ is a diffeomorphism between fibers). Since this leaf is diffeomorphic with N via $g^*\pi$, we can define $pt = \breve{g} \circ (g^*\pi|_L)^{-1}$. This map satisfies the given requirements.

That the integrability condition alone is not sufficient is shown in the following elementary example. We take $M = \mathcal{A}_0$ with the global even coordinate x and the trivial bundle $B = M \times \mathcal{A}_0 \equiv (\mathcal{A}_0)^2$ with the global even coordinates (x, y) and the obvious projection $\pi : B \to M$. On B we define the Ehresmann connection $\mathcal{H}$ by

$$\mathcal{H}_{(x,y)} = \{\lambda(\partial_x - y^2\partial_y) \mid \lambda \in \mathcal{A}\} \subset T_{(x,y)}B \ .$$

As curve we choose the canonical embedding $M \to M$, i.e., we see M itself as a curve in M. Now suppose that $pt : M \to B$ is a parallel transport map. Then it must be of the form $pt(x) = (x, f(x))$ for some smooth function f. And then the condition that Tpt maps T_xM into $\mathcal{H}_{pt(x)}$ implies that f must satisfy the condition $(\partial_x f)(x) = -f(x)^2$ because Tpt maps ∂_x to $\partial_x + (\partial_x f)(x)\,\partial_y$. This implies that f is of the form $f(x) = (x-c)^{-1}$ for some $c \in \mathbf{R}$. But such a map is not defined on the whole of M, and thus parallel transport along the whole of this curve does not exist (see also [5.10]).

2.7 Remark. In [Eh] C. Ehresmann introduced his general notion of a connection on an arbitrary fiber bundle. The definition he gave is slightly stronger than that of what here is called an Ehresmann connection. He added the requirement that parallel transport should always be defined. The underlying idea is that parallel transport provides an alternative way to define a connection. Since this approach becomes highly unwieldy in the case of $\mathcal{A}$-manifolds, we here only sketch the procedure in the case of $\mathbf{R}$-manifolds, neglecting all questions about smoothness. Let $g : [x, y] \to M$ be a curve, let $b \in \pi^{-1}(g(x))$ be arbitrary and let $\widetilde{g} : [x, y] \to M$ be a horizontal map satisfying $\pi \circ \widetilde{g} = g$ and $\widetilde{g}(x) = b$. Since $\widetilde{g}$ is uniquely determined by b, we obtain, by varying b, a well defined map $F_g : \pi^{-1}(g(x)) \to \pi^{-1}(g(y))$, $b = \widetilde{g}(x) \mapsto \widetilde{g}(y)$. Running through the curve g in the opposite direction shows that F_g is bijective. Taking the derivative with respect to y at $y = x$ gives us back the tangent vector of $\widetilde{g}$ at b. If parallel transport over all curves exists, we thus can recover the set of horizontal directions at b, i.e., we can recover the (Ehresmann) connection.

This analysis also shows that to any curve g we have associated a diffeomorphism F_g between the fibers over the endpoints. Since in general there is no canonical way to compare, in a fiber bundle, fibers over different points, these diffeomorphisms are a useful tool when one wants to do so. The idea of comparing different fibers in this way is one of the main motivations for the introduction of a connection in the form of parallel transport. In the context of $\mathcal{A}$-manifolds we have ignored this approach to a connection because not all points in a connected $\mathcal{A}$-manifold can be connected by a smooth curve.

2.8 Discussion/Definition. The maps γ_i in (2.4) depend, obviously, upon the trivializing set U. Most, if not all, types of connections are special cases of an Ehresmann connection, eventually in disguise. They use special features of the bundle to impose restrictions on the form of the maps γ_i. We will restrict our attention to one special form of these maps, a form that will be sufficiently universal to cover all our examples. The idea is quite simple. We have a typical fiber F with a (pseudo effective) left action Φ of the structure group G. We thus can require that all γ_i are fundamental vector fields associated to this action. More precisely, we define an *FVF connection* (for Fundamental Vector Field) to be an Ehresmann connection such that for each $m_o \in M$ there exists a trivializing coordinate chart (U, ψ) containing m_o such that the map $f \mapsto \gamma_i(m, f) \in T_f F$ is a fundamental vector field for all $m \in U$. The next results show that the notion of being a fundamental vector field has nice smoothness properties and is independent of the choices that can be made.

2.9 Lemma. *Let (U, ψ) be a trivializing coordinate chart with coordinates (x^i) such that the map $f \mapsto \gamma_i(m, f) \in T_f F$ is a fundamental vector field for all $m \in U$. Then there exists a unique map $A_i : U \to \mathfrak{g}$ such that $\gamma_i(m, f) = A_i(m)^F_f$. Moreover, the map A_i is smooth.*

Proof. By definition of a fundamental vector field, there exists an $A_i(m) \in \mathfrak{g}$ for each $m \in U$ such that $\gamma_i(m, f) = A_i(m)^F_f$. It thus remains to prove that the map A_i is unique and smooth, which will be a consequence of the fact that the structure group acts pseudo effectively. We start with smoothness.

In [2.3] we have seen that the maps $(m, f) \mapsto \gamma_i(m, f) \equiv A_i(m)^F_f$ are smooth. With respect to a basis (v_j) for $\mathfrak{g}$ the map A_i takes the form $A_i(m) = \sum_j c^j(m)\, v_j$ for functions

$c^j : U \to \mathcal{A}$. Since the map $T\Phi_f$ is left linear, we have $A_i(m)^F_f = \sum_j c^j(m)\,(v_j)^F_f$. With respect to a local coordinate system on F with coordinates (y,η), each fundamental vector field $(v_j)^F$ has the form $(v_j)^F_f = \sum_{k=1}^{\dim F} E^k_j(f)\,\partial_k$. Smoothness of A_i being equivalent to smoothness of the coefficients, we thus know that $\sum_j c^j(m)E^k_j(f)$ is smooth for all k, and we want to show that this implies that all c^j are smooth. Since the E^k_j are smooth, we have $E^k_j(y,\eta) = \sum_J E^k_{j,J}(y)\eta^J$ [III.3.17]. Taking derivatives with respect to the η's of the smooth functions $\sum_J(\sum_j c^j(m)E^k_{j,J}(y))\eta^J$, we obtain in particular that $\sum_j c^j(m)E^k_{j,J}(y)$ is smooth for all k and J. Following the proof of [VI.6.6] we introduce the functions $\phi^k_J(y) = \sum_j {}^j v \cdot E^k_{j,J}(y) \in {}^*\mathfrak{g}$ for real values of the (even) coordinates y. Varying also the local coordinate charts, we know from the proof of [VI.6.6] that there are $\ell \le \dim\mathfrak{g}$ independent elements $\phi^{k_1}_{J_1}(y_1), \ldots, \phi^{k_\ell}_{J_\ell}(y_\ell)$, such that all other elements $\phi^k_J(y)$ are linear combinations with real coefficients of these ℓ elements. Since these independent elements define $\mathfrak{g}^{\psi o}$, and since $\mathfrak{g}^{\psi o} = \{0\}$ [VI.6.10], we conclude that $\ell = \dim\mathfrak{g}$. Changing the basis of $\mathfrak{g}$ if necessary we may assume that the $\phi^{k_r}_{J_r}(y_r)$ form the left dual basis, i.e., $\phi^{k_r}_{J_r}(y_r) = {}^r v$. It follows that $E^{k_r}_{j,J_r}(y_r) = \delta^r_j$, and thus $c^r(m) = \sum_j c^j(m)E^{k_r}_{j,J_r}(y_r)$ is smooth.

To prove uniqueness of the A_i, suppose that A'_i is another solution. Then $A_i - A'_i$ is a smooth family for which the associated fundamental vector field is identically zero. Since $\mathfrak{g}^{\psi o} = \{0\}$, this implies that $A_i = A'_i$. QED

2.10 Lemma. *Let (U_a,ψ_a) be a trivializing coordinate chart with coordinates x^i and let $A^{a,x}_i : U \to \mathfrak{g}$ be smooth maps such that the Ehresmann connection $\mathcal{H}$ is spanned in the trivialization ψ_a by*

$$\partial_{x^i}|_m + A^{a,x}_i(m)^F_f \in T_mM \times T_fF \ .$$

If y^j is another system of coordinates on U_a, then $\mathcal{H}$ is also spanned in the trivialization ψ_a by

$$\partial_{y^j}|_m + A^{a,y}_j(m)^F_f \in T_mM \times T_fF \ ,$$

with $A^{a,y}_j : U \to \mathfrak{g}$ given by $A^{a,y}_j(m) = \sum_i(\partial_{y^j}x^i)(m)\cdot A^{a,x}_i(m)$.

If (U_b,ψ_b) is another trivialization with $U_a = U_b$, then $\mathcal{H}$ is spanned in the trivialization ψ_b by

$$\partial_{x^i}|_m + A^{b,x}_i(m)^F_f \in T_mM \times T_fF \ ,$$

with $A^{b,x}_i : U \to \mathfrak{g}$ given by

$$\begin{aligned} A^{b,x}_i(m) &= \mathrm{Ad}(\psi_{ba}(m))\left(A^{a,x}_i(m) - TL_{\psi_{ba}(m)^{-1}}\,T\psi_{ba}\,\partial_{x^i}|_m\right) \\ &= \mathrm{Ad}(\psi_{ab}(m)^{-1})\,A^{a,x}_i(m) + TL_{\psi_{ab}(m)^{-1}}\,T\psi_{ab}\,\partial_{x^i}|_m \ , \end{aligned} \tag{2.11}$$

where $\psi_{ba} : U_a = U_b \to G$ is the transition function related to the change of trivialization from ψ_a to ψ_b.

Proof. The first part is a direct consequence of the fact that the tangent map is left linear:

$$\frac{\partial}{\partial x^i}|_m + T\Phi_f\,A^{a,x}_i(m) = \sum_j \frac{\partial y^j}{\partial x^i}(m)\Big(\frac{\partial}{\partial y^j}|_m + T\Phi_f\Big(\sum_k \frac{\partial x^k}{\partial y^j}(m)\,A^{a,x}_k(m)\Big)\Big) \ ,$$

because $\sum_j \partial_{x^i} y^j \cdot \partial_{y^j} x^k = \delta_i^k$. Since the matrix $\partial_{x^i} y^j$ is invertible, the result follows.

To prove (2.11), we first note that $(\psi_b \circ \psi_a^{-1})(m, f) = (m, \Phi(\psi_{ba}(m), f))$ by definition of the transition function $\psi_{ba} : U \to G$. We then compute the image of the tangent vector $\partial_{x^i}|_m + A_i^{a,x}(m)_f^F$ under the map $\psi_b \circ \psi_a^{-1}$:

$$\begin{aligned} T(\psi_b \circ \psi_a^{-1})\,(\partial_{x^i}|_m + A_i^{a,x}(m)_f^F) &= \partial_{x^i}|_m + T\Phi_f\, T\psi_{ba}\, \partial_{x^i}|_m + T\Phi_{\psi_{ba}(m)}\, A_i^{a,x}(m)_f^F \\ &= \partial_{x^i}|_m + T\Phi_{\psi_{ba}(m)}\, \big(A_i^{a,x}(m)_f^F + T\Phi_f\, TL_{\psi_{ba}(m)^{-1}}\, T\psi_{ba}\, \partial_{x^i}|_m\big) \\ &= \partial_{x^i}|_m + T\Phi_{\psi_{ba}(m)}\, \big(A_i^{a,x}(m) - TL_{\psi_{ba}(m)^{-1}}\, T\psi_{ba}\, \partial_{x^i}|_m\big)_f^F \\ &= \partial_{x^i}|_m + \big(\mathrm{Ad}(\psi_{ba}(m))\, \big(A_i^{a,x}(m) - TL_{\psi_{ba}(m)^{-1}}\, T\psi_{ba}\, \partial_{x^i}|_m\big)\big)_{\Phi(\psi_{ba}(m),f)}^F , \end{aligned}$$

where the second equality follows from the equality $\Phi_f \circ L_{\psi_{ba}(m)} = \Phi_{\psi_{ba}(m)} \circ \Phi_f$ as maps from G to F. This proves the first equality of (2.11). The second equality is obtained by interchanging the roles of a and b and using that $\psi_{ba}(m) \cdot \psi_{ab}(m) = e$ for all $m \in U_a = U_b$. QED

2.12 Discussion. We learn from [2.9] that being an FVF connection can be expressed in terms of (local) smooth maps with values in $\mathfrak{g}$. And then [2.10] tells us that the notion of being a fundamental vector field on a trivialization is independent of the chosen trivialization as well as the chosen coordinate system. Moreover, the explicit dependence of $A_j^{a,y}(m)$ in terms of $A_i^{a,x}(m)$ also shows that the 1-form Γ^a with values in $\mathfrak{g}$ on U_a defined as

$$\Gamma^a(m) = \sum_i dx^i \otimes A_i^{a,x}(m) = \sum_j dy^j \otimes A_j^{a,y}(m)$$

is independent of the chosen coordinate system. This implies that if (U_a, ψ_a) is a local trivialization, there exists a $\mathfrak{g}$-valued 1-form Γ^a on U_a such that, *if* x^i are coordinates on (a part of) U_a, the local functions $A_i^{a,x}$ can be recovered from Γ^a by

$$A_i^{a,x}(m) = \iota(\partial_{x^i}|_m)\Gamma^a .$$

The existence of Γ^a is independent of whether there exists a global coordinate system on U_a or not, and $\mathcal{H}$ is given in the trivialization ψ_a by

$$\mathcal{H}_{(m,f)} = \{\, X_m + \big(\iota(X_m)\Gamma^a\big)_f^F \mid X_m \in T_mM \,\} .$$

2.13 Corollary. *Let $\pi : B \to M$ be a fiber bundle with typical fiber F and structure group G, and let* $=\{\, (U_a, \psi_a) \mid a \in I \,\}$ *be a trivializing atlas for B. If $\mathcal{H}$ is an FVF connection on B, there exist unique $\mathfrak{g}$-valued 1-forms Γ^a on U_a such that $\mathcal{H}$ is given in the trivialization ψ_a by*

$$\mathcal{H}_{(m,f)} = \{\, X_m + \big(\iota(X_m)\Gamma^a\big)_f^F \mid X_m \in T_mM \,\} . \tag{2.14}$$

Moreover, on overlaps $U_a \cap U_b$ the 1-forms Γ^a and Γ^b are related by

$$\Gamma^b(m) = \mathrm{Ad}(\psi_{ab}(m)^{-1}) \circ \Gamma^a(m) + (\psi_{ab}^* \Theta_{MC})|_m \,, \tag{2.15}$$

where Θ_{MC} is the Maurer-Cartan 1-form on G.

Conversely, if we have $\mathfrak{g}$-valued 1-forms Γ^a on U_a that are related on overlaps by (2.15), then (2.14) defines an FVF connection $\mathcal{H}$.

Proof. If $\mathcal{H}$ is given, (2.14) is a direct consequence of [2.12]. To prove (2.15) we choose (local) coordinates (x^i) on $U_a \cap U_b$. Using (2.11) and [1.3] we obtain

$$\begin{aligned}
\Gamma^b(m) &= \sum_i dx^i \otimes A_i^b(m) \\
&= \sum_i dx^i \otimes \big(\mathrm{Ad}(\psi_{ab}(m)^{-1}) A_i^a(m) + TL_{\psi_{ab}(m)^{-1}}\, T\psi_{ab}\, \partial_{x^i}|_m\big) \\
&= \mathrm{Ad}(\psi_{ab}(m)^{-1}) \circ \Gamma^a(m) + \sum_i dx^i \otimes TL_{\psi_{ab}(m)^{-1}}\, T\psi_{ab}\, \partial_{x^i}|_m \,.
\end{aligned}$$

To prove that the second term equals $\psi_{ab}^*\Theta_{MC}$, we first note that $T\psi_{ab}\partial_{x^i}|_m \in T_{\psi_{ab}(m)}G$. And then we use [1.5] to compute

$$\iota(\partial_{x^i}|_m)\psi_{ab}^*\Theta_{MC} = \iota(T\psi_{ab}\partial_{x^i}|_m)\Theta_{MC} = TL_{\psi_{ab}(m)^{-1}} T\psi_{ab}\partial_{x^i}|_m \,.$$

Since for any 1-form α we always have $\alpha = \sum_i dx^i \otimes \iota(\partial_{x^i})\alpha$, (2.15) follows.

To prove the converse, we first note that (2.14) obviously is the local expression of an FVF connection. It only remains to be shown that these local expressions coincide on overlaps $U_a \cap U_b$. But this is an immediate consequence of [2.10], (2.11). QED

2.16 Corollary. *Let $\pi : B \to M$ be a fiber bundle, let $\mathcal{H}$ be an FVF connection on B, and let $g : N \to M$ be a smooth map. Then the unique Ehresmann connection $g^*\mathcal{H}$ on the pull-back bundle g^*B such that for all $c \in g^*B$ we have $T\check{g}\,(g^*\mathcal{H})_c \subset \mathcal{H}_{\check{g}(c)}$ [2.5] is also an FVF connection. In particular, if $\mathcal{U}$ and $\mathcal{V}$ are (trivializing) atlases as in the proof of [2.5], then $g^*\mathcal{H}$ is determined by the local $\mathfrak{g}$-valued 1-forms $\check{\Gamma}^a = g^*\Gamma^{g(a)}$ [2.13].*

Proof. Since $\mathcal{H}$ is an FVF connection, the maps γ_i are given as $\gamma_i(m, f) = A_i(m)_f^F$ for $\mathfrak{g}$-valued functions A_i. Using the arguments and notation as in the proof of [2.5], $g^*\mathcal{H}$ is determined by the functions $\check{\gamma}_j$ given by

$$\check{\gamma}_j(n, f) = \sum_i \frac{\partial g^i}{\partial y^j}(n) \cdot A_i(g(n))_f^F = \Big(\sum_i \frac{\partial g^i}{\partial y^j}(n) \cdot A_i(g(n))\Big)_f^F \,.$$

This proves that $g^*\mathcal{H}$ is an FVF connection. In terms of the $\mathfrak{g}$-valued 1-forms associated to the trivializing atlases, we find:

$$\check{\Gamma}^a = \sum_j dy^j \otimes \sum_i \frac{\partial g^i}{\partial y^j}(n) \cdot A_i(g(n)) = \sum_i g^*dx^i \otimes A_i(g(n)) = g^*\Gamma^{g(a)} \,,$$

where $\Gamma^{g(a)} = \sum_i dx^i \otimes A_i(m)$ is the $\mathfrak{g}$-valued 1-form on $U_{g(a)}$ associated to the connection $\mathcal{H}$ on B. QED

2.17 Remark. In the context of general Ehresmann connections one could wonder why it is so easy to define a pull-back connection, because an Ehresmann connection is an object living on the tangent bundle, and for tangent vectors the notion of pull-back is not (directly) defined. A possible explanation is that the combination $\sum_i dx^i \otimes \gamma_i$ is independent of the chosen coordinate system, and for 1-forms we do have a natural notion of pull-back. But a precise definition of this object is not easy. However, in the context of FVF connections, the notion of a pull-back connection becomes natural. Such a connection is defined by local $\mathfrak{g}$-valued 1-forms. And, as we have seen, the pull-back of these 1-forms defines the pull-back connection.

2.18 Proposition. *Let $\pi : B \to M$ be a fiber bundle with typical fiber F and structure group G. Let $\mathcal{U} = \{U_a \mid a \in I\}$ be a trivializing atlas for B and let Γ^a be $\mathfrak{g}$-valued 1-forms on U_a defining an FVF connection $\mathcal{H}$ according to [2.13]. Then $\mathcal{H}$ is integrable if and only if (all) the local 2-forms*

$$d\Gamma^a + \tfrac{1}{2}[\Gamma^a \overset{\wedge}{,} \Gamma^a]$$

are identically zero.

Proof. We have to show that $\mathcal{H}$ is an involutive subbundle. Since the value of a commutator of two vector fields at a point depends only upon the behavior of the two vector fields in a neighborhood of the given point, it suffices to verify that $\mathcal{H}|_{\pi^{-1}(U_a)}$ is involutive for all $a \in I$. But on $\pi^{-1}(U_a)$ the connection $\mathcal{H}_{(m,f)}$ is spanned by the vector fields $\partial_{x^i}|_m + A_i^{a,x}(m)_f^F$ [2.12]. Using [V.1.19] it follows that it suffices to show that the commutator of two of these generating vector fields belongs to $\mathcal{H}$.

To compute such a commutator, we first compute the commutator $[\partial_{x^i}, A^F]$ for a smooth map $A : U_a \to \mathfrak{g}$. To do so, we choose a basis v_j for $\mathfrak{g}$ and (local) coordinates y^k on F. It follows that there exist smooth functions $A^j : U_a \to \mathcal{A}$ such that $A = \sum_j A^j v_j$. There also exist smooth functions $V_j^k : F \to \mathcal{A}$ such that $(v_j)^F = \sum_k V_j^k \, \partial_{y^k}$. And then we compute

$$[\partial_{x^i}, A^F] = \sum_{jk} [\partial_{x^i}, A^j V_j^k \, \partial_{y^k}] = \sum_{jk} (\partial_{x^i} A^j) \, V_j^k \, \partial_{y^k} = (\partial_{x^i} A)^F .$$

We thus find for the commutator of two generating vector fields:

$$[\partial_{x^i} + (A_i^{a,x})^F, \partial_{x^j} + (A_j^{a,x})^F] = (\partial_{x^i} A_j^{a,x})^F - (-1)^{\langle \varepsilon(x^i) | \varepsilon(x^j) \rangle} (\partial_{x^j} A_i^{a,x})^F + [A_i^{a,x}, A_j^{a,x}]^F ,$$

where we used [VI.5.2-iii]. Since this commutator is tangent to F, it projects to zero on U_a under the projection $T\pi$. Since $T\pi$ is a bijection from $\mathcal{H}_{(m,f)}$ onto $T_m M$, the condition that this commutator belongs to $\mathcal{H}$ becomes the condition that this commutator must be zero (for all i and j). Looking at the definition of $\mathfrak{g}^{\psi_o}$ [VI.6.2], it follows that the image of the smooth functions $F_{ij}^{a,x} : U_a \to \mathfrak{g}$ defined by

$$F_{ij}^{a,x}(m) = \partial_{x^i} A_j^{a,x} - (-1)^{\langle \varepsilon(x^i) | \varepsilon(x^j) \rangle} \partial_{x^j} A_i^{a,x} + [A_i^{a,x}, A_j^{a,x}]$$

belongs to $\mathfrak{g}^{\psi o}$. Since the action of G on F is pseudo effective, it follows from [VI.6.10] that $\mathcal{H}$ is integrable if and only if all functions $F_{ij}^{a,x}$ are identically zero. On the other hand, using [V.7.6], [1.4] and $\iota(\partial_{x^i})\Gamma^a = A_i^{a,x}$ [2.12], we find

$$-\iota(\partial_{x^i}, \partial_{x^j})\big(d\Gamma^a + \tfrac{1}{2}[\Gamma^a \wedge \Gamma^a]\big) = F_{ij}^{a,x} .$$

Since any 2-form β on U_a can be reconstructed from its contraction with the ∂_{x^i} by $\beta = \frac{1}{2}\sum_{ij} dx^j \wedge dx^i \cdot \iota(\partial_{x^i}, \partial_{x^j})\beta$, the result follows. $\boxed{QED}$

3. Connections on principal fiber bundles

In this section we show that an FVF connection on a principal fiber bundle can be described either as the kernel of a so called connection 1-form ω on the bundle or as an Ehresmann connection that is invariant under the right action of the structure group on the bundle. We also show how the connection 1-form ω can be reconstructed from the local 1-forms Γ^a defining the FVF connection. The description in terms of a connection 1-form allows us to prove quite easily that there always exist FVF connections on a principal fiber bundle.

3.1 Lemma. *Let $\pi : P \to M$ be a principal fiber bundle with structure group G and let $U \subset M$ be open. Let T denote the map which associates to each local trivialization $\psi : \pi^{-1}(U) \to U \times G$ of $\pi^{-1}(U)$ the local section $s \in \Gamma_U(P)$, $s(m) = \psi^{-1}(m, e)$ with $e \in G$ the identity element. Then T is a bijection between the set of all local trivializations of $\pi^{-1}(U)$ and $\Gamma_U(P)$. The inverse of T is given by the formula $\psi^{-1}(m, g) = s(m) \cdot g$.*

Proof. If s is given as $s(m) = \psi^{-1}(m, e)$, then by definition of the right action of G on P we have $\psi^{-1}(m, g) = s(m) \cdot g$. Hence the given formula is a left inverse for T. To show that it also is a right inverse and that it indeed defines a local trivialization, we suppose that $s \in \Gamma_U(P)$ is a local smooth section. We then define the smooth map $\Psi : U \times G \to \pi^{-1}(U)$ by $\Psi(m, g) = s(m) \cdot g$. If (V, χ) is any local trivializing chart for P, we obtain a map $s_\chi : U \cap V \to G$ such that $(\chi \circ s)(m) = (m, s_\chi(m))$ [IV.1.20]. It follows that $\chi \circ \Psi : U \cap V \times G \to U \cap V \times G$ is given by

$$(\chi \circ \Psi)(m, g) = (m, s_\chi(m) \cdot g) . \tag{3.2}$$

From this one deduces that Ψ is bijective, a bundle morphism and a local diffeomorphism. Hence $\psi = \Psi^{-1}$ is also smooth. And then (3.2) shows that ψ is compatible with the structure of the principal fiber bundle, i.e., (U, ψ) is a local trivialization. $\boxed{QED}$

3.3 Corollary. *A principal fiber bundle $\pi : P \to M$ is (isomorphic to) the trivial bundle if and only if $\Gamma(P)$ is not empty.*

Proof. This is the special case $U = M$ in [3.1]. $\boxed{QED}$

3.4 Discussion. In [3.1] we have established a bijection between local sections and local trivializations of a principal fiber bundle $\pi : P \to M$. We are thus allowed to speak of the (local) trivialization ψ determined by the (local) section s. Now suppose that $s_a \in \Gamma_{U_a}(P)$ and $s_b \in \Gamma_{U_b}(P)$ are two local sections determining two local trivializations ψ_a and ψ_b. These two trivializations determine a transition function $\psi_{ba} : U_a \cap U_b \to G$ by the formula $(\psi_b \circ \psi_a^{-1})(m, g) = (m, \psi_{ba}(m)g)$. Applying ψ_b^{-1} and substituting the definition of ψ_a and ψ_b in terms of s_a and s_b gives us

$$s_a(m) = s_b(m) \cdot \psi_{ba}(m) \ . \tag{3.5}$$

This formula gives us a way to determine the transition functions directly from the defining local sections: given s_a and s_b there exists for each $m \in U_a \cap U_b$ a unique $\psi_{ba}(m) \in G$ such that $s_a(m) = s_b(m) \cdot \psi_{ba}(m)$, just because the right action of G on the fibers of P is free and transitive. (3.5) then tell us that it must be the transition function determined by the two associated trivializations ψ_a and ψ_b.

3.6 Remark. In the physics literature a local section of a principal fiber bundle is often called a *(local) gauge* and changing a local section s to a local section $\hat{s}$ by the formula $\hat{s}(m) = s(m) \cdot \phi(m)$ is called a *(local) gauge transformation.*

3.7 Definitions. Let $\pi : P \to M$ be a principal fiber bundle with structure group G whose $\mathcal{A}$-Lie algebra is $\mathfrak{g}$. In a local trivialization $\pi^{-1}(U) \cong U \times G$ the vertical subbundle V is just the tangent space to the second factor TG. It follows that V is spanned by the left-invariant vector fields on G. Since the right action Φ of G on P corresponds to right multiplication on the second factor and since the fundamental vector fields of right multiplication are exactly the left-invariant vector fields [VI.5.4], we conclude that V is spanned by the fundamental vector fields of the right action of G on P, independent of the choice of a local trivialization. More precisely, $V_p = \{ x_p^P \mid x \in \mathfrak{g} \}$.

• An FVF connection on a principal fiber bundle P is called a *principal connection*, or simply a *connection*. Contrary to the vertical directions, the fundamental vector fields used in the definition of a principal connection on P are *right* invariant because the action of the structure group G on the typical fiber G is *left* multiplication.

• A *connection 1-form on the principal fiber bundle* P is an even $\mathfrak{g}$-valued 1-form ω on P satisfying the following two conditions.

(i) $\forall x \in \mathfrak{g} : \iota(x^P)\omega = x$.

(ii) $\forall g \in G : \Phi_g^*\omega = \mathrm{Ad}(g^{-1}) \circ \omega$.

Since the meaning of (i) is rather obvious, we concentrate on the precise meaning of (ii). On the left hand side $\Phi_g^*\omega$ indicates the generalized pull-back as defined in [V.7.23], [VI.8.12]. On the right hand side we have an example of the alternative notation [1.3] for the evaluation-wedge product of an $\mathrm{End}_R(\mathfrak{g})$-valued 0-form with a $\mathfrak{g}$-valued 1-form. Writing all definitions explicitly, condition (ii) says that for all $(p, g) \in P \times G$ and all $X_p \in T_pP$ we must have

$$\iota(\langle\!\langle (X_p, \underline{0}_g) \,\|\, T_{(p,g)}\Phi \rangle\!\rangle)\omega_{p\cdot g} = \mathrm{Ad}(g^{-1})\big(\iota(X_p)\omega_p\big) \ . \tag{ii}$$

3.8 Lemma. *An Ehresmann connection $\mathcal{H}$ on a principal fiber bundle $\pi : P \to M$ with structure group G is a principal connection on P if and only if $\mathcal{H}$ is invariant under the right action Φ of G on P, i.e., for all $g \in G : T\Phi_g \mathcal{H}_p = \mathcal{H}_{pg}$.*

Proof. In a local trivialization (U, ψ) a principal connection on P is given in terms of the maps γ_i by [2.9]

$$\gamma_i(m, g) = A_i(m)^G_g \ ,$$

for smooth maps $A_i : U \to \mathfrak{g}$. Since $A_i(m)^G_g = -TR_g A_i(m)$ [VI.5.1], we conclude that

$$TR_h(\partial_{x^i}|_m + A_i(m)^G_g) = \partial_{x^i}|_m + A_i(m)^G_{gh} \ ,$$

where now R_h denotes the natural right action of $h \in G$ on $U \times G$ [IV.2.12]. Since this action corresponds to the natural right action of G on P, we conclude that $T\Phi_h \mathcal{H}_p = \mathcal{H}_{ph}$.

Conversely, if $\mathcal{H}$ is invariant under the G-action, it follows that

$$TR_h(\partial_{x^i}|_m + \gamma_i(m, g)) = \partial_{x^i}|_m + \gamma_i(m, gh) \ .$$

In particular we must have $\gamma_i(m, g) = TR_g \gamma_i(m, e)$. Denoting $A_i(m) = -\gamma_i(m, e)$ we conclude that $\mathcal{H}$ is an FVF connection, i.e., a principal connection on P. QED

3.9 Lemma. *Let $\pi : P \to M$ be a principal fiber bundle with structure group G whose $\mathcal{A}$-Lie algebra is $\mathfrak{g}$. If ω is a connection 1-form on P, then* $\ker(\omega)$ *is a principal connection on P. Conversely, if $\mathcal{H}$ is a principal connection on P, then there exists a unique connection 1-form ω on P such that $\mathcal{H} = \ker(\omega)$.*

Proof. Note that $\ker(\omega)$ is defined as $\ker(\omega)_p = \{Y \in T_pP \mid \iota(Y)\omega_p = 0\}$. Our first goal will be to prove that $\ker(\omega)$ is a (smooth) subbundle of TP and a supplement to V. We can interpret ω as an even (because ω is even) left linear map from the bundle $TP \to P$ to the trivial bundle $P \times \mathfrak{g}$ by $Y \in T_pP \mapsto \iota(Y)\omega_p$. By condition (i) of a connection 1-form this map is surjective. We thus satisfy the conditions of [IV.3.18], proving that $\ker(\omega)$ is a subbundle of TP. For any $X \in V_p$ there exists $x \in \mathfrak{g}$ such that $X = x^P_p$. By condition (i) of a connection 1-form we have $\iota(X)\omega = x$, and thus $V \cap \ker(\omega) = \{0\}$. On the other hand, for any $Y \in T_pP$ we can define $x = \iota(Y)\omega \in \mathfrak{g}$, and then $Y - x^P_p \in \ker(\omega)_p$. This proves that $V \oplus \ker(\omega) = TP$. Condition (ii) of a connection 1-form implies that for a tangent vector $Y \in T_pP$ we have

$$\iota(T\Phi_g Y)\omega_{pg} = \mathrm{Ad}(g^{-1})\big(\iota(Y)\omega_p\big) \ .$$

It follows that $T\Phi_g(\ker(\omega)_p) \subset \ker(\omega)_{\Phi_g p}$. Since $T\Phi_g$ is a bijection, we conclude that $\ker(\omega)$ is invariant under the G-action and thus is a principal connection on P [3.8].

For the converse, let $\mathcal{H}$ be a principal connection on P. This implies that we have $T_pP = V_p \oplus \mathcal{H}_p$. Since V is generated by the fundamental vector fields of the right action of G on P, condition (i) of a connection 1-form together with $\mathcal{H} = \ker(\omega)$ completely determine ω, provided it exists. We thus *define* the 1-form ω by $\iota(\mathcal{H}_p)\omega_p = 0$

and for $x \in \mathfrak{g}$: $\iota(x_p^P)\omega_p = x$. In this way ω is a well defined $\mathfrak{g}$-valued 1-form on P satisfying condition (i) of a connection 1-form. To show that this ω is smooth, we invoke [IV.5.9], [IV.5.11] which says that ω is smooth if it maps smooth sections to smooth sections. Now any smooth section of TP is the sum of a smooth section of $\mathcal{H}$ and a smooth section of V. Our ω maps (smooth) sections of $\mathcal{H}$ to zero, which is smooth. If (v_i) is a basis for $\mathfrak{g}$, the corresponding v_i^P form a basis of the (globally) smooth sections of V. Since the v_i are real, their image under ω is smooth (a real constant). And thus ω is smooth. To prove condition (ii) of a connection 1-form, we first suppose that $X \in \mathcal{H}_p$, which implies (by invariance of $\mathcal{H}$ under the G-action) that $\iota(T\Phi_g X)\omega_{pg} = 0 = \mathrm{Ad}(g^{-1})\iota(X)\omega_p$. We then suppose that $X \in V_p$ and hence we have a unique $x \in \mathfrak{g}$ such that $X = x_p^P$. Since $T\Phi_g(x^P) = (\mathrm{Ad}(g^{-1})x)^P$ [VI.5.4], we have $\iota(T\Phi_g X)\omega_{pg} = \mathrm{Ad}(g^{-1})x = \mathrm{Ad}(g^{-1})(\iota(X)\omega_p)$. QED

3.10 Discussion. According to [3.8] we can define a principal connection either as being generated by the structure group or as being invariant under the (right) action of G on P. It is the latter definition which is usually given. However, the former is more intrinsic because it is a description which can be used for an arbitrary fiber bundle whereas the description in terms of the action of G on P is not easily generalized to arbitrary fiber bundles. According to [3.9] we can describe a principal connection also by a connection 1-form. Since it depends upon the context which description is the most useful, it is customary to mix both approaches and to switch between the description in terms of a subbundle of TP with its horizontal vectors and the description in terms of the 1-form. And in fact, both description can be given in terms of local $\mathfrak{g}$-valued 1-forms Γ^a satisfying (2.15). The subbundle $\mathcal{H}$ can be recovered via [2.12] and the 1-form ω via [3.11].

3.11 Proposition. *Let $\pi : P \to M$ be a principal fiber bundle with structure group G. Let $\mathcal{H}$ be a principal connection on P and let ω be the associated connection 1-form. Let $\{(U_a, \psi_a) \mid a \in I\}$ be a trivializing atlas for B and let $s_a : U_a \to P$ be the local section defining ψ_a [3.1]. Finally, let Γ^a be the $\mathfrak{g}$-valued 1-forms on U_a defining the principal connection $\mathcal{H}$ [2.13]. Then $s_a^*\omega = \Gamma^a$ and $\omega|_{\pi^{-1}(U_a)}$ can be reconstructed from Γ^a in the trivialization determined by s_a by*

$$\omega_{(m,g)} = \Theta_{MC}|_g + \mathrm{Ad}(g^{-1}) \circ \Gamma^a|_m \,. \tag{3.12}$$

Proof. Since ω is completely determined by the conditions $\ker(\omega) = \mathcal{H}$ and condition (i) of a connection 1-form, it suffices to verify these condition in the trivialization determined by s_a. In this trivialization the fundamental vector field x^P is given as $\vec{x}$ [3.7], and thus

$$\iota(\vec{x}_g)\big(\Theta_{MC}|_g + \mathrm{Ad}(g^{-1}) \circ \Gamma^a|_m\big) = x$$

by definition of the Maurer-Cartan 1-form. Introducing local coordinates x^i on U_a, we know that the connection $\mathcal{H}$ is generated by the vector fields $\partial_{x^i}|_m + A_i^{a,x}(m)_g^G$ with

$A_i^{a,x}(m) = \iota(\partial_{x^i}|_m)\Gamma^a$ [2.12]. Using the properties of the Maurer-Cartan 1-form [1.5] and the fact that $A_i^{a,x}(m)_g^G = -TR_g A_i^{a,x}(m)$ [VI.5.1], we then compute

$$\begin{aligned}\iota(\partial_{x^i}|_m + A_i^{a,x}(m)_g^G)(\Theta_{MC}|_g + \mathrm{Ad}(g^{-1}) \circ \Gamma^a|_m) &= \\ &= \mathrm{Ad}(g^{-1})\iota(\partial_{x^i}|_m)\Gamma^a|_m - \iota(TR_g A_i^{a,x}(m))\Theta_{MC}|_g \\ &= \mathrm{Ad}(g^{-1})A_i^{a,x}(m) - \iota(A_i^{a,x}(m))R_g^*\Theta_{MC} = 0 \ .\end{aligned}$$

We conclude that the given expression for ω in the local trivialization determined by s_a has the required properties and thus must coincide with ω. From this local expression for ω and the fact that in this trivialization the local section s_a is given as $s_a(m) = (m, e)$, it follows that $s_a^*\omega = \Gamma^a$. QED

3.13 Discussion. A natural question is whether there always exists a principal connection on a given principal fiber bundle $\pi : P \to M$. The answer is affirmative and relies on a partition of unity argument. Let $\mathcal{U} = \{(U_a, \psi_a) \mid a \in I\}$ be a trivializing atlas. For each $a \in I$ we choose the 1-form $\Gamma^a \equiv 0$. These choices do not (in general) satisfy (2.15), and thus do not define a global principal connection. But on the restriction $P|_{U_a} = \pi^{-1}(U_a)$ they *do* define a principal connection. According to [3.11] we thus have a connection 1-form $\omega_a = \Theta_{MC}$ on the local trivialization $\pi^{-1}(U_a) \cong U_a \times G$. Let $\{\rho_a \mid a \in I\}$ be a partition of unity associated to the open cover $\mathcal{U}$. For each $a \in I$ we then have the global 1-form $\rho_a\,\omega_a$. This global 1-form obviously satisfies condition (ii) of a connection 1-form, but condition (i) is replaced by $\iota(x^P)\rho_a\,\omega_a = \rho_a \cdot x$. It follows that $\sum_{a\in I} \rho_a\,\omega_a$ is a well defined global connection 1-form on the principal fiber bundle P.

4. The exterior covariant derivative and curvature

In this section we continue the study of FVF connections on principal fiber bundles. We define the exterior covariant derivative D and apply it to the connection 1-form ω to obtain the curvature 2-form $\Omega = D\omega$. We then prove the structure equations of Cartan $\Omega = d\omega + \frac{1}{2}[\omega \wedge \omega]$ and the Bianchi identities $d\Omega = [\Omega \wedge \omega]$. We also show that Ω is determined by the local 2-forms $d\,\Gamma^a + \frac{1}{2}[\,\Gamma^a \wedge \Gamma^a\,]$ if the FVF connection is determined by the local 1-forms Γ^a. Not surprisingly, we can prove that the FVF connection is integrable if and only if the curvature Ω is zero.

4.1 Definition. Let $\mathcal{H}$ be a principal connection on a principal fiber bundle $P \to M$ and let ω be the associated connection 1-form. For any k-form α on P with values in an $\mathcal{A}$-vector space E one defines the *exterior covariant derivative* $D\alpha$ (with respect to the connection 1-form ω) by the formula

$$\iota(X_1, \ldots, X_k)D\alpha = \iota(hX_1, \ldots, hX_k)d\alpha \ ,$$

where h denotes the projection on the horizontal part $h : T_pP \to \mathcal{H}_p$. In particular *the curvature 2-form* Ω is defined as the exterior covariant derivative of the connection 1-form $\omega : \Omega = D\omega$.

4.2 Lemma. *For any two vector fields X and Y we have $\iota(X,Y)\Omega = \iota([\,hX, hY\,])\omega$. As a consequence, the principal connection $\mathcal{H}$ on the principal fiber bundle $P \to M$ is integrable if and only if its curvature 2-form Ω is identically zero.*

Proof. Using (V.7.6) we have for homogeneous X and Y :

$$\begin{aligned} -\iota(X,Y)\Omega &= -\iota(hX, hY)d\omega \\ &= (hX)(\iota(hY)\omega) - (-1)^{\langle\varepsilon(X)|\varepsilon(Y)\rangle}(hY)(\iota(hX)\omega) - \iota([\,hX, hY\,])\omega \\ &= -\iota([\,hX, hY\,])\omega \,. \end{aligned}$$

According to Frobenius, the connection $\mathcal{H}$ is integrable if and only if for all $X, Y \in \mathcal{H}$ we have $[X, Y] \in \mathcal{H}$. Since $\mathcal{H} = \ker \omega$, this means that $\mathcal{H}$ is integrable if and only if have the implication $\iota(X)\omega = \iota(Y)\omega = 0 \Longrightarrow \iota([X, Y])\omega = 0$. Using the definition of the projection $h : TP \to \mathcal{H}$, this means that $\mathcal{H}$ is integrable if and only if for all vector fields X and Y we have $\iota([hX, hY])\omega = 0$. According to our previous computation, this is the case if and only if $\Omega = 0$. QED

4.3 Lemma. *Let ω be a connection 1-form on a principal fiber bundle $P \to M$ and let $\Omega = D\omega$ be its curvature 2-form. Then we have the following identities:*

$$D\omega \equiv \Omega = d\omega + \tfrac{1}{2}[\,\omega \wedge \omega\,] \qquad \textit{(the structure equations of Cartan)}$$

$$d\Omega = [\,\Omega \wedge \omega\,] \qquad \textit{(the Bianchi identities).}$$

Proof. Interpreting k-forms as skew-symmetric k-linear maps on smooth vector fields (IV.5.15), it suffices to evaluate these identities on smooth vector fields. Since they are 2- and 3-additive, it suffices to show that we have equality when evaluating on homogeneous vector fields. Moreover, since a vector field splits as a sum of a horizontal and a vertical vector field, we may restrict attention to smooth vector fields which are either horizontal or vertical. And finally, since k-forms are linear over $C^\infty(P)$ and since vector fields of the form x^P with $x \in \mathbf{B}\mathfrak{g}$ generate the module of smooth vertical vector fields [VI.1.19], [3.7], it suffices to use this kind of vertical vertical vector field.

We start with the structure equations of Cartan, for which we evaluate both sides on two homogeneous vectors X, Y. We distinguish three cases: both vertical, both horizontal, and X horizontal and Y vertical.

If both X and Y are horizontal, we have $\iota(X,Y)\Omega = \iota(X,Y)d\omega$. Since by [1.4] $\iota(X,Y)[\,\omega \wedge \omega\,] = 0$, we have equality for two horizontal vector fields.

If both X and Y are vertical, we may assume, as argued above, that $X = x^P$ and $Y = y^P$ for some $x, y \in \mathbf{B}\mathfrak{g}$. It follows that $[x^P, y^P] = [x, y]^P$ is vertical. By definition of Ω, $\iota(X,Y)\Omega = 0$. On the other hand we have

$$-\iota(x^P, y^P)d\omega = x^P(\iota(y^P)\omega) - (-1)^{\langle\varepsilon(x)|\varepsilon(y)\rangle} y^P(\iota(x^P)\omega) - \iota([x,y]^P)\omega = -[x,y]$$

and $\iota(x^P, y^P)[\omega \wedge \omega] = -2[\iota(x^P)\omega, \iota(y^P)\omega] = -2[x,y]$ [1.4], which shows that for two vertical vector fields we also have equality.

If X is horizontal and Y vertical, we may assume that $Y = y^P$ for a $y \in \mathbf{B}\mathfrak{g}$. By definition of a connection 1-form we have the equality $\Phi_g^*\omega = \mathrm{Ad}(g^{-1}) \circ \omega$. From [VI.8.20] we know that we can take the derivative of the left hand side in the direction of $\vec{y}$ and that at $g = e$ we obtain $\mathcal{L}(y^P)\omega$ ($\Phi_e^* = id$). To compute the derivative of the right hand side in the direction of $\vec{y}$ at $g = e$, we first note that it depends on g only via $\mathrm{Ad}(g^{-1})$. And then [VI.9.12] gives us $\vec{y}_e(\mathrm{Ad}(g^{-1}) \circ \omega) = -\mathrm{ad}_R(y) \circ \omega$, and thus $\mathcal{L}(y^P)\omega = -\mathrm{ad}_R(y) \circ \omega$. We then compute

$$\begin{aligned}\iota([X, y^P])\omega &= \iota(X)\mathcal{L}(y^P)\omega - (-1)^{\langle\varepsilon(X)|\varepsilon(y)\rangle}\mathcal{L}(y^P)\iota(X)\omega \\ &= -\iota(X)\,\mathrm{ad}_R(y) \circ \omega = -(-1)^{\langle\varepsilon(X)|\varepsilon(y)\rangle}\,\mathrm{ad}_R(y)(\iota(X)\omega) = 0\ ,\end{aligned}$$

because $\iota(X)\omega = 0$. In other words, if X is horizontal, then $[X, y^P]$ is also horizontal. Since $Y = y^P$ is vertical, $\iota(X,Y)\Omega = 0$. Moreover, again by [1.4], we have the equality $\iota(X,Y)[\omega \wedge \omega] = -2[\iota(X)\omega, \iota(Y)\omega] = 0$. Finally,

$$-\iota(X,Y)\omega = X(\iota(y^P)\omega) - (-1)^{\langle\varepsilon(X)|\varepsilon(Y)\rangle}Y(\iota(X)\omega) - \iota([X, y^P])\omega = 0\ ,$$

because $\iota(y^P)\omega$ is constant and X and $[X, y^P]$ are horizontal. We conclude that also in the third case we have equality, i.e., we have proven the structure equations of Cartan.

To prove the Bianchi identities, we first note that $d\Omega = \frac{1}{2}d[\omega \wedge \omega]$ according to the structure equations of Cartan. With respect to a basis (v_i) of $\mathfrak{g}$ we have ordinary 1-forms ω^i defined by $\omega = \sum_i \omega^i \otimes v_i$. And then we have $d\omega = \sum_i d\omega^i \otimes v_i$ and $[\omega \wedge \omega] = \sum_{ij} \omega^i \wedge \omega^j \otimes (-1)^{\langle\varepsilon(v_i)|\varepsilon(v_j)\rangle}[v_i, v_j]$, and thus:

$$\begin{aligned}d[\omega \wedge \omega] &= \sum_{ij}(d\omega^i \wedge \omega^j - \omega^i \wedge d\omega^j) \otimes (-1)^{\langle\varepsilon(v_i)|\varepsilon(v_j)\rangle}[v_i, v_j] \\ &= [d\omega \wedge \omega] - \sum_{ij}(-1)^{\langle\varepsilon(v_i)|\varepsilon(v_j)\rangle} d\omega^j \wedge \omega^i \otimes (-[v_j, v_i]) = 2[d\omega \wedge \omega]\ .\end{aligned}$$

Again using the structure equations of Cartan we thus have

$$d\Omega = [d\omega \wedge \omega] = [\Omega \wedge \omega] - \tfrac{1}{2}[[\omega \wedge \omega] \wedge \omega]$$

The result now follows because of the Jacobi identity , [1.4] and the following computation for three homogeneous vectors:

$$\begin{aligned}\iota(X,Y,Z)[\,[\,\omega \barwedge \omega\,] \barwedge \omega\,] &= [\,\iota(X,Y)[\,\omega \barwedge \omega\,], \iota(Z)\omega\,] \\ &\quad + (-1)^{\langle \varepsilon(X)|\varepsilon(Y)+\varepsilon(Z)\rangle}[\,\iota(Y,Z)[\,\omega \barwedge \omega\,], \iota(X)\omega\,] \\ &\quad + (-1)^{\langle \varepsilon(Z)|\varepsilon(X)+\varepsilon(Y)\rangle}[\,\iota(Z,X)[\,\omega \barwedge \omega\,], \iota(Y)\omega\,] \\ &= -2(-1)^{\langle \varepsilon(X)|\varepsilon(Z)\rangle}\Big((-1)^{\langle \varepsilon(X)|\varepsilon(Z)\rangle}[\,[\,\iota(X)\omega, \iota(Y)\omega\,], \iota(Z)\omega\,] \\ &\quad + (-1)^{\langle \varepsilon(X)|\varepsilon(Y)\rangle}[\,[\,\iota(Y)\omega, \iota(Z)\omega\,], \iota(X)\omega\,] \\ &\quad + (-1)^{\langle \varepsilon(Z)|\varepsilon(Y)\rangle}[\,[\,\iota(Z)\omega, \iota(X)\omega\,], \iota(Y)\omega\,]\Big) = 0\,. \end{aligned}$$

QED

4.4 Corollary. *On any principal fiber bundle with a connection 1-form ω we have the equality $D\Omega \equiv D^2\omega = 0$.*

Proof. Using [1.4] and the Bianchi identities [4.3] we have

$$\begin{aligned}\iota(X,Y,Z)D\Omega &= \iota(hX,hY,hZ)d\Omega = \iota(hX,hY,hZ)[\,\Omega \barwedge \omega\,] \\ &= [\,\iota(hX,hY)\Omega, \iota(hZ)\omega\,] + (-1)^{\langle \varepsilon(X)|\varepsilon(Y)+\varepsilon(Z)\rangle}[\,\iota(hY,hZ)\Omega, \iota(hX)\omega\,] \\ &\qquad + (-1)^{\langle \varepsilon(Z)|\varepsilon(X)+\varepsilon(Y)\rangle}[\,\iota(hZ,hX)\Omega, \iota(hY)\omega\,]\,.\end{aligned}$$

Since $\iota(hW)\omega$ is zero for all W, the result follows. QED

4.5 Remark. If G is an $\mathcal{A}$-Lie group, we can see it as a principal fiber bundle over a point $M = \{m_o\}$ with (trivial) projection $\pi : G \to \{m_o\}$. Since there is no non-zero 1-form on a zero dimensional $\mathcal{A}$-manifold, it follows from [3.11] that any connection 1-form ω on this principal fiber bundle necessarily is the Maurer-Cartan 1-form: $\omega = \Theta_{MC}$. And then the structure equations of Cartan [4.3] tell us that [1.6] can be interpreted as saying that the curvature of this connection is zero.

4.6 Discussion. Let ω be a connection 1-form on a principal fiber bundle $\pi : P \to M$, let $\mathcal{H}$ be the associated principal connection, let $\{(U_a, \psi_a) \mid a \in I\}$ be a trivializing atlas for P determined by local sections s_a, and let x^i be (local) coordinates on some U_a. In [3.11] we have seen that the local 1-forms $s_a^*\omega$ are the local 1-forms Γ^a defining $\mathcal{H}$, which are related on overlaps $U_a \cap U_b$ by (2.15). Moreover, the local functions $A_i^{a,x}$ describing the fundamental vector fields γ_i are given by $A_i^{a,x} = \iota(\partial_{x^i})s_a^*\omega$ [2.12].

We intend to give a similar description of the curvature 2-form $\Omega = D\omega$. We thus define the $\mathfrak{g}$-valued 2-forms $s_a^*\Omega$ on U_a and the homogeneous smooth functions $F_{ij}^{a,x} : U_a \to \mathfrak{g}$ by $F_{ij}^{a,x}(m) = -\iota(\partial_{x^i}, \partial_{x^j})(s_a^*\Omega)_m$ with parity $\varepsilon(F_{ij}^{a,x}) = \varepsilon(x^i) + \varepsilon(x^j)$. Since a 2-form is graded skew-symmetric in its entries, the functions $F_{ij}^{a,x}$ are graded skew-symmetric in their indices: $F_{ji}^{a,x} = -(-1)^{\langle \varepsilon(x_i)|\varepsilon(x_j)\rangle} F_{ij}^{a,x}$. With these functions the 2-form $s_a^*\Omega$ can be written as

$$(s_a^*\Omega)_m = -\tfrac{1}{2}\sum_{ij} dx^j \wedge dx^i \otimes F_{ij}^{a,x}(m)\,.$$

4.7 Remark. The minus sign in the definition of $F^{a,x}_{ij}$ is conventional. One could say that it is a consequence of our way to identify the dual of an exterior power [I.7.22], [V.7.14]. In the ungraded case it would allow us to write $s_a^*\Omega = \frac{1}{2}\sum_{ij} dx^i \wedge dx^j \otimes F^{a,x}_{ij}$ with the indices in the same order.

4.8 Proposition. *The 2-form $s_a^*\Omega$ is determined by the 1-form $s_a^*\omega = \Gamma^a$ as*

$$s_a^*\Omega = d(s_a^*\omega) + \tfrac{1}{2}[\, s_a^*\omega \mathbin{\overset{\wedge}{,}} s_a^*\omega\,] \equiv d\Gamma^a + \tfrac{1}{2}[\Gamma^a \mathbin{\overset{\wedge}{,}} \Gamma^a\,]\,.$$

In terms of the functions $A^{a,x}_i$ and $F^{a,x}_{ij}$ this equality is given as

$$F^{a,x}_{ij} = \partial_{x^i} A^{a,x}_j - (-1)^{\langle\varepsilon(x^i)|\varepsilon(x^j)\rangle}\partial_{x^j} A^{a,x}_i + [\,A^{a,x}_i, A^{a,x}_j\,]\,. \tag{4.9}$$

$\Omega|_{\pi^{-1}(U_a)}$ can be reconstructed from $s_a^*\Omega$ in the trivialization determined by s_a by

$$\Omega_{(m,g)} = \mathrm{Ad}(g^{-1})\circ(s_a^*\Omega)_m = -\tfrac{1}{2}\sum_{ij} dx^j \wedge dx^i \otimes \mathrm{Ad}(g^{-1})F^{a,x}_{ij}(m)\,. \tag{4.10}$$

On the intersection $U_a \cap U_b$ of two local trivializations we have

$$(s_b^*\Omega)_m = \mathrm{Ad}(\psi_{ab}(m)^{-1})\circ(s_a^*\Omega)_m\,. \tag{4.11}$$

Proof. The structure equations of Cartan tell us immediately that we have the equality $s_a^*\Omega = d(s_a^*\omega) + \frac{1}{2}[\, s_a^*\omega \mathbin{\overset{\wedge}{,}} s_a^*\omega\,]$. Substituting (3.12) in this equation gives us:

$$\begin{aligned}
s_a^*\Omega &= \sum_j d(dx^j \otimes A^{a,x}_j) + \tfrac{1}{2}\sum_{ij}[\, dx^j \otimes A^{a,x}_j(x) \mathbin{\overset{\wedge}{,}} dx^i \otimes A^{a,x}_i(x)\,] \\
&= -\sum_{ij} dx^j \wedge dx^i \otimes \partial_{x^i} A^{a,x}_j + \tfrac{1}{2}\sum_{ij} dx^j \wedge dx^i \otimes (-1)^{\langle\varepsilon(x^i)|\varepsilon(x^j)\rangle}[\,A^{a,x}_j, A^{a,x}_i\,] \\
&= -\tfrac{1}{2}\sum_{ij} dx^j \wedge dx^i \otimes \Big\{\partial_{x^i} A^{a,x}_j - (-1)^{\langle\varepsilon(x^i)|\varepsilon(x^j)\rangle}\partial_{x^j} A^{a,x}_i + [\,A^{a,x}_i, A^{a,x}_j\,]\Big\}\,.
\end{aligned}$$

From this (4.9) follows immediately.

To prove the local form of Ω, we first recall that the Adjoint representation is indeed a representation of $\mathfrak{g}$, i.e., $[\,\mathrm{Ad}(g)x, \mathrm{Ad}(g)y\,] = \mathrm{Ad}(g)[\,x, y\,]$. Using the local expression $\omega_{(m,g)} = \mathrm{Ad}(g^{-1})(s_a^*\omega)_m + \Theta_{MC}|_g$ and [1.6] we compute:

$$\begin{aligned}
\Omega_{(m,g)} &= d\omega + \tfrac{1}{2}[\,\omega \mathbin{\overset{\wedge}{,}} \omega\,] \\
&= d\big(\mathrm{Ad}(g^{-1})(s_a^*\omega)_m\big) + \tfrac{1}{2}[\,\mathrm{Ad}(g^{-1})(s_a^*\omega)_m \mathbin{\overset{\wedge}{,}} \mathrm{Ad}(g^{-1})(s_a^*\omega)_m\,] \\
&\qquad + \tfrac{1}{2}[\,\mathrm{Ad}(g^{-1})(s_a^*\omega)_m \mathbin{\overset{\wedge}{,}} \Theta_{MC}|_g\,] + \tfrac{1}{2}[\,\Theta_{MC}|_g \mathbin{\overset{\wedge}{,}} \mathrm{Ad}(g^{-1})(s_a^*\omega)_m\,] \\
&= \mathrm{Ad}(g^{-1})(s_a^*\Omega)_m + (d\,\mathrm{Ad}(g^{-1}))\wedge(s_a^*\omega)_m + [\,\mathrm{Ad}(g^{-1})(s_a^*\omega)_m \mathbin{\overset{\wedge}{,}} \Theta_{MC}|_g\,]\,.
\end{aligned}$$

The last line is a consequence of the proof of [1.4], from which one can deduce that for even $\mathfrak{g}$-valued 1-forms α and γ we have $[\alpha \wedge \gamma] = [\gamma \wedge \alpha]$. We thus have to show that the last two terms cancel. We will do this by evaluating on tangent vectors. Since each term is a product of a term which only acts on vectors in the M-direction and a term which only acts on vectors in the G-direction, we only have to show that for homogeneous $X \in T_m M$ and $Y \in T_g G$ we have

$$\iota(X,Y)\Big((d\,\mathrm{Ad}(g^{-1})) \wedge (s_a^*\omega)_m + [\,\mathrm{Ad}(g^{-1})(s_a^*\omega)_m \wedge \Theta_{MC}|_g\,]\Big) = 0\,.$$

For any $Y \in T_g G$ there exists $y \in \mathfrak{g}$ such that $Y = \vec{y}_g$. We then compute:

$$\iota(X,\vec{y})[\,\mathrm{Ad}(g^{-1})(s_a^*\omega)_m \wedge \Theta_{MC}|_g\,] = -[\,\mathrm{Ad}(g^{-1})\iota(X)(s_a^*\omega)_m\,,\, y\,]\,,$$

where the minus sign comes from the fact that $\iota(\vec{y})$ is a derivation and has to be commuted with the even 1-form $\mathrm{Ad}(g^{-1})(s_a^*\omega)_m$. Using [VI.9.12] we compute the second term:

$$\begin{aligned}
\iota(X,\vec{y})(d\,\mathrm{Ad}(g^{-1})) \wedge (s_a^*\omega)_m &= (-1)^{\langle\varepsilon(y)|\varepsilon(X)\rangle}\,(\vec{y}_g\,\mathrm{Ad}(g^{-1}))\cdot\iota(X)(s_a^*\omega)_m \\
&= -(-1)^{\langle\varepsilon(y)|\varepsilon(X)\rangle}\,\mathrm{ad}_R(y)\circ\mathrm{Ad}(g^{-1})\iota(X)(s_a^*\omega)_m \\
&= -(-1)^{\langle\varepsilon(y)|\varepsilon(X)\rangle}[\,y\,,\,\mathrm{Ad}(g^{-1})\iota(X)(s_a^*\omega)_m\,] \\
&= [\,\mathrm{Ad}(g^{-1})\iota(X)(s_a^*\omega)_m\,,\,y\,]\,.
\end{aligned}$$

This proves that the two terms cancel and thus we have proven (4.10).

To prove (4.11), we note that in the trivialization determined by s_a, the section s_b takes the form $s_b(m) = (m, \psi_{ab}(m))$ (a direct consequence of (3.5)). Using the local expression (4.10) we find immediately

$$s_b^*\Omega = \mathrm{Ad}(\psi_{ab}(m)^{-1})\circ(s_a^*\Omega)_m\,. \qquad \boxed{QED}$$

4.12 Proposition. *Let $\pi : P \to M$ be a principal fiber bundle with structure group G, and let ω be a connection 1-form on P and Ω its curvature 2-form.*

(i) *Let $U \subset M$ be open and $s \in \Gamma_U(P)$ a local section. Then s is a horizontal section if and only if $s^*\omega = 0$, and either of these two conditions implies that $\Omega|_{\pi^{-1}(U)} = 0$ and that the horizontal submanifolds in $\pi^{-1}(U) \cong U \times G$ are exactly the horizontal sets $U \times \{g\}$, $g \in \mathbf{B}G$ (see [2.1]).*

(ii) *$\Omega = 0$ on P if and only if there exists a trivializing atlas $\mathcal{U} = \{(U_a, \psi_a) \mid a \in I\}$ determined by local sections $s_a \in \Gamma_{U_a}(P)$ such that for all $a \in I$ we have $s_a^*\omega = 0$.*

Proof. The local section s is horizontal if and only if $Ts(X_m) \in \mathcal{H}_{s(m)}$ for all $m \in U$ and all $X_m \in T_m M$ [2.2]. Since $\mathcal{H}_{s(m)} = \ker\omega_{s(m)}$, this is the case if and only if $0 = \iota(\langle\!\langle X_m \| Ts \rangle\!\rangle)\omega_{s(m)} = \iota(X_m)(s^*\omega)_m$,which is the case if and only if $s^*\omega = 0$. Using the structure equations of Cartan, this implies directly that $s^*\Omega = 0$. And then

(4.10) tells us that $\Omega|_{\pi^{-1}(U)} = 0$. (3.12) shows that $\omega|_{\pi^{-1}(U)}$ is given in the trivialization determined by the local section s as Θ_{MC}. This implies that in this trivialization $\mathcal{H}_{(m,g)}$ is given as $T_mM \times \{0\}_g \subset T_{(m,g)}(U \times G)$. And the integral manifolds of this subbundle are obviously the subsets $U \times \{g\}$ as announced.

For (ii), if $\mathcal{U}$ exists, then by (i) $\Omega = 0$. We thus assume that $\Omega = 0$, which means that $\mathcal{H}$ is integrable [4.2]. Now let $p \in \mathbf{B}P$ be arbitrary, then through p passes a (unique) leaf (i, L) of the involutive subbundle $\mathcal{H} \subset TP$ [V.6.9]. Let $\ell \in L$ be the unique point such that $i(\ell) = p$. By definition of a leaf, $Ti : T_\ell L \to \mathcal{H}_p$ is an isomorphism, and by definition of a connection, $T\pi : \mathcal{H}_p \to T_{\pi(p)}$ is a bijection. Using [V.2.14] we deduce that there exist neighborhoods V of ℓ and U_m of $m = \pi(p)$ such that $\pi \circ i$ is a diffeomorphism from V to U_m. We then define $s : U_m \to P$ by $s = i \circ ((\pi \circ i)|_V)^{-1}$. Composing on the left with π shows that $\pi \circ s = id$, and thus s is a local section. Since (i, L) is a leaf, s is horizontal, i.e., $s^*\omega = 0$. Since $p \in \mathbf{B}P$ is arbitrary, we conclude that every $m \in \mathbf{B}M$ admits a neighborhood U_m on which there exists a local horizontal section s. Since for all $m \in M$ we have $m \in U_{\mathbf{B}m}$ by definition of the DeWitt topology, these neighborhoods cover M. The corresponding local sections thus define a trivializing atlas as desired. QED

5. FVF connections on associated fiber bundles

In this section we show that there is a natural way to introduce an FVF connection on an associated fiber bundle starting with one on the original bundle. Defining the structure bundle as the principal fiber bundle with the same structure group and transition functions as the original bundle allows us to show a close relationship between FVF connections on general fiber bundles and those on principal fiber bundles. Using this relationship, we show that a leaf of an integrable FVF connection is a covering space of the (connected) base space. In particular, a fiber bundle B over a simply connected base space admits an integrable FVF connection if and only if B is trivial. We end this section by giving an intrinsic description of an associated fiber bundle to a principal fiber bundle, which allows us to give alternative descriptions of some constructions concerning associated bundles.

5.1 Proposition. *Let $\pi : B \to M$ be a fiber bundle with typical fiber F and structure group G. Let H be another $\mathcal{A}$-Lie group with a pseudo effective action on an $\mathcal{A}$-manifold E. Let $\rho : G \to H$ be a morphism of $\mathcal{A}$-Lie groups. Finally, let $\mathcal{H}$ be an FVF connection on B defined by local $\mathfrak{g}$-valued 1-forms Γ^a relative to a trivializing atlas $\mathcal{U} = \{(U_a, \psi_a) \mid a \in I\}$ for B. Then the local $\mathfrak{h}$-valued 1-forms $\Gamma'^a = T_e\rho \circ \Gamma^a$ define an FVF connection $\mathcal{H}^\rho$ on the associated fiber bundle $B^{\rho,E} \to M$ [IV.2.1].*

Proof. Let us denote by ψ_{ba} the transition functions associated to the trivializing atlas $\mathcal{U}$. Then, according to its definition [IV.2.1], the transition functions of the associated bundle are given by $\psi'_{ba} = \rho \circ \psi_{ba}$. According to [2.13] we only have to prove that

$$\begin{aligned}\Gamma'^b &\equiv T_e\rho \circ \big(\mathrm{Ad}(\psi_{ab}(m)^{-1}) \circ \Gamma^a(m) + (\psi_{ab}^* \Theta_{MC}^G)|_m\big) \\ &= \mathrm{Ad}(\psi'_{ab}(m)^{-1}) \circ \Gamma'^a(m) + (\psi'^*_{ab} \Theta_{MC}^H)|_m \,.\end{aligned}$$

To prove this, we start with the equality $\rho \circ I_g = I_{\rho(g)} \circ \rho$ as maps from G to H, where I_g denotes, as in [VI.2.12], the map $x \mapsto gxg^{-1}$. Taking the tangent map at the identity and using the definition of the Adjoint representation gives us $T_e\rho \circ \mathrm{Ad}(g) = \mathrm{Ad}(\rho(g)) \circ T_e\rho$. This proves that $T_e\rho \circ \mathrm{Ad}(\psi_{ab}(m)^{-1}) \circ \Gamma^a(m) = \mathrm{Ad}(\psi'_{ab}(m)^{-1}) \circ \Gamma'^a(m)$. Next we note that $\rho \circ L_g = L_{\rho(g)} \circ \rho$ as maps from G to H. Taking the tangent map of this identity gives us the equality

$$T_e\rho \circ TL_{\psi_{ab}(m)^{-1}} \circ T\psi_{ab} = TL_{\rho(\psi_{ab}(m)^{-1})} \circ T_e\rho \circ T\psi_{ab} = TL_{\psi'_{ab}(m)^{-1}} \circ T\psi'_{ab} \,.$$

Comparing this with the equality $\psi_{ab}^*\Theta_{MC}^G = \sum_i dx^i \otimes TL_{\psi_{ab}(m)^{-1}}\, T\psi_{ab}\, \partial_{x^i}|_m$ given in the proof of [2.13], we can conclude that $T_e\rho \circ \psi_{ab}^*\Theta_{MC}^G = \psi'^*_{ab}\Theta_{MC}^H$. QED

5.2 Definition. In [IV.2.14] we have seen that any fiber bundle can be seen as being an associated fiber bundle to a principal fiber bundle. We formalize this by defining the *structure bundle* $\pi_{\mathcal{S}} : \mathcal{S}B \to M$ as being this principal fiber bundle, i.e., $\mathcal{S}B = B^{id,G}$ is the (principal) fiber bundle with typical fiber G and structure group G associated to the fiber bundle $\pi : B \to M$ with typical fiber F and structure group G by the identity representation $id : G \to G$. The underlying idea is that both B and $\mathcal{S}B$ are defined by the same transition functions ψ_{ab} associated to a trivializing atlas $\mathcal{U}$.

5.3 Corollary. *$\mathcal{S}B^{id,F} = B$ and if H is another $\mathcal{A}$-Lie group with a pseudo effective action on an $\mathcal{A}$-manifold E and if $\rho : G \to H$ is a morphism of $\mathcal{A}$-Lie groups, then $B^{\rho,E} = \mathcal{S}B^{\rho,E}$, i.e., constructing an associated fiber bundle from the original fiber bundle B or from its structure bundle yields the same result.*

5.4 Remark. The structure bundle is a generalization of the frame bundle for vector bundles. If $\pi : B \to M$ is a vector bundle with typical fiber the $\mathcal{A}$-vector space E, we can define for each $m \in M$ the set $\mathcal{F}_m$ of all bases, also called frames, of the $\mathfrak{A}$-graded $\mathcal{A}$-module B_m. Since two bases are related by an element of $\mathrm{Aut}(B_m)$ [II.2.6], the set $\mathcal{F}_m$ is isomorphic to $\mathrm{Aut}(B_m)$. In order to give the disjoint union $\mathcal{F}B = \amalg_{m\in M}\mathcal{F}_m$ the structure of a (principal) fiber bundle over M, we proceed as follows. We fix a basis (e_i) of E. In a local trivialization (U_a, ψ_a) we define a map $\{m\} \times \mathrm{Aut}(E) \to \mathcal{F}_m$ by $(m, V) \mapsto (\psi_a^{-1}(Ve_i))_{i=1}^{\dim E}$. This gives us an isomorphism χ_a^{-1} between $U_a \times \mathrm{Aut}(E)$ and $\amalg_{m\in U_a}\mathcal{F}_m$. To see how this isomorphism depends upon the chosen local trivialization, we choose another one (U_b, ψ_b), which gives us an isomorphism χ_b^{-1} between $U_b \times \mathrm{Aut}(E)$ and $\amalg_{m\in U_b}\mathcal{F}_m$ by $(m, W) \mapsto (\psi_b^{-1}(We_i))_{i=1}^{\dim E}$. For $m \in U_a \cap U_b$ we get the same basis of B_m if for all i we have $\psi_a^{-1}(Ve_i) = \psi_b^{-1}(We_i)$. Applying ψ_b and the definition of the transition function, this happens if and only if for all i we have $We_i = \psi_{ba}(m)Ve_i$, i.e., if and only if $W = \psi_{ba}(m) \circ V$. This implies that $\chi_b \circ \chi_a^{-1}$ is given by the map $(m, V) \mapsto (m, \psi_{ba}(m) \cdot V)$. We conclude that $\mathcal{F}B$ is a principal fiber bundle with structure group $\mathrm{Aut}(E)$ and *the same* transition functions as B. In other words, the frame bundle of the given vector bundle B is exactly the structure bundle as defined in [5.2].

5.5 Corollary. *Let $B \to M$ be any fiber bundle and let $\mathcal{S}B \to M$ be the associated structure bundle. Then there exists a canonical bijection between the set of FVF connections on B and the set of FVF/principal connections on $\mathcal{S}B$. In particular, any fiber bundle $B \to M$ admits an FVF connection.*

Proof. According to [IV.2.14] and [5.2] the fiber bundle $B \to M$ and the principal fiber bundle $\mathcal{S}B \to M$ are associated to each other by the identity representation. According to [5.1] an FVF connection on one of these two bundles defined by local 1-forms Γ^a determines an FVF connection on the other one by the same set of 1-forms. Finally, according to [3.13] any principal fiber bundle admits a principal/FVF connection. QED

5.6 Corollary. *Let $P \to M$ be a principal fiber bundle with structure group G, let H be an $\mathcal{A}$-Lie group with a pseudo effective action on F, let $\rho : G \to H$ be a morphism of $\mathcal{A}$-Lie groups, and let $B \equiv P^{\rho,F} \to M$ be the associated fiber bundle (associated to P by the representation ρ). If $\mathcal{H}^P$ is a principal/FVF connection on P with associated connection 1-form ω, and if $\mathcal{H}^B$ is the associated FVF connection on B given in [5.1], then $\mathcal{H}^B$ is integrable if and only if $T_e\rho \circ \Omega$ is zero, where Ω is the curvature 2-form on P.*

Proof. According to [2.18] $\mathcal{H}^B$ is integrable if and only if the local 2-forms

$$T_e\rho \circ \left(d\Gamma^a + \tfrac{1}{2}[\Gamma^a \wedge \Gamma^a]\right)$$

are zero (use that $[T_e\rho(x), T_e\rho(y)] = T_e\rho([x, y])$, i.e., $T_e\rho$ is a morphism of $\mathcal{A}$-Lie algebras). On the other hand, using (4.10), we find

$$(T_e\rho \circ \Omega)_{(m,g)} = T_e\rho \circ \mathrm{Ad}(g^{-1}) \circ (s_a^*\Omega)_m = \mathrm{Ad}(\rho(g)^{-1}) \circ T_e\rho \circ (s_a^*\Omega)_m \,.$$

Since $s_a^*\Omega = d\Gamma^a + \frac{1}{2}[\Gamma^a \wedge \Gamma^a]$, the result follows. QED

5.7 Corollary. *Let $\mathcal{H}^B$ be an FVF connection on a fiber bundle $B \to M$, let $\mathcal{H}^S$ be the associated FVF/principal connection on the structure bundle $\mathcal{S}B \to M$ [5.5], and let ω be its associated connection 1-form. Then the following four statements are equivalent:*

(i) *$\mathcal{H}^B$ is integrable;*
(ii) *$\mathcal{H}^S$ is integrable;*
(ii) *the curvature 2-form Ω on $\mathcal{S}B$ is zero;*
(iii) *there exists a trivializing atlas $\mathcal{U} = \{(U_a, \psi_a) \mid a \in I\}$ for B such that all local 1-forms Γ^a determining $\mathcal{H}^B$ are zero.*

Proof. The equivalence between (i) and (ii) has been shown in [4.2], and the equivalence between (i) and (iii) is an immediate consequence of [5.6] because $\mathcal{S}B$ is related to B by the identity representation. Since B and $\mathcal{S}B$ are determined by the same local 1-forms Γ^a, the equivalence between (iii) and (iv) is a direct consequence of [4.12]. QED

5.8 Proposition. *Let $\pi : B \to M$ be a fiber bundle with typical fiber F, let $\mathcal{H}$ be an integrable FVF connection on B and let (i, L) be a leaf of $\mathcal{H}$. Then $\pi(i(L))$ is open and closed in M and $\pi \circ i : L \to \pi(i(L))$ is a covering (§VI.7).*

Proof. Suppose that $m_o \in M$ belongs to the closure of $\pi(i(L))$. According to [5.7], there exists a local trivialization (U, ψ) containing m_o (part of a trivializing atlas) such that the local 1-form Γ on U is zero. Taking a smaller U if necessary, we may assume that U is connected. By definition of closure, there exists a point $m \in U \cap \pi(i(L))$. Let $\ell \in L$ be such that $\pi(i(\ell)) = m$, then $\pi(i(\mathbf{B}\ell)) = \mathbf{B}m \in U$ because $\pi \circ i$ is smooth and U open. Let $f \in F$ be such that $\psi(i(\ell)) = (m, f)$. Since the local 1-form Γ on U is zero, the local section $s : U \to B$, $s(m') = \psi^{-1}(m', \mathbf{B}f)$ is a smooth horizontal section. But then U is connected, s is tangent to $\mathcal{H}$ and $i(\mathbf{B}\ell) = \mathbf{B}i(\ell) \in s(U) \cap i(L)$ and thus by [V.6.9] $s(U) \subset i(L)$. Since s is a section, this implies that $U \subset \pi(i(L))$. We thus have shown that an arbitrary point m_o in the closure of $\pi(i(L))$ admits an open neighborhood U contained in $\pi(i(L))$, and thus $\pi(i(L))$ is open and closed.

To show that $\pi \circ i$ is a covering, we have to find for all $m \in \pi(i(L))$ an open neighborhood with certain properties [VI.7.1], [VI.7.3]. Since $\mathcal{H}$ is integrable, there exists (as above) a local trivialization (U, ψ) of B on which Γ is zero with U connected, contained in $\pi(i(L))$ and containing m. Let $(\pi \circ i)^{-1}(U) = \cup_{\alpha \in J} V_\alpha$ be the decomposition of $(\pi \circ i)^{-1}(U)$ in connected components. Then by definition the V_α are pairwise disjoint. Since $(\pi \circ i)^{-1}(U)$ is open in L, it follows from [III.1.3] that each V_α is open in L. Moreover, $\pi \circ i$ is smooth and $T_\ell(\pi \circ i)$ is a bijection for each $\ell \in L$ because $Ti : T_\ell L \to \mathcal{H}_{i(\ell)}$ is a bijection by definition of a leaf, and because $T\pi : \mathcal{H}_{i(\ell)} \to T_{\pi(i(\ell))}$ is a bijection by definition of a connection. If we can show that $\pi \circ i$ is a diffeomorphism from each V_α to U, we will have shown that $\pi \circ i$ is a covering map. Using the projection $\pi_F : U \times F \to F$, we define the map $g = \pi_F \circ \psi \circ i : \pi^{-1}(U) \to F$. Since i is tangent to $\mathcal{H}$ and since $\Gamma = 0$ on U, i.e., $T\psi(\mathcal{H}) = TU \times \{0\}$, it follows that $Tg = 0$. By [V.3.21] g is constant on each V_α. Since g is smooth, there thus exist $f_\alpha \in \mathbf{B}F$ such that $i(V_\alpha) \subset \psi^{-1}(U \times \{f_\alpha\})$. We now define the local smooth sections $s_\alpha(m') = \psi^{-1}(m', f_\alpha)$. Since $\Gamma = 0$ on U, this means (as before) that s_α is tangent to $\mathcal{H}$. Since $i(\ell) \in s_\alpha(U) \cap i(L)$ for any $\ell \in L$, it follows from [V.6.9] that there exist smooth maps $g_\alpha : U \to L$ such that $s_\alpha = i \circ g_\alpha$. Since $\pi \circ s_\alpha = id(U)$, $g_\alpha(U) \subset (\pi \circ i)^{-1}(U)$, and since U is connected, $g_\alpha(U)$ is contained in one of its connected components V_β. Moreover, again because $\pi \circ s_\alpha = id(U)$, $i(V_\beta) = \psi^{-1}(U \times \{f_\alpha\})$. Since i is an injective immersion, it follows that $i : V_\beta \to \psi^{-1}(U \times \{f_\alpha\})$ is a diffeomorphism (because L and U have the same dimension). But $i(V_\alpha) \subset \psi^{-1}(U \times \{f_\alpha\})$, and thus by injectivity of i we deduce that $\alpha = \beta$ and that $i : V_\alpha \to \psi^{-1}(U \times \{f_\alpha\})$ is a diffeomorphism. Since the same is true for $\pi : \psi^{-1}(U \times \{f_\alpha\}) \to U$, we have shown that $\pi \circ i : V_\alpha \to U$ is a diffeomorphism. QED

5.9 Corollary. *Let M be a simply connected $\mathcal{A}$-manifold and let $\pi_B : B \to M$ be a fiber bundle. Then B admits an integrable FVF connection if and only if it is trivial. Moreover, if B is a vector bundle, then triviality is as a vector bundle.*

Proof. If B admits an integrable FVF connection, then by [5.7] the corresponding FVF connection $\mathcal{H}^S$ on $\pi : \mathcal{S}B \to M$ is integrable. By [5.8] any leaf (i, L) of $\mathcal{H}^S$ is a covering of M. Since M is simply connected, this implies that $\pi \circ i : L \to M$ is a diffeomorphism. But then $i \circ (\pi \circ i)^{-1}$ is a global section of $\mathcal{S}B$ and thus $\mathcal{S}B$ is trivial. Adding this global trivialization of $\mathcal{S}B$ to a trivializing atlas, the corresponding trivializing atlas of the associated bundle B also contains a global trivialization, and thus B is trivial. In particular, if B is a vector bundle, this global trivialization is compatible with the vector bundle structure and thus B is trivial as a vector bundle.

Conversely, if $\psi : B \to M \times F$ is a global trivialization, then $(T\psi)^{-1}(TM \times \{0\})$ is an integrable FVF connection, determined by the (global) 1-form $\Gamma \equiv 0$. QED

5.10 Remark. Since any 1-dimensional connected $\mathcal{A}$-manifold N is simply connected (it is an interval [V.4.1]), it follows from [5.8] that any leaf of an integrable FVF connection on N is diffeomorphic to N via the projection map. A particular consequence is that parallel transport along an even curve always exists for FVF connections (see [2.6]).

5.11 Remark/Discussion. The definition of an associated fiber bundle and the construction of the FVF connection on an associated fiber bundle both use local trivializations. Even though this works quite well, one would like to have a more intrinsic/global description. Such a more global description can be given for arbitrary fiber bundles associated to a principal fiber bundle. The idea is as follows.

Let $\pi_P : P \to M$ be a principal fiber bundle with structure group G and let Φ^P denote the right action of G on P. Let H be an $\mathcal{A}$-Lie group with a pseudo effective action Φ^F on an $\mathcal{A}$-manifold F, and let $\rho : G \to H$ be an $\mathcal{A}$-Lie group morphism. These data allow us to define an associated fiber bundle $\pi_B : B \to M$ with structure group H and typical fiber F. The more intrinsic construction of B starts with the observation that the map $\Psi : G \times F \to F$ given by $\Psi(g, f) = \Phi^F(\rho(g), f)$ is a left action of G on F (not necessarily pseudo effective). With this action we define an effective right action $\Phi^\times$ of G on $P \times F$ by

$$\Phi^\times(p, f, g) = (\Phi^P(p, g), \Psi(g^{-1}, f)) .$$

The projection $\pi_P \circ \pi_1 : P \times F \to M$ (with $\pi_1 : P \times F \to P$ the projection on the first factor) is constant on G-orbits because π_P is constant on G-orbits in P. We thus have an induced map $\pi_\mathcal{O} : (P \times F)/G \to M$ from the orbit space $(P \times F)/G$ to M. We now define a map $\pi_\times : P \times F \to B$ in terms of the local trivializations ψ_a for P and $\widetilde{\psi}_a$ for B relative to the same trivializing atlas by

$$\pi_\times|_{\pi_P^{-1}(U_a) \times F} = \widetilde{\psi}_a^{-1} \circ (id_M \times \Psi) \circ (\psi_a \times id_F) ,$$

or equivalently, still in local trivializations, by

$$\pi_\times(m,g,f) = (m, \Psi(g,f)) \ .$$

We claim that this is a well defined map, independent of the chosen local trivialization, and that it is constant on G-orbits. Moreover, the induced map $\widetilde{\pi_\times} : (P \times F)/G \to B$ is a bijection verifying $\pi_B \circ \widetilde{\pi_\times} = \pi_{\mathcal{O}}$, i.e., we have the following commutative diagram

$$\begin{array}{ccccc} & & P \times F & & \\ & \swarrow & & \searrow^{\pi_\times} & \\ P \times F/G & & \xleftrightarrow{\ \widetilde{\pi_\times}\ } & & B \ . \\ & \pi_{\mathcal{O}}\searrow & & \swarrow \pi_B & \\ & & M & & \end{array}$$

The verification of these claims is straightforward and is left to the interested reader. We do not say that $\widetilde{\pi_\times}$ is a *diffeomorphism* between the orbit space $(P \times F)/G$ and B because we have not defined how to induce the structure of an $\mathcal{A}$-manifold on an orbit space (if possible at all; the only instance where we have defined the structure of an $\mathcal{A}$-manifold on an orbit space is for homogeneous $\mathcal{A}$-manifolds [VI.5.9]).

Either by using the bijection $\widetilde{\pi_\times}$ or by more direct means, one can give the orbit space $(P \times F)/G$ the structure of an $\mathcal{A}$-manifold and then the structure of a fiber bundle over M with structure group G and typical fiber F. Once we have this structure, the bijection $\widetilde{\pi_\times}$ becomes an isomorphism of fiber bundles. It follows that we can take the orbit space $(P \times F)/G$, for which one also finds the notation $P \times_G F$, as the definition of the associated fiber bundle B.

Once we have the description of the associated bundle B as the orbit space $(P \times F)/G$, we can give a global description of the induced FVF connection. To that end, let $\mathcal{H}^P$ be an FVF connection on the principal fiber bundle P. Then $\mathcal{H}^B \equiv T\pi_\times(H^P \times \{0\})$ is the FVF connection on the associated fiber bundle $B \equiv (P \times F)/G$ defined in [5.1]. Note that the principal fiber bundle P here plays the role of the bundle B of [5.1] and that the bundle B here is the associated fiber bundle, associated to P by the representation ρ of [5.1].

Another property of associated bundles that now can be given a more intrinsic description is the following. The construction of an associated bundle starts with a trivializing atlas for the initial bundle, and then the associated bundle has the same trivializing sets. Adding more elements to the original trivializing atlas, it follows that for each local trivialization (U, ψ) of the initial bundle there exists a corresponding local trivialization (U, ψ') for the associated bundle. If the initial bundle is a principal fiber bundle $\pi : P \to M$, we also know that a local trivialization of P is completely determined by a local section s. It follows that for each local section $s : U \to P$ there is a corresponding local trivialization (U, ψ_s) of B. In terms of the intrinsic description of B given above, this local trivialization is given as

$$\psi_s^{-1} : U \times F \to \pi_B^{-1}(U) \qquad , \qquad (m,f) \mapsto \pi_\times(s(m), f) \ .$$

6. The covariant derivative

In this section we introduce the notion of a covariant derivative on a vector bundle, which is a generalization of the derivative of a vector valued functions to sections of the vector bundle. We show that a covariant derivative is determined by local 1-forms Γ^a with values in $\mathrm{End}_R(E)$, the space of endomorphisms of the typical fiber E. We prove that these local 1-forms behave exactly as the local 1-forms defining an FVF connection, thus showing that a covariant derivative on a vector bundle is equivalent to an FVF connection on it.

6.1 Discussion. In [2.1] we discussed the idea of (locally) constant (local) sections of a fiber bundle in terms of Ehresmann connections. For vector bundles there is another approach in terms of a covariant derivative which is based more on the derivative of a function being zero than on the section being horizontal. Let M be an $\mathcal{A}$-manifold, E an $\mathcal{A}$-vector space and $f : M \to E$ a (smooth) function, corresponding to the section s of the (trivial) bundle $\pi : M \times E \to M$ given by $s(m) = (m, f(m))$. If X is a vector field on M, we have an action of X on f giving a new function $Xf : M \to E$ [V.1.24]. We can transform this action on functions into an action on sections by defining the section Xs as being given by $(Xs)(m) = (m, (Xf)(m))$. This is a simple transcription of the action of vector fields on E-valued functions to sections. We then can say that s is a constant section if and only if Xs is the zero section for all vector fields X on M.

If we try to generalize this to arbitrary vector bundles, we encounter a problem: a (global) section s is represented by local functions s_ψ, but there is no guarantee that the new local functions Xs_ψ glue together to form a new global section Xs. This corresponds of course to the fact that being horizontal in one trivialization does not necessarily correspond to being horizontal in another trivialization. The idea of a covariant derivative is to extract the essential features of the above definition of the action of vector fields on sections of a trivial bundle, and to use these to define something meaningful on an arbitrary vector bundle. In [6.18] and [7.2] we will see that the approach to define constant sections via a covariant derivative is equivalent to doing it via an FVF connection.

6.2 Definition. Let $\pi : B \to M$ be a vector bundle over M with typical fiber the $\mathcal{A}$-vector space E. Recall that $\Gamma(B)$ denotes the graded $C^\infty(M)$-module of smooth sections of the bundle B and that $\Gamma(TM)$ denotes the graded $C^\infty(M)$-module of all vector fields on M. *A covariant derivative ∇ on the bundle B* is a map $\nabla : \Gamma(TM) \times \Gamma(B) \to \Gamma(B)$ satisfying the following conditions.

(i) ∇ is bi-additive and even;

(ii) for all $f \in C^\infty(M)$, $X \in \Gamma(TM)$, $s \in \Gamma(B)$ we have $\nabla(fX, s) = f\nabla(X, s)$;

(iii) for homogeneous f, X, s we have

$$\nabla(X, fs) = (Xf)s + (-1)^{\langle \varepsilon(X) | \varepsilon(f) \rangle} f\nabla(X, s) ,$$

which can also be written as $\nabla(X, sf) = \nabla(X, s)f + (-1)^{\langle \varepsilon(s) | \varepsilon(f) \rangle}(Xf)s$.

Following custom, we will denote $\nabla(X, s)$ also as $\nabla_X s$; it is called *the covariant derivative of s in the direction of X.*

6.3 Example. Let M be an $\mathcal{A}$-manifold and E an $\mathcal{A}$-vector space. In the identification between E-valued functions on M and sections of the trivial bundle $M \times E$ (with its canonical trivialization), the action of vector fields on sections $(X, s) \mapsto Xs$ as in [6.1] is an example of a covariant derivative on this trivial bundle.

6.4 Proposition. *Let $\pi : B \to M$ be a vector bundle, ∇ a covariant derivative on B, and let $V \subset U \subset M$ be two open subsets.*

(i) *If $s, t \in \Gamma(B)$ are such that $s|_U = t|_U$, then $\nabla(X, s)|_U = (\nabla(X, t))|_U$ for any $X \in \Gamma(TM)$.*

(ii) *If $X, Y \in \Gamma(TM)$ are such that $X_m = Y_m$ for some $m \in M$, then for any $s \in \Gamma(B)$ we have $\nabla(X, s)(m) = \nabla(Y, s)(m)$.*

(iii) *There exists a unique covariant derivative ∇^U on $B|_U$ such that for all $s \in \Gamma(B)$ we have $\nabla(X, s)|_U = \nabla^U(X|_U, s|_U)$.*

(iv) $(\nabla^U)^V = \nabla^V$.

Proof. Recall first that $B|_U$ is the subbundle $\pi : B|_U \equiv \pi^{-1}(U) \to U$ [IV.1.13]. Apart from the difference coming from the presence of a bundle, the proof is a close copy of the proofs of [V.1.4], [V.1.5], [IV.5.8].

• (i) Without loss of generality we may assume that s and t are homogeneous of the same parity. For any $m \in U$, let ρ be a plateau function around m in U. It follows that $\rho(s - t) = 0$. Using the properties of a covariant derivative, we obtain

$$0 = \nabla(X, \rho(s - t)) = (X\rho)(s - t) + \rho\nabla(X, (s - t)) \qquad (\rho \text{ is even}).$$

Since $\rho(m) = 1$ this gives us $(\nabla_X s)(m) = (\nabla_X t)(m)$.

• (ii) Let (x^i) be local coordinates on a neighborhood W of m, then there exist functions X^i and Y^i on W such that $X|_W = \sum_i X^i\partial_{x^i}$ and $Y|_W = \sum_i Y^i\partial_{x^i}$. If ρ is a plateau function around m in W, then $\rho\partial_{x^i}$ is a global smooth vector field on M and ρX^i and ρY^i are global smooth functions. Since ρ is zero outside W, we have the global equalities $\rho^2 X = \sum_i(\rho X^i)\cdot(\rho\partial_{x^i})$ and $\rho^2 = \sum_i(\rho Y^i)\cdot(\rho\partial_{x^i})$. Using the properties of a covariant derivative we find

$$\rho^2\nabla(X, s) = \nabla(\rho^2 X, s) = \sum_i(\rho X^i)\cdot\nabla(\rho\partial_{x^i}, s)\ ,$$

and a similar equation with X replaced by Y. Evaluating these sections at m, and using $\rho(m) = 1$ and $X^i(m) = Y^i(m)$, we find

$$\begin{aligned}\nabla(X, s)(m) &= \sum_i X^i(m)\cdot\nabla(\rho\partial_{x^i}, s)(m) = \sum_i Y^i(m)\cdot\nabla(\rho\partial_{x^i}, s)(m)\\ &= \nabla(Y, s)(m)\ .\end{aligned}$$

• (iii) As for derivations, the main problem is that not every smooth section above U need be the restriction of a global smooth section. So let $t \in \Gamma_U(B)$, $Y \in \Gamma_U(TM)$ and $m \in U$ be arbitrary and let ρ be a plateau function around m in U. It follows that ρt is a well defined global smooth section of B and that ρY is a well defined global smooth vector field on M. Moreover, t and $(\rho t)|_U$ are two local sections above U that coincide in a neighborhood of m and Y and $(\rho Y)|_U$ are two vector fields on U such that $Y_m = ((\rho Y)|_U)_m$. Now, if ∇^U exists, we can combine (i) and (ii) with the defining property of ∇^U to obtain

$$\nabla^U(Y,t)(m) = \nabla^U(\rho Y)|_U(\rho t)|_U(m) = \nabla(\rho Y, \rho t)(m) \; . \tag{6.5}$$

This proves uniqueness of ∇^U, but we can also use (6.5) to *define* it. To see that (6.5) indeed produces a well defined ∇^U, independent of the choice for ρ, suppose $\hat\rho$ has the same properties as ρ. It follows that ρt and $\hat\rho t$ coincide in a neighborhood of m, and $(\rho Y)_m = (\hat\rho Y)_m$. And thus by the preceding result $\nabla(\rho Y, \rho t)(m) = \nabla(\hat\rho Y, \hat\rho t)(m)$, i.e., (6.5) is independent of the choice for ρ. Since s and $\rho(s|_U)$ coincide in a neighborhood of m, and $X_m = (\rho(X|_U))_m$, it follows that

$$\nabla(X,s)(m) = \nabla(\rho(X|_U), \rho(s|_U))(m) = \nabla^U(X|_U, s|_U)(m) \; ,$$

i.e., the covariant derivative defined by (6.5) has the desired property.

To prove that ∇^U is a covariant derivative, we first note that it is obviously bi-additive and even, since the same holds for ∇. Property (ii) of a covariant derivative is also a direct consequence of the corresponding property for ∇, simply because ρ is even and thus commutes with any function. To prove property (iii), we only need to add the argument that $\rho(m) = 1$.

• (iv) This is a direct corollary of the uniqueness in (ii). *QED*

6.6 Definition. The covariant derivative ∇^U on the subbundle $B|_U \equiv \pi^{-1}(U)$ is called the *induced covariant derivative*. As is customary, we will usually omit the superscript ^{U} and use the same symbol ∇ to denote the induced covariant derivative on the restriction to an open subset $U \subset M$. Worse, in most cases we will not even mention that we use the induced covariant derivative.

6.7 Discussion. If (U, ψ) is a local trivialization of B, then the structure of a free graded $\mathcal{A}$-module on each fiber is defined by declaring that the map $\psi : \pi^{-1}(U) \to U \times E$ is even and linear on each fiber. In other words, for $a, b \in \pi^{-1}(m)$ and $\lambda, \mu \in \mathcal{A}$, if $\psi(a) = (m, e)$ and $\psi(b) = (m, f)$ then $\psi(a\lambda + b\mu) = (m, e\lambda + f\mu)$. As in §IV.3 and §IV.5 we introduce the local sections $\underline{e}_i \in \Gamma_U(B)$ associated to a basis (e_i) of E by the formula

$$\underline{e}_i(m) = \psi^{-1}(m, e_i) \; .$$

In [IV.1.20] we have shown that there is a bijection between local sections $s \in \Gamma_U(B)$ and functions $s_\psi : U \to E$. Using (left) coordinates with respect to the basis (e_i), each (smooth) function s_ψ defines ordinary (smooth) functions $s^i : U \to \mathcal{A}$ by the equality $s_\psi(m) = \sum_i s^i(m) \cdot e_i$. Using the free graded $\mathcal{A}$-module structure on each fiber, the local sections $\underline{e}_i$, and the definition of the function s_ψ we thus have the equalities

$$s(m) = \psi^{-1}(m, s_\psi(m)) = \psi^{-1}(m, \sum_i s^i(m) \cdot e_i) = \sum_i s^i(m) \cdot \underline{e}_i(m) ,$$

valid for any local section $s : U \to B$. The (local) functions s^i thus define s and s_ψ by similar looking formulæ:

$$s_\psi(m) = \sum_i s^i(m) \cdot e_i \quad , \quad s(m) = \sum_i s^i(m) \cdot \underline{e}_i(m) . \tag{6.8}$$

An explicit example, though slightly hidden, of the use of the local sections $\underline{e}_i$ is given by the local vector fields $\partial_i \equiv \partial_{x^i}$ associated to local coordinates x^i on an $\mathcal{A}$-manifold [V.1.16]. As a consequence, even though it is not said explicitly, the local 1-forms dx^i, as well as the local k-forms $dx^{i_1} \wedge \cdots \wedge dx^{i_k}$ are examples of the use of the local sections $\underline{e}_i$.

6.9 Discussion. Let (U, ψ) be a local trivializing coordinate chart for B with coordinates x^i. Using the local sections $\underline{e}_i : U \to \pi^{-1}(U)$ introduced in [6.7] and the covariant derivative ∇ (officially we should say the induced covariant derivative ∇^U), we define homogeneous smooth functions $\Gamma_i{}^j{}_k : U \to \mathcal{A}$ of parity $\varepsilon(\Gamma_i{}^j{}_k) = \varepsilon(x^i) + \varepsilon(e_j) + \varepsilon(e_k)$ by

$$\nabla(\partial_{x^i}, \underline{e}_k)(m) = \sum_j \Gamma_i{}^j{}_k(m) \cdot \underline{e}_j(m) ,$$

where we used that the local sections $\underline{e}_i$ form a basis of the fiber at each point. We can put these functions together in homogeneous maps $\Gamma_i : U \to \mathrm{End}_{\mathbf{R}}(E)$ of parity $\varepsilon(\Gamma_i) = \varepsilon(x^i)$ by

$$\Gamma_i(m) = \sum_{jk} \Gamma_i{}^j{}_k \cdot e_j \otimes e^k ,$$

where as usual the e^k denote the right dual basis. In terms of the matrix representations given in [II.4.1] this means that we use left coordinates ${}^\bullet M_R$: $\Gamma_i{}^j{}_k = {}^\bullet M_R(\Gamma_i)^j{}_k$.

To show that Γ_i is independent of the choice of the basis (e_i) of E, let $(\widehat{e}_j)$ be another basis with the associated local sections $\widehat{\underline{e}}_j$ and functions $\widehat{\Gamma}_i{}^q{}_p$ defined by the equality $\nabla(\partial_{x^i}, \widehat{\underline{e}}_p)(m) = \sum_q \widehat{\Gamma}_i{}^q{}_p(m) \cdot \widehat{\underline{e}}_q(m)$. By definition the basis $(\widehat{e}_j)$ is related to the basis (e_i) by $\widehat{e}_j = \sum_i e_i a^i{}_j$ for some real valued matrix $(a^i{}_j)$ (real valued because we remain in the equivalence class). It follows that the right dual bases are related by $e^\ell = \sum_k a^\ell{}_k \widehat{e}^k$. From the definition of $\widehat{\underline{e}}_j$ we deduce the relation $\widehat{\underline{e}}_j = \sum_i \underline{e}_i\, a^i{}_j$ and thus, using [6.2-ii] and the fact that the $a^i{}_j$ are real constant, we obtain

$$\begin{aligned}\nabla(\partial_{x^i}, \widehat{\underline{e}}_p) &= \sum_q \widehat{\Gamma}_i{}^q{}_p \cdot \widehat{\underline{e}}_q = \sum_{jq} \widehat{\Gamma}_i{}^q{}_p \cdot \underline{e}_j \cdot a^j{}_q \\ &= \nabla(\partial_{x^i}, \sum_k \underline{e}_k\, a^k{}_p) = \sum_{jk} \Gamma_i{}^j{}_k \cdot \underline{e}_j \cdot a^k{}_p .\end{aligned}$$

Comparing coefficients of $\underline{e}_j$ gives us the equality $\sum_k \Gamma_i{}^j{}_k \cdot a^k{}_p = \sum_q \widehat{\Gamma}_i{}^q{}_p \cdot a^j{}_q$ with which we compute

$$\sum_{pq} \widehat{\Gamma}_i{}^q{}_p\, \widehat{e}_q \otimes \widehat{e}^p = \sum_{jpq} \widehat{\Gamma}_i{}^q{}_p \cdot a^j{}_q \cdot e_j \otimes \widehat{e}^p$$
$$= \sum_{jkp} \Gamma_i{}^j{}_k \cdot a^k{}_p e_j \otimes \widehat{e}^p = \sum_{jk} \Gamma_i{}^j{}_k e_j \otimes e^k \ ,$$

and thus $\Gamma_i = \widehat{\Gamma}_i$ is independent of the choice of a basis.

If we change the coordinates x^i to y^ℓ, we get new functions $\widehat{\Gamma}_\ell{}^j{}_k$ defined by the equality $\nabla(\partial_{y^\ell}, \underline{e}_k)(m) = \sum_j \widehat{\Gamma}_\ell{}^j{}_k(m) \cdot \underline{e}_j(m)$. Since $\partial_{y^\ell} = \sum_i(\partial_{y^\ell}\, x^i)\partial_{x^i}$, it follows from [6.2-ii] that these functions are related to the functions $\Gamma_i{}^j{}_k(m)$ by

$$\widehat{\Gamma}_\ell{}^j{}_k = \sum_i \frac{\partial x^i}{\partial y^\ell} \cdot \Gamma_i{}^j{}_k \ . \tag{6.10}$$

As a consequence, the maps $\widehat{\Gamma}_\ell \equiv \sum_{jk} \widehat{\Gamma}_\ell{}^j{}_k \cdot e_j \otimes e^k : U \to \mathrm{End}_R(E)$ are related to the Γ_i by $\widehat{\Gamma}_\ell = \sum_i(\partial_{y^\ell}\, x^i) \cdot \Gamma_i$. We conclude that the even $\mathrm{End}_R(E)$-valued 1-form Γ on U defined by

$$\Gamma = \sum_{ijk} dx^i \otimes \Gamma_i{}^j{}_k \cdot e^j \otimes e_k = \sum_i dx^i \otimes \Gamma_i = \sum_\ell dy^\ell \otimes \widehat{\Gamma}_\ell$$

is well defined. It follows that this 1-form exists on U, even if there does not exist a global coordinate system on U. With respect to the basis $(e_j \otimes e^k)$ of $\mathrm{End}_R(E)$, the $\mathrm{End}_R(E)$-valued 1-form Γ can be written as (see (VI.9.3))

$$\Gamma = \sum_{jk} \Gamma^j{}_k \otimes (e_j \otimes e^k) \tag{6.11}$$

with ordinary 1-forms $\Gamma^j{}_k = \sum_i dx^i \cdot \Gamma_i{}^j{}_k$. As for the $\mathrm{End}_R(E)$-valued 1-form Γ, it follows from (6.10) that these ordinary 1-forms are well-defined, independent of the chosen local coordinates.

6.12 Remark. In the particular case that the vector bundle $\pi : B \to M$ is the tangent bundle $B = TM$ and that the covariant derivative is derived from a metric, in that case the functions $\Gamma_i{}^j{}_k$ are called the*Christoffel symbols* associated to the metric.

6.13 Lemma. *Let (U, ψ) be a local trivialization, let $s \in \Gamma_U(B)$ be a local section with its associated local function s_ψ and let $X = \sum_i X^i\, \partial_{x^i}$ be a vector field on U. Then the local function $(\nabla_X s)_\psi$ associated to the local section $\nabla_X s$ is given by*

$$(\nabla_X s)_\psi = X s_\psi + \iota(X)\Gamma s_\psi \ . \tag{6.14}$$

Proof. Since (6.14) is additive in s, we may assume that s is homogeneous. The local section $\nabla_X s$ can then be written as

$$\nabla_X s = \sum_k (X s^k)\, \underline{e}_k + \sum_{ik} X^i (-1)^{\langle \varepsilon(s^k)|\varepsilon(x^i)\rangle} s^k \nabla(\partial_{x^i}, \underline{e}_k)$$
$$= \Big(\sum_j (X s^j) + \sum_{ijk} X^i (-1)^{\langle \varepsilon(s^k)|\varepsilon(x^i)\rangle} s^k\, \Gamma_i{}^j{}_k\Big)\, \underline{e}_j \,.$$

Using that $e^k(s_\psi) = e^k(\sum_p s^p e_p) = (-1)^{\langle \varepsilon(s^k)|\varepsilon(e_k)\rangle} s^k$, the function $(\nabla_X s)_\psi$ is given by

$$(\nabla_X s)_\psi = \Big(\sum_j (X s^j) + \sum_{ijk} X^i (-1)^{\langle \varepsilon(s^k)|\varepsilon(x^i)\rangle} s^k\, \Gamma_i{}^j{}_k\Big)\, e_j$$
$$= X\Big(\sum_j s^j\, e_j\Big) + \sum_{ijk} X^i\, \Gamma_i{}^j{}_k\, e_j\, (-1)^{\langle \varepsilon(s^k)|\varepsilon(e_k)\rangle} s^k$$
$$= X s_\psi + \sum_{ijk} X^i\, \Gamma_i{}^j{}_k\, e_j \cdot e^k(s_\psi) = X s_\psi + \sum_i X^i\, \Gamma_i \cdot s_\psi$$
$$= X s_\psi + \iota(X)\Gamma \cdot s_\psi \,.$$

QED

6.15 Lemma. *Let (U_a, ψ_a) and (U_b, ψ_b) be two local trivializations of B with transition function $\psi_{ab} : U_a \cap U_b \to \mathrm{Aut}(E)$ and let Γ^a and Γ^b be the $\mathrm{End}_R(E)$-valued 1-forms on U_a and U_b associated to the covariant derivative ∇. Then Γ^a and Γ^b are related on $U_a \cap U_b$ by*

$$\Gamma^b(m) = \psi_{ba}(m) \circ \Gamma^a(m) \circ \psi_{ba}(m)^{-1} + \psi_{ab}(m)^{-1} \circ d\psi_{ab}|_m \,. \tag{6.16}$$

Proof. Throughout this proof we will use left coordinates $^\bullet\mathcal{M}_R$ [II.4.1] for endomorphisms $A \in \mathrm{End}_R(E)$. The left coordinates $A^i{}_j \equiv {}^\bullet\mathcal{M}_R(A)^i{}_j$ can be defined by the formula $Ae_j = \sum_i A^i{}_j\, e_i$. This is compatible with the definition of the endomorphisms Γ_i in terms of the functions $\Gamma_i{}^j{}_k$ [6.9]. The transition function ψ_{ab} is defined by the equality $(\psi_a \circ \psi_b^{-1})(m, f) = (m, \psi_{ab}(m)\, f)$. Associated to each trivialization we have local sections $\underline{e}_i^a$ and $\underline{e}_j^b$. On $U_a \cap U_b$ they are related by:

$$\underline{e}_j^b(m) = \psi_b^{-1}(m, e_j) = (\psi_a^{-1} \circ \psi_a \circ \psi_b^{-1})(m, e_j) = \psi_a^{-1}(m, \psi_{ab}(m)\, e_j)$$
$$= \sum_i \psi_{ab}(m)^i{}_j \cdot \psi_a^{-1}(m, e_i) = \sum_i \psi_{ab}(m)^i{}_j \cdot \underline{e}_i^a(m) \,,$$

where the first equality of the second line follows from the fact that the ψ_a is compatible with the $\mathcal{A}$-module structure (actually, the $\mathcal{A}$-module structure on each fiber is defined in this way [IV.3.2]). Using coordinates x^i on $U_a \cap U_b$ we compute:

$$\sum_{pj} \Gamma_i^{bj}{}_k \cdot \psi_{ab}{}^p{}_j \cdot \underline{e}_p^a = \sum_j \Gamma_i^{bj}{}_k \cdot \underline{e}_j^b = \nabla(\partial_{x^i}, \underline{e}_k^b) = \sum_\ell \nabla(\partial_{x^i}, \psi_{ab}{}^\ell{}_k \cdot \underline{e}_\ell^a)$$
$$= \sum_p \frac{\partial \psi_{ab}{}^p{}_k}{\partial x^i} \cdot \underline{e}_p^a + \sum_{p\ell} (-1)^{\langle \varepsilon(x^i)|\varepsilon(e_\ell)+\varepsilon(e_k)\rangle} \psi_{ab}{}^\ell{}_k \cdot \Gamma_i^{ap}{}_\ell \cdot \underline{e}_p^a \,.$$

Comparing coefficients of $\underline{e}^a_p(m)$ and using that $\psi_{ba}(m)$ is the inverse of $\psi_{ab}(m)$ [IV.1.8] as well as (VI.1.22) for matrix multiplication in terms of left coordinates, we obtain the relation

$$\Gamma^{bj}_{i}{}_k(m) = \sum_p \frac{\partial \psi_{ab}{}^p{}_k}{\partial x^i} \cdot \psi_{ba}{}^j{}_p + \sum_{p\ell} (-1)^{\langle \varepsilon(x^i) | \varepsilon(e_\ell) + \varepsilon(e_k)\rangle} \psi_{ab}{}^\ell{}_k \cdot \Gamma^{ap}_{i}{}_\ell \cdot \psi_{ba}{}^j{}_p \,.$$

Substituting this in the definition of Γ^b_i and using (VI.1.21) several times for matrix multiplication in terms of left coordinates for homogeneous but not necessarily even endomorphisms, we compute:

$$\begin{aligned}
\Gamma^b &= \sum_{ijk} dx^i \otimes \Gamma^{bj}_{i}{}_k \cdot e_j \otimes e^k \\
&= \sum_{ijkp\ell} dx^i \otimes (-1)^{\langle \varepsilon(x^i) | \varepsilon(e_\ell) + \varepsilon(e_k)\rangle} \psi_{ab}{}^\ell{}_k \cdot \Gamma^{ap}_{i}{}_\ell \cdot \psi_{ba}{}^j{}_p \cdot e_j \otimes e^k \\
&\qquad + \sum_{ijkp} dx^i \otimes \frac{\partial \psi_{ab}{}^p{}_k}{\partial x^i} \cdot \psi_{ba}{}^j{}_p \cdot e_j \otimes e^k \\
&= \sum_{ijk\ell} dx^i \otimes (-1)^{\langle \varepsilon(x^i) | \varepsilon(e_\ell) + \varepsilon(e_k)\rangle} \psi_{ab}{}^\ell{}_k \cdot (\psi_{ba} \circ \Gamma^a_i)^j{}_\ell \cdot e_j \otimes e^k \\
&\qquad + \sum_{ijk} dx^i \otimes (\psi_{ba} \circ \frac{\partial \psi_{ab}}{\partial x^i})^j{}_k \cdot e_j \otimes e^k \\
&= \sum_{ijk} dx^i \otimes (\psi_{ba} \circ \Gamma^a_i \circ \psi_{ab})^j{}_k \cdot e_j \otimes e^k + \sum_i \psi_{ba} \circ (dx^i \otimes (\frac{\partial \psi_{ab}}{\partial x^i})) \\
&= \psi_{ba} \circ \Gamma^a \circ \psi_{ab} + \psi_{ba} \circ d\psi_{ab} \,,
\end{aligned}$$

where the last equality follows (among others) from the identity $df = \sum_i dx_i \cdot \partial_{x^i} f$, valid for any smooth function. Note that the composition symbol $\circ$ in the last line (and some in the line before that) is the alternative notation of the $\mathbin{\hat{\wedge}}$ symbol.

A faster proof, though less direct and demanding more explanation to justify all steps, is the following computation. Let s be a section, with local representative functions $s_a \equiv s_{\psi_a}$ and $s_b \equiv s_{\psi_b}$, and let X be a vector field. Then s_a and s_b are related by $s_a = \psi_{ab} s_b$. Since a similar relation holds for the local representatives for $\nabla_X s$, we have, using (6.14),

$$\begin{aligned}
\psi_{ab}(\nabla_X s)_b = (\nabla_X s)_a &= X s_a + \iota(X)\Gamma^a s_a = X(\psi_{ab} s_b) + \iota(X)\Gamma^a \psi_{ab} s_b \\
&= (X\psi_{ab}) s_b + \psi_{ab} X s_b + \iota(X)\Gamma^a \psi_{ab} s_b \\
&= \psi_{ab}\big(X s_b + \iota(X)(\psi_{ba} \circ \Gamma^a \circ \psi_{ab} + \psi_{ba} \circ d\psi_{ab}) s_b\big) \,.
\end{aligned}$$

Comparing this with the expression $(\nabla_X s)_b = X s_b + \iota(X)\Gamma^b s_b$ also gives the desired result. QED

6.17 Proposition. *Let $\mathcal{U} = \{(U_a, \psi_a)\}$ be a trivializing atlas for the vector bundle B and suppose that on each U_a we have an even $\mathrm{End}_R(E)$-valued 1-form Γ^a. If these Γ^a are*

related to each other by (6.16), then there exists a unique covariant derivative ∇ on B such that the Γ^a are determined by ∇ as in [6.9].

Proof. To prove existence, we construct ∇ as follows. For any $s \in \Gamma(B)$ and any vector field $X \in \Gamma(TM)$ we define the section $t = \nabla_X s$ piecewise on each trivializing chart U_a by (cf. (6.14))

$$t_a = Xs_a + \iota(X)\Gamma^a s_a \ .$$

These local functions glue together if and only if they satisfy $t_b(m) = \psi_{ba}(m)t_a(m)$ (IV.1.21). We thus compute:

$$\begin{aligned} t_b &= Xt_b + \iota(X)\Gamma^b t_b = X(\psi_{ba} s_a) + \iota(X)\big(\psi_{ba} \circ \Gamma^a \circ \psi_{ba}^{-1} + \psi_{ab}^{-1} \circ d\psi_{ab}\big)\psi_{ba} s_a \\ &= (X\psi_{ba})s_a + \psi_{ba} X s_a + \psi_{ba}\iota(X)\Gamma^a s_a + \psi_{ab}^{-1}(X\psi_{ab})\psi_{ba} s_a \\ &= \big((X\psi_{ba})\psi_{ab} + \psi_{ba} X\psi_{ab}\big)\psi_{ba} s_a + \psi_{ba}(Xs_a + \iota(X)\Gamma^a s_a) = \psi_{ba} t_a \ , \end{aligned}$$

where the last line follows because $\psi_{ba} \circ \psi_{ab} = id$, implying that $X(\psi_{ba} \circ \psi_{ab}) = 0$. This proves that t is indeed a well-defined section and thus that we have defined a map $\nabla : \Gamma(TM) \times \Gamma(B) \to \Gamma(B)$.

That this map has the properties of a covariant derivative follows from the fact that the local defining formula has these properties. Let us show this for the third condition, leaving the others to the reader. For X, f, and s as required we define sections $t = \nabla_X s$ and $u = \nabla_X fs$. In order to show that we have $u = (Xf)s + (-1)^{\langle \varepsilon(X)|\varepsilon(f)\rangle} ft$, we show this for all local representative functions, using that the local representative functions respect the $\mathcal{A}$-module structure:

$$\begin{aligned} u_\psi &= X(fs_\psi) + \iota(X)\Gamma f s_\psi = (Xf)s_\psi + (-1)^{\langle \varepsilon(X)|\varepsilon(f)\rangle} f(Xs_\psi + \iota(X)\Gamma s_\psi) \\ &= (Xf)s_\psi + (-1)^{\langle \varepsilon(X)|\varepsilon(f)\rangle} ft_\psi \ . \end{aligned}$$

To prove uniqueness, it suffices to note that the action of ∇ is completely determined by the local Γ via $(\nabla_X s)_\psi = Xs_\psi + \iota(X)\Gamma s$ (6.14). QED

6.18 Corollary. *Let $\pi_B : B \to M$ be a vector bundle with typical fiber E and let $\pi_S : SB \to M$ be the associated structure bundle. Then the following four objects*

(i) *an FVF connection on B,*
(ii) *a covariant derivative on B,*
(iii) *a (principal) connection on SB,*
(iv) *a connection 1-form on SB*

are four incarnations of a same concept: all four objects are determined by local even $\mathrm{End}_R(E)$-valued 1-forms Γ^a associated to a trivializing atlas and satisfying (2.15)/(6.16).

Proof. The only thing that has to be proven is that (2.15) and (6.16) are the same. But that is a direct consequence of [VI.2.13] and that $\Theta_{MC} = g^{-1} \circ dg$ on $\mathrm{Aut}(E)$ [1.5]. QED

6.19 Definition. Just as FVF connections on a principal fiber bundle have the special name principal connection, so are FVF connections on a vector bundle usually called *linear connections*. For a linear connection one can also find the name *affine connection* in the literature. However, since there is nothing really affine in such a connection (the 1-forms Γ^a take their values in the linear group $\mathrm{End}_R(E)$, not in the affine group of the typical fiber E), the name linear connection should be preferred.

7. More on covariant derivatives

In this section we study covariant derivatives on vector bundles in more detail. We start by showing that the link between a covariant derivative and an FVF connection is given by the fact that the covariant derivative measures how far a (local) section is away from being horizontal with respect to the FVF connection. Knowing that an FVF connection defines an FVF connection on a pull-back bundle, we show that the associated covariant derivative on a pull-back bundle again measures, in a sense to be made precise, how far away a lift is from being horizontal. In a similar way we can construct a covariant derivative on an associated vector bundle from one on the original bundle. Of this phenomenon we investigate several examples: the dual bundle, the bundle of homomorphisms and the second tensor power.

7.1 Discussion. In order to get a better understanding of the link between a covariant derivative and a linear connection, we have to delve deeper in the (coordinate) structure of a vector bundle. So let E be an $\mathcal{A}$-vector space, $e_1, \ldots, e_n$ a basis, and let $({}^ie)$ and (e^i) be the associated left and right dual bases. As an $\mathcal{A}$-manifold, E is modeled on $E^\sharp$ and it follows easily from [III.1.26] (see also [V.3.4]) that for any vector $v \in E$ its (left) coordinates are given by the values $y^i(v) = \langle\!\langle v_0 \,\|\, {}^ie \rangle\!\rangle$ and $\overline{y}^i(v) = \langle\!\langle v_1 \,\|\, {}^ie \rangle\!\rangle$, where $v = v_0 + v_1$ denotes the splitting into even and odd parts. And indeed we have

$$v = \sum_i \langle\!\langle v \,\|\, {}^ie \rangle\!\rangle \cdot e_i = \sum_i \langle\!\langle v_0 + v_1 \,\|\, {}^ie \rangle\!\rangle \cdot e_i = \sum_i (y^i(v) + \overline{y}^i(v))\, e_i \ .$$

To compute the fundamental vector field A^E on E associated to $A \in \mathrm{End}_R(E)$ ($\mathrm{End}_R(E)$ is the $\mathcal{A}$-Lie algebra of $\mathrm{Aut}(E)$), we have to be very careful in the use of coordinates. According to the definition of a fundamental vector field, we have $A^E_v = -\langle\!\langle A \,\|\, T\Phi_v \rangle\!\rangle$, where Φ denotes the left action of $\mathrm{Aut}(E)$ on E, and thus $\Phi_v(g) = g \cdot v$. According to [VI.1.20], on $\mathrm{Aut}(E)$ we have to use left coordinates $g^i{}_j = {}^\bullet M_R(g)^i{}_j$ [II.4.1] defined by $g = \sum_{ij} g^i{}_j\, e_i \otimes e^j$. And then the tangent vector $A \in \mathrm{End}_R(E) = T_e \,\mathrm{Aut}(E)$ is given as $A \cong \sum_{ij} A^i{}_j\, \partial_{g^i{}_j}$, when A itself is given as $A = \sum_{ij} A^i{}_j\, e_i \otimes e^j$. In terms of the coordinates $g^i{}_j$ on $\mathrm{Aut}(E)$ and $y^i, \overline{y}^i$ on $E \cong E^\sharp_0$, the coordinates of gv are given by (g is even)

$$y^i(gv) = \sum_j y^j(v)\, g^i{}_j = \sum_{pq} g^p{}_q \,\langle\!\langle e_p \otimes e^q(v_0) \,\|\, {}^ie \rangle\!\rangle$$

and similarly

$$\overline{y}^i(gv) = \sum_j \overline{y}^j(v)\, g^i{}_j = \sum_{pq} g^p{}_q \,\langle\!\langle e_p \otimes e^q(v_1) \,\|\, {}^ie \rangle\!\rangle$$

This gives us for the fundamental vector field the result

$$\begin{aligned} A_v^E &= -\sum_{pqi}\Big(A^p{}_q \,\langle\!\langle e_p \otimes e^q(v_0) \,\|\, {}^ie \rangle\!\rangle \,\frac{\partial}{\partial y^i}|_v + A^p{}_q \,\langle\!\langle e_p \otimes e^q(v_1) \,\|\, {}^ie \rangle\!\rangle \,\frac{\partial}{\partial \overline{y}^i}|_v \Big) \\ &= -\sum_i \Big(\langle\!\langle A(v_0) \,\|\, {}^ie \rangle\!\rangle \,\frac{\partial}{\partial y^i}|_v + \langle\!\langle A(v_1) \,\|\, {}^ie \rangle\!\rangle \,\frac{\partial}{\partial \overline{y}^i}|_v \Big) . \end{aligned}$$

Now let $\pi : B \to M$ be a vector bundle with typical fiber E, let $\mathcal{H}$ be a linear connection on B and let ∇ be a covariant derivative on B. Suppose furthermore that $\mathcal{H}$ and ∇ are determined by the same local $\mathrm{End}_R(E)$-valued 1-forms Γ^a associated to a trivializing atlas $\mathcal{U} = \{(U_a, \psi_a) \mid a \in I\}$. Then $\mathcal{H}$ is given in the trivialization ψ_a by (see [2.12])

$$\begin{aligned} \mathcal{H}_{(m,v)} = \{\, X - \sum_i \Big(&\langle\!\langle \iota(X)\Gamma^a(v_0) \,\|\, {}^ie \rangle\!\rangle \,\frac{\partial}{\partial y^i}|_v \\ &+ \langle\!\langle \iota(X)\Gamma^a(v_1) \,\|\, {}^ie \rangle\!\rangle \,\frac{\partial}{\partial \overline{y}^i}|_v \Big) \mid X \in T_mM \,\} . \end{aligned}$$

With these preparations we can state a more direct relationship between the linear connection $\mathcal{H}$ and the covariant derivative ∇.

7.2 Proposition. *Let $\pi : B \to M$ be a vector bundle with typical fiber E, let $\mathcal{H}$ be a linear connection on B and let ∇ be a covariant derivative on B. Suppose furthermore that $\mathcal{H}$ and ∇ are determined by the same local $\mathrm{End}_R(E)$-valued 1-forms Γ^a associated to a trivializing atlas $\mathcal{U} = \{(U_a, \psi_a) \mid a \in I\}$. Then for any section $s \in \Gamma(B)$ and any vector field X on M we have*

$$(\nabla_X s)(m) = \pi_\partial\Big(\langle\!\langle X_m \,\|\, Ts \rangle\!\rangle - \widetilde{X}^h_{s(m)} \Big) . \tag{7.3}$$

Roughly in words: $\nabla_X s$ measures how far away $\langle\!\langle X \,\|\, Ts \rangle\!\rangle$ is from being horizontal. Conversely, if X_m or s is homogeneous, then $\widetilde{X}^h_{s(m)}$ is determined by $(\nabla_X s)(m)$ via (7.3).

Proof. Let U_a be the trivializing set containing m and let $s_a : U_a \to E$ be the function determining the section s in the trivialization ψ_a, i.e., $\psi_a(s(m)) = (m, s_a(m))$. Let furthermore (y^i) be the (global) coordinates on E_0 with their associated coordinates $(\overline{y}^i)$ on E_1 (see [V.3.4]). Then $\widetilde{X}^h_{s(m)}$ is given in the trivialization ψ_a by

$$\begin{aligned} \widetilde{X}^h_{(m,s_a(m))} = X_m - \sum_i \Big(&\langle\!\langle \iota(X_m)\Gamma^a((s_a)_0) \,\|\, {}^ie \rangle\!\rangle \,\frac{\partial}{\partial y^i}|_{s_a(m)} \\ &+ \langle\!\langle \iota(X_m)\Gamma^a((s_a)_1) \,\|\, {}^ie \rangle\!\rangle \,\frac{\partial}{\partial \overline{y}^i}|_{s_a(m)} \Big) , \end{aligned}$$

and $\langle\!\langle X_m \| Ts \rangle\!\rangle$ is given in that same trivialization by (see [V.3.4])

$$\langle\!\langle X_m \| Ts \rangle\!\rangle = X_m + \sum_i \Big(\langle\!\langle X_m(s_a)_0 \| {}^i e \rangle\!\rangle \frac{\partial}{\partial y^i}|_{s_a(m)} + \langle\!\langle X_m(s_a)_1 \| {}^i e \rangle\!\rangle \frac{\partial}{\partial \overline{y}^i}|_{s_a(m)} \Big) .$$

It follows that $\langle\!\langle X_m \| Ts \rangle\!\rangle - \widetilde{X}^h_{s(m)}$ projects to zero under $T\pi$ and thus we can apply π_∂ [V.3.7]. This gives us in the trivialization determined by ψ_a

$$\begin{aligned}\pi_\partial(\langle\!\langle X_m \| Ts \rangle\!\rangle - \widetilde{X}^h_{s(m)}) &= \sum_i \Big(\langle\!\langle X_m(s_a)_0 \| {}^i e \rangle\!\rangle + \langle\!\langle X_m(s_a)_1 \| {}^i e \rangle\!\rangle + \\ &\qquad \langle\!\langle \iota(X_m)\Gamma^a((s_a)_0) \| {}^i e \rangle\!\rangle + \langle\!\langle \iota(X_m)\Gamma^a((s_a)_1) \| {}^i e \rangle\!\rangle \Big) e_i \\ &= X_m s_a + \iota(X_m)\Gamma^a s_a = (\nabla_X s)_a(m) ,\end{aligned}$$

where the last equality follows from [6.14]. Since this is the local expression of $(\nabla_X s)(m)$ in the trivialization determined by ψ_a, we have proven (7.3).

If we know $(\nabla_X s)(m)$, we know the coefficients $\langle\!\langle X_m s_a + \iota(X_m)\Gamma^a s_a \| {}^i e \rangle\!\rangle$, and thus we know the coefficients $\langle\!\langle \iota(X_m)\Gamma^a s_a \| {}^i e \rangle\!\rangle$, whereas for $\widetilde{X}^h_{s(m)}$ we need to know the values of $\langle\!\langle \iota(X_m)\Gamma^a((s_a)_\alpha) \| {}^i e \rangle\!\rangle$, for $\alpha = 0, 1$. If s is homogeneous, it is obvious that the former determines the latter. When it is X_m that is homogeneous, we can decompose $\langle\!\langle \iota(X_m)\Gamma^a s_a \| {}^i e \rangle\!\rangle$ in its homogeneous components which are given by the formula $(\langle\!\langle \iota(X_m)\Gamma^a s_a \| {}^i e \rangle\!\rangle)_{\varepsilon(e_i)+\varepsilon(X_m)+\alpha} = \langle\!\langle \iota(X_m)\Gamma^a((s_a)_\alpha) \| {}^i e \rangle\!\rangle$, and thus again we obtain the desired result. QED

7.4 Discussion. Let $\pi : B \to M$ be a vector bundle with typical fiber E and let ∇ be a covariant derivative on B. From [6.18] we know that ∇ is equivalent to a linear connection $\mathcal{H}$ on B. But for FVF connections we have two ways to construct new ones: on pull-back bundles [2.16] and on associated bundles [5.1]. It follows that we have two ways to create new covariant derivatives out of the given one. We first concentrate on pull-back bundles. We thus consider a smooth map $g : N \to M$ and the pull-back bundle $g^*\pi : g^*B \to N$ and the associated vector bundle map $\check{g} : g^*B \to B$ [IV.2.2]. Given the covariant derivative ∇ on B and the associated linear connection $\mathcal{H}$ [6.18], we apply [2.16] to obtain an induced linear connection $g^*\mathcal{H}$ on g^*B. And thus by [6.18] we have an induced covariant derivative $g^*\nabla$ on g^*B associated to this linear connection. If $\mathcal{U}$ and $\mathcal{V}$ are (trivializing) atlases as in the proof of [2.5] and if ∇ is determined by the local $\mathrm{End}_R(E)$-valued 1-forms Γ^a, then the induced covariant derivative $g^*\nabla$ is determined by the local $\mathrm{End}_R(E)$-valued 1-forms $\check{\Gamma}^a = g^*\Gamma^{g(a)}$.

The natural question that arises is whether there is another way to define $g^*\nabla$ which does not involve the local 1-forms Γ^a. According to the definition of a covariant derivative, if s is a section of g^*B and X a vector field on N, then we have to define $(g^*\nabla)_X s$ as a section of g^*B. The natural idea that comes to mind is the following. Find a vector field Y on M related to X by g, i.e., $\forall n \in N : Y_{g(n)} = \langle\!\langle X_n \| Tg \rangle\!\rangle$, and find a section t of B such that $t \circ g = \check{g} \circ s$. And then define $(g^*\nabla)_X s$ as

$$\check{g}\Big((g^*\nabla)_X s \Big) = (\nabla_Y s)(g(n)) ,$$

which is well defined because $\check{g}$ is a bijection between $(g^*\pi)^{-1}$ and $\pi^{-1}(g(n))$. For a generic smooth map g there is no hope that such Y and t exist, but perhaps we might circumvent this problem. For Y we indeed can: in view of [6.4-ii] we could use any Y such that, at a fixed point $n \in N$, we have $Y_{g(n)} = \langle\!\langle X_n \| Tg \rangle\!\rangle$ in order to define $(g^*\nabla)_X s$ at $n \in N$. How such a Y behaves elsewhere is of no importance for $(\nabla_Y s)(g(n))$. However, no such trick is available for the section t. If g is injective, t must be given on $g(N)$ by $t(g(n)) = \check{g}(s(n))$, and thus we have to extend this t outside the image $g(N)$. This might not be easy depending on $g(N)$. But when g is not injective, there is no hope to find a section t for a generic section s of g^*B. We thus have to be more subtle in our tentative to give a definition of $g^*\nabla$ without using the local 1-forms Γ^a.

7.5 Remark. In [6.4-iii] we have shown that there exists an induced covariant derivative on the restriction of the bundle to an open subset. But such a restriction is a particular case of a pull-back bundle [IV.2.3]. The fact that the proof of the existence of this induced covariant derivative is not immediate confirms that extending t outside of $g(N)$ (as described in [7.4]) will not be an easy task, if feasible at all.

7.6 Definition. Let $\pi : B \to M$ be a fiber bundle with typical fiber E and let $g : N \to M$ be a smooth map. A map $\sigma : N \to B$ will be called a *lift of g* if $\pi \circ \sigma = g$, i.e., if we have a commutative diagram

$$\begin{array}{ccc} & & B \\ & {}^{\sigma}\nearrow & \downarrow \pi \\ N & \underset{g}{\longrightarrow} & M\,. \end{array}$$

The set of all smooth lifts $\sigma : N \to B$ will be denoted as $\mathrm{Lift}_g(B)$. A particular case of a lift is a section: a section $s : M \to B$ is a lift of th identity map $id : M \to M$, and thus $\Gamma(B) = \mathrm{Lift}_{id}(B)$. If B is a vector bundle, the set $\mathrm{Lift}_g(B)$ is in a natural way a graded $C^\infty(N)$-module: for $f \in C^\infty(N)$ and $\sigma, \tau \in \mathrm{Lift}_g(B)$ we define $f \cdot \sigma + \tau \in \mathrm{Lift}_g(B)$ as

$$(f \cdot \sigma + \tau)(n) = f(n) \cdot \sigma(n) + \tau(n)\ .$$

7.7 Lemma. *Let $\pi : B \to M$ be a vector bundle with typical fiber E, let $g : N \to M$ be a smooth map, and let $g^*\pi : g^*B \to N$ be the pull-back bundle with associated vector bundle map $\check{g} : g^*B \to B$. Then the map $\mathfrak{I} : \Gamma(g^*B) \to \mathrm{Lift}_g(B)$ defined by $\mathfrak{I}(s) = \check{g} \circ s$ is an isomorphism of graded $C^\infty(N)$-modules.*

Proof. Let us first show that $\mathfrak{I}$ is a morphism of graded $C^\infty(N)$-modules. For that choose $f \in C^\infty(N)$ and $s, t \in \Gamma(g^*B)$. Then:

$$\begin{aligned} \mathfrak{I}(f \cdot s + t)(n) &= (\check{g} \circ (f \cdot s + t))(n) = \check{g}(f(n) \cdot s(n) + t(n)) \\ &= f(n) \cdot \check{g}(s(n)) + \check{g}(t(n)) = (f \cdot \mathfrak{I}(s) + \mathfrak{I}(t))(n)\ , \end{aligned}$$

where the second equality follows from the fact that $\check{g}$ is a vector bundle morphism [IV.4.19]. To prove that $\mathfrak{I}$ is injective, suppose $s, t \in \Gamma(g^*B)$ are different, i.e., $\exists n \in N : s(n) \neq t(n)$. Since both belong to the fiber $(g^*\pi)^{-1}(n)$ and since $\check{g}$ is an isomorphism when restricted to a fiber, $\check{g}(s(n)) \neq \check{g}(t(n))$, i.e., $\mathfrak{I}(s) \neq \mathfrak{I}(t)$.

Finally to prove surjectivity, let $\sigma \in \mathrm{Lift}_g(B)$ be arbitrary. Using the set theoretic definition of g^*B, we define $s : N \to g^*B$ by $s(n) = (n, \sigma(n))$, which indeed belongs to g^*B because $\pi(\sigma(n)) = g(n)$ by definition of a lift. Obviously $\check{g}(s(n)) = \sigma(n)$ by definition of $\check{g}$ in terms of the set theoretic definition of g^*B. To show that this s is smooth, we look at its representative in a local trivialization. Let $\mathcal{V}$ be a trivializing atlas for B and let $\mathcal{U}$ be an atlas for N such that for each $U_a \in \mathcal{U}$ there exists $V_{g(a)} \in \mathcal{V}$ such that $g(U_a) \subset V_{g(a)}$. According to [IV.2.2] we have a local trivialization of g^*B given by $\psi_a : (g^*\pi)^{-1}(U_a) \to U_a \times E$ of g^*B and $\chi_{g(a)} : \pi^{-1}(V_{g(a)} \to V_{g(a)} \times E$ of B. In terms of this trivialization the lift $\sigma|_{U_a}$ is given as $\sigma|_{U_a}(n) = (g(n), S(n))$ for some smooth function $S : U_a \to E$. And then $s|_{U_a}$ is given as $s|_{U_a}(n) = (n, S(n))$, which is obviously smooth. $\boxed{QED}$

7.8 Proposition. *Let $\pi : B \to M$ be a vector bundle with typical fiber E, let $g : N \to M$ be a smooth map, and let $g^*\pi : g^*B \to N$ be the pull-back bundle. Let furthermore ∇ be a covariant derivative on B and let $\mathcal{H}$ be the associated linear connection defined by the same local $\mathrm{End}_R(E)$-valued 1-forms. then the induced covariant derivative $g^*\nabla$ on g^*B is given by the following procedure. For X a (smooth) vector field on N and s a smooth section of g^*B, denote $\sigma = \mathfrak{I}(s) = \check{g} \circ s \in \mathrm{Lift}_g(B)$ and $\tau = \mathfrak{I}((g^*\nabla)(X, s))$. Then for any $n \in N$ we have*

$$\tau(n) = \pi_\partial\Big(\langle\!\langle X_n \| T\sigma \rangle\!\rangle - \widetilde{\langle\!\langle X_n \| Tg \rangle\!\rangle}^{h}_{\sigma(n)} \Big) . \tag{7.9}$$

Roughly in words: $\tau(n)$ measures how far $\langle\!\langle X_n \| T\sigma \rangle\!\rangle$ is from being horizontal.

Proof. We use notation as in the proof of [7.7]. According to that proof, we have (locally) $\psi_a(s(n)) = (n, s_a(n))$ for some smooth function $s_a : U_a \to E$ and simultaneously $\chi_{g(a)}(\sigma(n)) = (g(n), s_a(n))$. If $\Gamma^{g(a)}$ is the $\mathrm{End}_R(E)$-valued 1-form on $V_{g(a)}$ defining ∇ and $\mathcal{H}$, then $g^*\nabla$ is determined by $g^*\Gamma^{g(a)} \equiv \check{\Gamma}^a$ [2.16]. Thus in the trivialization $(V_{g(a)}, \chi_{g(a)})$ the local expression for $(g^*\nabla)(X, s)$ is given as

$$\big((g^*\nabla)(X, s)\big)_a(n) = X_n s_a + \iota(X_n)\check{\Gamma}^a \cdot s_a(n) = X s_a + \iota(\langle\!\langle X_n \| Tg \rangle\!\rangle)\Gamma^{g(a)} \cdot s_a(n) .$$

On the other hand, with notation as in the proof of [7.2], we have

$$\langle\!\langle X_n \| T\sigma \rangle\!\rangle = \langle\!\langle X_n \| Tg \rangle\!\rangle + \sum_i \Big(\langle\!\langle X_n (s_a)_0 \| {}^i e \rangle\!\rangle \frac{\partial}{\partial y^i}|_{s_a(n)} + \langle\!\langle X_n (s_a)_1 \| {}^i e \rangle\!\rangle \frac{\partial}{\partial \overline{y}^i}|_{s_a(n)} \Big)$$

and

$$\widetilde{\langle\!\langle X_n \| Tg \rangle\!\rangle}^{h}_{(g(n), s_a(n))} = \langle\!\langle X_n \| Tg \rangle\!\rangle - \sum_i \Big(\langle\!\langle \iota(\langle\!\langle X_n \| Tg \rangle\!\rangle)\Gamma^a((s_a)_0) \| {}^i e \rangle\!\rangle \frac{\partial}{\partial y^i}|_{s_a(m)}$$

$$+ \langle\!\langle \iota(\langle\!\langle X_n \| Tg \rangle\!\rangle)\Gamma^a((s_a)_1) \| {}^i e \rangle\!\rangle \frac{\partial}{\partial \overline{y}^i}|_{s_a(m)} \Big) ,$$

and then the result follows as in the proof of [7.2]. $\boxed{QED}$

7.10 Discussion. [7.7] tells us that the sections of the pull-back bundle are a natural generalization of sections of the original bundle. And then comparing (7.9) with (7.3) shows that the induced covariant derivative $g^*\nabla$ is the natural generalization of ∇ to these generalized sections.

Having treated induced covariant derivatives on pull-back bundles, we now turn our attention to covariant derivatives on associated bundles. As before we let $\pi : B \to M$ be a vector bundle with typical fiber E and ∇ a covariant derivative on B. But now we consider a representation, i.e., a homomorphism of $\mathcal{A}$-Lie groups, $\rho : \mathrm{Aut}(E) \to \mathrm{Aut}(F)$ for some $\mathcal{A}$-vector space F. If $\mathcal{H}$ is the linear connection on B having the same local $\mathrm{End}_R(E)$-valued 1-forms Γ^a as ∇, then we know from [5.1] that there exists a linear connection $\mathcal{H}^\rho$ on the associated vector bundle $\pi^\rho : B^{\rho,F} \to M$, associated to B by the representation ρ. And thus we have a corresponding covariant derivative ∇^ρ on $B^{\rho,F}$. Moreover, $\mathcal{H}^\rho$ and thus ∇^ρ is determined by the local $\mathrm{End}_R(F)$-valued 1-forms $\check{\Gamma}^a = T_e\rho(\Gamma^a)$ [5.1]. As for the situation with pull-back bundles, the natural question is whether there exists a way to define ∇^ρ which does not involve the local 1-forms Γ^a. The answer is positive for $\mathcal{A}$-vector spaces F that are constructed out of E by natural operations such as $F = E^*$ or $F = \mathrm{End}_R(E)$. However, the answer depends upon the representation ρ, so we will give several examples to show the idea.

7.11 Proposition. *If $F = \mathcal{A}$ and ρ the trivial representation, i.e., $\forall g \in \mathrm{Aut}(E)$: $\rho(g) = 1$, then $B^{\rho,\mathcal{A}}$ is the trivial bundle $B^{\rho,\mathcal{A}} = M \times \mathcal{A}$ and in the identification $\Gamma(B^{\rho,\mathcal{A}}) \cong C^\infty(M)$, the induced covariant derivative is given by $\nabla^\rho_X s = Xs$ (see [6.3]).*

7.12 Proposition. *If $F = E^*$ (the right dual of E) and ρ the natural representation, i.e., $\forall g \in \mathrm{Aut}(E) : \rho(g) = (g^{-1})^*$, then $B^{\rho,E^*} = B^*$ is the right-dual bundle and, using the identification $\Gamma(B^*) \cong \Gamma(B)^*$ [IV.5.14], the induced covariant derivative ∇^ρ is the unique covariant derivative on B^* such that for $s \in \Gamma(B)$ and for homogeneous $\alpha \in \Gamma(B^*)$, $X \in \Gamma(TM)$ we have*

$$(\nabla^\rho_X \alpha)(s) = X\big(\alpha(s)\big) - (-1)^{\langle \varepsilon(X) | \varepsilon(\alpha)\rangle} \alpha(\nabla_X s)\ . \tag{7.13}$$

7.14 Proposition. *If $F = \mathrm{End}_R(E)$ and $\rho = \mathrm{Ad}$ the adjoint representation, i.e., $\forall g \in \mathrm{Aut}(E)$, $\forall A \in \mathrm{End}_R(E) : \rho(g)(A) = \mathrm{Ad}(g)(A) = gAg^{-1}$ [VI.2.13], then $B^{\mathrm{Ad},\mathrm{End}_R(E)} = \mathrm{End}_R(B)$ is the bundle of right-linear endomorphisms of B and, using the identification $\Gamma(\mathrm{End}_R(B)) \cong \mathrm{End}_R(\Gamma(B))$ [IV.5.9], the induced covariant derivative ∇^ρ is the unique covariant derivative on $\mathrm{End}_R(B)$ such that for $s \in \Gamma(B)$ and for homogeneous $\phi \in \Gamma(\mathrm{End}_R(B))$, $X \in \Gamma(TM)$ we have*

$$(\nabla^\rho_X \phi)(s) = \nabla_X(\phi(s)) - (-1)^{\langle \varepsilon(X) | \varepsilon(\phi)\rangle} \phi(\nabla_X s)\ . \tag{7.15}$$

7.16 Proposition. *If $F = E \otimes E$ and ρ the natural representation, i.e., $\forall g \in \mathrm{Aut}(E)$: $\rho(g) = g \otimes g$, then $B^{\rho,E\otimes E} = B \otimes B$ is the tensor product of B with itself and, using the identification $\Gamma(B \otimes B) \cong \Gamma(B) \otimes \Gamma(B)$ [IV.5.17], the induced covariant derivative ∇^ρ is the unique covariant derivative on $B \otimes B$ such that for $t \in \Gamma(B)$ and for homogeneous $X \in \Gamma(TM)$, $s \in \Gamma(B)$ we have*

$$\nabla^\rho_X(s \otimes t) = (\nabla_X s) \otimes t + (-1)^{\langle \varepsilon(X)|\varepsilon(s)\rangle} s \otimes \nabla_X t \ . \tag{7.17}$$

Proofs. • [7.11] If $\mathcal{U}$ is any trivializing atlas for B (e.g., the full vector bundle structure), it also is a trivializing atlas for the associated bundle. But for the associated bundle all transition functions are constant 1, and thus we have a natural global trivialization of $B^{\rho,\mathcal{A}}$ as $B^{\rho,\mathcal{A}} \cong M \times \mathcal{A}$. Since $T_e\rho = 0$, it follows that all induced local 1-forms $\Gamma'^a = T_e\rho \circ \Gamma^a$ are identically zero. This shows that on any local trivializing chart (U, ψ) we have $(\nabla^\rho_X s)_\psi = X s_\psi$, and thus the result holds globally.

• [7.12] If the transition functions for B are given by ψ_{ab} with respect to some trivializing atlas, then the transition functions of B^{ρ,E^*} are given by $\rho(\psi_{ab}) = (\psi_{ab}^{-1})^*$, which are exactly the transition functions of the right-dual bundle B^* [IV.4.16].

To compute $T_e\rho$ we first note that we have identified $T_e\,\mathrm{Aut}(E)$ with $\mathrm{End}_R(E)$ and (thus) $T_e\,\mathrm{Aut}(E^*)$ with $\mathrm{End}_R(E^*)$. The representation ρ is the composition of the map $\mathit{Inv} : \mathrm{Aut}(E) \to \mathrm{Aut}(E)$ with taking the dual map: * : $\mathrm{End}_R(E) \to \mathrm{End}_L(E^*)$. We see that the natural target space of ρ is the space of *left* linear endomorphisms of E^*, not the *right* linear ones. However, since $g \in \mathrm{Aut}(E)$ is even, it is left and right linear and it is equal to its transpose (I.2.16), and thus we can as well define ρ as $\rho(g) = \mathfrak{T}(g^{-1})^*$. The map $A \mapsto \mathfrak{T}A^*$ is an even isomorphism from $\mathrm{End}_R(E)$ to $\mathrm{End}_R(E^*)$. Combining [VI.1.6] with [V.3.3] we find:

$$T_e\rho : A \mapsto -\mathfrak{T}A^* \ .$$

Using (6.14) we now compute the local expression of $\nabla^\rho_X\alpha$ in a trivialization (U, ψ):

$$\begin{aligned}(\nabla^\rho_X\alpha)_\psi &= X\alpha_\psi + \iota(X)(T_e\rho \circ \Gamma)\,\alpha_\psi = X\alpha_\psi + \big(T_e\rho(\iota(X)\Gamma)\big)(\alpha_\psi)\\ &= X\alpha_\psi - \big(\mathfrak{T}([\iota(X)\Gamma]^*)\big)(\alpha_\psi) = X\alpha_\psi - (-1)^{\langle\varepsilon(X)|\varepsilon(\alpha)\rangle}\langle\!\langle \alpha_\psi \,\|\, [\iota(X)\Gamma]^* \rangle\!\rangle \ ,\end{aligned}$$

where the last equality follows from (I.2.16) and the fact that Γ is even and thus that we have $\varepsilon(\iota(X)\Gamma) = \varepsilon(X)$. Since $C^\infty(U; E^*)$ is naturally isomorphic to $C^\infty(U; E)^*$, we can evaluate $(\nabla^\rho_X\alpha)_\psi \in C^\infty(U; E^*)$ on $s_\psi \in C^\infty(U; E)$, yielding

$$\begin{aligned}(\nabla^\rho_X\alpha)_\psi(s_\psi) &= (X\alpha_\psi)(s_\psi) - (-1)^{\langle\varepsilon(X)|\varepsilon(\alpha)\rangle}_\psi\langle\!\langle \alpha_\psi \,\|\, [\iota(X)\Gamma]^* \rangle\!\rangle(s_\psi)\\ &= (X\alpha_\psi)(s_\psi) - (-1)^{\langle\varepsilon(X)|\varepsilon(\alpha)\rangle}\alpha_\psi\big((\iota(X)\Gamma)(s_\psi)\big)\\ &= (X\alpha_\psi)(s_\psi) - (-1)^{\langle\varepsilon(X)|\varepsilon(\alpha)\rangle}\alpha_\psi\big((\nabla_X s)_\psi - X s_\psi\big)\\ &= (X\alpha_\psi)(s_\psi) + (-1)^{\langle\varepsilon(X)|\varepsilon(\alpha)\rangle}\alpha_\psi(X s_\psi) - (-1)^{\langle\varepsilon(X)|\varepsilon(\alpha)\rangle}\alpha_\psi((\nabla_X s)_\psi)\\ &= X\big(\alpha_\psi(s_\psi)\big) - (-1)^{\langle\varepsilon(X)|\varepsilon(\alpha)\rangle}\alpha_\psi((\nabla_X s)_\psi) \ ,\end{aligned}$$

where for the second equality we used [I.2.20]. Looking carefully at the identification $\Gamma(B)^*$ with $\Gamma(B^*)$ (see also [IV.4.5]) shows that we have $\alpha(s)|_U = \alpha_\psi(s_\psi) \in C^\infty(U)$. The above computation thus gives us the equality

$$(\nabla^\rho_X \alpha)(s)|_U = X(\alpha(s))|_U - (-1)^{\langle\varepsilon(X)|\varepsilon(\alpha)\rangle}\alpha(\nabla_X s)|_U \ .$$

Since U is an arbitrary trivializing chart,we have shown (7.13). Uniqueness of ∇^ρ follows immediately from the fact that via (7.13) we know the action of $\nabla^\rho_X \alpha$ on any $s \in \Gamma(B)$ and thus we know $\nabla^\rho_X \alpha$ itself.

• [7.14] The transition functions for B being ψ_{ba}, those for $B^{\mathrm{Ad},\mathrm{End}_R(E)}$ are given by $A \mapsto \psi_{ba} \cdot A \cdot \psi_{ba}^{-1}$. Comparing this with (IV.4.17) shows that these are exactly the transition functions of the bundle $\mathrm{End}_R(B)$ of right-linear endomorphisms of B.

Using (6.14) and [VI.2.15] we now compute the local expression of $\nabla^\rho_X \phi$ in a trivialization (U, ψ):

$$\begin{aligned}(\nabla^\rho_X \phi)_\psi &= X\phi_\psi + \iota(X)(T_e \,\mathrm{Ad} \circ \Gamma)\, \phi_\psi = X\phi_\psi + \mathrm{ad}_R(\iota(X)\Gamma)(\phi_\psi)\\ &= X\phi_\psi + \iota(X)\Gamma \circ \phi_\psi - (-1)^{\langle\varepsilon(X)|\varepsilon(\phi)\rangle}\phi_\psi \circ \iota(X)\Gamma \ .\end{aligned}$$

Since $C^\infty(U; \mathrm{End}_R(E))$ is naturally isomorphic to $\mathrm{End}_R(C^\infty(U; E))$, we can evaluate $(\nabla^\rho_X \phi)_\psi \in C^\infty(U; \mathrm{End}_R(E))$ on $s_\psi \in C^\infty(U; E)$, yielding

$$\begin{aligned}(\nabla^\rho_X \phi)_\psi(s_\psi) &= (X\phi_\psi)(s_\psi) + \iota(X)\Gamma(\phi_\psi(s_\psi)) - (-1)^{\langle\varepsilon(X)|\varepsilon(\phi)\rangle}\phi_\psi(\iota(X)\Gamma(s_\psi))\\ &= (X\phi_\psi)(s_\psi) + \iota(X)\Gamma(\phi_\psi(s_\psi)) - (-1)^{\langle\varepsilon(X)|\varepsilon(\phi)\rangle}\phi_\psi((\nabla_X s)_\psi - Xs_\psi)\\ &= X\big(\phi_\psi(s_\psi)\big) + \iota(X)\Gamma(\phi_\psi(s_\psi)) - (-1)^{\langle\varepsilon(X)|\varepsilon(\phi)\rangle}\phi_\psi((\nabla_X s)_\psi) \ .\end{aligned}$$

Analyzing the identification $\Gamma(\mathrm{End}_R(B)) \cong \mathrm{End}_R(\Gamma(B))$ shows that we have the equality $\phi_\psi(s_\psi) = (\phi(s))_\psi$ and thus the above computation gives us

$$\begin{aligned}((\nabla^\rho_X \phi)(s))_\psi &= X(\phi(s))_\psi + \iota(X)\Gamma(\phi(s))_\psi - (-1)^{\langle\varepsilon(X)|\varepsilon(\phi)\rangle}(\phi(\nabla_X s))_\psi\\ &= (\nabla_X(\phi(s)))_\psi - (-1)^{\langle\varepsilon(X)|\varepsilon(\phi)\rangle}(\phi(\nabla_X s))_\psi \ .\end{aligned}$$

Since this is valid for an arbitrary trivializing chart (U, ψ), we have shown (7.15). Uniqueness of ∇^ρ follows as for [7.12].

• [7.16] If the transition functions for B are given by ψ_{ab}, then the transition functions of $B^{\rho, E\otimes E}$ are given by $\rho(\psi_{ab}) = \psi_{ba} \otimes \psi_{ba}$, which are exactly the transition functions of the tensor product bundle $B \otimes B$ [IV.4.9].

Since the tensor product is indeed a product (as can be seen by computing explicitly the matrix elements of $\rho(g)$), the tangent map is given by

$$T_e\rho : A \mapsto A \otimes id + id \otimes A \ .$$

in a local trivializing chart (U, ψ) we thus obtain

$$\begin{aligned}(\nabla^\rho_X (s \otimes t))_\psi &= X(s \otimes t)_\psi + \iota(X)(T_e\rho \circ \Gamma)(s \otimes t)_\psi\\ &= X(s \otimes t)_\psi + (\iota(X)\Gamma \otimes id + id \otimes \iota(X)\Gamma)(s \otimes t)_\psi \ .\end{aligned}$$

The identification $\Gamma(B\otimes B)\cong\Gamma(B)\otimes\Gamma(B)$ is such that we have $(s\otimes t)_\psi = s_\psi\otimes t_\psi$, and thus we obtain:

$$\begin{aligned}(\nabla^\rho_X(s\otimes t))_\psi &= X(s_\psi\otimes t_\psi) + (\iota(X)\Gamma\otimes id + id\otimes\iota(X)\Gamma)(s_\psi\otimes t_\psi)\\ &= (Xs_\psi)\otimes t_\psi + (-1)^{\langle\varepsilon(X)|\varepsilon(s)\rangle} s_\psi\otimes Xt_\psi\\ &\qquad + (\iota(X)\Gamma s_\psi)\otimes t_\psi + (-1)^{\langle\varepsilon(X)|\varepsilon(s)\rangle} s_\psi\otimes\iota(X)\Gamma t_\psi\\ &= (\nabla_X s)_\psi\otimes t_\psi + (-1)^{\langle\varepsilon(X)|\varepsilon(s)\rangle} s_\psi\otimes(\nabla_X t)_\psi\\ &= ((\nabla_X s)\otimes t)_\psi + (-1)^{\langle\varepsilon(X)|\varepsilon(s)\rangle}(s\otimes\nabla_X t)_\psi\ .\end{aligned}$$

Since this is true for an arbitrary trivializing chart, we have shown (7.17). Uniqueness of ∇^ρ follows from the fact that the tensor products $s\otimes t$ generate $\Gamma(B\otimes B)$ (because of the isomorphism between $\Gamma(B\otimes B)$ and $\Gamma(B)\otimes\Gamma(B)$). QED

7.18 Remark. The representation Ad in [7.14] can also be written as $\mathrm{Ad}(g) = (g^{-1})^*\otimes g$ (see [IV.4.16]). Combining the ideas of the proofs of [7.12] and [7.16] yields for T_e Ad the formula

$$T_e\,\mathrm{Ad} : A\mapsto id\otimes A - \mathfrak{T}A^*\otimes id\ .$$

A careful study of the identification given in [I.7.12] for the right linear case shows that for homogeneous elements the action of $\mathfrak{T}A^*\otimes B$ on an endomorphism ϕ is given by

$$(\mathfrak{T}A^*\otimes B)(\phi) = (-1)^{\langle\varepsilon(A)|\varepsilon(B)+\varepsilon(\phi)\rangle} B\circ\phi\circ A\ .$$

We thus obtain

$$\begin{aligned}(\nabla^\rho_X\phi)_\psi &= X\phi_\psi + \iota(X)(id\otimes\Gamma - \mathfrak{T}\Gamma^*\otimes id)\,\phi_\psi = X\phi_\psi + \iota(X)(\Gamma\circ\phi_\psi - \phi_\psi\circ\Gamma)\\ &= X\phi_\psi + \iota(X)\Gamma\circ\phi_\psi - (-1)^{\langle\varepsilon(X)|\varepsilon(\phi)\rangle}\phi_\psi\circ\iota(X)\Gamma\ .\end{aligned}$$

This provides an alternative approach to the proof of [7.14].

8. Forms with values in a vector bundle

In previous sections we have introduced the exterior derivative of differential forms, the exterior covariant derivative on a principal fiber bundle and the covariant derivative on a vector bundle. In order to show how these objects are related, we have to generalize vector valued differential forms to differential forms with values in a vector bundle. This generalization is presented in this section.

8.1 Definition. Let $p : B \to M$ be a vector bundle with typical fiber E. A *k-form on M with values in B* is a section of the bundle $(\bigwedge^k {}^*TM) \otimes B$, just as an ordinary k-form is a section of the bundle $(\bigwedge^k {}^*TM)$. In analogy with ordinary k-forms and vector valued k-forms, the set of all smooth B-valued k-forms will be denoted as $\Omega^k(M; B)$, i.e.,

$$\Omega^k(M; B) \equiv \Gamma((\textstyle\bigwedge^k {}^*TM) \otimes B) .$$

As for ordinary differential forms, we define $\Omega(M; B)$ as the direct sum over all k :

$$\Omega(M; B) = \bigoplus_{k=0}^{\infty} \Omega^k(M; B) .$$

Let (e_i) be a basis for E and let (U, ψ) be a local trivializing chart for B (i.e., a chart for M with coordinates x^i and trivializing for B). According to [6.7] $dx^{i_1} \wedge \cdots \wedge dx^{i_k} \otimes \underline{e}_j$ forms a basis for a local section of the vector bundle $(\bigwedge^k {}^*TM) \otimes B$, i.e., a (smooth) k-form ω with values in B has the local expression

$$\omega|_U = \sum_{i_1,\dots,i_k,j} \omega^j_{i_1,\dots,i_k} \, dx^{i_1} \wedge \cdots \wedge dx^{i_k} \otimes \underline{e}_j = \sum_j \omega^j \otimes \underline{e}_j \, , \tag{8.2}$$

with $\omega^j = \sum_{i_1,\dots,i_k} \omega^j_{i_1,\dots,i_k} \, dx^{i_1} \wedge \cdots \wedge dx^{i_k}$ and where the $\omega^j_{i_1,\dots,i_k}$ are (smooth) functions on U. The parity of the B-valued k-form ω is determined in the usual way: $\varepsilon(\omega) = \varepsilon(\omega^j) + \varepsilon(e_j)$, where $\varepsilon(\omega^j)$ is the standard parity of the local k-form ω^j. Obviously the local k-forms ω^j on U are defined independently of the local coordinates x^i. However, they need not glue together to form global k-forms on M. More precisely, let (U_a, ψ_a) and (U_b, ψ_b) be two local trivializations with associated local sections $\underline{e}^a_i$ and $\underline{e}^b_i$ and associated local k-forms ω^j_a and ω^j_b. The transition function $\psi_{ba} : U_a \cap U_b \to \mathrm{Aut}(E)$ for the bundle B defines (left) matrix elements $\psi_{ba}(m)_i{}^j \equiv {}^\bullet M_R(\psi_{ba}(m))_i{}^j \in \mathcal{A}$ by the formula $\psi_{ba}(m)e_i = \sum_j \psi_{ba}(m)_i{}^j e_j$ [II.4.1], [VI.1.20]. Combining the definition of the transition function with that of the local sections $\underline{e}^a_i$ and $\underline{e}^b_j$ gives us the equality

$$\underline{e}^a_i(m) = \sum_j \psi_{ba}(m)_i{}^j \; \underline{e}^b_j(m) .$$

Since $\sum_i \omega^i_a \otimes \underline{e}^a_i$ and $\sum_j \omega^j_b \otimes \underline{e}^b_j$ both represent the same section on $U_a \cap U_b$, we must also have the equality

$$\omega^j_b|_m = \sum_i \omega^i_a|_m \; \psi_{ba}(m)_i{}^j . \tag{8.3}$$

This implies that if we want the local k-forms ω^j_a to glue together to form a global k-form ω^j, then the transition functions ψ_{ba} must be the identity. Said differently, they glue together to form a global k-form if the bundle B is trivial.

In the general case the (local) k-forms ω^i depend upon the local trivialization (U, ψ), but they also depend upon the choice for the basis (e_i) for E. If $\widehat{e}_1, \dots, \widehat{e}_n$ is another basis for E, it is related to the former basis by $\widehat{e}_i = \sum_j A_i{}^j e_j$ for (real) constants $A_i{}^j$. It immediately follows that $\widehat{\underline{e}}_i = \sum_j A_i{}^j \underline{e}_j$ and thus $\omega|_U = \sum_i \widehat{\omega}^i \otimes \widehat{\underline{e}}_i = \sum_j \omega^j \otimes \underline{e}_j$ with

$$\omega^j = \sum_i \widehat{\omega}^i A_i{}^j \ . \tag{8.4}$$

This formula looks exactly as (8.3) but its interpretation is completely different: (8.4) concerns a single trivialization (U, ψ) and two bases, whereas (8.3) concerns two trivializations and a single basis. Moreover, the matrix elements $\psi_{ba}(m)_i{}^j$ in (8.3) depend upon the choice of the basis for E, whereas the matrix elements $A_i{}^j$ in (8.4) do not depend upon the trivialization.

A way to hide the dependence of the local k-forms ω^j on the choice of a basis is to form the local E-valued k-form $\omega_\psi = \sum_j \omega^j \otimes e_j$, which is independent of such a choice. We thus have the similar looking formulæ

$$\omega|_U = \sum_j \omega^j \otimes \underline{e}_j \qquad , \qquad \omega_\psi = \sum_j \omega^j \otimes e_j \ , \tag{8.5}$$

where the first formula is a description of the B-valued k-form restricted to a local trivializing chart (U, ψ) with the trivialization hidden in the local sections $\underline{e}_j$, and where the second is a local E-valued k-form which depends upon the trivialization ψ, as indicated by the subscript ψ. This should be compared with (6.8), which can be seen as the special case of a 0-form.

Using the local E-valued k-forms ω_ψ we can give another interpretation of (8.3). If (U_a, ψ_a) and (U_b, ψ_b) are two local trivializations, we can form the local E-valued k-forms $\omega_a \equiv \omega_{\psi_a} = \sum_i \omega_a^i \otimes e_i$ and $\omega_b \equiv \omega_{\psi_b} = \sum_j \omega_b^j \otimes e_j$. We now interpret ψ_{ba} as an even $\mathrm{End}_R(E)$-valued 0-form on the intersection $U_a \cap U_b$ and we want to compute the wedge-dot product $\psi_{ba} \mathbin{\dot\wedge} \omega_a$ [1.3-ii]. Since the matrix elements $\psi_{ba}(m)_i{}^j$ are defined in such a way that we have $\psi_{ba} = \sum_{ij} \psi_{ba\,i}{}^j \, e_j \otimes e^i$, we compute according to (1.2):

$$\begin{aligned}
\psi_{ba} \mathbin{\dot\wedge} \omega_a &= \sum_{ijk} \psi_{ba\,i}{}^j \wedge \mathfrak{C}^{\varepsilon(e_j)+\varepsilon(e_i)}(\omega_a^k) \otimes (e_j \otimes e^i)(e_k) \\
&= \sum_{ijk} \psi_{ba\,i}{}^j \cdot \mathfrak{C}^{\varepsilon(e_j)+\varepsilon(e_i)}(\omega_a^k) \otimes e_j \, \delta_k^i = \sum_{ij} \omega_a^i \, \psi_{ba\,i}{}^j \otimes e_j \\
&= \sum_i \omega_a^i \otimes \psi_{ba} \, e_i \ .
\end{aligned}$$

To obtain the second equality we used that $\psi_{ba\,i}{}^j$ is a function (0-form) and thus the wedge product is just the ordinary product; to obtain the third equality we used that ψ_{ba} is even, and thus that $\varepsilon(\psi_{ba\,i}{}^j) = \varepsilon(e_j) + \varepsilon(e_i)$. Using the alternative notation for $\psi_{ba} \mathbin{\dot\wedge} \omega_a$ [1.3] we thus can rewrite (8.3) as

$$\omega_b = \psi_{ba} \circ \omega_a \equiv \psi_{ba} \mathbin{\dot\wedge} \omega_a \ . \tag{8.6}$$

We conclude that, given a trivializing atlas $\mathcal{U} = \{(U_a, \psi_a)\}$, the B-valued k-form ω is represented by a system of local E-valued k-forms ω_a on U_a satisfying the compatibility condition (8.6). It is not hard to show that conversely a system of local E-valued k-forms ω_a on U_a satisfying the compatibility condition (8.6) defines a (global) B-valued k-form ω. This description of B-valued k-forms should be compared with the description of sections of (vector) bundles given in [IV.1.20].

8.7 Corollary. *Given a trivializing atlas $\mathcal{U} = \{(U_a, \psi_a) \mid a \in I\}$ for the bundle B and a basis $e_1, \ldots, e_n$ for the typical fiber E, any B-valued k-form ω is represented by local k-forms ω_a^i, $1 \le i \le n$ on U_a which are related on overlaps by (8.3) and which change all at the same time by (8.4) in case we change the basis (e_i) to $(\widehat{e}_j)$ given by $\widehat{e}_i = \sum_j A_i{}^j e_j$.*

8.8 Discussion. In [VI.9.1] we have defined k-forms with values in an $\mathcal{A}$-vector space. This can be seen as a particular particular case of k-forms with values in a vector bundle as follows. We first construct the trivial bundle $B = M \times E$ with its canonical global trivialization $\psi : B \to M \times E$. In this (global) trivialization a B-valued k-form is given by n global k-forms ω^i as $\omega = \sum_i \omega^i \otimes \underline{e}_i$. From this we can obtain the global E-valued k-form $\sum_i \omega^i \otimes e_i$. Since they depend in the same way on the chosen basis for E, we get a bijection between B-valued k-forms and E-valued k-forms. In this sense an E-valued k-form is a particular case of a vector bundle valued k-form. However, whereas we are allowed to change the (global) trivialization when we consider B-valued k-forms, we are *not* allowed to change the trivialization for E-valued k-forms. E-valued k-forms correspond to B-valued k-forms in a given *fixed* global trivialization. Another way to interpret this particularity is to play with the various identifications as in [V.7.1]. This gives us that $\Gamma((\bigwedge^k {}^*TM) \otimes (M \times E))$ is isomorphic to $\mathrm{Hom}_L^{sk}({}^*\Gamma(TM)^k; \Gamma((M \times E)))$. Using a (global, fixed) trivialization ψ [IV.1.20] we obtain an isomorphism between $\Gamma(M \times E)$ and $C^\infty(M; E)$ given by $s \mapsto s_\psi$. Since the latter space represents the space of E-valued k-forms, we obtain an identification between k-forms with values in the trivial bundle $M \times E$ and E-valued k-forms. However, this identification obviously depends upon the chosen global trivialization ψ. Hence the statement that E-valued k-forms are a particular particular case of k-forms with values in a vector bundle.

8.9 Discussion. By definition, a B-valued 0-form is a section of the (vector) bundle $(\bigwedge^0 {}^*TM) \otimes B = (M \times \mathcal{A}) \otimes B$, where $\bigwedge^0 {}^*TM \cong M \times \mathcal{A}$ comes with a canonically defined trivialization. Using the natural isomorphism $\mathcal{A} \otimes E \cong E$ [I.4.10] we obtain a natural isomorphism $(M \times \mathcal{A}) \otimes B \cong B$ [IV.4.23]. It follows that a B-valued 0-form can be identified with a section of B. In terms of local representations this amounts to omitting the tensor product symbol: if the B-valued 0-form ω is locally given by $\omega|_U = \sum_j \omega^j \otimes \underline{e}_j$, then the ω^j are 0-forms, i.e., ordinary functions. It is identified with the (global) section of B whose local expression is given by $\sum_j \omega^j \cdot \underline{e}_j$.

Now suppose that B is the trivial bundle $B = M \times \mathcal{A}$ with its canonical trivialization. Using the same techniques as above, we obtain an identification between $\bigwedge^k {}^*TM$ and $(\bigwedge^k {}^*TM) \otimes (M \times \mathcal{A})$. It follows that k-forms with values in the trivial bundle $M \times \mathcal{A}$ are just ordinary k-forms, i.e., vector bundle valued k-forms are a generalization of ordinary k-forms.

8.10 More definitions. Playing around with the various identifications as in [V.7.1], we already argued that $\Gamma((\bigwedge^k {}^*TM) \otimes B)$ is isomorphic to $\mathrm{Hom}_L^{sk}({}^*\Gamma(TM)^k; \Gamma(B))$ [8.8]. In the same way we can show that it is isomorphic to $(\bigwedge^k {}^*\Gamma(TM)) \otimes \Gamma(B)$. The first isomorphism allows us to identify B-valued k-forms as skew-symmetric k-linear maps (over $C^\infty(M)$) from vector fields on M (sections of TM) to sections of B. The second isomorphism allows us to define the contraction of a B-valued k-form ω with a vector field X to yield a B-valued $(k-1)$-form $\iota(X)\omega$. There are at least two ways to define this contraction. The first is to apply [I.6.16], which does not yield a directly applicable formula. The second is to note that the contraction $\iota(X)\omega$ is such that, as a $(k-1)$-linear skew-symmetric map, this B-valued $(k-1)$-form is given by

$$\iota(Y_1, \ldots, Y_{k-1})(\iota(X)\omega) = \iota(Y_1, \ldots, Y_{k-1}, X)\omega \ .$$

However, the most useful formula to define this contraction is by using the local expression (8.2) and to note that the contraction is given by the local formula

$$(\iota(X)\omega)|_U = \sum_j \iota(X)\omega^j \otimes \underline{e}_j \ .$$

Using the compatibility condition (8.3), it is not hard to show that one can use the above local expression to *define* the global contraction $\iota(X)\omega$, without any reference to the various identifications described above. Moreover, using the local E-valued k-forms ω_ψ (8.5), it is immediate that we have $\iota(X)(\omega_\psi) = (\iota(X)\omega)_\psi$.

8.11 Remark. In [VI.9.1] we have defined the pull-back of a k-form with values in an $\mathcal{A}$-vector space. This can be seen as a special case of a more general notion of a pull-back of vector bundle valued k-forms involving the pull-back bundle. Since we will never need this more general notion, we do not go into the details of the more general notion.

8.12 Definition. In [1.1] we have defined the Φ-wedge product of vector valued differential forms. We now want to extend this definition to the case of arbitrary vector bundles. So let E, F, and G be three $\mathcal{A}$-vector spaces with homogeneous bases (e_i), (f_j), and (g_k) respectively, and let $\Phi : E \times F \to G$ be an even smooth bilinear map. Suppose furthermore that $B \to M$, $C \to M$, and $D \to M$ are vector bundles over M with typical fibers E, F, and G respectively. If α is a B-valued p-form and β a C-valued q-form, the *Φ-wedge product* $\alpha \wedge_\Phi \beta$ should be a D-valued $p+q$-form, all over M. The idea of the construction is as follows. In any common trivializing chart $U \subset M$ for the bundles B, C, and D, the forms α and β determine ordinary differential forms α^i and β^j on U according to

$$\alpha|_U = \sum_i \alpha^i \otimes \underline{e}_i \quad , \quad \beta|_U = \sum_j \beta^j \otimes \underline{f}_j \ .$$

We now introduce matrix elements for Φ by $\Phi(e_i, f_j) = \sum_k \Phi_{ij}^k\, g_k$. With these matrix elements, the D-valued $p+q$-form $\alpha \wedge_\Phi \beta$ should be defined on the trivializing chart U by

$$(\alpha \wedge_\Phi \beta)|_U = \sum_{i,j,k} \alpha^i \wedge \mathfrak{C}^{\varepsilon(e_i)}(\beta^j) \otimes \Phi_{ij}^k\, \underline{g}_k \,. \tag{8.13}$$

The main problem is that these local expressions need not glue together to form a globally well defined D-valued $p+q$-form.

8.14 Lemma. *Let $\mathcal{U} = \{U_a \mid a \in I\}$ be a joint trivializing atlas for the three bundles B, C, and D, and let ψ_{ba} be the transition functions for B, χ_{ba} those for C, and ϕ_{ba} those for D. If the map Φ intertwines these transition functions in the sense that for all $m \in U_a \cap U_b$ and all $e \in E$, $f \in F$ we have*

$$\Phi(\psi_{ba}(m)e, \chi_{ba}(m)f) = \phi_{ba}(m)\Phi(e,f) \,,$$

then the Φ-wedge product $\alpha \wedge_\Phi \beta$ is globally well defined by (8.13).

Proof. On two trivializing charts U_a and U_b we have, as in [8.1], local p-forms α_a^ℓ and α_b^i and local q-forms β_a^s and β_b^j. They satisfy the relations (8.3) $\alpha_b^i|_m = \sum_\ell \alpha_a^\ell|_m\, \psi_{ba}(m)_\ell{}^i$ and $\beta_b^j|_m = \sum_s \beta_a^s|_m\, \chi_{ba}(m)_s{}^j$. With these we compute:

$$\begin{aligned}
&\sum_{i,j,k} \alpha_b^i|_m \wedge \mathfrak{C}^{\varepsilon(e_i)}(\beta_b^j|_m) \otimes \Phi_{ij}^k\, g_k \\
&\quad = \sum_{i,j,\ell,s} \alpha_a^\ell|_m\, \psi_{ba}(m)_\ell{}^i \wedge \mathfrak{C}^{\varepsilon(e_i)}(\beta_a^s|_m\, \chi_{ba}(m)_s{}^j) \otimes \Phi(e_i, f_j) \\
&\quad = \sum_{i,j,\ell,s} \alpha_a^\ell|_m \wedge \mathfrak{C}^{\varepsilon(e_\ell)}(\beta_a^s|_m) \otimes \Phi(\psi_{ba}(m)_\ell{}^i\, e_i, \chi_{ba}(m)_s{}^j\, f_j) \\
&\quad = \sum_{\ell,s} \alpha_a^\ell|_m \wedge \mathfrak{C}^{\varepsilon(e_\ell)}(\beta_a^s|_m) \otimes \Phi(\psi_{ba}(m)e_\ell, \chi_{ba}(m)f_s) \\
&\quad = \sum_{\ell,s} \alpha_a^\ell|_m \wedge \mathfrak{C}^{\varepsilon(e_\ell)}(\beta_a^s|_m) \otimes \phi_{ba}(m)\Phi(e_\ell, f_s) \\
&\quad = \sum_{\ell,s,r} \alpha_a^\ell|_m \wedge \mathfrak{C}^{\varepsilon(e_\ell)}(\beta_a^s|_m) \otimes \Phi_{\ell s}^r\, \phi_{ba}(m) g_r \\
&\quad = \sum_{\ell,s,r,k} \alpha_a^\ell|_m \wedge \mathfrak{C}^{\varepsilon(e_\ell)}(\beta_a^s|_m) \otimes \Phi_{\ell s}^r\, \phi_{ba}(m)_r{}^k\, g_k \,.
\end{aligned}$$

To obtain the second equality, we have used, among others, that $\psi_{ba}(m)$ is even and thus that the parity of $\psi_{ba}(m)_\ell{}^i$ is $\varepsilon(e_i) + \varepsilon(e_\ell)$. Replacing g_k by $\underline{g}_k^b$ in the first and last line of this computation and noting that $\sum_k \phi_{ba}(m)_r{}^k \underline{g}_k^b = \underline{g}_r^a$, we may conclude that the local expressions on U_a and U_b coincide. QED

8.15 Discussion/Definition. If a given joint trivialization of the bundles B, C, and D satisfies the condition of [8.14], it is by no means guaranteed that this condition is satisfied for another joint trivialization. However, in the applications there usually is a natural choice for the joint trivializations such that the condition is automatically satisfied. (This is analogous to the situation in [IV.4.23].)

The first application we have in mind is the case that all three bundles are trivial: $B = M \times E, C = M \times F$, and $D = M \times G$, all three with a fixed trivialization. In other words, the case of k-forms with values in $\mathcal{A}$-vector spaces. In that case the local formula (8.13) becomes the global definition:

$$\Big(\sum_i \alpha^i \otimes e_i\Big) \wedge_\Phi \Big(\sum_j \beta^j \otimes f_j\Big) = \sum_{i,j} \alpha^i \wedge \mathfrak{C}^{\varepsilon(e_i)}(\beta^j) \otimes \Phi(e_i, f_j) .$$

Since the trivialization is fixed from the beginning, no problems arise here, no matter what map Φ we use. And we recover the previous definition for vector valued differential forms.

The second application is the case in which we start with a bundle $P \to M$ with typical fiber E, and then form the Hom-bundle $B = C = D = \mathrm{Hom}_R(P;P)$ with typical fiber $\mathrm{End}_R(E)$ [IV.4.16]. In that case the application Φ is just composition $\circ : \mathrm{End}_R(E) \times \mathrm{End}_R(E) \to \mathrm{End}_R(E)$, $(\phi, \chi) \mapsto \phi \circ \chi$. Since all three bundles are formed from one and the same bundle P, any trivializing atlas for P will do. Since the transition functions of the Hom-bundle are given by the formula (IV.4.17)

$$A \mapsto \psi_{ba}(m) \circ A \circ \psi_{ba}(m)^{-1} ,$$

it is immediate that composition intertwines these transition functions. And thus in this application the Φ-wedge product is well defined.

The third application is the case in which C is an arbitrary vector bundle with typical fiber E and D an arbitrary vector bundle with typical fiber F, but where B is the Hom-bundle $B = \mathrm{Hom}_R(C;D)$ with typical fiber $\mathrm{Hom}_R(E;F)$. In that case the map Φ is evaluation $\mathrm{Hom}_R(E;F) \times E \to F$, $(A, e) \mapsto Ae$. And again the particular form of the transition functions of the Hom-bundle guarantee that they intertwine with this Φ. And thus again the Φ-wedge product is well defined.

The fourth application is the case in which $\Phi : \mathcal{A} \times E \to E$ is left multiplication by elements of $\mathcal{A}$. Here the bundle B is the trivial bundle $B = M \times \mathcal{A}$ and $C = D$ is an arbitrary vector bundle. Using the canonical global trivialization for B and the same transition functions for C and D, it is immediate that left multiplication intertwines these transition functions, and thus the Φ-wedge product is, again, well defined.

8.16 Discussion. In the fourth application, the k-forms with values in the trivial bundle $B = M \times \mathcal{A}$ are just ordinary k-forms. We thus have defined a map

$$\Omega^k(M) \times \Omega^\ell(M;C) \to \Omega^{k+\ell}(M;C) .$$

We can extend this map by bilinearity to a map $\Omega(M) \times \Omega(M;C) \to \Omega(M;C)$. Since $\Omega(M)$ is a $\mathbf{Z} \times \mathbf{Z}_2$-graded commutative ring, the reader can easily establish the fact that this map gives $\Omega(M;C)$ the structure of a $\mathbf{Z} \times \mathbf{Z}_2$-graded $\Omega(M)$-module.

8.17 Notation. Each (admissible) map Φ has its associated Φ-wedge product which we denoted as $\wedge_\Phi$. For the maps Φ of the explicit applications we will use the same notation as for vector valued forms [1.3]: for the composition map in the second application we will denote the Φ-wedge product of α and β by $\alpha \mathring{\wedge} \beta$; for the evaluation map in the third application we will denote it by $\alpha \dot{\wedge} \beta$; for left multiplication in the fourth application we will denote it simply by $\alpha \wedge \beta$. And just as for vector valued forms, there are sometimes several ways to write the same object: if g is a 0-form on M, i.e., an ordinary function, and β a B-valued k-form, then the B-valued k-form $g \wedge \beta$ is the same as $g \cdot \beta$; if A is a $\mathrm{Hom}_R(C; D)$-valued 0-form and β an C-valued k-form, then it is customary to write $A \circ \beta$ instead of $A \dot{\wedge} \beta$; and if α is a $\mathrm{Hom}_R(C; D)$-valued k-form and s a C-valued 0-form, i.e., a section of C, then it is customary to write $\alpha(s)$ or $\alpha \cdot s$ for $\alpha \dot{\wedge} s$.

9. The covariant derivative revisited

In this section we show that a covariant derivative on a vector bundle can be extended to an operation on differential forms with values in the vector bundle, the original covariant derivative being the special case of 0-forms. Seen this way, the covariant derivative is a straightforward generalization of the exterior derivative as a covariant derivative on the trivial bundle $M \times \mathcal{A}$. At the end of this section we define the curvature tensor of a covariant derivative as the square of the covariant derivative and we prove that the FVF connection on the vector bundle associated to the covariant derivative is integrable if and only if this curvature tensor is zero.

9.1 Definition. In [VI.9.1] we defined the exterior derivative of a k-form ω with values in an $\mathcal{A}$-vector space E by the formula

$$d(\sum_i \omega^i \otimes e_i) = \sum_i d\omega^i \otimes e_i \ .$$

It is tempting to think that one can generalize this to vector bundle valued k-forms simply by using the local representation: if $\omega|_U = \sum_j \omega^j \otimes \underline{e}_j$, then $(d\omega)|_U = \sum_j (d\omega^j) \otimes \underline{e}_j$. However, M-dependent transition functions prevent these local definitions to glue together to a well defined global definition. Another idea might be to generalize the original definition of the exterior derivative for ordinary k-forms [V.7.3] in terms of contraction with vector fields and the action of vector fields on functions. However, when one tries to define the action of a vector field on a section of an arbitrary vector bundle, one encounters the same problem: local representations do not glue together to a global object. A solution to this problem, already discussed in [6.1], is given by the covariant derivative.

In order to prepare the actual generalization of the exterior derivative to bundle valued k-forms, we first note that the map $(X, f) \mapsto Xf$, which associates a vector valued function Xf to a vector field X and a vector valued function f, is an example of a covariant derivative [6.3]. We thus could write $\nabla_X f = Xf$. By definition of the exterior derivative, the action of X on f can also be written as $Xf = \iota(X)df$. We thus can write

$$\nabla_X f = \iota(X)df \ .$$

This analogy between covariant derivative and exterior derivative can be extended to arbitrary vector bundles as follows. Let $\pi : B \to M$ be a vector bundle with typical fiber E and let ∇ be a covariant derivative on B. For a fixed section $s \in \Gamma(B)$, the map $\nabla s : X \mapsto \nabla(X, s)$ from $\Gamma(TM)$ to $\Gamma(B)$ is linear over $C^\infty(M)$ [6.2]. According to [8.10] we thus can see it as a B-valued 1-form. It follows that we can see the covariant derivative ∇ as a map $\Omega^0(M; B) \to \Omega^1(M; B)$, $s \mapsto \nabla s$, just as we can see the exterior derivative as a map from functions to 1-forms.

With these analogies, we are ready to extend the notion of a covariant derivative to a map $\nabla : \Omega(M; B) \to \Omega(M; B)$ of parity $(1, 0)$ using a formula analogous to that for the exterior derivative (V.7.4) (recall that $\Omega(M; B)$ is a $\mathbf{Z} \times \mathbf{Z}_2$-graded $\Omega(M)$-module [8.16]). To that end we start with a B-valued k-form α on M, which we see as a k-linear graded skew-symmetric map over $C^\infty(M)$ from vector fields to sections of B [8.10]. We then define $\nabla\alpha$ as the B-valued $(k+1)$-form on M of the same parity as α, seen as a $(k+1)$-linear graded skew-symmetric map (over $C^\infty(M)$) from vector fields on M to sections of B, by the formula

$$(9.2)\quad (-1)^k \cdot \iota(X_0, \dots, X_k)\, \nabla\alpha =$$

$$= \sum_{0 \le i \le k} (-1)^{i + \sum_{p<i} \langle \varepsilon(X_p) | \varepsilon(X_i) \rangle} \nabla_{X_i}(\iota(X_0, \dots, X_{i-1}, X_{i+1}, \dots, X_k)\, \alpha)$$

$$+ \sum_{0 \le i < j \le k} (-1)^{j + \sum_{i<p<j} \langle \varepsilon(X_p) | \varepsilon(X_j) \rangle} \iota(X_0, \dots, X_{i-1}, [X_i, X_j], X_{i+1}, \dots$$

$$\dots, X_{j-1}, X_{j+1}, \dots, X_k)\, \alpha \ .$$

And as for the exterior derivative, this formula needs some explanation. The vector fields X_i are supposed to be homogeneous; taking the sum over homogeneous parts we obtain a $k+1$-additive function of vector fields on M to sections of B.

9.3 Lemma. *The map $\nabla\alpha$ given by (9.2) is $(k+1)$-linear over $C^\infty(M)$ and graded skew-symmetric. It thus defines a B-valued $(k+1)$-form on M.*

Proof. The proof is a direct copy of the corresponding proof for the exterior derivative [V.7.5] and is left to the reader. QED

9.4 Discussion. If $U \subset M$ is open, it is itself an $\mathcal{A}$-manifold. Restricting α and $\nabla\alpha$ to U gives us a $B|_U$-valued k-form $\alpha|_U$ and a $B|_U$-valued $(k+1)$-form $(\nabla\alpha)|_U$ on U. Denoting the induced connection on U by ∇^U [6.6], we also have the associated map ∇^U acting on $B|_U$-valued differential forms. As for the exterior derivative we are thus faced with the question whether $(\nabla\alpha)|_U$ equals $\nabla^U(\alpha|_U)$.

9.5 Lemma. *For any open $U \subset M$: $(\nabla\alpha)|_U = \nabla^U(\alpha|_U)$.*

Proof. The proof is left to the reader. It is a direct copy of the corresponding proof for the exterior derivative [V.7.7], now using [6.4] instead of [V.1.5]. QED

9.6 Discussion. Once we know that the induced map on B-valued differential forms commutes with restrictions, it is customary (again) not to attach the superscript ^{U} to the symbol ∇ when restricting to an open subset $U \subset M$. As for the exterior derivative, (9.2) is not meant for easy calculation. Using local trivializations, [9.7] gives us formulæ that do allow more explicit calculations.

9.7 Proposition. *Let α be a B-valued k-form β an ordinary ℓ-form on M, and let (U, ψ) be a local trivialization for the bundle $B \to M$. Then:*

(i) *denoting by Γ the local $\mathrm{End}_{\mathbb{R}}(E)$-valued 1-form on U defined by the covariant derivative ∇ [6.9], we have*

$$(\nabla\alpha)_\psi = d\alpha_\psi + \Gamma \mathbin{\dot\wedge} \alpha_\psi ;$$

(ii) $\nabla(\beta \wedge \alpha) = (d\beta) \wedge \alpha + (-1)^\ell \beta \wedge \nabla\alpha.$

Proof. • Both sides of the equality $(\nabla\alpha)_\psi = d\alpha_\psi + \Gamma \mathbin{\dot\wedge} \alpha_\psi$ are E-valued $(k+1)$-forms, which we interpret as $(k+1)$-linear skew-symmetric maps from vector fields to functions. We thus show that both sides give the same result when evaluated on $(k+1)$ vector fields. We first concentrate on the term $\Gamma \mathbin{\dot\wedge} \alpha_\psi$:

$$\begin{aligned}\iota(X_0, \ldots, X_k)(\Gamma \mathbin{\dot\wedge} \alpha_\psi) &= \iota(X_0, \ldots, X_{k-1})\big(\iota(X_k)\Gamma \cdot \alpha_\psi - \Gamma \mathbin{\dot\wedge} \iota(X_k)\alpha_\psi\big) \\ &= (-1)^{\sum_{p<k} \langle \varepsilon(X_p) | \varepsilon(X_k) \rangle} \cdot \iota(X_k)\Gamma \cdot \iota(X_0, \ldots, X_{k-1})\alpha_\psi \\ &\qquad - \iota(X_0, \ldots, X_{k-1})\big(\Gamma \mathbin{\dot\wedge} \iota(X_k)\alpha_\psi\big) ,\end{aligned}$$

where the first equality is a consequence of the fact that $\iota(X_k)$ is a derivation of degree $(-1, \varepsilon(X_k))$ [I.6.16] and that Γ is an even 1-form; the second equality a consequence of the fact that $\iota(X_k)\Gamma$ is a 0-form of parity $\varepsilon(X_k)$. By induction this gives us

$$(9.8) \quad (-1)^k \iota(X_0, \ldots, X_k)(\Gamma \mathbin{\dot\wedge} \alpha_\psi) =$$

$$\sum_{0 \le i \le k} (-1)^{i + \sum_{p<i} \langle \varepsilon(X_p) | \varepsilon(X_i) \rangle} \iota(X_i)\Gamma \cdot \iota(X_0, \ldots, X_{i-1}, X_{i+1}, \ldots, X_k)\alpha_\psi .$$

With this preparation, we compute:

$$(-1)^k \cdot \iota(X_0,\dots,X_k)(\nabla\alpha)_\psi =$$

$$= \sum_{0\le i\le k} (-1)^{i+\sum_{p<i}\langle \varepsilon(X_p)|\varepsilon(X_i)\rangle} \Big(\nabla_{X_i}\big(\iota(X_0,\dots,X_{i-1},X_{i+1},\dots,X_k)\,\alpha\big)\Big)_\psi$$
$$+ \sum_{0\le i<j\le k} (-1)^{j+\sum_{i<p<j}\langle \varepsilon(X_p)|\varepsilon(X_j)\rangle} \iota(X_0,\dots,X_{i-1},[X_i,X_j],X_{i+1},\dots$$
$$\dots,X_{j-1},X_{j+1},\dots,X_k)\,\alpha_\psi$$

$$= \sum_{0\le i\le k} (-1)^{i+\sum_{p<i}\langle \varepsilon(X_p)|\varepsilon(X_i)\rangle} \Big(X_i\big(\iota(X_0,\dots,X_{i-1},X_{i+1},\dots,X_k)\,\alpha_\psi\big)$$
$$+ \iota(X_i)\Gamma\cdot\iota(X_0,\dots,X_{i-1},X_{i+1},\dots,X_k)\,\alpha_\psi\Big)$$
$$+ \sum_{0\le i<j\le k} (-1)^{j+\sum_{i<p<j}\langle \varepsilon(X_p)|\varepsilon(X_j)\rangle} \iota(X_0,\dots,X_{i-1},[X_i,X_j],X_{i+1},\dots$$
$$\dots,X_{j-1},X_{j+1},\dots,X_k)\,\alpha_\psi$$

$$= (-1)^k\cdot\iota(X_0,\dots,X_k)(d\alpha_\psi) + (-1)^k\iota(X_0,\dots,X_k)(\Gamma \mathbin{\dot\wedge} \alpha_\psi)\ .$$

In this computation we used the general equality $\iota(X)(\omega_\psi) = (\iota(X)\omega)_\psi$ [8.10]. To obtain the second equality we used (6.14) and to obtain the last equality we used the definition of the exterior derivative (V.7.4) and (9.8). This proves (i).

• To prove (ii) we first note that β is a global 1-form and thus for a local trivialization (U,ψ) we have $\beta_\psi = \beta$. As a 1-form with values in the trivial bundle, we should officially write $\beta_\psi = \beta\otimes 1$ with 1 the canonical basis of $\mathcal{A}$. Applying the definition of $\beta\wedge\alpha$ [1.3] this means that if $\alpha_\psi = \sum_i \alpha^i\otimes e_i$, then $(\beta\wedge\alpha)_\psi = \sum_i(\beta\wedge\alpha^i)\otimes 1\cdot e_i$. Using the basis $(e_j\otimes e^k)$ of $\mathrm{End}_R(E)$, the local even 1-form Γ can be written as $\Gamma = \sum_{jk}\Gamma^j{}_k\, e_j\otimes e^k$ (6.11). Applying the definition of wedge-dot product we thus find:

$$\begin{aligned}\Gamma \mathbin{\dot\wedge} (\beta\wedge\alpha)_\psi &= \sum_{ijk}\Gamma^j{}_k \wedge \mathfrak{C}^{\varepsilon(e_j\otimes e^k)}(\beta\wedge\alpha^i)\otimes(e_j\otimes e^k)(e_i)\\ &= \sum_{ijk}(-1)^\ell\beta\wedge\Gamma^j{}_k\wedge\mathfrak{C}^{\varepsilon(e_j\otimes e^k)}(\alpha^i)\otimes(e_j\otimes e^k)(e_i)\\ &= (-1)^\ell\beta\wedge(\Gamma\mathbin{\dot\wedge}\alpha_\psi)\ ,\end{aligned}$$

where we used that $\Gamma^j{}_k$ is a 1-form of parity $\varepsilon(e_j\otimes e^k)$ to obtain the second equality. We now use (i) to compute $(\nabla(\beta\wedge\alpha))_\psi$ in an arbitrary local trivialization (U,ψ):

$$\begin{aligned}\big(\nabla(\beta\wedge\alpha)\big)_\psi &= d(\beta\wedge\alpha)_\psi + \Gamma\mathbin{\dot\wedge}(\beta\wedge\alpha)_\psi\\ &= (d\beta)\wedge\alpha_\psi + (-1)^\ell\beta\wedge d\alpha_\psi + (-1)^\ell\beta\wedge\Gamma\mathbin{\dot\wedge}\alpha_\psi\\ &= \big((d\beta)\wedge\alpha + (-1)^\ell\beta\wedge\nabla\alpha\big)_\psi\ ,\end{aligned}$$

where we used [V.7.9-iii] and (VI.9.4) to obtain the second equality. Since the trivialization was arbitrary, we have proven (ii). QED

9.9 Remark. Another approach to a covariant derivative consists in defining it as an **R**-linear map of parity $(1,0)$ on $\Omega(M;B)$ satisfying the Leibniz type rule [9.7-ii]. Starting from this definition, one can prove the existence of local 1-forms Γ such that [9.7-i] holds. In that way one recovers our original definition/construction of a covariant derivative. Apart from the minor difference that we use a bi-grading on $\Omega(M;B)$, this is the approach taken in [Qu].

9.10 Theorem. *Given a covariant derivative ∇ on a vector bundle $\pi : B \to M$, there exists a (unique) even $\mathrm{End}_R(B)$-valued 2-form R, such that for all B-valued k-forms we have*

$$\nabla(\nabla\alpha) = R \wedge \alpha \ .$$

In a local trivialization (U,ψ) for B with Γ the local $\mathrm{End}_R(E)$-valued 1-form on U defined by the covariant derivative ∇ [6.9], R is given by

$$R_\psi = d\Gamma + \tfrac{1}{2}[\Gamma \wedge \Gamma] = d\Gamma + \Gamma \wedge \Gamma \ .$$

Proof. If R exists, then in particular for any $s \in \Gamma(B)$, any two vector fields X, Y on M and a point $m \in M$ we must have

$$\langle\!\langle \iota(X_m, Y_m)R \| s(m) \rangle\!\rangle = (\iota(X,Y)\nabla(\nabla\alpha))(m) \ .$$

Since $\iota(X_m, Y_m)R \in \mathrm{End}_R(B_m)$ is uniquely determined by its action on elements in B_m, uniqueness of R follows. That it must be even follows from the fact that ∇ does not change parities.

To show existence, we proceed as follows. In any local trivialization (U,ψ) we have

$$\begin{aligned}(\nabla(\nabla\alpha))_\psi &= d(\nabla\alpha)_\psi + \Gamma \wedge (\nabla\alpha)_\psi = d(d\alpha_\psi + \Gamma \wedge \alpha_\psi) + \Gamma \wedge (d\alpha_\psi + \Gamma \wedge \alpha_\psi) \\ &= d(\Gamma \wedge \alpha_\psi) + \Gamma \wedge d\alpha_\psi + \Gamma \wedge (\Gamma \wedge \alpha_\psi) \\ &= d\Gamma \wedge \alpha_\psi - \Gamma \wedge d\alpha_\psi + \Gamma \wedge d\alpha_\psi + (\Gamma \wedge \Gamma) \wedge \alpha_\psi \\ &= (d\Gamma + \Gamma \wedge \Gamma) \wedge \alpha_\psi \ ,\end{aligned}$$

where the minus sign after the fourth equality comes from the fact that Γ is a 1-form. This computations shows that, if R exists, it is locally given by the announced formula $R_\psi = d\Gamma + \Gamma \wedge \Gamma$.

If we now take for α a B-valued 0-form s, i.e., a section of B, and if we fix two vector fields X, Y on M, we obtain a map $R(X,Y) : \Gamma(B) \to \Gamma(B)$,

$$s \mapsto \iota(X,Y)\nabla(\nabla s) \ .$$

The local expression for $\nabla(\nabla s)$ shows that this map is right linear over $C^\infty(M)$, i.e., $R(X,Y)$ is an element of $\text{End}_R(\Gamma(B))$. Since this space is isomorphic to $\Gamma(\text{End}_R(B))$ [IV.5.9], [IV.5.13] we obtain a map $R : \Gamma(TM)^2 \to \Gamma(\text{End}_R(B))$, $(X,Y) \mapsto R(X,Y)$. This map is obviously graded skew-symmetric and bilinear over $C^\infty(M)$. It thus defines an $\text{End}_R(B)$-valued 2-form R as announced. Coming back to the local expression for $\nabla(\nabla s)$ shows that the local expression of R is given by $R_\psi = d\Gamma + \Gamma \wedge \Gamma$. And then the local expression of $\nabla(\nabla\alpha)$ for an arbitrary B-valued k-form shows that we have the general result $\nabla(\nabla\alpha) = R \wedge \alpha$. QED

9.11 Definition/Discussion. The $\text{End}_R(B)$-valued 2-form R whose existence is guaranteed by [9.10] is called the *curvature tensor* of the covariant derivative ∇. We have defined it in terms of the extension of ∇ to an operator on $\Omega(M;B)$ as $R = \nabla^2$. The usual approach to the curvature tensor, which does not make use of this extension, starts from the observation that $\nabla_X s$ is the generalization of the derivative of a function in the direction of the vector field X. Since the commutator of two vector fields is defined in terms of the action of vector fields on functions, it is natural to ask to what extent the covariant derivative in the direction of a commutator corresponds to the commutator of the covariant derivatives. More precisely, if X and Y are two vector fields on M and s a section of B, one wants to know whether $[\nabla_X, \nabla_Y]s \equiv \nabla_X(\nabla_Y s) - (-1)^{\langle\varepsilon(X)|\varepsilon(Y)\rangle}\nabla_Y(\nabla_X s)$ is the same as $\nabla_{[X,Y]}s$. Looking at the official definition of ∇ applied to B-valued 1-forms we find:

$$\begin{aligned}
-\iota(X,Y)Rs &= -\iota(X,Y)\nabla(\nabla s) \\
&= \nabla_X(\iota(Y)\nabla s) - (-1)^{\langle\varepsilon(X)|\varepsilon(Y)\rangle}\nabla_Y(\iota(X)\nabla s) - \iota([X,Y])\nabla s \\
&= \nabla_X(\nabla_Y s) - (-1)^{\langle\varepsilon(X)|\varepsilon(Y)\rangle}\nabla_Y(\nabla_X s) - \nabla_{[X,Y]}s \\
&= [\nabla_X, \nabla_Y]s - \nabla_{[X,Y]}s\,.
\end{aligned}$$

We conclude that the difference between $[\nabla_X, \nabla_Y]s$ and $\nabla_{[X,Y]}s$ is linear over $C^\infty(M)$ in s and that it is measured by the curvature tensor. The usual definition of the curvature tensor takes the equation $-\iota(X,Y)Rs = [\nabla_X, \nabla_Y]s - \nabla_{[X,Y]}s$ as the defining equation for R.

9.12 Proposition. *Let ∇ be a covariant derivative on the vector bundle $\pi : B \to M$ and let $\mathcal{H}$ be the associated linear connection on B. The connection $\mathcal{H}$ is integrable if and only if the curvature tensor R of ∇ is (identically) zero.*

Proof. According to [9.10], R is (identically) zero if and only if the local expressions $R_{\psi_a} = d\Gamma^a + \frac{1}{2}[\Gamma^a \wedge \Gamma^a]$ are zero for a trivializing atlas $\mathcal{U} = \{(U_a, \psi_a) \mid a \in I\}$. According to [2.18] this is the case if and only if $\mathcal{H}$ is integrable. QED

10. PRINCIPAL FIBER BUNDLES VERSUS VECTOR BUNDLES

In this section we show that a special kind of differential forms on a principal fiber bundle are in 1–1 correspondence with differential forms with values in an associated vector bundle. Moreover, we prove that under this identification the exterior covariant derivative on the principal fiber bundle corresponds to the covariant derivative on the associated vector bundle.

10.1 Definition. Let $\pi : B \to M$ be a fiber bundle and let E be an $\mathcal{A}$-vector space. An E-valued k-form α on B is said to be *horizontal* if for all $b \in B$ and all $X \in V_b$ we have $\iota(X)\alpha = 0$, where $V \equiv \ker(T\pi) \subset TB$ denotes the vertical subbundle [2.2]; in words: α is horizontal if the contraction with a vertical tangent vector yields zero. In particular, any E-valued function on B (a 0-form) is horizontal.

10.2 Lemma. *Let $\pi : B \to M$ be a fiber bundle with typical fiber F, let (U, ψ) be a trivializing coordinate chart with coordinates x^i, let E be an $\mathcal{A}$-vector space, and let α be an E-valued k-form on B. The restriction $\alpha|_{\pi^{-1}(U)}$ is horizontal if and only if there exist E-valued functions $f_{i_1\ldots i_k}$ on $U \times F$ such that*

$$((\psi^{-1})^*\alpha)|_{(m,f)} = \sum_{i_1\ldots i_k} dx^{i_1} \wedge \cdots \wedge dx^{i_k} \otimes f_{i_1\ldots i_k}(m, f) \ . \tag{10.3}$$

Proof. If y^j are (local) coordinates on F, then a basis for the space of k-forms on $U \times F$ at $(m, f) \in U \times F$ is given by the set of k-forms

$$dy^{j_1} \wedge \cdots \wedge dy^{j_p} \wedge dx^{i_1} \wedge \cdots \wedge dx^{i_q}$$

with $p, q \in \mathbf{N}, p+q = k$ and $j_1, \ldots, j_p$ and $i_1, \ldots, i_q$ increasing sequences as in [II.3.16]. It follows that $((\psi^{-1})^*\alpha)|_{(m,f)}$ is a linear combination of these basis k-forms:

$$((\psi^{-1})^*\alpha)|_{(m,f)} =$$
$$\sum_{p=0}^{k} \sum_{j_1,\ldots,j_p} \sum_{i_1,\ldots,i_q} f^p_{j_1,\ldots,j_p;i_1,\ldots,i_q}(m, f) \cdot dy^{j_1} \wedge \cdots \wedge dy^{j_p} \wedge dx^{i_1} \wedge \cdots \wedge dx^{i_q} \ .$$

In the trivialization ψ, the vertical space V is spanned by the tangent vectors ∂_{y^j}. It follows immediately that if (10.3) is satisfied, then $\iota(X)\alpha = 0$ for all $X \in V_b$. We thus suppose that α is horizontal and we want to prove that all coefficients $f^p_{j_1,\ldots,j_p;i_1,\ldots,i_q}$ are zero for $p > 0$. Since α is horizontal, it follows in particular that all contractions $\iota(\partial_{y^j})((\psi^{-1})^*\alpha)|_{(m,f)}$ must be zero. For simplicity's sake we start considering contraction with ∂_{y^1}:

$$\iota(\partial_{y^1})dy^{j_1} \wedge \cdots \wedge dy^{j_p} \wedge dx^{i_1} \wedge \cdots \wedge dx^{i_q} = n_1\, dy^{j_2} \wedge \cdots \wedge dy^{j_p} \wedge dx^{i_1} \wedge \cdots \wedge dx^{i_q} \ ,$$

where the integer $n_1 \in \mathbf{N}$ is such that $1 = j_1 = \cdots = j_{n_1} < j_{n_1+1}$. Since the sequences $j_2, \ldots, j_p$ are still of the form given in [II.3.16], the $(k-1)$-forms $dy^{j_2} \wedge \cdots \wedge dy^{j_p} \wedge dx^{i_1} \wedge \cdots \wedge dx^{i_q}$ are independent. It follows that all coefficients $f^p_{j_1,\ldots,j_p;i_1,\ldots,i_q}$ with $j_1 = 1$ must be zero, i.e., we can restrict attention to increasing sequences $j_1, \ldots, j_p$ as in [II.3.16] with $j_1 \geq 2$. But then we contract with ∂_{y^2} to conclude that we also can discard sequences $j_1, \ldots, j_p$ starting at $j_1 = 2$. Continuing this way we conclude that horizontality of α implies that all coefficients $f^p_{j_1,\ldots,j_p;i_1,\ldots,i_q}$ with a sequence of j's must be zero. In other words, we have proven (10.3). QED

10.4 Definition. Let $\pi : P \to M$ be a principal fiber bundle with structure group G, let E be an $\mathcal{A}$-vector space, and let $\rho : G \to \mathrm{Aut}(E)$ be a representation. An E-valued k-form α on P is said to be *of type* ρ if for all $g \in G$ we have $\Phi_g^*\alpha = \rho(g^{-1}) \circ \alpha$, where Φ_g denotes the right action of $g \in G$ on P [IV.2.12]. The interpretation of this condition is analogous to that of condition (ii) of a connection 1-form: $\Phi_g^*\alpha$ denotes the generalized pull-back as defined in [V.7.23], [VI.8.12]. $\rho(g^{-1}) \circ \alpha$ is an example of the alternative notation [1.3] for the evaluation-wedge product of an $\mathrm{End}_R(E)$-valued 0-form with an E-valued k-form. More precisely, for $X_1, \ldots, X_k \in T_pP$ and $g \in G$ we must have

$$\begin{aligned}\iota(\langle\!\langle X_1 \,\|\, T\Phi_g \rangle\!\rangle, \ldots, \langle\!\langle X_k \,\|\, T\Phi_g \rangle\!\rangle)\alpha_{p\cdot g} &\equiv \iota(\langle\!\langle (X_1, 0) \,\|\, T\Phi \rangle\!\rangle, \ldots, \langle\!\langle (X_k, 0) \,\|\, T\Phi \rangle\!\rangle)\alpha_{p\cdot g} \\ &= \rho(g^{-1})\big(\iota(X_1, \ldots, X_k)\alpha_p\big) \ .\end{aligned}$$

In particular a function $f : P \to E$ is of type ρ if $f(pg) = \rho(g^{-1})(f(p))$.

10.5 Examples. • A connection 1-form ω on a principal fiber bundle is a $\mathfrak{g}$-valued 1-form of type Ad according to condition (ii) [3.7]. According to condition (i), it can never be horizontal. However, if ω_o is another connection 1-form on the principal fiber bundle P, then the difference $\omega - \omega_o$ is of type Ad and horizontal, because the fundamental vector fields x^P generate the vertical subspace V.

• The curvature 2-form $\Omega = D\omega$ of a connection 1-form ω is by definition horizontal. It is also of type Ad as can be seen directly from (4.10).

10.6 Discussion/Construction. If $\pi : B \to M$ is a fiber bundle and E an $\mathcal{A}$-vector space, then the set $\Omega^k(B; E)$ of all (smooth) E-valued k-forms is a graded $C^\infty(B)$-module. If $\alpha \in \Omega^k(B; E)$ is horizontal, its homogeneous components are also horizontal. It follows easily that the subset $\Omega^k_{hor}(B; E)$ of all horizontal E-valued k-forms is a graded submodule.

Now let $\pi : P \to M$ be a principal fiber bundle with structure group G. If an E-valued k-form $\alpha \in \Omega^k(P; E)$ is of type ρ for some representation $\rho : G \to \mathrm{Aut}(E)$, then the product $f\alpha$ of α with an arbitrary $f \in C^\infty(P)$ will not in general be again of type ρ. On the other hand, since π is surjective, the map $C^\infty(M) \to C^\infty(P)$, $f \mapsto f \circ \pi$ is an injective morphism of graded $\mathbf{R}$-algebras. We thus can interpret $C^\infty(M)$ as a subalgebra of $C^\infty(P)$ and we can see $\Omega^k(P;E)$ as a graded $C^\infty(M)$-module. Since $\pi \circ \Phi_g = \pi$, it follows easily that if α is of type ρ and $f \in C^\infty(M)$, then the product $f\alpha$ is again of type ρ. We conclude that the set $\Omega^k_\rho(P;E)$ of all (smooth) E-valued k-forms of type ρ is a graded $C^\infty(M)$-module. A particular consequence of the preceding discussion is that the intersection $\Omega^k_{hor,\rho}(P;E) = \Omega^k_\rho(P;E) \cap \Omega^k_{hor}(P;E)$ of all smooth horizontal E-valued k-forms of type ρ is a graded $C^\infty(M)$-module.

In order to lighten the notation a bit, we will denote, in the rest of this discussion, the vector bundle $P^{\rho,E}$ associated to P by the representation ρ as $B \equiv P^{\rho,E}$. Our purpose now is to construct an isomorphism $\mathfrak{I} : \Omega^k(M;B) \to \Omega^k_{hor,\rho}(P;E)$ between the graded $C^\infty(M)$-module $\Omega^k(M;B)$ of all (smooth) B-valued k-forms and the graded $C^\infty(M)$-module $\Omega^k_{hor,\rho}(P;E)$. The actual construction starts with the choice of a trivializing atlas $\mathcal{U} = \{(U_a,\psi_a) \mid a \in I\}$ for P and the associated trivializing atlas $\{(U_a,\widetilde{\psi}_a) \mid a \in I\}$ for B [IV.2.1]. Now if $\widetilde{\alpha} \in \Omega^k(M;B)$ is a B-valued k-form on M, then it is represented by local k-forms $\widetilde{\alpha}^i_a$ as $\widetilde{\alpha}|_{\pi^{-1}(U_a)} = \sum_i \widetilde{\alpha}^i_a \otimes \underline{e}_i$ (8.2). If ψ_{ba} is a transition function for P, then the corresponding transition function $\widetilde{\psi}_{ba}$ for B is given by $\widetilde{\psi}_{ba} = \rho \circ \psi_{ba}$ [IV.2.1]. It follows that the local k-forms $\widetilde{\alpha}^i_a$ are related on overlaps $U_a \cap U_b$ by (8.3):

$$\widetilde{\alpha}^j_b|_m = \sum_i \widetilde{\alpha}^i_a \, \rho(\psi_{ba}(m))_i{}^j \ .$$

On $U_a \times G$ we now define the E-valued k-form α_a by

$$\alpha_a|_{(m,g)} = \rho(g^{-1}) \circ \sum_i \widetilde{\alpha}^i_a|_m \otimes e_i \ . \tag{10.7}$$

Obviously α_a is horizontal and $\alpha_a|_{(m,gh)} = \rho(h^{-1}) \circ \alpha_a|_{(m,g)}$, i.e., α_a is horizontal and of type ρ. Moreover,

$$\begin{aligned}
\alpha_b|_{(m,\psi_{ba}(m)\cdot g)} &= \rho(g^{-1}) \circ \rho(\psi_{ba}(m)^{-1}) \circ \Big(\sum_j \widetilde{\alpha}^j_b|_m \otimes e_j\Big) \\
&= \rho(g^{-1}) \circ \rho(\psi_{ba}(m)^{-1}) \circ \Big(\sum_{i,j} \widetilde{\alpha}^i_a|_m \, \rho(\psi_{ba}(m))_i{}^j \otimes e_j\Big) \\
&= \rho(g^{-1}) \circ \rho(\psi_{ba}(m)^{-1}) \circ \Big(\sum_i \widetilde{\alpha}^i_a|_m \otimes \rho(\psi_{ba}(m))(e_i)\Big) \\
&= \rho(g^{-1}) \circ \Big(\sum_i \widetilde{\alpha}^i_a|_m \otimes e_i\Big) = \alpha_a|_{(m,g)} \ .
\end{aligned}$$

Since $(\psi_b \circ \psi_a^{-1})(m,g) = (m, \psi_{ba}(m)\cdot g)$, we thus have shown that $(\psi_b \circ \psi_a^{-1})^*\alpha_b = \alpha_a$. In other words, the E-valued k-forms $\psi_a^*\alpha_a$ and $\psi_b^*\alpha_b$ coincide on overlaps $U_a \cap U_b$. It follows that there exists a well defined global E-valued k-form $\alpha = \mathfrak{I}(\widetilde{\alpha})$ on P such that $\alpha|_{\pi^{-1}(U_a)} = \psi_a^*\alpha_a$. Moreover, this E-valued k-form is horizontal and of type ρ, i.e., $\mathfrak{I}(\widetilde{\alpha}) \equiv \alpha \in \Omega^k_{hor,\rho}(P;E)$.

A priori the construction of α depends upon the trivializing atlas $\mathcal{U}$. A way to make it manifestly independent of such a choice would be to take for $\mathcal{U}$ the whole fiber bundle structure. However, since any chart added to $\mathcal{U}$ is compatible with the elements of $\mathcal{U}$, this will not change α. And thus, even though the construction of α from $\widetilde{\alpha}$ depends upon an atlas, the result does not.

10.8 Proposition. *The map* $\mathfrak{I} : \Omega^k(M; P^{\rho,E}) \to \Omega^k_{hor,\rho}(P; E)$, $\widetilde{\alpha} \mapsto \mathfrak{I}(\widetilde{\alpha}) \equiv \alpha$ *is an isomorphism of graded* $C^\infty(M)$*-modules.*

Proof. Replacing $\widetilde{\alpha} \in \Omega^k(M; P^{\rho,E})$ by a sum of two elements, or multiplying it by a function on M will replace α by the corresponding sum or multiply the result by the given function. Hence the map $\mathfrak{I}$ is a morphism of graded $C^\infty(M)$-modules. An E-valued k-form α on P is zero if and only if all local k-forms $\alpha_a = (\psi_a^{-1})^*\alpha$ are zero, which is the case if and only if all $\widetilde{\alpha}_a^i$ are zero (10.7). But this happens if and only if $\widetilde{\alpha}$ is zero, showing that $\mathfrak{I}$ is injective.

To prove surjectivity, let α be an E-valued k-form on P which is horizontal and of type ρ. Since it is horizontal, there exist E-valued functions $f^a_{i_1\dots i_k}$ on $U_a \times G$ such that $\alpha_a \equiv ((\psi_a^{-1})^*\alpha)|_{(m,g)} = \sum_{i_1\dots i_k} dx^{i_1} \wedge \cdots \wedge dx^{i_k} \otimes f^a_{i_1\dots i_k}(m,g)$ [10.2]. Since α is of type ρ, we must have $f^a_{i_1\dots i_k}(m,g) = \rho(g^{-1})f^a_{i_1\dots i_k}(m,e)$. Decomposing $f^a_{i_1\dots i_k}(m,g)$ with respect to a basis (e_j) of E as $f^a_{i_1\dots i_k}(m,g) = \sum_j f^{a,j}_{i_1\dots i_k}(m,g)\cdot e_j$, we can define the local k-forms $\widetilde{\alpha}^j_a$ on U_a by

$$\widetilde{\alpha}^j_a|_m = \sum_{i_1\dots i_k} dx^{i_1} \wedge \cdots \wedge dx^{i_k} \cdot f^{a,j}_{i_1\dots i_k}(m,e) \ .$$

Since $\rho(g)$ is even, we thus have by construction $\alpha_a|_{(m,g)} = \rho(g^{-1}) \circ \sum_j \widetilde{\alpha}^j_a|_m$. Comparing this with the construction of $\mathfrak{I}$, we see that if the local k-forms $\sum_j \widetilde{\alpha}^j_a \otimes \underline{e}_j$ glue together to form a global k-form $\widetilde{\alpha}$, then the α we started with is the one constructed from $\widetilde{\alpha}$, i.e., $\mathfrak{I}$ is surjective.

To prove that the local k-forms $\sum_j \widetilde{\alpha}^j_a \otimes \underline{e}_j$ glue together, we note that, by construction, we have $(\psi_b \circ \psi_a^{-1})^*\alpha_b = \alpha_a$. Since $(\psi_b \circ \psi_a^{-1})(m,g) = (m, \psi_{ba}(m)\cdot g)$, this implies that the local functions $f^a_{i_1\dots i_k}$ must satisfy the compatibility condition

$$f^b_{i_1\dots i_k}(m, \psi_{ba}(m)\cdot g) = f^a_{i_1\dots i_k}(m,g) \ .$$

Together with the type ρ condition, this implies that we have

$$f^a_{i_1\dots i_k}(m,e) = \rho(\psi_{ba}(m)) f^b_{i_1\dots i_k}(m,e) \ ,$$

and thus $\widetilde{\alpha}^j_b|_m = \sum_i \widetilde{\alpha}^i_a|_m \ \rho(\psi_{ba}(m))_i{}^j$. By [8.7] this means that the $\widetilde{\alpha}_a$ glue together to form a global $P^{\rho,E}$-valued k form on M. QED

10.9 Remark. As before we use the abbreviation $B = P^{\rho,E}$. For 0-forms the identification $\mathfrak{I} : \Omega^0(M;B) \to \Omega^0_{hor,\rho}(P;E)$, which is an identification between $\Gamma(B)$ and $C^\infty(P;E)$, can be stated in a more intrinsic way using the description [5.11] of the associated bundle. Let $f : P \to E$ be a function of type ρ (it is automatically horizontal), i.e., $f(pg) = \rho(g^{-1})f(p)$, and let $s = \mathfrak{I}^{-1}(f) \in \Gamma(B)$ be the associated section of B. Then we have or all $p \in P$ the equality

$$\pi_\times(p, f(p)) = s(\pi(p)) \ . \tag{10.10}$$

This formula can also be used to construct the correspondence $f \leftrightarrow s$ as follows. If the function f is horizontal and of type ρ, then $(pg, f(pg)) = (pg, \rho(g^{-1})f(p)) = (p, f(p))\cdot g$, i.e., the map $p \mapsto \pi_\times(p, f(p))$ is constant on the fibers $\pi^{-1}(m)$. We thus can define the section $s : M \to B$ by $s(m) = \pi_\times(p, f(p))$ for an arbitrary $p \in \pi^{-1}(m)$. Conversely, if $s : M \to B$ is a section, we claim that there is a unique function $f : P \to E$ of type ρ satisfying (10.10). Uniqueness of f follows from the fact that if $\pi_\times(p, e) = \pi_\times(p, e')$ then $e = e'$ (the pre-images of $\pi_\times$ are G-orbits in $P \times E$). Existence follows from the surjectivity of $\pi_\times$. Since $\pi_\times(p, f(p)) = \pi_\times(pg, \rho(g^{-1})f(p))$ it follows that f satisfies $f(pg) = \rho(g^{-1})f(p)$, i.e., f is of type ρ.

10.11 Examples. • Let $\pi : P \to M$ be a principal fiber bundle with structure group G. Denoting by $\mathrm{Conn}(P)$ the set of all connection 1-forms on P, we know it is not empty [10.5]. The difference $\omega - \omega_o$ of two elements $\omega_o, \omega \in \mathrm{Conn}(P)$ being horizontal and of type Ad [10.5], we can define a map $\mathrm{Conn}(P) \to \Omega^1_{hor,\mathrm{Ad}}(P; \mathfrak{g})$ by $\omega \mapsto \omega - \omega_o$ for a given fixed $\omega_o \in \mathrm{Conn}(P)$. This map is obviously injective; it is also surjective because if δ is a horizontal $\mathfrak{g}$-valued 1-form of type Ad, then $\omega = \omega_o + \delta$ satisfies the conditions of a connection 1-form. We conclude that the space of all connection 1-forms (the space of all principal connections) on P is an affine space modeled on the graded $C^\infty(M)$-module $\Omega^1_{hor,\mathrm{Ad}}(P; \mathfrak{g})$. Since this space is isomorphic to $\Omega^1(M; P^{\mathrm{Ad},\mathfrak{g}})$, the difference $\omega - \omega_o$ of two connection 1-forms can be seen as a $P^{\mathrm{Ad},\mathfrak{g}}$-valued 1-form. We could have shown this result directly by combining [2.13] and [3.11]: $\omega - \omega_o$ is determined by local $\mathfrak{g}$-valued 1-forms $\Gamma^a - \Gamma^a_o$ which, according to (2.15), are related on overlaps by

$$(\Gamma^b - \Gamma^b_o) = \mathrm{Ad}(\psi_{ab}) \circ (\Gamma^a - \Gamma^a_o) \ .$$

Here we have deduced this result from the more general identification given in [10.8]. The presence of the term $\psi^*_{ab}\Theta_{MC}$ in (2.15), which contains derivatives of the transition function ψ_{ab} prevents the individual connection 1-forms to be interpreted as sections of some associated bundle.

Since the space of FVF connections on a fiber bundle only depends upon the structure group [5.5], we conclude that the set of all FVF connections on a fiber bundle $\pi : B \to M$ is an affine space modeled on a graded $C^\infty(M)$-module of sections of a vector bundle (determined by B).

• Coming back to a principal fiber bundle $\pi : P \to M$ with structure group G and a connection 1-form ω, we have seen that the curvature 2-form $\Omega = D\omega$ is horizontal and of type Ad [10.5]. We thus can see Ω as a $P^{\mathrm{Ad},\mathfrak{g}}$-valued 2-form. And, in a similar way as for the difference of two connection 1-forms, this result could have been deduced directly from (4.11).

• Twisting our point of view, we now start with a vector bundle $\pi : B \to M$ with typical fiber E. Combining [5.3] with the first part of [7.14] then shows that we have $\mathrm{End}_{\mathbb{R}}(B) = \mathcal{S}B^{\mathrm{Ad},\mathrm{End}_{\mathbb{R}}(E)}$. Now suppose that ∇ is a covariant derivative on B. Then it can also be seen as an FVF connection on B or as a connection 1-form ω on $\mathcal{S}B$ [6.18]. When we see it as a connection 1-form, we have the associated curvature 2-form

Ω, which in turn corresponds to an $\mathcal{S}B^{\mathrm{Ad},\mathrm{End}_R(E)}$-valued 2-form, i.e., an $\mathrm{End}_R(B)$-valued 2-form via the identification $\mathfrak{I} : \Omega^2(M; \mathrm{End}_R(B)) \to \Omega^2_{hor,\mathrm{Ad}}(\mathcal{S}B; E)$. If we now compare [4.8] with this construction (in particular with (10.7)) and with the local description of the curvature tensor R of the covariant derivative ∇ as given in [9.10], we see that $\mathfrak{I}(R) = \Omega$, i.e., the incarnation of the curvature 2-form Ω on the structure bundle $\mathcal{S}B$ as an $\mathrm{End}_R(B)$-valued 2-form is exactly the curvature tensor R of the covariant derivative.

10.12 Proposition. *Let $\pi_P : P \to M$ be a principal fiber bundle with structure group G, let $\rho : G \to \mathrm{Aut}(E)$ be a representation and let $B \equiv P^{\rho,E}$ be the vector bundle associated to P by the representation ρ. Let furthermore ω be a connection 1-form on P and let $\alpha \in \Omega^k_{hor,\rho}(P; E)$ be a horizontal E-valued k-form on P of type ρ. Then the exterior covariant derivative $D\alpha$ is also horizontal and of type ρ and is given by the formula*

$$D\alpha = d\alpha + (T_e\rho \circ \omega) \wedge \alpha \,. \tag{10.13}$$

Moreover, if $\mathfrak{I}_k : \Omega^k(M; B) \to \Omega^k_{hor,\rho}(P; E)$ is the identification between B-valued k-forms on M and horizontal E-valued k-forms on P of type ρ and if ∇ is the covariant derivative on B associated to the connection ω, then we have for $\sigma \in \Omega^k(M; B)$ the equality

$$\mathfrak{I}_{k+1}(\nabla\sigma) = D(\mathfrak{I}_k(\sigma)) \,, \tag{10.14}$$

i.e., the identification $\mathfrak{I}$ intertwines the covariant derivative ∇ and the exterior covariant derivative D.

Proof. To prove (10.13) we proceed as in the proof of [4.3]: we show that we have equality when evaluating on $k+1$ smooth homogeneous vector fields $X_0, \ldots, X_k$ that are either horizontal or vertical, and using only vertical vector fields of the form y^P, $y \in \mathbf{B}\mathfrak{g}$.

If all X_i are horizontal, contraction of the left gives by definition of the exterior covariant derivative the value $\iota(X_0, \ldots, X_k)d\alpha$. Contraction of the Φ-wedge product $(T_e\rho \circ \omega) \wedge \alpha$ with $X_0, \ldots, X_k$ yields a sum of terms, each involving the contraction of ω with some X_i, which is zero. We conclude that the equality holds when evaluating on $k+1$ horizontal vector fields.

Whenever one of the X_i is vertical, the left hand side of (10.13) is zero by definition. If at least two are vertical, contraction of $(T_e\rho \circ \omega) \wedge \alpha$ with $X_0, \ldots, X_k$ yields a sum of terms, each involving the contraction of α with k vector fields among $X_0, \ldots, X_k$. Since among these k at least one is vertical, the result is zero. Still assuming that at least two among the X_i are vertical, contraction of $d\alpha$ with $X_0, \ldots, X_k$ gives a sum of two terms, the first being a single sum, the second being a double sum (V.7.4). Each summand in the first term contains the contraction of α with k elements among $X_0, \ldots, X_k$ and thus is zero. Each summand in the second term contains the contraction of α with $k-1$ elements among $X_0, \ldots, X_k$ as well as contraction with the commutator of the remaining two. If the remaining two are not both vertical, at least one among the $k-1$ is vertical and thus the full contraction is zero. If the remaining two are both vertical, i.e., of the form x^P and y^P for $x, y \in \mathbf{B}\mathfrak{g}$, then their commutator $[x^P, y^P] = [x,y]^P$ is also vertical, and thus again the full contraction is zero. We conclude that if at least two among the X_i are vertical, then contraction of the right hand side with $X_0, \ldots, X_k$ yields zero, and thus in that case too we have equality.

Remains the case with only one vertical vector field. By skew-symmetry we may assume that it is the first: $X_0 = y^P$ for some $y \in \mathbf{B}\mathfrak{g}$ and X_i is horizontal for $i \geq 1$. If X is any horizontal vector field and β any (suitable) ℓ-form, then

$$\begin{aligned}\iota(X)\big((T_e\rho\circ\omega) \mathbin{\text{\AA}} \beta\big) &= \big(\iota(X)(T_e\rho\circ\omega)\big) \mathbin{\text{\AA}} \beta - (T_e\rho\circ\omega) \mathbin{\text{\AA}} \iota(X)\beta \\ &= -(T_e\rho\circ\omega) \mathbin{\text{\AA}} \iota(X)\beta\ ,\end{aligned}$$

where the minus sign after the first equality comes from the fact that $T_e\rho\circ\omega$ is a 1-form, and where the second equality comes from the fact that X is horizontal. Since we have $\iota(y^P, X_1, \ldots, X_k) = \iota(y^P)\circ\iota(X_1)\circ\cdots\circ\iota(X_k)$, we obtain the equality

$$\begin{aligned}\iota(y^P, X_1, \ldots, X_k)\big((T_e\rho\circ\omega) \mathbin{\text{\AA}} \alpha\big) &= (-1)^k\ (T_e\rho\ \iota(y^P)\omega)\cdot\iota(X_1,\ldots,X_k)\alpha \\ &= (-1)^k\ T_e\rho(y)\cdot\iota(X_1,\ldots,X_k)\alpha\ ,\end{aligned}$$

where the right hand side should be interpreted as the action of $T_e\rho(y) \in \mathrm{End}_R(E)$ on the E-valued function $\iota(X_1,\ldots,X_k)\alpha$. As said before, contraction of $d\alpha$ with the vector fields $y^P, X_1, \ldots, X_k$ yields a sum of two terms, the first a single summation and the second a double summation. In the single summation, only the term in which α is not contracted with y^P remains (all other being zero) and gives $y^P(\iota(X_1,\ldots,X_k)\alpha)$. Whenever, in the double summation, the vector field y^P does not appear in the commutator, contraction with α yields zero, and thus only a single summation remains and we find:

$$\begin{aligned}(-1)^k\cdot\iota(y^P, X_1, \ldots, X_k)d\alpha &= y^P(\iota(X_1,\ldots,X_k)\alpha) \\ &\quad + \sum_{0<j\leq k}(-1)^{j+\sum_{0<p<j}\langle\varepsilon(X_p)|\varepsilon(X_j)\rangle}\ \iota([y^P, X_j], \\ &\qquad\qquad X_1,\ldots,X_{j-1},X_{j+1},\ldots,X_k)\alpha \\ &= \mathcal{L}(y^P)(\iota(X_1,\ldots,X_k)\alpha) \\ &\quad - \sum_{0<j\leq k}(-1)^{\sum_{0<p<j}\langle\varepsilon(X_p)|\varepsilon(y)\rangle}\ \iota(X_1,\ldots,X_{j-1}, \\ &\qquad\qquad [y^P, X_j], X_{j+1},\ldots,X_k)\alpha \\ &= (-1)^{\sum_{1\leq p\leq k}\langle\varepsilon(X_p)|\varepsilon(y)\rangle}\ \iota(X_1,\ldots,X_k)(\mathcal{L}(y^P)\alpha)\ ,\end{aligned}$$

where the last equality follows from [V.7.17]. Since α is of type ρ, we have the equality $\Phi_g^*\alpha = \rho(g^{-1})\circ\alpha$. Taking the derivative of this equation in the direction of $\vec{y}$ at $g = e$

as in the proof of [4.3] we obtain on the left $\mathcal{L}(y^P)\alpha$. On the right hand side we have to compute $\vec{y}_e(\rho(g^{-1})\circ\alpha)$. Taking into account that $T_e Inv = -id$ we obtain $\vec{y}_e(\rho(g^{-1})\circ\alpha) = -(T_e\rho(y))\circ\alpha$, and thus $\mathcal{L}(y^P)\alpha = -(T_e\rho(y))\circ\alpha$. Substituting this in the previous result, we obtain

$$\iota(y^P, X_1, \ldots, X_k)d\alpha = -(-1)^k\, T_e\rho(y)\cdot\iota(X_1,\ldots,X_k)\alpha\ .$$

Combining this with the contraction of $(T_e\rho\circ\omega)\mathbin{\wedge}\alpha$ with $y^P, X_1, \ldots, X_k$, we see that contraction of the right hand side of (10.13) with $y^P, X_1, \ldots, X_k$ yields zero, and thus we have equality also in this last case with just one vertical vector field, i.e., we have proven (10.13).

To show (10.14), we choose a local trivialization (U,ψ) of P and we denote by $(U,\widehat{\psi})$ the associated local trivialization of $B \equiv P^{\rho,E}$. For ease of notation we introduce the functions $I_\rho : G \to \mathrm{End}_R(E)$, $g \mapsto \rho(g^{-1})$ and $I_{\mathrm{Ad}} : G \to \mathrm{End}_R(\mathfrak{g})$, $g \mapsto \mathrm{Ad}(g^{-1})$. Given the B-valued k-form σ and the local trivialization $(U,\widehat{\psi})$, we have an E-valued k-form $\sigma_{\widehat{\psi}}$ on U (8.5). According to (10.7), the local expression $(\psi^{-1})^*(\mathfrak{J}_k\sigma)$ of $\mathfrak{J}_k\sigma$ as E-valued k-form on $U\times G$ is given by the formula

$$(\psi^{-1})^*(\mathfrak{J}_k\sigma) = I_\rho\circ\sigma_{\widehat{\psi}}\ .$$

Using (10.13) and the local expression (3.12) of ω, we compute:

$$\begin{aligned}(\psi^{-1})^*(D(\mathfrak{J}_k\sigma)) &= (\psi^{-1})^*(d(\mathfrak{J}_k\sigma) + (T_e\rho\circ\omega)\mathbin{\wedge}\mathfrak{J}_k\sigma)\\ &= d(I_\rho\circ\sigma_{\widehat{\psi}}) + (T_e\rho\circ(\Theta_{MC} + I_{\mathrm{Ad}}\circ\Gamma))\mathbin{\wedge}(I_\rho\circ\sigma_{\widehat{\psi}})\\ &= (dI_\rho)\mathbin{\wedge}\sigma_{\widehat{\psi}} + I_\rho\circ d\sigma_{\widehat{\psi}} + ((T_e\rho\circ\Theta_{MC})\circ I_\rho)\mathbin{\wedge}\sigma_{\widehat{\psi}}\\ &\qquad + ((T_e\rho\circ I_{\mathrm{Ad}}\circ\Gamma)\circ I_\rho)\mathbin{\wedge}\sigma_{\widehat{\psi}}\ ,\end{aligned}$$

where the reader should be aware that all composition operators $\circ$ are in fact alternative notations for wedge products, either $\wedge$ or $\wedge$. We now claim that we have the equalities

$$dI_\rho = -(T_e\rho\circ\Theta_{MC})\circ I_\rho \qquad\text{and}\qquad (T_e\rho\circ I_{\mathrm{Ad}}\circ\Gamma)\circ I_\rho = I_\rho\circ(T_e\rho\circ\Gamma)\ .$$

Using this, [10.7], and [9.7-i] we obtain:

$$(\psi^{-1})^*(D(\mathfrak{J}_k\sigma)) = I_\rho\circ\left(d\sigma_{\widehat{\psi}} + (T_e\rho\circ\Gamma)\mathbin{\wedge}\sigma_{\widehat{\psi}}\right) = I_\rho\circ(\nabla\sigma)_{\widehat{\psi}} = (\psi^{-1})^*(\mathfrak{J}_{k+1}(\nabla\sigma))\ .$$

Since the trivializing chart (U,ψ) is arbitrary, we obtain the global result (10.14). Moreover, this also shows that $D\alpha$ is horizontal and of type ρ whenever α is because: $\mathfrak{J}_k$ is bijective, so there exist σ such that $\alpha = \mathfrak{J}_k(\sigma)$, and thus $D\alpha = \mathfrak{J}_{k+1}(\nabla\sigma)$ is in the image of $\mathfrak{J}_{k+1}$ and thus also horizontal and of type ρ.

To finish we now prove the two claims. Contracting dI_ρ with a left-invariant vector field $\vec{x}$ yields:

$$\iota(\vec{x}_g)dI_\rho = T_gI_\rho(\vec{x}_g) = T_gI_\rho(TL_gx) = T_e(I_\rho\circ L_g)(x)\ .$$

Now $I_\rho \circ L_g : G \to \mathrm{End}_R(E)$ is the map $h \mapsto \rho(h^{-1}) \circ \rho(g^{-1})$, and thus, taking into account that $T_e Inv(x) = -x$, we find

$$\begin{aligned}\iota(\vec{x}_g) dI_\rho &= -(T_e\rho(x)) \circ \rho(g^{-1}) = -(T_e\rho(\iota(\vec{x}_g)\Theta_{MC})) \circ I_\rho(g) \\ &= -\iota(\vec{x}_g)(T_e\rho \circ \Theta_{MC}) \circ I_\rho) .\end{aligned}$$

Since $x \in \mathfrak{g}$ is arbitrary, we have shown the first of our claims. To prove the second one, we start by studying the map $f : G \to \mathrm{End}_R(E)$ defined as

$$f(h) = \rho(g^{-1}hg) \circ \rho(g^{-1}) = \rho(g^{-1}) \circ \rho(h) .$$

Evaluating $T_e f$ on $x \in \mathfrak{g}$ we thus find the equality

$$(T_e\rho(\mathrm{Ad}(g^{-1})x)) \circ \rho(g^{-1}) = \rho(g^{-1}) \circ T_e\rho(x) .$$

Substituting for x the value $x = \iota(Y_{(m,g)})\Gamma$ for some $Y_{(m,g)} \in T_{(m,g)}(U \times G)$ we find

$$\begin{aligned}\iota(Y_{(m,g)})((T_e\rho \circ I_{\mathrm{Ad}} \circ \Gamma) \circ I_\rho) &= (T_e\rho(\mathrm{Ad}(g^{-1})(x))) \circ \rho(g^{-1}) = \rho(g^{-1}) \circ T_e\rho(x) \\ &= \iota(Y_{(m,g)})(I_\rho \circ (T_e\rho \circ \Gamma)) .\end{aligned}$$

Since $Y_{(m,g)}$ is arbitrary, we have also shown the second claim. *QED*

10.15 Remark. If α is horizontal and of type ρ, we used in the proof of [10.12] an indirect way to prove that $D\alpha$ is again horizontal and of type ρ. By definition of D it is obvious that $D\alpha$ is horizontal; this is true even if α itself is not horizontal. However, a direct proof that $D\alpha$ is again of type ρ involves the exterior derivative of a generalized pull-back. Correctly defining this notion involves a lot of technical problems that were avoided in our indirect proof.

10.16 Corollary. *On a principal fiber bundle $\pi : P \to M$ with connection 1-form ω and curvature 2-form Ω, the Bianchi identities are equivalent to the statement that $D\Omega = 0$.*

Proof. In [4.4] we have seen that the Bianchi identities imply $D\Omega = 0$. From (10.13) we learn that $D\Omega = d\Omega + (T_e \,\mathrm{Ad} \circ \omega) \mathbin{\mathbin{\wedge}\!\!\!\cdot}\, \Omega = d\Omega + (\mathrm{ad}_R \circ \omega) \mathbin{\wedge}\!\!\!\cdot\, \Omega$. In order to show the converse, it thus is sufficient to show that $(\mathrm{ad}_R \circ \omega) \mathbin{\wedge}\!\!\!\cdot\, \Omega = -[\Omega \,\overset{\wedge}{,}\, \omega]$. Writing $\omega = \sum_i \omega^i \otimes v_i$ and $\Omega = \sum_j \Omega^j \otimes v_j$ with respect to a basis (v_i) of $\mathfrak{g}$, we compute

$$\begin{aligned}(\mathrm{ad}_R \circ \omega) \mathbin{\wedge}\!\!\!\cdot\, \Omega &= \sum_{ij} (\mathrm{ad}_R \circ (\omega^i \otimes v_i)) \mathbin{\wedge}\!\!\!\cdot\, (\Omega^j \otimes v_j) = \sum_{ij} (\omega^i \otimes \mathrm{ad}_R(v_i)) \mathbin{\wedge}\!\!\!\cdot\, (\Omega^j \otimes v_j) \\ &= \sum_{ij} \omega^i \wedge \Omega^j \otimes (-1)^{\langle \varepsilon(v_i) | \varepsilon(v_j)\rangle} \,\mathrm{ad}_R(v_i)(v_j) \\ &= -\sum_{ij} \Omega^j \wedge \omega^i \otimes (-1)^{\langle \varepsilon(v_i) | \varepsilon(v_j)\rangle} [v_j, v_i] = -[\Omega \,\overset{\wedge}{,}\, \omega] ,\end{aligned}$$

where the third equality follows from the fact that ω and Ω are even and thus that we have $\varepsilon(\omega^i) = \varepsilon(v_i) = \varepsilon(\Omega^i)$. The fourth equality follows from [I.6.9], [I.5.9] and graded skew-symmetry of the bracket $[\_, \_]$. *QED*

10.17 Corollary. *Let $\pi : B \to M$ be a vector bundle with typical fiber E and ∇ a covariant derivative on it. If $R \in \mathrm{End}_R(B)$ is the curvature tensor of ∇ then $\nabla^{\mathrm{Ad}} R = 0$.*

Proof. The covariant derivative ∇ on B is equivalent to a connection 1-form ω on the structure bundle $\mathcal{S}B$ [6.18]. The representation Ad defines the associated vector bundle $\mathcal{S}B^{\mathrm{Ad},\mathrm{End}_R(E)} \equiv \mathrm{End}_R(B)$ and an induced covariant derivative ∇^{Ad} on $\mathrm{End}_R(B)$ [5.3], [7.14]. According to [10.11], $\mathfrak{I}(R) = \Omega$, and then, combining [10.12] with [4.4] we compute $\mathfrak{I}(\nabla^{\mathrm{Ad}} R) = D\Omega = 0$. Since $\mathfrak{I}$ is an isomorphism, the result follows. QED

10.18 Corollary. *Let $\pi : P \to M$ be a principal fiber bundle with structure group G and let ω be a connection 1-form on P with associated curvature 2-form Ω. Let furthermore $\rho : G \to \mathrm{Aut}(E)$ be a representation and let $\alpha \in \Omega^k_{hor,\rho}(P; E)$ be a horizontal E-valued k-form on P of type ρ. Then $D^2\alpha$ is given by the formula (compare with [9.10])*

$$D^2\alpha = (T_e\rho \circ \Omega) \mathbin{\bar{\wedge}} \alpha .$$

Proof. Using [10.12] we compute:

$$\begin{aligned} D^2\alpha &= D(d\alpha + (T_e\rho \circ \omega) \mathbin{\bar{\wedge}} \alpha) \\ &= d(d\alpha + (T_e\rho \circ \omega) \mathbin{\bar{\wedge}} \alpha) + (T_e\rho \circ \omega) \mathbin{\bar{\wedge}} (d\alpha + (T_e\rho \circ \omega) \mathbin{\bar{\wedge}} \alpha) \\ &= d((T_e\rho \circ \omega) \mathbin{\bar{\wedge}} \alpha) + (T_e\rho \circ \omega) \mathbin{\bar{\wedge}} d\alpha + (T_e\rho \circ \omega) \mathbin{\bar{\wedge}} ((T_e\rho \circ \omega) \mathbin{\bar{\wedge}} \alpha) \\ &= (T_e\rho \circ d\omega) \mathbin{\bar{\wedge}} \alpha + \big((T_e\rho \circ \omega) \mathbin{\bar{\wedge}}_{\circ} (T_e\rho \circ \omega)\big) \wedge \alpha , \end{aligned}$$

where for the fourth equality we used [V.7.9-iii] and the fact that $T_e\rho$ is constant and thus $d(T_e\rho \circ d\omega) = T_e\rho \circ d\omega$. Since $T_e\rho$ is an $\mathrm{End}_R(E)$-valued 1-form, we can apply [1.4] to get $(T_e\rho \circ \omega) \mathbin{\bar{\wedge}}_{\circ} (T_e\rho \circ \omega) = \frac{1}{2}[\,T_e\rho \circ \omega \mathbin{\bar{\wedge}} T_e\rho \circ \omega\,]$. But $T_e\rho$ is a morphism of $\mathcal{A}$-Lie algebras and thus $[\,T_e\rho \circ \omega \mathbin{\bar{\wedge}} T_e\rho \circ \omega\,] = T_e\rho \circ [\,\omega \mathbin{\bar{\wedge}} \omega\,]$. Plugging this in the final equality for $D^2\alpha$ above gives us $D^2\alpha = (T_e\rho \circ (d\omega + \frac{1}{2}[\,\omega \mathbin{\bar{\wedge}} \omega\,])) \mathbin{\bar{\wedge}} \alpha$, which is the desired result by [4.3]. QED

References

Some other books/papers on supermanifolds (in square brackets the page(s) where the work is cited)

[BBH-R] C. Bartocci, U. Bruzzo, and D. Hernández-Ruipérez, *The Geometry of Supermanifolds*, MIA 71, Kluwer Acad. Publ., Dordrecht, 1991.

[Ber] F.A. Berezin, *Introduction to superanalysis*, Mathematical Physics and Applied Mathematics vol. 19, Reidel, Dordrecht, 1987.

[DW] B. DeWitt, *Supermanifolds*, Cambridge UP, Cambridge, 1984, [p. 271].

[Fre] P.G.O. Freund, *Introduction to Supersymmetry*, Cambridge UP, Cambridge, 1986.

[Ko] B. Kostant, *Graded manifolds, graded Lie theory, and prequantization*, Differential geometric methods in mathematical physics (K. Bleuler & A. Reetz, eds.), Proceedings Conference, Bonn 1975. LNM 570, Springer-Verlag, Berlin, 1977, pp. 177–306.

[Le] D.A. Leites, *Introduction to the theory of supermanifolds*, Russian Math. Surveys **35** (1980), 1–64.

[Ma] Y. I. Manin, *Gauge field theory and complex geometry*, Grundlehren der Mathematischen Wissenschaften, 289, Springer-Verlag, Berlin, 1997.

General references (in square brackets the page(s) where the work is cited)

[Ba] M. Batchelor, *The structure of supermanifolds*, Trans. Amer. Math. Soc. **253** (1979), 329–338, [p. 196].

[Bo] N. Bourbaki, *Groupes et algèbres de Lie, Ch 4–6*, Éléments de mathématique, Fascicule XXXIV, Hermann, Paris, 1968; Masson, Paris, 1981, [p. 23].

[Br] K.S. Brown, *Cohomology of Groups*, GTM 87, Springer-Verlag, Berlin, New York, 1982, [p. 325].

[É.Ca] É. Cartan, *La topologie des espaces représentatifs de groupes de Lie*, Œuvres I (part 2), Gauthier-Villars, Paris, 1952, pp. 1307–1330, [p. 325].

[H.Ca] H. Cartan, *Notions d'algèbre différentielle; application aux groupes de Lie et aux variétés où opère un groupe de Lie*, CBRM Colloque de Topologie, Bruxelles 5–8 juin 1950, George Thone, Liège, 1950, pp. 15–27, [p. 248].

[Du] J. Dugundji, *Topology*, Allyn and Bacon, Inc, Boston, 1966, [p. 93].

[Eh] C. Ehresmann, *Les connexions infinitésimales dans un espace fibré différentiable*, CBRM Colloque de Topologie, Bruxelles 5–8 juin 1950, George Thone, Liège, 1950, pp. 29–55, [p. 344].

[GHV] W. Greub, S. Halperin, & R. Vanstone, *Connections, Curvature, and Cohomology*, vol. I, Academic Press, Inc., San Diego, London, etc., 1972, [pp. 145, 195].

[God] C. Godbillon, *Éléments de topologie algébrique*, Hermann, Paris, 1971, [p. 327].

[Gor] V.V. Gorbatsevich, *The construction of a simply connected Lie group with a given Lie algebra*, Russian Math. Surveys **41** (1986), 207–208, [p. 325].

[Hi] F. Hirzebruch, *Topological Methods in Algebraic Geometry*, Springer, Berlin, Heidelberg, New York, 1978, [p. 145].

[HS] M. W. Hirsch & Stephen Smale, *Differential Equations, Dynamical Systems, and Linear Algebra*, Academic Press, Inc., San Diego, London, etc., 1974, [p. 233].

[Hu] D. Husemoller, *Fibre Bundles*, GTM 20, Springer, Berlin, New York, 1994, [p. 145].

[Iw] K. Iwasawa, *On some types of topological groups*, Annals of Mathematics **50** (1949), 507–558, [p. 327].

[Lie] S. Lie, *Theorie der Transformationsgruppen Vol III*, Teubner, Leipzig, 1893; Reprinted, Chelsea Publishing Company, New York, NY, 1970, [p. 325].

[MS-V] J. Monterde & A.O. Sánchez-Valenzuela, *Existence and uniqueness of solutions to superdifferential equations*, J. Geom. Phys. **10** (1993), 315–343, [p. 334].

[Pij] H.G.J. Pijls, *The Yang-Mills equations*, Proceedings Seminar 1981–1982 Mathematical Structures in field theories (E.M. de Jager & H.G.J. Pijls, eds.), CWI Syllabus 2, CWI, Amsterdam, 1984, pp. 119–182, [p. xii].

[Qu] D. Quillen, *Superconnections and the Chern character*, Topology **24** (1985), 89-95, [p. 393].

[Ra] R.M. Range, *Holomorphic functions and integral representations in several complex variables*, GTM 108, Springer-Verlag, Berlin, New York, 1986, [p. 95].

[Sc] L. Schwartz, *Les Tenseurs*, Hermann, Paris, 1975, [p. xii].

[So] J.-M. Souriau, *Structure des Systèmes Dynamiques*, Dunod, Paris, 1969; English translation, *Structure of Dynamical Systems, A Symplectic View of Physics*, PM 149, Birkhäuser, Boston, Basel, 1997, [p. 332].

[Spa] E.H. Spanier, *Algebraic Topology*, Springer-Verlag, Berlin, New York, 1966, [p. 312].

[Spi] M. Spivak, *A Comprehensive Introduction to Differential Geometry*, vol. I, Publish or Perish, Berkeley, 1970, [pp. xii, 123, 135].

[Sta] J.D. Stasheff, *Continuous Cohomology of Groups and Classifying Spaces*, Bull. Amer. Math. Soc. **84** (1978), 513–530, [p. 325].

[Ste] N. Steenrod, *The Topology of Fibre bundles*, Princeton University Press, Princeton, NJ, 1951, [p. 145].

[Tu1] G.M. Tuynman, *An elementary proof of Lie's third theorem*, Publications IRMA **34** (1994), X.1–X.4, Internal publication of the Department of Mathematics of the University of Lille, [p. 325].

[Tu2] G.M. Tuynman, *Functions of class C^k without derivatives*, Publicacions Matemàtiques **41** (1997), 417–435, [p. 97].

[TW] G.M. Tuynman & W.A.J.J. Wiegerinck, *Central Extensions in Physics*, J. Geom. Phys. **4** (1987), 207–258, [p. 325].

[vE] W.T. van Est, *Une démonstration de E. Cartan du troisième théorème de Lie*, Actions hamiltoniennes de groupes — Troisième théorème de Lie, Séminaire Sud-Rhodanien de Géométrie VIII (may 1986) — Travaux en cours 27, Hermann, Paris, 1988, pp. 83–96, [p. 325].

[Wa] F.W. Warner, *Foundations of Differentiable Manifolds and Lie Groups*, GTM 94, Springer-Verlag, Berlin, New-York, 1983, [pp. xii, 300, 300].

Index of Notation

Index

CPSIA information can be obtained
at www.ICGtesting.com
Printed in the USA
LVHW042210120420
653147LV00008BB/1076

9 781402 022968